MICRO- AND NANOSCALE FLUID MECHANICS: TRANSPORT IN MICROFLUIDIC DEVICES

This text describes the physics of fluid transport in microfabricated and nanofabricated liquid-phase systems, with consideration of particles and macromolecules. This text brings together fluid mechanics, electrodynamics, and interface science with a focused goal of preparing the modern microfluidics researcher to analyze and model continuum fluid mechanical systems encountered when working with micro- and nanofabricated devices. This text is designed for classroom instruction and also serves as a useful reference for practicing researchers. Worked sample problems are inserted throughout to assist the student, and exercises are included at the end of each chapter to facilitate use in classes.

Brian J. Kirby currently directs the Micro/Nanofluidics Laboratory in the Sibley School of Mechanical and Aerospace Engineering at Cornell University. He joined the school in August 2004. Previously, he was a Senior Member of the Technical Staff in the Microfluidics Department at Sandia National Laboratories in Livermore, California. He was educated at Stanford University and The University of Michigan. Professor Kirby has received numerous research and teaching awards, including the Presidential Early Career Award for Scientists and Engineers (PECASE) and the Mr. and Mrs. Robert F. Tucker Excellence in Teaching Award.

Micro- and Nanoscale Fluid Mechanics

TRANSPORT IN MICROFLUIDIC DEVICES

Brian J. Kirby

Cornell University

CAMBRIDGE UNIVERSITY PRESS
Cambridge, New York, Melbourne, Madrid, Cape Town,
Singapore, São Paulo, Delhi, Mexico City

Cambridge University Press
32 Avenue of the Americas, New York, NY 10013-2473, USA

www.cambridge.org
Information on this title: www.cambridge.org/9781107617209

First published 2010
Reprinted 2011, 2012, 2013
First paperback edition 2013

A catalog record for this publication is available from the British Library.

Library of Congress Cataloging in Publication Data

Kirby, Brian (Brian J.)
Micro- and nanoscale fluid mechanics : transport in microfluidic devices / Brian Kirby.
p. cm.
Includes bibliographical references and index.
ISBN 978-0-521-11903-0 (hardback)
1. Microfluidic devices. 2. Microfluidics. 3. Nanofluids. I. Title.
TJ853.4.M53K57 2010
620.1′064–dc22 2009053537

ISBN 978-0-521-11903-0 Hardback
ISBN 978-1-107-61720-9 Paperback

Additional resources for this publication at www.cambridge.org/**kirby**

Contents

Preface

This text focuses on the physics of liquid transport in micro- and nanofabricated systems. It evolved from a graduate course I have taught at Cornell University since 2005, titled "Physics of Micro- and Nanoscale Fluid Mechanics," housed primarily in the Mechanical and Aerospace Engineering Department but attracting students from Physics, Applied Physics, Chemical Engineering, Materials Science, and Biological Engineering. This text was designed with the goal of bringing together several areas that are often taught separately – namely, fluid mechanics, electrodynamics, and interfacial chemistry and electrochemistry – with a focused goal of preparing the modern microfluidics researcher to analyze and model continuum fluid-mechanical systems encountered when working with micro- and nanofabricated devices. It omits many standard topics found in other texts – turbulent and transitional flows, rheology, transport in gel phase, Van der Waals forces, electrode kinetics, colloid stability, and electrode potentials are just a few of countless examples of fascinating and useful topics that are found in other texts, but are omitted here as they are not central to the fluid flows I wish to discuss.

Although I hope that this text may also serve as a useful reference for practicing researchers, it has been designed primarily for classroom instruction. It is thus occasionally repetitive and discursive (where others might state results succinctly and only once) when this is deemed useful for instruction. Worked sample problems are inserted throughout to assist the student, and exercises are included at the end of each chapter to facilitate use in classes. Solutions for qualified instructors are available from the publisher at http://www.cambridge.org/kirby. This text is *not* a summary of current research in the field and omits any discussion of microfabrication techniques or any attempt to summarize the technological state of the art.

The text considers, in turn, (a) low-Reynolds-number fluid mechanics and hydraulic circuits; (b) outer solutions for microscale flow, focusing primarily on the unique aspects of electroosmotic flow outside the electrical double layer; (c) inner solutions for microscale flow, focusing on sources of interfacial charge and modeling of electrical double layers; and (d) unsteady and nonequilibrium solutions, focusing on nonlinear electrokinetics, dynamics of electrical double layers, electrowetting, and related phenomena. In each case, several applications are selected to motivate the presentation, including microfluidic mixing, DNA and protein separations, microscale fluid velocity measurements, dielectrophoretic particle manipulation, electrokinetic pumps, and the like.

I select notation with the goal of helping students new to the field and with the understanding that this (on occasion) leads to redundant or unwieldy results. I minimize use of one symbol for multiple different variables, so the radius in spherical coordinates (r) is typeset with a symbol different from the radius in cylindrical coordinates ($\mathscr{r}$) and the colatitudinal angle ϑ in spherical coordinates is distinguished from the polar coordinate in cylindrical coordinates (θ). Because I teach from this text using a chalkboard, I use symbols that I can reproduce on a chalkboard – thus I avoid the use of the Greek letter ν for the kinematic viscosity $\nu = \eta/\rho$, because I am utterly unable to make it distinguishable from the y velocity v. Vectors, though they are placed in boldface to make them stand out, are also written with (admittedly redundant) superscripted arrows to match the chalkboard presentation.

This material is used for a semester-long graduate course at Cornell. Chapters 1, 2, 5, 7, and 8, as well as the appendices, are not covered in class as they are considered review or supplementary material. The remainder of the text is covered in approximately forty-two 50-minute classroom sessions.

I would like to acknowledge a number of people who helped with various aspects of this text. In particular, Dr. Elizabeth Strychalski and Professors Stephen Pope and Claude Cohen at Cornell, Professor Shelley Anna of Carnegie-Mellon University, Professor Kevin Dorfman of the University of Minnesota, Professor Nicolas Green of the University of Southampton, Donald Aubrecht of Harvard University, Professor Sumita Pennathur of UCSB, and Professor Aaron Wheeler of the University of Toronto were kind enough to offer useful suggestions. Professor Amy Herr of the University of California, Berkeley, used a draft of this text for her class during spring 2009; her insight and the feedback from her students were both immensely helpful. Professor Martin Bazant of the Massachusetts Institute of Technology provided materials helpful in completing the bibliography for several of the chapters. The students that have taken my class since 2004 have all contributed to this text in some way, but I would like to thank my student researchers Alex Barbati, Ben Hawkins, Sowmya Kondapalli, and Vishal Tandon in particular for their input, and my student Michael Allen for careful proofreading. Ben Hawkins and Dr. Jason Gleghorn contributed a number of the figures and helped to write material that was included in the chapters on Stokes flow and dielectrophoresis. David J. Griffiths (Reed College) provided files that assisted with typesetting. Gabe Terrizzi created many of the figures; his contributions were immensely helpful. Greg Parker (gparker@chorus.net) designed the cover.

Although many people assisted with review of this text, I am solely responsible for any errors, and I hope that readers will notify me or the publisher of those that they find. Errata will be maintained at http://www.cambridge.org/kirby.

Brian J. Kirby
Ithaca, NY
May 2010

Nomenclature

Symbol	Meaning	Page of first use or definition
A	area	61
$\mathcal{A}$	Helmholtz free energy	324
α	coefficient	112
α	phase lag angle	69
α	rotation angle	158
α	thermal diffusivity	80
a	acceleration	255
a	particle radius	171
a_i	activity	413
β	compressibility	75
β	coefficient	236
b	slip length	31
$\vec{\boldsymbol{B}}$	applied magnetic field	391
$\mathcal{B}$	Brillouin function	104
c_i	species molar concentration	407
c_p	specific heat	80
c	passive scalar	80
C	capacitance	117
C	constant of integration	43
C_h	compliance	66
$\underset{\sim}{C}$	complex number	465
C_D	drag coefficient	188
Γ	2D vortex strength	163
Γ	circulation	13
Γ	surface chemical site density	229
Γ	magnitude of injected sample	90
γ	surface tension	20
γ_i	natural logarithm of species concentration	259
χ	electrokinetic coupling matrix	65
χ_e	electric susceptibility	100
χ_m	magnetic susceptibility	98
d	depth	140
d	diameter	22
D	scalar diffusivity	80
D_i	species diffusivity	252
$\vec{\boldsymbol{D}}$	electric displacement	100

Symbol	Meaning	Page of first use or definition
Du	Dukhin number	263
δ	Dirac delta function	458
$\overleftrightarrow{\boldsymbol{\delta}}$	identity tensor	16
∇	del operator	426
e	eccentricity	188
e	fundamental charge	201
e_1	singlet potential	475
e_2	pair potential	476
e_{mf}	potential of mean force	227
$\vec{\boldsymbol{E}}$	electric field	97
ε	electrical permittivity	98
$\underset{\sim}{\varepsilon}$	complex electrical permittivity for sinusoidal fields	113
ε_{S}	Stern layer permittivity	360
ε_0	electrical permittivity of free space	100
ε_{r}	relative permittivity, i.e., dielectric constant	101
ε'	reactive permittivity	115
ε''	dissipative permittivity	115
$\varepsilon_{\mathrm{LJ}}$	potential well depth	477
$\overleftrightarrow{\boldsymbol{\varepsilon}}$	strain rate tensor	10
$\frac{\partial \varepsilon}{\partial c}$	dielectric increment	413
F	Faraday constant	99
$\vec{\boldsymbol{F}}$	force	108
$\vec{\boldsymbol{f}}$	force per unit volume	6
$\underset{\sim}{f_{\mathrm{CM}}}$	Clausius–Mossotti factor	393
f_{ad}	adjusted distribution function	480
f_{d}	distribution function	217
f_{dc}	direct correlation function	482
f_{tc}	total correlation function	482
f_{M}	Mayer f function	480
f_0	Henry's function	288
f	electrophoretic correction factor	287
ϕ	electric potential	97
φ	electric potential difference from bulk	133
φ_0	total potential drop across the double layer	133
ϕ_{v}	velocity potential	153
$\underset{\sim}{\phi_{\mathrm{v}}}$	complex velocity potential	158
φ	azimuthal coordinate	419
Φ	cross-correlation	189
ζ	zeta potential	139
G	Gibbs free energy	20
G	electrical conductance	117
G_{s}	excess surface conductance	262
$\vec{\boldsymbol{g}}$	gravitational acceleration	6
g_i	chemical potential	227
$\overline{g_i}$	electrochemical potential	227

Symbol	Meaning	Page of first use or definition
$\vec{\vec{G}}$	hydrodynamic interaction tensor	187
$\vec{\vec{G}}_0$	Oseen–Burgers tensor	187
H	capillary height	23
$\vec{H}$	induced magnetic field	98
h	height	43
η	dynamic viscosity	17
$\vec{i}$	current density	110
i_0	exchange current density	112
I	current	64
I	second moment of area	309
I_c	ionic strength	408
j	square root of minus 1	157
$\vec{j}$	scalar flux density	80
$\mathcal{J}$	Joukowski transform	171
k	spring constant	325
k	chemical reaction rate	409
k_{ve}	viscoelectric coefficient	235
k_B	Boltzmann constant	104
K_a	acid dissociation constant	409
K_{eq}	equilibrium constant	409
K_{sp}	solubility product	412
κ	2D doublet strength	165
κ	Debye screening parameter	288
Λ	molar conductivity	256
Λ	2D source strength	160
λ_B	Bjerrum length	478
λ_D	Debye length	202
λ_{HS}	hard-sphere packing length	213
λ_S	Stern layer thickness	360
ℓ_c	polymer contour length	301
ℓ_e	polymer end-to-end length	303
ℓ_K	polymer Kuhn length	312
ℓ_p	polymer persistence length	299
L	length	61
L	electrical inductance	117
L	depolarization factor	384
m	mass	184
$\vec{M}$	magnetization	98
μ	viscous mobility	252
μ_{DEP}	dielectrophoretic mobility	374
μ_{EK}	electrokinetic mobility	265
μ_{EO}	electroosmotic mobility	138
μ_{EP}	electrophoretic mobility	252
μ_{mag}	magnetic permeability	98
$\mu_{mag,0}$	magnetic permeability of free space	98
N_A	Avogadro's number	112

Symbol	Meaning	Page of first use or definition
N_{bp}	number of base pairs in DNA molecule	301
n	normal coordinate	106
p	pressure	6
$\vec{\boldsymbol{p}}$	dipole moment	104
pK_a	negative logarithm of acid dissociation constant	410
pH	negative logarithm of molar proton concentration	410
pOH	negative logarithm of molar hydroxyl ion concentration	411
pzc	point of zero charge	230
$\mathcal{P}$	perimeter	63
$\mathcal{P}$	probability density function	313
Pe	mass transfer Peclet number	79
ϖ	dummy frequency integration variable	115
ψ	stream function	8
ψ_S	Stokes stream function	9
ψ_e	electric stream function	469
$\vec{\boldsymbol{P}}$	electric polarization	100
$\bar{\boldsymbol{P}}$	pressure interaction tensor	187
Q	volumetric flow rate	60
q	electric charge	97
q''	electric areal charge density	359
ρ	fluid density	6
ρ_E	net charge density	99
r	radial coordinate – spherical coordinates	418
r_h	hydraulic radius	63
Δr	radial distance – spherical coordinates	98
$\vec{\boldsymbol{\Delta r}}$	distance vector	98
$\imath$	radial coordinate – cylindrical coordinates	418
$\Delta \imath$	radial distance – cylindrical coordinates	157
Re	Reynolds number	442
R	universal gas constant	112
R	electrical resistance	117
R	radius of channel	47
R	radius of curvature	21
R	separation resolution	267
R_h	hydraulic resistance	61
$\langle r_g \rangle$	radius of gyration	303
s	arc length	302
$\mathcal{S}$	entropy	324
$\mathcal{S}$	Schwarz–Christoffel transform	473
σ	conductivity	110
σ_{LJ}	Lennard–Jones "bond length"	477
$\underset{\sim}{\sigma}$	complex electrical conductivity	114
σ_s	effective surface conductivity	210

Symbol	Meaning	Page of first use or definition
Sk	Stokes number	186
St	Strouhal number	442
t	time	7
T	Kelvin temperature	20
$\vec{\boldsymbol{T}}$	torque	109
$\vec{\vec{\boldsymbol{T}}}$	Maxwell stress tensor	107
$\vec{\vec{\boldsymbol{\tau}}}$	stress tensor	15
τ	characteristic time	103
θ	polar coordinate – cylindrical coordinates	418
θ	contact angle	21
θ_0	corner angle	170
ϑ	colatitude coordinate – spherical coordinates	418
$\Delta\theta$	polar coordinate of distance vector	157
$\vec{\boldsymbol{u}}$	velocity vector	7
$\underset{\sim}{u}$	complex velocity	159
u_{EK}	electrokinetic velocity	269
u_{EO}	electroosmotic velocity	140
u_{EP}	electrophoretic velocity	255
$u_{\mathscr{r}}$	radial velocity – cylindrical coordinates	8
u_r	radial velocity – spherical coordinates	9
u_θ	circumferential velocity – cylindrical coordinates	8
u_ϑ	circumferential velocity – spherical coordinates	9
$\mathcal{U}$	molecular internal energy	324
V	voltage	106
$\mathcal{V}$	volume	66
ω	angular frequency	49
$\vec{\boldsymbol{\omega}}$	vorticity	12
$\vec{\vec{\boldsymbol{\omega}}}$	rotation rate tensor	11
w	width	90
x	x coordinate	418
ξ	hard-sphere packing parameter	215
ξ	thermodynamic efficiency	143
y	y coordinate	418
Y	Young's modulus	309
z	z coordinate	418
z	valence magnitude for symmetric electrolytes	203
z_i	species valence	99
Z	partition function	326
$\underset{\sim}{Z}$	impedance	119
$\underset{\sim}{Z}_{\mathfrak{h}}$	hydraulic impedance	69

Subscript	Example	Meaning
0	p_0	phasor or sinusoid magnitude
0	w_0	value at reference state
∞	$c_{i,\infty}$	value in freestream or in bulk
bend	$\mathcal{U}_{\mathrm{bend}}$	bending
conv	$\vec{\boldsymbol{j}}_{\mathrm{conv},i}$	convective
diff	$\vec{\boldsymbol{j}}_{\mathrm{diff},i}$	diffusive
edl	q''_{edl}	electrical double layer
eff	ζ_{eff}	effective
ext	$\vec{\boldsymbol{E}}_{\mathrm{ext}}$	extrinsic
H	u_H	high
L	u_L	low
m	$\underset{\sim}{\varepsilon}_{\mathrm{m}}$	suspending medium
n	E_n	normal
p	ρ_{p}	particle
pre	$\vec{\vec{\boldsymbol{\tau}}}_{\mathrm{pre}}$	isotropic (pressure) components
str	I_{str}	streaming
t	u_{t}	tangential
visc	$\vec{\vec{\boldsymbol{\tau}}}_{\mathrm{visc}}$	deviatoric (viscous) components
w	ρ_{w}	water

Superscript, accent	Example(s)	Meaning
$\circ$	g_i°	value at reference condition
$'$	φ', y'	dummy integration variable
$'$	F', I'	per unit length
$''$	F'', q''	per unit area
$'$, $''$	f', f''	derivatives of functions
$'$	ε'	reactive component
$''$	ε''	dissipative component
$-$	$\overline{u}$	spatially averaged
$\sim$	$\underset{\sim}{Z}$	analytic representation of real parameters
$\vec{}$	$\vec{\boldsymbol{u}}$, $\vec{\boldsymbol{T}}$	vector or pseudovector
$\vec{\vec{}}$	$\vec{\vec{\boldsymbol{\tau}}}$, $\vec{\vec{\boldsymbol{\varepsilon}}}$	rank 2 tensor
$\hat{}$	$\hat{\boldsymbol{x}}$, $\hat{\boldsymbol{\vartheta}}$	unit vector
$\hat{}$	$\hat{e}_1$	molar value
$*$	d^*, p^*	nondimensionalized quantity
$\langle\rangle$	$\langle\ell_{\mathrm{e}}\rangle$, $\langle r_{\mathrm{g}}\rangle$	time- or ensemble-averaged property
Δ	Δp, Δx	difference in property

Introduction

Micro- and nanofabricated devices have led to revolutionary changes in our ability to manipulate tiny volumes of fluid or micro- and nanoparticles contained therein. This has led to countless applications for chemical and particulate separation and analysis, biological characterization, sensors, cell capture and counting, micropumps and actuators, high-throughput design and parallelization, and system integration, to name a few areas. Because biological and chemical analysis is typically concerned with molecules and bioparticles with small dimensions (some examples are shown in Fig. 0.1), the tools used to manipulate these objects are naturally of a similar scale, and the developments in micro- and nanofabrication in recent decades has brought engineering tools to a scale that easily matches these objects.

From a fluid-mechanical standpoint, our ability to manufacture micro- and nanoscale devices creates a number of challenges and provides matching opportunities, some of which are denoted schematically in Fig. 0.2. If we focus on liquid-phase devices, which have dominated most bioanalytical applications, shrinking the length scales makes interfacial phenomena and electrokinetic phenomena much more important, and reduces the importance of gravity and pressure. The no-slip boundary condition, safely assumed for macroscopic flows, can be inaccurate when the length scale is small. Although the low-Reynolds-number characteristic of most of these flows eliminates the challenges of nonlinearity in the convective term and the associated difficulty in modeling turbulent flows, we are instead forced to consider the nonlinearity of the source term in the Poisson–Boltzmann equation, nonlinearity of the coupling of electrodynamics with fluid flow, and uncertainty in predicting electroosmotic boundary conditions. Often, the microfluidics researcher worries not about *how* to solve the relevant governing equations and boundary conditions, but rather *what* those equations and boundary conditions are and how his or her analytical goals can be reconciled with fabrication

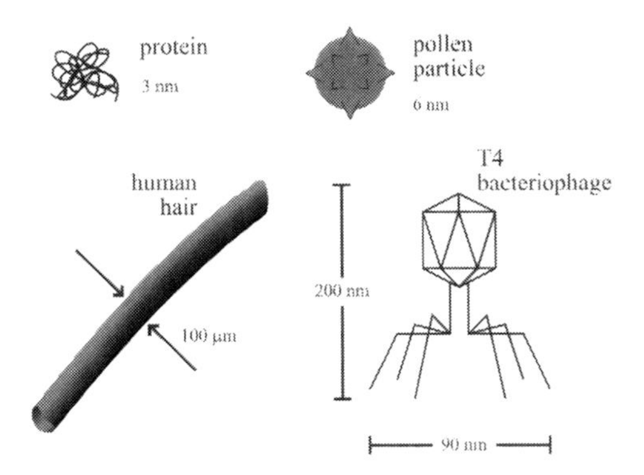

Fig. 0.1 Length scales of some biological objects ranging clockwise from top left from nano- to microscale.

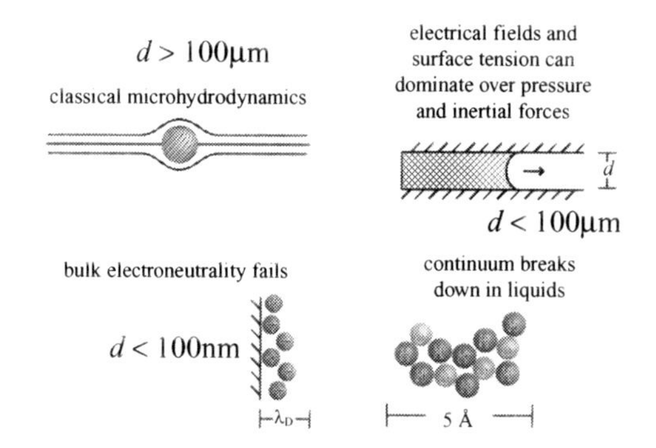

Fig. 0.2 Some fundamental changes in fluid physics and the length scales at which this occurs.

and experimentation concerns. Microscale problems are approached by use of continuum modeling and analysis, but the continuum governing equations must be modified to account for forces that are routinely ignored in macroscale systems. Figure 0.3, for example, highlights the general structure of a microfluidic device and some of the parametric inputs and fluid issues attendant with their use. Nanoscale problems are approached with a mix of continuum and atomistic approaches, depending on the the problem.

General properties of micro- and nanoscale flows. This text considers flow in micro- and nanofabricated devices, typically fabricated by photolithographic patterning combined with etching or molding processes evolved from microcircuit fabrication processes. The nature of the geometries, length scales, and materials used in these processes leads to a specialized set of physical phenomena and flow regimes, which have their own interesting properties and applications. Microscale flows are typically laminar owing to the short length scales, but can have large mass transfer Peclet numbers owing to the low diffusivity of macromolecules and particles of interest. These flows can be driven with

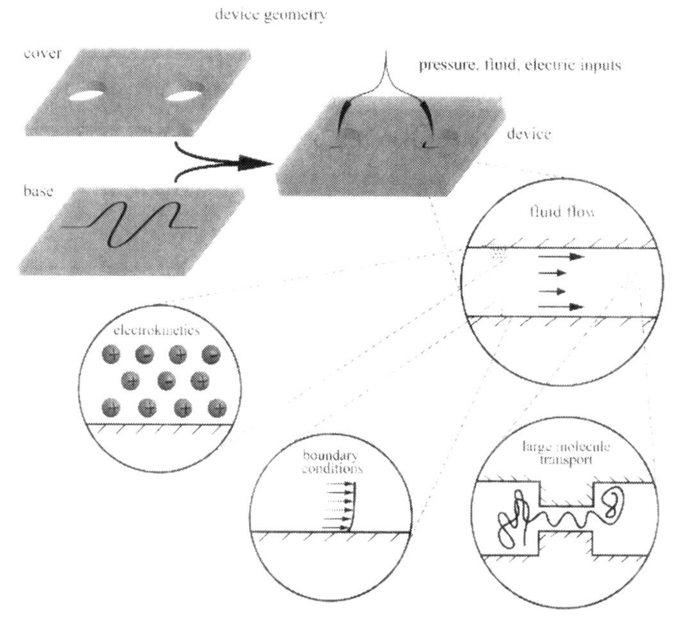

Fig. 0.3 A microfluidic device, its inputs, and some aspects of the fluid and analyte flow therein.

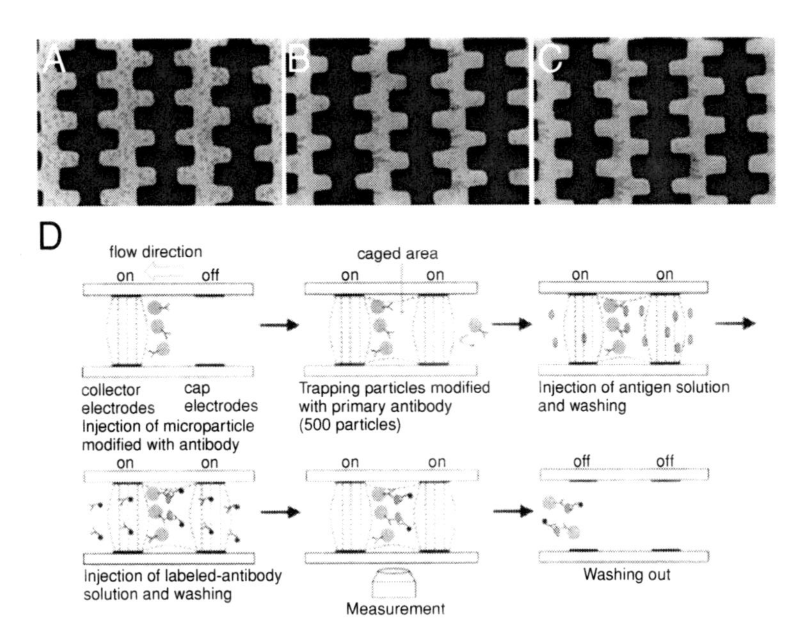

Fig. 0.4 (a) Patterned interdigitated electrodes used [1] to isolate leukemia cells from diluted blood. (b) A dilute blood suspension flows through a channel, and dielectrophoresis (Chapter 17) is used to capture both leukemia cells and erythrocytes (red blood cells) while (c) subsequently the frequency applied to the electrodes was decreased to retain the leukemia cells and elute the erythrocytes. (d) Microsystems are also used as [2] "microbeakers" whereby particles are caged from a flow stream by use of dielectrophoresis while chemistry and analysis are conducted by flowing various solutions over the captured particles on a microscope stage. Used with permission.

pressure, but applied electric fields are often more convenient or elegant to actuate these systems. Even if not applied, intrinsic electric fields exist at interfaces in all cases, driven usually by chemical reaction. Thus electrodynamics, chemistry, and fluid mechanics are inextricably intertwined, so that electric fields can create fluid flow and fluid flow can create electric fields, with a degree of coupling driven by the surface chemistry. The flow coupling is described by electrostatic source terms in the Navier–Stokes equations or particle transport equations. Many useful tools arise from these forces, such as electrokinetic pumps and dielectrophoretic manipulation of cells.

Boundary conditions become much more of an issue in microsystems, owing to high surface area–volume ratios. Boundary conditions that are taken for granted at the macroscale (e.g., the no-slip condition) can often fail in these systems. Further, microscale fluid mechanics is often closely related to chemical issues at surfaces. Multiphase implementations, designed to optimize certain aspects of transport, lead to additional interfacial concerns. Here, our boundary conditions must vary based on electric fields or chemistry.

Our ability to engineer and model microdevices is often limited by fabrication issues and instrumentation. Typical geometries resulting from microfabrication influence our discussion of transport issues in these systems – for example, most micro- and nanodevices are quasi-2D, and thus many of our analytical techniques will be used with quasi-2D structures in mind. The small scale requires specific instrumentation and techniques that facilitate inquiry into microscale flows. Our microfluidic devices are often implemented to study bioanalytes such as proteins or DNA, and bioparticles such as cells (Fig. 0.4) or virions. Often, we need to work in non-Newtonian systems, which requires a modification of the constitutive relation used in the Navier–Stokes equations.

Outline of the material covered in this text. To a great extent, the material in this text is specified by the incompressibility condition for mass conservation, the Navier–Stokes equations for fluid flow and momentum conservation, the Nernst–Planck equation for species transport, and the Poisson equation for electrostatics, combined with boundary conditions for microchannel walls, inlets and outlets, particle/droplet/bubble interfaces, and electrodes. We combine these equations in various ways throughout, starting with low-Reynolds-number fluid mechanics described by incompressibility and Navier–Stokes, and then folding in the effects of electrodynamics first with equilibrium systems (defined by the Poisson equation and Boltzmann statistics) and later with nonequilibrium systems (defined by the Poisson equation and the Nernst–Planck equation). By their nature, the results of these analyses lend themselves naturally to the use of boundary-layer theory and matched asymptotic techniques (for the electrical double layer) and a coupling matrix formulation for electromechanical coupling (owing to the linearity of the Stokes equations). Throughout, additional results of condensed matter physics and chemistry are included as needed to describe unique aspects of these flows.

The text begins with low-Reynolds-number fluid mechanics, to ensure that the treatment stands on its own and to put these classical topics in a micro- or nanoscale context. For those cases in which the Reynolds number is low, boundary conditions are classical, solutes are small, and no electric fields are applied, classical undergraduate fluid-mechanical tools apply. Chapters 1 and 2 represent primarily classical material, with the notable exception being the discussion of Navier slip models in Chapter 1 – although this model is itself classical, experimental measurements of slip lengths have mostly been performed in the last twenty years. Chapter 3 on hydraulic circuits has focused relevance owing to the prevalence of long narrow channels in microdevices and the utility of circuit analysis in designing massively parallel microfluidic circuits. This chapter also leads into Chapter 4, which combines standard undergraduate mass transfer with the generally graduate level topic of the kinematics of mixing and chaos. It also highlights the particular importance of the low-*Re*, high-*Pe* limit found in many microfluidic devices, including both its detrimental effect on mixing and its benefits to laminar-flow-patterning devices.

The text then addresses the effect of electric fields on flow far from walls, with particular attention given to electroosmosis. Chapter 5 includes an elementary treatment of electrostatics and electrodynamics, which allows Chapter 6 to present an integral analysis of electroosmosis that highlights flow-current similitude outside the electrical double layer. Because purely electroosmotic flows with thin double layers are potential flows, Chapter 7 provides a discussion of potential flow. Because Chapters 3, 5, and 7 all use complex numbers, Appendix G provides a reference for key concepts. Chapter 8 presents Stokes flow relations, with specific attention to the motion of small particles. Although the flows in Chapter 8 are not primarily driven by electric fields, the material in this chapter builds on analytical techniques from Chapter 7 and is thus positioned immediately after it. Readers at this point are directed to Appendix F, which introduces the multipolar theory for the Laplace and Stokes equations, and prepares for later subjects, in particular multipolar models of dielectrophoretic forces. Electrokinetic pumps (Section 6.5) are discussed as an early application, as much of the work on electrokinetic pumps requires only the material in these introductory chapters.

The attention then turns to the boundary layer close to micro- and nanodevice interfaces. With background information on electrolyte solution properties from Appendix B, Chapter 9 introduces the details of the Gouy–Chapman electrical double layer, as well as modified Poisson–Boltzmann equations. Appendix H provides background on interaction potentials and facilitates expansion on these ideas. Chapter 10 summarizes experimental observations of the surface potential that is the boundary condition that drives the EDL models. With this description, the text proceeds to nonequilibrium description of species and charge distributions (Chapter 11) with applications to microchip separations, particularly protein separations (Chapter 12). These skills enable discussion of electrophoresis of small particles (Chapter 13), which is the first case in which the double-layer thickness plays a primary role and the fluid mechanics and the ion distribution in the double layer exhibit two-way coupling. Macromolecule transport follows, building on the descriptions of small ion and microparticle transport – we choose to discuss this by using DNA (Chapter 14). This segues into nanofluidics, in particular a discussion of electrokinetic effects with full two-way coupling (Chapter 15).

Finally, we explore solutions for which interfacial charge is no longer in equilibrium – here, the dynamics of charge caused by interfacial potential or interfacial discontinuities in current are critical. We discuss, in turn, the dynamics of electrical double layers at electrodes and polarizable materials (Chapter 16) and nonlinear electrokinetic manipulation of particles or droplets by using dielectrophoresis, magnetophoresis, and electrowetting (Chapter 17).

Supplementary reading. Throughout the text, supplementary reading is provided in each chapter to expand on the material of that chapter. By necessity, some topics have been omitted, and the reader is pointed here to excellent source material for a few of these topics. This text omits gas-phase microfluidic flows, for which Karniadakis [3] is a thorough source. We also omit microfabrication; general microfabrication details can be found in [4, 5], and treatments with a focus on microfluidics can be found in [6, 7]. This text is primarily focused on analytical techniques and avoids numerical simulation approaches. Some useful sources for numerical techniques include [8, 9, 10, 11, 12, 13, 14, 15]; Refs. [12, 13, 14] have stressed accessibility and are recommended for those seeking an introduction to numerical work.

1 Kinematics, Conservation Equations, and Boundary Conditions for Incompressible Flow

This text describes liquid flow in microsystems, primarily flow of water and aqueous solutions. To this end, this chapter describes basic relations suitable for describing the flow of water. For flows in microfluidic devices, liquids are well approximated as *incompressible*, i.e., having approximately uniform density, so this text describes incompressible flow exclusively.

This chapter describes the kinematics of flow fields, which describes the *motion* and *deformation* of fluids. As part of this process, key concepts are introduced, such as streamlines, pathlines, streaklines, the stream function, vorticity, circulation, and strain rate and rotation rate tensors. These concepts provide the language used throughout the text to communicate the modes of fluid motion and deformation. We discuss conservation of mass and momentum for incompressible flows of Newtonian fluids. Finally, we discuss boundary conditions for the governing equations, including solid and free interfaces with surface tension, and in particular we give attention to the no-slip condition and its applicability in micro- and nanoscale devices. This chapter assumes familiarity with vector calculus, which is reviewed in Appendix C. Importantly, Appendix C also covers the notation and coordinate systems used throughout.

We define a fluid as a material that deforms continuously when experiencing a nonuniform stress of any magnitude. We are primarily interested in a *continuum* description of the fluid flow, meaning that we are interested in the macroscopic manifestation of the motions of the individual molecules that make up the fluid, i.e., the velocity and the pressure of the fluid as a function of time and space. We also consider continuum field properties such as the temperature of the fluid or concentrations of chemical species in solution.

1.1 FLUID STATICS

In the case in which the fluid is assumed motionless, the equilibrium of fluid is determined by the interplay between the fluid pressure and the body forces:

$$\nabla p = \sum_i \vec{f}_i , \qquad (1.1)$$

where p is the pressure [Pa] and $\vec{f}_i$ is a body force per unit volume [N/m^3]. Thus, in the presence of body forces, the fluid-static equations predict that pressure varies spatially. The most common fluid-static result relates to the pressure in a static column of liquid. For liquid in a gravitational field, $\vec{f}_i = -\rho g \hat{z}$, where $g = 9.8$ m/s^2. Integrating this in the z direction, we obtain

$$p - p_0 = -\rho g z , \qquad (1.2)$$

where p_0 is the pressure at $z = 0$. Similar relations can be determined for fluid in other potential fields, for example, charged fluids in an electric field.

1.2 KINEMATICS OF A FLUID VELOCITY FIELD

If we consider fluids *in motion*, we benefit from *kinematic relations*, which describe motion and deformation of a fluid. Because kinematics describes fluid motion but not the forces that generate that motion, kinematic relations are properties of the continuum velocity field alone. This velocity field gives the velocity at any point in space and time and is denoted by $\vec{u}(\vec{r}, t)$, where t is time and $\vec{r}$ is a position vector specifying a location in space. Kinematics provides language that helps us understand velocity fields as well as the mathematical relationships that frame the physics of the system. Kinematic relations and definitions that classify types of flows often provide insight into which governing equation should be used.

1.2.1 Important geometric definitions

This subsection defines a number of curves that relate to the velocity field, namely *pathlines*, *streaklines*, *streamlines*, and *material lines*, each of which can facilitate our analysis. Streamlines are the most common tool used analytically to understand flows, because streamlines are analytically simple to generate and provide a clear image of the instantaneous velocity in a system. Pathlines and streaklines, in contrast, are straightforward to reproduce in the laboratory and are thus the most common experimental tools used for visualization. In particular, compelling simplifications for two-dimensional (2D) flows are achieved with the *stream function*, which is related to the streamlines in the system. Two-dimensional flows with plane symmetry are often relevant in microfluidic devices, because the devices created with lithography and etching often have a uniform depth. These lines describe the motion of a *fluid particle*, i.e., a point that moves in a fluid flow with a velocity equal to the local fluid velocity.

PATHLINES

Pathlines are the loci of points traced out by the motion through the flow of a fluid particle that was at location $\vec{r}_0$ at time t_0 (see Fig. 1.1). We can envision a pathline by imagining inserting a small fluorescent particle into a point in a fluid flow and then taking a long exposure of the particle as a function of time. Pathlines experimentally are a temporal record of the path of a point marker. The starting point and time of the fluid particle as well as the time history of the velocity field influence the resulting pathlines.

STREAKLINES

Streaklines are the loci of fluid particles that have passed through a point $\vec{r}_0$. We can envision a streakline by imagining inserting a small tube into a point in a fluid flow, releasing fluorescent dye from this tube starting at t_0, and then taking a snapshot of the dye at a later time. In this sense, streaklines experimentally are an instantaneous record of a curvilinear marker. As was the case with pathlines, the starting point and time of the dye release as well as the time history of the velocity field influence the resulting pathlines. Figure 1.2 contrasts pathlines and streaklines.

STREAMLINES

Streamlines are lines that are everywhere tangent to the instantaneous velocity. Unlike pathlines and streaklines, streamlines are properties of the *instantaneous* velocity field.

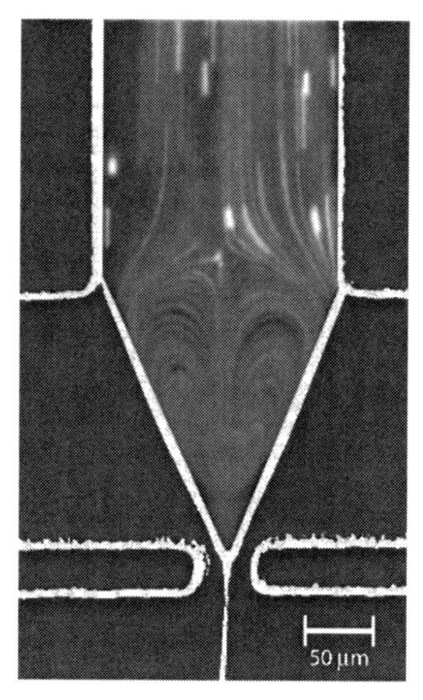

Fig. 1.1 Microparticle pathlines for a multiphase oil–water flow. Water inflow is from top, oil inflow from left and right. The resulting tipstreaming flow is used to generate micron-sized oil droplets. Fluorescent particles in the aqueous phase form pathlines on the camera, illustrating the recirculation induced by the fast oil flow on the slower water flow (Reproduced with permission from [16]).

For steady flow, all particles that pass through $\vec{r}_0$ follow the same trajectory. This trajectory is always tangent to the local velocity, and thus streamlines are identical to pathlines and streaklines for steady flows.[1]

Importantly, for 2D flows, we can define a scalar function that defines these streamlines without integration. We define the stream function (which we denote as ψ for plane-symmetric flows and ψ_S for axially symmetric flows) such that its isocontours are always tangent to the velocity.[2] The stream function used depends on the nature of the symmetry in the flow, whereas the mathematical form of the velocity–stream function relationship is dependent on the coordinate system. In a plane 2D flow defined with Cartesian variables, we define the stream function by using

$$u = \frac{\partial \psi}{\partial y}, \tag{1.3}$$

$$v = -\frac{\partial \psi}{\partial x}. \tag{1.4}$$

The same plane 2D flow and stream function just described can be written in terms of cylindrical coordinates as follows:

$$u_r = \frac{1}{r}\frac{\partial \psi}{\partial \theta} \tag{1.5}$$

and

$$u_\theta = -\frac{\partial \psi}{\partial r}. \tag{1.6}$$

[1] For all of these curves, the integration terminates at stagnation points (points of zero velocity). For flows with stagnation points, generating complete streamlines sometimes requires that integration commence from multiple starting points.

[2] In fact, the preceding velocity–stream function relation satisfies two requirements: (1) that isocontours of the stream function are streamlines, and (2) that any velocity field specified by the stream function also satisfies the conservation of mass equation (1.21) discussed in Section 1.3. Using ψ or ψ_S to satisfy conservation of mass can greatly simplify fluid problems.

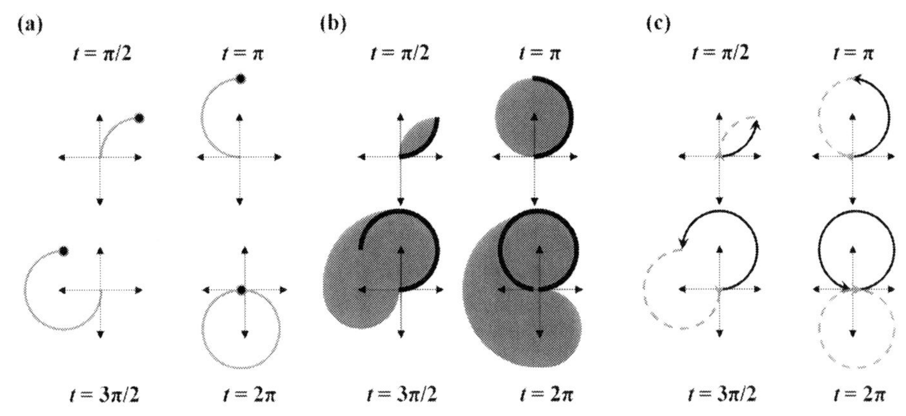

Fig. 1.2 Streaklines and pathlines. A particle released from the origin at time $t = 0$ is denoted in black, and a stream of dye released continuously from the origin starting at $t = 0$ is denoted in gray. The flow is the time-varying uniform flow $u = \cos t$, $v = \sin t$, which causes fluid particles to follow circular orbits. (a): location of the particle and dye stream at four instants (as would be visualized experimentally by a short-exposure image). The instantaneous location of the dye stream is the streakline associated with the origin and $t > 0$. (b): locus of particle and dye stream locations for $t > 0$ (as would be visualized experimentally by a long exposure starting at time $t = 0$). The time history of the particle traces out a pathline associated with the origin and $t = 0$. (c): pathlines (solid black lines) and streaklines (dashed gray lines) are shown by extracting the instantaneous dye contour from (a) and the particle time history from (b). Streamlines at any instant (not shown) are all straight lines aligned at an angle $\theta = t$ with respect to the x axis.

For axisymmetric flows, we define the *Stokes stream function* ψ_S, and the velocity–stream function relationships are written in cylindrical coordinates as

$$u_\imath = \frac{1}{\imath}\frac{\partial \psi_S}{\partial z}, \tag{1.7}$$

$$u_z = -\frac{1}{\imath}\frac{\partial \psi_S}{\partial \imath}, \tag{1.8}$$

whereas in spherical coordinates, these relationships are given by

$$u_r = \frac{1}{r^2 \sin\vartheta}\frac{\partial \psi_S}{\partial \vartheta}, \tag{1.9}$$

$$u_\vartheta = -\frac{1}{r \sin\vartheta}\frac{\partial \psi_S}{\partial r}. \tag{1.10}$$

The stream function for plane flows has units of square meters per second and is different from the Stokes stream function for systems with axial symmetry, which has units of cubic meters per second. The volumetric flow rate between two streamlines is related to the difference between the stream functions of the two streamlines. For plane-symmetric flow, the difference in stream function between two streamlines is equal to the volumetric flow per unit depth; for axisymmetric flow, the difference in the Stokes stream function between two streamlines is equal to the volumetric flow per radian.

MATERIAL LINES

Material lines trace the location of a curve in a flow field at specific instants in time. We can envision them by considering small fluorescent lines embedded in a fluid flow. Given a curve C_0 defined at a time t_0, the material line C as a function of time is simply the curve through the fluid particles that C_0 originally comprised.

1.2.2 Strain rate and rotation rate tensors

For our purposes, a fluid is a material that responds to forces by deforming at a measurable rate, and a fluid's unloaded state is defined by motionlessness but not by a specific configuration. This is in contrast to a solid, which responds to an applied force by acquiring a finite deformation from its unloaded state. For fluids the stresses (forces per unit area) in the material are related to the *rate of strain*. The rate of strain or *strain rate* of a fluid flow at a point is a measure of the velocity gradients at that point or, equivalently, the rate at which fluid elements are being deformed by a flow. The response of fluids is inherently *viscous*. For solids, the stresses (forces per unit area) in the material are related to the *strain*. The strain of a solid material is a measure of its static deformation from its unloaded state. In this sense, the response of solids is inherently *elastic*.

This subsection shows that the velocity gradients (expressed through the velocity gradient tensor) can be rewritten in terms of a *strain rate tensor* and a *rotation rate tensor*. Section 1.4 shows that the viscous forces or stresses in a Newtonian fluid (such as water or air) are linearly proportional to the strain rates as expressed by the strain rate tensor. The magnitude of the vorticity or, equivalently, the magnitude of the rotation rate tensor dictates what analytical tools we use to treat a specific flow problem.

STRAIN RATE FOR UNIDIRECTIONAL FLOWS

Consider a unidirectional flow $u = u(y)$ moving in the x direction. In this simple case, the scalar strain rate magnitude $\dot{\gamma}\,[\mathrm{s}^{-1}]$ is given by

$$\dot{\gamma} = \frac{1}{2}\frac{\partial u}{\partial y} . \tag{1.11}$$

This unidirectional flow is simple, and this result is not general; however, it does illustrate two basic ideas – that the strain rate is a measure of how rapidly the fluid elements are deformed and that this property is related to the local velocity gradients. If the velocity is uniform, fluid elements are not distorted. If u varies spatially, then a fluid element is sheared, extended, or both.

GENERAL STRAIN RATE FOR THREE-DIMENSIONAL FLOWS

Equation (1.11) is simple and gives a scalar that measures the strain rate of a unidirectional flow; however, its result is not general. For a general three-dimensional (3D) flow, a scalar is not enough information to describe the deformation of fluid flow. The strain rate tensor $\vec{\vec{\varepsilon}}\ [\mathrm{s}^{-1}]$ is a convenient way to record the detailed structure of the instantaneous fluid deformation. It also classifies two ways that a flow deforms: by extension and by shear.

In Cartesian coordinates, the strain rate tensor is defined as

strain rate tensor, Cartesian coordinates

$$\vec{\vec{\varepsilon}} = \begin{bmatrix} \varepsilon_{xx} & \varepsilon_{xy} & \varepsilon_{xz} \\ \varepsilon_{yx} & \varepsilon_{yy} & \varepsilon_{yz} \\ \varepsilon_{zx} & \varepsilon_{zy} & \varepsilon_{zz} \end{bmatrix} = \begin{bmatrix} \frac{\partial u}{\partial x} & \frac{1}{2}\left(\frac{\partial u}{\partial y} + \frac{\partial v}{\partial x}\right) & \frac{1}{2}\left(\frac{\partial u}{\partial z} + \frac{\partial w}{\partial x}\right) \\ \frac{1}{2}\left(\frac{\partial u}{\partial y} + \frac{\partial v}{\partial x}\right) & \frac{\partial v}{\partial y} & \frac{1}{2}\left(\frac{\partial v}{\partial z} + \frac{\partial w}{\partial y}\right) \\ \frac{1}{2}\left(\frac{\partial u}{\partial z} + \frac{\partial w}{\partial x}\right) & \frac{1}{2}\left(\frac{\partial v}{\partial z} + \frac{\partial w}{\partial y}\right) & \frac{\partial w}{\partial z} \end{bmatrix} . \tag{1.12}$$

We determine both the strain rate tensor and the rotation rate tensor by splitting $\nabla\vec{u}$ into its symmetric and antisymmetric parts.[3] To do this, we start with $\nabla\vec{u}$:

$$\nabla\vec{u} = \begin{bmatrix} \frac{\partial u}{\partial x} & \frac{\partial u}{\partial y} & \frac{\partial u}{\partial z} \\ \frac{\partial v}{\partial x} & \frac{\partial v}{\partial y} & \frac{\partial v}{\partial z} \\ \frac{\partial w}{\partial x} & \frac{\partial w}{\partial y} & \frac{\partial w}{\partial z} \end{bmatrix}. \tag{1.13}$$

We find the symmetric part by averaging the velocity gradient tensor with its transpose[4] to get $\vec{\vec{\varepsilon}} = \frac{1}{2}(\nabla\vec{u} + \nabla\vec{u}^{\mathrm{T}})$, and we find the antisymmetric part by averaging the velocity gradient tensor with its antitranspose to get $\vec{\vec{\omega}} = \frac{1}{2}(\nabla\vec{u} - \nabla\vec{u}^{\mathrm{T}})$:

$$\nabla\vec{u} = \tfrac{1}{2}\nabla\vec{u} + \tfrac{1}{2}\nabla\vec{u},$$

$$\nabla\vec{u} = \tfrac{1}{2}\nabla\vec{u} + \tfrac{1}{2}\nabla\vec{u}^{\mathrm{T}} + \tfrac{1}{2}\nabla\vec{u} - \tfrac{1}{2}\nabla\vec{u}^{\mathrm{T}},$$

$$\nabla\vec{u} = \begin{bmatrix} \frac{\partial u}{\partial x} & \frac{1}{2}\left(\frac{\partial u}{\partial y} + \frac{\partial v}{\partial x}\right) & \frac{1}{2}\left(\frac{\partial u}{\partial z} + \frac{\partial w}{\partial x}\right) \\ \frac{1}{2}\left(\frac{\partial v}{\partial x} + \frac{\partial u}{\partial y}\right) & \frac{\partial v}{\partial y} & \frac{1}{2}\left(\frac{\partial v}{\partial z} + \frac{\partial w}{\partial y}\right) \\ \frac{1}{2}\left(\frac{\partial w}{\partial x} + \frac{\partial u}{\partial z}\right) & \frac{1}{2}\left(\frac{\partial w}{\partial y} + \frac{\partial v}{\partial z}\right) & \frac{\partial w}{\partial z} \end{bmatrix} + \begin{bmatrix} 0 & \frac{1}{2}\left(\frac{\partial u}{\partial y} - \frac{\partial v}{\partial x}\right) & \frac{1}{2}\left(\frac{\partial u}{\partial z} - \frac{\partial w}{\partial x}\right) \\ \frac{1}{2}\left(\frac{\partial v}{\partial x} - \frac{\partial u}{\partial y}\right) & 0 & \frac{1}{2}\left(\frac{\partial v}{\partial z} - \frac{\partial w}{\partial y}\right) \\ \frac{1}{2}\left(\frac{\partial w}{\partial x} - \frac{\partial u}{\partial z}\right) & \frac{1}{2}\left(\frac{\partial w}{\partial y} - \frac{\partial v}{\partial z}\right) & 0 \end{bmatrix},$$

$$\nabla\vec{u} = \vec{\vec{\varepsilon}} + \vec{\vec{\omega}}. \tag{1.16}$$

We call the symmetric part the strain rate tensor $\vec{\vec{\varepsilon}}$, and we call the antisymmetric part the rotation rate tensor[5] $\vec{\vec{\omega}}$.

EXTENSIONAL STRAIN RATE

Extensional strain rate is a measure of how much the flow *stretches or squeezes* fluid elements along the coordinate axes (see top diagram in Fig. 1.3). Extensional strain rate is characterized by terms such as $\frac{\partial u}{\partial x}$, $\frac{\partial v}{\partial y}$, and $\frac{\partial w}{\partial z}$. Stretching is indicated when these terms are positive, and squeezing is indicated when these terms are negative.

---

[3] The elements of a symmetric tensor are the same if the row number and the column number of the element are switched. The elements of an antisymmetric tensor are equal in magnitude but have opposite signs if the row number and the column number of the element are switched.

[4] The transpose of this tensor is the same as the transpose of a matrix – i.e., we obtain it by switching rows and columns. So if $\nabla\vec{u}$ is given by

$$\nabla\vec{u} = \begin{bmatrix} \frac{\partial u}{\partial x} & \frac{\partial u}{\partial y} & \frac{\partial u}{\partial z} \\ \frac{\partial v}{\partial x} & \frac{\partial v}{\partial y} & \frac{\partial v}{\partial z} \\ \frac{\partial w}{\partial x} & \frac{\partial w}{\partial y} & \frac{\partial w}{\partial z} \end{bmatrix}, \tag{1.14}$$

then its *transpose* $\nabla\vec{u}^{\mathrm{T}}$ is given by

$$\nabla\vec{u}^{\mathrm{T}} = \begin{bmatrix} \frac{\partial u}{\partial x} & \frac{\partial v}{\partial x} & \frac{\partial w}{\partial x} \\ \frac{\partial u}{\partial y} & \frac{\partial v}{\partial y} & \frac{\partial w}{\partial y} \\ \frac{\partial u}{\partial z} & \frac{\partial v}{\partial z} & \frac{\partial w}{\partial z} \end{bmatrix}. \tag{1.15}$$

The *antitranspose* is just the transpose multiplied by minus 1.

[5] Some authors define the rotation rate tensor as *twice* the antisymmetric component of the velocity gradient tensor.

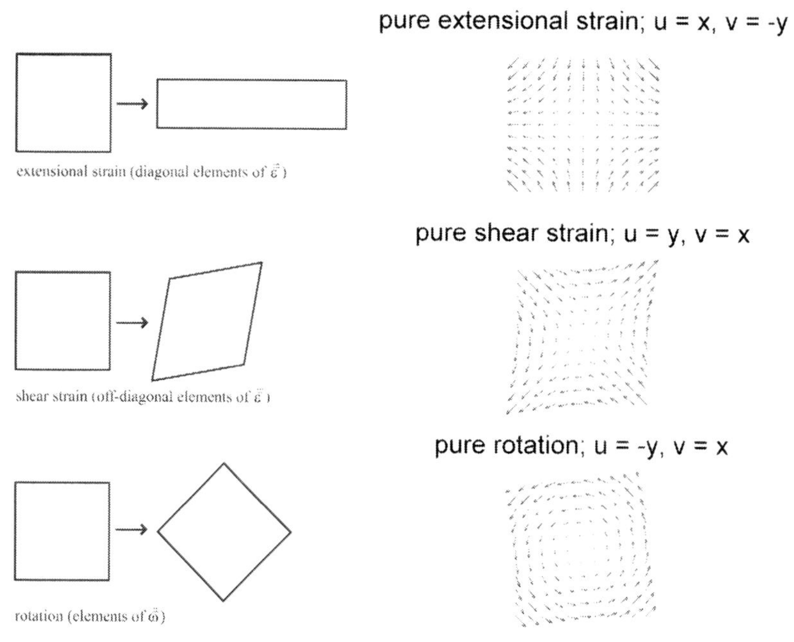

Fig. 1.3 Three examples of how components of a velocity field deform fluid elements. Fluid elements are shown as squares at left; the deformed shape shows the shape the fluid element would take after deformation by fluid motion. At right, the velocity field that would generate this deformation is shown in terms of velocity vectors. Top: The diagonal elements of $\vec{\vec{\varepsilon}}$ show extensional strain (stretching and squeezing). After application of pure extensional strain, a rectangle remains a rectangle. Here the extensional strain is positive in the x direction and negative in the y direction. Middle: The off-diagonal elements of $\vec{\vec{\varepsilon}}$ show shear strain (skewing). After application of shear strain, a rectangle becomes a parallelogram. Bottom: The elements of $\vec{\omega}$ show solid-body rotation. The square remains a square, but its orientation with respect to the coordinate axes changes.

These terms are the diagonal components of the strain rate tensor $\vec{\vec{\varepsilon}}$. The sum of the diagonal elements of the strain rate tensor is zero for an incompressible flow, as shown in Section 1.3, owing to conservation of mass. An example of a flow that leads to extensional strain rates is the stagnation flow in Exercise 1.18(a). That flow squeezes the fluid in the y direction ($\frac{\partial v}{\partial y} < 0$) and stretches the fluid in the x direction ($\frac{\partial u}{\partial x} > 0$), as can be seen in the components of $\vec{\vec{\varepsilon}}$.

SHEAR STRAIN RATE

Shear strain rate is a measure of how much the flow skews fluid elements (see the middle diagram in Fig. 1.3). Shear strain is characterized by terms such as $\frac{1}{2}(\frac{\partial u}{\partial y} + \frac{\partial v}{\partial x})$. An example of a flow with shear strains is the simple shear flow in Exercise 1.18(b).

The use of the modifiers *extensional* and *shear* to modify strain rate expresses the relation of the strain to the chosen coordinate system. For an anisotropic material, the relevant coordinate system is related to the principal axes of the material. Simple fluids such as water are isotropic, and the choice of coordinate axes for isotropic materials is arbitrary – by changing coordinate systems, we can convert an extensional strain to a shear strain and vice versa.

VORTICITY

The vorticity $\vec{\omega}$ is defined as $\vec{\omega} = \nabla \times \vec{u}$ and has units of radian per second. Its magnitude is equal to twice the rate of solid-body rotation of a point in a flow, and its direction defines the axis of this rotation. The vorticity pseudovector has three components – these components contain the minimum amount of information required for describing the magnitude and the direction of the solid-body rotation. The rotation rate tensor $\vec{\vec{\omega}}$ is closely related to the vorticity $\vec{\omega}$ – in fact, if we write the components of the vorticity vector as $\vec{\omega} = (\omega_1, \omega_2, \omega_3)$, then the rotation rate tensor can be written as

rotation rate tensor in terms of vorticity

$$\vec{\vec{\omega}} = \begin{pmatrix} 0 & -\frac{1}{2}\omega_3 & \frac{1}{2}\omega_2 \\ \frac{1}{2}\omega_3 & 0 & -\frac{1}{2}\omega_1 \\ -\frac{1}{2}\omega_2 & \frac{1}{2}\omega_1 & 0 \end{pmatrix} . \tag{1.17}$$

The rotation rate tensor is a true antisymmetric rank 2 tensor, whereas the vorticity is a pseudovector. A flow field is termed *irrotational* if the vorticity is zero everywhere in the flow. Often, the term *irrotational* is used also for flows whose vorticity is present exclusively in point singularities, such as flows with irrotational vortexes. Chapter 7 exhibits the considerable mathematical simplifications obtained when $\vec{\omega}$ is zero.

CIRCULATION

The circulation Γ around a closed contour is defined as

$$\Gamma = \int_C \vec{u} \cdot \hat{t}\, ds = \int_S \vec{\omega} \cdot \hat{n}\, dA \tag{1.18}$$

and defines the total net vorticity or total net solid-body rotation within an enclosed area. The integral $\int_S dA$ is the integral over a specified surface; this area is alternately specified by a closed, bounding path defined by $\int_C ds$, where ds is a differential distance along the contour, with the local contour direction defined by $\hat{t}$. The dot product $\vec{u} \cdot \hat{t}$ defines the velocity component tangent to ds. Circulation is important in irrotational flows (Chapter 7) – in particular, the fluid-mechanical equations in some cases allow for an infinite number of solutions corresponding to varying levels of circulation. For the electroosmotic flows considered starting in Chapter 6, we resolve this ambiguity by selecting the solution for which the circulation around enclosed objects is zero.

1.3 GOVERNING EQUATIONS FOR INCOMPRESSIBLE FLOW

A fluid flow is termed incompressible if the density gradients are small enough such that the density can be assumed uniform to a first approximation. Incompressibility leads to a significant simplification of the fluid flow equations, because the density can be assumed uniform and the transfer of energy from kinetic energy (velocity) to internal energy (temperature) can be ignored.[6] The governing equations for incompressible laminar fluid flow include continuity (conservation of mass) and the Navier–Stokes equations (conservation of momentum).

1.3.1 Conservation of mass: continuity equation

Conservation of mass is specified by the integral relation

$$\frac{\partial}{\partial t} \int_V \rho\, dV = -\int_S (\rho \vec{u}) \cdot \hat{n}\, dA, \tag{1.19}$$

[6] Technically, viscous stresses do lead to irreversible transfer from kinetic to internal energy. However, we generally ignore such effects in this text, as they are insignificant for more microscale flows.

where $\hat{\boldsymbol{n}}$ is a unit outward normal along the surface $\mathcal{S}$, t is time [s], and ρ is the fluid density [kg/m^3]. This relation states that the change in mass within a control volume denoted by $\mathcal{V}$ is given by the surface integral of the flux of mass crossing the surface of the volume. This conservation of mass equation is also called the *continuity equation*. For an incompressible fluid, we assume that ρ is a constant and thus the mass within the control volume does not change with time. This, combined with the divergence theorem, means that

$$\nabla \cdot (\rho \vec{\boldsymbol{u}}) = 0 \,, \tag{1.20}$$

and because ρ is uniform for incompressible flow, this simplifies to

conservation of mass, incompressible fluid

$$\nabla \cdot \vec{\boldsymbol{u}} = 0 \,. \tag{1.21}$$

EXAMPLE PROBLEM 1.1

Show that any velocity field in 2D Cartesian space defined by the stream function and the equations

$$u = \frac{\partial \psi}{\partial y} \tag{1.22}$$

and

$$v = -\frac{\partial \psi}{\partial x} \tag{1.23}$$

automatically solves the conservation of mass equation.

SOLUTION: Conservation of mass in two dimensions requires

$$\frac{\partial u}{\partial x} + \frac{\partial v}{\partial y} = 0 \,, \tag{1.24}$$

thus

$$\frac{\partial^2 \psi}{\partial y \partial x} - \frac{\partial^2 \psi}{\partial x \partial y} = 0 \,. \tag{1.25}$$

From equality of mixed partials, the left-hand side is identically zero, and thus this stream function guarantees conservation of mass.

1.3.2 Conservation of momentum: the Navier–Stokes equations

The conservation of momentum equation for a continuum is given by the integral relation

$$\frac{\partial}{\partial t} \int_{\mathcal{V}} \rho \vec{\boldsymbol{u}} \, dV = -\int_{\mathcal{S}} (\rho \vec{\boldsymbol{u}} \vec{\boldsymbol{u}}) \cdot \hat{\boldsymbol{n}} \, dA + \int_{\mathcal{S}} \vec{\vec{\boldsymbol{\tau}}} \cdot \hat{\boldsymbol{n}} \, dA + \int_{\mathcal{V}} \sum_i \vec{\boldsymbol{f}}_i \, dV \,, \tag{1.26}$$

where $\hat{\boldsymbol{n}}$ is again a unit outward normal, dA is a differential area upon the surface $\mathcal{S}$, $\vec{\vec{\boldsymbol{\tau}}}$ is the stress tensor, and $\vec{\boldsymbol{f}}_i$ indicates one of perhaps several body forces (which can include, for example, gravity or Coulomb forces). The $\vec{\boldsymbol{u}}\vec{\boldsymbol{u}}$ term is a dyadic tensor; see Subsection C.2.6 in Appendix C.

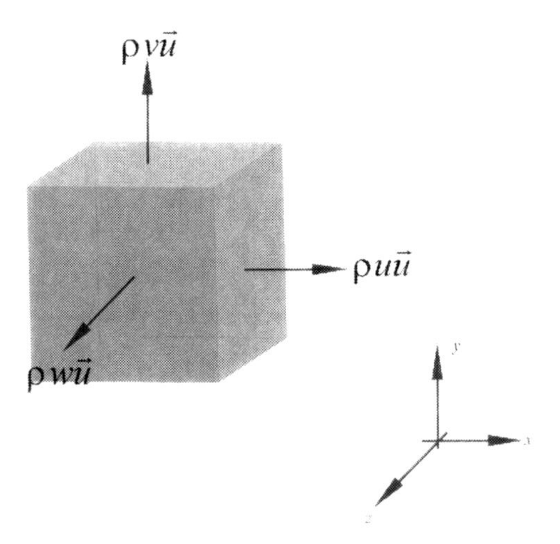

Fig. 1.4 Convective momentum fluxes for a differential Cartesian control volume.

For an incompressible fluid, we assume that the density ρ is uniform. Applying the divergence theorem and using the conservation of mass relation, we can write the conservation of momentum equation in differential form as

Cauchy momentum equation

$$\rho \frac{\partial \vec{u}}{\partial t} + \rho \vec{u} \cdot \nabla \vec{u} = \nabla \cdot \vec{\vec{\tau}} + \sum_i \vec{f}_i \,. \tag{1.27}$$

In this general form, the conservation of momentum equation is referred to as the *Cauchy momentum equation*. The Cauchy momentum equation states that temporal changes in momentum ($\rho \frac{\partial \vec{u}}{\partial t}$) are a result of the net momentum convected out of the control volume by the fluid flow ($\rho \vec{u} \cdot \nabla \vec{u}$), the net volumetric force caused by stresses applied to the control surface ($\nabla \cdot \vec{\vec{\tau}}$), and the net body force per unit volume ($\vec{f}_i$). In microfluidic systems, the most common body force term is the Coulomb force on a fluid with net charge density in an electric field. Gravitational forces, which are commonly discussed in the literature regarding macroscopic fluid mechanics, are often negligible in microscale liquid flows.

The forces and momentum fluxes are often well illustrated by differential control volumes. A schematic of the convective fluxes for a differential Cartesian control volume is shown in Fig. 1.4. The velocity normal to each face (u, v, or w) carries the vector fluid momentum ($\rho \vec{u}$) through that boundary. These contributions lead to a net outflow of momentum equal to $\rho \vec{u} \cdot \nabla \vec{u}$, as shown in Exercise 1.10. The surface stresses have two components: *pressure forces*, which are always normal to the surfaces and are independent of the velocity field, and *viscous forces*, which in general have components normal to and tangent to the surface and *are* dependent on the velocity field. We can write the stress tensor $\vec{\vec{\tau}}$ as a sum of these two components:

stress tensor decomposition into pressure and viscous forces

$$\vec{\vec{\tau}} = \vec{\vec{\tau}}_{\text{pre}} + \vec{\vec{\tau}}_{\text{visc}} \,, \tag{1.28}$$

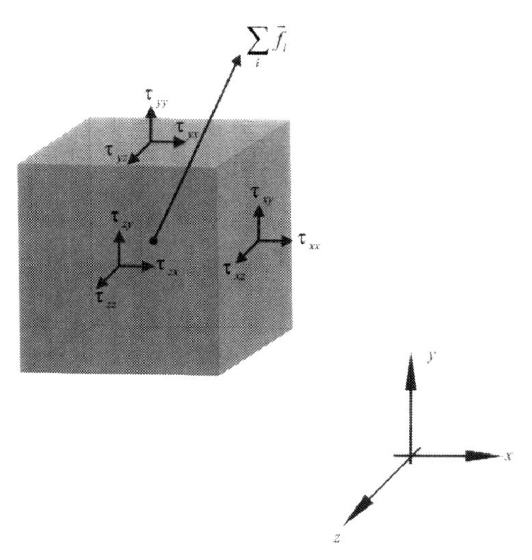

Fig. 1.5 Surface momentum fluxes and body force for a Cartesian control volume.

where the stresses from the pressure are captured in $\vec{\vec{\tau}}_{\text{pre}}$, written for Cartesian systems as

pressure stress tensor

$$\vec{\vec{\tau}}_{\text{pre}} = -p\vec{\vec{\delta}} = \begin{bmatrix} -p & 0 & 0 \\ 0 & -p & 0 \\ 0 & 0 & -p \end{bmatrix}, \tag{1.29}$$

and the identity tensor $\vec{\vec{\delta}}$ is defined for Cartesian systems as

$$\vec{\vec{\delta}} = \begin{bmatrix} 1 & 0 & 0 \\ 0 & 1 & 0 \\ 0 & 0 & 1 \end{bmatrix}. \tag{1.30}$$

The rest of the stresses are captured in $\vec{\vec{\tau}}_{\text{visc}}$. Because $\nabla \cdot (-p\vec{\vec{\delta}}) = -\nabla p$, we can write the Cauchy momentum equation as

Cauchy momentum equation

$$\rho \frac{\partial \vec{u}}{\partial t} + \rho \vec{u} \cdot \nabla \vec{u} = -\nabla p + \nabla \cdot \vec{\vec{\tau}}_{\text{visc}} + \sum_i \vec{f}_i. \tag{1.31}$$

To define $\vec{\vec{\tau}}_{\text{visc}}$ in terms of the velocities in the system, we need a *constitutive relation*. Section 1.4 discusses constitutive relations for fluids in detail, which describe $\vec{\vec{\tau}}_{\text{visc}}$ in terms of the velocity field and the viscosity of the fluid.

A schematic of the surface stresses and the body forces on a differential Cartesian control volume is shown in Fig. 1.5. Divergence of these surface stresses leads to a net force on the control volume equal to $\nabla \cdot \vec{\vec{\tau}}_{\text{visc}}$.

1.4 CONSTITUTIVE RELATIONS

Constitutive relations are the link between the microscopic states of matter (i.e., the details of how the molecules interact with each other) and the macroscopic states (e.g., velocity and pressure). Constitutive relations are different from *conservation equations* in that conservation equations, as used in this text, are strictly continuum descriptions, whereas constitutive relations are continuum descriptions of underlying atomistic properties. For fluid flow, the key constitutive relation needed is a relation between the velocities and the stress tensor. In this section, the viscous stresses are defined in terms of the strain rate tensor, and thus the momentum equation can then be written strictly in terms of velocities, pressures, and fluid parameters.

1.4.1 Relation between strain rate and stress

The relation between $\vec{\vec{\varepsilon}}$ and $\vec{\vec{\tau}}_{\text{visc}}$ *does not* stem from any of the conservation relations previously discussed. Rather, constitutive models come from a microscopic description of how molecular collisions lead to momentum transfer. Thus continuum fluid mechanics postulates the constitutive relations that cannot be derived.

NEWTONIAN RELATION FOR SHEAR STRESS IN A ONE-DIMENSIONAL FLOW

The fundamental postulate of the Newtonian model is that fluids have a property that we call *viscosity*, which relates the strain rate linearly to the stress. The Newtonian model assumes that the viscosity itself is independent of the strain rate or of any other velocity parameter. Fluids well described by this model are called *Newtonian fluids.* Air and water are Newtonian fluids, so the great majority of the flows of interest to engineers involve Newtonian fluids. In microfluidic systems, departures from the Newtonian approximation are observed when the fluid contains large polymer molecules or when a complex colloidal system (e.g., blood) is modeled as a simple fluid.

Consider a unidirectional flow $u = u(y)$ of a Newtonian fluid in the x direction and consider a rectangular control volume embedded in the flow field. In this simple case, the viscous stress τ_{xy} in the y direction on the x face of the control volume is given by

$$\tau_{xy} = 2\eta\dot{\gamma} = \eta\frac{\partial u}{\partial y}\,, \tag{1.32}$$

where $\dot{\gamma}$ is the local strain rate and η is the *viscosity* of the fluid [Pa s].[7] The viscosity is thus the fundamental link between velocity gradients and the surface stresses that result from that velocity gradient. The Newtonian model postulates that η is a property of the fluid, but not a property of the local strain rate.

GENERAL NEWTONIAN RELATION FOR SHEAR STRESS IN A 3D FLOW

For Newtonian fluids, the shear stress tensor and the viscous component of the strain rate tensor are linearly related:

Newtonian postulate for viscous stress tensor in terms of strain rate tensor

$$\vec{\vec{\tau}}_{\text{visc}} = 2\eta\vec{\vec{\varepsilon}}\,. \tag{1.33}$$

[7] Viscosity is often written with the symbol μ; however, we reserve μ for the viscous mobility and several other mobilities, such as the electroosmotic mobility μ_{EO} and the electrophoretic mobility μ_{EP}.

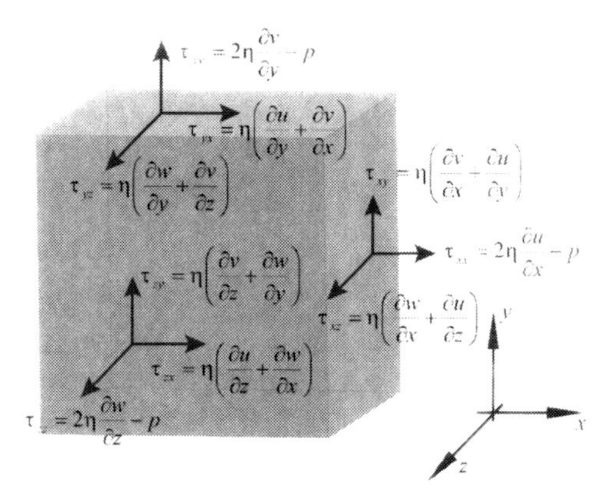

Fig. 1.6 Surface stresses for a Newtonian fluid for a Cartesian control volume.

This equation has a similar form to that of Eq. (1.32) but is now a tensor equation that gives a description of the nine components of surface stress. Noting that

$$\vec{\vec{\tau}}_{\text{pre}} = -p\vec{\vec{\delta}}\,, \tag{1.34}$$

we find that the complete surface stress tensor is given by

surface stress tensor for a Newtonian fluid

$$\vec{\vec{\tau}} = 2\eta\vec{\vec{\varepsilon}} - p\vec{\vec{\delta}}\,. \tag{1.35}$$

A schematic of the surface stresses for a Newtonian fluid is shown in Fig. 1.6. These relations dictate that *viscous stresses are proportional to strain rates*, and the constant of proportionality is 2η:

$$\vec{\vec{\tau}}_{\text{visc}} = \begin{bmatrix} \tau_{xx} & \tau_{xy} & \tau_{xz} \\ \tau_{yx} & \tau_{yy} & \tau_{yz} \\ \tau_{zx} & \tau_{zy} & \tau_{zz} \end{bmatrix} = \begin{bmatrix} 2\eta\frac{\partial u}{\partial x} & \eta\left(\frac{\partial u}{\partial y} + \frac{\partial v}{\partial x}\right) & \eta\left(\frac{\partial u}{\partial z} + \frac{\partial w}{\partial x}\right) \\ \eta\left(\frac{\partial v}{\partial x} + \frac{\partial u}{\partial y}\right) & 2\eta\frac{\partial v}{\partial y} & \eta\left(\frac{\partial v}{\partial z} + \frac{\partial w}{\partial y}\right) \\ \eta\left(\frac{\partial w}{\partial x} + \frac{\partial u}{\partial z}\right) & \eta\left(\frac{\partial w}{\partial y} + \frac{\partial v}{\partial z}\right) & 2\eta\frac{\partial w}{\partial z} \end{bmatrix}. \tag{1.36}$$

If a fluid is Newtonian, the surface stress term $\nabla \cdot \vec{\vec{\tau}}$ from the Navier–Stokes equation simplifies. The pressure component $\nabla \cdot -p\vec{\vec{\delta}}$ is equal to $-\nabla p$, and $\nabla \cdot 2\eta\vec{\vec{\varepsilon}}$ is equal to $\nabla \cdot \eta\nabla\vec{u}$. In this case, the Cauchy momentum equations become

Navier–Stokes equations, Newtonian fluid

$$\rho\frac{\partial \vec{u}}{\partial t} + \rho\vec{u}\cdot\nabla\vec{u} = -\nabla p + \nabla\cdot\eta\nabla\vec{u}\,. \tag{1.37}$$

Equation (1.37) lists the *Navier–Stokes equations* for a Newtonian fluid. If the viscosity can be assumed uniform, then the Navier–Stokes equations become

Navier–Stokes equations, Newtonian fluid, uniform fluid properties

$$\rho \frac{\partial \vec{u}}{\partial t} + \rho \vec{u} \cdot \nabla \vec{u} = -\nabla p + \eta \nabla^2 \vec{u} . \quad (1.38)$$

Equations (1.37) and (1.38) are *vector equations* – each term is a vector with three components, and we use the vector notation to collapse the three equations (for x, y, and z momentum) into one equation. We can write these equations out term by term, in which case we obtain three scalar equations. For Cartesian coordinate systems, we obtain

Navier–Stokes equations, Cartesian coordinates

$$\begin{aligned}
\rho \tfrac{\partial u}{\partial t} + \rho u \tfrac{\partial u}{\partial x} + \rho v \tfrac{\partial u}{\partial y} + \rho w \tfrac{\partial u}{\partial z} &= -\tfrac{\partial p}{\partial x} + \eta \tfrac{\partial^2 u}{\partial x^2} + \eta \tfrac{\partial^2 u}{\partial y^2} + \eta \tfrac{\partial^2 u}{\partial z^2} , \\
\rho \tfrac{\partial v}{\partial t} + \rho u \tfrac{\partial v}{\partial x} + \rho v \tfrac{\partial v}{\partial y} + \rho w \tfrac{\partial v}{\partial z} &= -\tfrac{\partial p}{\partial y} + \eta \tfrac{\partial^2 v}{\partial x^2} + \eta \tfrac{\partial^2 v}{\partial y^2} + \eta \tfrac{\partial^2 v}{\partial z^2} , \quad (1.39) \\
\rho \tfrac{\partial w}{\partial t} + \rho u \tfrac{\partial w}{\partial x} + \rho v \tfrac{\partial w}{\partial y} + \rho w \tfrac{\partial w}{\partial z} &= -\tfrac{\partial p}{\partial z} + \eta \tfrac{\partial^2 w}{\partial x^2} + \eta \tfrac{\partial^2 w}{\partial y^2} + \eta \tfrac{\partial^2 w}{\partial z^2} .
\end{aligned}$$

The Navier–Stokes equations are presented in several coordinate systems in Appendix D.

1.4.2 Non-Newtonian fluids

The Newtonian description listed is immensely successful in describing flows of air and water, among many examples. The linear relation between stress and strain rate links kinematic concepts such as the strain rate tensor directly to the stress tensor, which plays a role in conservation of momentum. However, many fluids exist for which the Newtonian formulation is inaccurate. For these *non-Newtonian fluids*, in which the stress-strain rate relation is not linear, other models must be developed.

Two basic forms of non-Newtonian fluids are *shear-thinning fluids* and *shear-thickening fluids*. Shear-thinning fluids have effective viscosities that decrease as the strain rate increases. This behavior is common in fluids made of long polymeric molecules, which align when they are sheared and slide along one another more easily at high strain rate. Some household examples include most salad dressings and sweet chili sauce – all of which typically contain cellulose gum or xanthan gum, which are large polysaccharide polymers. In microfluidic devices, we use long polymeric materials in separation media for DNA (Chapter 14) and protein (Chapter 12) separations. We also often work with colloidal systems (one example is blood) – the rheology of these systems is rarely well described by a Newtonian model. Shear-thickening fluids, which are much less common, have effective viscosities that increase as the strain rate increases. Silly Putty is one example.

Another non-Newtonian fluid type (which we do not discuss in any detail) is a *viscoelastic fluid* (e.g., egg whites), which has a shear stress that is a function of the time history of the strain rate. In this sense, viscoelastic fluids combine a viscous (fluid) response with an elastic (solid) response. Fluids of this type exhibit some of the properties we typically associate with solids, for example, the ability to spring back into a

Table 1.1 **Surface tensions for some liquids with air at 20°C.**

Interface	Surface tension, [mN/m]
Mercury–air	484
Water–air	73
Ethanol–air	23

preferred configuration. In microfluidic devices, viscoelasticity is again observed primarily in entangled polymer solutions. *Bingham plastics* are fluids that behave as solids at low stresses and fluids at high stresses. This is common for granular media, for example, toothpaste.

1.5 SURFACE TENSION

Surface tension is a critical parameter whose influence increases as length scales decrease. Surface tension affects our ability to fill microchannels and, if controlled, can actuate droplets in microsystems. Surface tension is closely related to boundary conditions for fluid flow, described in Section 1.6, and the material in this section prepares for that discussion.

1.5.1 Definition of surface tension and interfacial energy

We can define the surface tension γ as follows: Given a system that consists of two phases (1 and 2) separated by an interface, the surface tension γ_{12} can be defined as

$$\gamma_{12} = \left(\frac{\partial G}{\partial A}\right)_{\text{constant } T,p}, \tag{1.40}$$

where G is the Gibbs free energy of the entire system (phase 1 plus phase 2 plus interface) and A is the area of the interface. Thus the surface tension is a measure of the additional energy associated with the interface.

The extra Gibbs free energy at the interface exists because the molecules in this interfacial region are in a state different from either bulk state. The system as a whole has a higher free energy if some of the molecules are in this in-between state. The surface tensions for several interfaces are listed in Table 1.1.

Because interfaces between phases create added potential energy, the surface tension can be thought of as a force per unit length that tends to minimize the surface area of an interface – analogous to the tension in a membrane that has been stretched.

1.5.2 Young–Laplace equation

Equilibrium relations at an interface dictate the shape of the interface and the forces at the interface. These relations lead to the Young–Laplace equation. The Young–Laplace equation states that any curved surface at equilibrium separating phase 1 from phase 2 maintains a pressure drop across the surface:

Young–Laplace equation for pressure drop across a curved interface

$$p_1 - p_2 = \gamma_{12}\left(\frac{1}{R_1} + \frac{1}{R_2}\right), \tag{1.41}$$

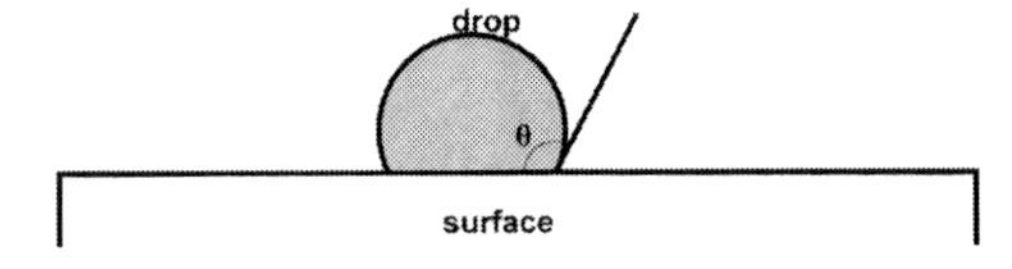

Fig. 1.7 Schematic of contact angle.

where R_1 and R_2 are the radii of curvature of the surface in two orthogonal directions, with the sign defined as positive if the center of curvature is in domain 1. For a gas bubble in liquid, which takes a spherical shape, this becomes

Young–Laplace equation for pressure drop across an isotropic curved interface

$$\Delta p_{lg} = 2\gamma_{lg}/R, \tag{1.42}$$

where γ_{lg} is the surface tension between liquid and gas, and R is the radius of the bubble. The sign of Δp_{lg} is defined such that the bubble is at a higher pressure than the liquid.

1.5.3 Contact angle

One can imagine that measuring the radius of a bubble as well as the pressure difference would allow calculation of the surface tension; however, this measurement is difficult. In contrast, the *contact angle* does not give the surface tension of an interface directly (it is a function of several different surface tensions, as specified by Young's equation) but is easily measured. Consider a liquid drop on a solid surface surrounded by a gas (Fig. 1.7). At equilibrium, the contact angle is the angle the interface makes with the solid surface, as measured through the dense medium. This can be interpreted in two ways–with a force equilibrium at the triple point and with minimization of the Gibbs function at the equilibrium condition.

Force equilibrium at the triple point. One way to derive Young's equation is to equate forces at the triple point. Because the surface tension can be considered as a force per unit length, we can simply show these forces all acting at the triple point (Fig. 1.8). At equilibrium these forces sum to zero:

$$\gamma_{lg} \cos \theta + \gamma_{sl} = \gamma_{sg}, \tag{1.43}$$

leading to Young's equation:

Young's equation for contact angle

$$\cos \theta = \frac{\gamma_{sg} - \gamma_{sl}}{\gamma_{lg}}. \tag{1.44}$$

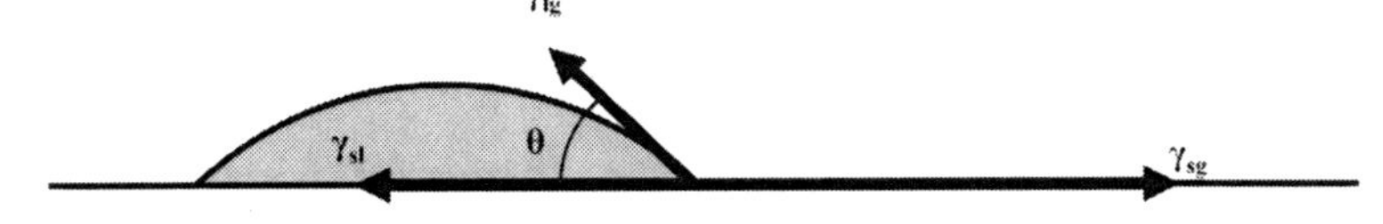

Fig. 1.8 Schematic of surface tension forces acting at the triple point.

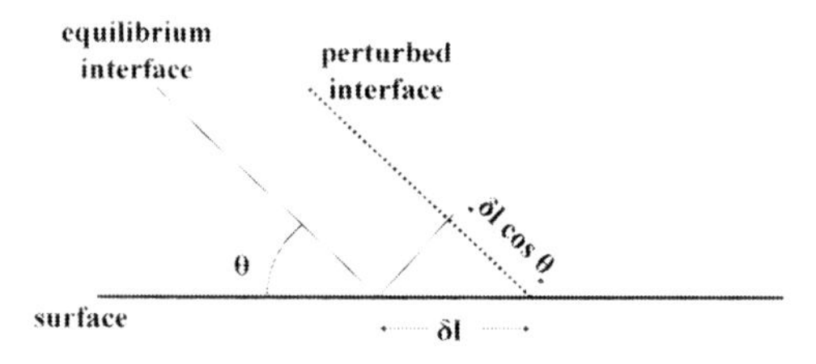

Fig. 1.9 Schematic of surface tension Gibbs energies created by triple-point motion away from equilibrium.

Thermodynamic equilibrium. Alternatively, we can derive Young's equation by considering thermodynamic equilibrium. Because the state is at equilibrium, the Gibbs free energy is at its minimum, and thus the derivative of the Gibbs free energy with respect to any system parameter must be zero. If we define l as the x location of the triple point, then $\partial G/\partial l = 0$. Considering a 2D case, we can similarly say that the Gibbs free energy per unit length G' satisfies the same condition: $\partial G'/\partial l = 0$. Now consider the equilibrium point and a perturbation away from that equilibrium point that involves moving the triple point by a distance δl (Fig. 1.9).[8] The change in Gibbs energy per unit length is

$$\delta G' = -\gamma_{sg}\delta l + \gamma_{sl}\delta l + \gamma_{lg}\delta l \cos\theta \,, \tag{1.45}$$

and because $\delta G'$ must be zero, this again leads to Young's equation:

Young's equation for contact angle

$$\cos\theta = \frac{\gamma_{sg} - \gamma_{sl}}{\gamma_{lg}} \,. \tag{1.46}$$

The contact angles for water drops in air on several solid surfaces are listed in Table 1.2.

1.5.4 Capillary height

Consider a small channel with circular cross section and diameter d in a reservoir of fluid at equilibrium (Fig. 1.10). Assume that interface shapes are unaffected by gravity.[9]

At equilibrium, the pressure drop across the interface must be equal to the change in hydrostatic head, as defined by Eq. (1.2). Given a radius of curvature R at the interface, the Young–Laplace equation gives

$$\Delta p_{lg} = \frac{2\gamma_{lg}}{R} = \rho g H \,, \tag{1.47}$$

where H is the equilibrium height, g is the gravitational acceleration, and ρ is the density of the liquid. Because the interface is spherical, we can argue geometrically that

$$R = \frac{d}{2\cos\theta} \,, \tag{1.48}$$

[8] The location of the interface far away from the solid surface is tacitly fixed. So this is actually a *rotation* of the interface. We assume that θ can be modeled as constant, but that the rotation requires that an extra $\delta l \cos\theta$ of surface be created.

[9] We can show this rigorously by showing that gravity can be neglected if the channel size is less than $\sqrt{\gamma_{lg}/\rho g}$, which for water on Earth is about 3 mm.

Table 1.2 **Contact angles for water drops in air on several solid surfaces at 20°C.**

Surface	Contact angle [deg]
Glass	0–30
poly(methyl methacrylate)	40–70
poly(dimethylsiloxane)	50–105
Teflon	105–120

and from Eqs. (1.47) and (1.48) along with Young's equation, we can show that

equilibrium capillary height relation for a circular tube

$$H = \frac{4}{\rho g d}\left(\gamma_{sg} - \gamma_{sl}\right) . \tag{1.49}$$

EXAMPLE PROBLEM 1.2

Derive the capillary height equation (1.49) by using the interfacial stress equality in Eq. (1.47), Young's equation Eq. (1.46), and geometric arguments from Eq. (1.48).

SOLUTION: Equating hydrostatic head and Laplace pressure, we have

$$\Delta p_{lg} = \frac{2\gamma_{lg}}{R} = \rho g H . \tag{1.50}$$

We substitute in $R = \frac{d}{2\cos\theta}$ to get

$$\Delta p_{lg} = \frac{4\gamma_{lg}}{d}\cos\theta = \rho g H , \tag{1.51}$$

substitute $\cos\theta = \frac{\gamma_{sg}-\gamma_{sl}}{\gamma_{lg}}$ to get

$$\Delta p_{lg} = \frac{4\gamma_{lg}}{d}\frac{\gamma_{sg}-\gamma_{sl}}{\gamma_{lg}} = \rho g H , \tag{1.52}$$

and rearrange to get

$$H = \frac{4}{\rho g d}\left(\gamma_{sg} - \gamma_{sl}\right) . \tag{1.53}$$

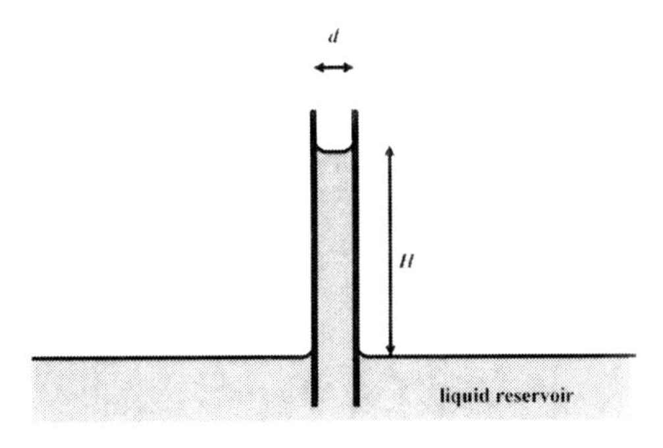

Fig. 1.10 Capillary height due to surface tension in a solvophilic system.

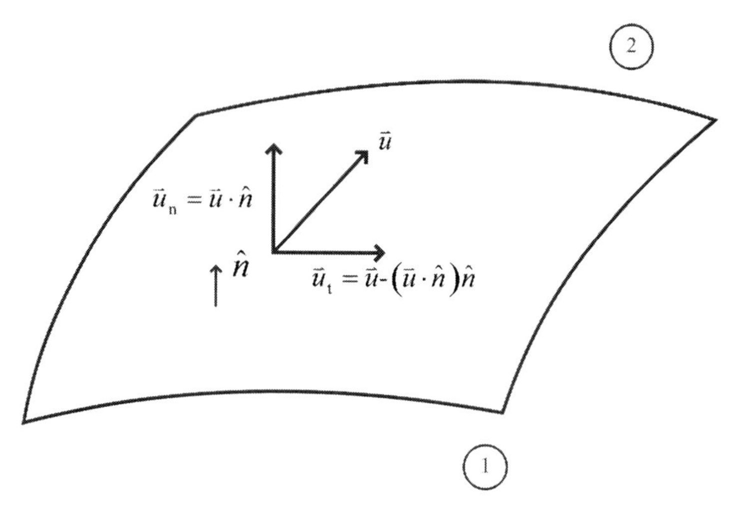

Fig. 1.11 Decomposition of interfacial velocity into normal and tangential components.

1.5.5 Dynamic contact angle

Contact angle is most effective when describing static systems. The use of a single contact angle fails when the contact line on a surface is moving – the contact angles show *hysteresis*, in that the contact angle is a function of how the triple point is moving. This phenomenon cannot be captured by a single surface tension value alone. Thus the dynamics of how a bubble or droplet on a surface moves requires additional analysis.

1.6 VELOCITY AND STRESS BOUNDARY CONDITIONS AT INTERFACES

The Navier–Stokes equations prescribe solutions of fluid flow problems as long as initial and boundary conditions are specified.[10] This section details the velocity boundary conditions for fluid flow, which include velocity conditions and stress conditions. For all interfaces, two velocity boundary conditions are required, one each for the normal and tangential velocities (this decomposition is shown in Fig. 1.11). For deforming interfaces, two additional boundary conditions are required for the normal and tangential stresses, each of which are related to the velocity gradients. For stationary (solid–liquid) interfaces, these conditions can be ignored, because the stresses are not related to velocities in the solid.

1.6.1 Kinematic boundary condition for continuity of normal velocity

An interface or wall defining the boundary between phases by its nature prescribes a boundary condition for the velocity normal to an interface. Unless the interface has chemical reactions or physical state changes, or a wall is permeable, the interface or wall denotes a surface through which mass does not cross. Because of this, the normal velocity distribution is continuous through the interface. This property is fundamentally a *kinematic* property of the fluid velocity, as it is a description of the deformation of the interface.

[10] No existence proof exists for solutions of the Navier–Stokes equations; however, the Navier–Stokes equations can be solved for well-posed fluid flow problems.

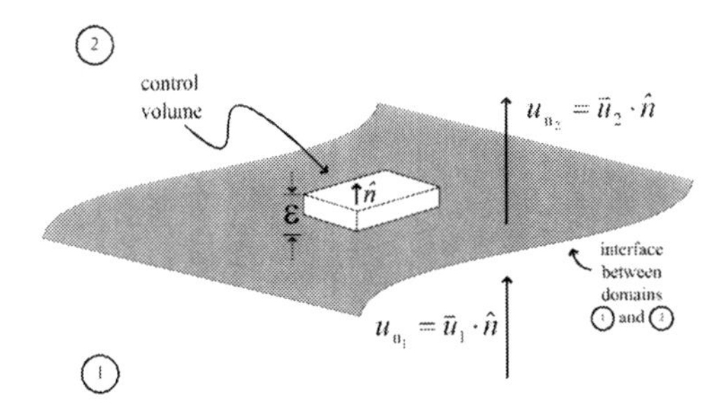

Fig. 1.12 Control volume used to derive the kinematic boundary condition for flow normal to an interface.

We start by describing the general kinematic boundary condition for normal velocities, which applies to both fluid–solid and fluid–fluid interfaces, and then we simplify this for the most common boundary in microdevices – a motionless, impermeable, solid wall.

Consider the interface between two domains, labeled 1 and 2 in Fig. 1.12, and use subscripts 1 and 2 to denote properties on either side of the interface. We define the unit normal to the interface $\hat{\boldsymbol{n}}$ pointing from domain 1 to domain 2, and we define the magnitude of the fluid velocity in the normal direction as $u_{\mathrm{n}} = \vec{\boldsymbol{u}} \cdot \hat{\boldsymbol{n}}$. The general kinematic condition for the velocity normal to the boundary between two domains 1 and 2 is given by

$$u_{\mathrm{n},1} = u_{\mathrm{n},2} \,. \tag{1.54}$$

The normal velocity is continuous across the interface – we can derive Eq. (1.54) by applying the conservation of mass equation (assuming no chemical reaction or phase change) to a flat control volume of infinitesimal thickness aligned normal to the surface (see Fig. 1.12). Equation (1.54) is directly applicable to fluid–fluid interfaces, for example, the interface between oil and water, for which a nonzero normal velocity is possible. If one of the domains is assumed to be a motionless, impermeable solid, then its velocity is by definition zero, and the boundary condition becomes

no-penetration condition

$$\vec{\boldsymbol{u}} \cdot \hat{\boldsymbol{n}} = u_{\mathrm{n}} = 0 \,. \tag{1.55}$$

The $u_{\mathrm{n}} = 0$ condition is termed the *no-penetration condition* for fluid flow at an impermeable wall.

1.6.2 Dynamic boundary condition for continuity of tangential velocity

We start by describing the general dynamic boundary condition (the *no-slip* condition) for tangential velocities, which applies to both fluid–solid and fluid–fluid interfaces, and then we simplify this for the most common implementation in microdevices – the case of a motionless wall. Unlike the condition for the normal velocity, the condition for the tangent velocity is not one of *kinematics* but rather one of the *dynamics* of momentum transport across interfaces.

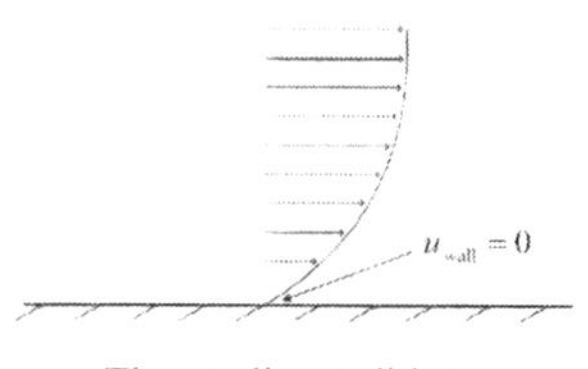

Fig. 1.13 The no-slip condition.

Consider the same interface between the two domains described in the preceding subsection. We find the fluid velocity in the tangent direction $\vec{u}_t$ by subtracting the normal component of the fluid velocity from the total:

$$\vec{u}_t = \vec{u} - (\vec{u} \cdot \hat{n})\,\hat{n}\,. \tag{1.56}$$

A schematic of the normal and tangent components of velocity is shown in Fig. 1.11. With this definition, the tangential direction can be anywhere in the plane of the interface. We typically *assume* that the dynamic boundary condition is that the tangential velocity is continuous through the interface:

$$\vec{u}_{t,1} = \vec{u}_{t,2}\,. \tag{1.57}$$

Unlike the kinematic condition, this boundary condition cannot be derived from a conservation relation or any other continuum argument. Like the viscous constitutive relation for the shear–strain rate relation for a Newtonian fluid, Eq. (1.57) is a postulate that is consistent with experiment for most systems and is supported by microscopic models (e.g., the kinetic theory or atomistic simulations) of momentum transport in fluids. It does not always hold, as described in Subsection 1.6.4.

For fluid–fluid interfaces, Eq. (1.57) is applicable and nonzero tangential velocities can be observed. If one of the domains is assumed to be a motionless solid, then its velocity is by definition zero, and the boundary condition becomes

no-slip condition

$$\vec{u}_t = 0\,. \tag{1.58}$$

This boundary condition (see Fig. 1.13) is termed the *no-slip condition* for fluid flow at a motionless wall.

1.6.3 Dynamic boundary conditions for stresses

For motionless interfaces, such as a motionless solid wall, the velocity continuity boundary conditions in Eqs. (1.55) and (1.58) are sufficient to solve the fluid-mechanical equations. However, if the fluid interface can move, then we must also satisfy dynamic boundary conditions regarding the motion of the interface.

The force equilibrium on a surface is a balance of the *fluid* stresses (evaluated at the surface) combined with the *interface* stresses, that is, the surface tension. The fluid stress at the surface is given by $\overline{\overline{\tau}} \cdot \hat{n}$, which can also be written for a Newtonian fluid as

$$\overline{\overline{\tau}} \cdot \hat{n} = -p\hat{n} + \eta\frac{\partial \vec{u}}{\partial n} + \eta\nabla u_n\,, \tag{1.59}$$

where the n coordinate is the normal coordinate, i.e., the coordinate in the direction of $\hat{n}$. The net interfacial stress owing to surface tension has two parts: one tangential stress

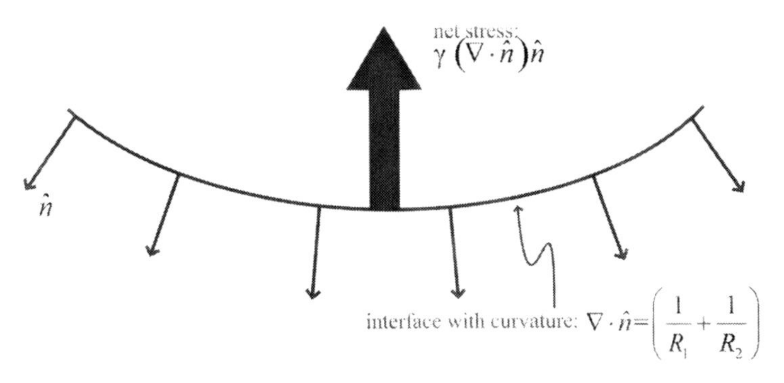

Fig. 1.14 Net normal stress from surface tension at a curved interface.

stemming from spatial gradients of the surface tension γ, and one normal stress from the curvature of the interface. The dynamic boundary condition is given by

$$\overline{\overline{\boldsymbol{\tau}}}_2 \cdot \hat{\boldsymbol{n}} - \overline{\overline{\boldsymbol{\tau}}}_1 \cdot \hat{\boldsymbol{n}} = (\gamma \nabla \cdot \hat{\boldsymbol{n}})\, \hat{\boldsymbol{n}} - [\nabla \gamma - \hat{\boldsymbol{n}}\, (\hat{\boldsymbol{n}} \cdot \nabla \gamma)] \,, \tag{1.60}$$

where $\overline{\overline{\boldsymbol{\tau}}}_1$ and $\overline{\overline{\boldsymbol{\tau}}}_2$ refer to the fluid stresses in domains 1 and 2, respectively. The normal stress from surface tension is proportional to the curvature of the interface, given by the sum of the two principal radii $\nabla \cdot \hat{\boldsymbol{n}} = \frac{1}{R_1} + \frac{1}{R_2}$ (Fig. 1.14). The tangential stress from surface tension is proportional to the component of the gradient of γ that is in the plane of the interface, which is shown in Fig. 1.15 and is expressed by $[\nabla \gamma - \hat{\boldsymbol{n}}\, (\hat{\boldsymbol{n}} \cdot \nabla \gamma)]$. The expression for the gradient in the interfacial plane takes the gradient $\nabla \gamma$ and subtracts away its normal component $\hat{\boldsymbol{n}}\, (\hat{\boldsymbol{n}} \cdot \nabla \gamma)$ to leave only the component in the interfacial plane (Fig. 1.16).

We now split the interfacial stress condition into normal and tangential components, and we examine normal and tangential stresses separately. The normal component of Eq. (1.60) is given by

$$\left[\overline{\overline{\boldsymbol{\tau}}}_2 \cdot \hat{\boldsymbol{n}} - \overline{\overline{\boldsymbol{\tau}}}_1 \cdot \hat{\boldsymbol{n}}\right] \cdot \hat{\boldsymbol{n}} = \gamma \nabla \cdot \hat{\boldsymbol{n}} \,. \tag{1.61}$$

Using the constitutive relation for fluid flow, rewriting the normal stress in terms of the derivative of the velocity with respect to the normal coordinate, and rewriting the surface tension term using the radii of curvature, we can manipulate Eq. (1.61) to obtain

normal stress condition for a moving interface

$$p_1 - p_2 + 2\eta_2 \frac{\partial u_{n,2}}{\partial n} - 2\eta_1 \frac{\partial u_{n,1}}{\partial n} = \gamma \left(\frac{1}{R_1} + \frac{1}{R_2}\right) . \tag{1.62}$$

The symbols R_1 and R_2 denote the two radii of curvature of the interface in any two orthogonal planes containing $\hat{\boldsymbol{n}}$; as used in Eq. (1.62), these radii are defined as positive if $\hat{\boldsymbol{n}}$ points toward the center of curvature (Fig. 1.17).

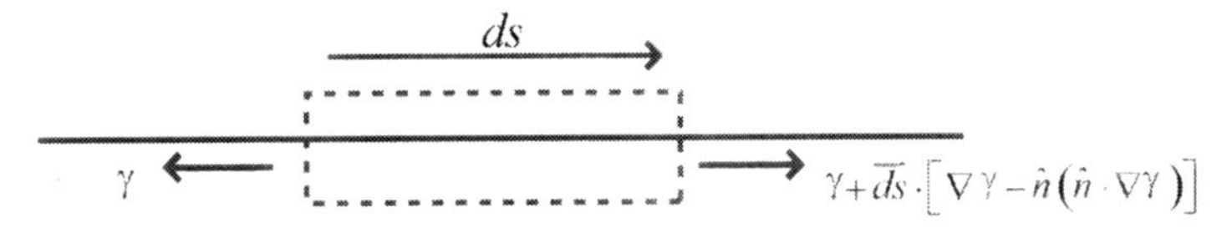

Fig. 1.15 Net tangent stress from surface tension along an interface. The differential distance along the interface is denoted as ds.

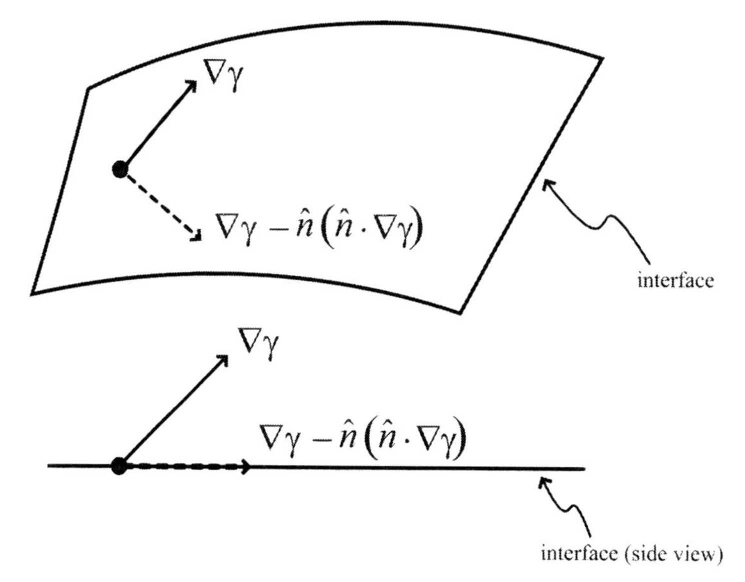

Fig. 1.16 Definition of gradient along an interface.

Unlike the boundary conditions for velocities, Eq. (1.62) shows that the normal stress is *discontinuous* owing to the interfacial tension. For a quiescent ($\vec{u} = 0$) system, the only contribution to the stress vector $\vec{\tau}$ is the pressure, and the boundary condition becomes

Young–Laplace equation for pressure drop across a curved interface

$$p_1 - p_2 = \gamma \left(\frac{1}{R_1} + \frac{1}{R_2} \right). \tag{1.63}$$

Equation (1.63) is the *Young–Laplace equation* for the pressure difference across a static interface; it states that the pressure must be larger inside a curved interface because

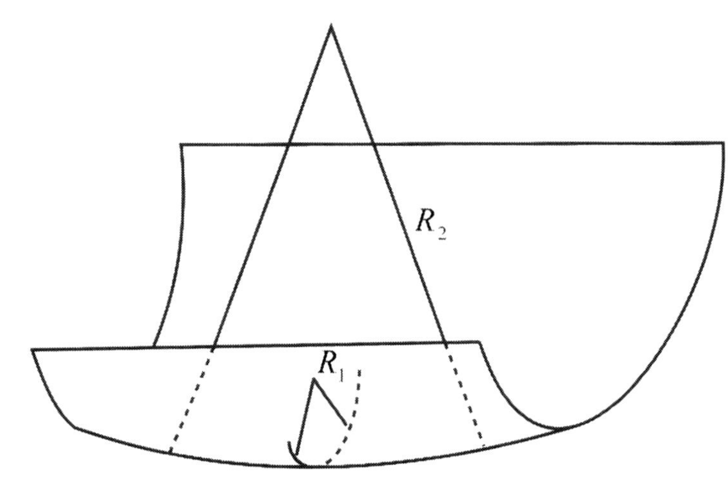

Fig. 1.17 Principal radii of curvature at an interface.

the internal pressure must balance out the interfacial tension's tendency to contract the interface. For a spherical interface, $R = R_1 = R_2$, and this becomes

Young–Laplace equation for pressure drop across an isotropic curved interface

$$p_1 - p_2 = \frac{2\gamma}{R}. \tag{1.64}$$

Equations (1.63) and (1.64) recapitulate Eqs. (1.41) and (1.42). In fact, the only stable configuration for a motionless interface that satisfies the Young–Laplace equation is a sphere – any other configuration either has nonuniform curvature (requiring viscous stresses and thus fluid motion) or is unstable to perturbations (and thus breaks apart, for example, as a cylinder of water from a faucet does). Flows generated by capillary force imbalances are termed *capillary flows*.

Examining the tangential stresses of Eq. (1.60), we can show that the tangential stresses come from two sources: the component of the surface fluid stresses tangent to the interface, which is given by $(\bar{\bar{\tau}} \cdot \hat{\boldsymbol{n}}) \cdot \hat{\boldsymbol{t}}$; and the tangent stress caused by surface tension, which is given by $[\nabla\gamma - \hat{\boldsymbol{n}}(\hat{\boldsymbol{n}} \cdot \nabla\gamma)]$. Thus the dynamic boundary condition for normal stresses is given by

$$\left(\bar{\bar{\tau}}_1 \cdot \hat{\boldsymbol{n}}\right) \cdot \hat{\boldsymbol{t}} - \left(\bar{\bar{\tau}}_2 \cdot \hat{\boldsymbol{n}}\right) \cdot \hat{\boldsymbol{t}} = [\nabla\gamma - \hat{\boldsymbol{n}}(\hat{\boldsymbol{n}} \cdot \nabla\gamma)] \cdot \hat{\boldsymbol{t}} \tag{1.65}$$

Because $(\bar{\bar{\tau}} \cdot \hat{\boldsymbol{n}}) \cdot \hat{\boldsymbol{t}} = \eta\frac{\partial u_n}{\partial t} + \eta\frac{\partial u_t}{\partial n}$ for a Newtonian fluid, this can be rewritten as

dynamic BC for Marangoni flows

$$\eta_1\frac{\partial u_{t,1}}{\partial n} + \eta_1\frac{\partial u_{n,1}}{\partial t} - \eta_2\frac{\partial u_{t,2}}{\partial n} - \eta_2\frac{\partial u_{n,2}}{\partial t} = -\left[\nabla\gamma - \hat{\boldsymbol{n}}(\hat{\boldsymbol{n}} \cdot \nabla\gamma)\right], \tag{1.66}$$

where u_t is the component of the velocity tangent to the interface. The tangential stress is discontinuous if the surface tension is changing along the interface. Surface tension can be variable at an interface if the surface contains a gradient of surfactant concentration, interfacial electric charge density, or temperature. Flows generated by variations in surface tension are generally called *Marangoni flows* or, for thermal surface tension variations, *thermocapillary flows*. If surface tension is uniform, then the tangential stress is continuous, and

stress continuity relation for interfaces with uniform surface tension

$$\eta_1\frac{\partial u_{t,1}}{\partial n} + \eta_1\frac{\partial u_{n,1}}{\partial t} = \eta_2\frac{\partial u_{t,2}}{\partial n} + \eta_2\frac{\partial u_{n,2}}{\partial t}. \tag{1.67}$$

Other surface forces can play a role in determining the interfacial condition, for example, net electric charge density at an interface in the presence of an electric field.

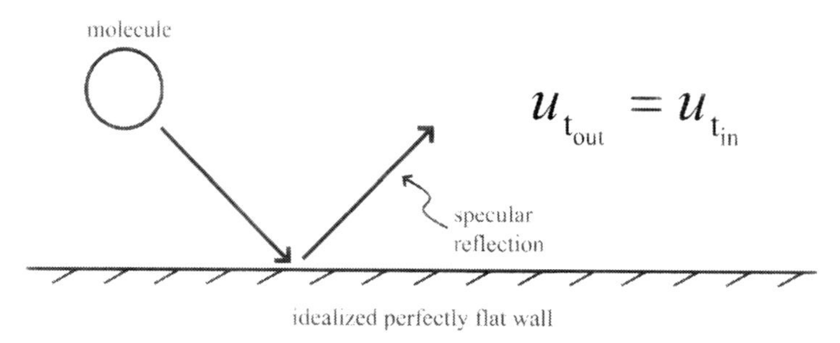

Fig. 1.18 Specular reflections of an isolated molecule off of a perfectly flat surface.

1.6.4 The physics of the tangential velocity boundary condition

As seen in the previous subsection, the stress boundary conditions and the normal velocity boundary condition follow from conservation of momentum or mass, but the tangent velocity continuity boundary condition is not guaranteed by any conservation equation.

The no-slip boundary condition is consistent with observations of macroscale flows and works well for most micro- and nanoscale flows as well. The no-slip condition, however, can be inaccurate in certain special cases, as subsequently described.

SLIP OF AN ISOLATED FLUID MOLECULE AT AN IDEAL, PERFECTLY FLAT SOLID SURFACE

Consider an ideal, perfectly rigid, perfectly flat surface. At an atomistic level, atoms and molecules interact with this wall primarily because of the electrostatic repulsion between atomic or molecular orbitals; in this sense, then, this ideal surface would be a uniform plane with a repulsive force that prevents atoms or molecules from penetrating the wall.

Now consider noninteracting fluid molecules in contact with this wall. A molecule moving in isolation that is incident with this wall reflects off of the wall specularly – the momentum normal to the wall is reversed, and the momentum tangent to the wall is unaffected (Fig. 1.18). Thus *an interface consisting of isolated fluid molecules (i.e., an ideal gas) in contact with a mathematically flat wall leads to a full-slip condition at the surface.* In this hypothetical case, the boundary condition at the wall is

$$\left.\frac{\partial u_t}{\partial n}\right|_{\text{wall}} = 0 . \tag{1.68}$$

This result is never observed in practice, simply because no surface is mathematically flat, not even a perfect crystal plane. Isolated molecules incident upon real surfaces do not reflect specularly, because the surface is not smooth at an atomic level. An imprecise, but useful, analogy is to consider a ball thrown against a smooth wall as compared with one thrown against a surface with large-scale roughness. The ball bounces off of a smooth surface in a predictable and approximately specular fashion, but the ball's reflection off of the rough surface is unpredictable and not specular. Because the reflection off of a rough surface is not specular, the tangential momentum of the ball reflecting off a rough wall (or a molecule reflecting off of any real surface) is, on average, reduced (Fig. 1.19). Thus, on average, a real surface reduces the tangential momentum of a molecule that collides with it.

We can further consider the effect of the number density and therefore the frequency of interaction of the molecules. If the number density of the molecules is low, the molecules travel a long distance in between collisions (their so-called mean free path

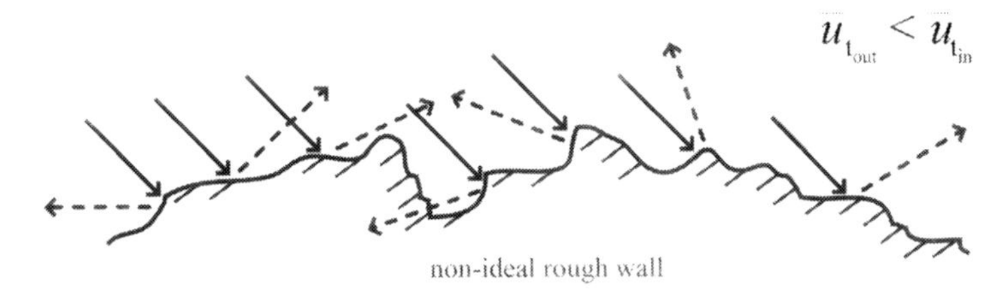

Fig. 1.19 Nonspecular reflections of impacts of an isolated molecule off of a rough surface, leading to a reduction (on average) in tangential velocity.

is large), and molecules near the wall collide with the wall only rarely. If, on the other hand, the molecules are dense, molecules that collide with the wall immediately collide with other molecules, and the molecules bounce back and forth against the wall over and over, quickly removing (on average) any tangential momentum. Two schematics denoting rarefied and dense molecules are shown in Fig. 1.20.

The preceding arguments are the basis for slip models for *gases*. The no-slip condition for gases is an accurate approximation if the characteristic length of the flow is much bigger than the mean free path of the gas. For example, the mean free path of air at room temperature and atmospheric pressure is approximately 100 nm, meaning that the molecules in air travel 100 nm on average before each collision. If the characteristic size of the flow is much larger than 100 nm, then the fluid near the wall exchanges tangential momentum with the wall more than it exchanges tangential momentum with the bulk fluid, and the tangential velocity of the fluid near the wall is well approximated as zero.

The preceding arguments provide an initial framework for examining the physics of tangential velocities at walls; however, in this text we are interested in liquids, so isolated fluid molecules do not accurately represent the fluids of relevance here. For liquids such as water, the mean free path is of the order of 0.2 Å, and thus the molecules cannot be described as noninteracting. To describe slip in liquid systems, a detailed study of the molecular-scale interactions is required. Most often, slip is measured experimentally and characterized by the Navier slip length (see Fig. 1.21).

THE NAVIER SLIP BOUNDARY CONDITION

A more general continuum description of the tangential stress condition at an interface uses a *slip length* that is a property of the fluid–solid interface:

definition of slip length b

$$u_t = b \left. \frac{\partial u_t}{\partial n} \right|_{\text{wall}} , \qquad (1.69)$$

where n is the coordinate normal to the interface and b is the slip length. The no-slip condition is defined by $b = 0$. A nonzero slip length b implies that fluids always have a nonzero velocity at the wall if shear is present. Typical values postulated for the slip length are typically of the order of 1 nm and occasionally of the order of 1 μm. To put

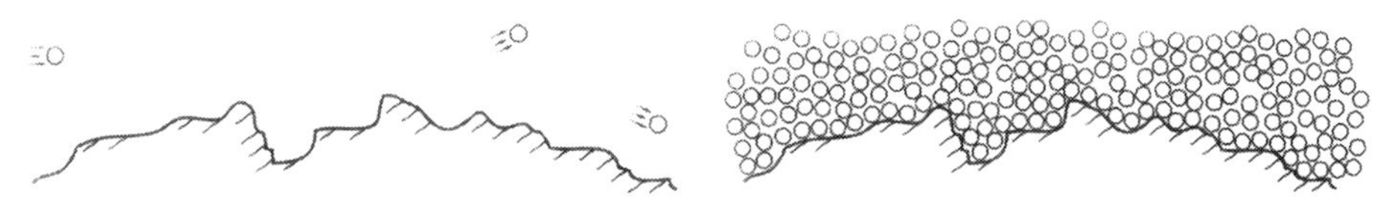

Fig. 1.20 Schematic depictions of (left) rarefied and (right) dense molecules at a surface.

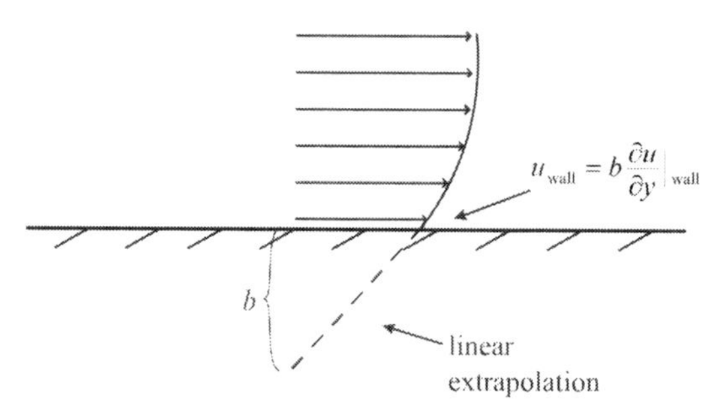

Fig. 1.21 The Navier slip model.

the effect of slip length in macroscopic flows in perspective, flow of water through a 1-cm Teflon pipe at 1 m/s might be expected to have a slip velocity of 100 nm/s, which is approximately 1×10^{-7} of the typical velocities in the system and thus safely ignored in any experiment or computation. The relative magnitude of the velocity at the wall $u_{\text{wall}}/u_{\text{bulk}}$ is roughly on the order of b/d, where d is the characteristic length over which velocities change (roughly the diameter of the system).

The fluid velocity in electroosmotic flow (presented in Chapter 6) varies spatially from the value at the wall (assumed zero in our previous discussions) to the bulk velocity over a distance of the order of the Debye length λ_{D}, so the velocity at the wall is roughly $u_{\text{bulk}}b/\lambda_{\text{D}}$. Because λ_{D} can be nanometers in length (or less), small slip lengths can have enormous effects on the wall slip velocity in an electroosmotic system.

The slip length is an imperfect description of the interfacial condition; however, it does provide a simple and effective framework for incorporating interfacial slip.

1.7 SOLVING THE GOVERNING EQUATIONS

The techniques used to solve the Navier–Stokes solutions include analytical techniques in simple geometries and numerical solutions for complex geometries. For example, we use direct integration when the geometry is simple enough to do so, or for slightly more complex boundary conditions we use separation of variables and eigenfunction expansions (usually in finite domains) or similarity transforms (usually in infinite domains). For complex geometries that do not allow analytical solution, our approach is to use numerical techniques, i.e., discretize the equations and solve a set of linear algebraic equations that approximate the nonlinear partial differential equation. Figure 1.22 gives a pictorial view of these approaches for solving the Navier–Stokes equations.

In addition to solving the Navier–Stokes equations directly, we often find that the system under study allows us to simplify the Navier–Stokes equations (Fig. 1.23) and use the equations in different form – these different forms are chosen because they are easier to solve mathematically. Specifically, for certain flows near a surface, we can ignore diffusion of momentum in the direction parallel to the surface – this is the *boundary-layer approximation* and simplifies the Navier–Stokes equations to the boundary-layer equations. This approach is used in Chapter 6. For irrotational flow (i.e., flows with $\vec{\omega} = 0$), the kinematic constraint and the conservation of mass relation can be combined to form the Laplace equation, a linear equation that can be solved for the velocity potential, which in turn specifies the velocity field. This approach is described in Chapter 7. For inertialess flow (i.e., flow at low Reynolds number, with the Reynolds number defined in Subsection E.2.1), the unsteady and convective terms of the Navier–Stokes equations can be ignored, leaving the *Stokes equations*. This approach is used in Chapter 8.

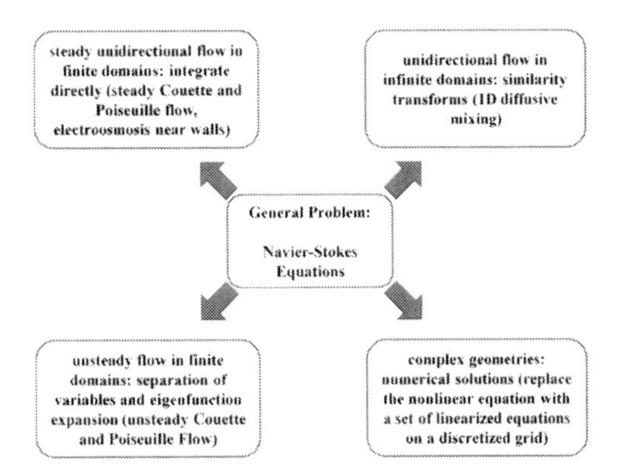

Fig. 1.22 Techniques for solving the Navier–Stokes equations.

1.8 FLOW REGIMES

The different regimes of fluid flow can be grossly summarized by the terms *laminar*, *transitional*, and *turbulent*. The term *laminar flow* comes from the observation that, for viscous-dominated flows, the fluid moves in orderly sheets or lamina, although this is insufficient to precisely define laminar flow. Laminar flow specifically implies that, for specified boundary conditions, a single flow is experimentally observed, and that this solution is *stable to perturbations*.[11] Laminar flows with steady boundary conditions are steady. *Transitional* and *turbulent flow* both refer to flows that have an inherent instability. In these cases, flows exist that satisfy the Navier–Stokes equations but nonetheless are not observed in practice, because these flows are unstable and flow perturbations are enough to disrupt and change the flow. *Turbulent flow* implies that the structure of the flow is dominated by fluid motions that are random and are not microscopically predictable from the boundary conditions. Turbulent flow is always 3D, even if the boundary conditions have an axis or plane of symmetry, and turbulent flow is always unsteady, even when the boundary conditions are steady. *Transitional flow* is a catchall for flow that is no longer laminar (solutions exist that are unstable and not observed), but is not fully dominated by random vortical structures and cannot be described as fully turbulent.

The Reynolds number is used to characterize the transition between these flow regimes, and, for a specific geometry, points of transition can be defined with the Reynolds number. For example, laminar flow is normally observed in long channels of circular cross section for $Re < 2300$. Fully turbulent flow is observed in these flows for $Re > 4000$. For comparison, the Reynolds number for flow of water in a 100 μm channel at a mean velocity of 100 μm/s is $Re = 0.01$. This illustrates that most flows in microscale channels are laminar. Transitional or turbulent flows in microchannels are difficult to achieve and typically require large pressure gradients.

1.9 A WORD ON TERMINOLOGY AND THE MICROFLUIDICS LITERATURE

Fluid mechanics uses a wealth of terminology that has precise meaning in the fluid mechanics community; this terminology can easily be confused and is often used incorrectly in the microfluidics literature. Terms such as *turbulent*, *unsteady*, *chaotic*, and

[11] Laminar flow can also be used to describe the cases with multiple (but finite) stable solutions or a well-defined oscillation.

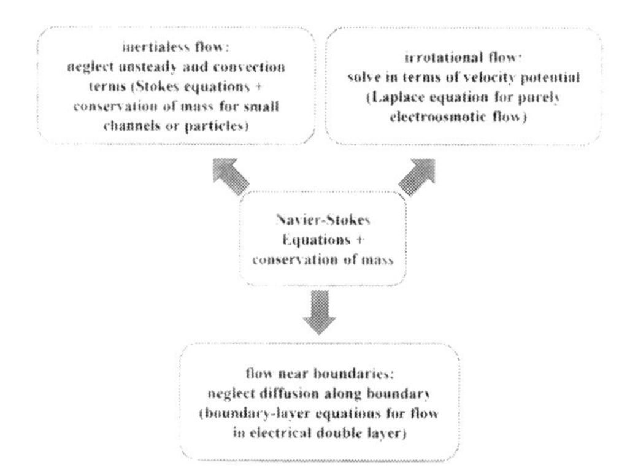

Fig. 1.23 Three approaches for simplifying the Navier–Stokes equations into a more easily solved equation or set of equations.

recirculatory are related but have different meanings. This section endeavors to bring the reader's attention to terms whose misuse has permeated the largely interdisciplinary literature focused on flow in microscale devices.

We use terms *unsteady* and *steady* to denote whether a flow (i.e., the pressure and velocity fields) varies with time. We use *rotational* and *irrotational* to denote whether the magnitude of the vorticity or the rotation rate tensor is zero – irrotational flows imply that $\vec{\omega}$ and $\vec{\vec{\omega}}$ are zero, whereas rotational flows imply that $\vec{\omega}$ and $\vec{\vec{\omega}}$ are nonzero. *Recirculatory flows* imply flows for which some regions of the flow include velocity in a direction opposite the main flow. Recirculatory flows are often rotational but need not be. *Turbulent flows* imply that the laminar flow is unstable and not observed. Turbulent flows are always unsteady and rotational, but unsteady flows and rotational flows are not necessarily turbulent. *Chaotic flows* imply flows for which streamlines separate from each other exponentially as a function of time. Such streamline separation is observed in turbulent flows, but chaotic flows need not be turbulent.

1.10 SUMMARY

This chapter summarizes kinematics, conservation equations, and boundary conditions for incompressible flow. For incompressible flow, conservation of mass is given by

$$\nabla \cdot \vec{\boldsymbol{u}} = 0 . \tag{1.70}$$

For a Newtonian fluid with a spatially uniform viscosity, the Navier–Stokes equations for conservation of momentum are given by

$$\rho \frac{\partial \vec{\boldsymbol{u}}}{\partial t} + \rho \vec{\boldsymbol{u}} \cdot \nabla \vec{\boldsymbol{u}} = -\nabla p + \eta \nabla^2 \vec{\boldsymbol{u}} + \sum_i \vec{\boldsymbol{f}}_i . \tag{1.71}$$

We derive this from Eq. (1.27) by postulating the Newtonian relation between surface stresses and fluid velocities:

$$\vec{\vec{\tau}} = 2\eta\vec{\vec{\varepsilon}} - p\vec{\vec{\delta}}\,, \tag{1.72}$$

This model is excellent for air and water and many other fluids. To write this equation, we develop kinematic definitions of the strain rate tensor as the symmetric part of the velocity gradient tensor:

$$\vec{\vec{\varepsilon}} = \begin{bmatrix} \varepsilon_{xx} & \varepsilon_{xy} & \varepsilon_{xz} \\ \varepsilon_{yx} & \varepsilon_{yy} & \varepsilon_{yz} \\ \varepsilon_{zx} & \varepsilon_{zy} & \varepsilon_{zz} \end{bmatrix} = \begin{bmatrix} \frac{\partial u}{\partial x} & \frac{1}{2}\left(\frac{\partial u}{\partial y} + \frac{\partial v}{\partial x}\right) & \frac{1}{2}\left(\frac{\partial u}{\partial z} + \frac{\partial w}{\partial x}\right) \\ \frac{1}{2}\left(\frac{\partial u}{\partial y} + \frac{\partial v}{\partial x}\right) & \frac{\partial v}{\partial y} & \frac{1}{2}\left(\frac{\partial v}{\partial z} + \frac{\partial w}{\partial y}\right) \\ \frac{1}{2}\left(\frac{\partial u}{\partial z} + \frac{\partial w}{\partial x}\right) & \frac{1}{2}\left(\frac{\partial v}{\partial z} + \frac{\partial w}{\partial y}\right) & \frac{\partial w}{\partial z} \end{bmatrix}. \tag{1.73}$$

The kinematic definitions of $\vec{\vec{\varepsilon}}$ and $\vec{\vec{\omega}}$ (as well as $\nabla\vec{u}$) relate the stretching, skewing, and rotation of fluid elements in terms of components of these tensors.

The boundary conditions for flow at a solid, impermeable wall are the no-penetration condition

$$u_{\mathrm{n}} = 0\,, \tag{1.74}$$

and the no-slip condition

$$u_{\mathrm{t}} = 0\,. \tag{1.75}$$

The no-slip condition is typically suitable for describing tangential velocities at solid walls but can fail especially if the solid wall is hydrophobic and the liquid is water. In this case, a Navier slip model is typically used to describe the boundary condition:

$$u_{\mathrm{t}} = b\left.\frac{\partial u_{\mathrm{t}}}{\partial n}\right|_{\mathrm{wall}}. \tag{1.76}$$

For moving interfaces with uniform surface tension separating Newtonian fluids, the tangential stress is matched on either side of the interface:

$$\eta_1 \frac{\partial u_{t,1}}{\partial n} + \eta_1 \frac{\partial u_{n,1}}{\partial t} = \eta_2 \frac{\partial u_{t,2}}{\partial n} + \eta_2 \frac{\partial u_{n,2}}{\partial t} . \tag{1.77}$$

1.11 SUPPLEMENTARY READING

Modern introductory texts that cover the basic fluid mechanical equations include Fox, Pritchard, and McDonald [17], Munson, Young, and Okiishi [18], White [19], and Bird, Stewart, and Lightfoot [20]. These texts progress through this material more methodically, and are a good resource for those with minimal fluids training. More advanced treatment can be found in Panton [21], White [22], Kundu and Cohen [23], or Batchelor [24]. Batchelor provides a particularly lucid description of the Newtonian approximation, the fundamental meaning of pressure in this context, and why its form follows naturally from basic assumptions about isotropy of the fluid. Texts on kinetic theory [25, 26] provide a molecular-level description of the foundations of the viscosity and Newtonian model.

Although the classical fluids texts are excellent sources for the governing equations, kinematic relations, constitutive relations, and classical boundary conditions, they typically do not treat slip phenomena at liquid–solid interfaces. An excellent and comprehensive review of slip phenomena at liquid–solid interfaces can be found in [27] and the references therein. Slip in gas–solid systems is discussed in [3].

The treatment of surface tension in this chapter is similar to that found in basic fluids texts [21, 24] but omits many critical topics, including surfactants. Reference [28] covers these topics in great detail and is an invaluable resource. References [29, 30] cover flows owing to surface tension gradients, e.g., thermocapillary flows. A detailed discussion of boundary conditions is found in [31].

Although porous media and gels are commonly used in microdevices, especially for chemical separations, this text focuses on bulk fluid flow in micro- and nanochannels and omits consideration of flow through porous media and gels. Reference [29] provides one source to describe these flows. Another fascinating rheological topic that is largely omitted here is the flow of particulate suspensions and granular systems, with blood being a prominent example. Discussions of biorheology can be found in [23], and colloid science texts [29, 32] treat particulate suspensions and their rheology.

1.12 EXERCISES

1.1 In general, the sum of the extensional strains ($\varepsilon_{xx} + \varepsilon_{yy} + \varepsilon_{zz}$) in an incompressible system always has the same value. What is this value? Why is this value known?

1.2 For a 2D flow (no z velocity components and all derivatives with respect to z are zero), write the components of the strain rate tensor $\vec{\vec{\varepsilon}}$ in terms of velocity derivatives.

1.3 Given the following strain rate tensors, draw a square-shaped fluid element and then show the shape that fluid element would take after being deformed by the fluid flow.

(a) $\vec{\vec{\varepsilon}} = \begin{bmatrix} 1 & 0 & 0 \\ 0 & -1 & 0 \\ 0 & 0 & 0 \end{bmatrix}$.

(b) $\vec{\vec{\varepsilon}} = \begin{bmatrix} -1 & 0 & 0 \\ 0 & 1 & 0 \\ 0 & 0 & 0 \end{bmatrix}$.

(c) $\vec{\vec{\varepsilon}} = \begin{bmatrix} 0 & 1 & 0 \\ 1 & 0 & 0 \\ 0 & 0 & 0 \end{bmatrix}$.

1.4 The following strain rate tensor is not valid for incompressible flow. Why?

$$\vec{\vec{\varepsilon}} = \begin{bmatrix} 1 & 1 & 1 \\ 1 & 1 & 1 \\ 1 & 1 & 1 \end{bmatrix}. \quad (1.78)$$

1.5 Could the following tensor be a strain rate tensor? If yes, explain the two properties that this tensor satisfies that make it valid. If no, explain why this tensor could not be a strain rate tensor.

$$\vec{\vec{\varepsilon}} = \begin{bmatrix} 1 & 1 & 1 \\ 1 & 0 & -1 \\ -1 & 1 & -1 \end{bmatrix}. \quad (1.79)$$

1.6 Consider an incompressible flow field in cylindrical coordinates with axial symmetry (for example, a laminar jet issuing from a circular orifice). The axial symmetry implies that the flow field is a function of r and z but not θ. Can a stream function be derived for this case? If so, what is the relation between the derivatives of the stream function and the r and z velocities?

1.7 Consider the following two velocity gradient tensors:

(a) $\nabla\vec{u} = \begin{pmatrix} 0 & 1 & 0 \\ 1 & 0 & 0 \\ 0 & 0 & 0 \end{pmatrix}$.

(b) $\nabla\vec{u} = \begin{pmatrix} 1 & 0 & 0 \\ 0 & -1 & 0 \\ 0 & 0 & 0 \end{pmatrix}$.

Draw the streamlines for each velocity gradient tensor. With respect to the coordinate axes, identify which of these exhibits extensional strain and which exhibits shear strain. Following this, redraw the streamlines for each on axes that have been rotated 45° counterclockwise, using $x' = x/\sqrt{2} + y/\sqrt{2}$ and $y' = -x/\sqrt{2} + y/\sqrt{2}$. Are your conclusions about extensional and shear strain the same for the flow once you have rotated the axes? Do the definitions of extensional and shear strain depend on the coordinate system?

1.8 Write out the components of $\rho\vec{u} \cdot \nabla\vec{u}$ by using Cartesian derivatives. Is $\rho\vec{u} \cdot \nabla\vec{u}$ a scalar, vector, or second-order tensor?

1.9 Give an example of a 3D velocity gradient tensor that corresponds to a purely rotational flow.

1.10 Consider a differential Cartesian control volume and examine the convective momentum fluxes into and out of the control volume. Show that, for an incompressible fluid, the net convective outflow of momentum per unit volume is given by $\rho\vec{u} \cdot \nabla\vec{u}$.

1.11 Consider a differential Cartesian control volume and examine the viscous stresses on the control volume. Do not use a particular model for these viscous stresses; simply assume that $\overline{\overline{\tau}}_{\text{visc}}$ is known. Show that the net inflow of momentum from these forces is given by $\nabla \cdot \overline{\overline{\tau}}_{\text{visc}}$.

1.12 Show that $\nabla \cdot 2\eta\overline{\overline{\varepsilon}} = \eta\nabla^2\vec{u}$ if the viscosity is uniform and the fluid is incompressible.

1.13 For a one dimensional flow given by $\vec{u} = u(y)$, the strain rate magnitude is given by $\frac{1}{2}\frac{\partial u}{\partial y}$ and the vorticity magnitude is given by $\frac{\partial u}{\partial y}$. Are the strain rate and vorticity proportional to each other in general? If not, why are they proportional in this case?

1.14 Write out the Navier–Stokes equations in cylindrical coordinates (see Appendix D). Simplify these equations for the case of plane symmetry.

1.15 Write out the Navier–Stokes equations in cylindrical coordinates (see Appendix D). Simplify these equations for the case of axial symmetry.

1.16 Write out the Navier–Stokes equations in spherical coordinates (see Appendix D). Simplify these equations for the case of axial symmetry.

1.17 For each of the following Cartesian velocity gradient tensors, (1) calculate the strain rate tensor, (2) calculate the rotation rate tensor, and (3) sketch the streamlines for the flow:

(a) $\nabla u = \begin{bmatrix} 0 & 1 & 0 \\ 1 & 0 & 0 \\ 0 & 0 & 0 \end{bmatrix}.$

(b) $\nabla u = \begin{bmatrix} -1 & 0 & 0 \\ 0 & 1 & 0 \\ 0 & 0 & 0 \end{bmatrix}.$

(c) $\nabla u = \begin{bmatrix} 0 & 1 & 0 \\ -1 & 0 & 0 \\ 0 & 0 & 0 \end{bmatrix}.$

(d) $\nabla u = \begin{bmatrix} 0 & 1 & 0 \\ 0 & 0 & 0 \\ 0 & 0 & 0 \end{bmatrix}.$

1.18 Consider the 2D flows defined by the following stream functions. The symbols A, B, C, and D denote constants.

(a) $\psi = Axy$.

(b) $\psi = \frac{1}{2}By^2$.

(c) $\psi = C \ln\left(\sqrt{x^2 + y^2}\right)$.

(d) $\psi = -D\left(x^2 + y^2\right)$.

For the flow field denoted by each of the preceding stream functions, execute the following:

(a) Show that the flow field satisfies conservation of mass.

(b) Derive the four components of the Cartesian strain rate tensor $\overline{\overline{\varepsilon}}$.

(c) Plot streamlines for these flows in the regions $-5 < x < 5$ and $-5 < y < 5$.

(d) Assume that the pressure field $p(x, y)$ is known for each flow. Derive the four components of the Cartesian stress tensor $\bar{\bar{\tau}}$, assuming that the fluid is Newtonian.

(e) Imagine that a 5×5 grid of lines (see Fig. 1.24) is visualized by instantaneously making a grid of tiny bubbles in a flow of water. Sketch the result if the grid were convected in the specified flow field starting at time $t = 0$ and a picture of the deformed grid was taken at a later time.

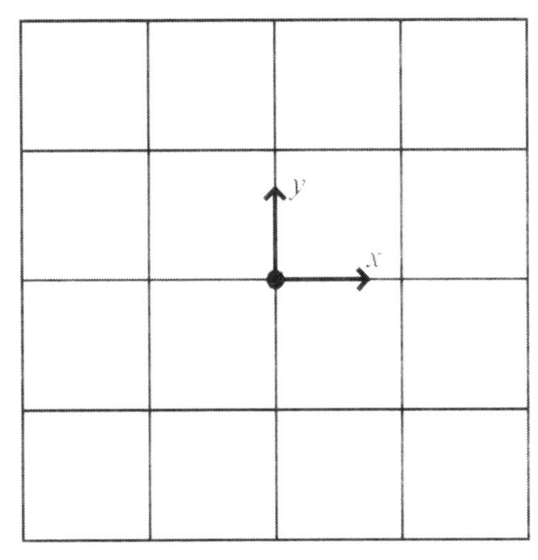

Fig. 1.24 A grid that can be used to visualize how a flow deforms.

1.19 Create an infinitesimal control volume in cylindrical coordinates with edge lengths $d\imath$, $\imath d\theta$, and dz. Use the integral equation for conservation of mass:

$$\frac{\partial}{\partial t}\int_{\mathcal{V}} \rho dV = -\int_{S} (\rho \vec{u}) \cdot \hat{n} \, dA, \tag{1.80}$$

where V is the volume of the control volume, to derive the incompressible continuity equation in cylindrical coordinates.

1.20 Using thermodynamic arguments, derive the Young–Laplace equation Eq. (1.41).

1.21 Use trigonometric and geometric arguments to derive Eq. (1.48).

1.22 Consider a capillary of diameter d oriented along the y axis and inserted into a reservoir of a fluid. Assume the surface tension of the liquid–gas interface is given by γ_{lg}. At the interface, the radius of curvature R can be assumed uniform everywhere in the xz plane (i.e., the interface is spherical) if the variations in the local pressure drop across the interface are small compared with the nominal value of the pressure drop across the interface.

(a) Write a relation for the pressure drop across the interface as a function of γ_{lg} and R.

(b) As a function of θ, evaluate the difference in height of the fluid at the center of the capillary with respect to the fluid at the outside edge of the capillary, and thus evaluate the difference in hydrostatic head between the center and edge of the capillary.

(c) The criterion for approximating the interface as spherical is that the pressure drop variations from center to edge are small compared with the pressure drop itself. Determine the maximum value for d for which the interface can be assumed spherical. Your result will be of the order of $\sqrt{\gamma_{lg}/\rho g}$.

1.23 Show that the Euclidean norm of the rotation rate tensor is equal to $\sqrt{2}$ times the solid-body rotation rate of a point in a flow.

1.24 Draw a flat control volume at an interface between two domains and derive the general kinematic boundary condition for the normal velocities that is given in Eq. (1.54).

2 Unidirectional Flow

Although the Navier–Stokes solutions cannot be solved analytically in the general case, we can still obtain solutions that guide engineering analysis of fluid systems. If we make certain geometric simplifications, specifically that the flow is unidirectional through a channel of infinite extent, the Navier–Stokes equations can be simplified and solved by direct integration. The key simplification enabled by this assumption is that the convective term of the Navier–Stokes equations can be neglected, because the fluid velocity and the velocity gradients are orthogonal. The solutions in this limit include laminar flow between two moving plates (Couette flow) and pressure-driven laminar flow in a pipe (Poiseuille flow). These flows are simultaneously the simplest solutions of the Navier–Stokes equations and the most common types of flows observed in long, narrow channels. Many microchannel flows are described by these solutions, their superposition, or a small perturbation of these flows. This chapter presents these solutions and interprets these solutions in terms of flow kinematics, viscous stresses, and Reynolds number.

2.1 STEADY PRESSURE- AND BOUNDARY-DRIVEN FLOW THROUGH LONG CHANNELS

The Navier–Stokes equations can be simplified when flow proceeds through an infinitely long channel of uniform cross section, owing to the geometric elimination of the nonlinear term in the equation. Couette and Poiseuille flows, in turn, describe flow driven by boundary motion or pressure gradients.

2.1.1 Couette flow

Couette flow is the flow of fluid in between two infinite parallel flat plates driven by the motion of one or more of the plates. Consider steady flow between two infinite parallel moving plates with a uniform pressure field. Consider plates located at $y = \pm h$. The top plate moves in the x direction with speed u_H, and the other plate moves in the x direction with speed u_L. A schematic of this flow is shown in Fig. 2.1. The governing equations are the Navier–Stokes equations without body forces:

$$\rho \frac{\partial \vec{u}}{\partial t} + \rho \vec{u} \cdot \nabla \vec{u} = -\nabla p + \eta \nabla^2 \vec{u} . \tag{2.1}$$

Simplified equation. We solve this flow by making two simplifying approximations. The resulting solution satisfies the boundary condition and governing equation and thus validates these approximations. We begin by assuming that the fluid motion is in the x direction only. Because the flow is in the x direction only, we can replace $\vec{u}$ with u, resulting in

$$\rho \frac{\partial u}{\partial t} + \rho u \frac{\partial u}{\partial x} = -\frac{\partial p}{\partial x} + \eta \frac{\partial^2 u}{\partial x^2} + \eta \frac{\partial^2 u}{\partial y^2} + \eta \frac{\partial^2 u}{\partial z^2} . \tag{2.2}$$

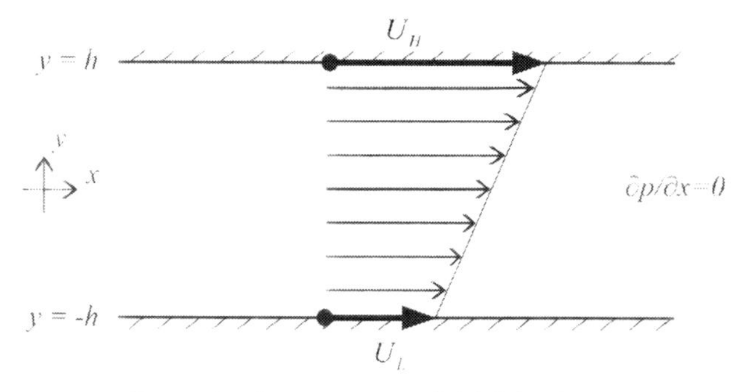

Fig. 2.1 Couette flow between two parallel, infinite plates.

Our next simplifying approximation is that the velocity profile is independent of x. The first three terms of the equation can therefore be assumed zero, in turn, because (a) the flow is assumed steady, (b) u is assumed independent of x, and (c) the pressure is assumed uniform. Thus the simplified governing equation for this flow is

$$0 = \eta \frac{\partial^2 u}{\partial y^2} \,. \tag{2.3}$$

In this case, we have chosen a problem statement and geometry that allow much of the Navier–Stokes equation to be ignored; in so doing, we have a system that is much simpler mathematically.

EXAMPLE PROBLEM 2.1

Show that, if η is uniform and flow is in the x direction only such that $\vec{\boldsymbol{u}} = (u, 0, 0)$, then $\nabla \cdot \eta \nabla \vec{\boldsymbol{u}} = \left[\eta\left(\frac{\partial^2 u}{\partial x^2} + \frac{\partial^2 u}{\partial y^2} + \frac{\partial^2 u}{\partial z^2}\right), 0, 0\right]$.

SOLUTION: Evaluate as follows:

$$\nabla \vec{\boldsymbol{u}} = \begin{bmatrix} \frac{\partial u}{\partial x} & \frac{\partial u}{\partial y} & \frac{\partial u}{\partial z} \\ \frac{\partial v}{\partial x} & \frac{\partial v}{\partial y} & \frac{\partial v}{\partial z} \\ \frac{\partial w}{\partial x} & \frac{\partial w}{\partial y} & \frac{\partial w}{\partial z} \end{bmatrix} = \begin{bmatrix} \frac{\partial u}{\partial x} & \frac{\partial u}{\partial y} & \frac{\partial u}{\partial z} \\ 0 & 0 & 0 \\ 0 & 0 & 0 \end{bmatrix} , \tag{2.4}$$

$$\eta \nabla \vec{\boldsymbol{u}} = \begin{bmatrix} \eta\frac{\partial u}{\partial x} & \eta\frac{\partial u}{\partial y} & \eta\frac{\partial u}{\partial z} \\ 0 & 0 & 0 \\ 0 & 0 & 0 \end{bmatrix} . \tag{2.5}$$

Recall that

$$\nabla = \left(\frac{\partial}{\partial x}, \frac{\partial}{\partial y}, \frac{\partial}{\partial z}\right) , \tag{2.6}$$

leading to

$$\boxed{\nabla \cdot \eta \nabla \vec{\boldsymbol{u}} = \left[\eta\left(\frac{\partial^2 u}{\partial x^2} + \frac{\partial^2 u}{\partial y^2} + \frac{\partial^2 u}{\partial z^2}\right), 0, 0\right] .} \tag{2.7}$$

Solution. Equation (2.3) can be integrated directly to give a linear distribution:

$$u = C_1 y + C_2 \,. \tag{2.8}$$

Applying boundary conditions at the top plate,

$$u(y = h) = u_H \,, \tag{2.9}$$

as well as at the bottom,

$$u(y = -h) = u_L \,, \tag{2.10}$$

we find

Couette flow between infinite parallel plates at $y = \pm h$

$$u = \frac{u_H + u_L}{2} + \frac{u_H - u_L}{2}\left(\frac{y}{h}\right) . \tag{2.11}$$

Physical interpretation. The Couette flow has no acceleration, no net pressure forces, and no net convective transport of momentum. Because of this, the governing equation also says that the net viscous force on any control volume is also zero. For uniform viscosity, this means that the concavity (i.e., the second derivative with respect to y) of the flow distribution is zero. If we draw a control volume around the flow, we can see that the top surface applies a force per unit area to the fluid given by $\eta\frac{u_H - u_L}{2h}$, and the bottom surface applies an equal and opposite force per unit area. The velocity profile of a Couette flow is not a function of the viscosity, but the force required to move the plates is.

Flow kinematics and viscous stresses. The Couette flow has a simple functional form. Although a tensor description is not necessary for this flow, our Couette flow is a good opportunity to use the tensor description in a simple case.

Evaluating the strain rate and viscous stress tensor, we obtain

$$\vec{\vec{\tau}}_{\text{visc}} = \begin{bmatrix} \tau_{xx} & \tau_{xy} & \tau_{xz} \\ \tau_{yx} & \tau_{yy} & \tau_{yz} \\ \tau_{zx} & \tau_{zy} & \tau_{zz} \end{bmatrix} = 2\eta\vec{\vec{\varepsilon}} = \begin{bmatrix} 2\eta\varepsilon_{xx} & 2\eta\varepsilon_{xy} & 2\eta\varepsilon_{xz} \\ 2\eta\varepsilon_{yx} & 2\eta\varepsilon_{yy} & 2\eta\varepsilon_{yz} \\ 2\eta\varepsilon_{zx} & 2\eta\varepsilon_{zy} & 2\eta\varepsilon_{zz} \end{bmatrix} \tag{2.12}$$

or

$$\vec{\vec{\tau}}_{\text{visc}} = \begin{bmatrix} 2\eta\frac{\partial u}{\partial x} & \eta\left(\frac{\partial u}{\partial y} + \frac{\partial v}{\partial x}\right) & \eta\left(\frac{\partial u}{\partial z} + \frac{\partial w}{\partial x}\right) \\ \eta\left(\frac{\partial u}{\partial y} + \frac{\partial v}{\partial x}\right) & 2\eta\frac{\partial v}{\partial y} & \eta\left(\frac{\partial v}{\partial z} + \frac{\partial w}{\partial y}\right) \\ \eta\left(\frac{\partial u}{\partial z} + \frac{\partial w}{\partial x}\right) & \eta\left(\frac{\partial v}{\partial z} + \frac{\partial w}{\partial y}\right) & 2\eta\frac{\partial w}{\partial z} \end{bmatrix} = \begin{bmatrix} 0 & \eta\frac{u_H - u_L}{2h} & 0 \\ \eta\frac{u_H - u_L}{2h} & 0 & 0 \\ 0 & 0 & 0 \end{bmatrix} . \tag{2.13}$$

The only nonzero viscous stress terms are the τ_{xy} and τ_{yx} terms, each of which has a value equal to the viscosity times the velocity gradient:

viscous stresses for Couette flow

$$\tau_{xy} = \tau_{yx} = \eta\frac{u_H - u_L}{2h} . \tag{2.14}$$

This expression has no x or y dependence. The viscous stress is uniform, consistent with the governing equation, which can also be written as $\frac{\partial \tau_{xy}}{\partial y} = 0$. This flow exhibits both shear strain (off-diagonal elements of $\vec{\vec{\varepsilon}}$) and rotation (off-diagonal elements of $\vec{\vec{\omega}}$).

Reynolds number. The Reynolds number is defined as $Re = \rho U\ell/\eta$, where U and ℓ are the characteristic velocity and length, respectively, which come from the boundary conditions (see Appendix E, Subsection E.2.1). By *characteristic*, we mean that the velocity and length are representative of (a) the inertial forces, as represented by the dynamic pressure $\frac{1}{2}\rho U^2$, and (b) the viscous forces, as represented by the viscous stress $\eta U/\ell$. In both cases, we care most precisely about velocity *differences*.

For Couette flow, we define the characteristic velocity U as $U = |u_H - u_L|$, which is a measure of the total difference between the fastest part of the flow and the slowest part of the flow. The dynamic pressure $\frac{1}{2}\rho U^2$ tells us the amount the pressure would increase if the fast fluid were slowed down along an irrotational streamline path to the speed of the slow fluid, and thus gives a measure of the inertia of the fast fluid relative to that of the slow fluid.

For Couette flow, we define the characteristic length ℓ as $\ell = 2h$, because the viscous stresses are $\tau_{xy} = \eta\frac{|u_H - u_L|}{2h}$ everywhere in the flow.[1]

Substituting these into the definition of the Reynolds number, we get that the Reynolds number for Couette flows is

$$Re_{\text{Cou}} = \frac{2\rho\,|u_H - u_L|\,h}{\eta}. \tag{2.15}$$

For Couette flows, the Reynolds number indicates whether the laminar solution just derived is observed experimentally or if a turbulent solution is observed. For $Re < 100$, the laminar solution derived in this subsection is the one observed. In microsystems, this is typically satisfied, owing to the small h values used.

EXAMPLE PROBLEM 2.2

Calculate the rotation rate tensor and vorticity of the Couette flow described in Subsection 2.1.1.

SOLUTION: The Couette solution is

$$u = \frac{u_H + u_L}{2} + \frac{u_H - u_L}{2}\left(\frac{y}{h}\right). \tag{2.16}$$

The rotation rate tensor is written in matrix form for Cartesian coordinates by

$$\vec{\vec{\omega}} = \begin{bmatrix} 0 & \frac{u_H - u_L}{4h} & 0 \\ -\frac{u_H - u_L}{4h} & 0 & 0 \\ 0 & 0 & 0 \end{bmatrix}, \tag{2.17}$$

and the vorticity is given by

$$\vec{\omega} = -\frac{u_H - u_L}{2h}\hat{\mathbf{z}}. \tag{2.18}$$

[1] When we define nondimensional parameters, the numerical prefactor is arbitrary. We define these numbers to match convention, to make the bookkeeping easy, or to make the number more effective for comparing different types of flows. We use $2h$ to match convention.

EXAMPLE PROBLEM 2.3

A drop of water ($\rho = 1000$ kg/m^3; $\eta = 1$ mPa s) is placed between two flat $1'' \times 3''$ glass coverslides that are pressed together to make a $d = 10$ μm film of water. One glass slide is moved with respect to the other at a velocity of 200 μm/s. Model the flow between these plates as a Couette flow.

1. What is *Re* for this flow?
2. What is the total force required to move the slide?

SOLUTION: Defining $y = 0$ at the bottom coverslide, which we model as motionless, we find that the velocity profile is given by

$$u = 200\frac{y}{d}\ \mu\text{m/s}. \tag{2.19}$$

The Reynolds number is given by

$$Re = \frac{\rho u_H d}{\eta} = \frac{1 \times 10^3\ \text{kg/m}^3 \times 200 \times 10^{-6}\ \text{m/s} \times 10 \times 10^{-6}\ \text{m}}{1 \times 10^{-3}\ \text{Pa s}} = 2 \times 10^{-3}. \tag{2.20}$$

The force is given by the shear stress multiplied by the area:

$$F = \eta\frac{\partial u}{\partial y}A = 1\,\text{mPa s} \times 20\,\text{s}^{-1} \times 2.54 \times 10^{-2}\,\text{m} \times 7.62 \times 10^{-2}\,\text{m} = 38.7 \times 10^{-6}\,\text{N} \tag{2.21}$$

EXAMPLE PROBLEM 2.4

Consider a flow of water ($\eta = \eta_1$) and oil ($\eta = \eta_2$) between two infinite plates with no pressure gradient. Assume that the oil-water interface is the $y = 0$ plane, and the plates are located at the $y = \pm h$ planes. The oil is above $y = 0$ and the water is below $y = 0$. The upper plate is moving in the x direction with a velocity $u = u_H$. The lower plate is motionless. Derive $u(y)$ for this steady flow, and calculate the distribution for $\eta_2 = 6\eta_1$. How does the shear stress vary with y?

SOLUTION: The governing equation is $\frac{\partial}{\partial y}\eta\frac{\partial u}{\partial y} = 0$, with boundary conditions $u = u_H$ at $y = h$, $u = 0$ at $y = -h$, $u_{\text{water}} = u_{\text{oil}}$ at $y = 0$, and $\eta_1\frac{\partial u}{\partial y}|_{\text{water}} = \eta_2\frac{\partial u}{\partial y}|_{\text{oil}}$ at $y = 0$. The dynamic boundary condition is simplified because the flat interface means that $\nabla \cdot \hat{\boldsymbol{n}} = 0$ on the surface. In each domain, the velocity solution can be written as a linear distribution, meaning that four parameters must be determined, and four boundary conditions are available to specify these. Solution of four algebraic equations in four unknowns leads to a solution in the water ($y < 0$) given by

$$u = \frac{\eta_2}{\eta_1 + \eta_2} u_H + \frac{\eta_2}{\eta_1 + \eta_2} u_H \frac{y}{h}, \tag{2.22}$$

and a solution in the oil ($y > 0$) given by

$$u = \frac{\eta_2}{\eta_1 + \eta_2} u_H + \frac{\eta_1}{\eta_1 + \eta_2} u_H \frac{y}{h}. \tag{2.23}$$

Given the specified values, a numerical result can be calculated. For $y < 0$,

$$u = \frac{6}{7} u_H + \frac{6}{7} u_H \frac{y}{h}, \tag{2.24}$$

and for $y > 0$,

$$u = \frac{6}{7} u_H + \frac{1}{7} u_H \frac{y}{h}. \tag{2.25}$$

The shear stress is uniform throughout the domain and is given by $\tau_{xy} = \frac{\eta_1 \eta_2}{\eta_1 + \eta_2} \frac{u_H}{h}$.

2.1.2 Poiseuille flow

Poiseuille flow describes the flow resulting from pressure gradients in a tube, and Hagen–Poiseuille flow refers specifically to pressure-driven flow through a tube of circular cross section. This type of flow is characteristic of pressure-driven flow through channels in microdevices, and this solution (Fig. 2.2) is the fundamental source for all of the results in Chapter 3.

To solve for the Hagen–Poiseuille flow, we assume that all flow is in the z direction ($u_r = 0$, $u_\theta = 0$) and that velocity gradients in the θ and z directions are zero. The Navier–Stokes equations are

$$\rho \frac{\partial \vec{u}}{\partial t} + \rho \vec{u} \cdot \nabla \vec{u} = -\nabla p + \eta \nabla^2 \vec{u}. \tag{2.26}$$

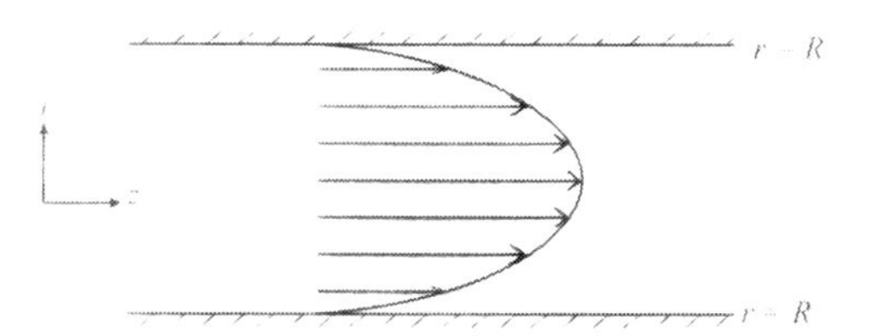

Fig. 2.2 Poiseuille flow in a circular tube.

Simplified equation. Because of the geometric simplifications, the convection and unsteady terms are all zero, leaving:

$$\nabla p = \eta \nabla^2 \vec{u} . \tag{2.27}$$

For radially symmetric flow, we write this as:

$$\frac{\partial p}{\partial z} = \eta \frac{1}{r} \frac{\partial}{\partial r} r \frac{\partial u_z}{\partial r} . \tag{2.28}$$

Solution. To solve, we assume that $\frac{\partial p}{\partial z}$ is uniform, and integrate the r terms:

$$\frac{\partial p}{\partial z} \frac{1}{2} r^2 + C_1 = r \eta \frac{\partial u_z}{\partial r} , \tag{2.29}$$

rearrange,

$$\frac{\partial p}{\partial z} \frac{r}{2\eta} + \frac{C_1}{\eta r} = \frac{\partial u_z}{\partial r} \tag{2.30}$$

and integrate again to find

$$u_z = \frac{\partial p}{\partial z} \frac{r^2}{4\eta} + \frac{C_1}{\eta} \ln r + C_2 . \tag{2.31}$$

By requiring that (a) u_z be bounded at $r = 0$ and (b) $u_z = 0$ at $r = R$, we solve and find

Poiseuille solution for flow in a tube of radius R

$$u_z = -\frac{1}{4\eta} \frac{\partial p}{\partial z} \left(R^2 - r^2 \right) . \tag{2.32}$$

We can integrate this distribution spatially to find the volumetric flow rate Q:

flow rate in a tube of radius R caused by a pressure gradient

$$Q = -\frac{\pi R^4}{8\eta} \frac{\partial p}{\partial z} , \tag{2.33}$$

and normalize by the area to generate the area-averaged velocity $\overline{u_z}$:

mean velocity for flow in a tube of radius R caused by a pressure gradient

$$\overline{u_z} = \left(-\frac{\partial p}{\partial z} \right) \frac{R^2}{8\eta} . \tag{2.34}$$

Physical interpretation. The steady Poiseuille flow has no convective transport of momentum, nor acceleration. The governing equation highlights the balance between the net pressure forces and the net viscous force. Because the pressure gradient (and thus the net pressure force) is uniform, the derivative of the shear stress is uniform, and thus the concavity (i.e., 2nd derivative with respect to y) of the distribution is uniform. If we draw a control volume around the fluid in its entirety, we can see that the net pressure force on the control volume is $\left(-\frac{\partial p}{\partial z}dz(\pi R^2)\right)$, whereas the viscous force per unit area that the wall applies to the fluid is $\frac{R}{2}\frac{\partial p}{\partial z}$, and the net viscous force in total is $\frac{\partial p}{\partial z}dz(\pi R^2)$.

Flow kinematics and viscous stresses. There are only two nonzero viscous stress terms for Poiseuille flow – the τ_{rz} and τ_{zr} terms. Each is given by the viscosity times the velocity gradient:

$$\tau_{zr} = \tau_{rz} = \eta \frac{\partial u_z}{\partial r}. \tag{2.35}$$

Evaluating the derivative, we obtain

shear stress for Poiseuille flow in a tube

$$\tau_{zr} = \tau_{rz} = \frac{r}{2}\frac{\partial p}{\partial z}. \tag{2.36}$$

For Poiseuille flows, the viscous stress varies linearly, with a minimum of zero at $r = 0$ and a maximum of $\frac{R}{2}\frac{\partial p}{\partial z}$ at $r = R$. The velocity distribution of a Poiseuille flow is parabolic.

EXAMPLE PROBLEM 2.5

Calculate the vorticity of the Poiseuille flow described in Subsection 2.1.2.

SOLUTION: The velocity is

$$\vec{u} = -\frac{1}{4\eta}\frac{\partial p}{\partial z}\left(R^2 - r^2\right)\hat{z}, \tag{2.37}$$

and the vorticity is given by $\vec{\omega} = \nabla \times \vec{u}$:

$$\vec{\omega} = -\frac{1}{2\eta}\frac{\partial p}{\partial z} r\hat{\theta}. \tag{2.38}$$

Reynolds number. As was the case with the Couette flow, we need a characteristic velocity and a characteristic length to use to define the Reynolds number.

Unlike for the Couette flow, the problem statement does not prescribe any velocities; rather, it specifies a pressure gradient. We therefore use the velocities from the solution to guide definition of the Reynolds number. We typically define $U = |\overline{u_z}|$ for Poiseuille flow and define $\ell = 2R$ for this flow. With these definitions,

definition of Reynolds number for Poiseuille flow

$$Re_{\text{Poi}} = \frac{2\rho\,|\overline{u_z}|\,R}{\eta} = \frac{\rho R^3 \left|\frac{\partial p}{\partial z}\right|}{4\eta^2}. \tag{2.39}$$

As was the case for Couette flow, *Re* for Poiseuille flow tells us whether the laminar flow distribution derived above is observed – for $Re < 2300$, the flow is laminar.

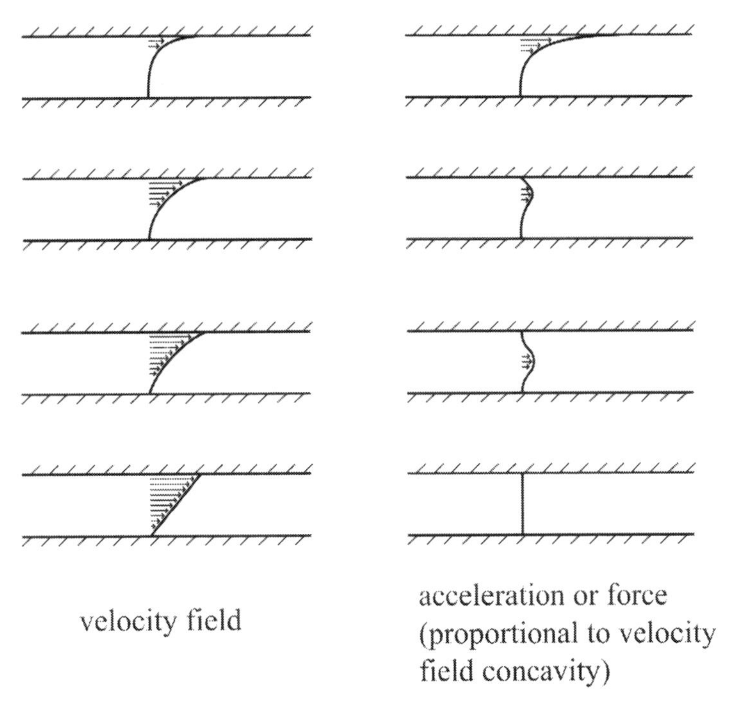

Fig. 2.3 Startup of a Couette flow. Elapsed time increases from top to bottom.

2.2 STARTUP AND DEVELOPMENT OF UNIDIRECTIONAL FLOWS

The solutions above are for *steady* flows through infinitely long channels. We are also interested in flows with unsteady boundary conditions, for example the *startup* of the flow or the flow in response to oscillatory boundary conditions. For example, startup of a Couette flow corresponds to the case where two plates and the fluid are motionless for time $t < 0$, but the plates move for $t > 0$. Figure 2.3 shows the startup of Couette flow, both in terms of velocity as well as local force and acceleration. The startup and steady-state of a Couette flow is a useful illustration of the forces, velocities, and accelerations for a simple flow. Consider first the steady-state solution. Because this is at equilibrium, the net force on a control volume is zero. In fact, each term in the Navier–Stokes equation is zero. In contrast, upon startup, the fluid near the top wall feels a net viscous force pulling it forward, and thus the fluid near the wall accelerates. The viscous force and thus the acceleration diffuses down toward the other plate, until eventually the steady-state solution is reached. Similarly, the startup of Poiseuille flow (acceleration from rest of quiescent fluid caused by a pressure gradient applied at $t = 0$) initially involves uniform acceleration as the pressure gradient is applied. This is then counteracted by a viscous force, first at the walls, and then throughout the fluid, until steady-state is reached. The startup of a Poiseuille flow is shown in Fig. 2.4. Startup of Couette and Poiseuille flows (and, in fact, all time-varying Couette and Poiseuille flows) can be solved analytically with separation of variables, with solutions given by eigenfunction expansions – sine series for plane-symmetric flow and Bessel series for axisymmetric flow. This is possible in part because the governing equation for each flow is linear, because the (nonlinear) convection term is zero for these flows. We ignore startup effects for flow through a channel of radius or half-depth R at all elapsed times $t \gg \frac{\rho R^2}{\eta}$ or for changes with a characteristic frequency $\omega \ll \frac{\eta}{\rho R^2}$.

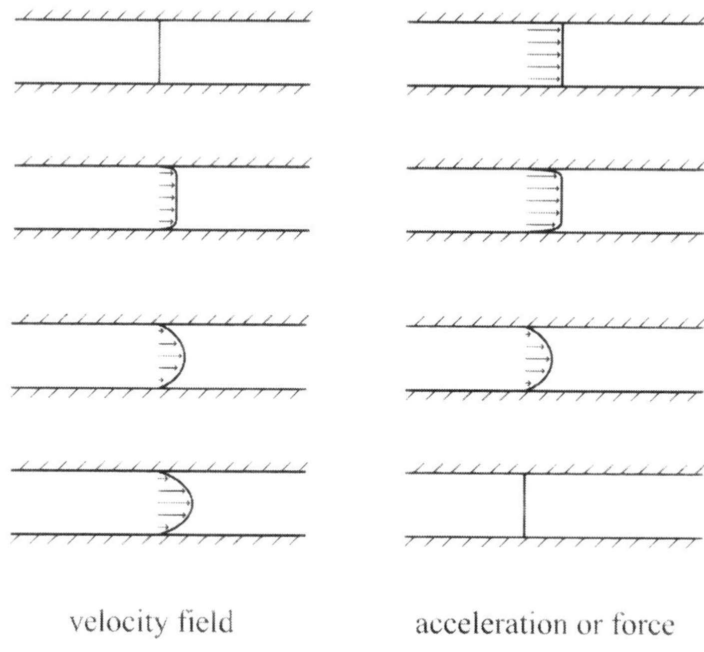

Fig. 2.4 Startup of a Poiseuille flow. Elapsed time increases from top to bottom.

In contrast to startup, which refers to a temporal change in a fluid system, *development* of a flow refers to the spatial evolution of the flow profile from an inlet profile to the final profile that the channel would have if it were infinitely long. Near the inlet of a channel, we cannot assume that the velocity gradients are normal to the flow direction, and thus the convective terms are nonzero and the Navier–Stokes equations cannot be solved analytically. Solutions for the flow near an inlet to a channel are typically performed numerically. We refer to a region of a flow as *fully developed* if the velocity profile in that region is equal to that which would be observed if the channel were infinitely long. The laminar flow in a long, narrow channel of radius or half-depth R can be assumed fully developed when the distance from the inlet ℓ satisfies the relations $\ell/R \gg Re$ and $\ell/R \gg 1$. When this is satisfied, we can use the results from this chapter directly to analyze flows in channels of finite length.

2.3 SUMMARY

The Navier–Stokes equations can be solved analytically if certain simplifications are made. In particular, the convection term $\vec{u} \cdot \nabla \vec{u}$ is the most mathematically difficult term in the Navier–Stokes equations to handle, and this chapter examines flows for which the geometry is so simple that we can assume away any spatial dependence that leads to nonzero net convective flux. This approach leads to steady solutions for Couette flow, i.e., flow between two infinite parallel plates:

$$u = \frac{u_H + u_L}{2} + \frac{u_H - u_L}{2}\left(\frac{y}{h}\right), \tag{2.40}$$

and Hagen–Poiseuille flow, i.e., pressure-driven flow through a circular channel

$$u_z = -\frac{1}{4\eta}\frac{\partial p}{\partial z}\left(R^2 - r^2\right) . \tag{2.41}$$

The *startup* of these flows highlights differences among forces, velocities, and accelerations. During startup, the acceleration is proportional to the force, which is a sum of the pressure and viscous forces. At equilibrium, the acceleration is by definition zero, and the concavity of the velocity distribution (which is proportional to the viscous force) is proportional to the local pressure gradient. Thus steady Couette flow, which has no pressure gradient, has no concavity in the velocity distribution, whereas steady Poiseuille flow, which has a uniform pressure gradient, has uniform concavity in the velocity profile. Development of these flows includes nonzero convective terms, which are nonlinear and preclude general analytical solution.

2.4 SUPPLEMENTARY READING

Couette and Poiseuille flows are the most basic fluid flows. Steady solutions are covered in introductory fluids or transport texts such as Fox, Pritchard, and McDonald [17], Munson, Young, and Okiishi [18], White [19], and Bird, Stewart, and Lightfoot [20]. More advanced texts [22, 30] cover unsteady solutions by use of separation of variables. Asymptotic approximations used to describe perturbations to these flows are discussed in Bruus [33] and Leal [30], in addition to Van Dyke's classic monograph [34].

2.5 EXERCISES

2.1 Show that, if $\vec{u} = (u, 0, 0)$, i.e., flow is in the x direction only, then $\vec{u} \cdot \nabla\vec{u} = (u\frac{\partial u}{\partial x}, 0, 0)$.

2.2 Consider the following two cases:

(a) A Newtonian fluid.

(b) A power-law fluid, i.e., a fluid for which $\tau_{xy} = K\frac{du}{dy}|\frac{du}{dy}|^{n-1}$. You may simplify your math by assuming that $\frac{du}{dy}$ is positive.

Consider laminar flow between two infinite parallel plates, each aligned with the xz plane. The plates are located at $y = \pm h$. There are no applied pressure gradients. Assume the top plate moves in the x direction with velocity u_H and the bottom plate moves in the x direction with velocity u_L. For each case,

(a) Solve for the flow between the plates as a function of y.

(b) Derive relations for $u(y)$, $\vec{\vec{\tau}}(y)$, and $\vec{\vec{\varepsilon}}(y)$. Note that $\vec{\vec{\tau}}$ and $\vec{\vec{\varepsilon}}$ should both be second-order tensors. For each of these three parameters, comment on how the result is influenced by the magnitude of the viscosity as well as by its strain rate dependence.

(c) Evaluate the force per unit area that each surface must apply *to the fluid* to maintain this flow. Comment on how the result is influenced by the magnitude of the viscosity as well as by its strain rate dependence.

2.3 Consider two flat, parallel, infinite plates located at $y = h$ and $y = -h$ separated by a Newtonian fluid with viscosity η. Assume the lower plate is motionless. Assume that the top plate is actuated in the positive x direction with a force per unit area given by F''. In terms of the given parameters, determine an expression for the velocity distribution between the plates.

2.4 Consider a microchannel of circular cross section with length $L = 5$ cm and radius $R = 20$ μm connected to two large reservoirs with $d = 4$ mm. One reservoir contains a column of water of height 1 cm, and the other contains a column of water of height 0.5 cm. Assume that the flow through the channel can be approximated for all times as quasi-steady Hagen–Poiseuille flow, and describe the velocity in the microchannel as a function of r and t. Given the time variation of the flow, confirm your assumption about the quasi-steady nature by evaluating the characteristic time of the flow changes and comparing that with $\frac{\rho R^2}{\eta}$.

2.5 Integrate Eq. (2.32) to derive Eq. (2.33).

2.6 Average Eq. (2.33) to derive Eq. (2.34).

2.7 Complete the following first for a Newtonian fluid and then for a power-law fluid. Define a power-law fluid as one whose viscous stress is given by ($\tau_{rz} = K\frac{du_z}{dr}|\frac{du_z}{dr}|^{n-1}$). You may simplify your math by assuming that $\frac{du_z}{dr}$ is positive. Consider laminar flow through a circular pipe whose axis is along the z axis. The pipe has radius R. Assume a pressure gradient $\frac{dp}{dz}$ is applied.

(a) Solve for the flow in the tube as a function of r.

(b) Derive relations for $u_z(r)$, $\vec{\vec{\tau}}(r)$, and $\vec{\vec{\varepsilon}}(r)$. Note that $\vec{\vec{\tau}}$ and $\vec{\vec{\varepsilon}}$ should both be second-order tensors. For each of these three parameters, comment on how your result is influenced by the magnitude of the viscosity as well as by its strain rate dependence.

(c) Define Q as the total flow rate through the tube. Derive a relation for Q. What is the relation between Q and $\frac{dp}{dz}$? Is it linear?

(d) Plot the velocity distribution for the following cases. Put all three graphs on one plot. Plot u_z on the abscissa and plot r on the ordinate. Assume $\frac{dp}{dz} = -1$ N/m^3, $R = 1$ m:

i. $n = 1$, $K = 1$ kg/m s (Newtonian).
ii. $n = 0.25$, $K = 0.517$ kg/m s$^{1.75}$ (shear thinning),
iii. $n = 4$, $K = 18.36$ kg s^2/m (shear thickening).

These n and K values are chosen to ensure that the total flow rate through the system is the same in all cases. Note that the units of K depend on the value of n.

2.8 Consider a Newtonian fluid inside a pipe with a circular cross section and a radius R. Assume the fluid is motionless at $t = 0$. A pressure gradient is applied at $t = 0$. As a function of r and other fluid and system parameters, what is the instantaneous fluid acceleration immediately after the pressure gradient is applied?

2.9 Consider steady developing pipe flow of a Newtonian fluid at the inlet of a long pipe with circular cross section and radius R. Assume the velocity at the inlet is given by $u(r) = U$. Simplify the Navier–Stokes equations that govern the flow *at the inlet* (not necessarily anywhere else in the flow, just at the inlet). Which terms in the Navier–Stokes equations are nonzero?

2.10 Consider the fluid-static problem of water in between two stationary infinite plates located at $\pm d$. Assume all fluid is motionless. Assume that the fluid (water) has a density of 1000 kg/m^3 and assume that gravity acts on the water with an downward gravitational acceleration $g = -9.8$ m/s^2. Assume that the pressure at $z = 0$ is given by p_0.

Given that the 1D Navier–Stokes equations for the z velocity w in a system with these geometric simplifications but nonzero gravitational body forces is

$$0 = -\frac{\partial p}{\partial z} + \eta \frac{\partial^2 w}{\partial x^2} - \rho g \,, \tag{2.42}$$

solve for the pressure distribution.

2.11 Consider 1D steady flow in the z direction between two stationary infinite plates located at $\pm d$. Assume that there are no pressure gradients. Assume that all flow is in the z direction and all velocity gradients are in the x direction. Assume that the fluid (water) has a density of 1000 kg/m^3, and assume that gravity acts on the water with an downward gravitational acceleration $g = -9.8$ m/s^2.

Given that the 1D Navier–Stokes equations for the z velocity w in a system with these geometric simplifications but nonzero gravitational body forces is

$$0 = \eta \frac{\partial^2 w}{\partial x^2} - \rho g \,, \tag{2.43}$$

solve for the steady-state velocity distribution between the plates.

2.12 Consider 1D steady flow in the z direction between two stationary infinite plates located at $x = \pm d$. Assume that the pressure in the system is given by $p = p_0 - \rho g z$, where ρ is the nominal density of the fluid and p_0 is by definition the pressure at $z = 0$. Assume that all flow is in the z direction and all velocity gradients are in the x direction.

Given that the 1D Navier–Stokes equations for the z velocity w in a system with these geometric simplifications but nonzero gravitational body forces is

$$0 = -\frac{\partial p}{\partial z} + \eta \frac{\partial^2 w}{\partial x^2} - \rho g \,, \tag{2.44}$$

complete the following exercises:

(a) Show that the steady-state velocity distribution between the plates is zero, i.e., the fluid is motionless.

(b) Perturb this system by changing the density of the fluid as a function of x. Assume that the density averaged over x is unchanged, but now assume that the fluid at the center of the channel is made denser by adding a large number of metal microparticles that are heaver than water, and assume that the fluid at the edges of the channel is made less dense by adding a large number of oil microdroplets. Assume that the effect of these local changes in density is that there is a local body force per unit volume:

$$\text{body force per volume} = \begin{cases} f & \text{for all } \; |x| > d\frac{\alpha-2}{\alpha} \\ -f\frac{2}{\alpha-2} & \text{for all } \; |x| < d\frac{\alpha-2}{\alpha} \end{cases} , \tag{2.45}$$

where $2 < \alpha < \infty$ is a geometric parameter that describes the distribution of the buoyancy perturbation ($\alpha \to \infty$ implies that the positive buoyancy is all concentrated near the wall, and $\alpha \to 2$ implies that the negative buoyancy is all concentrated in the center). Assume that this body force is constant with time. Confirm that the force definition in Eq. (2.45) leads to no net force by integrating this distribution over $-d < x < d$ and showing that the integral is equal to zero.

(c) Assume that the flow remains 1D, and solve for the velocity distribution in this system and show that it is given by

$$w = \begin{cases} \frac{f}{\eta}\left(\frac{x^2}{\alpha-2} - \frac{d^2}{\alpha}\right) & \text{for all} \quad |x| < d\frac{\alpha-2}{\alpha} \\ -\frac{f}{2\eta}(x-d)^2 & \text{for all} \quad |x| > d\frac{\alpha-2}{\alpha} \end{cases} \tag{2.46}$$

What is the centerline velocity?

(d) Plot your velocity distribution (normalized by fd/η) versus x/d for $\alpha = \frac{1}{0.04}, \frac{1}{0.13}, \frac{1}{0.22}, \frac{1}{0.31}, \frac{1}{0.40}, \frac{1}{0.49}$.

(e) Draw a control volume around a rectangular section of the flow spanning over $-d < x < d$ and over a finite range in z. What is the net body force? The net convection of momentum through the top and bottom faces? The net momentum transfer at the walls owing to viscous forces? How would the net convective and diffusive transfer of momentum be different if there were a net body force?

(f) Qualitatively describe how a motionless fluid would accelerate to this steady-state solution if the gravitational acceleration were "turned on" at time $t = 0$.

(g) This flow has no net body force, yet has nonzero local and averaged z velocities. Explain how the *distribution* of body forces in this system creates a flow even though the net body force is zero.

2.13 Consider 1D steady flow in the z direction between two stationary infinite plates located at $x = \pm d$. Assume that the pressure in the system is given by $p = p_0 - \rho g z$, where ρ is the nominal density of the fluid and p_0 is by definition the pressure at $z = 0$. Assume all flow is in the z direction and all velocity gradients are in the x direction.

Given that the 1D Navier–Stokes equations for the z velocity w in a system with these geometric simplifications but nonzero gravitational body forces is

$$0 = -\frac{\partial p}{\partial z} + \eta\frac{\partial^2 w}{\partial x^2} - \rho g\,, \tag{2.47}$$

complete the following exercises:

(a) Show that the steady-state velocity distribution between the plates is zero, i.e., the fluid is motionless.

(b) Now perturb this system by changing the density of the fluid as a function of x. Assume that the fluid near the walls is made denser by adding a large number of metal microparticles that are heaver than water. Assume that the concentration of microparticles is largest at the wall and decays exponentially with the distance from the wall (i.e., $d - |x|$) with a characteristic length denoted by λ (assume λ is small relative to d), so that the net body force on the fluid is given by

$$\text{body force per unit volume} = -\frac{fd}{\lambda}\exp\left(-\frac{d-|x|}{\lambda}\right). \tag{2.48}$$

Show that the total body force per height per depth (including effects at both walls) is $-2fd$.

(c) Assume that the flow remains 1D, and solve for the velocity distribution in this system and show that it is given by

$$w = -\frac{fd^2}{\eta}\frac{\lambda}{d}\left[1 - \exp\left(-\frac{d - |x|}{\lambda}\right)\right]. \tag{2.49}$$

What is the centerline velocity?

(d) Plot the velocity distribution (normalized by fd^2/η) for the case in which $\lambda = d/100$.

(e) Draw a control volume around a rectangular section of the flow spanning over $-d < x < d$ and over a finite range in z. What is the net body force? The net convection of momentum through the top and bottom faces? The net momentum transfer at the walls owing to viscous forces? Are these results a function of λ?

(f) This flow has a net force that is independent of λ but a velocity distribution that is linearly proportional to λ (at least far from the wall). Explain how the *distribution* of body forces in this system affects the flow velocity even though the net body force is independent of λ.

2.14 Consider the dynamic problem of a capillary inserted into a reservoir such that the height of the water in the capillary at time $t = 0$ is equal to the height outside. This is similar to the capillary height problem, but we are searching for the time history of the solution.

(a) As we have shown in the text, the gravity forces balance the capillary forces at equilibrium (when the height $h = H$). What is the net force for the general case of a height $h \neq H$?

(b) Assume that the liquid rises in response to this force. If flow through the capillary is modeled as Poiseuille flow through a circular capillary, what is the flow rate and the average velocity u_{avg} in terms of the equilibrium height H, the current height h, and geometric and fluid properties?

(c) Given that $u_{avg} = dh/dt$, write an ordinary differential equation (ODE) for $h(t)$.

(d) Your expression for dh/dt should have two terms, one that dominates as $t \to 0$ and one that dominates as $t \to \infty$. Solve the ODE in each limit.

2.15 Consider a microchannel connected at each end to a macroscopic reservoir, one of which is full (Fig. 2.5). Neglect gravity (i.e., assume the filled reservoir is large enough to serve as a reservoir of fluid but small enough that hydrostatic head can be neglected). The liquid–air interface is right at the entrance of the microchannel proper. If the surface is solvophilic ($\theta < 90$), the liquid will be pulled into the channel owing to capillary forces. This problem is equivalent to the capillary height problem, but with gravity forces neglected. Assume that the microchannel is much wider than it is deep, i.e., model the microchannel flow as 2D infinite plate flow in a channel of depth d. Define l as the distance the interface has moved down the channel, and take $l = 0$ at $t = 0$. Solve for $l(t)$.

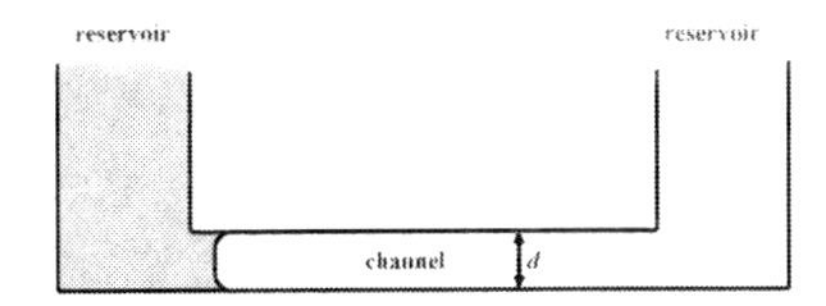

Fig. 2.5 Schematic of a microchip that is about to be filled by capillary action.

As can be seen from the result, there is *no* equilibrium solution in this case, and the interface will move through the microchannel until it is filled.

2.16 Consider two motionless infinite flat plates located at $y = 0$ and $y = 2h$ and separated by a Newtonian fluid. Assume that the pressure everywhere is uniform.

Assume that, at $t = 0$, the instantaneous velocity between the plates is given by

$$u_0(y) = u(y, t = 0) = A \sin\left(\frac{\pi y}{2h}\right), \tag{2.50}$$

where A is a constant [m/s]. This is clearly *not* an equilibrium solution, and the velocity will change with time.

(a) Rearrange the Navier–Stokes equations in vector form,

$$\rho \frac{\partial}{\partial t}\vec{\boldsymbol{u}} + \rho \vec{\boldsymbol{u}} \cdot \nabla \vec{\boldsymbol{u}} = -\nabla p + \eta \nabla^2 \vec{\boldsymbol{u}}, \tag{2.51}$$

to solve for $\frac{\partial}{\partial t}\vec{\boldsymbol{u}}$ as a function of other parameters.

(b) Assume that $v = w = 0$ and $\frac{\partial u}{\partial x} = \frac{\partial u}{\partial z} = 0$ and simplify your vector equation to show that

$$\frac{\partial u}{\partial t} = \frac{\eta}{\rho}\frac{\partial^2 u}{\partial y^2}. \tag{2.52}$$

(c) Given $u_0(y)$, evaluate the viscous term at $t = 0$ and compare the viscous term with $u_0(y)$. Show that the viscous term is directly proportional to u at $t = 0$. Although you have not proved so yet, this proportionality holds for all t. Given that this holds for all t, solve for $u(y, t)$.

(d) Will this technique work for other initial conditions? For example, would this work if the initial condition were a parabolic relation such as

$$u_0(y) = Ay(2h - y)? \tag{2.53}$$

2.17 Consider two infinite flat plates located at $y = 0$ and $y = 2h$ separated by a uniform-pressure Newtonian fluid.

Consider the case in which the system has reached a steady state with the top plate motionless and the bottom ($y = 0$) plate moving in the x direction with velocity U, i.e., the velocity is given by $u(t = 0) = U(1 - \frac{y}{2h})$. At $t = 0$, the bottom plate is stopped, and both plates are motionless for all times $t > 0$. This flow is no longer steady, and the velocity will change with time.

(a) This problem has homogeneous boundary conditions and can be solved by use of separation of variables and linear superposition of solutions. We proceed first without paying attention to the initial conditions, derive a general solution, and then apply the initial condition.

Assume that the total solution $u(y, t)$ can be written as a sum of *modes* u_n:

$$u = \sum_{n=1}^{\infty} u_n, \tag{2.54}$$

Where each u_n can be written as the product of two functions, one of which is a function of y only and one of which is a function of t only:

$$u_n(y, t) = Y_n(y) T_n(t) \tag{2.55}$$

Substitute Eq. (2.55) into Eq. (2.52) and rearrange the resulting equation into a left-hand side (LHS) that is a function of *only* t and a right-hand side (RHS) that is a function of *only* y. Explain why both the LHS and the RHS of this equation should be equal to a constant, which you can call $-k_n$.

(b) Set the RHS of the equation you derived equal to $-k_n$ and identify the form the Y_n solutions must take if they are to satisfy the boundary conditions as well as the equation you just wrote. Only a specific set of k_n values will satisfy the boundary conditions – what are these values?

(c) Given the known k_n values, solve for T_n by setting the LHS of the separated equation equal to k_n.

(d) Write the complete solution for $u(y, t)$ and simplify the solution at $t = 0$ to evaluate $u(y, t = 0)$.

(e) Now we return to the initial condition specified originally. Conveniently, u_0 can be written as a Fourier sine series,

$$u(y, t = 0) = U(1 - \frac{y}{2h}) = U \sum_{n=1}^{\infty} \frac{2}{n\pi} \sin \frac{n\pi y}{2h}, \tag{2.56}$$

which directly specifies the solution because the initial condition and the general solution can be matched term by term. Write out this solution. Evaluate and plot this solution (u on the abscissa; y on the ordinate) for the case in which $\eta = 1 \times 10^{-3}$ Pa s, $\rho = 1000$ kg/m^3, $U = 4$ cm/s, and $h = 1$ mm. Evaluate the sum from $n = 1$ to $n = 100$. Note that the Fourier series approximation of the initial conditions will be oscillatory at $t = 0$ regardless of how many terms you sum. On this plot, graph the distribution of u for $t = 0$, $t = 0.001$ s, $t = 0.01$ s, and every 0.05 s from 0.05 s to 1 s.

2.18 Stokes' first problem considers a motionless infinite flat plate in an infinite quiescent fluid. At time $t = 0$, the flat plate is instantly accelerated to a speed U. The problem is to derive the velocity field $u(y, t)$ for all $t > 0$. Assuming the flow is in the x direction only and velocity gradients are in the y direction only, the simplified equation is

$$\frac{\partial u}{\partial t} = \frac{\eta}{\rho} \frac{\partial^2 u}{\partial y^2}. \tag{2.57}$$

(a) We solve this problem by using a *similarity solution*. Similarity solutions are often appropriate when problems have infinite or semi-infinite domains (in contrast, separation of variables often works well on problems with finite domains).

Similarity solutions attempt to convert partial differential equations (PDEs) to ODEs by changing variables. For this problem, we will find that if we recast the problem in terms of a different variable (rather than y or t), the PDE can be converted to an ODE.

We can achieve similarity solutions by *guessing* a functional form for the so-called similarity variable η and by *guessing* the relation between η and the velocity or the stream function.[2]

Postulate that u can be written as

$$u = Bt^q f(\eta), \tag{2.58}$$

[2] In contrast, for separation of variables we guess that the velocity can be written as the product of two functions.

where B and q are parameters we can choose arbitrarily, and η is given by

$$\eta = Ayt^p , \tag{2.59}$$

where A and p are also parameters we can choose.

Given the forms for η and u, substitute into the governing equation and derive an equation that links f, f', and f''. Here f' implies $df/d\eta$ and f'' implies $d^2 f/d\eta^2$.

(b) If the relation is to be an ODE, it can have f, f', f'', η, p, q, and the like, but it cannot have y or t in it. Eliminate y, as needed, by replacing y with η/At^p and simplify the equation so that only one term has t in it. What value(s) of p and q (which are otherwise arbitrary) allow t and y to both be eliminated from this equation?

(c) Simplify the equation with the correct choices of p and q, and assign $A = \frac{1}{2\sqrt{\eta/\rho}}$. Note that p is dictated by the need to make the resulting equation an ODE, and the parameter q is dictated by the steady boundary condition.

Make note of the fact that the *error function* (erf) is proportional to the area under a Gaussian curve from 0 to x. It is defined as

$$\text{erf}(x) = \frac{2}{\sqrt{\pi}} \int_0^x \exp\left(-x'^2\right) dx' , \tag{2.60}$$

and the *complementary error function* (erfc) gives one minus the error function:

$$\text{erfc}(x) = 1 - \text{erf}(x) = 1 - \frac{2}{\sqrt{\pi}} \int_0^x \exp\left(-x'^2\right) dx' . \tag{2.61}$$

Integrate the resulting ODE analytically to find f and show that the solution for u is

$$u = U \,\text{erfc}\left(\frac{y}{2\sqrt{\eta t/\rho}}\right) . \tag{2.62}$$

2.19 We see in Chapter 6 that electric-field-driven fluid flow in microchannels can often be approximated as flow driven by a surface that is effectively moving. Because of this, the startup of electroosmotic flow is effectively the startup of a Couette flow. Thus the startup of a Couette flow is not just a classical separation of variables problem; it is immediately relevant for any microfluidic experiment in which an electric field is turned on.

Consider two parallel plates separated by a distance $2d$, with plates located at $y = \pm d$. Assume that, for $t < 0$, the fluid and plates are motionless. Assume that, at time $t = 0$, both plates are instantaneously accelerated to velocity U. Using separation of variables, calculate $u(y, t)$.

2.20 Consider flow in an infinite circular tube of uniform circular cross section driven by a sinusoidally varying pressure gradient:

$$\frac{\partial p}{\partial z} = -A\left(1 + \alpha \sin \omega t\right) . \tag{2.63}$$

Using separation of variables, solve for $u_z(\imath, t)$.

2.21 The pressure-driven flow through a circular pipe of uniform cross section is relatively straightforward to solve and is often used as an approximation of flow through microfluidic channels of a variety of geometries. However, flow through channels of varying cross section or that are not straight is of course not correctly described by this approximation. One situation in which flow through a more complicated structure can be better approximated is the case of a slightly curved tube. Such structures have been used for separation channels in a variety of configurations and have also been used with a view toward how the secondary flow leads to unique flow structures.

Consider a pipe of uniform circular cross section with radius R. Assume that the axis of this pipe is curved with radius R_0, where $R_0 \gg R$. Consider fully developed flow, which is independent of distance along the pipe. Write the solution as a perturbation expansion in R/R_0, and calculate the solution to first order in the perturbation parameter. The secondary flow of this problem is called *Dean flow*.

2.22 Microfabrication techniques make it rather straightforward to create surfaces with controlled texture or raised surfaces. These surfaces have been used to generate novel flow patterns.

Consider flow between two infinite parallel plates that are separated by a distance d. Assume that the plates are not perfectly flat – in fact, they have a sinusoidal variation that makes the surface grooved. Specifically, assume that the walls are located at

$$y = \pm\frac{d}{2}\left(1 + \alpha \sin\frac{2\pi x}{L}\right). \tag{2.64}$$

Assume that $\alpha \ll 1$ and use a first-order perturbation expansion in α to predict the flow distribution if a pressure gradient $\frac{\partial p}{\partial z}$ is specified and the flow is parallel to the grooves.

2.23 Consider flow between two infinite parallel plates that are separated by a distance d. Assume that the plates are not perfectly flat – in fact, they have a sinusoidal variation that makes the surface grooved. Specifically, assume that the walls are located at

$$y = \pm\frac{d}{2}\left(1 + \alpha \sin\frac{2\pi x}{L}\right). \tag{2.65}$$

Assume that $\alpha \ll 1$ and use a first-order perturbation expansion in α to predict the flow distribution if a pressure gradient $\frac{\partial p}{\partial x}$ is specified and the flow is perpendicular to the grooves.

3 Hydraulic Circuit Analysis

The previous chapters introduce the governing equations for fluid flow and provide solutions for simple unidirectional flows. Although we can only rarely use these simple solutions to *exactly* describe real flows, these unidirectional flows provide a framework for generating *engineering estimates* for many flows, and these estimation techniques reduce complex systems with large numbers of components into a relatively simple, approximate, linear description. This is relevant for flow in microdevices both (1) because microdevices can straightforwardly be made with large numbers of microchannels and (2) because those microchannels often have long, thin geometries that lead to largely unidirectional flows that are well approximated by the solutions presented in Chapter 2. Our approach, called *hydraulic circuit analysis*, involves assuming that the Poiseuille flow derived in Chapter 2 provides a sound engineering estimate of the pressure drops and flow rates through long, straight channels, even if the cross section is not exactly circular and the channels are neither perfectly straight nor infinite in extent. We thus write an approximate *linear* relation between pressure drops and flow rates through these channels – the *Hagen–Poiseuille law*. This law, combined with conservation of mass, approximately prescribes fluid flow through complex rigid networks by solution of sets of algebraic equations. By adding a hydraulic capacitance or *compliance*, this linearized analysis also allows prediction of unsteady flow through channels with a finite flexibility.

3.1 HYDRAULIC CIRCUIT ANALYSIS

Recall from Chapter 2 that the solution for steady flow of a Newtonian fluid through a channel with uniform circular cross section aligned in the z direction and defined by a radius R is given by

Poiseuille flow for tube of radius R

$$u_z = -\frac{1}{4\eta}\frac{\partial p}{\partial z}\left(R^2 - r^2\right) . \qquad (3.1)$$

From this velocity distribution, the volumetric flow rate Q in the positive z direction is given by

volumetric flow rate for Poiseuille flow through tube of radius R

$$Q = -\frac{\pi R^4}{8\eta}\frac{\partial p}{\partial z} . \qquad (3.2)$$

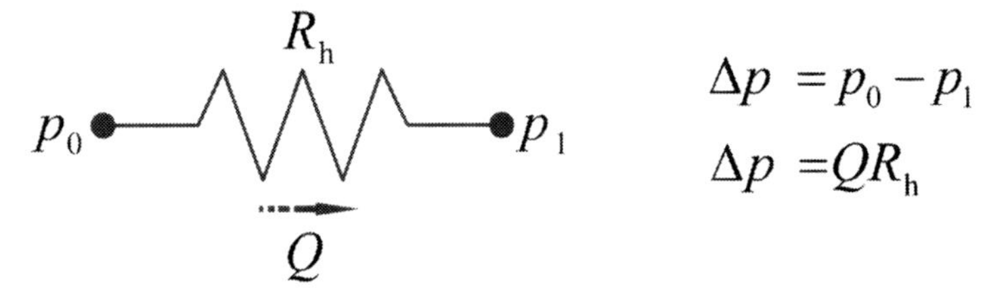

Fig. 3.1 Symbol and pressure–flow-rate relation for a tube with steady flow.

Strictly speaking, this result applies only for a channel that is perfectly straight and infinitely long. Let us, though, neglect entrance effects and simply use the infinite-tube result as an approximation of the flow through a finite tube. This approximation is good if $\frac{R}{L} \ll 1$ and $\frac{R}{L} \ll 1/Re$, in which case we can assume that the pressure gradient is uniform along the length of the tube. Further, if the flow is axial and the pressure gradient is uniform, we need pay attention to the magnitude of the pressure gradient alone, rather than its orientation with respect to a coordinate system. Thus we replace $-\frac{\partial p}{\partial z}$ with $\Delta p/L$, where Δp is the difference between the pressure at the inlet and that at the outlet, and L is the length of the tube. With this approximation, we have

$$Q = \frac{\pi R^4}{8\eta L}\Delta p\,, \tag{3.3}$$

where Q is defined as positive for flow from inlet to outlet. This can be rewritten in terms of the cross-sectional area as

$$Q = \frac{AR^2}{8\eta L}\Delta p\,. \tag{3.4}$$

If we now define the *hydraulic resistance* R_h as

$$R_h = \frac{8\eta L}{AR^2}\,, \tag{3.5}$$

we can write Eq. (3.2) as

Hagen–Poiseuille law

$$Q = \frac{\Delta p}{R_h}\,. \tag{3.6}$$

This relation is the *Hagen–Poiseuille law*, with the hydraulic resistance in Eq. (3.5) defined for a channel of circular cross section and radius R. This relation is approximately correct for long channels ($\frac{R}{L} \ll 1$ and $\frac{R}{L} \ll 1/Re$), for which this relation's errors (which are primarily at the entrance to the tube) can be neglected, and furthermore requires that the flow rate be laminar ($Re < 2300$). Equation (3.6) is analogous to Ohm's law for electrical circuits ($I = \Delta V/R$), which is presented in Chapter 5, and we adopt notation and symbolism similar to that for electrical circuits (Fig. 3.1) when denoting a channel in a channel network. In fact, all of the relations for hydraulic circuits are analogous to electrical circuit relations. Analogies between fluid flow and current flow are described in Table 3.1.

Table 3.1 Fluid-mechanical analogs for electrical circuit elements. Because these relations assume $\frac{d\Delta p}{dt} \simeq \frac{dp}{dt}$, the analogies between capacitance and compliance are approximate.

Electric circuit property or component	Fluid-mechanical analog
Battery or power supply	Pump
Voltage drop ΔV	Pressure drop Δp
Current I	Volumetric flow rate Q
Current density i	Fluid velocity u
Resistance R	Hydraulic resistance R_h
Capacitance C	Compliance C_h
Inductance L	Inertia (which is negligible in microfluidic systems)
Resistor and capacitor in parallel	Channel–fluid system that stores fluid by channel expansion, bubble compression, or fluid compression (very common)
Resistor and capacitor in series	Two channels in series separated by a fixed, compressible, impermeable object (relatively uncommon)
Current through resistor	Spatially averaged flow rate through a channel
Current through capacitor	$Q_{\text{inlet}} - Q_{\text{outlet}}$; volume stored in channel expansion, bubble compression, or fluid compression
Total current through circuit	Sum of spatially averaged flow rate through a channel plus fluid stored in channel

Now consider several tubes that intersect at a rigid node or junction. Conservation of mass requires that the net flow into or out of the junction must be zero:

conservation of mass at the node of a fluid network

$$\sum_{\text{channels}} Q = 0\,, \tag{3.7}$$

where positive Q denotes flow rate *into* the node.

The Hagen–Poiseuille law and conservation of mass together form the basis for steady-state hydraulic circuit analysis. Because both of these relations are linear relations, the pressures and flow rates in a fluid network can be written symbolically (see Fig. 3.2) and solved by use of systems of algebraic equations. In the sections below, we use this approach to describe flow through channels of a variety of cross sections and rigidities.

3.2 HYDRAULIC CIRCUIT EQUIVALENTS FOR FLUID FLOW IN MICROCHANNELS

The Hagen–Poiseuille relation is a good approximation of steady flow through long, narrow, rigid, microchannels with circular cross section, for which end effects can be ignored. However, microchannels rarely have circular cross section – the fabrication techniques used to make microchannels typically lead to rectangular, trapezoidal, or semicircular channel shapes. Thus the engineering application of hydraulic circuit analysis requires that we develop a more general model for the hydraulic resistance of a channel. Further, microchannels are often made of elastomers and are therefore not rigid; even rigid channels often behave as if they were not rigid, owing to bubbles trapped in the

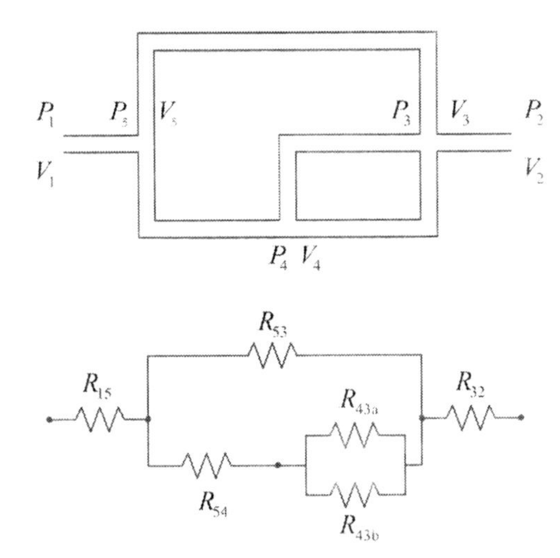

Fig. 3.2 A simple microfluidic device and its hydraulic circuit analog.

channel. Because of these factors, we define the hydraulic resistance and compliance of channels as a means for simplifying our analysis of these systems.

Hydraulic resistance. The preceding description for a circular microchannel can be generalized (at least approximately) by replacing the tube radius R with the hydraulic radius r_h. This *hydraulic radius* is a tool for estimating the flow resistance of a channel without solving the fluid flow equations exactly. In this form, our 1D model of a microchannel has hydraulic resistance R_h approximated by

hydraulic resistance of a circular microchannel

$$R_h \simeq \frac{8\eta L}{r_h^2 A}, \tag{3.8}$$

where r_h is the *hydraulic radius* of the channel,[1] which is given by

definition of hydraulic radius

$$r_h = \frac{2A}{\mathcal{P}}, \tag{3.9}$$

where A is the cross-sectional area of the channel, and $\mathcal{P}$ is the length of the perimeter of the channel. If the channel cross section is circular, r_h is equal to the radius of the circle. Relation (3.8) is derived from a conservation of momentum analysis on a control volume (see Exercise 3.1) and, owing to the approximations used in its derivation, is exact only for infinitely long, circular channels. For finite channels or noncircular shapes, using the hydraulic radius to predict hydraulic resistance is typically a good approximation (within 20%) of the correct result.

[1] R_h and r_h are different parameters – the former is a measure of the pressure–flow relation of a tube and the latter is an equivalent length that describes the geometry of a tube.

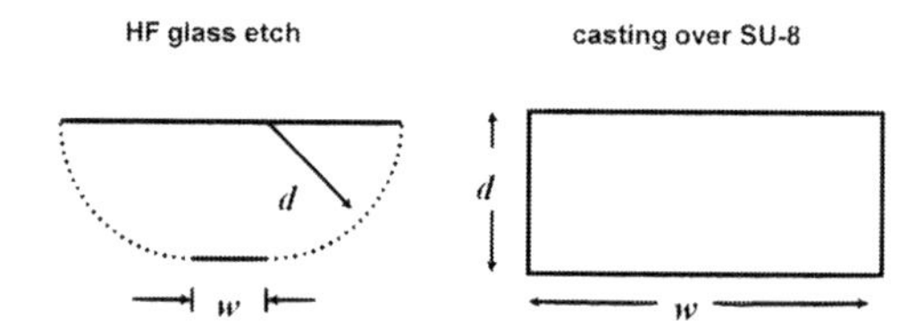

Fig. 3.3 Shapes of channels based on (a) wet etching of glass by HF and (b) casting over an SU-8 template.

EXAMPLE PROBLEM 3.1

Calculate the hydraulic radius of

1. a glass channel made by use of photoresist lithographically patterned to width $w = 240\ \mu$m and etched isotropically in HF to a depth of $d = 30\ \mu$m,
2. a channel cast in poly(dimethylsiloxane) (PDMS) made by patterning SU-8 molds of width $w = 40\ \mu$m and depth $d = 30\ \mu$m and casting the PDMS over the SU-8 mold.

Isotropic etches lead to microchannel cross sections consisting of circular side walls and flat top and bottom surfaces, whereas casting over SU-8 makes roughly rectangular channel cross sections (see Fig. 3.3).

SOLUTION: For the HF etch, we find

$$r_{\mathrm{h}} = \frac{2A}{\mathcal{P}} = 2\frac{240 \times 30 + 30^2\pi/4 + 30^2\pi/4}{240 + 300 + 30\pi/2 + 30\pi/2} \simeq 25\ \mu\mathrm{m}\,, \tag{3.10}$$

and for the SU-8 mold, we get

$$r_{\mathrm{h}} = \frac{2 \times 40 \times 30}{40 + 30 + 40 + 30} =\simeq 17\ \mu\mathrm{m}\,. \tag{3.11}$$

Electrokinetic coupling matrix. The relations between hydraulic circuits and electrical circuits are discussed throughout this text, both (a) because the systems are analogous in isolation and (b) because fluid flow and current become coupled whenever we consider a system with electrically charged walls. We use an *electrokinetic coupling equation*, which relates fluid flow and current flow to applied pressure fields and electric fields. If we ignore the effects of charged surfaces, this electrokinetic coupling equation is given (for a channel aligned in the x direction) by

$$\begin{bmatrix} Q/A \\ I/A \end{bmatrix} = \begin{bmatrix} r_{\mathrm{h}}^2/8\eta & 0 \\ 0 & \sigma \end{bmatrix} \begin{bmatrix} -\frac{dp}{dx} \\ E \end{bmatrix}, \tag{3.12}$$

where Q and I are the volumetric flow rate [m^3/s] and current [A], respectively, A is the channel area [m^2], σ is the electrical conductivity [S/m], and E is the magnitude of the electric field in the x direction [V/m]. The first of these equations is the Hagen–Poiseuille

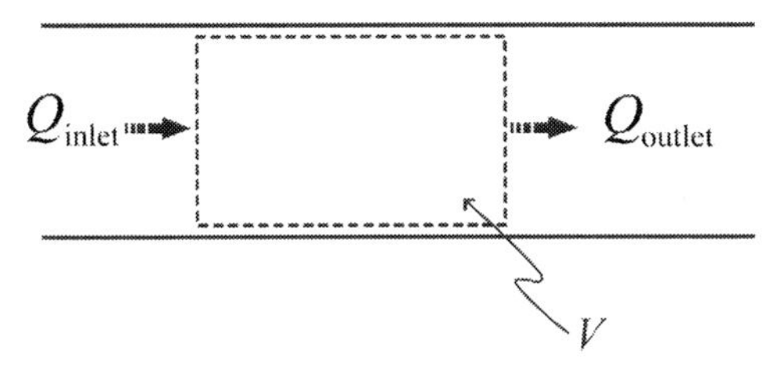

Fig. 3.4 Control volume for Eq. (3.15). Differences in inlet and outlet flow rates must lead to expansion of the control volume by expansion of the channel cross-sectional area.

law (3.6), and the second is Ohm's law for current density (5.69), to be discussed in Chapter 5.

Following Eq. (3.12), the *electrokinetic coupling matrix* χ is denoted as

$$\chi = \begin{bmatrix} \chi_{11} & \chi_{12} \\ \chi_{21} & \chi_{22} \end{bmatrix} = \begin{bmatrix} r_h^2/8\eta & 0 \\ 0 & \sigma \end{bmatrix}. \tag{3.13}$$

This electrokinetic coupling matrix becomes useful for describing the coupling of charged surfaces (electroosmosis, electrokinetic pumping, streaming current, and streaming potential), as well as the way these phenomena change when the channels under discussion become small. At present, we focus on the Hagen–Poiseuille law and use only the first equation described by this coupling matrix.

Hydraulic capacitance: system compliance. So far, we have assumed that all solid surfaces were infinitely rigid when we considered the fluid flow therein. If this is the case, then the control volume we use to analyze the tube is stationary, and the mass in the tube is constant with time. In this case, the Hagen–Poiseuille relation is a complete (albeit approximate) description of the flow–pressure relation in long narrow tubes. However, if the walls are not rigid, then the control volume we use to analyze the tube no longer has stationary boundaries, and pressure can cause the cross-sectional area of a microchannel to contract or expand. Thus the mass (or volume) inside the tube is no longer constant with time. Consider, for example, a small-diameter tube made out of the same material used to make a balloon. If we apply a low pressure to one end, the tube will not expand, and the flow–pressure relation is given by the Hagen–Poiseuille law. If we apply a high pressure to one end, the tube expands to a larger diameter, and the governing flow–pressure relation for the tube is the Hagen–Poiseuille law, but with a larger channel radius. If we apply a temporally varying pressure, the Hagen–Poiseuille law is insufficient to describe the pressure–flow relation in the tube, because the tube volume varies with time. We consider the conservation of mass relation for a control volume, Eq. (1.19):

$$\frac{\partial}{\partial t}\int_{\mathcal{V}} \rho \, dV = -\int_{S} (\rho \vec{u}) \cdot \hat{n} \, dA. \tag{3.14}$$

For incompressible, unidirectional flow, we can draw a simple control volume (Fig. 3.4) and simplify this relation to

$$\frac{\partial}{\partial t}\mathcal{V} = Q_{\text{inlet}} - Q_{\text{outlet}}. \tag{3.15}$$

If the volume of a tube changes, the time change in the tube volume is equal to the difference in the inlet and outlet volumetric flow rates. For a rigid tube, $\frac{\partial}{\partial t}\mathcal{V} = 0$ and $Q_{\text{inlet}} = Q_{\text{outlet}}$.

We describe the ability of a channel–fluid system to store fluid when pressurized as its *compliance* or hydraulic capacitance C_h [m^3/Pa]. The compliance of a system is given by the increase in effective stored fluid volume $\mathcal{V}$ per change in pressure,

definition of compliance, i.e., hydraulic capacitance

$$C_h = \frac{d\mathcal{V}}{dp}. \tag{3.16}$$

The fluid can be stored in an increase in the volume accessible to the fluid (via channel expansion or bubble compression) or via compression of the fluid itself. The effective volumetric rate at which fluid is stored by a compliant system ($Q_C = Q_{inlet} - Q_{outlet}$) is given by $Q_C = C_h \frac{dp}{dt}$. By substituting in the definition of compliance, this relation gives $Q_C = C_h \frac{dp}{dt} = \frac{d\mathcal{V}}{dp}\frac{dp}{dt} = \frac{d\mathcal{V}}{dt}$, which recapitulates the conservation of mass relation in Eq. (3.14).

The compliance is related to both the compressibility of the liquid as well as to the physical rigidity of its encapsulating material. Both factors are positive contributions to compliance and lead to outlet flow that is lower than the inlet flow when the pressure is increasing. For a working liquid with compressibility β, the compliance from fluid compressibility[2] is $\beta\mathcal{V}$. In microfluidic systems with water as the working fluid, the compression of the water itself is usually safely ignored. If we consider an incompressible fluid in a compliant channel, an increase in pressure pushes out on the material that is encapsulating the fluid, and thus the volume of the working fluid in the channel increases. However, the flexibility of the material encapsulating the fluid is widely variable depending on the microdevice material. Water in an infinitely rigid channel has low compliance, because the water is hard to compress and the channel is impossible to expand. Water enclosed within a flexible polymer channel has a large compliance, because the compliant channel inflates like a balloon when the pressure is increased, so the volume of water stored in the channel when the pressure is increased. Also, if a liquid system has bubbles, the bubbles play the role of a flexible encapsulating material, and, from the standpoint of water flow, a microdevice with bubbles inside it has large compliance. Overall, then, the effective fluid volume stored in the channel expands when the pressure increases.

As mentioned earlier, the most compliant component of a microdevice is usually any bubbles that have accidentally worked their way into the system, owing to the compliant nature of an ideal gas. The compliance of rigid microchannel substrates can often be ignored (for example, for glass, silicon, Zeonor, polycarbonate, and Plexiglas). However, the compliance of PDMS can typically not be ignored. Figure 3.5 illustrates the mechanical compliance of rigid and nonrigid walls and the effect of bubbles in these systems.

We typically represent compliance with a hydraulic capacitor with the symbol and pressure–flow-rate relation shown in Fig. 3.6. A channel that deforms in response to

[2] The compressibility $\beta = -\frac{1}{\mathcal{V}_m}\frac{d\mathcal{V}_m}{dp}$ of a fluid describes the change in volume $\mathcal{V}_m$ of a finite mass of fluid when pressure is changed. In contrast, the compliance of a channel describes the change in the amount of fluid stored in the channel. Thus the fluid compressibility describes how fluid shrinks in volume upon pressurization, allowing more fluid to be stored in the same volume, and the resulting compliance describes how the compression allows a net inflow of fluid into the system while this occurs. This explains how channel *expansion* and fluid *compression* both lead to compliance with the same sign.

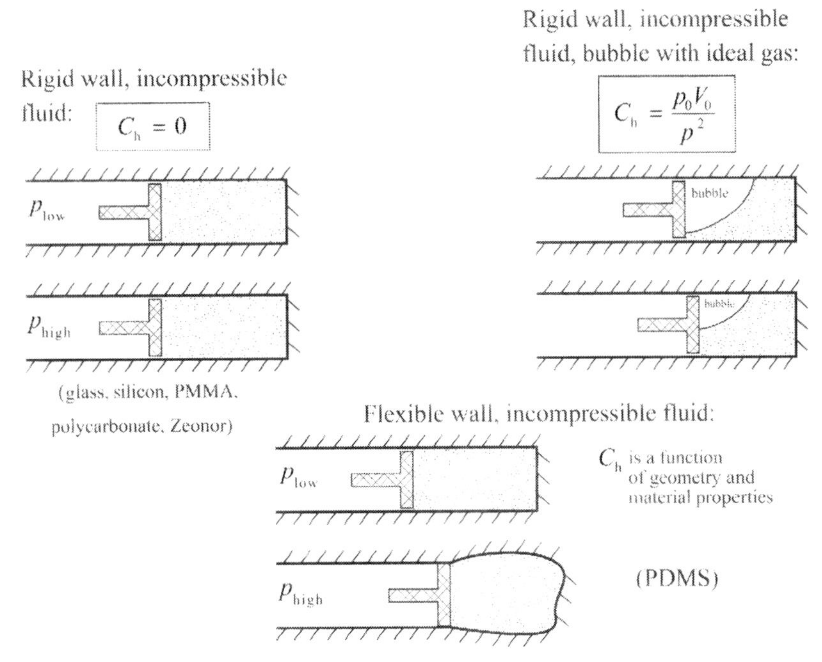

Fig. 3.5 Mechanical compliance of different types of microchannels. Top left: a rigid wall encapsulating an incompressible fluid has no compliance, and the hydraulic impedance of a channel is purely real. Top right: a rigid wall encapsulating both an incompressible fluid and a bubble has a large compliance, because the squeezing of the air caused by an applied pressure allows the volume available to the water in the channel to expand. Bottom: a compliant wall leads to a large compliance, because applied pressure causes the channel to expand, and thus the volume available to the water increases. PMMA, poly(methyl methacrylate).

pressure is most accurately represented with distributed resistances combined with distributed capacitances tied to a reference pressure. When we treat a channel as a single element, each channel can be approximated as a single resistor–capacitor pair. This simplification reduces the accuracy because the volume storage of the channel is discretized and localized to the capacitor, artificially introducing inlet–outlet asymmetry because the capacitor has to be on one side of the resistor. If we are willing to tolerate the artificial asymmetry of this mathematical model, such a description can properly model flow into and out of channels. However, this description is still not amenable to a convenient mathematical description such as the hydraulic impedance, described in Section 3.2.2. The mathematical complication is that the volume stored is proportional to absolute pressure changes, whereas the volume traversing the channel is proportional to the relative

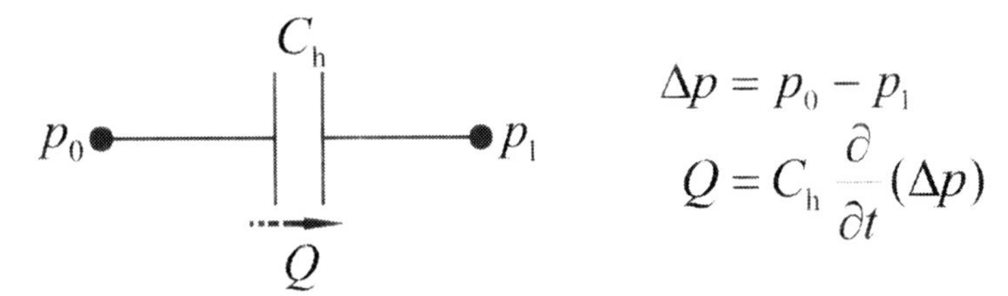

Fig. 3.6 Symbol and pressure–flow-rate relation for a hydraulic capacitor. Although the fluid storage in a compliant channel is physically related to the absolute pressure rather than to the pressure drop across the channel, the temporal changes of the two are often closely related, and we therefore write the system response approximately in terms of the pressure drop.

pressure drop across a channel. If the pressure fluctuations in a system are sinusoidal and relatively small, we can approximate $\frac{dp}{dt}$ with $\frac{d\Delta p}{dt}$, and we can treat the capacitance as being parallel to the resistance. The effect of this is that the volume stored in the channel is a function of the pressure *difference* rather than the absolute pressure. This is a major simplification analytically, because all equations are in terms of Δp and the impedance techniques in Section 3.2.2 can be used. The key conceptual weakness of this approach is that the equivalent circuit no longer properly distinguishes between the flow *into* a channel and *out of* the channel. Rather, the flow rate through the hydraulic resistor now corresponds to an inlet-to-outlet spatial average of the flow rate over the length of the tube, and the flow through the hydraulic capacitor now corresponds to an inlet-to-outlet integral of the fluid stored in the control volume expansion of fluid compression.

In this text, compliant channels are represented as a hydraulic resistor and hydraulic capacitor in parallel, and it is understood that this analysis quantitatively handles only small pressure fluctuations. Treating the compliance as a capacitor in parallel is mathematically convenient and appropriate if the goal is to determine inlet-to-outlet spatial average quantities.

3.2.1 Analytic representation of sinusoidal pressures and flow rates

If the time dependences of our pressures or flow rates are sinusoidal, the relations between flow properties and their derivatives take on a special form. These harmonic functions (sines and cosines) are also closely related to complex exponentials. Because of this, we use the *analytic* or *complex representation* of real quantities like pressure and flow rate, discussed in detail in Section G.3 in Appendix G.

EXAMPLE PROBLEM 3.2

Consider a flexible microchannel with a hydraulic resistance R_h and a compliance C_h in parallel. A pressure drop $\Delta p = \Delta p_0 \cos \omega t$ with $\Delta p > 0$ is applied across this circuit. Assume that $\frac{dp}{dt}$ is well approximated by $\frac{d\Delta p}{dt}$.

Using the relations $Q_R = \Delta p / R_h$ and $Q_C = C_h \frac{d\Delta p}{dt}$ and conservation of mass, solve for the spatially averaged volume flow rate traveling through the channel, the rate of fluid storage owing to channel expansion, and the total volume traveling through or stored by the circuit. Write your answers in terms of cosine functions with phase lags.

SOLUTION: For the spatially averaged volume flow rate traveling through the channel,

$$Q_R = \frac{\Delta p}{R_h} = \frac{\Delta p_0}{R_h} \cos \omega t \,. \tag{3.17}$$

The rate at which fluid is stored by expansion of the channel volume is $Q_C = C_h \frac{d\Delta p}{dt} = -C_h \, \Delta p_0 \, \omega \, \sin \omega t$ or

$$Q_C = C_h \, \Delta p_0 \, \omega \, \cos(\omega t + \pi/2) \,. \tag{3.18}$$

From conservation of mass, the total flow rate traveling through or stored by the channel is $Q = \frac{\Delta p_0}{R_h} \cos \omega t - C_h \, \Delta p_0 \, \omega \sin \omega t$. Using the trigonometric identity $a \sin x + b \cos x = \sqrt{a^2 + b^2} \cos(x + \alpha)$, where $\alpha = \text{atan2}(a, b)$, this becomes $Q = \frac{\Delta p_0}{R_h} \sqrt{1 + \omega^2 R_h^2 C_h^2} \cos[\omega t + \text{atan2}(-C_h \, \Delta p_0 \, \omega, \frac{\Delta p_0}{R_h})]$. The final solution for the flow rate traveling through or stored by the channel is

$$Q = Q_R + Q_C = \frac{\Delta p_0}{R_h} \sqrt{1 + \omega^2 R_h^2 C_h^2} \cos\left[\omega t + \tan^{-1}(-\omega R_h C_h) + 2\pi\right] . \tag{3.19}$$

3.2.2 Hydraulic impedance

The impedance of a hydraulic element is a complex quantity that extends the Hagen–Poiseuille law ($\Delta p = Q R_h$) to handle sinusoidally time-varying fluid flows, as long as the flow is always well approximated by the steady-state, infinite-tube Poiseuille pressure–flow-rate relationship[3] and if $\frac{d\Delta p}{dt}$ is a good approximation for $\frac{dp}{dt}$. We can write sinusoidal pressures and flow rates as follows (note Re means the real part of):

$$\Delta p = \text{Re}(\underset{\sim}{\Delta p}) = \text{Re}\left\{\Delta p_0 \exp[j(\omega t + \alpha_p)]\right\} = \text{Re}\left(\underset{\sim}{\Delta p_0} \exp j\omega t\right) , \tag{3.20}$$

$$Q = \text{Re}(\underset{\sim}{Q}) = \text{Re}\left\{Q_0 \exp[j(\omega t + \alpha_Q)]\right\} = \text{Re}\left(\underset{\sim}{Q_0} \exp j\omega t\right) . \tag{3.21}$$

In this case, we have defined the analytic representations $\underset{\sim}{\Delta p}$ and $\underset{\sim}{Q}$ of the pressure and flow rate, respectively, using phasors $\underset{\sim}{\Delta p_0}$ and $\underset{\sim}{Q_0}$. The undertilde describes a complex quantity, with properties described in Appendix G. With these definitions, for circuit elements we can define a *hydraulic impedance* $\underset{\sim}{Z_h}$, which describes the pressure–flow-rate relationship through the complex equation $\underset{\sim}{\Delta p} = \underset{\sim}{Q}\underset{\sim}{Z_h}$ or phasor equation $\underset{\sim}{\Delta p_0} = \underset{\sim}{Q_0}\underset{\sim}{Z_h}$. Each hydraulic circuit element has a complex impedance corresponding to its circuit properties, both hydraulic resistors,

hydraulic impedance contribution from hydraulic resistance

$$\underset{\sim}{Z_h} = R_h , \tag{3.22}$$

and hydraulic capacitors,

hydraulic impedance contribution from hydraulic capacitance, i.e., compliance

$$\underset{\sim}{Z_h} = \frac{1}{j\omega C_h} = \frac{1}{\omega C_h} \exp(-j\frac{\pi}{2}) . \tag{3.23}$$

[3] Here we are neglecting the variation in flow distribution caused by the acceleration. The unsteady flow distribution can be solved using separation of variables and is an infinite sum of Bessel functions. The assumption to neglect this variation is sound if $\omega \ll \frac{\eta}{\rho R^2}$.

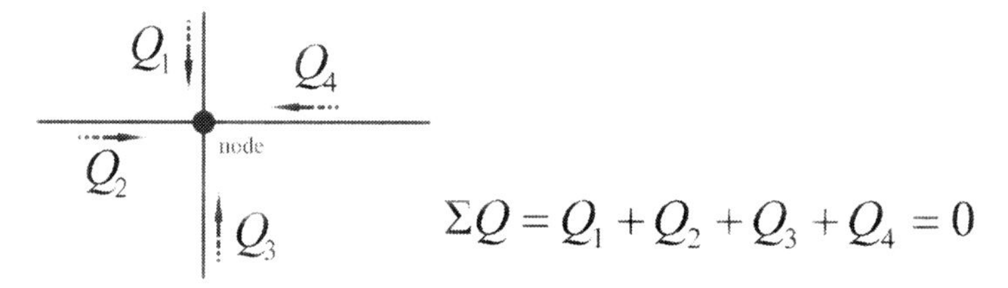

Fig. 3.7 Implementation of conservation of mass at a node linking four circuit elements.

3.2.3 Hydraulic circuit relations

The Hagen–Poiseuille law and the relations for hydraulic impedance describe the volumetric flow rate through an element or the pressure drop across that element. For a circuit that is composed of a network of these elements, mass conservation links the channel flow rate relations to specify the system solution. Conservation of mass requires that the net volumetric flow into or out of a node is zero (Fig. 3.7):

conservation of mass at the node of a fluid network

$$\sum_{\text{channels}} Q = 0 , \tag{3.24}$$

where Q in this case is defined as positive *into* the node. This, combined with the Hagen–Poiseuille law and hydraulic impedance, prescribes the approximate flow solution with a system of algebraic equations.

3.2.4 Series and parallel component rules

Conservation of mass is all that is needed to calculate the flow rate and pressures in a general network of circuit elements. In a number of special cases, though, we have developed rules of thumb by applying conservation of mass to fundamental network components.

Series hydraulic circuit rules. The hydraulic resistance of two hydraulic resistors in series is equal to the sum of the hydraulic resistances,

series relation for hydraulic resistors

$$R_h = R_{h,1} + R_{h,2} ; \tag{3.25}$$

the reciprocal of the compliance of two hydraulic capacitors in series is equal to the sum of the reciprocals of the compliances,

series relation for hydraulic capacitors

$$\frac{1}{C_h} = \frac{1}{C_{h,1}} + \frac{1}{C_{h,2}} ; \tag{3.26}$$

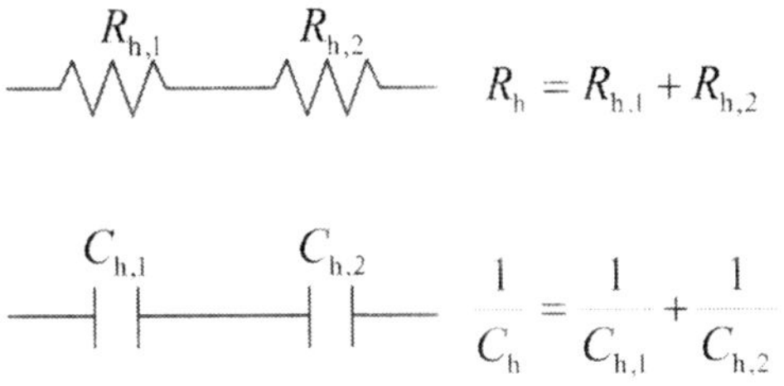

Fig. 3.8 Series component relations for hydraulic circuit elements.

and the hydraulic impedance of two hydraulic impedances in series is equal to the sum of the hydraulic impedances,

series relation for hydraulic impedances

$$\underset{\sim}{Z}_h = \underset{\sim}{Z}_{h,1} + \underset{\sim}{Z}_{h,2} \,. \tag{3.27}$$

Series component relations for hydraulic circuits are depicted in Fig. 3.8.

Parallel hydraulic circuit rules. The reciprocal of the hydraulic resistance of two hydraulic resistors in parallel is equal to the sum of the reciprocals of the hydraulic resistances,

parallel relation for hydraulic resistances

$$\frac{1}{R_h} = \frac{1}{R_{h,1}} + \frac{1}{R_{h,2}} \,; \tag{3.28}$$

the compliance of two hydraulic capacitors in parallel is equal to the sum of the compliances,

parallel relation for hydraulic capacitances

$$C_h = C_{h,1} + C_{h,2} \,; \tag{3.29}$$

and the reciprocal of the hydraulic impedance of two hydraulic impedances in parallel is equal to the sum of the reciprocals of the hydraulic impedances:

parallel relation for hydraulic impedances

$$\frac{1}{\underset{\sim}{Z}_h} = \frac{1}{\underset{\sim}{Z}_{h,1}} + \frac{1}{\underset{\sim}{Z}_{h,2}} \,. \tag{3.30}$$

Parallel component relations are depicted in Fig. 3.9.

EXAMPLE PROBLEM 3.3

Consider a flexible microchannel that can be modeled with a hydraulic resistance R_{h} and a compliance C_{h} in parallel. A pressure drop $\Delta p = \Delta p_0 \cos \omega t$ with $\Delta p > 0$ is applied across this circuit. Assume that $\frac{dp}{dt}$ is well approximated by $\frac{d\Delta p}{dt}$.

Write the analytic representation of the applied pressure drop and solve for the analytic representation and phasor representation of the total flow rate traveling through or stored by the system by using $Q = \underset{\sim}{\Delta p}/\underset{\sim}{Z_{\text{h}}}$ and the parallel circuit rules for complex hydraulic impedance. Use a similar relation to determine the analytic representation and phasor representation of the separate flow rates traveling through or stored by the channel. Once these values are determined, write the real representation of these flow rates.

Consult Appendix G as needed for complex relations.

SOLUTION: The impedance of the channel resistance is $\underset{\sim}{Z_R} = R_{\text{h}}$ and the impedance of the channel compliance is $\underset{\sim}{Z_C} = 1/j\omega C_{\text{h}}$. Using parallel circuit rules, we find that the total impedance is given by $1/\underset{\sim}{Z} = 1/R_{\text{h}} + j\omega C_{\text{h}}$ or $\underset{\sim}{Z} = \frac{R_{\text{h}}}{1+j\omega R_{\text{h}} C_{\text{h}}}$.

The analytic representation of the applied pressure drop is $\underset{\sim}{\Delta p} = \Delta p_0 \exp j\omega t$ and the phasor representation is $\underset{\sim}{\Delta p_0} = \Delta p_0$. The Hagen–Poiseuille law for the analytic representations is $Q = \underset{\sim}{\Delta p}/\underset{\sim}{Z_{\text{h}}}$ and for the phasors is $\underset{\sim}{Q_0} = \underset{\sim}{\Delta p_0}/\underset{\sim}{Z_{\text{h}}}$. Using the phasor equation, the phasor for the spatially averaged flow rate traveling through the channel is $\underset{\sim}{Q_0} = \Delta p_0/R_{\text{h}}$. The magnitude of this phasor is $\frac{\Delta p_0}{R_{\text{h}}}$ and the angle of this phasor is $\text{atan2}(0, \Delta p_0/R_{\text{h}}) = 0$. Similarly, the phasor for the rate at which fluid is stored by channel expansion is $\underset{\sim}{Q_0} = \Delta p_0 j\omega C_{\text{h}}$. The magnitude of this phasor is $\Delta p_0 \omega C_{\text{h}}$ and the angle of this phasor is $\text{atan2}(\Delta p_0 \omega C_{\text{h}}, 0) = \pi/2$. The phasor for the total flow rate (flow and storage) is $\underset{\sim}{Q_0} = \Delta p_0 \frac{1+j\omega R_{\text{h}} C_{\text{h}}}{R_{\text{h}}}$. The magnitude of this phasor is $\frac{\Delta p_0}{R_{\text{h}}}\sqrt{1+\omega^2 R_{\text{h}}^2 C_{\text{h}}^2}$. The angle of this phasor is $\text{atan2}(\frac{\Delta p}{R_{\text{h}}}\omega R_{\text{h}} C_{\text{h}}, \frac{\Delta p}{R_{\text{h}}}) = \tan^{-1}(-\omega R_{\text{h}} C_{\text{h}}) + 2\pi$. Returning to real quantities, the spatially averaged flowrate through the channel is $Q_R = \Delta p_0 R_{\text{h}} \cos \omega t$, the rate at which fluid is stored by channel expansion is $Q_C = \Delta p_0 \omega C_{\text{h}} \cos(\omega t + \pi/2)$, and the total flow traveling through or stored by the channel is $Q = \frac{\Delta p_0}{R_{\text{h}}}\sqrt{1+\omega^2 R_{\text{h}}^2 C_{\text{h}}^2} \cos(\omega t + \tan^{-1}(\omega R_{\text{h}} C_{\text{h}}) + 2\pi)$.

3.3 SOLUTION TECHNIQUES

Hydraulic circuit systems lead to algebraic systems of equations that can be solved directly with symbolic or matrix manipulation software. Given the hydraulic impedances

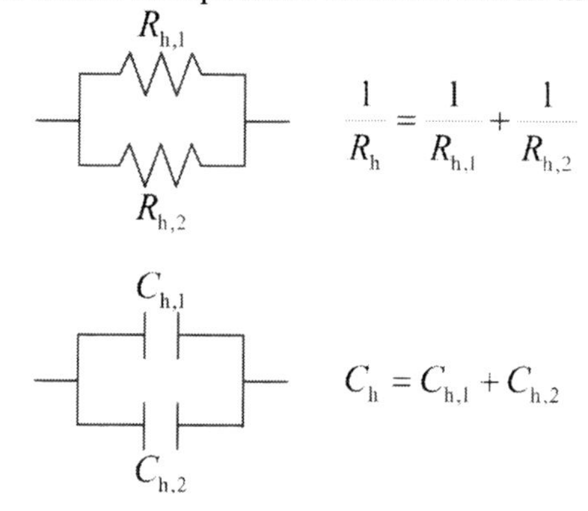

Fig. 3.9 Parallel component relations for hydraulic circuit elements.

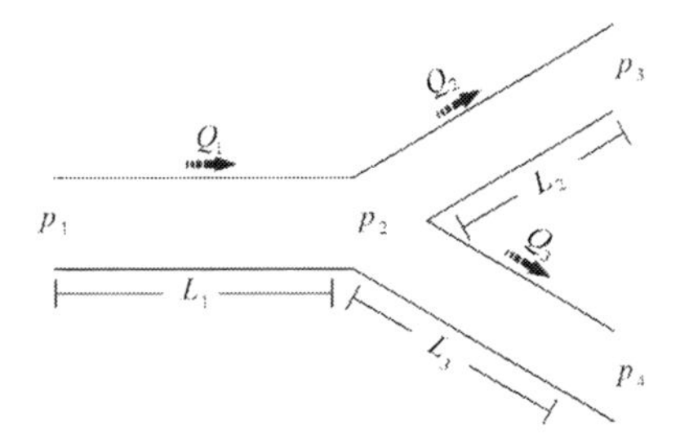

Fig. 3.10 A simple fluid circuit. All channels are circular in cross section with radius $R = 20\ \mu$m. Lengths are specified by $L_1 = 0.5$ cm, $L_2 = 1$ cm, and $L_3 = 2$ cm. Pressures at the ports are $p_1 = 0.11$ MPa, $p_3 = 0.1$ MPa, and $p_4 = 0.1$ MPa. The fluid is water, with viscosity 1×10^{-3} Pa s.

of the channels, the unknowns are the pressures at the N nodes and the flow rates in the M channels. The N equations from conservation of mass at each node and M equations from the Hagen–Poiseuille law for each channel define the system. Thus an $N + M \times N + M$ matrix equation can be written and inverted to obtain the solution. If the channels are all purely resistive, then the pressure and flow-rate phasors are real. If the channels include compliance, then the pressure and flow-rate phasors are complex and the angle of the complex numbers describes the phase lag of the response.

EXAMPLE PROBLEM 3.4

Consider the fluid circuit shown in Fig. 3.10. Write the seven equations in the unknowns p_1, p_2, p_3, p_4, Q_1, Q_2, Q_3. Three of these equations are given by the problem statement, one is conservation of mass, and three are Hagen–Poiseuille relations. Write these equations as a matrix equation. Invert the matrix to solve the system for the seven unknowns.

SOLUTION: First we must solve for the hydraulic resistance of the three channels, using $R_h = 8\eta L/\pi R^4$. This gives hydraulic resistances of 7.96×10^{13}, 1.59×10^{14}, and 3.18×10^{14} N/s m^5 for the three channels.

The seven equations are

$$\begin{aligned} p_1 &= 0.11\ \text{MPa}, & p_1 - p_2 &= Q_1 R_{h,1}, \\ p_3 &= 0.1\ \text{MPa}, & p_2 - p_3 &= Q_2 R_{h,2}, \\ p_4 &= 0.1\ \text{MPa}, & p_2 - p_4 &= Q_3 R_{h,3}, \\ & & Q_1 &= Q_2 + Q_3 . \end{aligned} \tag{3.31}$$

In matrix form, this is

$$\begin{bmatrix} 1 & 0 & 0 & 0 & 0 & 0 & 0 \\ 0 & 0 & 1 & 0 & 0 & 0 & 0 \\ 0 & 0 & 0 & 1 & 0 & 0 & 0 \\ 1 & -1 & 0 & 0 & -R_{h,1} & 0 & 0 \\ 0 & 1 & -1 & 0 & 0 & -R_{h,2} & 0 \\ 0 & 1 & 0 & -1 & 0 & 0 & -R_{h,3} \\ 0 & 0 & 0 & 0 & 1 & -1 & -1 \end{bmatrix} \begin{bmatrix} p_1 \\ p_2 \\ p_3 \\ p_4 \\ Q_1 \\ Q_2 \\ Q_3 \end{bmatrix} = \begin{bmatrix} 0.11\ \text{MPa} \\ 0.1\ \text{MPa} \\ 0.1\ \text{MPa} \\ 0 \\ 0 \\ 0 \\ 0 \end{bmatrix} . \tag{3.32}$$

The solution is $p_2 = 0.1057$ MPa, $Q_1 = 54 \times 10^{-12}$ m^3/s, $Q_2 = 36 \times 10^{-12}$ m^3/s, and $Q_3 = 18 \times 10^{-12}$ m^3/s. The matrix is ill-conditioned in SI units (because the R_h values are many orders of magnitude larger than the other matrix elements), and numerical inversion works best if a non-SI unit system is chosen so that the elements of the matrix are of the same order.

3.4 SUMMARY

This chapter presents relations for flow through hydraulic circuits. Hydraulic circuit analysis uses the infinite-tube Poiseuille flow solution to approximate the pressure–flow relations in long channels, using the Hagen–Poiseuille relation,

$$\Delta p = Q R_h \,, \tag{3.33}$$

as well as the definition of hydraulic resistance,

$$R_h = \frac{8\eta L}{A r_h^2} \,. \tag{3.34}$$

The mechanical compliance of a channel is described for small pressure fluctuations by use of a hydraulic capacitance or compliance, and the net response of a channel to sinusoidal forcing is given by its hydraulic impedance:

$$\underset{\sim}{Z_h} = R_h + \frac{1}{j\omega C_h} \,. \tag{3.35}$$

To consider time-varying pressures and flow rates, we adopt analytic representations of these properties and use the complex hydraulic impedance of the circuit elements in the equations, leading to the complex form of the Hagen–Poiseuille law,

$$\underset{\sim}{\Delta p} = \underset{\sim}{Q} \underset{\sim}{Z_h} \,. \tag{3.36}$$

The resulting equations are analogous to electrical circuit relations, in that the Hagen–Poiseuille law and continuity replace Ohm's law and Kirchoff's law. Although the analogy between Poiseuille flow rates through a long channel and current through a resistor is excellent, the analogy between fluid storage in a flexible channel and current through a capacitor is inexact, and the hydraulic capacitance as implemented in this chapter is effective for estimating only channel-averaged responses to small, sinusoidal perturbations.

Physically, the hydraulic system has pumps, channels, and microchannel compliance in the same role that voltage sources, resistors, and capacitors play in electrical circuits, and pressure, volumetric flow rate, and hydraulic resistance replace voltage, current, and electrical resistance from electrical circuit modeling. These hydraulic circuit analysis techniques provide useful engineering approximations to flow rates in channel networks, which are common in microfluidic systems.

3.5 SUPPLEMENTARY READING

Bruus [33] goes into significant detail about hydraulic models for flow through tubes and equivalent circuits, with specific attention to Stokes flow criteria for the applicability of circuit models. White [22] details errors of hydraulic radius calculations.

3.6 EXERCISES

3.1 Consider steady Poiseuille flow through a control volume with a circular cross section with radius R, perimeter $\mathcal{P}$, cross-sectional area A, and differential length dz. Assume a pressure gradient with magnitude $\frac{\partial p}{\partial z}$ is present. Write the net pressure on the control volume in terms of A, dz, and $\frac{\partial p}{\partial z}$. Write the wall shear stress in terms of R and $\frac{\partial p}{\partial z}$. The sum of these forces is zero at equilibrium. Given this, write the relation between R, A, and $\mathcal{P}$.

For a circle, the expression relating R, A, and $\mathcal{P}$ follows directly from geometry, and the preceding analysis was not necessary. However, for a channel of unknown geometry but known A and $\mathcal{P}$, the preceding analysis allows us to derive the hydraulic radius r_h.

Repeat the above analysis for a cross section of unknown geometry but known A and $\mathcal{P}$. To do this analysis, you will have to *assume* that the surface stress is uniform, and define it as $\tau_{\perp z} = \frac{r_h}{2}\frac{\partial p}{\partial z}$. What is the value of r_h in terms of A and $\mathcal{P}$?

This analysis shows that the key approximation associated with using the hydraulic radius is the assumption that the wall shear stress is uniform along the perimeter of the channel. For what geometries is this assumption good? For what geometries is this assumption bad?

3.2 Consider water ($\eta = 1 \times 10^{-3}$ Pa s) in an infinitely rigid tube of circular cross section with radius $R = 10\ \mu$m and length $L = 10$ cm. The compressibility of water at standard temperature and pressure is a thermodynamic property and is roughly equal to 5×10^{-5}atm^{-1}. Calculate the hydraulic resistance and the compliance of the tube–water system.

3.3 Consider flow of an incompressible fluid through a long, narrow microchannel made of a flexible material. Model this channel as a rigid tube with hydraulic resistance R_h with two identical balloons attached to it – one attached to the inlet and one attached to the outlet. The balloons each have a compliance $C_h/2$ such that the conservation of mass relation at the inlet and outlet is described by

$$\Delta Q = -C_h \frac{d\mathcal{V}}{dt} . \tag{3.37}$$

Assume that the pressure at the inlet is given by $\frac{1}{2} p \cos \omega t$ and the pressure at the outlet is given by $-\frac{1}{2} p \cos \omega t$. All other values will also be sinusoidal, with phase lags that can be determined as part of the solution.

(a) Determine the flow rate through the tube, in the inlet, and out the outlet. Determine also the rate at which volume is stored in each balloon.

(b) Solve the system by treating this system as a hydraulic resistor with hydraulic resistance R_h and a hydraulic capacitor with compliance C_h in parallel. How do your results compare?

3.4 Consider a dead-end microchannel that is filled with water except for a volume of air $\mathcal{V}$ at the end of the channel. The column of water has a hydraulic resistance of R_h. Model this system as a hydraulic resistor and a hydraulic capacitor in series and predict $\mathcal{V}(t)$ if a sinusoidal pressure signal of $p = p \cos \omega t$ is applied.

3.5 In circuit analysis, we often consider a voltage source connected to a series *RC* circuit; however, a *current* source connected to a parallel *RC* circuit is also important owing to its analogy with fluid flow through a compliant system.

Consider an electrical circuit composed of a resistor with resistance R and a capacitor with capacitance C in parallel (see Chapter 5 for electrical circuit relations).

(a) Given a steady current source of 1 μA, how much current goes through the resistor? The capacitor?

(b) Given an oscillating current source of magnitude $I = I_0 \cos \omega t$ with $I_0 = 1\ \mu$A, what is the time-dependent current through the resistor? What is the current into the capacitor?

Now consider a circular microchannel with length 2 cm and radius 10 μm that is filled with water with $\eta = 1$ mPa s. Halfway down the channel, the microchannel goes through a 120° turn, as shown in Fig. 3.11. When the channel is initially filled, a bubble with volume $(200\ \mu\text{m})^3$ at 1 atm is trapped at the turn (perhaps owing to a fabrication error). The pressure at the outlet is 1 atm. Assume the water is incompressible and the channel is not compliant.

(a) Using the ideal gas law relation for the bubble's volume as a function of pressure, calculate the mechanical compliance of the bubble at 1 atm.

(b) Assuming, for simplicity, that the mechanical compliance of the bubble does not change when it is compressed (this is valid only for small pressure perturbations), model the bubble as a capacitor in parallel with the channels (resistors). Evaluate the volumetric flow rate out the outlet for a constant input flow rate Q_0.

(c) Under similar conditions, evaluate the volumetric flow rate out the outlet for $Q_{\text{inlet}} = Q_0 \cos \omega t$. At what frequency does the bubble damp the magnitude of the oscillation of the outlet flow rate by a factor of two with respect to the inlet oscillation?

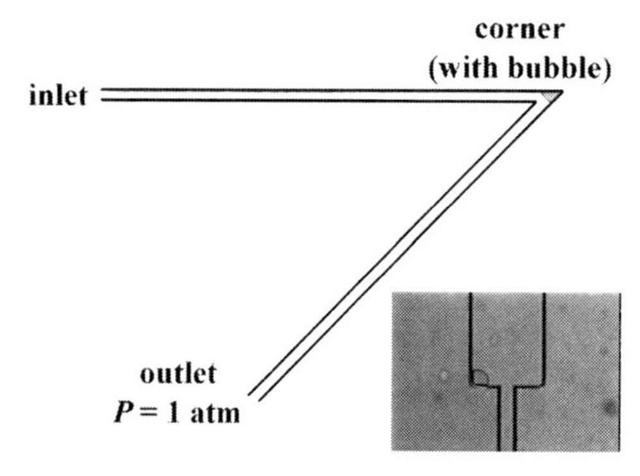

Fig. 3.11 Schematic of a microchannel with a bubble trapped at a turn. Lower right: a micrograph of an experimental realization of such a bubble.

3.6 Consider a microfluidic device consisting of four microchannels arranged in a cross configuration (Fig. 3.12). Assume all of the microchannels can be approximated as having a circular cross section and assume that all flow is laminar and described by Poiseuille flow relations. The viscosity is 1 mPa s and the channel radius is 2 μm.

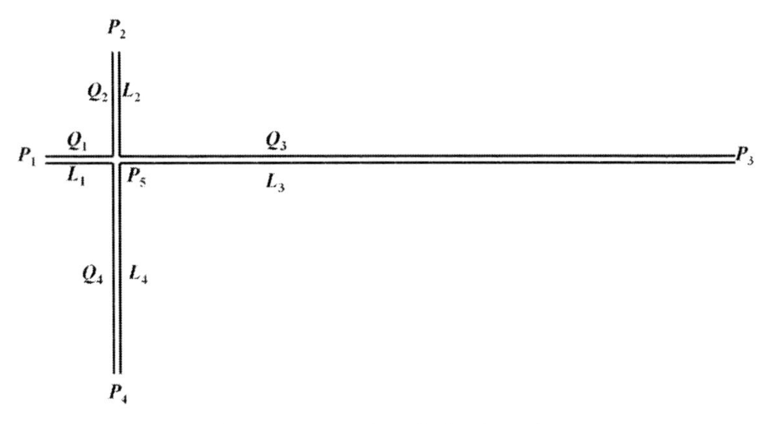

Fig. 3.12 Four-port, cross-shaped microchannel.

Outlet 4 is at $p_4 = 100{,}000$ Pa. Pumps are attached to inlets 1, 2, and 3. The pump attached to inlet 1 is at $p_1 = 400{,}000$ Pa. The four channels have lengths $L_1 = 1$ cm, $L_2 = 1.5$ cm, $L_3 = 9$ cm, and $L_4 = 3$ cm.

What must the pressures be at inlets 2 and 3 so that the volumetric flow rate through each inlet is the same? Given those pressures, what is the pressure p_5 at the intersection of the channels?

What is the velocity distribution $u(\imath)$ in each channel? For what regions of the microdevice is this velocity a good approximation of the real velocity distribution?

We solve this most easily by noting that the nine unknowns are the five pressures p_1–p_5 and the four volumetric flow rates Q_1–Q_4. The nine equations come from (a) the four $\Delta p = QR_h$ equations for the four channels, (b) one conservation of mass equation at the intersection, and (c) four more bits of information specified by the problem (p_1, p_4, the relation between Q_1 and Q_2, and the relation between Q_1 and Q_3). Thus a set of nine equations for nine unknowns can be written. This set of nine equations can be written as a matrix equation $Ax = b$, where x is a 1×9 column vector with the nine unknowns, A is the 9×9 multiplication matrix, and b is a 1×9 column vector with nine constants. Numerically, this inversion will work best if the matrix A is well conditioned, i.e., if the components of the matrix are all of the same order. This will be achieved if the problem is solved with pressures in pascals, flow rates in cubic micrometers per second, and hydraulic resistances in pascal-seconds per cubic micrometer.

3.7 Assume that you are designing a microfluidic device to study the effect of shear rates on endothelial cells growing in several chambers in a microfluidic device. Your goal is to design a device that houses four 300 μm $\times$ 1000 μm chambers in which the cells grow, and to design a channel network such that flow introduced in one inlet passes through each of these four chambers in parallel. To make the fabrication straightforward, the depth of all channels in the device will be 40 μm. Design a 2D microchannel geometry such that equivalent circuit analysis can be used to model the system and such that the shear rates through the four chambers are in the ratio 1:3:10:30. Many geometries are possible.

3.8 Consider the microfluidic channel in Fig. 3.13.

(a) A pressure of 2 atm is applied at the inlet and a pressure of 1 atm is applied at the outlet. The channels are semicircular in cross section, with a 10-μm radius. Each of the twelve sections is 1 mm long. The working fluid is water. What is the magnitude of the volumetric flow rate through each of the twelve channels?

(b) Assume channel 4 becomes clogged and the flow rate in channel 4 goes to zero, but all other channels are unchanged. What are the new flow rates?

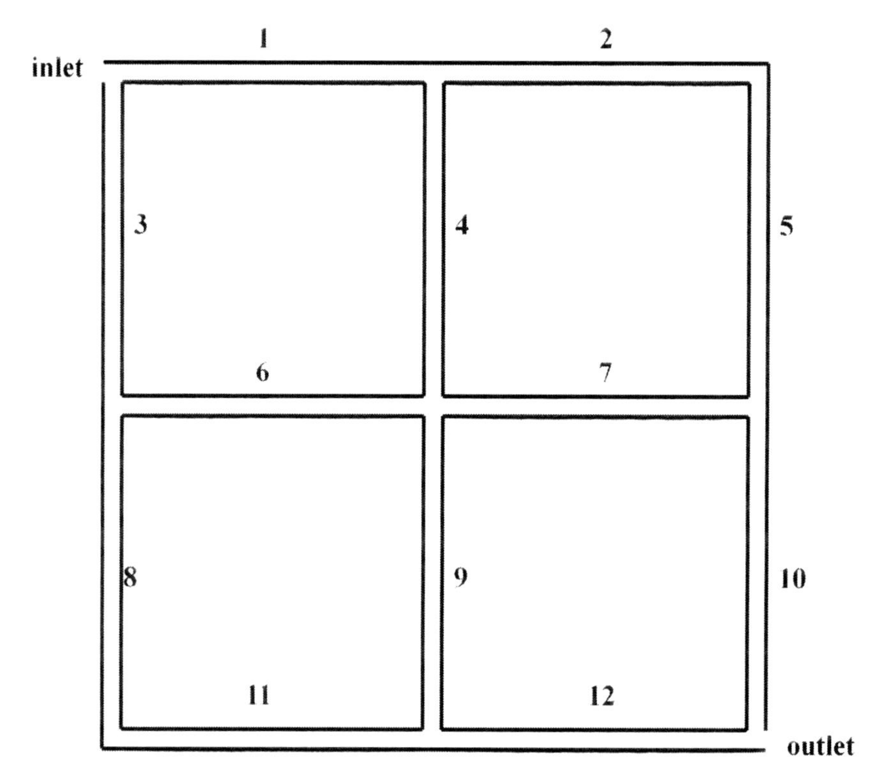

Fig. 3.13 A schematic of a microfluidic device with one inlet, one outlet, and twelve channels.

3.9 Using equivalent circuit techniques, design a microdevice geometry with one input and one output. For a given set of pressures (one at the input and one at the output – your specification of these pressures should be part of your answer), the input should split into four separate channels that each experience four different cross-section–averaged velocities: 1 μm/s, 10 μm/s, 100 μm/s, and 1 mm/s. Your design should include specifications of the pressures at input and output, as well as enough information about the microdevice geometry to ensure the specified velocities are observed. Assume no electric fields exist in this device.

4 Passive Scalar Transport: Dispersion, Patterning, and Mixing

Many microfluidic systems are used to manipulate the distribution of chemical species. Chemical separations, for example, physically separate components of a multispecies mixture so that the quantities of each component can be analyzed or so that useful species can be concentrated or purified from a mixture. Many biochemical assays, for example DNA microarrays, require that a reagent be brought into contact with the entirety of a functionalized surface, i.e., that the reagents in the system be well mixed. Studies of homogeneous kinetics in solution require that a system become well mixed on a time scale faster than the kinetics of the reaction. In contrast to these, extracting functionality from a spatial variation of surface chemistry often depends on the ability to pattern surface chemistry with flow techniques, which requires that components of the solution remain *unmixed*.

These topics all motivate discussion of the *passive scalar transport equation*. This convection–diffusion equation governs the transport of any conserved property that is carried along with a fluid flow, moves with the fluid, and does not affect that fluid flow. Chemical species and temperature are two examples of properties that can be handled in this way, as long as (1) the chemical concentration or temperature variations are low enough that transport properties such as density or viscosity can safely be assumed uniform, and (2) we neglect electric fields, which can cause migration of chemical species relative to the fluid.

We start by introducing the scalar convection–diffusion equation, which describes species transport, and discuss the physics of mixing. We then note that, owing to the nature of microfabrication techniques and the species of interest in biochemical analysis systems, many microfluidic species transport systems reside in the limit of low Reynolds number (laminar) but high Peclet number (minimal diffusion). This limit makes it straightforward to isolate chemical species in microdevices, enabling *laminar flow-patterning* techniques, which can be analyzed with simple 1D arguments that are a minor extension of the hydraulic circuit analyses presented in Chapter 3. Unfortunately, this same situation leads to challenges when species must be mixed, leading to the so-called *microfluidic mixing* problem. The challenges of mixing in these systems has given rise to interest in *chaotic advection*, which uses flow fields with special properties that can exponentially increase the scalar mixing by using a deterministic flow field to amplify the random effects of species diffusion. In chemical separation systems, we are interested in the mixing of species in the axial direction, because mixing in this direction decreases the resolution of a chemical separation. This motivates our study of *Taylor–Aris dispersion*, in which the dispersive nature of the flow leads to a significant increase in the effective diffusion of the scalar in the axial direction.

4.1 PASSIVE SCALAR TRANSPORT EQUATION

The following sections describe the sources of passive scalar fluxes and show how these fluxes, when applied to a control volume, lead to the conservation equations for scalars.

4.1.1 Scalar fluxes and constitutive properties

The two mechanisms that lead to flux of scalars into or out of a control volume are diffusion and convection. Diffusion refers to the net migration of a fluid property owing to random thermal fluctuations in the fluid system. Although we typically treat the fluid as a continuum and ignore any properties on the atomistic scale, the molecules nonetheless exhibit extensive motion on the molecular scale, and the essentially random nature of this motion leads to random fluctuations in the distribution of a passive scalar. The net effect is a flux of the scalar in the direction opposite to the local scalar gradient.

In the ideal solution limit (which is applicable for temperature and most chemical species at the conditions used for most microfluidic devices; see Subsection B.3.5), Fick's law is a constitutive relation that links the flux density of a scalar to both the gradient of the scalar and the *diffusivity* of the scalar in the solvent:

Fick's law for scalar flux density

$$\vec{j}_{\text{diff}} = -D\nabla c\,, \tag{4.1}$$

where $\vec{j}_{\text{diff}}$ is the diffusive scalar flux density (i.e., the net amount of the scalar diffusing across a surface per unit area per time), D [m^2/s] is the diffusivity of the scalar in the fluid, and c is the scalar, which in this text is often the concentration c_i of a chemical species. Fick's law is a macroscopic representation of the summed effect of the random motion of species owing to thermal fluctuations. Fick's law is analogous to the Fourier law for thermal energy flux caused by a temperature gradient as well as to the Newtonian model for momentum flux induced by a velocity gradient; the species diffusivity D is analogous to the thermal diffusivity $\alpha = k/\rho c_p$ [m^2/s] and the momentum diffusivity η/ρ [m^2/s].

In addition to the random fluctuations of a scalar that are due to thermal motion, the deterministic transport of the scalar that is due to fluid convection also leads to a convective species flux:

convective scalar flux density

$$\vec{j}_{\text{conv}} = \vec{u}c\,, \tag{4.2}$$

where $\vec{j}_{\text{conv}}$ is the convective scalar flux density (i.e., the net amount of the scalar convecting across a surface per unit area per time) and $\vec{u}$ is the velocity of the fluid.

4.1.2 Scalar conservation equation

Given the preceding fluxes, the conservation equation for a scalar c can be written as

$$\frac{\partial}{\partial t}\int_{\mathcal{V}} c\,dV = -\int_{S} \vec{j}\cdot\hat{\boldsymbol{n}}\,dA, \tag{4.3}$$

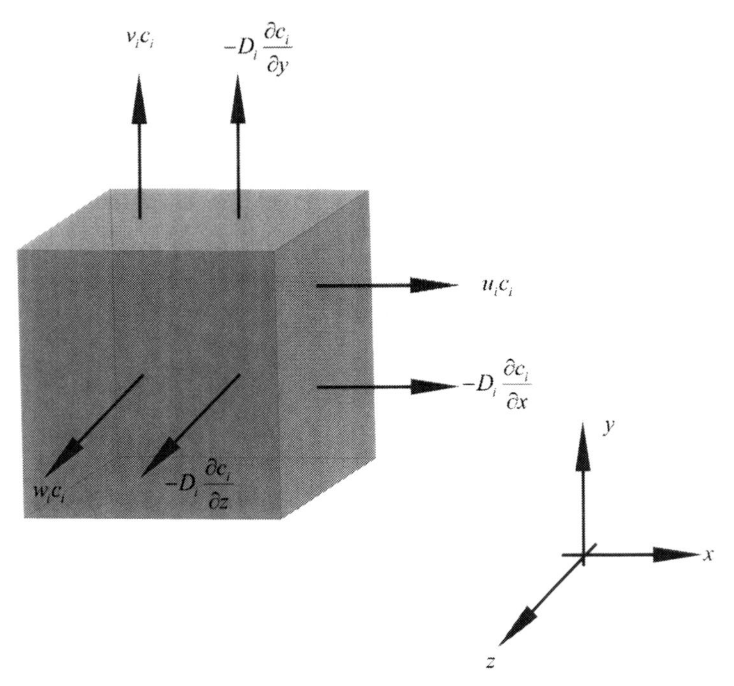

Fig. 4.1 Species fluxes for a Cartesian control volume.

where $\mathcal{V}$ is a control volume with differential element dV, $\mathcal{S}$ is its control surface with differential element dA, $\hat{\boldsymbol{n}}$ is a unit outward normal vector, and $\vec{\boldsymbol{j}}$ is the total scalar flux density owing both to diffusion and convection. Application of the fluxes just described to a differential control volume (such as the Cartesian control volume shown in Fig. 4.1) leads to the differential form of the scalar convection–diffusion equation, written for uniform D as

passive scalar convection–diffusion equation, uniform fluid properties

$$\frac{\partial c}{\partial t} + \vec{\boldsymbol{u}} \cdot \nabla c = D\nabla^2 c. \tag{4.4}$$

Compared with the Navier–Stokes equations for momentum transport, the passive scalar transport equation is simpler owing to its linear dependence on $\vec{\boldsymbol{u}}$ and the absence of the pressure term. Nondimensionalization of this equation for a flow with steady boundary conditions (see Subsection E.2.2) leads to the following form:

nondimensional scalar convection–diffusion equation, uniform fluid properties

$$\frac{\partial c^*}{\partial t^*} + \vec{\boldsymbol{u}}^* \cdot \nabla^* c^* = \frac{1}{Pe}\nabla^{*2} c^*, \tag{4.5}$$

in which starred properties have been nondimensionalized and the Peclet number $Pe = U\ell/D$. This nondimensional form highlights that the relative magnitude of the convective fluxes compared with the diffusive fluxes is proportional to the Peclet number. Thus

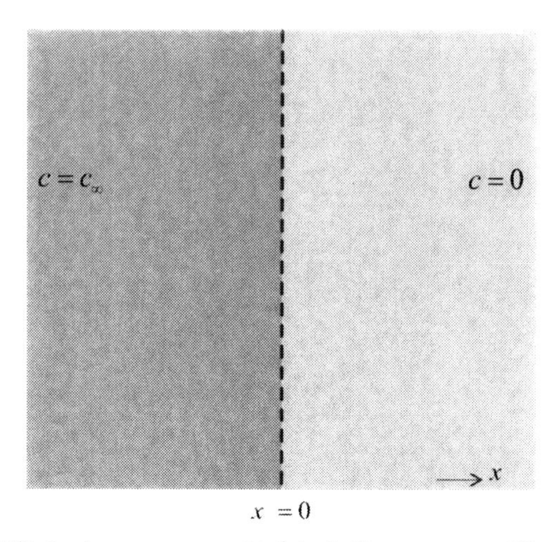

Fig. 4.2 Initial condition for 1D diffusion between two semi-infinite half-spaces at two different scalar concentrations. Grayscale indicates initial concentration.

a system with high Peclet number has negligible diffusion and scalars move about primarily by fluid convection, whereas a system with low Peclet number has a large amount of diffusion and the scalar distribution is spread out quickly by diffusive processes.

4.2 PHYSICS OF MIXING

We find it useful to discuss the mixing process in terms of each of the two secondary processes that are governed by the passive scalar convection–diffusion equation. Although these phenomena are inseparable, we find we gain considerable insight by isolating these processes and examining their functional dependences. We thus decompose mixing into two processes: (1) diffusion across scalar gradients and (2) shortening of diffusion length scales by motion of the fluid. These phenomena are the diffusive and convective actions of the governing equation.

Diffusion across scalar gradients. Consider an infinite 1D domain, where the scalar for all $x < 0$ at $t = 0$ is given by $c = c_\infty$, and the scalar for all $x > 0$ at $t = 0$ is given by $c = 0$. Consider quiescent fluid (Fig. 4.2). The governing equation for this passive scalar diffusion problem is given by

$$\frac{\partial c}{\partial t} = D\frac{\partial^2 c}{\partial x^2}, \tag{4.6}$$

and the solution, achieved by similarity transform (see Exercise 4.5), is

$$c = \frac{1}{2}c_\infty \operatorname{erfc}\left(\frac{x}{2\sqrt{Dt}}\right). \tag{4.7}$$

This solution illustrates the action of diffusion in eliminating scalar gradients. For $t \to \infty$, the solution approaches $c = c_\infty/2$ everywhere. An illustration of this solution is shown in Fig. 4.3. The distance $\ell = \sqrt{Dt}$ is the distance at which the solution has diffused to $\frac{1}{2}c_\infty \operatorname{erfc}(\frac{1}{2}) \simeq \frac{1}{4}$, which is approximately halfway toward the equilibrium solution. The expression $\ell = \sqrt{Dt}$ is commonly used as the *diffusion length scale* of this system. The diffusion length scale characterizes how far into the domain the species has diffused as a function of time. For a given time t, $\ell_{\text{diff}} = \sqrt{Dt}$ denotes the characteristic length over

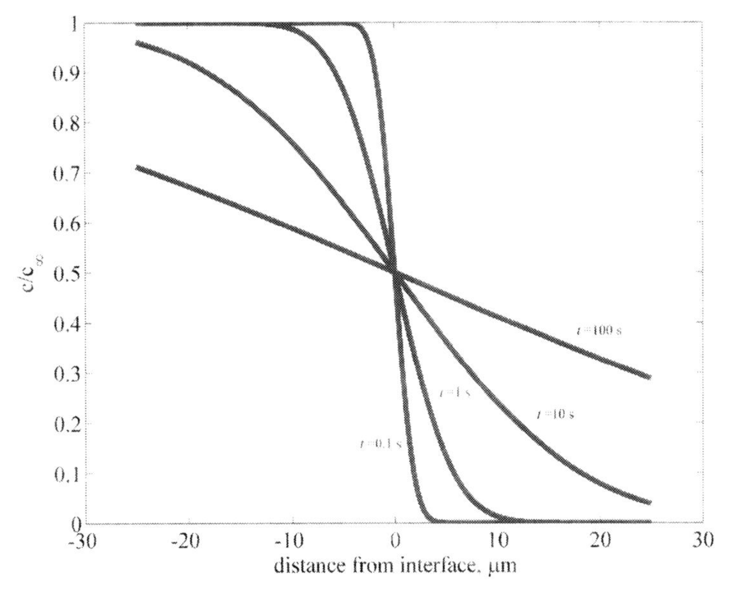

Fig. 4.3 Illustration of pure 1D diffusion of a passive scalar with diffusivity 1×10^{-11} m^2/s.

which diffusion has occurred. Similarly, for a given reservoir size R, the time required for diffusion to mix two components in that reservoir is proportional to $t_{\text{diff}} = R^2/D$. In microscale systems, this time can be long – for example, the time required for a solution of bovine serum albumin (a protein found in cow blood) to diffuse across a quiescent 100 μm channel is about 2 min, and if we were somehow able to measure the time required for a dilute suspension of neutrally buoyant 10-μm cells to diffuse itself across a quiescent 100-μm channel, we would measure approximately 30 years.

This 1D diffusion problem relates directly to microscale flows of interest. Consider the fluid flow in Fig. 4.4, in which two fluids are brought into contact and diffusion transverse to flow occurs while the fluids are convected downstream. In the limit where the channels are shallow relative to their width, the transverse distribution of depth-averaged species in this case is identical to the 1D diffusion problem previously specified,

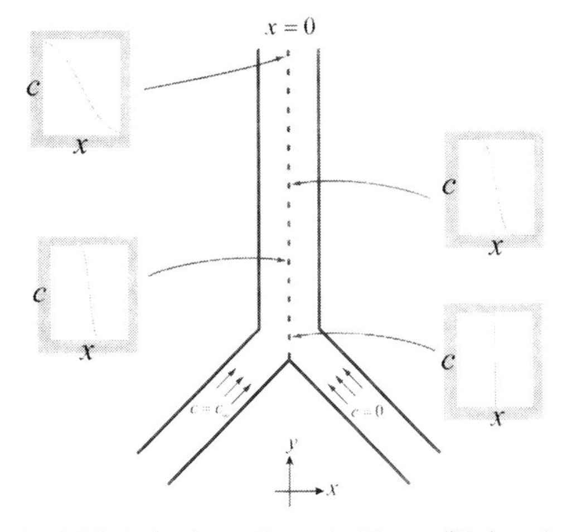

Fig. 4.4 A two-input, one-output microfluidic device that replicates the 1D pure diffusion solution from Section 4.2. The channel depth is assumed small relative to the width.

with only one difference – this is a *steady* flow with a steady species distribution, and the distribution varies with y/U (the time since the fluid entered the channel) rather than with the time since the experiment started.

Shortening diffusion length scales. Solutions to the 1D diffusion equation show that species mix over a distance $\ell_{\text{diff}} = \sqrt{Dt}$, which requires that the time for diffusion to mix scalars over a length R be proportional to R^2/D, and this time for many practical microfluidic systems is quite long. The system shown in Fig. 4.4 has convection, but this convection is normal to the scalar gradients and does not affect the transverse diffusion process. Thus the geometry in Fig. 4.4 is ideal if the mixing is to be *minimized* so as to keep the two fluid components separate. If mixing is to be *maximized*, the convection must somehow enhance the process by shortening of diffusion length scales. Although convection and diffusion are inseparable, we nonetheless find it useful to describe mixing as a two-step process consisting of (1) convective stirring of fluid over a time t until the spatial separation between components 1 and 2 is reduced to a characteristic length ℓ, and (2) diffusive spreading of the scalar field over a length scale $\sqrt{Dt}$. The characteristic time to mix is then given by the time t such that the length ℓ created by the stirring is of the same order as the diffusive length scale $\sqrt{Dt}$. In the absence of convection, the time is given by R^2/D. If convection actively stirs the fluid and quickly reduces the length scales over which diffusion must occur, the mixing time is much less than R^2/D. This is why we stir things to mix them up, for example, stirring cream added to coffee or chocolate syrup added to milk. The fluid flow generated by the spoon tends to shorten the diffusion length scales, and makes the mixing occur much more quickly. Two types of kinematic structures that shorten diffusion length scales include the baker's transformation (stretch and fold), denoted schematically in Fig. 4.5, and a twist map or vortex, denoted schematically in Fig. 4.6 and demonstrated experimentally in Fig. 4.7. In both cases, the motion of fluid reduces the characteristic size of the scalar domains. These flow structures occur naturally in high-*Re* flows but are absent in many low-*Re* flows. Thus mixing is often a slow process in microfluidic devices unless system geometries are designed specifically to generate flow structures that shorten diffusion length scales.

4.3 MEASURING AND QUANTIFYING MIXING AND RELATED PARAMETERS

Quantifying mixing typically involves some characterization of the spatial inhomogeneity of a scalar distribution. We typically measure the effectiveness of a mixing scheme by monitoring the mixing time or the mixing distance, depending on the application, and evaluating how this mixing time or distance depends on the Peclet number.

Diffusive mixing regime. As an example, consider a flow of two miscible streams of fluid through a channel of width L at a characteristic velocity U. The Peclet number for this flow is UL/D. The characteristic time for diffusive mixing (as illustrated in the solution of the preceding diffusion equation) is L^2/D or $Pe\frac{L}{U}$ and the characteristic length is UL^2/D or $Pe\ L$. Thus, for a purely diffusive mixing process, the characteristic length or time for mixing is proportional to the Peclet number. For example, for a solution of bovine serum albumin ($D = 1 \times 10^{-10}$) flowing at $\overline{u} = 100\ \mu$m/s through a channel of width 100 μm, the Peclet number $Pe = 100$, which means that the fluid remains largely unmixed until the flow has traveled a distance 100 times the width of the channel, i.e., 1 cm.

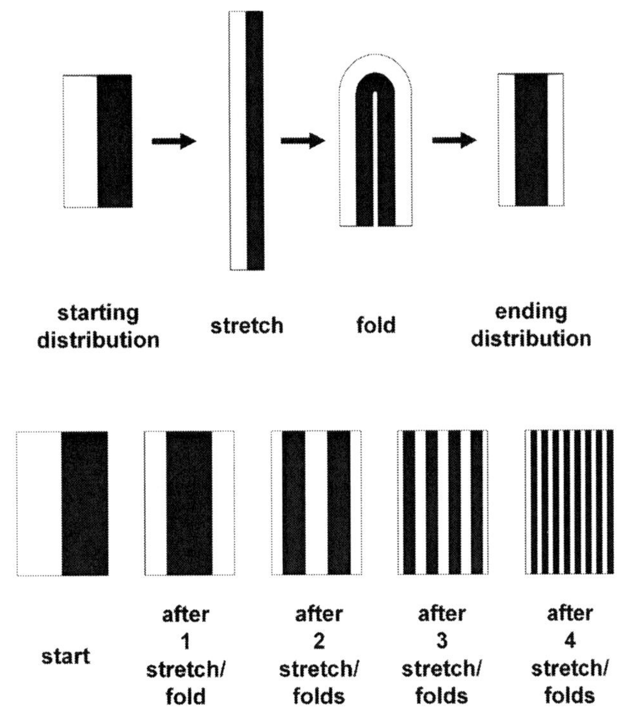

Fig. 4.5 Schematic of the baker's transformation and its role in shortening the length scales over which diffusion must act. Note how repeated stretching (extensional strain) and folding (rotation) leads to narrow sheets. Diffusion need act only over these short length scales to mix these flows.

Chaotic mixing regime. *Chaotic mixing* is a term commonly used in the low-*Re* mixing literature and indicates mixing processes with flows that lead to an exponential decay of the characteristic length over which diffusion must act. The term *chaotic mixing* implies that (1) trajectories in the flow become separated by a distance that grows exponentially with time, or, alternatively, that (2) that the interfacial area between two fluids grows exponentially with time. This, in turn, implies that the net effect of diffusion, which is inherently random on a macroscopic scale, is deterministically amplified by the fluid flow. Thus a deterministic *fluid flow* can lead to a chaotic *mixing result* if the fluid flow

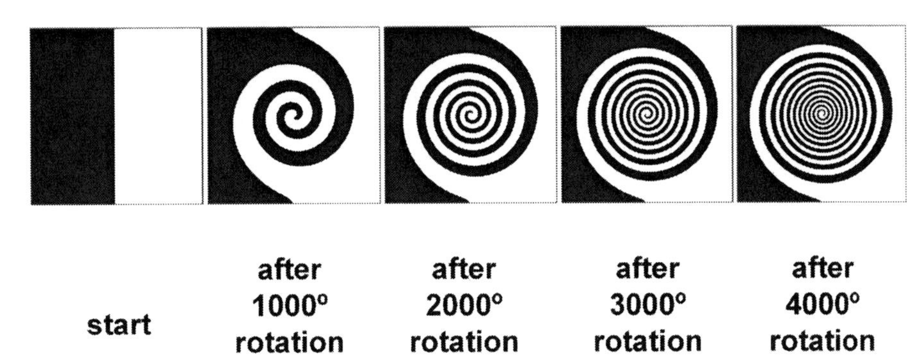

Fig. 4.6 Schematic of a vortex or whorl and its role in shortening the length scales over which diffusion must act. This sort of flow is often described mathematically by twist maps. Note again the presence of narrow sheets over which diffusion must act.

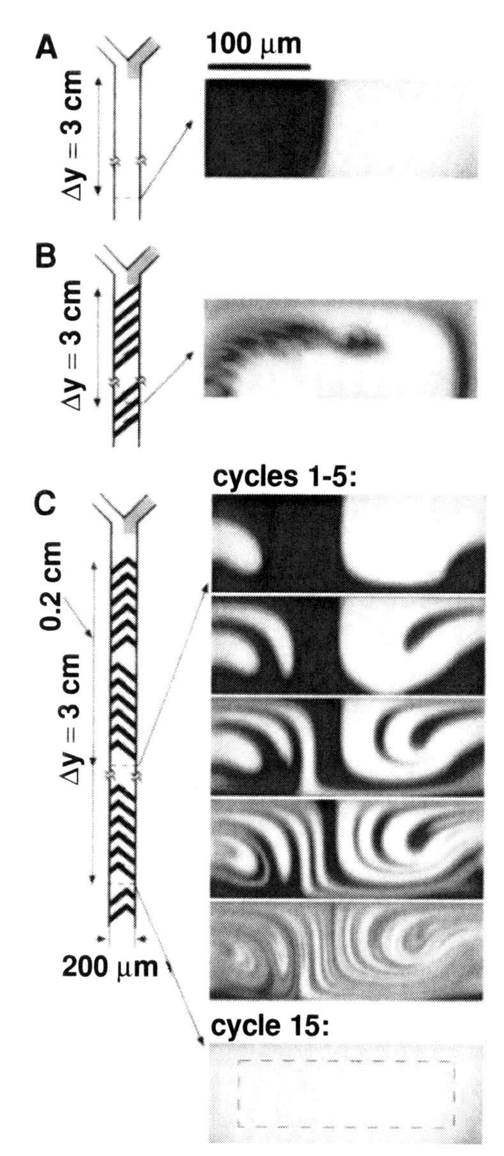

Fig. 4.7 Cross sections of the dye distribution in a microfluidic channel designed to create staggered, time-dependent whorls or twist maps. From [35], used with permission.

amplifies the random aspect of the molecular diffusion. For chaotic mixing processes, the characteristic mixing time or length is proportional to ln *Pe*. This regime is possible only far away from walls, and thus this scaling is observable only if the majority of the observed mixing is happening far from walls.

Chaotic Batchelor regime. The chaotic Batchelor regime implies the situation in which the flow is partially chaotic but the mixing is eventually limited by the nonchaotic flow near the wall. In this limit, the characteristic mixing time or length is proportional to $Pe^{\frac{1}{4}}$. Although the difference between the characteristic times of pure diffusion and chaotic mixing (with or without boundaries) is enormous, the effect of boundaries on the Peclet

Table 4.1 **Diffusivities for dilute analytes in water at 25°C. Particle diffusivities calculated from the Stokes–Einstein relation $D = k_B T/6\pi\eta a$, where a is the particle radius.**

Analyte	D
Na^+	1×10^{-9}
bovine serum albumin, 66 kDa	1×10^{-10}
10 nm particle	1×10^{-14}
1 μm particle	1×10^{-16}
10 μm particle	1×10^{-17}

number dependence of chaotic mixing is important from an engineering standpoint only if the Peclet number range under investigation is of the order of 1×10^4 or more.

4.4 THE LOW-REYNOLDS-NUMBER, HIGH-PECLET-NUMBER LIMIT

From the previous section, we see that mixing can be efficient if Pe is low, in which case diffusion occurs quickly, or if the kinematics of the flow field leads to a repetitive process that shortens the length scales over which diffusion must act. However, if Pe is high and the kinematic structure of the flow field does little to shorten diffusion length scales, then mixing is negligible. In low-Re systems with simple, steady boundary conditions, the laminar flow kinematically does not shorten diffusion length scales. Thus flows in most microfluidic devices do not lead to a shortening of diffusion length scales unless specifically designed to do so.

4.4.1 The high-Peclet-number limit

We find from Subsection E.2.2 that the *mass transfer Peclet number* for a dilute species i is given by $Pe = U\ell/D_i$, where D_i is the binary diffusivity of the species i in the solvent, ℓ is a characteristic length, and U is a characteristic velocity. Here, ℓ should characterize the length over which species must diffuse, and U should characterize the velocity transverse to this diffusion. Table 4.1 shows diffusivities for some example ions, molecules, and particles – from this table, we can see that the diffusivities of particles and macromolecules can be quite small, and because many of the species of interest are large and slowly diffusing, the Peclet number is often large for these flows.

4.4.2 The low-Reynolds-number limit

The results of the previous section all are relevant if the flow has a characteristic direction and diffusion acts transverse to that to distribute chemical species. This is typically true if flow moves in an orderly, laminar fashion through long, narrow channels (as are typical of many microfluidic chips). In this case, flow is unidirectional and the Peclet number governs diffusion. However, if the flow has no characteristic direction, but rather turns about, especially if the flow varies with time, the transport of chemical species is a function of both (1) diffusion, as specified by the mass transfer Peclet number, and (2) the flow itself (as specified by the Reynolds number). Note from Appendix E that the *Reynolds number* is given by $Re = \rho U l/\eta$, where U is the velocity, ℓ is a characteristic length, and η and ρ are the dynamic viscosity and density of the fluid. The flow is almost

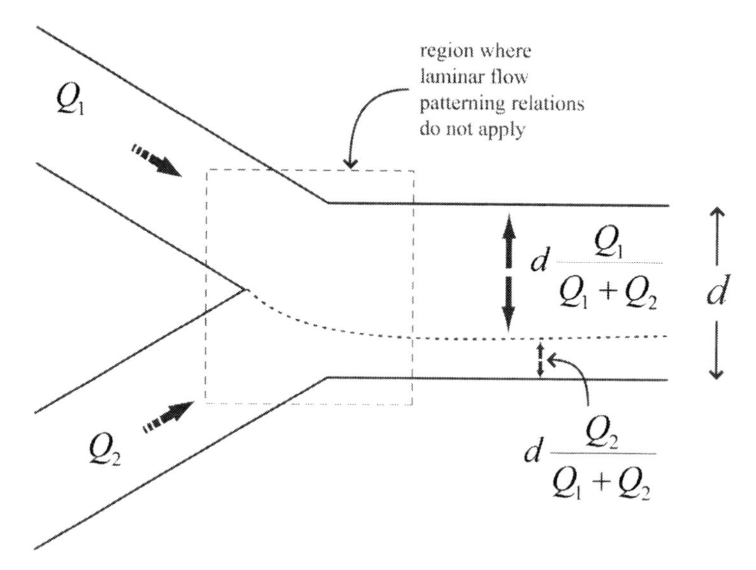

Fig. 4.8 Domain geometries as predicted by inlet flow rates.

always in the low-*Re* limit in microfluidic devices, and thus in the laminar regime – this means the flow is stable and any shortening of diffusion length scales must come from complex geometric boundary conditions, and does not occur naturally from flow instability as is the case in large, high-*Re* systems.

4.5 LAMINAR FLOW PATTERNING IN MICRODEVICES

In the high-*Pe*, low-*Re* limit, laminar flow patterning can be used to control the spatial position of chemical species because multiple solutions can be brought into contact without their chemical components mixing. Simple arguments regarding the flow rate of each solution dictate the area occupied by each component when traveling through a microfluidic channel. This is common for microscale flows. For nanoscale flows, *Pe* and *Re* are often both small.

Consider a microchannel system as a resistor network with nodes, as described in Chapter 3. Consider two channels that generate miscible input flows into a node, the first with flow rate Q_1 of a solution of species A and the second with flow rate Q_2 of a solution of species B. If the mixing between these two species is slow and the channel is shallower than it is wide, then there will be a clear interface between the two streams, and the location of the interface between these two solutions can be predicted with simple flow arguments.

Far from the channel junction or node, the depth-averaged velocity is uniform across the width of the channel (but of course varies strongly along the depth axis). In this case, we can use conservation of species to infer what the cross-sectional areas of each flow is. From this argument, we can show that the fraction of the channel filled with species A is given by $Q_1/(Q_1 + Q_2)$. Similar relations can be derived for the other species, or for each species in a multicomponent system. An example configuration is shown in Fig. 4.8.

Practically speaking, this result means that, if the device and fluids are designed properly, meaning that the channels are wider than they are deep, the Reynolds number is low, and the mass transfer Peclet number is high, then we can control the distribution of the species in a channel simply by controlling the input flow rates of each, either through control of channel depths, channel widths, or input pressures.

EXAMPLE PROBLEM 4.1

Consider a 100-μm-wide channel with a depth of 10 μm. Two inlets into the channel allow influx of flow. Inlet 1 has a solution of deionized water, and inlet 2 has a solution of a chemical species with diffusivity $D = 1 \times 10^{-11}$ m^2/s in water. The mean velocity in the main channel is $\overline{u} = 100$ μm/s. Define coordinate x as the distance along the main channel. Estimate for what x can one assume that the system is mostly mixed.

SOLUTION: The Peclet number for this system is given by $\overline{u}\ell/D$, where the proper ℓ is the length over which species must mix, i.e., 100 μm. Thus $Pe = 1000$. The system is mixed for $x \gg Pe\,\ell$, or $x \gg 10$ cm. This length is longer than the size of most microfluidic chips.

EXAMPLE PROBLEM 4.2

Calculate Pe and Re for flow through a circular microchannel with diameter $d =$ 10 μm. Consider water as the liquid ($\eta/\rho = 1 \times 10^{-6}$ m^2/s) and assume that the mean velocity is 100 μm/s. Assume the chemical species is at low concentration and has a binary diffusion coefficient in water of $D = 1 \times 10^{-12}$ m^2/s.

SOLUTION: $Pe = 1 \times 10^3$, $Re = 1 \times 10^{-3}$. This flow is in the high-Pe, low-Re limit.

The microfluidics community often uses the terms "laminar flow" or "the technique of laminar flow" to imply laminar flow patterning – the control of species distributions in a channel in the limit in which flow rates directly control interfacial positioning. This is common shorthand that has permeated the community, though it obscures an important distinction – *laminar flow* implies a specific flow regime at low Re characterized by stable sheetlike flow structures, whereas *laminar flow patterning* implies a technique for controlling the location of fluids in long, narrow channels at low Re and high Pe.

4.6 TAYLOR–ARIS DISPERSION

The flow in Fig. 4.4 illustrates a 1D diffusion process across gradients transverse to the flow direction. However, many microfluidic devices manipulate boluses of fluid, for example, for chemical separation. In this case, we wish to explore the *axial* diffusion and dispersion of an isolated bolus of fluid as the flow moves through a long narrow microchannel.

We are concerned with the effects of cross-sectional variations in velocity and how these variations affect the measured cross-sectionally averaged concentration. This averaged concentration, for example, is measured by the detector in an electrophoretic separation apparatus. Taylor–Aris dispersion describes how axial convection, axial diffusion, and transverse diffusion combine to control analyte transport in pressure-driven flow through a microchannel. An illustration of how pressure-driven flow leads to dispersion is shown in Fig. 4.9.

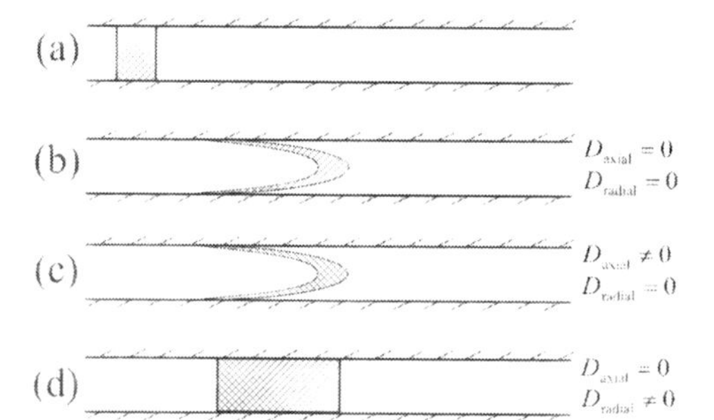

Fig. 4.9 Schematic of Taylor–Aris dispersion in a Poiseuille flow: (a) initial bolus of fluid, (b) fluid after elapsed time t in the absence of diffusion, (c) fluid after elapsed time t with a finite axial diffusion. A slight broadening of the distribution is seen, but the effect of diffusion is minor compared with the dispersive effect of the flow. (d) Fluid after elapsed time t with finite radial and transverse diffusion. The radial diffusion reduces the dispersive effects by causing the scalar to sample both slow- and fast-moving regions of the flow.

Consider Poiseuille flow through a circular microchannel of length L and radius R. The two governing parameters for this flow are the Peclet number $Pe = \overline{u}R/D$ and the length ratio L/R. The symbol $\overline{u}$ denotes the average velocity in the Poiseuille flow, which is given (for a circular microchannel) by

$$\overline{u} = -\frac{R^2}{8\eta}\frac{dp}{dx}. \tag{4.8}$$

The Peclet number characterizes the relative importance of species convection to species diffusion, and the length ratio characterizes the relative importance of radial diffusion to axial diffusion.

Pure Convection. If $Pe \gg L/R$, diffusion can be ignored, as illustrated in Fig. 4.9(b). In this case, which is uncommon in microfluidic devices, the width w of a thin injected sample bolus containing Γ moles per unit area can be shown to grow linearly with time, and the averaged concentration within the bolus decreases inversely with time:

$$w = 2\overline{u}t, \tag{4.9}$$

and

$$\overline{c} = \frac{\Gamma}{w}. \tag{4.10}$$

These relations are valid only for large times. This fluid flow is inherently dispersive – the transverse variation of velocity leads to a spreading of the cross-section-average scalar distribution.

Convection–diffusion. If $Pe \ll L/R$, we can solve the 2D convection–diffusion problem and show that an averaged 1D equation can be written:

$$\frac{\partial \overline{c}}{\partial t} + \overline{u}\frac{\partial \overline{c}}{\partial x} = D_{\text{eff}}\frac{\partial^2 \overline{c}}{\partial x^2}, \tag{4.11}$$

in which the effective diffusivity D_{eff} is given by

effective diffusivity in Taylor–Aris dispersion limit

$$D_{\text{eff}} = D\left(1 + \frac{Pe^2}{A}\right), \tag{4.12}$$

where A is a constant that depends on the geometry and flow boundary conditions. This constant is given by $A = 48$ for Hagen–Poiseuille flow and $A = 30$ for Couette flow, and, although it varies with channel geometry, this constant is of the order of 50 for all standard microchannel geometries. The effective diffusivity leads to an effective diffusive growth of the bandwidth with a $w \propto t^{\frac{1}{2}}$ dependence:

temporal dependence of bandwidth for dispersive transport

$$w = 4\sqrt{\ln 2}\sqrt{D_{\text{eff}}t}\,. \tag{4.13}$$

The 1 in Eq. (4.12) comes from the diffusive component, and the other term is a dispersive component. Diffusion plays two roles in this equation. Axial diffusion leads to the unity term in this equation and tends to increase the effective diffusion (albeit significantly only if Pe is small). Radial diffusion leads to the use of the Taylor–Aris dispersion relation and a $w \propto t^{\frac{1}{2}}$ rather than a linear dependence. Whereas axial diffusion causes band broadening, radial diffusion *minimizes* band broadening. Radial diffusion causes analyte molecules to sample both fast-moving and slow-moving parts of the Poiseuille flow, so each molecule sees an average velocity rather than the widely varying, radially dependent x velocities.

4.7 SUMMARY

In this chapter, we present the passive scalar transport equation:

$$\frac{\partial c}{\partial t} + \vec{u} \cdot \nabla c = D\nabla^2 c\,, \tag{4.14}$$

which can be nondimensionalized to the following form:

$$\frac{\partial c^*}{\partial t^*} + \vec{u}^* \cdot \nabla^* c^* = \frac{1}{Pe}\nabla^{*2} c^*\,. \tag{4.15}$$

This nondimensional form highlights the Peclet number $Pe = U\ell/D$, which indicates the relative magnitude of the convective fluxes compared with the diffusive fluxes in the system.

Mixing is driven by diffusion processes, which proceed over a characteristic length proportional to $\sqrt{Dt}$. Flow processes can facilitate mixing by changing the characteristic length over which this diffusive mixing must occur. Microfluidic devices often use flows with low Re and high Pe, leading to slow mixing. The net effect is that scalars in microfluidic systems are often unmixed over time scales relevant to experiments. This attribute facilitates laminar flow patterning but interferes with processes that require mixing. The situation is often different in nanoscale devices, in which Pe is often also small. When mixing is slow, chaotic advection facilitates mixing by shortening the required diffusion length scales, but this sort of advection occurs only if specifically designed for with carefully crafted microfluidic geometries.

When scalars are transported along circular microchannels, Taylor–Aris dispersion controls the effective axial diffusivity, leading to an effective diffusivity given by

$$D_{\text{eff}} = D\left(1 + \frac{Pe^2}{48}\right). \tag{4.16}$$

4.8 SUPPLEMENTARY READING

This chapter presumes the existence of a diffusivity but largely ignores its atomistic foundations. The physics of random walk processes [33, 36] are illustrative for those who want to build their macroscopic picture of diffusivity from a microscopic foundation. The diffusivity itself and Fick's law are an appropriate model for diffusion if the system is in the dilute solution limit and if thermal gradients are low. Diffusion becomes more complicated if (a) concentration gradients exist simultaneously with thermal gradients, in which case thermodiffusion or Soret effects occur; or (b) the system is a dense solution, in which case the diffusivity is a function of the concentrations of all species components, rather than being a binary property of the species and the solvent. The Maxwell–Stefan formulation, which treats these details, is described in [37]. Species transport can also be driven by electric fields, and directed species transport motivates use of the Nernst–Planck transport equations. These issues are ignored in this chapter but discussed in detail in Chapter 11.

An extensive literature on the physics of mixing exists, and detailed terminology exists to describe chaotic advection and its application to mixing. Two examples of important terminology that have been omitted here include Poincaré maps and Lyapunov exponents, which characterize the space explored by chaotic trajectories and the exponential rate at which trajectories separate. Ottino [38] is the standard reference in the area of kinematics of mixing, and covers these terms and other key mixing concepts. Strogatz [39] is an excellent general reference for those interested broadly in the chaos of nonlinear systems.

This chapter largely ignores description of modern microfluidic mixing geometries, which are described in other texts [3, 6, 7] and in reviews [40, 41, 42], but the reader would benefit from examining papers on microfluidic mixing [35, 43, 44] and laminar flow patterning [45, 46] directly. Transport in DNA microarrays is an area in which mixing issues are of current interest [47, 48, 49, 50, 51] and that links the material in this chapter with that of Chapter 14.

This chapter's description of laminar flow patterning description is a special case of a Hele-Shaw flow analysis, which is described in a more general sense in Chapter 8. Taylor–Aris dispersion is a classical topic that is covered in many texts, including Probstein [29] and Chang and Yeo [52]. The Lévêque problem of diffusion in shear flow, largely omitted here except for the exercises, is discussed in [53, 54].

4.9 EXERCISES

4.1 Calculate Pe for the following sets of characteristic parameters:

(a) $U = 100\ \mu\text{m/s}$, $\ell = 10\ \mu\text{m}$, and $D = 1e - 9\ \text{m}^2/\text{s}$.

(b) $U = 100\ \mu\text{m/s}$, $\ell = 10\ \mu\text{m}$, and $D = 1e - 11\ \text{m}^2/\text{s}$.

(c) $U = 100\ \mu\text{m/s}$, $\ell = 10\ \mu\text{m}$, and $D = 1e - 13\ \text{m}^2/\text{s}$.

4.2 Consider flow of water ($\rho = 1000$ kg/m^3; $\eta = 1$ mPa s) through a circular microchannel of length 5 cm. What is the largest magnitude pressure drop that can be applied while maintaining strictly laminar flow if the microchannel diameter is

(a) 5 μm,

(b) 50 μm,

(c) 50 nm.

4.3 Read Refs. [35, 43, 45]. Answer the following questions:

(a) Consider laminar flow. In terms of its proportionality to *Pe* and ℓ, what is the distance along a channel two streams must travel before they are mixed?

(b) Stroock et al. [35] plot Δy_{90} vs. ln *Pe* in Figure 3E. Why are these data plotted this way? Why is Δy_{90} not plotted vs. *Pe*?

(c) Why do Song et al. [43] plot t_{mix} vs. $\frac{w}{U}$ log *Pe*? Is this the same or different from Stroock's approach?

(d) Given the diagrams and size and flow specifications in the paper by Takayama et al. [45] combined with information you can glean from the text, your independent research, or both, what are the relevant *Re* and *Pe* for this flow? You may have to make engineering estimates for some parameters that are not well specified in the paper.

4.4 Exercise 2.18 used a similarity transform to solve for the time dependence of flow near an impulsively started infinite flat plate. Consider a mathematically identical problem related to passive scalar diffusion. Consider a quiescent fluid containing a passive scalar initially at $c = 0$ for time $t < 0$. For time $t \geq 0$, the boundary condition $c = \frac{1}{2}c_\infty$ is applied at $x = 0$. The boundary condition at $x = \infty$ is that $c = 0$.

(a) Using an approach analogous to Exercise 2.18, solve for the scalar distribution as a function of time.

(b) Show that the concentration at $x = \ell = \sqrt{Dt}$ always has a scalar value of $\text{erfc}(\frac{1}{2})$ times the boundary condition.

4.5 Consider a quiescent fluid containing a passive scalar initially at $c = 0$ for time $t < 0$. For time $t \geq 0$, the boundary condition $c = \frac{1}{2}c_\infty$ is applied at $x = 0$. The fluid region is of finite x dimension d, where the boundary condition at $x = d$ is given by $\frac{\partial c}{\partial x} = 0$. This will require using separation of variables and solving for the distribution in terms of a Fourier series. The result will be quite similar to that of Exercise 2.17.

4.6 Consider a microfluidic device consisting of four microchannels arranged in a cross configuration (Fig. 4.10). Assume all of the microchannels can be approximated as being much wider than they are deep. Assume that all flow is laminar and described by Poiseuille flow relations. The viscosity of all fluids is 1 mPa s and the channel depth is 2 μm. The width of channels 1 and 3 is 20 μm, and the width of channels 2 and 4 is 200 μm.

Outlet 4 is at $p_4 = 100{,}000$ Pa. Pumps are attached to inlets 1, 2, and 3. The pump attached to inlet 1 is at $p_1 = 150{,}000$ Pa. The four channels have lengths $L_1 = L_2 = L_3 =$ 1 cm and $L_4 = 3$ cm. The fluid injected into inlets 1 and 3 is water, and the fluid injected into inlet 2 is a dye with binary diffusivity $D = 2$ $(\mu\text{m})^2$/s. What pressures at inlet 2 and inlet 3 are required to ensure that, in channel 4, the flow consists of a narrow, 6-μm

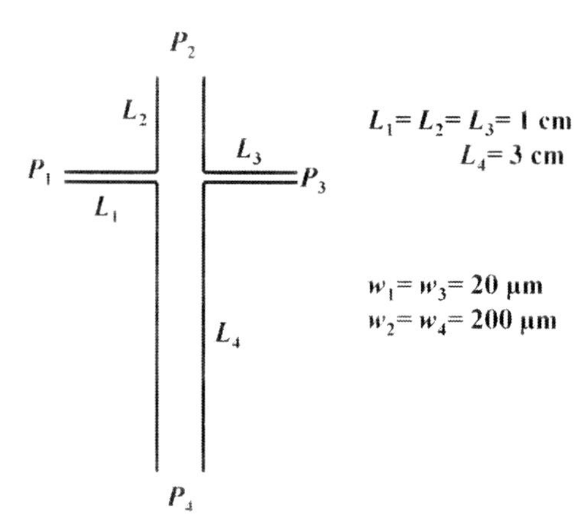

Fig. 4.10 Four-port, cross-shaped microchannel.

stream of dye whose center is located 40 μm left of the centerline of the channel? Estimate *Pe* for the flow through channel 4 and determine if this stream of dye will diffuse quickly or not.

We solve this most easily by noting that the nine "unknowns" are the five pressures p_1–p_5 and the four volumetric flow rates Q_1–Q_4. The nine equations come from (a) the four $\Delta p = QR_\text{h}$ equations for the four channels, (b) one conservation of mass equation at the intersection, and (c) four more bits of information specified by the problem (p_1, p_4, the relation between Q_1 and Q_2, and the relation between Q_1 and Q_3). Thus, a set of nine equations for nine unknowns can be written. This set of nine equations can be written as a matrix equation $Ax = b$, where x is a 1×9 column vector with the nine unknowns, A is the 9×9 multiplication matrix, and b is a 1×9 column vector with nine constants. Numerically, this inversion will work best if the matrix A is well-conditioned, i.e., if the components of the matrix are all of the same order. This will be achieved if the problem is solved with units of pascals for pressures, cubic micrometers per second for flow rates, and pascal-seconds per cubic micrometer for hydraulic resistances.

4.7 Consider the microchannel system viewed from the top (Fig. 4.11). Assume the depths of the channels are uniform throughout. Assume that the distances between the "intersections" (i.e., places where the geometry changes abruptly) are large.

(a) If the flow is pressure driven and the depth of the channel is *much* less than the width, this is what is called a *Hele-Shaw* system. In this case, the streamlines (i.e., lines everywhere normal to the instantaneous flow velocity) for this flow are the same as the streamlines for potential flow through this system.[1] Draw (or calculate and plot) the streamlines for $Q_\text{A} = Q_\text{B}$ and a channel depth much less than the width. You may choose w_0, w_1, and w_2 as you wish for these purposes.

(b) Continue assuming that the channel depth is much less than the width. In the absence of diffusion, the "interfacial" streamline labeled in Fig. 4.11 separates the flow of fluid A from the flow of fluid B. Far from the intersections, calculate the positions y_1 and y_2 of the interfacial streamlines, given w_0, w_1, and w_2 as well as Q_A and Q_B.

(c) Now consider a fluid element located at an arbitrary height y_0 at left. Given Q_A and Q_B as well as the channel geometry, what will this fluid element's y position be in regions 1 and 2? Again, consider positions far from the intersections.

[1] Note, however, that the velocity magnitudes and pressure distributions are *not* the same.

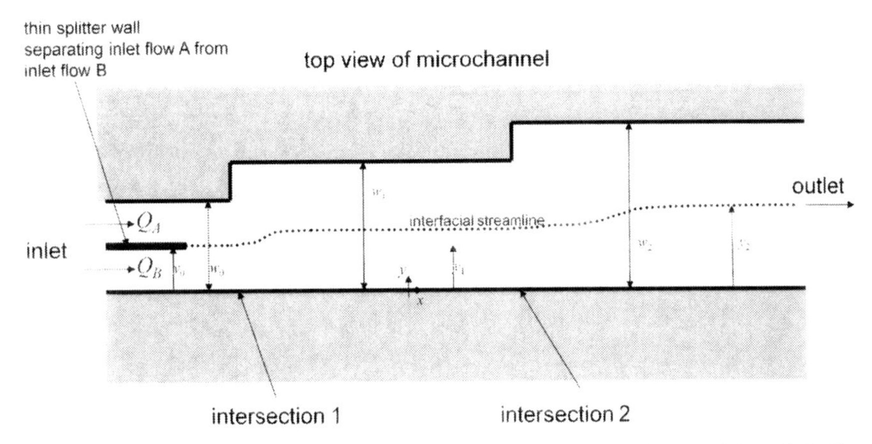

Fig. 4.11 A microchannel system design. Interfacial streamline is shown for a specific set of values of Q_A and Q_B. Example shown for $w_0 < w_1 < w_2$, although this relation is not required. Not to scale – the lengths between intersections are *large* relative to the channel widths. Inlet splitter plate is at centerline of inlet region.

(d) Continue to ignore diffusion. Assume that fluid A consists of a suspension of mammalian cells randomly distributed in space, with a distribution of radii ranging from 5 μm to 20 μm. Assume that particles follow the streamlines (i.e., they are Lagrangian fluid tracers) *except* that the cell walls are rigid; thus the center of a cell with radius a cannot approach the side of the wall more closely than a distance a.

Design a cell size sorter that uses this topology, i.e., design Q_A, Q_B, w_0, w_1, and w_2 such that the y location y_2 of a cell at the outlet is dependent primarily on the cell radius and *not* on the location of the cell in the inlet flow. You need not limit yourself to geometries for which the channel width is much larger than the depth, though you may do so. Complete the outlet geometry so that, rather than one large outlet as shown in Fig. 4.11, sixteen outlets are at the end of the device, each collecting cells in a 1-μm range (i.e., outlet 1 takes only 5-μm particles, outlet 2 takes only 6-μm particles, etc.). *Quantify* and comment in detail on the following issues:

i. How sensitive is your design to cell size, i.e., by how much will cells of different sizes be spatially separated?
ii. How sensitive is your design to the accuracy with which Q_A, Q_B, w_0, w_1, and w_2 can be specified? Presume that w_1 and w_2 can each be specified with ± 1 μm accuracy (owing to fabrication limitations), and Q_A and Q_B can each be specified with $\pm 10\%$ accuracy (owing to uncertainties in linear displacement pumps at low flow rates).
iii. What is the dilution of the particle suspension?

The optimal design will be a sensitive sorter that is largely independent of input errors and dilutes the cell suspension as little as possible. Optimizing all of these simultaneously will not be possible. Justify your design decisions with brief comments about how you prioritized these issues.

4.8 Consider 2D flow between two infinite parallel plates separated by 50 μm. Assume all features are infinite in the z direction. Assume that the plate at the bottom is moving at 100 μm/s and the top plate is moving at 200 μm/s. Assume that a bolus of a chemical species at concentration $c = 1 \times 10^{-3}$ M and $D = 1 \times 10^{-10}$ m^2/s with width 50 μm is injected in between the plates.

(a) Estimate the times when (a) axial and (b) transverse diffusion are negligible.

(b) Consider the early-time solution for which the dispersion is primarily due to convection; solve for the mean concentration $\overline{c}(t)$.

(c) Neglect axial diffusion and calculate the dispersion coefficient.

(d) Calculate the dispersion coefficient if axial diffusion is not ignored.

4.9 The floor of a 1000-μm-wide, 50-μm-deep channel is lined with cells attached to the bottom surface. The two flow inputs into this system are designed to create a region in which the concentration of a soluble chemical species varies approximately linearly with space, and the migration of the cells in response to this chemical gradient will be observed. Where should observation happen?

4.10 Consider the diffusion of a scalar near a wall, in the region where the fluid velocity is linearly proportional to the distance from the wall. Such diffusion is relevant for the mixing near the top and bottom of the microchannel in Fig. 4.4 if the microchannel is no longer assumed shallow relative to its width. Using the coordinate system from Fig. 4.4, show that the width of the mixing region between the two scalar domains is proportional to $\sqrt[3]{y/\overline{u}}$. Compare this with the result expected for a shallow channel.

4.11 Laminar flow-patterning concepts are routinely used in microchannels to shape flows for a variety of purposes. For example, flow cytometry, a technique for counting and characterizing the cells or particles in a fluid suspension, often uses sheath flows to ensure that cells under test always reside at the same (central) location in a microchannel cross section. Mixing studies have used similar concepts to create a thin stream of fluid that quickly mixes diffusively.

Consider a microfluidic flow cytometry device with four channels that meet at a central node. Each channel is long relative to its depth and width. The depths of all channels are equal. The widths of all channels are equal. The lengths of all channels are equal. Assume that flow enters through ports 1–3 and exits through port 4. Inlet flow from ports 1 and 3 consists of fluid only (i.e., the sheath flow), and the inlet flow in port 2 contains a suspension of cells to be analyzed. Assume that the Peclet number based on the diffusivity of the cells is infinite. Assume that the channel exiting in port 4 is 100 μm in width and that the flow is interrogated with a laser whose beam is focused to a 10-μm width that is centered in this channel. The cell suspension must pass through this central 10-μm region to be measured. To maximize the rate at which cells can be analyzed, the cell suspension should completely fill the 10-μm region.

(a) Given that the input volumetric flow rate into port 2 is 2×10^5 $\mu m^3/s$, calculate the flow rates that must be used in ports 1 and 3 to ensure that all cells pass through the interrogation region.

(b) How would the sheath flow rates (i.e., flow rates through ports 1 and 3) need to be changed if the laser focus were misaligned, resulting in an interrogation region 15 μm in width, located 25 μm from one edge?

5 Electrostatics and Electrodynamics

This text is primarily concerned with the behavior of *fluids* in micro- and nanofabricated systems; however, the ubiquity of diffuse charge in solution and the routine use of applied electric fields requires that the electrodynamic equations be solved simultaneously with the fluid equations, leading to a body force term that modifies the fluid velocity fields. This chapter summarizes the fundamental equations of electrostatics and electrodynamics with specific focus on aqueous solutions with boundary conditions typical of microfluidic devices. A description of electrical circuits, which describe the electrostatics and electrodynamics of discretized elements, is also presented.

5.1 ELECTROSTATICS IN MATTER

Electrostatics describes the effects caused by stationary *source* charges or static electric fields on other charges, termed *test charges*. The electrostatic limit applies when all charges are stationary and the current is zero. The equations and boundary conditions of electrostatics are all derived from Coulomb's law.

We study charges *in matter*, usually in an aqueous solution (we also call this an electrolyte solution) or in a metal conductor. In matter, it becomes unwieldy to keep track of all of the electrostatic interactions. To simplify things, we distinguish between *free charge* and *bound charge* and keep detailed track of only the free charge. Free charge implies a charge that is mobile over distances that are large relative to atomic length scales. Free charge typically comes from electrons (in a metal) or ions (in an aqueous solution). We treat free charges specifically in the sections to follow. Bound charge implies charges of equal magnitude but opposite signs that are held in close proximity and are free to move only atomic distances (roughly 1 Å or less). Examples of bound charge include the positive charge of an atomic nucleus and the negative charge of its associated electron cloud, the uneven partial charges in a heteronuclear covalent bond, or the bound ions in a solid crystal. Rather than calculate all of the details of bound charge with detailed electrostatic equations, we replace the detailed effects of bound charge by use of the continuum *electrical permittivity*.

Because microscale flows are often driven by voltages specified by power supplies, we begin our description with the electrical potential and the electric field.

5.1.1 Electrical potential and electric field

We consider a point charge q and the force it feels owing to the presence of other charges. We define the *electrical potential* ϕ (also called the *voltage*) at a point in space such that the electrostatic potential energy that a point charge has with respect to a reference position is given by $q\phi$. We define the *electric field* $\vec{E}$ such that the force on a point charge q is given by $\vec{F} = q\vec{E}$:

definition of electric field

$$\vec{E} = -\nabla\phi. \tag{5.1}$$

5.1.2 Coulomb's law, Gauss's law for electricity in a material, curl of electric field

Given a static source charge q [C] embedded in a linear, instantaneously responsive, isotropic material of electric permittivity ε [C/V m], Coulomb's law describes the electric field induced in the material by this charge:

Coulomb's law in a material, uniform fluid properties

$$\vec{E} = \frac{q}{4\pi\varepsilon}\frac{\vec{\Delta r}}{\Delta r^3} = \frac{q}{4\pi\varepsilon}\frac{\hat{r}}{\Delta r^2}, \tag{5.2}$$

where $\vec{\Delta r}$ is the distance vector from the point charge to the location in question, $\hat{r}$ is the unit vector in this direction ($\hat{r} = \vec{\Delta r}/\Delta r$), and Δr is the magnitude of this vector. We can similarly write the electrical potential caused by a point charge in matter:

$$\phi = \frac{q}{4\pi\varepsilon\Delta r}. \tag{5.3}$$

Because these relations are for a charge embedded in matter, they take into account both the source charge *and* the response of all of the bound charge in the material. In treating the medium as linear and isotropic, we are neglecting nonlinear effects seen at high fields, as well as anisotropy observed, for example, in some crystals. Equation (5.2) implicitly treats the material as responding instantaneously. Finite response times are addressed later in the chapter when the frequency dependence of the quasielectrostatic permittivity is described.

The flux of the electric field caused by an ensemble of point charges can be integrated over a surface, deriving *Gauss's law for electricity*[1]:

---

[1] Gauss's law is one of the four equations commonly referred to as Maxwell's equations. The other Maxwell's equations are as follows:

1. Gauss's law for magnetism, either in integral form,

$$\int_S \vec{B} \cdot \hat{n}\, dA = 0, \tag{5.4}$$

where $\hat{n}$ is a unit outward normal on the surface S and dA is an infinitesimal area on that surface; or in differential form,

$$\nabla \cdot \vec{B} = 0. \tag{5.5}$$

The applied magnetic field is denoted by $\vec{B}$. The applied magnetic field is related to the magnetization and the induced magnetic field by the magnetic permeability:

$$\vec{B} = \mu_0(\vec{M} + \vec{H}) = \mu_0(1 + \chi_m)\vec{H} = \mu_{mag}\vec{H}, \tag{5.6}$$

where $\vec{H}$ is the induced magnetic field and $\vec{M}$ is the resulting magnetization of the material. The magnetic permeability of free space $\mu_{mag,0}$ is $4\pi \times 10^{-7}$ H/m. The unit of inductance H is termed a Henry and is equal to 1 Vs/A. The magnetic permeabilities [Vs/A] of other materials are given by $\mu_{mag} = \mu_{mag,0}(1 + \chi_m)$, where χ_m is the magnetic susceptibility of the material. Gauss's law for magnetism takes the same form as Gauss's law for electricity, but the RHS is zero because magnetic monopoles are absent.

2. Faraday's law of induction:

$$\nabla \times \vec{E} = -\frac{\partial}{\partial t}\vec{B}. \tag{5.7}$$

3. Ampere's law:

$$\nabla \times \vec{H} = \vec{i} + \frac{\partial}{\partial t}\vec{D}. \tag{5.8}$$

Gauss's law for electricity

$$\int_S \varepsilon \vec{E} \cdot \hat{n} \, dA = \sum q \,. \tag{5.9}$$

Here, the integral is performed over a closed surface S, $\hat{n}$ is a unit outward normal along this surface, dA is a differential area along this surface, ε is the electrical permittivity of the material, and $\sum q$ is the sum of the charges [C] inside the surface. By application of the divergence theorem, this integral relation can be put in differential form:

Poisson's equation, i.e., Gauss's law in differential form

$$\nabla \cdot \varepsilon \vec{E} = \rho_E \,, \tag{5.10}$$

where ρ_E is the volumetric net free charge density in the material [C/m^3]. We refer to ρ_E simply as the *charge density* in solution.

CHARGE DENSITY OF AN ELECTROLYTE SOLUTION

We are concerned with the local net free charge density of the fluids we are studying, which are typically aqueous solutions. Ignoring the solvent, which is typically neutral, we can relate the net free charge density ρ_E to species concentrations by

$$\rho_E = \sum_i c_i z_i F \,, \tag{5.11}$$

where c_i is the molar concentration of species i, z_i is the valence (charge normalized by the elementary charge) of species i, and F is Faraday's constant, equal to $F = eN_A = 96485$ C/mol. The free charge density thus corresponds (in liquid systems) to the density of charge of mobile ions. The bound charge has been accounted for by the continuum concept of the electrical permittivity, as discussed earlier. In the absence of external forces, the potential energy of any system of ions is minimized if the net charge is distributed evenly. Thus most media are electroneutral, i.e., the net charge density at all points within domains is zero – any charge is located only at surfaces. This is true for perfect conductors. In the fluid systems we are considering, which have finite conductivity and are also affected by diffusive forces stemming from the thermal energy of the solvent molecules, the fluid has a nonzero net charge density within a finite region near walls, specified by the Debye length λ_D, defined in Chapter 9.

By calculating the line integral of the electric field around a closed contour, we can also show that

$$\int_C \vec{E} \cdot \hat{t} \, ds = 0 \,, \tag{5.12}$$

where ds is a differential line element. This can be converted to differential form with Stokes' theorem, leading to

electric fields are irrotational

$$\nabla \times \vec{E} = 0 \,. \tag{5.13}$$

Equation (5.13) highlights the irrotational property of the electric field. This property follows from the definition of the electric field, because the electric field is defined in Eq. (5.1) in terms of the gradient of a scalar potential and the curl of the gradient of a scalar is zero by definition.

Gauss's law can also be written as

Gauss's law in terms of electric displacement

$$\nabla \cdot \vec{D} = \rho_E \,, \tag{5.14}$$

where $\vec{D}$ is the *electric displacement* or *electric flux density* [C/m^2], defined as

definition of electric displacement

$$\vec{D} = \varepsilon \vec{E} \,, \tag{5.15}$$

and, with this definition, Eq. (5.2) can be written as

$$\vec{D} = \frac{q}{4\pi} \frac{\hat{r}}{\Delta r^2} \,. \tag{5.16}$$

Although $\vec{E}$ and $\vec{D}$ can be converted back and forth quite easily, their analytical purposes are different. The effects *induced by a source charge*, including polarization and electric field, are described by the electric displacement (through $\nabla \cdot \vec{D} = \rho_E$). The effect *experienced by a test charge* is described by the electric field (through $\vec{F} = q\vec{E}$). In vacuum, the two properties have the same value. In matter, the two values are different.

5.1.3 Polarization of matter and electric permittivity

A source charge in vacuum induces an electric field, as defined by Coulomb's law and the permittivity of free space ε_0. The *electrical permittivity of free space* ε_0 is a fundamental constant and is given by $\varepsilon_0 = 8.85 \times 10^{-12}$ C/Vm.

When a source charge is embedded in matter, we divide the effects of the source charge into two parts: (a) the polarization of the medium, which we treat as a continuum, and (b) the residual electric field that exists despite the polarization of the matter. Thus we can think of all electric fields in matter as being *residual* electric fields, the left-over field after some fraction of the effects of the source charge are canceled by the polarization of the matter. Gauss's law for electricity in matter lumps the two effects of the source charge (polarization and residual electric field) into one term, the electric displacement.

The electric polarization $\vec{P}$ [C/m^2] in matter is the dipole moment per unit volume, and measures the degree to which bound charge is polarized. The polarizability of a medium is described by the electric susceptibility χ_e[unitless], which is a function of the atom-scale structure of the material as well as the temperature:

$$\vec{P} = \varepsilon_0 \chi_e \vec{E} \,. \tag{5.17}$$

The electrical permittivity [C/Vm] of a material reflects the sum total of the polarization response of the matter plus the residual electric field and is given by $\varepsilon = \varepsilon_0(1 + \chi_e)$.

Table 5.1 Electric susceptibilities and electric permittivities for several materials.

Material	χ_e	ε_r
Vacuum	0	1
Dry air	5×10^{-4}	1.0005
Dodecane	1	2
Glass	5	6
Silicon	11	12
Isopropanol	17	18
Ethanol	23	24
Methanol	32	33
Acetonitrile	36	37
Water	79	80

Electrical permittivities for several materials are given in Table 5.1. The *relative permittivity* or *dielectric constant*, defined as

$$\varepsilon_r = \varepsilon/\varepsilon_0 , \tag{5.18}$$

is used to denote the ratio of the total response (both polarization and residual electric field) of a particular medium to the electric field alone.

The terminology here is subtle. The *electric susceptibility* indicates how much a medium polarizes in response to an electric field. Free space has nothing to polarize, and its electric susceptibility is zero, whereas water has a large dipole, and its electric susceptibility is large. The *electric permittivity* indicates how much electric field is caused by a source charge and is useful for relating fields to charge with Gauss's law. As compared with vacuum, the dielectric constant of a medium indicates how much smaller the electric fields are in a medium if the same source charge is used.

A common misconception is that the dielectric constant indicates how much more a medium polarizes than a vacuum – this is incorrect, because a vacuum cannot polarize. Water's dielectric constant is 80 *not* because water polarizes 80 times more than space; rather, the dielectric constant is 80 because, when water polarizes, the electric field that is left over is 1/80 as big as it would have been if the water remained unpolarized. The electric susceptibility, similarly, does not indicate that water polarizes 79 times more than space – it indicates that the field caused by the polarized water is 79 times bigger than the residual electric field.

To illustrate polarization of matter, consider a sodium ion and a chloride ion dissolved in water. As seen in Fig. 5.1, the sodium ion causes the polar water molecules to orient themselves in response to the electric field and would cause an attractive force on a negatively charged ion (for example, a chloride ion). The presence of the water molecules tends to decrease the electric field caused by the ions, because the polarized water molecules generate their own electric field in the opposite direction. The net effect is that the test charge (the chloride ion) feels an attractive force, but a much smaller one than would be felt if the ions were in a vacuum. As we have written it, Coulomb's law treats the *free* charges (sodium, chloride) on an individual, atomistic level, but subsumes the effects of *bound* charges (water OH bonds) into a continuum with a continuum property ε. Subsection 5.3.1 extends this to consider finite response times of materials and the attendant frequency dependence of the permittivity.

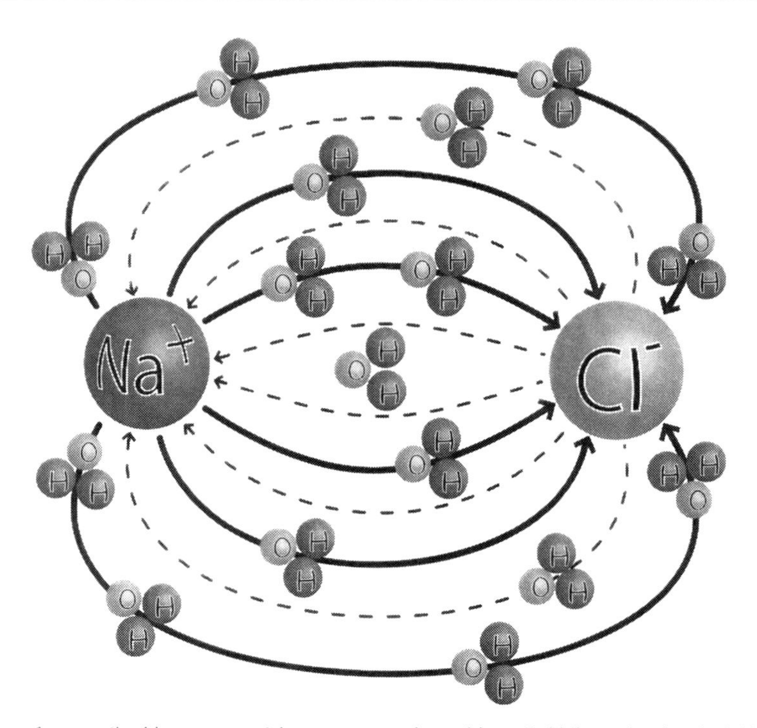

Fig. 5.1 Polarization of matter (in this case, water) in response to charged ions. Bold lines: the electric field lines that the sodium and chloride ions would have generated if they were in a vacuum. The polar water molecules orient in response to the ions, with the oxygen (carrying a partial negative charge) toward the sodium atom and the hydrogens (carrying partial positive charges) toward the chloride atom. The orientation of the polar water molecules causes its own electric field (dashed lines), which cancels out most ($\frac{\varepsilon_r-1}{\varepsilon_r}$) of the electric field that would exist if the ions were in a vacuum. The resulting net electric field is well approximated by the field given by Coulomb's law for charges in matter, Eq. (5.2). Ions and molecules are not drawn to scale. The extent of the polarization is exaggerated in this figure – in fact, the linear material assumption in Eq. (5.15) describes matter response to electric fields only if the induced polarization is a small perturbation of the state of the material.

5.1.4 Material, frequency, and electric-field dependence of electrical permittivity

Materials experience an electric polarization in an electric field, as described by Eq. (5.17). The electric polarization is a measure of the degree to which the bound charge in a medium polarizes, with positive components being pulled in the direction of the electric field and negative components being pulled in the direction opposite to the electric field. Multiple polarization mechanisms contribute to this phenomenon, and these mechanisms each have a different characteristic time. In the presence of an electric field, the electron cloud of an atom is displaced with respect to the nucleus, leading to *electronic polarization*, and atomic bonds stretch, leading to *atomic polarization*. At 100 V/cm, this deflection is roughly 1 billionth of an atomic radius. These phenomena are rapid, occurring over characteristic times in the range 1×10^{-15}–1×10^{-18} s. Molecules with permanent dipoles orient rotationally (owing to the torque; see Subsection 5.1.10), leading to a net *orientational polarization*. This contribution is what makes water's permittivity so large compared with other liquids. Although atomic polarization requires motion on subatomic length scales (much smaller than 1 Å), dipole orientation requires motion on atomic length scales (1 Å or higher), which takes longer. The effective rotational relaxation time for water at room temperature, for example, is about 8 ps. Orientational polarization occurs only if the frequency of the electric field is below 1×10^{12} Hz.

Debye relaxation model for frequency dependence of quasielectrostatic permittivity. Equation (5.15) assumes that the response of the medium to an applied field is instantaneous. However, the polarization of any real medium requires a finite time. In order to use electrostatic equations to describe systems whose polarization cannot be assumed infinitely fast, we define a frequency-dependent quasielectrostatic permittivity $\varepsilon(\omega)$, which describes the magnitude of the medium's polarization response to a sinusoidal applied electric field. In defining the quasielectrostatic permittivity, we obviate an electrodynamic description of moving bound charge by use of a frequency-dependent electrostatic parameter. For applied fields whose frequency is much higher than the characteristic frequency of the polarization process, the polarization process is largely ineffective, and thus that process does not contribute to the permittivity at that frequency. For those applied fields whose frequency is much lower than the characteristic frequency of the polarization process, the polarization process contributes to the permittivity at that frequency. We thus think of $\varepsilon(\omega)$ in terms of the sum of contributions from these different physical processes. Each process has a characteristic time τ that signifies the speed of the process – for example, τ for atomic polarization is approximately 1.0×10^{-15} s and τ for orientational polarization is approximately 1.2×10^{-11} s. In the *Debye model* of permittivity, the contribution to the electric susceptibility from each polarization mechanism i takes the form

$$\chi_{e,i}(\omega) = \frac{\chi_{e,i}(0)}{1 + (\omega\tau_i)^2}, \tag{5.19}$$

where ω is the frequency of the applied field, $\chi_{e,i}(0)$ is the contribution of mechanism i to the polarization if the applied field is steady, and τ_i is the characteristic time of polarization mechanism i. This expression is analogous to the expression for the frequency dependence of the power transmitted through a capacitor in an *RC* circuit. The total electric susceptibility for a material is thus given by

$$\chi_e(\omega) = \sum_i \frac{\chi_{e,i}(0)}{1 + (\omega\tau_i)^2}, \tag{5.20}$$

and the permittivity is given by

$$\varepsilon(\omega) = \varepsilon_0 \left[1 + \sum_i \frac{\chi_{e,i}(0)}{1 + (\omega\tau_i)^2} \right]. \tag{5.21}$$

For water at room temperature, for example, $\chi_{e,i}$ for atomic polarization is approximately equal to 5, whereas $\chi_{e,i}$ for orientational polarization is approximately equal to 72, so a Debye-model approximation for the permittivity of water at room temperature is

$$\varepsilon \simeq \varepsilon_0 \left[1 + \frac{5}{1 + (\omega \times 1 \times 10^{-15}\ \mathrm{s})^2} + \simeq \frac{72}{1 + (\omega \times 1.2 \times 10^{-11}\ \mathrm{s})^2} \right]. \tag{5.22}$$

The Debye model assumes that the response of the medium is made up of noninteracting dipoles with a single response time for orientation. This model is a good approximate model but typically must be replaced with a model that is more sophisticated or has more free parameters, if experimental data are to be well predicted. For orientational polarization, the Debye model does not account for interactions between molecules and the attendant spread of response times; for atomic and electronic polarization, the Debye model is inaccurate because the dipoles are created by the field rather than oriented by the field. Thus the predictions of Rel. (5.22) should be considered qualitative.

Electric field dependence of permittivity. The use of a permittivity as previously described implies that the medium is linear, meaning that double the electric field leads to double the polarization. This is correct for electronic and atomic polarization of water. This approximation is correct for orientation of water only if the orientational energy of the inherent dipole moment of each molecule in the field is small compared with $k_B T$, meaning that the statistical orientation of the molecules is minor. Here, k_B is the Boltzmann constant (1.38×10^{-23}J/K). For water, when the applied electric field is of the order of 1×10^9 V/m, the orientational energy $p_w E \simeq k_B T$ is of the same order as the thermal energy of the system, and the water is largely aligned. Once the alignment is significant, the linear model for the response of the material is no longer accurate. An approximate description of the permittivity of water as a function of the electric field uses the Brillouin function to approximate the statistical vector average of the dipole moment of the water molecules as a function of applied field:

$$\langle p \rangle = p_w \mathcal{B}\left(\frac{p_w E}{k_B T}\right), \tag{5.23}$$

where p_w is the magnitude of the dipole moment of an individual water molecule (2.95 D), $\langle p \rangle$ denotes the statistical vector average dipole moment of an ensemble of water molecules, E is the magnitude of the applied electric field, and $\mathcal{B}$ denotes the Brillouin function for dipole orientation with two possible states, i.e., $\mathcal{B}(x) = 2\coth 2x - \coth x$. From Eq. (5.23), the ensemble-averaged dipole moment $\langle p \rangle$ of water molecules varies linearly with E at low fields, because $\mathcal{B}(x) \simeq (x)$ for small x, but approaches $\langle p \rangle = p_w$ as the electric field E approaches infinity. Because the ensemble-averaged dipole moment cannot increase beyond p_w, the permittivity is proportional to $1/E$ as E approaches infinity. The approximate description of the permittivity of water is thus given by

$$\varepsilon_r(E) = 6 + 72\frac{k_B T}{p_w E}\mathcal{B}\left(\frac{p_w E}{k_B T}\right), \tag{5.24}$$

where the values 6 and 72 are approximate values for water at room temperature.[2] The constant value (6) includes atomic and electronic polarization, and the frequency-dependent term (proportional to 72) corresponds to the orientational polarization. This relation is approximate and ignores other nonlinear effects such as electrostriction.

5.1.5 Poisson and Laplace equations

The Poisson and Laplace equations are simplified forms of Gauss's law that apply if no time-varying magnetic fields are present. From Subsection 5.1.2, Gauss's law for electricity is

$$\nabla \cdot \vec{D} = \rho_E \,. \tag{5.25}$$

Considering irrotational electric fields ($\nabla \times \vec{E} = 0$) and writing the electric field as the gradient of the electrical potential ($\vec{E} = -\nabla\phi$), we get the Poisson equation:

$$-\nabla \cdot \varepsilon \nabla \phi = \rho_E \,, \tag{5.26}$$

[2] Although $\lim_{x \to 0} \mathcal{B}(x) = x$, it is challenging numerically to evaluate $\mathcal{B}(x)$ in the $x \to 0$ limit.

written here for nonuniform ε. For uniform ε, this can be rewritten as

Poisson equation, uniform fluid properties

$$-\varepsilon\nabla^2\phi = \rho_E . \tag{5.27}$$

The Laplace equation is derived from the Poisson equation if ρ_E is assumed to be zero:

Laplace equation for electroneutral material

$$\nabla^2\phi = 0 , \tag{5.28}$$

which is the case for electrolyte solutions far away from solid walls.

5.1.6 Classification of material types

We classify materials into several categories based on their response to electric fields. *Ideal dielectrics* or, equivalently, *perfect insulators* are materials that have no free charge – their charge is all bound at the atomic level. These materials cannot conduct current ohmically, so their conductivity $\sigma = 0$, but they do polarize in response to electric fields and thus they have an electrical permittivity ε. *Perfect conductors* or *ideal conductors* have charges (typically electrons) that are assumed to be free to move in the material. The electric field and charge density in a perfect conductor are both zero, and thus the conductor is an equipotential and is electroneutral. Regardless of any externally applied electric field, the charge in a conductor aligns itself at the edges of the conductor to perfectly cancel out this field, and all of the net charge in a conductor is located *on the surface* of the conductor. Perfect conductors have $\sigma = \infty$, and, because they experience no electric field, the permittivity of perfect conductors is undefined. In between these two limits, materials that carry finite current and support a finite electric field are called *weak conductors* or *lossy dielectrics*. These materials have finite conductivity σ and finite permittivity ε. In microfluidic systems that use electric fields, the device substrate is typically an insulator (e.g., glass or polymer) or has an insulating coating (such as a silicon oxide coating on a silicon microdevice), electrodes are made from conducting materials, and the working fluid is typically an aqueous solution that we treat as a lossy dielectric. Thus microsystem boundaries are usually perfect insulators or perfect conductors, whereas the governing equations are used to study weakly conducting fluids.

5.1.7 Electrostatic boundary conditions

For electrostatic systems (i.e., systems in which all charges are static), Gauss's law can be applied to the boundary between two domains to determine the boundary conditions for the normal and tangential components of the electric field. Given an interface with a charge per unit area given by q'', separating two domains labeled 1 and 2 with permittivities of ε_1 and ε_2, respectively, the boundary condition for the normal component of the electric field is given by

$$\varepsilon_2\vec{\boldsymbol{E}}_2 \cdot \hat{\boldsymbol{n}} - \varepsilon_1\vec{\boldsymbol{E}}_1 \cdot \hat{\boldsymbol{n}} = q'' . \tag{5.29}$$

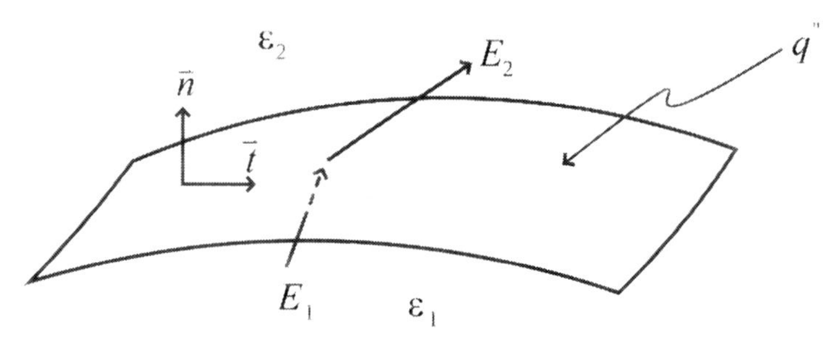

Fig. 5.2 The boundary between two domains. The normal vector $\hat{\boldsymbol{n}}$ points, by convention, from domain 1 to domain 2. The tangential boundary condition applies for any $\hat{\boldsymbol{t}}$ vector that is normal to $\hat{\boldsymbol{n}}$.

The direction of $\hat{\boldsymbol{n}}$ points from domain 1 to domain 2, as shown in Fig. 5.2. A line integral over a loop at the boundary gives the boundary condition for any tangent component of the electric field:

$$\vec{\boldsymbol{E}}_1 \cdot \hat{\boldsymbol{t}} = \vec{\boldsymbol{E}}_2 \cdot \hat{\boldsymbol{t}} . \tag{5.30}$$

Eq. (5.30) holds for any unit vector tangent to the surface.

Boundary conditions at microdevice walls and inlets. We are generally concerned with the boundary conditions for the electric field for a weakly conducting fluid at a solid surface, either an insulating surface (such as a polymer or glass) or a conducting surface (such as a metal electrode). Because the fluid channels in a microdevice are usually connected to large reservoirs that are, in turn, connected to electrode wires, we often need to treat the inlets and outlets of a device as boundaries in an analytical or computational treatment of the electrostatic equations.

For an insulating wall that has a surface charge density q'', Eq. (5.29) is simplified because the electric field in the insulating wall ($\vec{\boldsymbol{E}}_1$) normal to the surface is approximately zero. Thus the normal boundary condition is a relation between the surface charge density and the potential gradient normal to the surface:

$$\frac{\partial \phi}{\partial n} = -\frac{q''_{\text{wall}}}{\varepsilon} , \tag{5.31}$$

where n is the coordinate normal to the surface, pointing into the weakly conducting fluid.

For a conducting wall such as a metal electrode, the metal surface provides a constant-potential boundary condition:

$$\phi_{\text{wall}} = V_{\text{electrode}} . \tag{5.32}$$

Rigorous treatment of inlets and outlets to microchannels requires specification of the entire reservoir geometry. However, because microdevice reservoirs are typically enormous compared with the microchannels themselves, we neglect any voltage drop between the electrode and the inlet to the microchannel. Thus, in the electrostatic case,

$$\phi_{\text{inlet}} = V_{\text{inlet}} \tag{5.33}$$

and

$$\phi_{\text{outlet}} = V_{\text{outlet}} . \tag{5.34}$$

Figure 5.3 depicts these boundary conditions for a sample microdevice.

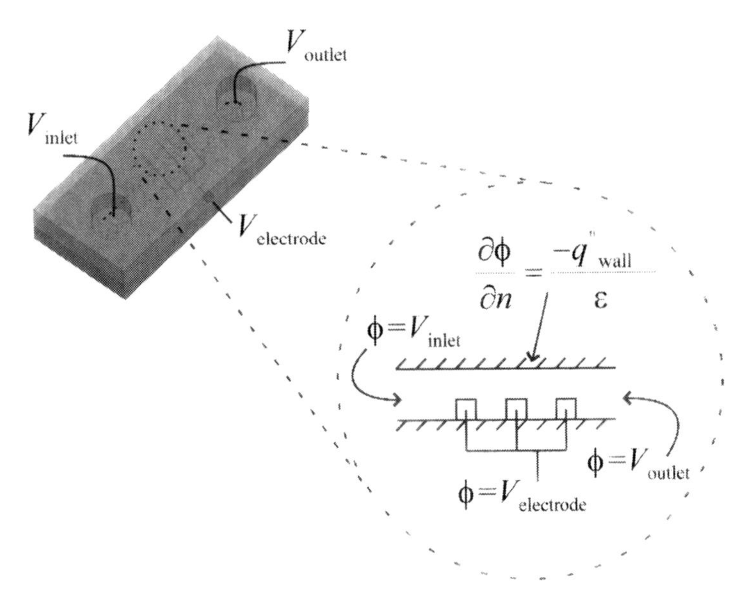

Fig. 5.3 Upper left: schematic of a glass microdevice with inlet and outlet reservoirs with potentials specified by electrical connections made with platinum wires. A microchannel connects the inlet and outlet, and micropatterned electrodes on the bottom of the microchannel are independently addressed. Lower right: schematic of the microchannel fluid domain and the electrostatic boundary conditions in the limit without current or ion motion. The glass surface is treated as an insulating surface with a surface charge density, the electrode is treated as a conductor and therefore an isopotential, and the inlet and outlet are approximated as being equal to the potential applied to the reservoir.

5.1.8 Solution of electrostatic equations

The electrostatic equations are typically solved numerically. For systems that can be approximated as having charge only at interfaces, the Laplace equation applies and thus conformal mapping is applicable. For microchannel systems that can be approximated as having polygonal cross sections that vary slowly along their axis, Schwarz–Christoffel transforms enable rapid analytical solution (Appendix G).

5.1.9 Maxwell stress tensor

Fundamentally, if magnetic fields are steady and if the volume is not moving quickly, the electrical forces on a particle are simply the sum of all Coulomb interactions between charges and fields. Because Maxwell's equations relate charges and fields, the force on a control volume can be put into a form that is a function of the fields only. This is particularly useful when we know the electric fields along the surface of a control volume better than we know the distribution of charge inside the volume. To calculate the force on a volume, we introduce the Maxwell stress tensor $\vec{\vec{T}}$:

definition of Maxwell stress tensor

$$\vec{\vec{T}} = \varepsilon \vec{E}\vec{E} + \frac{1}{\mu_{\text{mag}}}\vec{B}\vec{B} - \frac{1}{2}\vec{\vec{\delta}}\left(\varepsilon \vec{E}\cdot\vec{E} + \frac{1}{\mu_{\text{mag}}}\vec{B}\cdot\vec{B}\right), \tag{5.35}$$

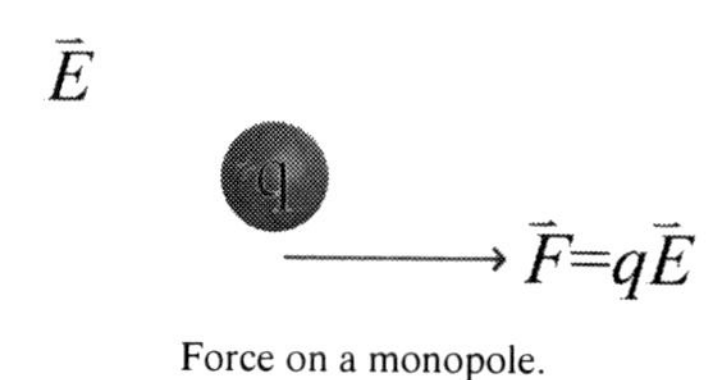

Fig. 5.4 Force on a monopole.

which, in the absence of magnetic fields, is given by

definition of Maxwell stress tensor, zero magnetic fields

$$\vec{\vec{T}} = \varepsilon \vec{E}\vec{E} - \frac{1}{2}\vec{\vec{\delta}}\left(\varepsilon \vec{E} \cdot \vec{E}\right). \tag{5.36}$$

With these definitions, the force on a volume can be written in terms of the Maxwell stress tensor:

force on a volume in terms of the Maxwell stress tensor

$$\vec{F} = \int_S (\vec{\vec{T}} \cdot \hat{n})\, dA. \tag{5.37}$$

The Maxwell stress tensor is a dyadic tensor; see Subsection C.2.6. The relation between the force and the Maxwell stress tensor is used in Chapter 17 for determining the force on a microparticle as a function of the electric fields applied to it.

5.1.10 Effects of electrostatic fields on multipoles

The *multipolar theory* is used to describe electricity and magnetism for both mathematical and physical reasons. Mathematically, multipolar solutions (described in Appendix F) arise naturally from use of separation of variables to analytically solve the Laplace equation. Physically, some of these solutions correspond to physical objects that play a central role in electrodynamic systems, such as point charges and magnetic dipoles.

Here the relations for forces and torques on monopoles (point charges) and dipoles are summarized. In all cases, the force is written in terms of the applied electric field at the location of the multipole. By *applied* electric field, we imply the field that would have been at that point if the multipole were absent.

ELECTRICAL FORCE ON A MONOPOLE (I.E., POINT CHARGE)

The electrical force on a point charge is given by

force on a point charge

$$\vec{F} = q\vec{E}. \tag{5.38}$$

This is shown schematically in Fig. 5.4.

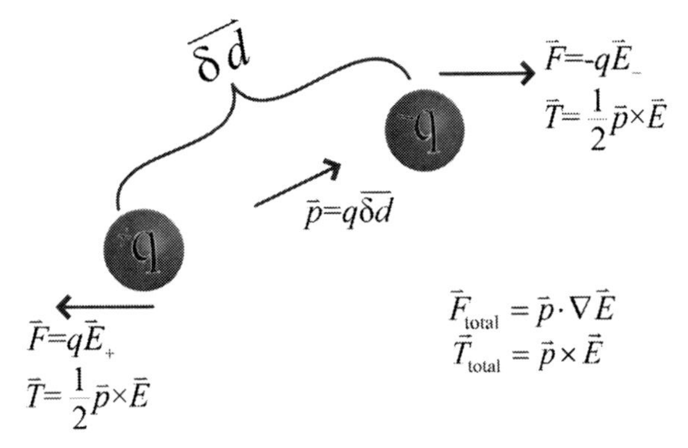

Fig. 5.5 Force on a dipole.

ELECTRICAL FORCE ON A DIPOLE

We can find the electrical force on a dipole in a nonuniform field by summing the forces on its constituent monopoles. This is shown schematically in Fig. 5.5. The result can be written in terms of the dipole moment $\vec{\boldsymbol{p}}$ [C m]:

force on an electrical dipole

$$\vec{\boldsymbol{F}} = (\vec{\boldsymbol{p}} \cdot \nabla)\, \vec{\boldsymbol{E}}. \tag{5.39}$$

The dipole moment is measured in units of Coulomb-meters or debyes (where 1 D = 3.33×10^{-30} C m).

ELECTRICAL TORQUE ON A DIPOLE

We can find the electrical torque on a dipole in a nonuniform field by summing the forces on its constituent monopoles (Fig. 5.5):

torque on an electrical dipole

$$\vec{\boldsymbol{T}} = \vec{\boldsymbol{p}} \times \vec{\boldsymbol{E}}. \tag{5.40}$$

5.2 ELECTRODYNAMICS

Electrodynamics describes the physics of moving charges, which is critical because when we apply fields to aqueous solutions, the charged ions in the solution move, carrying a current. Fortunately, we need consider only the simplest aspects of electrodynamics, i.e., current owing to mobile dissolved ions. As mentioned earlier, we assume all magnetic fields are constant because time-varying magnetic fields are rarely used in microfluidic systems. Because the current in aqueous solutions is carried by ions, which move slowly, we typically ignore the electrodynamic force ($\vec{\boldsymbol{F}} = q\vec{\boldsymbol{u}}_{\rm ion} \times \vec{\boldsymbol{B}}$) on a moving charge as well. Because of this, this chapter treats only the charge conservation equation and omits much of the field of electrodynamics.

5.2.1 Charge conservation equation

The charge conservation equation, written in integral form as

$$\frac{\partial \sum q}{\partial t} + \int_S \vec{i} \cdot \vec{n}\, dA = 0 \tag{5.41}$$

and in differential form (upon application of the divergence theorem) as

$$\frac{\partial \rho_E}{\partial t} + \nabla \cdot \vec{i} = 0\,, \tag{5.42}$$

describes the *motion* of charge and is thus a statement of electrodynamics rather than of electrostatics. The symbol $\vec{i}$ denotes the current density (or, equivalently, the charge flux density), which in an aqueous solution includes both ohmic current stemming from the electromigration of ions with respect to the fluid as well as net convective and diffusive charge flux. The ohmic current flux (i.e., the current flux from motion of free charge relative to the fluid) is given by

$$\vec{i} = \sigma \vec{E}\,, \tag{5.43}$$

where σ is the *conductivity* of the medium. In most regions of microfluidic systems, this ohmic flux dominates over charge transport by fluid convection or net diffusion of ions.

5.2.2 Electrodynamic boundary conditions

The electrodynamic boundary conditions for our systems come from a combination of Gauss's law and the charge conservation equation applied to the interface. The electrostatic condition for the normal field still applies:

$$\varepsilon_2 \vec{E}_2 \cdot \hat{n} - \varepsilon_1 \vec{E}_1 \cdot \hat{n} = q''\,, \tag{5.44}$$

as does the electrostatic condition for the tangent field:

$$\vec{E}_1 \cdot \hat{t} = \vec{E}_2 \cdot \hat{t}\,. \tag{5.45}$$

We add to this the charge conservation equation applied across the interface

dynamic condition for charge at interface

$$\frac{\partial q''}{\partial t} = \vec{i}_1 \cdot \hat{n} - \vec{i}_2 \cdot \hat{n}\,, \tag{5.46}$$

which, if the current flux is strictly ohmic (that is, if ion migration is the only source of current, and ion diffusion and fluid flow can be ignored), implies

$$\frac{\partial q''}{\partial t} = \sigma_1 \vec{E}_1 \cdot \hat{n} - \sigma_2 \vec{E}_2 \cdot \hat{n}\,. \tag{5.47}$$

Thus Gauss's equation relates the normal fields to the interfacial charge density, whereas the charge conservation equation relates the normal fields to the time derivative of the interfacial charge density.

EXAMPLE PROBLEM 5.1

Consider an interface between two materials with permittivities and conductivities given by ε_1, ε_2, σ_1, and σ_2. Assume the system is in steady state.
Applying Gauss's law at an interface leads to the boundary condition for the electric displacement normal to the interface:

$$\varepsilon_2 \vec{E}_2 \cdot \hat{n} - \varepsilon_1 \vec{E}_1 \cdot \hat{n} = q'' . \tag{5.48}$$

If ion diffusion is ignored, the ohmic current is given by $\vec{i} = \sigma \vec{E}$. Applying the charge conservation equation at an interface at steady state leads to

$$\sigma_1 \vec{E}_1 \cdot \hat{n} - \sigma_2 \vec{E}_2 \cdot \hat{n} = 0 . \tag{5.49}$$

Thus, if current passes through an interface, the equilibrium solution will include a charge density that must reside at the surface. Write an expression for this interface charge density in terms of the material parameters and $\vec{E}_2 \cdot \hat{n}$.

SOLUTION: From charge conservation at equilibrium,

$$\sigma_1 \vec{E}_1 \cdot \hat{n} = \sigma_2 \vec{E}_2 \cdot \hat{n} , \tag{5.50}$$

from which we can write

$$\vec{E}_1 \cdot \hat{n} = \frac{\sigma_2}{\sigma_1} \vec{E}_2 \cdot \hat{n} . \tag{5.51}$$

From Gauss's law,

$$\varepsilon_2 \vec{E}_2 \cdot \hat{n} - \varepsilon_1 \vec{E}_1 \cdot \hat{n} = q'' . \tag{5.52}$$

Substituting in for $\vec{E}_1 \cdot \hat{n}$, we get

$$\varepsilon_2 \vec{E}_2 \cdot \hat{n} - \varepsilon_1 \frac{\sigma_2}{\sigma_1} \vec{E}_2 \cdot \hat{n} = q'' . \tag{5.53}$$

Simplifying, we obtain

$$q'' = \left(\varepsilon_2 - \varepsilon_1 \frac{\sigma_2}{\sigma_1} \right) \vec{E}_2 \cdot \hat{n} . \tag{5.54}$$

Current boundary conditions at electrodes. A voltage applied to a conducting metal electrode induces a current to pass through the interface. The interface, though, is complicated by the fact that the current in metals is carried by electrons, whereas the current in an aqueous solution is carried by ions. Thus, for a current to pass through an electrode, an electrochemical reaction must occur. This reaction is driven by a potential difference between the electrode and the solution in contact with it, as denoted in Fig. 5.6. The current density is defined by the Butler–Volmer equation:[3]

$$\vec{i}_1 \cdot \hat{n} = i_0 \left[\exp\left(\frac{-\alpha n F(\phi_2 - \phi_1)}{RT} \right) - \exp\left(\frac{(1-\alpha) n F(\phi_2 - \phi_1)}{RT} \right) \right] , \tag{5.55}$$

[3] This form of the equation is valid in the limit in which mass exchange is infinitely fast, and the concentration of ions at the electrode surface is at equilibrium with the bulk solution. If this is not the case, then the mass transfer must also be accounted for.

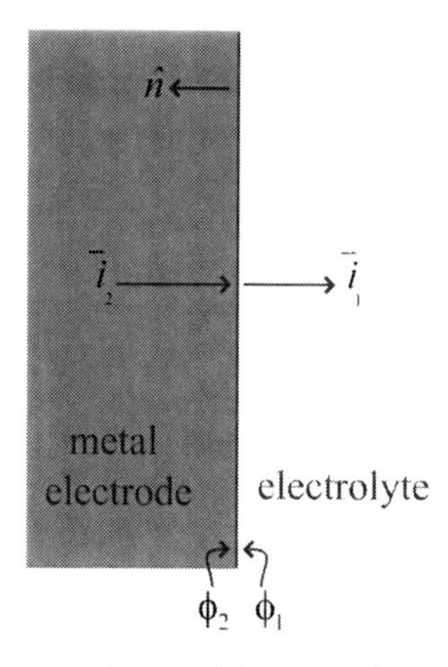

Fig. 5.6 Current and potential drop at an electrode.

where n is the number of electrons transferred in the chemical reaction, F is the Faraday constant, ϕ_1 and ϕ_2 are the electrical potentials on the solution and the electrode sides of the interface, respectively, $R = k_B N_A$ is the universal gas constant, T is temperature, i_0 is a constant referred to as the exchange current density,[4] and α is a parameter between 0 and 1 that denotes the sensitivity of chemical transition states to ϕ_1 and ϕ_2. Lacking detailed information about the transition state, we often assume that α can be approximated as $\frac{1}{2}$, indicating that the chemical transition state is equally sensitive to both potentials.

If the potential drop is small relative to RT/nF (approximately 25 mV at room temperature for a single-electron reaction), then the first-order Taylor series expansion of the Butler–Volmer equation gives

$$\vec{i}_1 \cdot \hat{n} = i_0 \frac{nF(\phi_1 - \phi_2)}{RT} . \tag{5.56}$$

In this limit, the interface has an effective resistance of $RT/i_0 nF$. As the current increases and the exponential term grows, the effective resistance drops.

5.2.3 Field lines at substrate walls

Perfect insulators and perfect conductors in contact with weakly conducting liquids lead to well-defined geometries of field lines at the interface, as denoted in Fig. 5.7. Because the tangent electric field inside a conductor is zero, the tangent electric field just outside the conductor is also zero, and thus electrostatics dictates that the electric field just outside a perfect conductor is *normal* to the surface. Similarly, because the conductivity through a perfect insulator is zero, the current normal to the surface must be zero, and electrodynamics dictates that the electric field just outside a perfect insulator be *tangent* to the surface.

5.3 ANALYTIC REPRESENTATIONS OF ELECTRODYNAMIC QUANTITIES: COMPLEX PERMITTIVITY AND CONDUCTIVITY

Section G.3 notes that, if the time dependences of the voltages are sinusoidal, the relations between the voltages and their derivatives take on a special form. We take

[4] The exchange current density is the current density at zero potential drop.

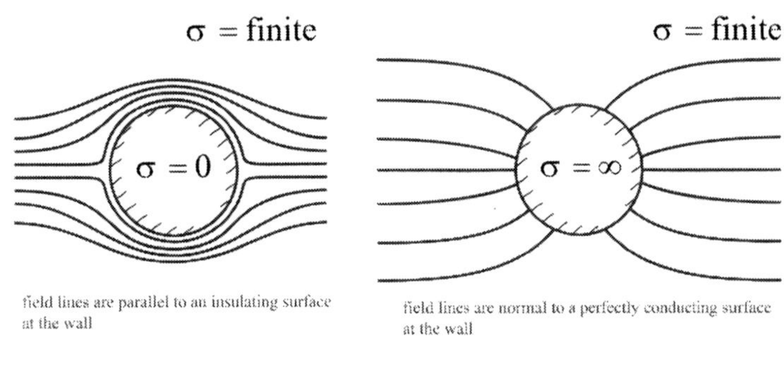

$$\frac{\partial \phi}{\partial n} = 0 \qquad\qquad \frac{\partial \phi}{\partial t} = 0$$

Fig. 5.7 Boundary conditions for the electric field lines at perfect insulators and conductors in contact with weakly conducting liquids.

advantage of this by using the *analytic representation* of the sinusoidal signals, in which case we exchange the real description of a physical quantity with a closely related complex function. This approach makes analysis of system responses to sinusoidal fields enormously simpler. For sinusoidal fields, we can replace the real electric field $\vec{E} = \vec{E}_0 \cos \omega t$ with its analytic representation $\underset{\sim}{\vec{E}} = \vec{E}_0 \exp j\omega t$ and replace linear, real equations of electrodynamics with equivalent analytic representations. For these linear systems, for which all parameters vary at the driving frequency, we can furthermore drop the $\exp j\omega t$ from each term and write the result strictly in terms of the phasors. For example, we can replace $\vec{D} = \varepsilon \vec{E}$ with

$$\underset{\sim}{\vec{D}} = \varepsilon \underset{\sim}{\vec{E}}, \tag{5.57}$$

where $\underset{\sim}{\vec{D}}$ is the analytic representation of the electrical displacement, or

$$\underset{\sim}{\vec{D}_0} = \varepsilon \underset{\sim}{\vec{E}_0}, \tag{5.58}$$

where $\underset{\sim}{\vec{D}_0}$ and $\underset{\sim}{\vec{E}_0}$ are the displacement and electric field phasors.

For electrodynamic systems, the analytic representation allows combination of the quasielectrostatic relation for the electric displacement with the electrodynamic relation for ohmic current. In so doing, the two sources of current in a stationary medium are combined into one complex equation,

complex sum of displacement and ohmic current

$$\underset{\sim}{\vec{D}} + \frac{\underset{\sim}{\vec{i}}}{j\omega} = \varepsilon \underset{\sim}{\vec{E}} + \frac{\sigma \underset{\sim}{\vec{E}}}{j\omega} = \underset{\sim}{\varepsilon} \underset{\sim}{\vec{E}}, \tag{5.59}$$

where the *complex permittivity* $\underset{\sim}{\varepsilon}$ is defined as

definition of complex permittivity

$$\underset{\sim}{\varepsilon} = \varepsilon + \frac{\sigma}{j\omega}. \tag{5.60}$$

(a) ohmic current: motion of free charge

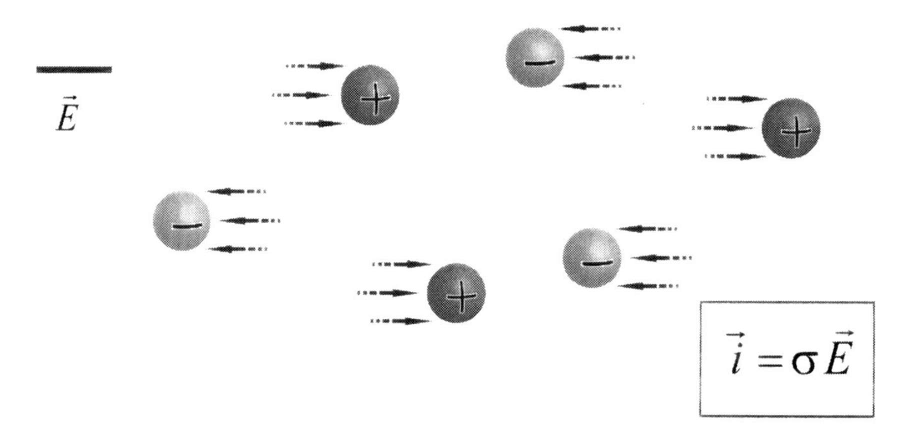

(b) displacement current: motion of bound charge

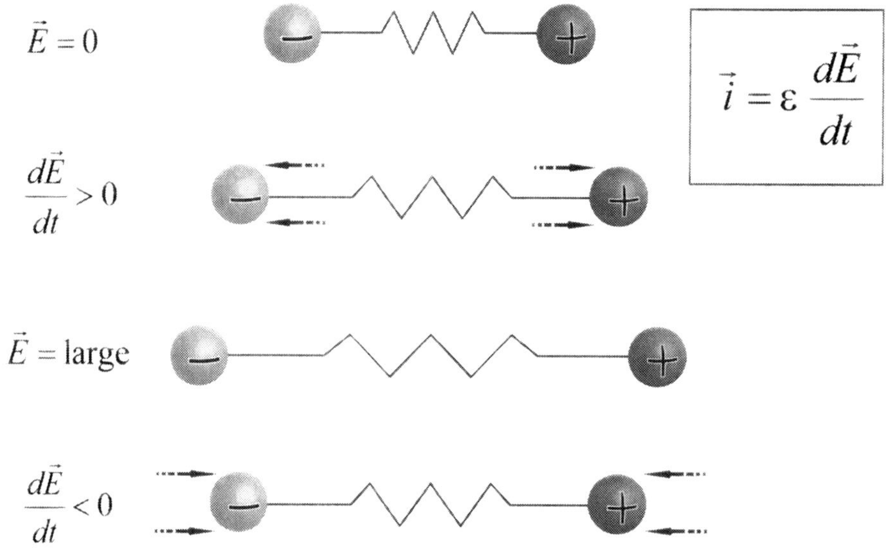

Fig. 5.8 Ohmic and displacement current. (a) Ohmic current in an electrolyte describes the motion of free charges (ions) in response to electric field. The current is proportional to the electric field and the conductivity σ is the constitutive constant. (b) Displacement current describes the motion of bound charges (partial charges of covalent bonds, atomic nuclei and electron orbitals, etc.) in response to this electric field. The displacement current is proportional to the time derivative of the electric field, and the electrical permittivity ε is the constitutive constant.

The two current modes (ohmic current and displacement current) are shown in Fig. 5.8. Equivalently, one could write Eq. (5.59) as

$$\underset{\sim}{\vec{\boldsymbol{i}}} + j\omega\underset{\sim}{\vec{\boldsymbol{D}}} = \underset{\sim}{\sigma}\underset{\sim}{\vec{\boldsymbol{E}}}, \tag{5.61}$$

where the *complex conductivity* $\underset{\sim}{\sigma}$ is given by

$$\underset{\sim}{\sigma} \;=\; j\omega\underset{\sim}{\varepsilon} \;=\; \sigma + j\omega\varepsilon\,. \tag{5.62}$$

In this description, one complex parameter (typically $\underset{\sim}{\varepsilon}$) describes both the ohmic current (due to motion of free charge) and displacement current (due to polarization of bound charge). Although the undertilde is used to describe the parameters $\underset{\sim}{\varepsilon}$ and $\underset{\sim}{\sigma}$ because they are complex, $\underset{\sim}{\varepsilon}$ and $\underset{\sim}{\sigma}$ must not be confused with analytic representations of sinusoidal functions. They are time-independent complex parameters used to combine the equations relating analytical representations of the electric displacement and the ohmic current.

5.3.1 Complex description of dielectric loss

The preceding description and the description in Subsection 5.1.4 are both incomplete. Taken together, these two descriptions account for both an electrodynamic description of ohmic current and a quasielectrostatic description of lossless displacement current. Real systems also experience dielectric loss, in which the energy of oscillating bound charge dipoles is absorbed and converted to heat. It is customary to use ε' to denote the quasielectrostatic polarization response of the material (termed the *reactive permittivity*), which is in phase with the applied field. It is customary also to use ε'' to denote the conversion of oscillating polarization to heat (termed the *dissipative permittivity* or *dielectric loss*) – this conversion is in phase with the *derivative* of the polarization and is thus 90° out of phase with the applied field. For sinusoidal fields in materials with finite dielectric loss, the complex permittivity $\underset{\sim}{\varepsilon}$ becomes

$$\underset{\sim}{\varepsilon} = \varepsilon' + \frac{\varepsilon''}{j} + \frac{\sigma}{j\omega} = \varepsilon' - j\left(\varepsilon'' + \frac{\sigma}{\omega}\right) . \tag{5.63}$$

For the *Debye model* of material response, ε' is given by

$$\varepsilon'(\omega) = \varepsilon_0 \left[1 + \sum_i \frac{\chi_{e,i}(0)}{1 + (\omega\tau_i)^2} \right] , \tag{5.64}$$

and ε'' is given by

$$\varepsilon''(\omega) = \varepsilon_0 \left[\sum_i \frac{\chi_{e,i}(0)\omega\tau_i}{1 + (\omega\tau_i)^2} \right] . \tag{5.65}$$

The Debye model treats the material response as the orientation of dipoles – this model is a good match for the orientational polarization of water.

From these relations, we see that ε' is a measure of how much the medium polarizes in response to an electric field, and ε'' is the product of the relative medium response and $\omega\tau_i$. The energy dissipation is low at high frequency because the medium has no time to polarize and no energy is transferred to the medium from the field. The energy dissipation is low at low frequency because, though the medium polarizes during each cycle, the number of cycles is low. However, at frequencies near $1/\tau_i$, the medium is polarized quite strongly (one half of the polarization at steady state) and the number of cycles is high. Thus significant energy is pumped into the medium for each cycle, and there are many cycles per second. The total energy conversion to heat is thus proportional to $\sigma/\omega + \varepsilon''$.

Although we never use it directly in this text, the complex quantity $\varepsilon' + \frac{1}{j}\varepsilon''$ is an analytic function,[5] and ε' and ε'' are related by the Kramers–Krönig relations, discussed

[5] Some authors refer to $\varepsilon' + \frac{1}{j}\varepsilon''$ as the complex permittivity, but we reserve that term for the quantity $\underset{\sim}{\varepsilon} = \varepsilon' + \frac{1}{j}\varepsilon'' + \frac{\sigma}{j\omega}$.

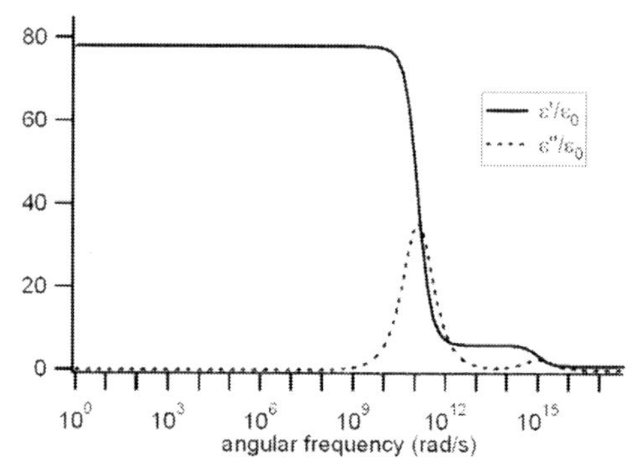

Fig. 5.9 Frequency dependence of electric permittivity and dielectric loss calculated using the Debye model for all modes.

in Appendix G:

$$\varepsilon'(\omega) = \frac{2}{\pi} \int_0^\infty \frac{\varpi \varepsilon''(\varpi)}{\varpi^2 - \omega^2} d\varpi , \tag{5.66}$$

$$\varepsilon''(\omega) = -\frac{2}{\pi} \int_0^\infty \frac{\omega \varepsilon'(\varpi)}{\varpi^2 - \omega^2} d\varpi . \tag{5.67}$$

Although these relations are exact, they do not immediately lend themselves to an intuitive interpretation. For the sigmoidal functions that describe the permittivity in the Debye model, the following relation holds, which is exact at $\omega = 1/\tau_i$ and approximate for frequencies near $\omega = 1/\tau_i$:

$$\varepsilon'' \simeq -\frac{1}{\tau_i} \frac{\partial \varepsilon'}{\partial \omega} \simeq -\omega \frac{\partial \varepsilon'}{\partial \omega} = -\frac{\partial \varepsilon'}{\partial (\ln \omega)} . \tag{5.68}$$

If plotted on a semilogarithmic axis, the dissipative response is approximately proportional to the derivative of the reactive response. Thus energy dissipation coincides with those frequencies where the permittivity is decreasing with ω. A graph of the frequency dependences of the reactive permittivity and the dissipative permittivity are shown in Fig. 5.9.

5.4 ELECTRICAL CIRCUITS

In electrical circuits, we consider idealized elements and simple relations that relate voltage and current. In the subsections to follow, we list circuit components, discuss

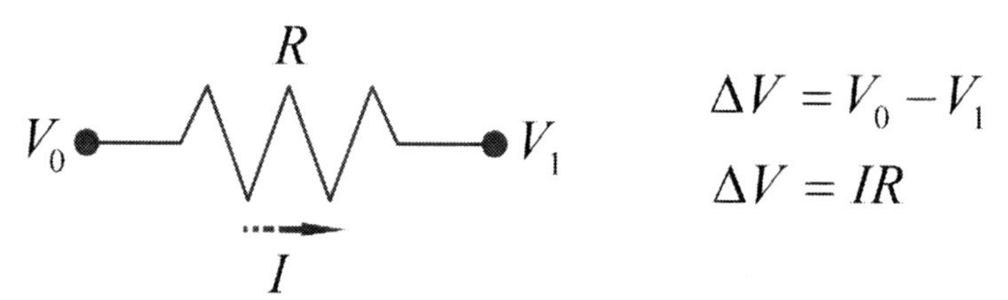

Fig. 5.10 Symbol and voltage–current relation for a resistor.

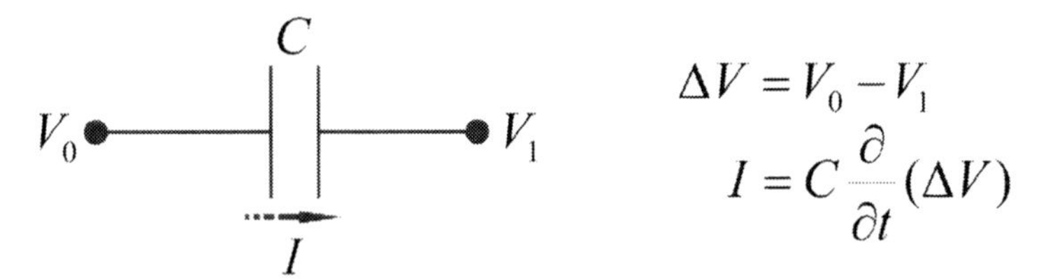

Fig. 5.11 Symbol and voltage–current relation for a capacitor.

notational concepts such as impedance and phasors, and summarize circuit relations. This discussion is a description of electrostatics and electrodynamics for discretized elements with well-defined electrical properties.

5.4.1 Components and properties

Key idealized circuit components include *voltage sources*, *wires*, *resistors*, *capacitors*, and *inductors*. Voltage sources are objects that specify a voltage V [V] at a given point. Ideal wires link together regions of space that have the same voltage V or electrical potential.[6] Ideal resistors (Fig. 5.10) are objects with a finite resistance to the motion of charge, defined by the resistance[7] R [Ω] and a voltage–current relation given by

$$\Delta V = IR. \tag{5.69}$$

Ideal capacitors (Fig. 5.11) are physical systems with a finite capacity to carry charge, denoted by a capacitance C [F] and a voltage–current relation given by $I = C\frac{d}{dt}\Delta V$.[8] An ideal capacitor consists of two conductors separated by a dielectric material – the two conductors each are at a specific voltage, and the polarization of the dielectric causes a net charge on the surface of the conductors.[9] Although electrons do not pass through a capacitor directly, the relative motion of electrons on either side of a capacitor does carry current. Ideal inductors (Fig. 5.12) have an inductance L [H] and a voltage–current relation given by $\Delta V = L\frac{d}{dt}I$.[10] We construct electrical circuits by wiring together voltage sources with these resistors, capacitors, and inductors.

6 We often use V to describe voltage at nodes in electrical circuits, but we use ϕ to denote the electrical potential when discussing its spatial distribution in a continuum.

7 The symbol R is used for both resistance and radius in this text. *Conductance* G is defined as the reciprocal of resistance: $G = 1/R$.

8 In this text, C denotes capacitance, and c denotes molar concentration.

9 The definitions for a capacitor as they relate to the definition of capacitance are important and less obvious than for other circuit elements. This is important in Chapter 16 when we define equivalent capacitances of more complicated systems. For the purposes of evaluating the capacitance of a capacitor, the voltage drop across a capacitor is defined as the potential of the more positive conductor minus the more negative one, and the charge stored by a capacitor is defined as the charge stored at the more negative conductor. Both of these properties are positive, and the capacitance (which is defined as the ratio of the stored charge to the voltage drop) is by definition positive. We often speak of the capacitance of a single conductor, in which case the (imagined) second conductor is an imaginary spherical shell of infinite radius at zero voltage. The charge stored is still defined as the charge on the negative conductor, so if we apply a positive voltage to a single conductor, the charge stored by the capacitor is actually the opposite of the charge at that conductor.

10 We rarely discuss inductors in this text, and L is instead used to denote length.

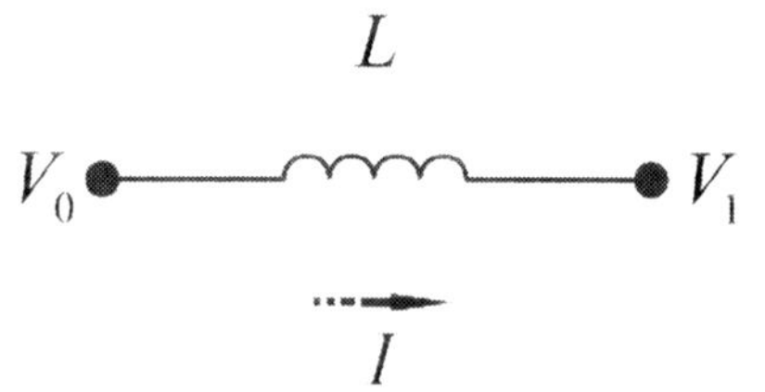

L

V_0 V_1

I

$$\Delta V = V_0 - V_1$$
$$\Delta V = L \frac{\partial}{\partial t} I$$

Fig. 5.12 Symbol and voltage–current relation for an inductor.

EXAMPLE PROBLEM 5.2

Consider a circuit with a resistor with resistance R and a capacitor with capacitance C in parallel. A voltage drop $\Delta V = \Delta V_0 \cos \omega t$ with $\Delta V > 0$ is applied across this circuit.

Using conservation of current and the relations $I_R = \Delta V / R$ and $I_C = C \frac{d\Delta V}{dt}$, solve for the current through the resistor, the current through the capacitor, and the total current passing through the circuit. Write your answers in terms of cosine functions with phase lags.

SOLUTION: For the current through the resistor,

$$I_R = \frac{\Delta V}{R} = \frac{\Delta V_0}{R} \cos \omega t \,. \tag{5.70}$$

The current through the capacitor is $I_C = C \frac{d\Delta V}{dt} = -C \, \Delta V_0 \, \omega \, \sin \omega t$ or

$$I_C = C \, \Delta V_0 \, \omega \, \cos(\omega t + \pi/2) \,. \tag{5.71}$$

From conservation of current, the total current through the circuit is $I = \frac{\Delta V_0}{R} \cos \omega t - C \, \Delta V_0 \, \omega \, \sin \omega t$. Using the trigonometric identity $a \sin x + b \cos x = \sqrt{a^2 + b^2} \cos(x + \alpha)$, where $\alpha = \text{atan2}(a, b)$, this becomes $I = \frac{\Delta V_0}{R} \sqrt{1 + \omega^2 R^2 C^2} \cos\left[\omega t + \text{atan2}(-C \, \Delta V_0 \, \omega, \frac{\Delta V_0}{R})\right]$ and therefore

$$I = \frac{\Delta V_0}{R} \sqrt{1 + \omega^2 R^2 C^2} \cos\left[\omega t + \tan^{-1}(-\omega RC)\right] \,. \tag{5.72}$$

Per our definition of atan2, the phase lag is given by $\tan^{-1}(-\omega RC) + 2\pi$ because the first argument is negative; we neglect the 2π in writing Eq. (5.72).

5.4.2 Electrical impedance

The impedance of a circuit element is a complex quantity that extends Ohm's law ($V = IR$) to AC circuits. If the voltages and currents are sinusoidal and we use their analytic representations,

$$V = \mathrm{Re}(\underset{\sim}{V}) = \mathrm{Re}\left\{V_0 \exp[j(\omega t + \alpha_V)]\right\} = \mathrm{Re}\left(\underset{\sim}{V}_0 \exp j\omega t\right), \tag{5.73}$$

$$I = \mathrm{Re}(\underset{\sim}{I}) = \mathrm{Re}\left\{I_0 \exp[j(\omega t + \alpha_I)]\right\} = \mathrm{Re}\left(\underset{\sim}{I}_0 \exp j\omega t\right), \tag{5.74}$$

then for circuit elements we define an *impedance* $\underset{\sim}{Z}$, which describes the voltage–current relationship through the equation $\underset{\sim}{V} = \underset{\sim}{I}\underset{\sim}{Z}$ or $\underset{\sim}{V}_0 = \underset{\sim}{I}_0\underset{\sim}{Z}$. As is the case for $\underset{\sim}{\varepsilon}$ and $\underset{\sim}{\sigma}$, the electrical impedance $\underset{\sim}{Z}$ is a complex parameter that combines the ohmic and capacitive current responses of a circuit. Each circuit element has a complex impedance corresponding to its circuit properties, including resistors,

impedance of an electrical resistor

$$\underset{\sim}{Z} = R, \tag{5.75}$$

capacitors,

impedance of an electrical capacitor

$$\underset{\sim}{Z} = \frac{1}{j\omega C} = \frac{1}{\omega C}\exp(-j\frac{\pi}{2}), \tag{5.76}$$

and inductors,

impedance of an electrical inductor

$$\underset{\sim}{Z} = j\omega L = \omega L \exp(j\frac{\pi}{2}). \tag{5.77}$$

5.4.3 Circuit relations

Ohm's law and the relations for impedance describe the current through an element or the voltage drop across that element. For a circuit that is composed of a network of these elements, Kirchoff's current law links these network elements by using the conservation of current relation (Fig. 5.13):

Kirchoff's law for conservation of current

$$\sum_{\text{channels}} I = 0, \tag{5.78}$$

where I in this case is defined as positive *into* the node. Circuit networks can be solved as systems of algebraic equations constructed from Ohm's law and circuit element impedance relations.

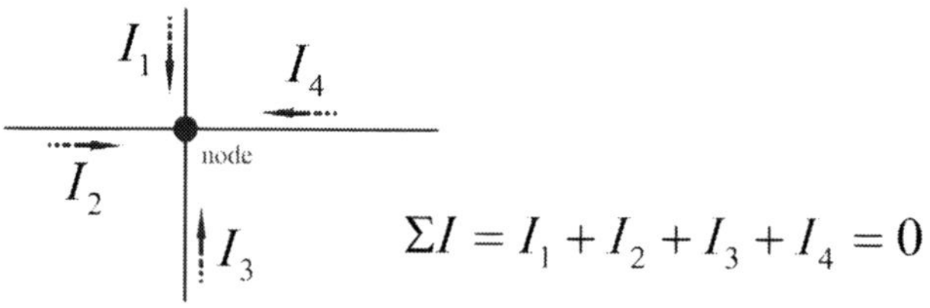

I_1 I_4 node I_2 I_3

$$\Sigma I = I_1 + I_2 + I_3 + I_4 = 0$$

Fig. 5.13 Implementation of Kirchoff's law at a node linking four circuit elements.

5.4.4 Series and parallel component rules

The results of Kirchoff's law for parallel and series circuits can be used to determine the response of combinations of circuit components.

Series circuit rules. The resistance of two resistors in series is equal to the sum of the resistances:

series circuit rule for resistors

$$R = R_1 + R_2\,; \tag{5.79}$$

the reciprocal of the capacitance of two capacitors in series is equal to the sum of the reciprocals of the capacitances:

series circuit rule for capacitors

$$\frac{1}{C} = \frac{1}{C_1} + \frac{1}{C_2}\,; \tag{5.80}$$

the inductance of two inductors in series is equal to the sum of the inductances:

series circuit rule for inductors

$$L = L_1 + L_2\,; \tag{5.81}$$

and the impedance of two impedances in series is equal to the sum of the impedances:

series circuit rule for impedances

$$\underset{\sim}{Z} = \underset{\sim}{Z}_1 + \underset{\sim}{Z}_2\,. \tag{5.82}$$

These series component relations are depicted in Fig. 5.14.

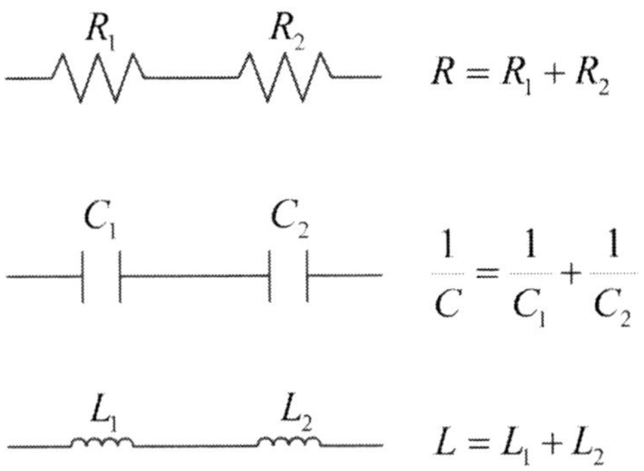

$R = R_1 + R_2$

$\frac{1}{C} = \frac{1}{C_1} + \frac{1}{C_2}$

$L = L_1 + L_2$

Fig. 5.14 Series component relations for circuit elements.

Parallel circuit rules. The reciprocal of the resistance of two resistors in parallel is equal to the sum of the reciprocals of the resistances:

parallel circuit rule for resistances

$$\frac{1}{R} = \frac{1}{R_1} + \frac{1}{R_2}; \quad (5.83)$$

the capacitance of two capacitors in parallel is equal to the sum of the capacitances:

parallel circuit rule for capacitors

$$C = C_1 + C_2; \quad (5.84)$$

the reciprocal of the inductance of two inductors in parallel is equal to the sum of the reciprocals of the inductances:

parallel circuit rule for inductors

$$\frac{1}{L} = \frac{1}{L_1} + \frac{1}{L_2}; \quad (5.85)$$

and the reciprocal of the impedance of two impedances in parallel is equal to the sum of the reciprocals of the impedances:

parallel circuit rule for impedances

$$\frac{1}{\underset{\sim}{Z}} = \frac{1}{\underset{\sim}{Z}_1} + \frac{1}{\underset{\sim}{Z}_2}. \quad (5.86)$$

Parallel component relations are depicted in Fig. 5.15.

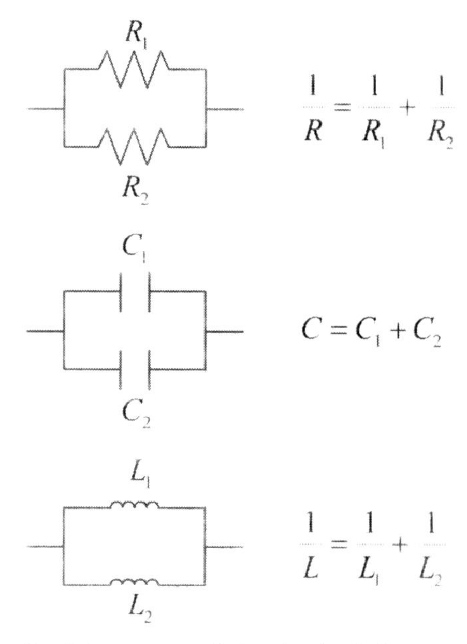

Fig. 5.15 Parallel component relations for circuit elements.

5.5 EQUIVALENT CIRCUITS FOR CURRENT IN ELECTROLYTE-FILLED MICROCHANNELS

Although a complete description of fluid and charge transport in a microchannel typically involves a relatively complicated analysis, 1D models often provide an approximate solution that guides intuition or system design. We treat microchannels as 1D elements and their intersections as nodes, and we model both the electrical properties and the fluid flow properties with circuitlike elements.

This approach is well suited for analyzing microfluidic devices for several reasons. First, microchannels have lengths much longer than their diameters, so the fully developed flow approximation is accurate. Second, complicated microchannel networks are ubiquitous in microsystems.

5.5.1 Electrical circuit equivalents of hydraulic components

To describe the current and voltage in a microsystem by using 1D models, we describe microchannels in terms of equivalent circuit elements and replace the microchannel system with an electrical circuit. We model microchannels as resistors, junctions as nodes, and electrodes as capacitors. The governing equation is Ohm's law ($\Delta V = IR$ or $\Delta \underset{\sim}{V} = \underset{\sim}{I}\,\underset{\sim}{Z}$) and the boundary conditions are given by Kirchoff's current law at each node.[11]

Electrical circuit equivalent of microchannels. Our 1D model of a microchannel is that of a resistor with resistance R, specified both by the geometry and the fluid conductivity σ:

bulk fluid resistance of a microchannel

$$R = \frac{L}{\sigma A}, \tag{5.87}$$

[11] We define $\Delta V = V_{\text{inlet}} - V_{\text{outlet}}$ and define I as positive for current flow from inlet to outlet.

where L is the length of the channel, σ is the bulk conductivity of the fluid, and A is the cross-sectional area. The SI units for conductivity are inverse ohm-meters [$\Omega^{-1}\text{m}^{-1}$], but it is more commonly reported in terms of microsiemens per centimeter [μS/cm] or millisiemens per meter [mS/m], where a siemens is an inverse ohm ($1\ \text{S} = 1\ \Omega^{-1}$).[12]

For microchannels with discrete changes in cross-sectional area, we can describe them with several discrete resistors. For systems with cross-sectional areas or conductivities that vary continuously and slowly, we can use a 1D integral relation:

$$R = \int_{x_1}^{x_2} \frac{1}{\sigma(x)A(x)} dx\,, \tag{5.88}$$

where x_1 and x_2 denote the locations of the beginning and end of the microchannel ($L = x_2 - x_1$). Here, σ and A are assumed to be functions of x.

Electrical circuit equivalent of microchannel intersections. In a 1D model, we treat the intersection between channels as a node and apply Kirchoff's law:

Kirchoff's law for conservation of current

$$\sum_{\text{channels}} I = 0\,, \tag{5.89}$$

where I denotes current *into* the node.

Electrical circuit equivalent of electrodes. *Electrodes* denote the interface between a solid conductor and an electrolyte solution. Fields are applied to our microchannel system by connecting voltage sources to metal wires inserted into reservoirs or micropatterned metal electrodes on the surfaces of the microdevices.[13] The interface (described in more detail in Chapter 16) has a capacitance, which can be roughly approximated by

capacitance of a metal electrode with an electrical double layer

$$C = \frac{\varepsilon A_{\text{electrode}}}{d_{\text{electrode}}} = \frac{\varepsilon A_{\text{electrode}}}{\lambda_D}\,. \tag{5.90}$$

Here C is the capacitance, $A_{\text{electrode}}$ is the *area of the electrode surface*, ε is the electrical permittivity of the fluid, and λ_D is the *Debye length* of the electrolyte solution, which is defined in Chapter 9 and typically has a magnitude between 1 and 100 nm. The reactions that take place at the electrode – consuming or creating electrons in the electrode through chemical reactions in the electrolyte that consume or create ions – lead to an effective resistance as well, which can be predicted with electrode kinetics models and is a function of the catalytic behavior of the electrode (Subsection 5.2.2). For microfluidic systems, which have high resistances, the resistance of the electrode can often be neglected.

[12] Equation (5.87) ignores the effect of surfaces, which add an additional conductance component. See Exercise 11.9 for more info on this topic.

[13] Not all electrodes need be metal, but metal electrodes are the most common.

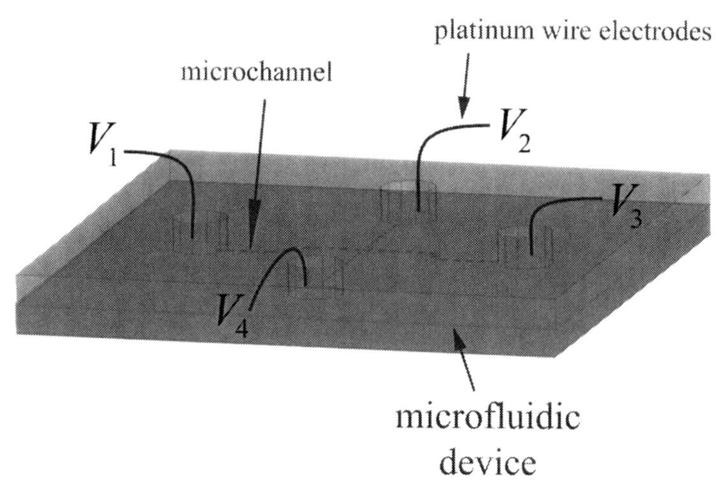

Fig. 5.16 A simple microfluidic device.

From the preceding relations, we see that potentials applied to microchannels can be modeled (if the microchannel is well approximated by a 1D model) as if they were being applied to resistor–capacitor systems. An example of this for a four-port microfluidic device is shown in Figs. 5.16 and 5.17.

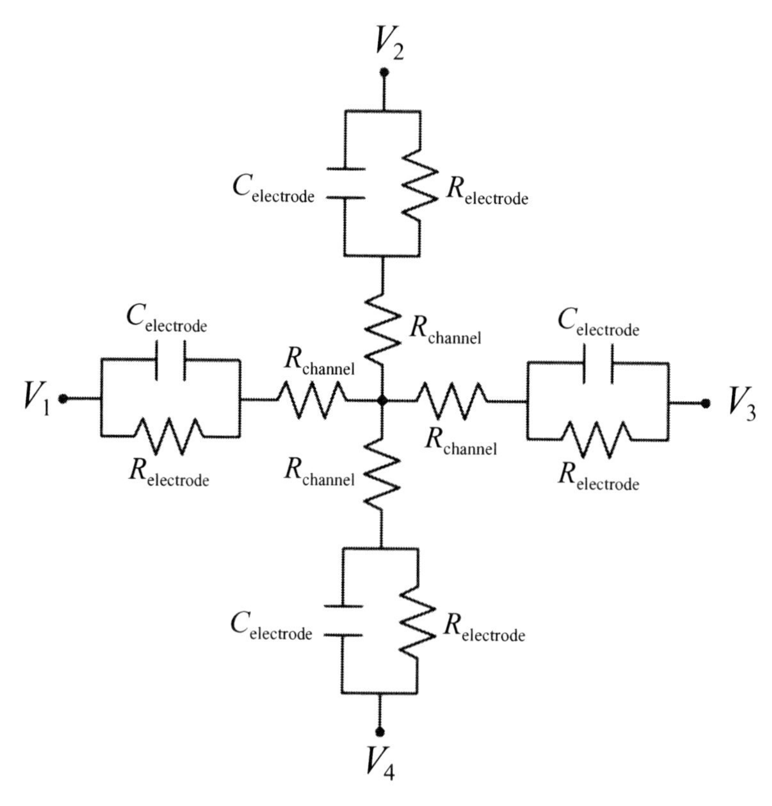

Fig. 5.17 The circuit analog for the device in Fig. 5.16.

EXAMPLE PROBLEM 5.3

Consider a circuit with a resistor with resistance R and a capacitor with capacitance C in parallel. A voltage drop $\Delta V = \Delta V_0 \cos \omega t$ with $\Delta V > 0$ is applied across this circuit.

Write the analytic representation of the applied voltage drop and solve for the analytic representation and phasor representation of the total current through the circuit by using $\underset{\sim}{I} = \Delta \underset{\sim}{V} / \underset{\sim}{Z}$ and the parallel circuit rules for complex impedance. Use a similar relation to determine the analytic representation and phasor representation of the current through each component. Once these values are determined, write the real representation of these currents. Consult Appendix G as needed for complex relations.

SOLUTION: The impedance of the resistor is $\underset{\sim}{Z}_R = R$ and the impedance of the capacitor is $\underset{\sim}{Z}_C = 1/j\omega C$. By use of parallel circuit rules, the total impedance is given by $1/\underset{\sim}{Z} = 1/R + j\omega C$ or $\underset{\sim}{Z} = \frac{R}{1+j\omega RC}$.

The analytic representation of the applied voltage drop is $\Delta \underset{\sim}{V} = \Delta V_0 \exp j\omega t$ and the phasor representation is $\Delta \underset{\sim}{V}_0 = \Delta V_0$. Ohm's law for the analytic representations is $\underset{\sim}{I} = \Delta \underset{\sim}{V} / \underset{\sim}{Z}$ and for the phasors is $\underset{\sim}{I}_0 = \Delta \underset{\sim}{V}_0 / \underset{\sim}{Z}$. Using the phasor equation, the phasor for the current through the resistor is $\underset{\sim}{I}_0 = \Delta V_0 / R$. The magnitude of this phasor is $\frac{\Delta V_0}{R}$ and the angle of this phasor is atan2$(0, \Delta V_0 / R) = 0$. Similarly, the phasor for the current through the capacitor is $\underset{\sim}{I}_0 = \Delta V_0 j\omega C$. The magnitude of this phasor is $\Delta V_0 \omega C$ and the angle of this phasor is atan2$(\Delta V_0 \omega C, 0) = \pi/2$. The phasor for the total current through the circuit is $\underset{\sim}{I}_0 = \Delta V_0 \frac{1+j\omega RC}{R}$. The magnitude of this phasor is $\frac{\Delta V_0}{R}\sqrt{1 + \omega^2 R^2 C^2}$, and the angle of this phasor is atan2$(V_0 \omega C, V_0/R) = \tan^{-1}(\omega RC)$. Returning to real quantities, the current through the resistor is $I_R = \Delta V_0 R \cos \omega t$, through the capacitor is $I_C = \Delta V_0 \omega C \cos(\omega t + \pi/2)$, and through the circuit in total is $I = \frac{\Delta V_0}{R}\sqrt{1 + \omega^2 R^2 C^2} \cos\left[\omega t + \tan^{-1}(\omega RC)\right]$.

EXAMPLE PROBLEM 5.4

Consider a parallel-plate capacitor in which the plates are separated by a medium with a permittivity ε and conductivity σ. Assume a sinusoidal electric field with frequency ω is applied. Model this as a perfect capacitor in parallel with a resistor, where the capacitor properties are defined by the permittivity of the medium and the resistor properties are defined by the conductivity of the medium.

For this electric field, write the complex impedance of the capacitor in terms of ω, the capacitor area A, the plate separation d, the permittivity ε, and the conductivity σ. Rewrite this complex impedance in terms of an effective permittivity (i.e., the permittivity that would make the complex impedance for a pure capacitor equal to the impedance for this capacitor–resistor system) and show that this effective permittivity is given by the complex permittivity $\underset{\sim}{\varepsilon} = \varepsilon + \sigma/j\omega$.

SOLUTION: The capacitance of a parallel-plate capacitor is $C = \frac{\varepsilon A}{d}$.

The resistance of the parallel-plate system is $R = d/\sigma A$.

The complex impedance of the capacitor is $\underset{\sim}{Z} = 1/j\omega C$. The complex impedance of the resistor is $\underset{\sim}{Z} = d/\sigma A$.

Thus the complex impedance is given by

$$\frac{1}{\underset{\sim}{Z}} = \frac{j\omega\varepsilon A}{d} + \frac{\sigma A}{d}, \tag{5.91}$$

$$\frac{1}{\underset{\sim}{Z}} = \frac{j\omega\underset{\sim}{\varepsilon}A}{d} = \frac{j\omega A}{d}(\varepsilon + \sigma/j\omega), \tag{5.92}$$

and the effective permittivity is

$$\underset{\sim}{\varepsilon} = \varepsilon + \sigma/j\omega. \tag{5.93}$$

5.6 SUMMARY

This chapter presents basic relations for electrostatic and electrodynamics in electrolyte solutions. The most important electrostatic relation is Gauss's law for electricity:

$$\nabla \cdot \vec{\boldsymbol{D}} = \rho_E, \tag{5.94}$$

which relates the divergence of the electric displacement to the instantaneous charge density. Gauss's law is the source of the Poisson equation as well as of the electrostatic boundary conditions. The key electrodynamic relation is the charge conservation equation:

$$\frac{\partial \rho_E}{\partial t} + \nabla \cdot \vec{\boldsymbol{i}} = 0, \tag{5.95}$$

which describes the motion of charge and its relation to the net charge density. These two equations prescribe the solution of the electrical component of microfluidic problems. We can solve for the electric field distribution caused by charges, either net areal charge densities on surfaces (which are observed on most microfluidic substrates and at all electrode surfaces) or net volumetric charge densities in the fluid (which are observed near the substrate surface and throughout nanofluidic systems) with the relations in this chapter. In many microfluidic systems, electric fields play a central role through the effect of Coulomb forces on the net charge density in the fluid.

This chapter also presents electrical circuit analysis, which simplifies our analytical task when the electrical components can be discretized. The key governing equations for electrical circuits are Ohm's law,

$$\Delta\underset{\sim}{V} = \underset{\sim}{I}\underset{\sim}{Z}, \tag{5.96}$$

combined with Kirchoff's law,

$$\sum_{\text{channels}} I = 0\,. \tag{5.97}$$

These relations combined with the voltage–current relationships for specific circuit elements cast the physical system of an electrical circuit in terms of a set of algebraic equations. We introduce analytic representations of the voltage and current because the complex mathematics simplifies analysis for sinusoidal signals.

5.7 SUPPLEMENTARY READING

The material in this chapter is classical physics and is covered thoroughly in electricity and magnetism texts. Griffiths [55] is an excellent introductory text, whereas Jackson [56] is more advanced. Electronic versions of Haus and Melcher [57] are in the public domain. The electrochemical relations for electrodes are generally outside the focus of electricity and magnetism texts, and so texts focusing on electrochemistry [58, 59] are the most useful. Morgan and Green's AC electrokinetics text [60] and Jones's electromechanics of particles text [61] are focused primarily on electromagnetic effects on particles, and both include coverage of material closely related to this chapter. Pethig [62] discusses electrical properties of many materials of interest. Models for the permittivity of water as a function of the electric field are found in [26]. For electrical circuits, an introductory electronic circuits text, e.g., [63] is useful.

Other chapters in this textbook build on the results in this chapter. It is shown in Chapter 6 that purely electroosmotic flows can be approximated by solutions to the Laplace equation. Further, the charge conservation equation is important in a number of applications, notably the dynamics of the electrical double layer (Chapter 16), which lead to induced-charge flows (Section 16.3), and dielectrophoretic manipulation of particles (Chapter 17). This chapter focuses on ohmic current when we are considering the charge conservation equation. The effects of convection and diffusion are considered in detail in the context of ion transport in Subsection 11.3.1.

5.8 EXERCISES

5.1 Consider a dipole of strength $\vec{p} = q\vec{d}$, consisting of a negatively charged monopole with charge $-q$ and a positively charged monopole with charge q. The vector $\vec{d}$ indicates the vector distance from the negative charge to the positive charge. Assume this dipole resides in a spatially varying electric field $\vec{E}$, where $\vec{E}$ denotes the electric field that would exist if the dipole were absent.

Derive the relation $\vec{F} = (\vec{p} \cdot \nabla)\,\vec{E}$ by evaluating the forces on two monopoles if the monopole distance is small enough that the spatially varying electric field can be linearized around the centroid of the dipole.

5.2 Consider a dipole of strength $\vec{p} = q\vec{d}$, consisting of a negatively charged monopole with charge $-q$ and a positively charged monopole with charge q. The vector $\vec{d}$ indicates the vector distance from the negative charge to the positive charge. Assume this dipole resides in a spatially varying electric field $\vec{E}$, where $\vec{E}$ denotes the electric field that would exist if the dipole were absent.

Derive the relation $\vec{T} = \vec{p} \times \vec{E}$ by evaluating the forces and resulting torques on two monopoles if the monopole distance is small enough that the spatially varying electric field can be linearized around the centroid of the dipole. Note that torque can be written as $\vec{T} = \vec{r} \times \vec{F}$.

5.3 Calculate the electric field of a positive point charge at the origin. Plot the electric field isocontours in the $z = 0$ plane.

5.4 Calculate and plot (using a quiver plot) the 2D electric field of a positive line charge with magnitude 1 C/m at $x = 0.5$ m and a negative line charge with magnitude 1 C/m at $x = -0.5$ m.

5.5 Calculate the electric field induced by a dipole of strength $p = 1$. Plot isocontours of the electric field as well as arrows to denote the electric field direction and magnitude in the $z = 0$ plane.

5.6 Consider a parallel-plate capacitor in which the plates are separated by a perfect dielectric (i.e., zero conductivity) with permittivity ε. Assume a sinusoidal electric field with frequency ω is applied. For this electric field, write the complex impedance of the capacitor in terms of ω, the capacitor area A, the plate separation d, and the permittivity ε.

5.7 Consider a resistor in which two conducting plates are separated by a medium with conductivity σ. Assume a sinusoidal electric field with frequency ω is applied. For this electric field, write the complex impedance of the resistor in terms of ω, the cross-sectional area A, the plate separation d, and the conductivity σ.

5.8 Consider a resistor in which two conducting plates are separated by a medium with a permittivity ε and conductivity σ. Assume a sinusoidal electric field with frequency ω is applied. Model this as a resistor in parallel with a perfect capacitor, for which the resistor properties are defined by the conductivity of the medium and the capacitor properties are defined by the permittivity of the medium.

For this electric field, write the complex impedance of the resistor in terms of ω, the resistor area A, the plate separation d, the permittivity ε, and the conductivity σ. Rewrite this complex impedance in terms of an effective conductivity (i.e., the conductivity that would make the complex impedance for a resistor equal to the impedance for this resistor–capacitor system), and show that this effective conductivity is given by the complex conductivity $\underset{\sim}{\sigma} = \sigma + j\omega\varepsilon$.

5.9 Microwave ovens heat water by applying electric fields to the water, typically at a frequency of 2.45 GHz. Given that the characteristic orientational relaxation time of water is approximately 8 ps, estimate the magnitude of the dielectric loss at the oven frequency. Why might 2.45 GHz work better than 2.45 MHz? Why might 2.45 GHz work better than 125 GHz?

5.10 Gauss's law in differential form relates the electric displacement $\vec{D}$ to the net charge density:

$$\nabla \cdot \vec{D} = \rho_E \,, \tag{5.98}$$

where $\vec{D} = \varepsilon \vec{E}$.

Similarly, the conservation of charge relation for ohmic current relates the ohmic current density $\vec{i}$ to the net charge density:

$$\nabla \cdot \vec{i} = -\frac{\partial \rho_E}{\partial t} \,, \tag{5.99}$$

where $\vec{i} = \sigma \vec{E}$ if charge diffusion and fluid convection are ignored.

For sinusoidal fields, consider the complex quantity $\vec{\underset{\sim}{D}}+\vec{\underset{\sim}{i}}/j\omega$. This complex quantity is analogous to the complex permittivity $\underset{\sim}{\varepsilon}$.

(a) What is the relation between the electric field $\vec{E}$ and the complex quantity $\vec{\underset{\sim}{D}}+\vec{\underset{\sim}{i}}/j\omega$?

(b) What is the divergence of $\vec{\underset{\sim}{D}}+\vec{\underset{\sim}{i}}/j\omega$?

5.11 Consider an interface between two media with conductivities and permittivities given by σ_1, σ_2, ε_1, and ε_2. Assume that a sinusoidal field is applied to the system. Assume that the net charge at the interface is also sinusoidal, i.e., assume that any net charge at the interface is induced by the sinusoidal electric field. Use the integral form of Gauss's equation with a complex electric displacement to show that the boundary condition is given by

$$\underset{\sim}{\varepsilon}_1 \frac{\partial \underset{\sim}{\phi}_1}{dn} = \underset{\sim}{\varepsilon}_2 \frac{\partial \underset{\sim}{\phi}_2}{dn}, \tag{5.100}$$

where n is a coordinate normal to the interface.

5.12 Consider a rectangular microchannel with 50-μm depth and 100-μm width. Two electrodes are patterned on the floor of this microchannel, each 20 μm wide, each offset 25 μm from the centerline. All geometries are constant along the axis of the channel, such that the system can be assumed 2D. The rightmost electrode has an applied voltage of 1 V whereas the other electrode is grounded. Use a Schwarz–Christoffel mapping technique (see Section G.5) to solve for the electric potential in the microchannel.

5.13 Consider an insulating boundary with a thin layer of thickness d and net volumetric charge density q''/d. Integrate Poisson's equation over this layer to determine the potential gradient at the edge of the charged layer. Now take the limit of this system as $d \to 0$ and in so doing derive Eq. (5.31).

5.14 Show that the contribution of a polarization mechanism is one-half as large at $\omega = 1/\tau_i$ as it is at $\omega = 0$.

5.15 Show that the maximum dielectric loss for a polarization mechanism that undergoes Debye relaxation occurs at $\omega = 1/\tau_i$.

5.16 Consider an ion with valence z. Considering the electric field generated by this ion in a vacuum and the permittivity of water as a function of electric field, derive an expression for the permittivity of water in the vicinity of this ion. Calculate and plot the permittivity as a function of radial distance from the ion.

5.17 Consider a microchannel fabricated with a uniform channel depth of 10 μm and cross-sectional area of 20 μm^2. Four inlets are connected at a single junction, and the overall microsystem is cross shaped (Fig. 5.18). The lengths of the four channels are denoted L_1, L_2, L_3, and L_4. An electrode is placed at each inlet, and voltages V_1, V_2, V_3, and V_4 are applied at the inlets. The fluid is of uniform conductivity given by $\sigma =$ 100 μS/cm.

For each of the following cases, calculate the voltage at the junction and the current *into* the junction from each of the four inlets. Of course, current will flow out of the junction in some directions; this should be denoted with a negative current.

(a) $L_1 = L_2 = L_3 = 1$ cm; $L_4 = 3$ cm; $V_2 = 1$ kV; $V_3 = 1$ kV; $V_1 = V_4 = 0$ V.

(b) $L_1 = L_2 = L_3 = 1$ cm; $L_4 = 3$ cm; $V_2 = 1$ kV; $V_3 = 0$ kV; $V_1 = V_4 = 0$ V.

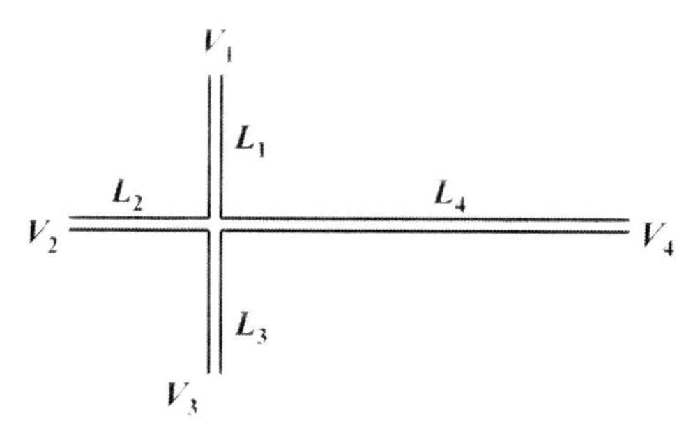

Fig. 5.18 Schematic of top view of a cross-shaped microchannel with four inlets that converge at a single junction.

5.18 Using equivalent circuit techniques, design a microdevice geometry with one input and one output. For a given set of voltages (one at the input and one at the output – your specification of these voltages should be part of your answer), the input should split into four separate channels that each experience four different electric fields: 1 V/cm, 10 V/cm, 100 V/cm, and 1000 V/cm. Your design should include specifications of the voltages at input and output, as well as enough information about the microdevice geometry to ensure the specified fields are observed. Assume that no pressure gradients exist in this device.

6 Electroosmosis

When electric fields are applied across capillaries or microchannels, bulk fluid motion is observed. The velocity of this motion is linearly proportional to the applied electric field, and dependent on both (a) the material used to construct the microchannel and (b) the solution in contact with the channel wall. This motion is referred to as *electroosmosis* and stems from electrical forces on ions in the *electrical double layer* or EDL, a thin layer of ions that is located near a wall exposed to an aqueous solution. If the fluid velocity is interrogated at micrometer resolution, for example, by observation of the fluid flow with a light microscope, the fluid flow in a channel of uniform cross section appears to be uniform. If the fluid velocity is interrogated with nanometer resolution (which is experimentally difficult), we find that the fluid velocity is uniform far from the wall but decays to zero at the wall over a length scale λ_D ranging from approximately 0.5 to 200 nm. Figure 6.1 illustrates the velocity profile in an electroosmotic flow.

This fluid flow can be immensely useful in microfluidic systems, because it is often much more straightforward experimentally to address voltage signals sent to electrodes rather than to implement and control a miniaturized mechanical pressure pump. This flow, however, comes with its own complications: Its velocity distribution is different from pressure-driven flow, it is sensitive to chemical features at the interface, and the act of applying electric fields can also move particles relative to the fluid or cause Joule heating (i.e., resistive heating) throughout the fluid.

Just as our impressions of the fluid flow vary depending on the spatial resolution of our velocity observations, our mathematical descriptions of this system can take different forms, depending on what level of detail is required. A full description of the flow requires that the Navier–Stokes equations be combined with a Coulomb body force term caused by the net charge density near the wall, and that the Poisson equation be solved in conjunction with an equilibrium Boltzmann distribution of ions to determine the spatial variation of this body force. This treatment describes the fluid velocity in complete detail. However, this level of detail is not required for describing electroosmotic flows at micrometer resolution, and we therefore start by describing electroosmosis with simpler relations that are dependent on properties of the wall but ignore the details of the flow near the wall. In developing simpler relations, we use an integral analysis of the flow near the wall to derive a result for the *outer* solution of this flow, i.e., the solution far from the wall. In this chapter, many details of the *inner* solution of this flow, i.e., the solution close to the wall, are left unspecified.

Our approach in this chapter is to use a 1D integral analysis of the EDL to relate the *electroosmotic mobility* of a solid–liquid interface to the electrical potential at the wall (called the *surface potential*), which is a chemical property of both the wall material and the electrolyte solution. This electroosmotic mobility describes the *apparent* slip at the wall if the observer ignores the portion of the fluid velocity distribution proximal to the wall.

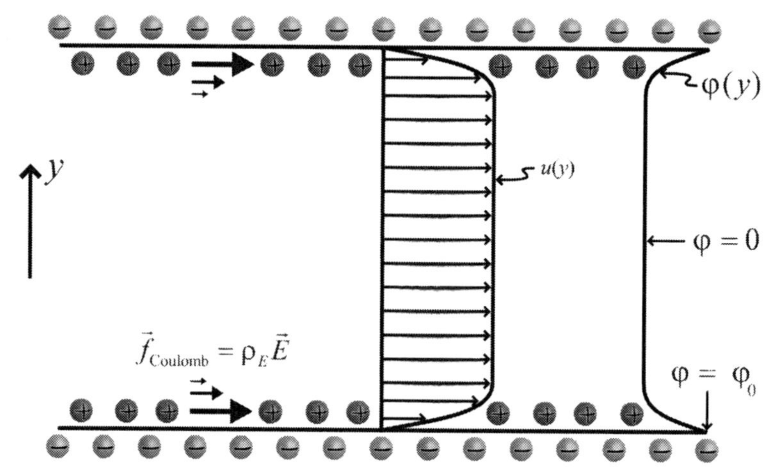

Fig. 6.1 Cartoon of EDL and electroosmotic flow. A negatively charged wall in this figure coincides with a thin, positively charged EDL. Coulomb forces in the EDL induce a fluid flow that is approximately uniform outside the EDL. The (typically nanoscale) thickness of the EDL is exaggerated for the purposes of the figure. The difference between the local electrical potential and that of the bulk fluid is denoted by φ; this value is zero far from the wall but is nonzero near the wall.

The solutions presented in this chapter provide no information about the size of the EDL, its physical underpinnings, or its structure, though these solutions do require that we assume that the double layer is a *boundary layer*, i.e., it is relatively thin compared with the dimensions of the flowfield. However, despite these omissions, the 1D analysis leads to several important conclusions about purely electroosmotically driven flows: (1) The velocity field near the wall is proportional to the voltage difference between that point and the wall; (2) the velocity at the edge of the boundary layer is proportional to the local electric field and the voltage difference between the bulk fluid and the wall; and (3) if the fluid conductivity and electroosmotic mobility are uniform, the velocity at any point far from the wall is proportional to the local electric field, and thus the velocity field far from the wall is irrotational.

6.1 MATCHED ASYMPTOTICS IN ELECTROOSMOTIC FLOW

Because the thin-EDL assumption implies scale separation between the EDL and the bulk flow field, we often find it useful to analyze electroosmotic flow in terms of matched asymptotics. In this case, we find two asymptotic solutions: (1) the *inner* solution, corresponding to the EDL, in which we keep track of the details of the Coulomb forces and the resulting velocity gradients and vorticity, but we assume that the extrinsic electric field is uniform, and (2) the *outer* solution, for which we assume that the fluid is electroneutral and irrotational, but allow the extrinsic electric field to vary spatially (Fig. 6.2). The inner solution is valid in the EDL, but gives incorrect results far from the wall if the extrinsic field varies with distance from the wall. The outer solution is valid outside the EDL, but gives incorrect results near the wall and violates the no-slip condition.

6.2 INTEGRAL ANALYSIS OF COULOMB FORCES ON THE EDL

A variety of processes (for example, ion adsorption and acid–base reactions) lead solid surfaces to acquire a nonzero surface charge density in an aqueous solution, and this

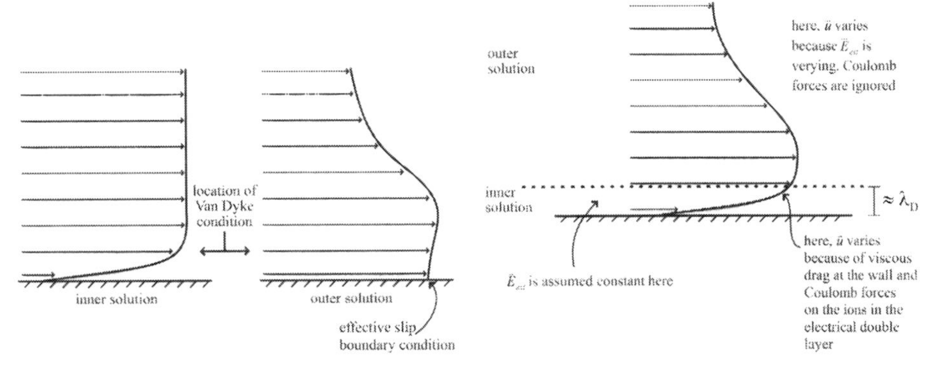

Fig. 6.2 Left: Inner and outer solutions for electroosmotic flow. The inner solution is valid in the EDL, but is generally invalid in the bulk. The outer solution is valid in the bulk, but invalid in the EDL and at the wall. Right: The composite solution smoothly transitions from the inner solution to the outer solution.

charge makes the electrical potential at the wall different from that in the bulk. Near the surface, the charge of the wall is counteracted by a thin cloud of oppositely charged ions, which is contained in what we call the *electrical double layer*. Unlike the bulk solution, which is *electroneutral*, meaning that its net charge density ρ_E is zero, the solution in the EDL has a nonzero net charge density. This EDL has a finite thickness, which we denote approximately as λ_D. Electroosmosis stems from the net Coulomb force felt by the fluid near the wall because of the cloud of ions.

To describe this phenomenon, we start by defining a *double-layer potential* $\varphi = \phi - \phi_{bulk}$ so that φ is, by definition, zero in the bulk.[1] The value of φ thus specifies how a potential differs from that in the bulk far from walls. The potential difference between the wall and the bulk solution is denoted by φ_0. If an *extrinsic* electric field[2] is applied, the force that this field applies to the ions near the wall induces fluid flow in the system. In this section, we show that all we need to predict the bulk fluid flow is the potential change between the bulk fluid and the wall.

We can describe the fluid flow induced by an extrinsic electric field by using an integral treatment of the Navier–Stokes equations with an electrostatic source term $\rho_E \vec{E}_{ext}$. The domain we consider is shown in Fig. 6.3 – the limits of integration are the wall and a point in the fluid at which the potential can be approximated by $\varphi = 0$, typically a point that is much more than λ_D away from the wall. We assume that three quantities are known: φ at the top ($\varphi = 0$) and bottom ($\varphi = \varphi_0$) surfaces, as well as the velocity at the bottom surface ($u = 0$). We also assume that this domain is small enough that we can assume the extrinsic electric field $\vec{E}_{ext}$ is uniform. The integral analysis then determines the velocity at the top surface.

[1] This definition is convenient when the electrical double layer is thin and the bulk is well defined. For cases in which the bulk is not well defined, we must define the bulk as a hypothetical location at which the electrochemical potential of an ion is independent of its charge.

[2] By extrinsic, we mean this field is applied externally with a power supply. This distinguishes this field from the *intrinsic* electric field, which is caused by the charge at the wall. The intrinsic field is coincident with the presence of charge in the double layer – if there were no intrinsic field, there would be no ions and no net Coulomb force on the double layer. The extrinsic and the intrinsic fields are separable in this simple geometry because the fields are normal to each other and of very different magnitudes.

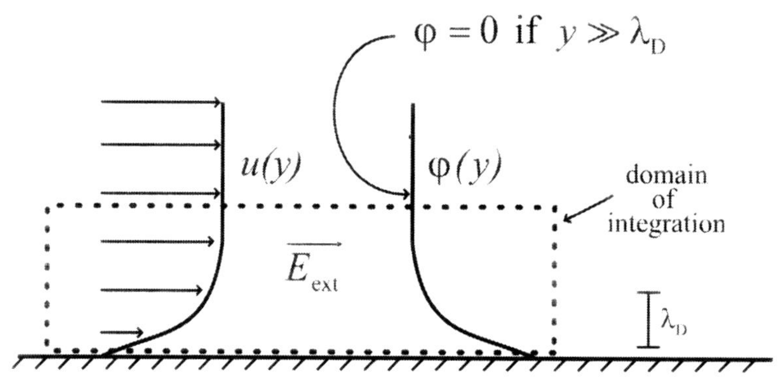

Fig. 6.3 Domain of integration for the integral analysis of the Coulomb forces on the EDL.

We write the uniform-viscosity Navier–Stokes equations for a thin region near the wall and include the electrostatic body force term $\rho_E \vec{E}_{ext}$:

Navier–Stokes equation with electrostatic source term

$$\rho \frac{\partial \vec{u}}{\partial t} + \rho \vec{u} \cdot \nabla \vec{u} = -\nabla p + \eta \nabla^2 \vec{u} + \rho_E \vec{E}_{ext} \,. \tag{6.1}$$

Here we assume that $\vec{E}_{ext}$ is caused by an external power supply and is uniform within the EDL, with a value equal to $\vec{E}_{ext,wall}$. The intrinsic electric field $E_{int} = -\frac{\partial \varphi}{\partial y}$ is caused by the chemistry at the surface and is nonuniform.

If we consider steady isobaric flow strictly along a wall aligned in the x direction, with velocity and potential gradients *only* in the y direction, Eq. (6.1) reduces to a simple conservation of x-momentum equation:

$$0 = \eta \frac{\partial^2 u}{\partial y^2} + \rho_E E_{ext,wall} \,. \tag{6.2}$$

In this simplified geometry, $\vec{E}_{ext,wall}$ is tangent to the wall. Recall next the uniform-permittivity Poisson equation (5.26):

$$-\varepsilon \nabla^2 \phi = \rho_E \,. \tag{6.3}$$

Using Eq. (6.3) to substitute for ρ_E and retaining only y gradients, we find

$$0 = \eta \frac{\partial^2 u}{\partial y^2} - \varepsilon \frac{\partial^2 \varphi}{\partial y^2} E_{ext,wall} \,. \tag{6.4}$$

The intrinsic field impacts Eq. (6.4) through the y gradients in velocity and electrical potential, whereas the extrinsic field impacts Eq. (6.4) explicitly. Rearranging, we obtain

$$\varepsilon E_{ext,wall} \frac{\partial^2 \varphi}{\partial y^2} = \eta \frac{\partial^2 u}{\partial y^2} \,. \tag{6.5}$$

Now we integrate from the wall ($y = 0$) to a point outside the double layer ($y \gg \lambda_D$, at which $\varphi = 0$ by definition). We find from this integral that

$$\eta u = \varepsilon E_{ext,wall} \varphi + C_1 y + C_2 \,. \tag{6.6}$$

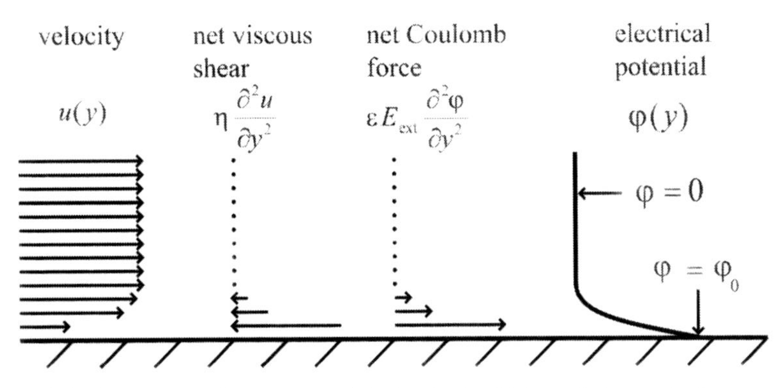

Fig. 6.4 Velocity, shear, Coulomb force, and electrical potential in an electroosmotic flow.

Applying the no-slip boundary condition and forcing the velocity to be bounded at $y = \infty$, we obtain

$$u = \frac{\varepsilon E_{ext,wall}}{\eta}(\varphi - \varphi_0), \tag{6.7}$$

where φ_0 is defined as the value of φ at the wall, i.e., at $y = 0$.

Inner solution. Equation (6.7) is the *inner* solution for electroosmotic flow, and we can write this solution with a subscript for clarity:

1D inner solution for electroosmotic flow, uniform fluid properties

$$u_{inner} = \frac{\varepsilon E_{ext,wall}}{\eta}(\varphi - \varphi_0). \tag{6.8}$$

The inner solution is valid only near the wall, where it is correct to assume that $E_{ext,wall}$ is uniform.

The velocity and the intrinsic potential are *similar* (i.e., proportional to each other) inside the EDL. Thus the spatial variation in the electrical potential is the same as the spatial variation in the velocity. The spatial distribution of φ is as yet underived, and thus Eq. (6.8) succeeds only in describing one unknown (the velocity) in terms of another unknown (the electrical potential). However, qualitative schematics of the velocity distribution and associated potential distribution, shear, and Coulomb forces predicted for EDLs are shown in Fig. 6.4. This equilibrium velocity distribution involves counterbalanced (a) viscous shear forces and (b) Coulomb forces near the wall. The viscous shear forces are related to velocity gradients, which decrease as the distance from the wall increases; similarly, the net Coulomb forces are induced by large net charge densities, which also decrease as the distance from the wall increases.

6.3 SOLVING THE NAVIER–STOKES EQUATIONS FOR ELECTROOSMOTIC FLOW IN THE THIN-EDL LIMIT

The previous section derived the *inner* solution for electroosmotic flow in terms of the electrical potential distribution, and built intuition about the nature of the flow in the EDL. The *outer* solution describes the flow outside the EDL, which is the part of the flow that we readily observe under a microscope. In both cases, a basic assumption of this analysis is that the EDL thickness is small relative to the characteristic size of the microchannel.

6.3.1 Outer solution

In the part of the flow where the distance from the wall is much greater than λ_D (where φ becomes zero by definition), substituting $\varphi = 0$ into Eq. (6.8) leads to the *Helmholtz–Smoluchowski equation*:

Helmholtz–Smoluchowski solution for electroosmotic flow far from a flat wall

$$u = -\frac{\varepsilon \varphi_0}{\eta} E_{\text{ext,wall}} \,. \tag{6.9}$$

Equation (6.9) is the exterior boundary condition of the inner solution. We find the outer solution for electroosmotic flow by considering the Navier–Stokes solutions for the bulk fluid, for which the net charge density is zero, and by using Eq. (6.9) as the interior boundary condition. Thus the governing equation for the outer flow is

Navier–Stokes equation, uniform fluid properties, electroneutral solution

$$\rho \frac{\partial \vec{u}_{\text{outer}}}{\partial t} + \rho \vec{u}_{\text{outer}} \cdot \nabla \vec{u}_{\text{outer}} = -\nabla p + \eta \nabla^2 \vec{u}_{\text{outer}} \,, \tag{6.10}$$

and the interior boundary condition is

effective slip condition used to describe the outer solution for electroosmotic flow

$$u_{\text{outer}}(y = 0) = -\frac{\varepsilon \varphi_0}{\eta} E_{\text{ext,wall}} \,. \tag{6.11}$$

Equation (6.11) is an example of a *Van Dyke matching condition* [34]. The idea is that the inner solution describes the flow in the EDL, the outer solution describes the flow outside the double layer, and the Van Dyke matching condition matches the two at a point in the middle. The outer solution is invalid in the EDL, and violates the no-slip condition at the wall, but properly describes the velocity distribution outside the double layer. For a straight channel of uniform cross section, the velocity profile outside the EDL is everywhere equal to this effective slip velocity.

The preceding outer solution gives the electroosmotic flow distribution outside the EDL, which is the most important region of the flow in most cases, because the EDL is thin and difficult to interrogate. In this section, we discuss how general electroosmotic flow problems can be simplified and solved with relatively simple equations.

6.3.2 Replacing the EDL with an effective slip boundary condition

The preceding matched asymptotic approach subsumes the net charge density of the EDL into a boundary condition for flow outside the EDL. The Van Dyke condition in Eq. (6.11) dictates that the limit of the velocity as the outer solution approaches the wall be a function of the wall potential, fluid permittivity, fluid viscosity, and the electric field magnitude near the wall. Thus, if we wish to solve only for the outer solution (in which the fluid is electroneutral), we can describe the flow by solving the Laplace equation for the electrical potential,

Laplace equation for electric potential in the electroneutral bulk solution

$$\nabla^2 \phi = 0 \,, \tag{6.12}$$

and by solving the Navier–Stokes equations *with no source term*:

Navier–Stokes equation for electroneutral fluid, uniform fluid properties

$$\rho \frac{\partial \vec{u}_{\text{outer}}}{\partial t} + \rho \vec{u}_{\text{outer}} \cdot \nabla \vec{u}_{\text{outer}} = -\nabla p + \eta \nabla^2 \vec{u}_{\text{outer}} \,, \tag{6.13}$$

combined with the effective electroosmotic slip velocity:

electroosmotic slip velocity

$$\vec{u}_{\text{outer,wall}} = -\frac{\varepsilon \varphi_0}{\eta} \vec{E}_{\text{ext,wall}} \,. \tag{6.14}$$

The electroosmotic slip velocity requires knowledge of the electric field at the wall, which is a result of the solution for ϕ. So, to solve this system, we solve the Laplace equation, use the electric field solution to specify the effective slip boundary conditions, and then use the effective slip boundary conditions to solve the Navier–Stokes equations.

6.3.3 Replacing the Navier–Stokes equations with the Laplace equation: flow–current similitude

If a number of conditions are met, the outer solution for the velocity is irrotational, in which case we need solve only the Laplace equation to determine the flow field, rather than solve the conservation equations for mass and momentum. These conditions (see Exercise 6.5) are satisfied for most microscale devices made from a single material as long as a pressure gradient is not applied to the system. This is an enormous simplification that greatly changes our ability to study electroosmotic flows. In these systems, *the velocity far away from any wall is proportional to the local electric field multiplied by the electroosmotic mobility* (Fig. 6.5):

$$\vec{u} = -\frac{\varepsilon \varphi_0}{\eta} \vec{E}_{\text{ext}} \,. \tag{6.15}$$

This is a potential flow and the velocity potential and electric potential both satisfy the Laplace equation. It satisfies the no-penetration condition at the wall because the electric field solution satisfies the no-current condition at an insulating wall. It also satisfies the electroosmotic slip condition at the wall and applies to any geometry.

Equation (6.15) is hugely important, because electroosmotic flows can be calculated rather easily in this limit, and analytical solutions for potential flows can give immediate

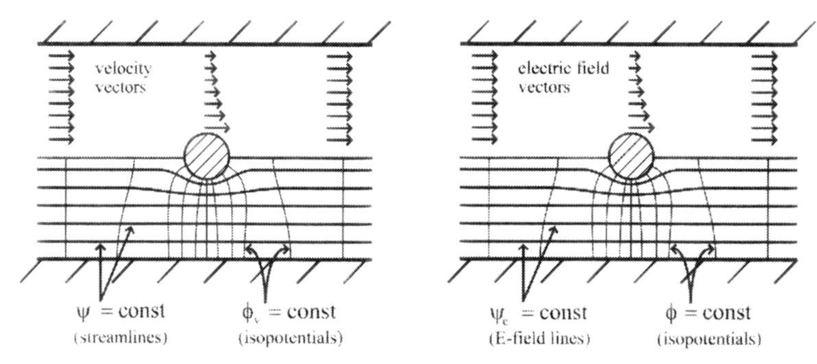

Fig. 6.5 Flow-current similitude for electroosmotic flow in a channel with a circular insulating wall. If the conductivity and surface potential φ_0 are uniform, the flow velocity is proportional to the local electric field, and thus (a) electric field lines are the same as streamlines and (b) velocity potential isocontours are the same as electrical potential isocontours.

physical insight into the flow distributions in an electroosmotically driven system. These potential flows are discussed in detail in Chapter 7.

Experimental applicability. Experiments are often designed to satisfy the assumptions required for Eq. (6.15) to apply, because system design is then more straightforward. We can achieve uniform conductivity by ensuring that the fluid is well mixed and that Joule heating (that is, fluid heating owing to the current conducted through the fluid) is minimal. We can achieve uniform surface potential by making a device from one material and ensuring that the surfaces are clean. Removing applied pressure fields typically requires monitoring the hydrostatic head and the interface curvature at the reservoirs. If care is taken to satisfy these requirements, the flow generated in these systems is easy to model and often advantageous for a variety of applications. For example, the potential flow described here can lead to dispersionless transport, which leads to optimal performance for capillary electrophoresis separations.

6.3.4 Reconciling the no-slip condition with irrotational flow

Electroosmosis outside thin double layers is described by an irrotational flow with a velocity proportional everywhere to the local electric field. Vorticity is generated at the wall, but the electrostatic force on the charge density in the double layer generates a vorticity source term (Fig. 6.6) that, when integrated over the double layer, is of precisely the same magnitude but of opposite sign to that of the vorticity generated at the wall. Thus the vorticity *inside* the double layer is large, but the vorticity *outside* the double layer is identically zero.

6.4 ELECTROOSMOTIC MOBILITY AND THE ELECTROKINETIC POTENTIAL

The effective slip boundary condition observed for electroosmosis can be *phenomenologically* described by use of an electroosmotic mobility:

definition of electroosmotic mobility

$$\vec{u}_{\text{outer,wall}} = \mu_{\text{EO}} \vec{E}_{\text{ext,wall}} . \tag{6.16}$$

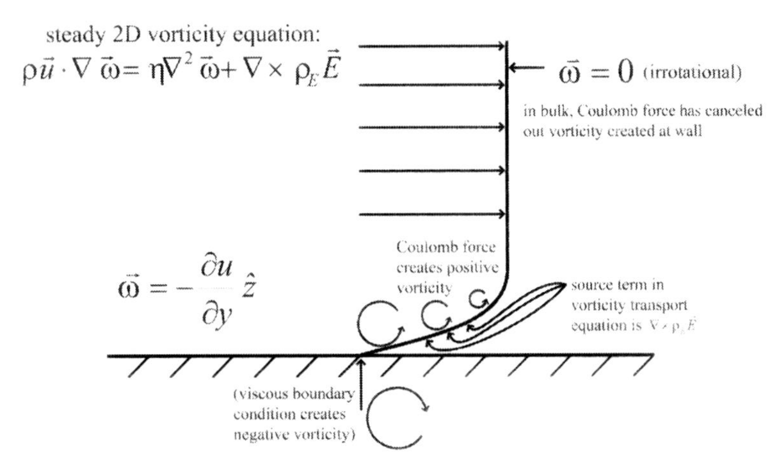

Fig. 6.6 Vorticity generation and cancellation in EDLs.

A typical magnitude observed for μ_{EO} in aqueous systems might be of the order of 1×10^{-8} m^2/V s (see Table 6.1, which lists typical approximate values of electroosmotic mobilities). By comparing Eq. (6.16) with Eq. (6.14), we can see that, for a system with uniform fluid properties,

electroosmotic mobility, uniform fluid properties, simple interface

$$\mu_{EO}=-\frac{\varepsilon\varphi_0}{\eta}. \tag{6.17}$$

It is common to report the *zeta potential* or *electrokinetic potential* inferred from the electroosmotic mobility:

definition of electrokinetic potential for electroosmosis

$$\zeta=-\frac{\mu_{EO}\eta_{bulk}}{\varepsilon_{bulk}}. \tag{6.18}$$

ζ is an experimentally observed quantity that has units of volts. If the fluid has uniform ε and η, then the measured ζ is equal to φ_0. If the fluid permittivity or viscosity varies in the EDL, then a different integral analysis must be performed to relate ζ to φ_0.

Table 6.1 **Typical electroosmotic mobilities.**

Wall material	Solution	μ_{EO} [m^2/V s]
Glass	pH7, 1 mM NaCl	3×10^{-8}
Glass	pH5, 1 mM NaCl	1×10^{-8}
Silicon	pH7, 1 mM NaCl	3×10^{-8}
PDMS	pH7, 1 mM NaCl	1.5×10^{-8}
Polycarbonate	pH7, 1 mM NaCl	2×10^{-8}

Note: pH and concentrations are defined in Appendix B.

6.4.1 Electrokinetic coupling matrix representation of electroosmosis

For the analysis presented in this chapter (thin double layers, uniform fluid properties), the bulk velocity result $u = -\frac{\varepsilon\zeta}{\eta}E$ implies that the mean velocity through a tube with an electrokinetic potential ζ is given by $\overline{u} = -\frac{\varepsilon\zeta}{\eta}E$. Recalling the electrokinetic coupling matrix, first presented in Chapter 3,

$$\begin{bmatrix} Q/A \\ I/A \end{bmatrix} = \chi \begin{bmatrix} -\frac{dp}{dx} \\ E \end{bmatrix}, \tag{6.19}$$

we see that electroosmosis in channels with thin double layers is described by setting $\chi_{12} = -\frac{\varepsilon\zeta}{\eta}$.

6.5 ELECTROKINETIC PUMPS

Unlike Poiseuille flow (Chapter 2), for which the mean velocity for a given pressure gradient decreases as the channel dimension decreases, the mean electroosmotic velocity is independent of length scale as long as the thin-EDL approximation applies. This means that, as length scales are reduced, electric fields become much more effective at generating fluid flow than pressure gradients are. This property is central to the performance of an electroosmotic or *electrokinetic pump*, in which an electric field is applied along a capillary to generate flow and pressure. This is a simple flow case, but one that shows compellingly how length scale can dictate the relative importance of flow phenomena.

6.5.1 A planar electrokinetic pump

Consider a system in which an electric field $E = \Delta V/L$ is applied across an open rectangular microchannel with depth $2d$, width w, length L, and cross-sectional area $A = wd$. Assume that $\lambda_D \ll d \ll w \ll L$, and no pressure gradient is applied. As we have discussed previously, the boundary condition can be modeled with the equation $\vec{\boldsymbol{u}}_{\text{wall}} = \mu_{\text{EO}}\vec{\boldsymbol{E}}$. If the microchannel area is uniform, the electroosmotic velocity in the channel is uniform (this is a degenerate case of a Couette flow, in which both surfaces move with the same velocity). Summarizing the relations for the pressure drop across the capillary (Δp), the velocity distribution across the capillary cross section, and the total flow rate in this system (Q_{tot}), we have

$$\Delta p = 0, \tag{6.20}$$

$$u = \mu_{\text{EO}}E, \tag{6.21}$$

$$Q_{\text{tot}} = uA. \tag{6.22}$$

Now suppose the downstream port of this capillary is closed, and we attach a pressure transducer to this port. For simplicity (and because $L \gg w$), we consider only the region of the capillary that is far from the ends, and we assume all flow is strictly in the x direction. In this case, the net flow through any cross section must be zero because the port is closed. The electroosmotic flow equation still holds, but the assumption that the pressure gradient is zero does not. In fact, an adverse pressure gradient and a reverse Poiseuille flow must coincide with the electroosmotic flow to satisfy mass conservation.

For a 2D system (plates of width w separated by $2d$, where $w \gg d$), the electroosmotic and pressure-driven contributions to the flow are

$$Q_{\mathrm{EOF}} = w \int_{-d}^{d} u_{\mathrm{wall}}\, dy\,, \tag{6.23}$$

$$Q_{\mathrm{PDF}} = w \int_{-d}^{d} \frac{1}{2\eta}\left(-\frac{\partial p}{\partial x}\right)(d^2 - y^2)\, dy\,. \tag{6.24}$$

Integrating and setting the net flow rate equal to zero, we find that the pressure difference generated by the flow is given by

pressure generated by a parallel-plate electroosmotic pump, thin-EDL approximation

$$\frac{\Delta p}{\Delta V} = -\frac{3\mu_{\mathrm{EO}}\eta}{d^2}\,. \tag{6.25}$$

In the thin-EDL limit, an electric field applied across a closed microchannel generates a downstream pressure proportional to the applied voltage and inversely proportional to the channel depth squared. A similar relation can be derived for a circular capillary of radius R:

pressure generated by a circular capillary electroosmotic pump, thin-EDL approximation

$$\frac{\Delta p}{\Delta V} = -\frac{8\mu_{\mathrm{EO}}\eta}{R^2}\,. \tag{6.26}$$

EXAMPLE PROBLEM 6.1

Consider a system in which an electric field is applied across an open rectangular capillary with depth $2d$, width w, length L, and cross-sectional area $A = wd$. Assume that $r \ll w \ll L$, and no pressure gradient is applied. If the wall electroosmotic mobility is given, derive the pressure per voltage achievable by the pump.

SOLUTION: The electroosmotic and pressure-driven flow rates are

$$Q_{\mathrm{EOF}} = w \int_{-d}^{d} u_{\mathrm{wall}}\, dy\,, \tag{6.27}$$

$$Q_{\mathrm{PDF}} = w \int_{-d}^{d} \frac{1}{2\eta}\left(-\frac{\partial p}{\partial x}\right)(d^2 - y^2)\, dy\,. \tag{6.28}$$

Integrating, we find

$$Q_{\mathrm{EOF}} = 2wu_{\mathrm{wall}}d\,, \tag{6.29}$$

$$Q_{\mathrm{PDF}} = \frac{2w}{3\eta}\left(-\frac{\partial p}{\partial x}\right)d^3\,. \tag{6.30}$$

Setting $Q_{\mathrm{tot}} = Q_{\mathrm{EOF}} + Q_{\mathrm{PDF}} = 0$, we find

$$2wu_{\mathrm{wall}}d + \frac{2w}{3\eta}\left(-\frac{\partial p}{\partial x}\right)d^3 = 0\,. \tag{6.31}$$

Defining Δp as $p_{\text{inlet}} - p_{\text{outlet}}$ and ΔV as $V_{\text{inlet}} - V_{\text{outlet}}$, and replacing u_{wall} with $\mu_{\text{EO}} E$, E with $\Delta V / L$, and $\partial p / \partial x$ with $-\Delta p / L$, we find

$$2w \frac{\Delta V}{L} \mu_{\text{EO}} d + \frac{2w}{3\eta} \left(-\frac{\partial p}{\partial x} \right) d^3 = 0 , \tag{6.32}$$

and

$$\Delta p = -\frac{3 \mu_{\text{EO}} \eta \Delta V}{d^2} . \tag{6.33}$$

From this, we see that the generated pressure is linear with ΔV, and we can define the pressure generated per volt, which is given by

$$\frac{\Delta p}{\Delta V} = -\frac{3 \mu_{\text{EO}} \eta}{d^2} . \tag{6.34}$$

We have thus derived two limiting cases: an open capillary, for which the generated flow rate is maximum, but the pressure generated is zero, and a closed microchannel, for which the generated flow rate is zero, but the pressure generated is maximum. In the general case for which Q_{tot} is nonzero and Δp is nonzero, the solution is a linear combination of these two solutions, with an example flow field shown in Fig. 6.7. We can write for the planar electrokinetic pump,

$$Q = 2w u_{\text{wall}} d + \frac{2w}{3\eta} \left(-\frac{dp}{dx} \right) d^3 , \tag{6.35}$$

and we can rearrange this as

$$Q = Q_{\max} \left(\frac{\Delta p_{\max} - \Delta p}{\Delta p_{\max}} \right) , \tag{6.36}$$

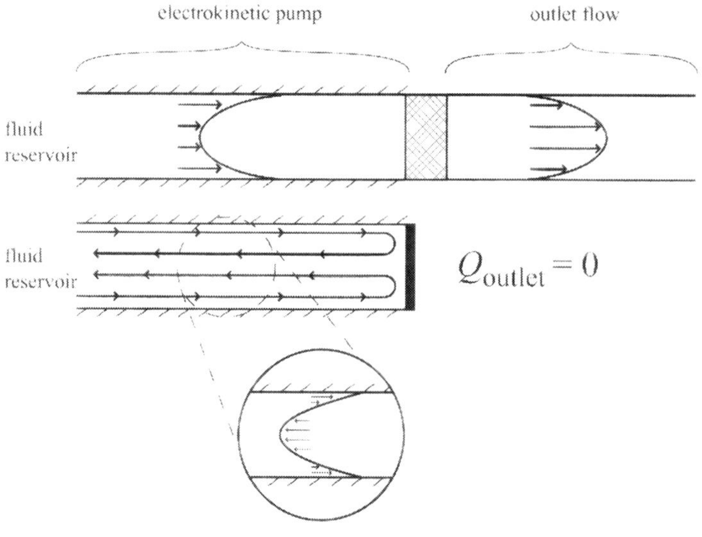

Fig. 6.7 Flow field inside an electrokinetic pump.

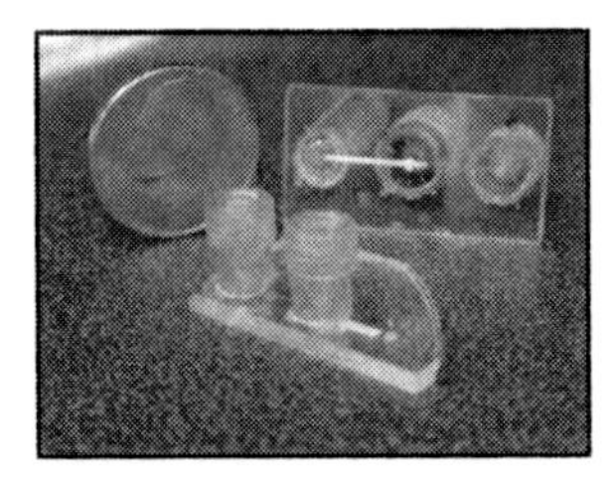

Fig. 6.8 Images of electrokinetic pumps made from (left) particle-packed glass chips and (right) porous polymer monoliths. Courtesy Sandia National Labs.

where Q_{max} is the maximum flow rate (at $\Delta p = 0$), given by

$$Q_{max} = 2wd\mu_{EO}\frac{\Delta V}{L}, \tag{6.37}$$

and Δp_{max} is given by

$$\Delta p_{max} = -\frac{3\mu_{EO}\eta\Delta V}{d^2}. \tag{6.38}$$

Because the power generated by a pump is given by ΔpQ, Eq. (6.36) can be used to derive the maximum power point. Because the power input to an electrokinetic pump is given by the current–voltage product ΔVI, the thermodynamic efficiency can be defined as

thermodynamic efficiency of an electroosmotic pump

$$\xi = \frac{\Delta pQ}{\Delta VI}, \tag{6.39}$$

where the current I is given by $I = \Delta V/R$ and the resistance R is given by $R = L/\sigma A$ in the bulk conductivity limit.[3] Efficiencies of electrokinetic pumps range approximately from 0.01% to 5%.

6.5.2 Types of electrokinetic pumps

The preceding analysis above is for a single capillary, but these results can be extended analytically to other geometries. Geometries used have included single capillaries [64], arrays of capillaries, capillaries packed with microparticles [65, 66, 67], planar microfabricated structures, and porous polymer monoliths. Two examples of these are shown in Fig. 6.8. These devices have been used for chemical separations [67] and many other applications.

Experimentally, we can evaluate these pumps by monitoring flow rates and pressures as applied voltages are varied.

[3] Here σ is the bulk conductivity, A is the cross-sectional area of the channels, and L is the length of the device. This relation holds only if no interface phenomena affect the system conductivity.

EXAMPLE PROBLEM 6.2

Consider a microchannel with cross-sectional area equal to 200 μm^2 and a surface electroosmotic mobility equal to 2×10^{-8} m^2/V s. If an electric field of 100 V/cm is applied to this microchannel, what is the resulting fluid velocity and the resulting total volumetric flow rate?

SOLUTION: The fluid velocity is given by

$$u = \mu_{EO} E, \tag{6.40}$$

or

$$u = 2 \times 10^{-8}\, m^2/V\,s \times 1 \times 10^4\, V/m = 2 \times 10^2\, \mu m/s. \tag{6.41}$$

The volumetric flow rate is thus

$$Q = 2 \times 10^2\, \mu m/s \times 200\, \mu m^2 = 4 \times 10^4\, \mu m^3/s. \tag{6.42}$$

EXAMPLE PROBLEM 6.3

Consider a microdevice consisting of two infinite parallel plates separated by a distance $2h$ and filled with an aqueous solution. Assume that the top plate is glass and has an electroosmotic mobility of 4×10^{-8} m^2/V s, and that the bottom plate is Teflon and has an electroosmotic mobility of 2×10^{-8} m^2/V s. Derive the resulting velocity distribution if an electric field of 150 V/cm is applied. How does the velocity distribution for this flow relate to the velocity distribution in Eq. (2.40)?

SOLUTION: The velocity at the top plate $y = h$ is given by

$$u = 4 \times 10^{-8}\, m^2/V\,s \times 150\, V/cm = 6 \times 10^2\, \mu m/s, \tag{6.43}$$

and the velocity at the bottom plate $y = h$ is given by

$$u = 2 \times 10^{-8}\, m^2/V\,s \times 150\, V/cm = 3 \times 10^2\, \mu m/s. \tag{6.44}$$

This flow is precisely a Couette flow, as described in Eq. (2.40):

$$u = \left(4.5 \times 10^2 + \frac{y}{h} 1.5 \times 10^2\right)\, \mu m/s. \tag{6.45}$$

6.6 SUMMARY

This chapter describes electroosmotic flow in the thin-EDL limit, in which we are largely unconcerned with the details of the flow inside the EDL, and care only about the net effect of the EDL on the bulk flow. This is often applicable for microscale devices because the EDL is typically only nanometers in thickness. If the double layer is thin, we can solve the system by using a matched asymptotic approach, in which the extrinsic electric field is assumed uniform inside the EDL, and Coulomb forces are ignored outside the double layer. In this case, the EDL can be replaced with an effective slip condition defined by the electroosmotic mobility:

$$\vec{u}_{\text{outer,wall}} = \mu_{\text{EO}} \vec{E}_{\text{ext,wall}} \,, \tag{6.46}$$

which, for uniform fluid properties and simple interfaces, is given by

$$\mu_{\text{EO}} = -\frac{\varepsilon\varphi_0}{\eta} \,. \tag{6.47}$$

For flows with uniform interfacial properties, uniform fluid properties, and no applied pressure gradients, the fluid flow in a purely electroosmotic flow is a potential flow in which the fluid velocity is proportional to the local electric field:

$$\vec{u} = -\frac{\varepsilon\varphi_0}{\eta} \vec{E}_{\text{ext}} \,. \tag{6.48}$$

These relations link the interfacial potential to the fluid velocity in the system; fluid flow in microdevices can be calculated with relatively straightforward simulations by use of effective slip boundary conditions.

6.7 SUPPLEMENTARY READING

The integral analysis that leads to the description of the bulk electroosmotic velocity as a function of the wall potential can be found in many sources [29, 68, 69, 70, 71, 72]. Li [72] gives a detailed treatment of several electrokinetic flows. A number of researchers have discussed [73, 74, 75, 76, 77] similitude between the electric field and the fluid velocity in a more rigorous mathematical fashion, as well as the dynamics of electroosmosis with unsteady electric fields [78].

Relevant sources include discussion of electrokinetically generated pressures that are due to nonuniform zeta potential [79], electrokinetic pumps [64, 65], the effect of permittivity on electrokinetic pumps [66], detailed calculations of electrokinetic pumps without the many simplifying approximations used in this chapter [80, 81], and discussion of electrical double layers with wall curvature [52].

It is critical to highlight what we have and have not accomplished. We have assumed that we knew the electrical potential at the wall, in which case we can evaluate the velocity far from the wall with an integral analysis. The magnitude of the velocity of the flow far from the wall is dependent on the total variation in electrical potential between the wall and the bulk fluid, i.e., φ_0, but is independent of the spatial dependence of φ. However, many important issues are left unresolved. The chemistry of microdevice walls typically leads to a surface charge density q''_{wall}. Knowledge of this surface charge density would help us know the electrical force on the wall and the net body force on the fluid as well. However, we cannot evaluate any of these parameters without a detailed picture of the ion distribution near a charged wall, which is modeled in Chapter 9. Our study of distributed body forces in 1D flow systems (for example, Exercise 2.12), though, has shown us that velocities in these systems are functions of the *distribution* of the body forces. We were able to get an outer velocity solution without knowledge of the details of the inner solution *only* because of a coincidence – namely, that the mathematical form of the bulk velocity (a second spatial integral of the Coulomb force on a net charge density) and the mathematical form of the wall potential (a second spatial integral of the net charge density) are the same for this system. In our result for the inner solution, we have simply written one unknown (u_{inner}) in terms of another (φ); however, this relation predicts purely electroosmotic flow fields once we make experimental measurements of φ_0.

Although the integral description used here to link surface potential to fluid velocity is a straightforward and intuitive one, its ability to match experimental observations is a matter of debate, owing to questions about the validity of the assumption that the fluid properties are uniform and the interface is simple. This is discussed in more detail in Chapter 10.

6.8 EXERCISES

6.1 Consider a 1D case in which the cross-sectional area of a polycarbonate microchannel with $\mu_{\text{EO}} = 2.6 \times 10^{-8}$ m^2/V s changes instantaneously between two regions, which we refer to as region 1 and region 2. The conductivity σ is uniform throughout. The magnitude of the electric field in region 1 is denoted by E_1 and the magnitude of the electric field in region 2 is denoted by E_2, the cross-sectional area in region 1 is given by $A_1 = 200\ \mu\text{m}^2$ and the cross-sectional area in region 2 is given by $A_2 = 400\ \mu\text{m}^2$, and the length of region 1 is given by $L_1 = 1$ cm and the length of region 2 is given by $L_2 = 1.5$ cm. A potential difference of $\Delta V = 300$ V is applied across the channel.

(a) Using the techniques of Chapter 3, determine the resistance and potential drop in each region in terms of the input parameters. Determine E_1 and E_2.

(b) Determine the magnitudes of the electric fields E_1 and E_2 in the two regions, and calculate the velocity magnitude in each region.

(c) Does this velocity solution satisfy conservation of mass? What is the relation between conservation of mass and conservation of current in this flow?

6.2 Consider flow of an aqueous solution ($\varepsilon = 80\varepsilon_0$, $\eta = 1 \times 10^{-3}$ Pa s) through a $L = 1$ cm microchannel made of glass, as shown in Fig. 6.9. Assume that the glass microchannel has uniform cross section and is much longer than it is wide, and thus assume that the system is approximately 1D. Assume that the glass is an insulator, and that the interface between the solution and the glass has a surface potential that is 70 mV lower than the potential of the bulk solution, as well as a surface charge density of $q''_{\text{wall}} = -0.05$ C/m^2.

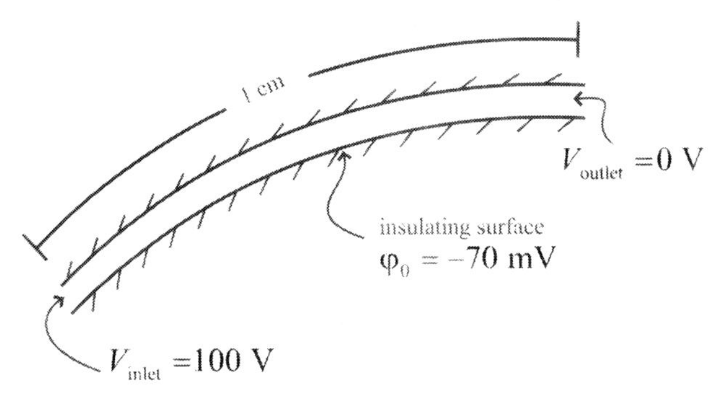

Fig. 6.9 Schematic of a microchannel.

Assume that the voltage at the inlet is $V_1 = 100$ V and the voltage at the outlet is $V_2 = 0$ V.

(a) Formulate the governing equations and boundary conditions for a complete solution of this problem, including electrical potential and fluid velocity both near to and far from the wall. Do not attempt to solve this system of equations.

(b) Execute a 1D integral analysis of the boundary layer near the wall and determine the effective slip boundary condition that describes flow outside the EDL.

(c) Given your effective slip boundary condition, reformulate the problem as follows:

  i. Define the governing equation for the outer solution for fluid flow in this system.
  ii. Prescribe the boundary conditions for the outer solution for the fluid flow in terms of the electric field.
  iii. Prescribe the governing equations and boundary conditions required to solve for the electric field distribution in the system.
  iv. Solve for the outer solution for the velocity distribution for this system.

6.3 In deriving the effective slip owing to electroosmosis, we supposed that the surface *potential* was known, leading to a prediction of the effective slip of the outer velocity solution. If we presumed, instead, that the surface *charge density* was known but the potential was not, could the effective slip be predicted with the 1D integral analysis presented in this chapter?

6.4 Draw a control volume around the boundary layer near a charged wall. Given the velocity distribution derived in this chapter, evaluate the drag force per unit length at the wall in terms of q''_{wall}. Integrate the Coulomb force in the control volume to evaluate the total electrostatic force per unit length on the ions in the double layer. What does this tell you about the relation between the viscous drag force and the total Coulomb force? What does this tell you about the relation between the areal charge density on the wall and the volumetric charge density in the boundary layer?

6.5 The conditions required for guaranteeing that the outer solution for electroosmotic flow in a channel with impermeable, insulating walls is irrotational are

(a) the flow is quasi-steady (i.e., changes in the applied electric field or the chemical properties of the surface are slow compared with the characteristic frequency $\eta/\rho R^2$),

(b) fluid properties are uniform,

(c) the surface electroosmotic mobility is uniform,

(d) the EDL is thin,

(e) stagnation pressures are equal at all inlets and outlets,

(f) flows at inlets and outlets are given by $\vec{u} = \mu_{EO}\vec{E}$, and

(g) the Reynolds number is small.

For each condition, describe how a flow that violates the condition can lead to nonzero vorticity in the outer solution.

6.6 In deriving relations for an electrokinetic pump at the maximum pressure condition, the flow rate is assumed zero. Prove that the flow rate is zero by applying conservation of mass to a carefully chosen control volume within the pump geometry.

6.7 Calculation of the solution of the Laplace equation in a microchannel with a turn highlights the nonuniformity of the resulting flow field, as the electric field at the inside edge of a turn is higher than that at the outside edge. Presume that you have been asked to design a microdevice to generate a turn in a channel that leads to electroosmotic flow that is piecewise uniform (i.e., a uniform inlet flow is changed to a uniform outlet flow in a different direction, and this change occurs at a well-defined interface). You have decided to achieve this by varying the depth (in the z axis) of the channel so that it is d_1 on one side of the interface and d_2 on the other side. By combining the change of depth with proper control of the channel geometry in the xy plane, you hope to achieve a piecewise-uniform flow.

Consider the channel geometry specified in Fig. 6.10. We will assume that this device works (i.e., that the flow is piecewise uniform) and determine the relations required to make this so.

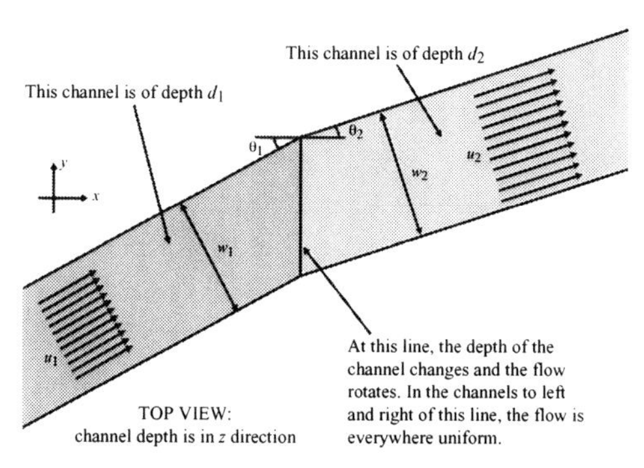

Fig. 6.10 Schematic (top view) of a channel that incorporates a depth change and a turn to lead to piecewise-uniform electroosmotic flow.

(a) Given that the flow at left is u_1 and the flow at right is u_2, write a mass continuity equation that relates the velocities, depths, and widths of channels 1 and 2.

(b) The widths w_1 and w_2 must be related to the angles θ_1 and θ_2 because the two channels share an interface. Write an equation to relate these.

(c) The flow entering and exiting the interface can be decomposed into components normal and tangential to the interface. The tangential velocity is continuous through the interface. Write an equation relating u_1, u_2, θ_1, and θ_2 that ensures continuity of tangential velocity.

(d) Combine these relations to show that the flow is piecewise uniform if these two relations are satisfied:

$$\frac{\tan\theta_1}{d_1} = \frac{\tan\theta_2}{d_2}, \tag{6.49}$$

and

$$u_1 \sin\theta_1 = u_2 \sin\theta_2 . \tag{6.50}$$

(e) Given that d_1/d_2 is specified, graph θ_2 as a function of θ_1 for $\theta_1 = 0$ to $\theta_1 = \pi/2$. Use values of d_1/d_2 equal to 1, 2, 4, 8, 16, 32, and 64.

(f) Given that d_1/d_2 is specified, graph u_2/u_1 as a function of the turning angle (i.e., $\theta_1 - \theta_2$) for $\theta_1 = 0$ to $\theta_1 = \pi/2$. Use values of d_1/d_2 equal to 1, 2, 4, 8, 16, 32, and 64. Note that this function is dual valued.

6.8 A fundamental challenge for electrophoretic separations in microchips is in eliminating dispersion, which is due to fluid flow. Electroosmosis in a straight channel is dispersion free, but electroosmosis through turns generates large dispersion, which is due to the different travel times for different regions of the flow. Thus any attempt to efficiently fabricate long channels on a microchip by simply turning the channel back and forth will, in general, lead to poor separation performance.

Recall that geometries can be specified such that electroosmosis is piecewise uniform if the channel depth is allowed to change across an interface and the angles that the inlet and outlet channels make with the interface satisfy the appropriate relations. Piecewise-uniform flow has no dispersion while the fluid is traveling through the channels; however, the interface itself will rotate an analyte band with respect to the fluid velocity. Thus an analyte band that is oriented normal to the fluid flow will *not* be normal to the fluid flow upon exiting the interface. This rotation will also effectively broaden the analyte band.

In this problem, you will design a way to have a microfluidic channel with a turn that creates *no* dispersion and *does not* rotate the analyte band with respect to the fluid velocity. This will require that the depth be changed and that a number of angles be designed so that the no-dispersion solution is obtained. In theory, this design will allow for a "perfect" microfluidic turn.

(a) Consider the first interface in Fig. 6.11. This interface will rotate a material line in the inlet channel (region 1) to a different angle in the outlet channel (region 2), as shown in Exercise 6.7. Thus a material line perpendicular to the flow in the inlet channel will not be perpendicular to the flow in the outlet channel. As a function of θ_1 and the depth ratio d_1/d_2, derive the angle the material line in region 2 makes with a line perpendicular to the flow.

(b) Consider a channel with two interfaces, i.e., add a region 3 with depth equal to that of region 1. This is shown in Fig. 6.11. For the second interface, define θ_3 and θ_4 as the angles the inlet and outlet make with the normal to the interface. As a function of θ_1 and the depth ratio d_1/d_2, derive the two values of θ_3 that will allow a material line perpendicular to the flow in region 1 to be perpendicular to the flow when exiting in region 3. One of these solutions will lead to outlet flow that is parallel to the inlet flow. This solution is of no practical use for designing a turn. The other solution leads to outlet flow that is *not* parallel to the inlet flow. This solution can be used to

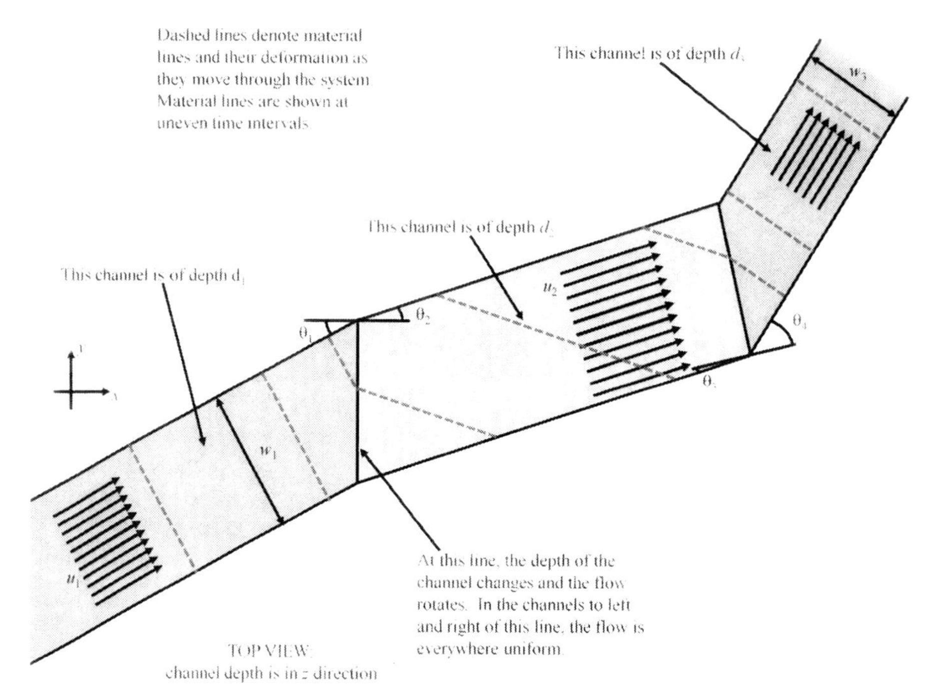

Fig. 6.11 Schematic of a channel turn that incorporates a depth change and two interfaces to lead to piecewise-uniform electroosmotic flow and *no* rotation of material lines with respect to the fluid velocity vector.

design a microfluidic turn. For the case that turns the flow, what is the net turning angle α as a function of θ_1 and the depth ratio?

(c) Assume $d_1/d_2 = 12$. Calculate the angles $\theta_1, \theta_2, \theta_3, \theta_4$ required for the design of a channel with two interfaces that rotates an electroosmotic flow by 45° and maintains the perpendicularity of material lines to the flow. There are actually four possible sets of solutions. Draw or plot one of these microchannel geometries. What is the width ratio w_3/w_1?

(d) Assume $d_1/d_2 = 12$. Calculate the angles $\theta_1, \theta_2, \theta_3, \theta_4, \theta_5, \theta_6, \theta_7, \theta_8$ required for the design of a channel with four interfaces that rotates an electroosmotic flow by 90°, maintains the perpendicularity of material lines to the flow, and *does not* change the width (i.e., $w_5/w_1 = 1$). Draw or plot this geometry. By definition $d_1 = d_3 = d_5$ and $d_2 = d_4$.

(e) Read one of Refs. [82, 83]. Both of these papers propose low-dispersion microfluidic turns. This problem has modeled a turn with *no* dispersion. Comment on the applicability of the turns designed in the references compared with the turn designed in this exercise. How are the modeling techniques in Refs. [82, 83] and in your solution limited? Which solution is better? How would your interpretation of the relative merits of the solutions change with more detailed modeling?

6.9 Consider the significance of viscous dissipation, i.e., the conversion of kinetic energy to heat that is due to viscosity, compared with Joule heating in an electroosmotic system of unspecified geometry but characteristic length L. Here we are modeling electroosmosis as a slip velocity at the wall, and we are paying no attention to what goes on close to the wall.

Assume fields with magnitude of the order of E are applied and the fluid has conductivity and viscosity of σ and η, respectively. Assume that the electroosmotic mobility of the system is μ_{EO}. Derive an approximate scaling relation (make suitable approximations to make this derivation straightforward and algebraic) that gives the ratio of the total viscous dissipation in the channel to the total Joule heating. To do this, note that the local viscous dissipation is given by $2\eta\left|\nabla\vec{\boldsymbol{u}}\right|^2$.

Evaluate this scaling relation for a system with $\eta = 1 \times 10^{-3}$ Pa s, $\sigma = 100\ \mu$S/cm, $L = 20\ \mu$m, and $\mu_{EO} = 4 \times 10^{-8}$ m^2/V s.

6.10 Consider the startup of electroosmotic flow, shown for flow between two plates in Fig. 6.12. Consider two parallel plates separated by a distance $2d$, with plates located at $y = \pm d$. Assume that, for $t < 0$, the fluid and plates are motionless. Assume that, at time $t = 0$, an electric field E is applied. Using separation of variables, calculate the outer flow solution $u_{\mathrm{outer}}(y, t)$.

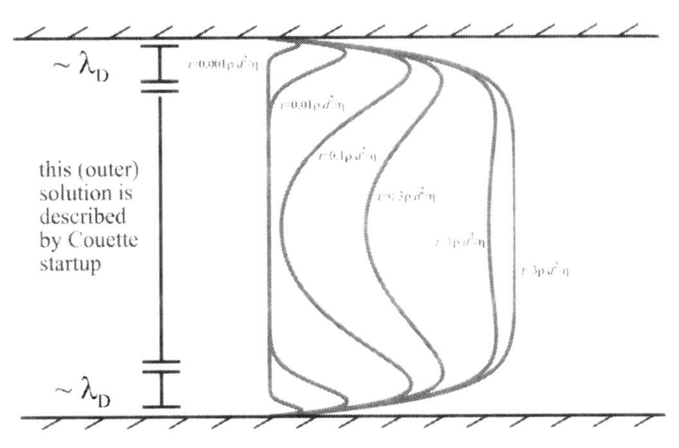

Fig. 6.12 Startup of electroosmotic flow.

6.11 Consider oscillatory electroosmotic flow. Consider two parallel plates separated by a distance $2d$, with plates located at $y = \pm d$. Assume that an electric field equal to $E \cos \omega t$ is applied. Using separation of variables, calculate $u_{\mathrm{outer}}(y, t)$.

6.12 Derive the zero-mass-flow electrokinetic pump relation in Eq. (6.26).

6.13 Derive the maximum efficiency of a planar (2D) electrokinetic pump as a function of the electroosmotic mobility and fluid properties such as the viscosity, conductivity, and permittivity. Assume the conductivity of the system is given by the bulk conductivity of the fluid (i.e., ignore surface conductivity).

6.14 The electrical permittivity of the electrolyte affects the thermodynamic efficiency of an electrokinetic pump owing to the dependence of electroosmotic flow on the permittivity. Read Ref. [66] and summarize

(a) the dependence of the flow rate on permittivity,

(b) the dependence of the generated pressure on permittivity, and

(c) the scaling of the thermodynamic efficiency with electrical permittivity.

Also, answer the following questions:

(a) How can the permittivity of a solution be changed? List at least two ways. One is summarized in Ref. [66].

(b) What other properties of a solution will also be changed if these techniques are implemented? Will these other effects tend to increase or decrease the thermodynamic efficiency owing to these ancillary effects?

(c) What key concepts or symbols or definitions are in this paper that are not covered in the treatment in this chapter?

6.15 The simple relations derived in Section 6.5 do not completely describe the thermodynamic efficiency of an electrokinetic pump. Read Ref. [81] and provide answers to the following questions:

(a) What are the key assumptions used in the simple relations written in Section 6.5?

(b) When do these assumptions break down? Write your answers in terms of electrolyte concentration, system conductivity, and characteristic system size.

(c) Do the simple relations overpredict or underpredict performance? How significant are the changes when a more thorough modeling approach is used?

6.16 Consider the electrokinetic pumps described in Refs. [65, 66]. How do these physical systems deviate from the idealized geometry used to derive the simple relations in Section 6.5? What experimental parameters are measured in Ref. [65] to facilitate the use of simple models to predict the performance of the system?

6.17 Numerically simulate flow through a long 2D channel with electroosmotic slip conditions on the side walls and closed, no-slip end walls. Plot streamlines for this flow.

6.18 Given a planar electrokinetic pump with fluid flow between plates at $y = \pm d$, find the y location(s) at which the fluid velocity is zero.

7 Potential Fluid Flow

This chapter discusses the physical relevance of potential fluid flow to flows in microfluidic devices and describes analytical tools for creating potential flow solutions. This chapter also focuses on the use of complex mathematics for 2D potential problems with plane symmetry. These plane-symmetric flows are relevant for microsystems, because microchannels are often shallower than they are wide and thus depth-averaged properties are often well approximated by 2D analysis.

In particular, we want to retain perspective on the engineering importance of these flows as well as on the relative importance of analysis versus numerics. The Laplace equation is rather straightforward to solve numerically, and therefore numerical simulation is a suitable approach for most Laplace equation systems. For example, simulation of the electroosmotic flow within a microdevice with a complicated geometry would be simulated, because analytical solution would be impossible. Despite the importance of numerics, the analytical solutions are important because they lend physical insight and because simple analytical solutions for important cases (for example, the potential flow around a sphere) facilitate expedient solutions of more complicated problems. For example, the study of electrophoresis of a suspension of charged spheres is typically analyzed with techniques informed by the analytical solution for potential flow around a sphere and *not* with detailed and extensive numerical solutions of the Laplace equation.

7.1 APPROACH FOR FINDING POTENTIAL FLOW SOLUTIONS TO THE NAVIER–STOKES EQUATIONS

For irrotational flows, we can define velocity fields in terms of the gradient of a *velocity potential*. This leads to a simple approach (Fig. 7.1) for satisfying conservation of mass and conservation of momentum, but problems when we try to satisfy boundary conditions.

We can satisfy conservation of mass by solving the Laplace equation given the boundary or freestream conditions, which is numerically and analytically easy compared with solving the Navier–Stokes equations. The boundary condition we use at walls is the no-penetration condition, i.e., that the velocity normal to a solid boundary is zero. In the freestream or at inlets, the boundary condition is a specified velocity. Solution of the Laplace equation specifies the velocity field everywhere in the flow, including at the boundary. Because the velocity field is then known, the conservation of momentum equations (Navier–Stokes equations) can be satisfied by rearranging the Navier–Stokes equations to solve for the pressure as a function of space, though the pressure is of little use in microscale irrotational flows and is rarely calculated.

Unfortunately, although this solution satisfies the no-penetration condition at the walls, it fails in general to satisfy the no-slip condition at the walls. The Laplace equation solution (combined with the appropriate p field) thus satisfies mass conservation, the

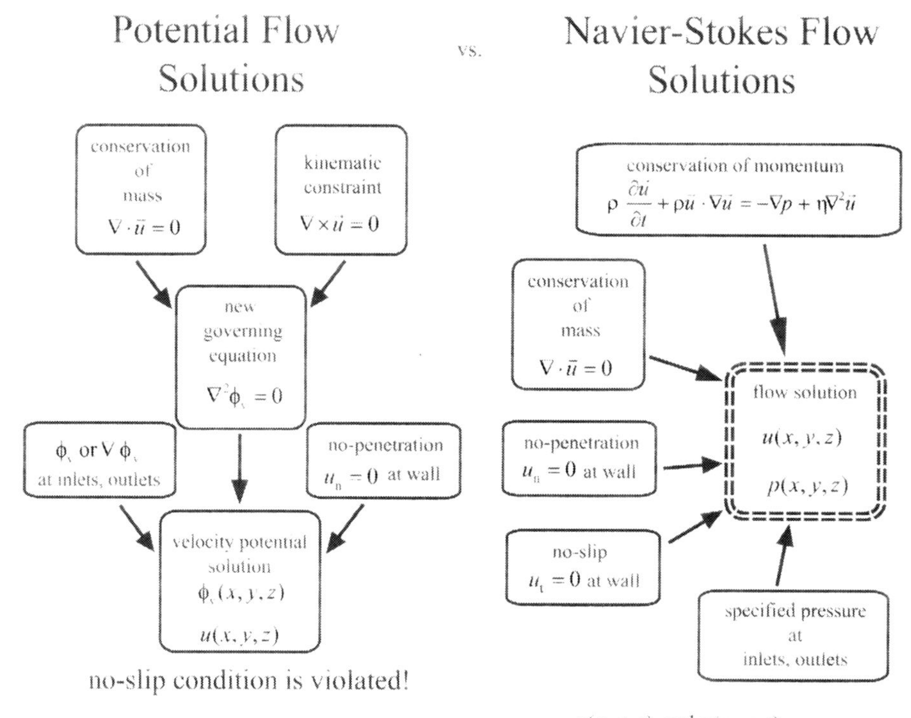

Fig. 7.1 Approaches for solving fluid flows with (a) potential flow compared with (b) Navier–Stokes solutions. The Navier–Stokes approach is primarily a mass and momentum conservation approach, whereas potential flow is primarily a kinematic approach – for potential flow, conservation of momentum is used only after the Laplace equation is solved, and only to determine the pressure field.

Navier–Stokes equations and *one* of the boundary conditions (the no-penetration condition), but it fails to satisfy the other boundary condition (the no-slip condition). Potential flow solutions are therefore useful for describing flow reasonably far from walls, and these solutions are relevant when the regions of vorticity are localized to a region we are willing to ignore or treat separately. Chapter 6 illustrates an important example of such flows, because the vorticity is contained inside a thin EDL.

7.2 LAPLACE EQUATION FOR VELOCITY POTENTIAL AND STREAM FUNCTION

This section illustrates how solutions of the Laplace equation can solve the fluid flow equations by use of the velocity potential or the stream function.

7.2.1 Laplace equation for the velocity potential

If the region of fluid flow we are considering is irrotational and therefore the velocity can be defined in terms of the gradient of a scalar,[1]

$$\vec{u} = \nabla \phi_v \,, \tag{7.3}$$

[1] Many parameters are defined as gradients of a scalar. One example is the velocity potential:

$$\vec{u} = \nabla \phi_v \,. \tag{7.1}$$

Another example is the electrical potential:

$$\vec{E} = -\nabla \phi \,. \tag{7.2}$$

The sign conventions in these two cases are opposite.

then conservation of mass is satisfied when Eq. (7.3) is inserted into the continuity equation ($\nabla \cdot \vec{u} = 0$), giving

$$\nabla \cdot \nabla \phi_v = 0 \tag{7.4}$$

or

Laplace equation for velocity potential

$$\nabla^2 \phi_v = 0 . \tag{7.5}$$

Equation (7.5) is the Laplace equation for the velocity potential.

LAPLACE EQUATION: BOUNDARY CONDITIONS AT WALLS

The fluid boundary condition used at a wall is the no-penetration condition, i.e., $u_n = 0$ at the wall, where u_n is the velocity normal to the wall. Cast in terms of the velocity potential (ϕ_v) and the coordinate normal to the wall (n), we obtain

$$\frac{\partial \phi_v}{\partial n} = 0 . \tag{7.6}$$

LAPLACE EQUATION: BOUNDARY CONDITIONS AT INLETS OR IN THE FREESTREAM

In the freestream or at inlets, the boundary condition is typically a specified velocity:

$$\nabla \phi_v|_{\text{inlet}} = \vec{u}_{\text{inlet}} . \tag{7.7}$$

Equations (7.5)–(7.7) constitute a complete specification of the problem for ϕ_v.

Alternatively, we can define the stream function ψ such that

definition of stream function for Cartesian coordinates, part I

$$u = \frac{\partial \psi}{\partial y} \tag{7.8}$$

and

definition of stream function for Cartesian coordinates, part II

$$v = -\frac{\partial \psi}{\partial x} . \tag{7.9}$$

This satisfies continuity automatically because $\frac{\partial u}{\partial x} + \frac{\partial v}{\partial y} = 0$ given this definition.

Now, to satisfy irrotationality, we set $\nabla \times \vec{u} = 0$, which in two dimensions means $\frac{\partial v}{\partial x} - \frac{\partial u}{\partial y} = 0$, so

$$\frac{\partial}{\partial x}\left(-\frac{\partial \psi}{\partial x}\right) - \frac{\partial}{\partial y}\frac{\partial \psi}{\partial y} = 0 \tag{7.10}$$

or

$$\frac{\partial^2}{\partial x^2}\psi + \frac{\partial^2}{\partial y^2}\psi = 0 . \tag{7.11}$$

Just as before, we can now solve the Laplace equation (this time for the stream function) and obtain a result that then leads directly to a velocity field.

The two approaches just described are equivalent. The velocity potential is more general and applies in both two and three dimensions. The stream function is typically used only in 2D flows, but it leads conveniently to streamlines (isocontours of the stream function), which are very useful.

SOLVING THE LAPLACE EQUATION

The Laplace equation is quite straightforward to solve numerically, and solutions can easily be achieved with differential equation solvers. The key limitation of a numerical approach is that, for 2D flows around objects, the solution of the Laplace equation is not unique, and an infinite number of solutions are possible, each with a different level of circulation.[2] Thus the circulation must be specified before the Laplace equation can be solved. In 2D aerodynamic theory (the most common fluid-mechanical application of 2D potential flow), the circulation is specified by enforcement of the Kutta condition, the viscous condition that a streamline must emanate from a sharp trailing edge. In microfluidic flows driven by electroosmosis, the flow is proportional to the electric field, which has no circulation, so electroosmotic potential flows have precisely zero circulation. Fortunately, the zero-circulation solution is the one typically found by numerical solvers, because their initial conditions are usually circulation-free.

Although analytical techniques work only for special geometries, they lead to powerful results that are furthermore analogous to many electric field solutions of engineering significance. Superposition of solutions can be used because the Laplace equation is a linear equation. We can superpose a set of solutions to satisfy specific boundary conditions or circulation constraints. These solutions (sources, doublets, vortices, and uniform flows) are discussed in this chapter. We focus on plane 2D flows, because the plane-symmetric case leads to an elegant means for writing the solutions and the efficient use of complex algebra.

7.2.2 No-slip condition

Unfortunately, although the solution to the Laplace equation for ϕ_v or ψ gives a velocity field that satisfies the Navier–Stokes equations and the no-penetration condition at the walls, it fails to satisfy the no-slip condition at the walls. This solution is thus *not* a correct description of the flow field at the wall. A boundary-layer theory must be added to a potential flow solution if the flow near the wall is to be specified; the boundary-layer theory delineates the spatial range of applicability of the potential flow solution. For electroosmotic flows, this boundary layer is the electrical double layer.

7.3 POTENTIAL FLOWS WITH PLANE SYMMETRY

Potential flow with plane symmetry has a number of applications, including airfoil theory and prediction of flow in shallow microfluidic devices. In both cases, the variations in one dimension (e.g., the depth of a microchannel) are assumed minor compared with the variations in the other direction. When this is the case, the flow may be approximated as 2D. For 2D irrotational flows, the velocity potential and stream function are orthogonal harmonic functions – isopotential contours are orthogonal to streamlines, and these two functions can be combined and manipulated by use of complex algebra.

[2] The circulation Γ around a closed contour is defined as $\Gamma = \int_C \vec{u} \cdot \hat{t}\, ds = \int_S \vec{\omega} \cdot \hat{n}\, dA$ and defines the total net solid-body rotation within an enclosed area. See Subsection 1.2.1.

Table 7.1 Analogies between the role of complex algebra in manipulating sinusoidal functions and the role of complex algebra in treating plane-symmetric potential flow.

Parameter	Sinusoidal functions	Plane-symmetric potential flow
Key functions	$\sin t$, $\cos t$	$\phi_v(x, y)$, $\psi(x, y)$
Relation between functions	$\sin t = \cos(t - \pi/2)$; functions are 90° out of phase	Isocontours of ϕ_v and ψ are spatially orthogonal to each other – rotated 90° in the xy plane.
Equations the functions satisfy	$f'' + f = 0$; second-order homogeneous ODE	$\nabla^2 f = 0$; second-order homogeneous PDE
Role of complex algebra in simplifying the functions	$\exp j\omega t = \cos t + j \sin t$; function not simplified	$\underset{\sim}{\phi_v} = \phi_v + j\psi$; two functions combined into one
Role of complex algebra in simplifying derivatives	$\frac{\partial}{\partial t}(\exp j\omega t) = j\omega \exp j\omega t$; differential equation converted to algebraic equation	derivatives not simplified
Role of complex algebra in simplifying solutions	Solutions to the equations are *always* of the form $A \exp j(\omega t + \alpha)$	Solutions to the equations are *always* of the form $\underset{\sim}{\phi_v} = f(\underset{\sim}{z}$ only)

In this section, we create flow solutions from basic elements, such as uniform flows, vortexes, sources, sinks, and doublets. Sources, sinks, and doublets are components of a multipolar expansion of the flow solution, whose general solution is discussed in Appendix F. First, we discuss how complex algebra is used for bookkeeping, then we discuss basic elements, and finally, we discuss how these elements can be superposed. This is a Green's function approach to the solution of the governing equation.

7.3.1 Complex algebra and its use in plane-symmetric potential flow

We use complex algebra in this subsection to simplify the mathematics for plane-symmetric potential flows. The algebra of complex variables is convenient for describing velocity potential and stream function, for reasons analogous to those for circuit analysis with sinusoidal functions, as summarized in Table 7.1.

Complex distance. Given two points in the xy plane separated by an x distance $\Delta x = x_2 - x_1$ and a y distance $\Delta y = y_2 - y_1$, we can define a *complex distance* $\underset{\sim}{z}$:

definition of complex distance in Cartesian coordinates

$$\underset{\sim}{z} = \Delta x + j \Delta y , \tag{7.12}$$

where $j = \sqrt{-1}$. This complex distance is a complex number that contains both x and y distance information. The undertilde is used to denote a complex representation of two real, physical quantities. This distance can also be written in terms of one length and one angle, in a manner analogous to cylindrical coordinates:

definition of complex distance in cylindrical coordinates

$$\underset{\sim}{z} = \Delta r \exp j \Delta\theta , \tag{7.13}$$

where $\Delta r = \sqrt{\Delta x^2 + \Delta y^2}$ and $\Delta\theta = \text{atan2}(\Delta y, \Delta x)$.[3] Here, Δr is the distance between the points, and $\Delta\theta$ is the angle the line from point 1 to point 2 makes with the x axis. The value of atan2 as used in this text[4] is understood to fall between 0 and 2π, $\cos\,\Delta\theta$ is equal to $\frac{\Delta x}{\sqrt{\Delta x^2+\Delta y^2}}$, and $\sin\,\Delta\theta$ is equal to $\frac{\Delta y}{\sqrt{\Delta x^2+\Delta y^2}}$. Also, recall Euler's formula ($\exp j\alpha = \cos\alpha + j\sin\alpha$).

Rotation. An immediate example of the utility of complex descriptions of distances in the xy plane is the ease with which a distance can be rotated. We can rotate a complex distance $\underset{\sim}{z}$ by an angle α by multiplying it by $\exp[j\alpha]$. To rotate a flow by the angle α, we replace $\underset{\sim}{z}$ with $\underset{\sim}{z}\exp[-j\alpha]$ in the formulas for $\underset{\sim}{\phi_v}$.

EXAMPLE PROBLEM 7.1

Use complex algebra to show that a vector in the x direction can be rotated into the y direction by rotating it by an angle $\pi/2$. Specifically, show that the complex distance $\underset{\sim}{z} = 1 + 0j$, when multiplied by $\exp(j\pi/2)$, is equal to $0 + 1j$.

SOLUTION: Start by multiplying $\underset{\sim}{z}$ by $\exp(j\pi/2)$ to get

$$(1 + 0j)\exp\left(j\frac{\pi}{2}\right), \tag{7.15}$$

and multiply out to get

$$\exp\left(j\frac{\pi}{2}\right). \tag{7.16}$$

We then use Euler's formula to write the result as

$$\cos\frac{\pi}{2} + j\sin\frac{\pi}{2}, \tag{7.17}$$

from which we obtain

$$0 + 1j. \tag{7.18}$$

Complex velocity potential. We define also a *complex velocity potential* $\underset{\sim}{\phi_v}$:

definition of complex velocity potential

$$\underset{\sim}{\phi_v} = \phi_v + j\psi, \tag{7.19}$$

[3] The symbols Δr and $\Delta\theta$ denote radial and angular coordinates of a *distance*. They are different from r and θ, which denote radial and angular coordinates of a point with respect to the origin.

[4] Here atan2 is the four-quadrant inverse tangent, whose value depends on the signs of Δy and Δx, as well as the value of $\Delta y/\Delta x$. Specifically, if $\tan^{-1}$ results in a value between $-\pi/2$ and $\pi/2$, atan2 is given by

$$\text{atan2}(\Delta y, \Delta x) = \begin{cases} \tan^{-1}(\Delta y/\Delta x) & \text{if } \Delta x > 0, \Delta y > 0 \\ \tan^{-1}(\Delta y/\Delta x) + \pi & \text{if } \Delta x < 0 \\ \tan^{-1}(\Delta y/\Delta x) + 2\pi & \text{if } \Delta x > 0, \Delta y < 0 \\ \pi/2 & \text{if } \Delta x = 0, \Delta y > 0 \\ 3\pi/2 & \text{if } \Delta x = 0, \Delta y < 0 \end{cases} \tag{7.14}$$

where ϕ_v is the velocity potential and ψ is the stream function. This combines the velocity potential and the stream function into one complex function. Just as the x direction and y direction are orthogonal and conveniently represented with a complex variable $\underset{\sim}{z}$, ϕ_v and ψ are orthogonal and conveniently represented with a complex value $\underset{\sim}{\phi_v}$.

Complex velocity. Define also a *complex velocity* $\underset{\sim}{u}$:

definition of complex velocity in Cartesian coordinates

$$\underset{\sim}{u} = u + jv \,, \tag{7.20}$$

where u and v are the velocity components in the x and y directions.[5]

When a flow is centered on the origin, it is useful to describe flow in terms of radial ($u_{\imath}$) and counterclockwise circumferential (u_θ) velocities. In this case, we can equivalently define the complex velocity as

definition of complex velocity in cylindrical coordinates

$$\underset{\sim}{u} = (u_{\imath} + ju_\theta)\exp j\theta \,. \tag{7.21}$$

The velocity components $u_{\imath}$ and u_θ are defined with respect to the origin, as are the distances $\imath$ and θ. The distances $\Delta\imath$ and $\Delta\theta$, in contrast, are measured with respect to a point of interest.

Expression relating complex velocity, distance, and velocity potential. The expression relating the complex velocity, complex potential, and complex distance is analogous to the relation $\vec{u} = \nabla\phi_v$:

relation between complex potential and complex velocity

$$\underset{\sim}{u}^* = u - jv = \frac{\partial \underset{\sim}{\phi_v}}{\partial \underset{\sim}{z}} \,, \tag{7.22}$$

where the asterisk denotes the complex conjugate. The derivative with respect to $\underset{\sim}{z}$ can also be written as

$$\frac{\partial \underset{\sim}{\phi_v}}{\partial \underset{\sim}{z}} = \frac{\partial \underset{\sim}{\phi_v}}{\partial x} = -j\frac{\partial \underset{\sim}{\phi_v}}{\partial y} \,. \tag{7.23}$$

See Appendix G for more detailed information about complex differentiation.

Solutions to the Laplace equation. The use of complex variables has an even more elegant result – *any well-behaved (i.e., suitably differentiable) complex velocity potential* $\underset{\sim}{\phi_v} =$

[5] Many authors define the complex velocity as $u - jv$, not $u + jv$. We use $u + jv$ to keep a definition analogous to the complex velocity potential and complex distance. This definition requires that Eq. (7.22) use the complex conjugate of the complex velocity.

$\underset{\sim}{\phi_v}(\underset{\sim}{z})$ *that is a function of* $\underset{\sim}{z}$ *alone automatically satisfies the Laplace equation and is thus a solution of the fluid-mechanical equations.* This mathematical concept is discussed in Subsection G.1.2, but is stated here as well. *Any* $\underset{\sim}{\phi_v}(\underset{\sim}{z})$ satisfies the Laplace equation, so $\underset{\sim}{\phi_v} = \underset{\sim}{z}$, $\underset{\sim}{\phi_v} = \underset{\sim}{z}^3$, $\underset{\sim}{\phi_v} = \ln \underset{\sim}{z}$, and $\underset{\sim}{\phi_v} = \sqrt[14]{\underset{\sim}{z}}$ all satisfy the Laplace equation where they are differentiable. This *is not* true if we specify any $\phi_v(x, y)$ – for example, $\phi_v = x^2$, $\phi_v = y^{-1}$, $\phi_v = x^3 + y^3$, and $\phi_v = yx \ln x$ all fail to satisfy the Laplace equation. The reason the complex variable technique works is because the use of a differentiable function of a complex variable enforces a specific spatial relation between the derivatives with respect to x and y.

Thus a Green's function-type approach (i.e., first finding solutions to the governing equations, then combining them to solve boundary conditions) becomes simple – any complex velocity potential $\underset{\sim}{\phi_v}(\underset{\sim}{z})$ is already a solution, so all that is required is to combine functions to match the intended boundary conditions, rather than to solve the Laplace equation directly. Because the Laplace equation is linear and homogeneous, we can superpose solutions at will to match boundary conditions. Some of these solutions are described in the following subsections.

7.3.2 Monopolar flow: plane-symmetric (line) source with volume outflow per unit depth Λ

A plane-symmetric source with volume outflow per unit depth Λ located at the origin is denoted by a velocity potential of $\phi_v = \frac{\Lambda}{2\pi} \ln \imath$ and a stream function of $\psi = \frac{\Lambda}{2\pi}\theta$ and leads to an exclusively radial velocity of $u_\imath = \frac{\Lambda}{2\pi \imath}$. A source is a *monopole*, i.e., a point singularity that induces a flow in the field around it (see Subsection F.1.2).

More generally, the complex velocity potential,

complex velocity potential for flow induced by a point source of strength Λ

$$\underset{\sim}{\phi_v} = \frac{\Lambda}{2\pi} \ln \underset{\sim}{z}, \tag{7.24}$$

corresponds to the flow induced by a point source of fluid, where Λ is the *source strength*[6] or volume outflow rate per unit depth [m²/s] and $\underset{\sim}{z}$ is the complex distance measured from the location of the source. After complex differentiation of $\underset{\sim}{\phi_v}$, we examine the real part to find

x velocity induced by a point source of strength Λ

$$u = \frac{\Lambda}{2\pi} \frac{\Delta x}{(\Delta x^2 + \Delta y^2)} \tag{7.25}$$

and examine the imaginary part to find

y velocity induced by a point source of strength Λ

$$v = \frac{\Lambda}{2\pi} \frac{\Delta y}{(\Delta x^2 + \Delta y^2)} . \tag{7.26}$$

[6] A source with a negative Λ is referred to as a *sink*.

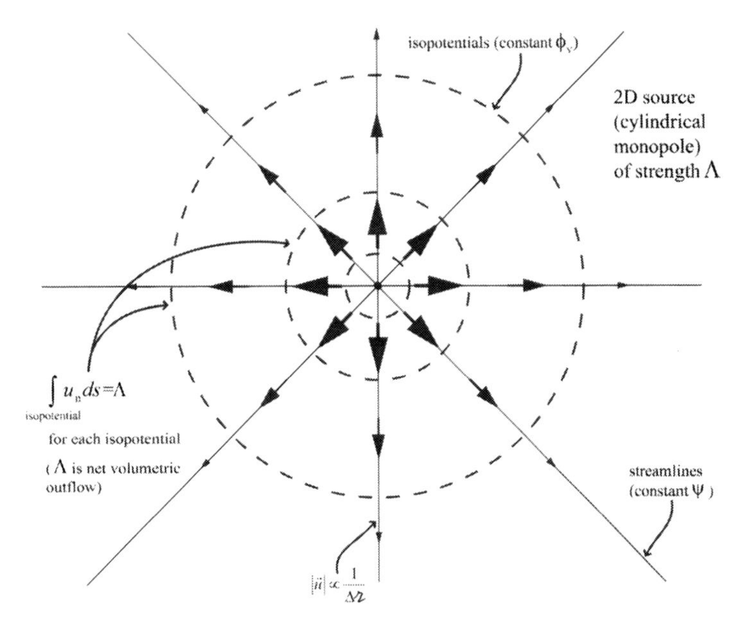

Fig. 7.2 Isopotentials, streamlines, velocity magnitudes, and net volumetric outflow for a 2D point source with strength Λ.

This corresponds to a flow radiating outward from the location of the source. The velocity magnitude scales inversely with the distance from the source (as is the case for all monopolar flows) and is equivalent in all directions. The velocity potential is

$$\phi_V = \frac{\Lambda}{2\pi} \ln(\Delta r) = \frac{\Lambda}{2\pi} \ln\sqrt{\Delta x^2 + \Delta y^2}\,, \tag{7.27}$$

and the stream function is

$$\psi = \frac{\Lambda}{2\pi} \Delta\theta = \frac{\Lambda}{2\pi} \text{atan2}(\Delta y, \Delta x)\,. \tag{7.28}$$

Thus the the isopotentials are concentric circles around the source, and the streamlines are lines radiating out from the source (Fig. 7.2).

EXAMPLE PROBLEM 7.2

Evaluate the velocity components of a plane-symmetric source by complex differentiation of $\underset{\sim}{\phi_V} = \frac{\Lambda}{2\pi} \ln \underset{\sim}{z}$.

SOLUTION: The source is straightforward to characterize by differentiation of $\underset{\sim}{\phi_V}$. By definition,

$$\underset{\sim}{u}^* = \frac{\partial \underset{\sim}{\phi_V}}{\partial \underset{\sim}{z}}\,. \tag{7.29}$$

We differentiate $\underset{\sim}{\phi_V} = \frac{\Lambda}{2\pi} \ln \underset{\sim}{z}$ to find

$$\underset{\sim}{u}^* = \frac{\Lambda}{2\pi \underset{\sim}{z}}\,. \tag{7.30}$$

We then multiply by $\underset{\sim}{z}^*$ to make the denominator real:

$$\underset{\sim}{u}^* = \frac{\Lambda \underset{\sim}{z}^*}{2\pi \underset{\sim}{z}\underset{\sim}{z}^*} = \frac{\Lambda \underset{\sim}{z}^*}{2\pi \left|\underset{\sim}{z}\right|^2}, \tag{7.31}$$

and take the complex conjugate of both sides:

$$\underset{\sim}{u} = \frac{\Lambda \underset{\sim}{z}}{2\pi \left|\underset{\sim}{z}\right|^2}. \tag{7.32}$$

Examining the real part, we find

$$u = \frac{\Lambda}{2\pi} \frac{\Delta x}{(\Delta x^2 + \Delta y^2)}, \tag{7.33}$$

and, examining the imaginary part, we find

$$v = \frac{\Lambda}{2\pi} \frac{\Delta y}{(\Delta x^2 + \Delta y^2)}. \tag{7.34}$$

Plane-symmetric source located at the origin. For the special case of a source at the origin, $\Delta x = x$, $\Delta y = y$, $\Delta r = r$, and $\Delta\theta = \theta$. Thus $\left|\underset{\sim}{z}\right|^2 = r^2$ and $\underset{\sim}{z} = r \exp j\theta$. Given that

$$\underset{\sim}{u} = \frac{\Lambda \underset{\sim}{z}}{2\pi \left|\underset{\sim}{z}\right|^2}, \tag{7.35}$$

we can write

$$\underset{\sim}{u} = \frac{\Lambda}{2\pi r^2} r \exp j\theta = \frac{\Lambda}{2\pi r} \exp j\theta, \tag{7.36}$$

And from $\underset{\sim}{u} = (u_r + j u_\theta) \exp j\theta$, we find

radial velocity induced by a point source of strength Λ

$$u_r = \frac{\partial \phi_v}{\partial r} = \frac{1}{r}\frac{\partial \psi}{\partial \theta} = \frac{\Lambda}{2\pi r} \tag{7.37}$$

and

circumferential velocity induced by a point source of strength Λ

$$u_\theta = -\frac{\partial \psi}{\partial r} = \frac{1}{r}\frac{\partial \phi_v}{\partial \theta} = 0. \tag{7.38}$$

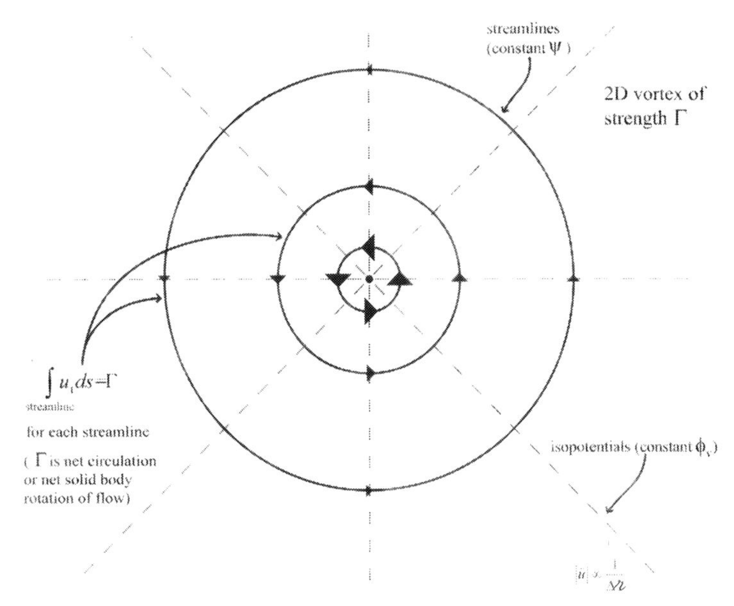

Fig. 7.3 Isopotentials, streamlines, velocity magnitudes, and net solid-body rotation for a 2D point vortex with strength Γ.

7.3.3 Plane-symmetric vortex with counterclockwise circulation per unit depth Γ

A plane-symmetric vortex with counterclockwise circulation per unit depth Γ located at the origin is denoted by a velocity potential of $\phi_v = \frac{\Gamma}{2\pi}\theta$ and a stream function of $\psi = -\frac{\Gamma}{2\pi} \ln \imath$ and leads to an exclusively circumferential velocity of $u_\theta = \frac{\Gamma}{2\pi \imath}$.

More generally, the complex velocity potential,

complex velocity potential induced by a point vortex of strength Γ

$$\underset{\sim}{\phi_v} = -j\frac{\Gamma}{2\pi} \ln \underset{\sim}{z}, \tag{7.39}$$

corresponds to the flow induced by a plane-symmetric point vortex, where Γ is the *vortex strength* or circulation per unit depth [m²/s] and $\underset{\sim}{z}$ is measured from the location of the vortex. The 2D vortex (Fig. 7.3) is given by an expression analogous to that of the complex potential for a 2D source, but multiplied by $1/j$. In this sense, *a vortex is a monopole with an imaginary strength.*

The flow induced by a plane-symmetric vortex can be determined by differentiation of $\underset{\sim}{\phi_v}$ followed by an examination of the real and imaginary parts of the result, leading to

x velocity induced by a point vortex of strength Γ

$$u = \frac{\Gamma}{2\pi} \frac{-\Delta y}{(\Delta x^2 + \Delta y^2)} \tag{7.40}$$

and

y velocity induced by a point vortex of strength Γ

$$v = \frac{\Gamma}{2\pi} \frac{\Delta x}{(\Delta x^2 + \Delta y^2)} . \tag{7.41}$$

This corresponds to a flow rotating counterclockwise (for positive Γ) around the location of the vortex. The velocity magnitude scales inversely with the distance from the vortex and is equivalent in all directions.

The velocity potential is given by

$$\phi_v = \frac{\Gamma}{2\pi} \Delta\theta = \frac{\Gamma}{2\pi} \text{atan2}(\Delta y, \Delta x) , \tag{7.42}$$

and the stream function is given by

$$\psi = -\frac{\Gamma}{2\pi} \ln(\Delta r) = -\frac{\Gamma}{2\pi} \ln \sqrt{\Delta x^2 + \Delta y^2} . \tag{7.43}$$

From this, we see the structure of isopotentials (lines radiating from the vortex) and streamlines (concentric circles around the vortex).

Plane-symmetric vortex located at the origin. For the case of a vortex at the origin, $\Delta x = x$, $\Delta y = y$, $\Delta r = r$, and $\Delta\theta = \theta$. Thus $|\underset{\sim}{z}|^2 = r^2$ and $\underset{\sim}{z} = r \exp j\theta$. Given that

$$\underset{\sim}{u} = j \frac{\Gamma \underset{\sim}{z}}{2\pi |\underset{\sim}{z}|^2} , \tag{7.44}$$

we can write

$$\underset{\sim}{u} = j \frac{\Gamma}{2\pi r^2} r \exp j\theta = j \frac{\Gamma}{2\pi r} \exp j\theta , \tag{7.45}$$

And from $\underset{\sim}{u} = (u_r + j u_\theta) \exp j\theta$,

radial velocity induced by a point vortex of strength Γ

$$u_r = \frac{\partial \phi_v}{\partial r} = \frac{1}{r} \frac{\partial \psi}{\partial \theta} = 0 \tag{7.46}$$

and

circumferential velocity induced by a point vortex of strength Γ

$$u_\theta = -\frac{\partial \psi}{\partial r} = \frac{1}{r} \frac{\partial \phi_v}{\partial \theta} = \frac{\Gamma}{2\pi r} . \tag{7.47}$$

EXAMPLE PROBLEM 7.3

Evaluate the velocity induced by a plane-symmetric vortex by evaluating the stream function or velocity potential directly.

SOLUTION: The complex velocity potential is

$$\underset{\sim}{\phi}_{\mathrm{v}} = -j\frac{\Gamma}{2\pi}\ln \underset{\approx}{z} = -j\frac{\Gamma}{2\pi}\ln\left(\Delta r \exp j\Delta\theta\right), \tag{7.48}$$

$$\underset{\sim}{\phi}_{\mathrm{v}} = -j\frac{\Gamma}{2\pi}\ln\left(\Delta r\right) - j\frac{\Gamma}{2\pi}\ln\left(\exp j\Delta\theta\right), \tag{7.49}$$

$$\underset{\sim}{\phi}_{\mathrm{v}} = \frac{\Gamma}{2\pi}\Delta\theta - j\frac{\Gamma}{2\pi}\ln\left(\Delta r\right). \tag{7.50}$$

Thus, for the velocity potential, we get

$$\phi_{\mathrm{v}} = \frac{\Gamma}{2\pi}\Delta\theta = \frac{\Gamma}{2\pi}\mathrm{atan2}\left(\Delta y, \Delta x\right), \tag{7.51}$$

and for the stream function we get

$$\psi = -\frac{\Gamma}{2\pi}\ln\left(\Delta r\right) = -\frac{\Gamma}{2\pi}\ln\sqrt{\Delta x^2 + \Delta y^2}. \tag{7.52}$$

From this, we can see that

$$u = \frac{\partial\phi_{\mathrm{v}}}{\partial x} = \frac{\partial\psi}{\partial y} = -\frac{\Gamma}{2\pi}\frac{\Delta y}{\left(\Delta x^2 + \Delta y^2\right)}, \tag{7.53}$$

$$v = \frac{\partial\phi_{\mathrm{v}}}{\partial y} = -\frac{\partial\psi}{\partial x} = \frac{\Gamma}{2\pi}\frac{\Delta x}{\left(\Delta x^2 + \Delta y^2\right)}. \tag{7.54}$$

7.3.4 Dipolar flow: plane-symmetric doublet with dipole moment κ

A plane-symmetric doublet is mathematically equivalent to a source and a sink (both of equal and infinite strength) separated by an infinitesimal distance.[7] We define the vector dipole moment of the doublet in the direction from the source to the sink. The doublet is a *dipole*, which is the mathematical limit of a combination of two oppositely signed, infinitely strong monopoles separated by an infinitesimal distance (see Subsection F.1.2).

A plane-symmetric doublet with *doublet strength* or *dipole moment* κ located at the origin and aligned along the x axis is denoted by a velocity potential of $\phi_{\mathrm{v}} = \frac{\kappa x}{2\pi(x^2+y^2)}$ and a stream function of $\psi = -\frac{\kappa y}{2\pi(x^2+y^2)}$. It leads to a radial velocity of $u_r = -\frac{\kappa}{2\pi r^2}\cos\theta$ and a circumferential velocity of $u_\theta = -\frac{\kappa}{2\pi r^2}\sin\theta$.

[7] The doublet can also be created mathematically by two vortexes of opposite strength infinitesimally far apart from each other and aligned 90° away from the dipole moment, i.e., two monopoles of imaginary magnitude separated by an imaginary distance.

The general complex velocity potential for a plane-symmetric doublet is given by

complex velocity potential induced by a point doublet of strength κ

$$\underset{\sim}{\phi_{\mathrm{v}}} = \frac{\kappa}{2\pi \underset{\sim}{z} \exp(-j\alpha)} = \frac{\kappa}{2\pi \underset{\sim}{z}} \exp(j\alpha), \tag{7.55}$$

where κ is the dipole moment per unit depth [m^3/s]. Here $\underset{\sim}{z}$ is measured from the location of the source and α indicates the angle of the vector dipole moment measured counterclockwise with respect to the positive x axis. The dipole moment thus has components of $\kappa \cos\alpha$ and $\kappa \sin\alpha$. We characterize the flow induced by a plane-symmetric doublet by differentiation of $\underset{\sim}{\phi_{\mathrm{v}}}$. The real component of the result is

x velocity induced by a point doublet of strength κ

$$u = \frac{-\kappa}{2\pi\,(\Delta x^2 + \Delta y^2)} \left[\cos\alpha \left(\frac{\Delta x^2 - \Delta y^2}{\Delta x^2 + \Delta y^2}\right) + \sin\alpha \left(\frac{2\Delta x \Delta y}{\Delta x^2 + \Delta y^2}\right)\right], \tag{7.56}$$

and the imaginary part is

y velocity induced by a point doublet of strength κ

$$v = \frac{-\kappa}{2\pi(\Delta x^2 + \Delta y^2)} \left[-\sin\alpha \left(\frac{\Delta x^2 - \Delta y^2}{\Delta x^2 + \Delta y^2}\right) + \cos\alpha \left(\frac{2\Delta x \Delta y}{\Delta x^2 + \Delta y^2}\right)\right]. \tag{7.57}$$

This corresponds (for positive κ) to a flow emanating from a point just left of the doublet, circling around, and returning to a point just right of the doublet. The velocity magnitude scales inversely with the squared distance from the doublet (this is the case for all dipolar flows) and is equivalent in all directions.

The velocity potential for this flow is

$$\phi_{\mathrm{v}} = \frac{\kappa}{2\pi(\Delta x^2 + \Delta y^2)} (\Delta x \cos\alpha + \Delta y \sin\alpha)\,, \tag{7.58}$$

and the stream function is

$$\psi = \frac{-\kappa}{2\pi(\Delta x^2 + \Delta y^2)} (-\Delta x \sin\alpha + \Delta y \cos\alpha)\,. \tag{7.59}$$

The streamlines of the doublet flow are circles tangent to the axis of the doublet, and the isopotentials are circles tangent to the normal to the axis of the doublet (Fig. 7.4).

Plane-symmetric doublet located at the origin with dipole moment aligned along the positive x axis. For the case of a doublet at the origin with the dipole moment aligned with the positive x axis, $\Delta x = x$, $\Delta y = y$, $\Delta \imath = \imath$, $\Delta\theta = \theta$, and $\alpha = 0$. Thus $|\underset{\sim}{z}|^2 = \imath^2$ and $\underset{\sim}{z} = \imath \exp j\theta$. Given that

$$\underset{\sim}{u} = \frac{-\kappa}{2\pi \Delta \imath^2} \exp[j(-\alpha + 2\Delta\theta)]\,, \tag{7.60}$$

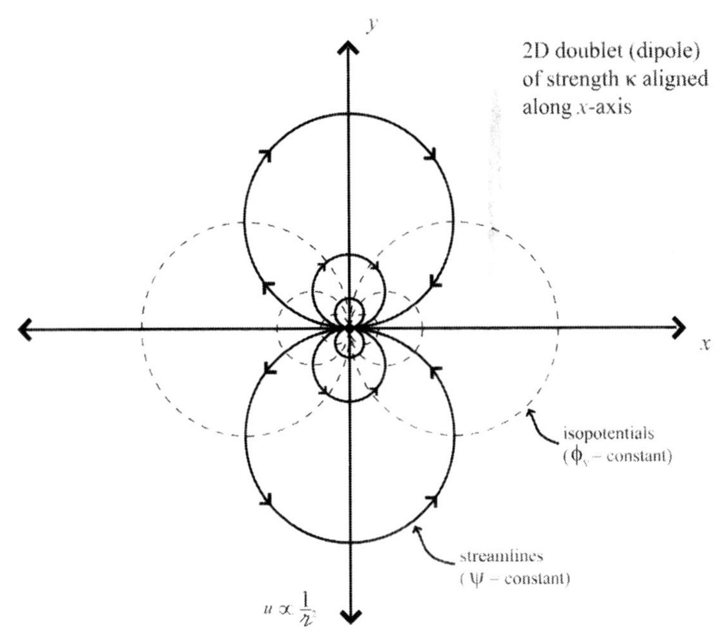

Fig. 7.4 Isopotentials, streamlines, and velocity magnitudes for a 2D doublet with strength κ.

we set $\alpha = 0$, $\Delta r = r$, and $\Delta\theta = \theta$:

$$\underset{\sim}{u} = \frac{-\kappa}{2\pi r^2} \exp j2\theta\,. \tag{7.61}$$

Because this flow is centered on the origin, cylindrical coordinates are convenient, and we look for velocity in the form of $\underset{\sim}{u} = (u_r + ju_\theta)\exp j\theta$:

$$\underset{\sim}{u} = \frac{-\kappa}{2\pi r^2}(\cos\theta + j\sin\theta)\exp j\theta\,, \tag{7.62}$$

from which we see that

radial velocity induced by a point doublet of strength κ

$$u_r = \frac{\partial\phi_v}{\partial r} = \frac{1}{r}\frac{\partial\psi}{\partial\theta} = \frac{-\kappa}{2\pi r^2}\cos\theta \tag{7.63}$$

and

circumferential velocity induced by a point doublet of strength κ

$$u_\theta = -\frac{\partial\psi}{\partial r} = \frac{1}{r}\frac{\partial\phi_v}{\partial\theta} = \frac{-\kappa}{2\pi r^2}\sin\theta\,. \tag{7.64}$$

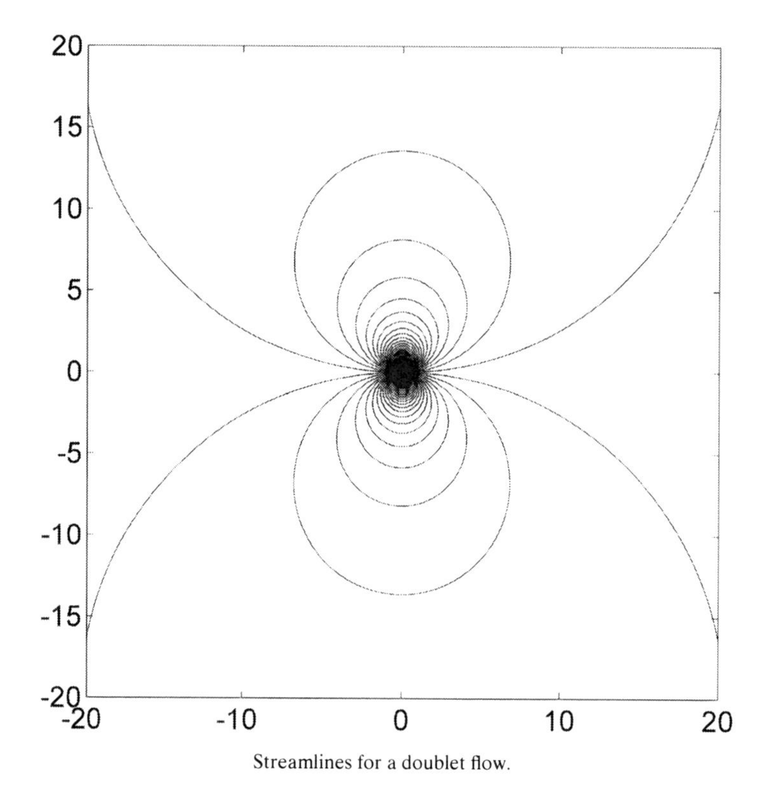

Fig. 7.5 Streamlines for a doublet flow.

EXAMPLE PROBLEM 7.4

Consider 2D potential flow. In the domain $-20 < x < 20$, $-20 < y < 20$, plot the streamlines for a doublet located at the origin with strength 50π and dipole moment aligned in the x direction.

SOLUTION: See Fig. 7.5.

7.3.5 Uniform flow with speed U

A uniform flow moving with speed U in the x direction is given simply by $\phi_v = Ux$ and $\psi = Uy$. Uniform flow is the degenerate case of the Legendre polynomial solution to Laplace's equation, where $A_0 = U$.

More generally, the complex velocity potential,

complex velocity potential corresponding to a uniform flow at velocity U aligned α with respect to the x axis

$$\underset{\sim}{\phi_v} = U\underset{\sim}{z}\exp[-j\alpha], \tag{7.65}$$

corresponds to a uniform flow at speed U in a direction rotated α counterclockwise from the positive x axis, i.e., a flow with x- and y-velocity components $U\cos\alpha$ and $U\sin\alpha$. This flow is by definition uniform in space, so here $\underset{\sim}{z}$ can be measured relative to any arbitrary point. Usually $\underset{\sim}{z}$ is measured from the origin.

Uniform Flow: velocity by complex differentiation. The uniform flow is easy to characterize by differentiation of $\underset{\sim}{\phi}_v$, from which we can see that

x velocity corresponding to a uniform flow at velocity U aligned α with respect to the x axis

$$u = U \cos \alpha \tag{7.66}$$

and

y velocity corresponding to a uniform flow at velocity U aligned α with respect to the x axis

$$v = U \sin \alpha \,. \tag{7.67}$$

The velocity potential is given by

$$\phi_v = U \left(\Delta x \cos \alpha + \Delta y \sin \alpha \right) , \tag{7.68}$$

and the stream function is given by

$$\psi = U \left(\Delta y \cos \alpha - \Delta x \sin \alpha \right) . \tag{7.69}$$

The streamlines for this flow are aligned at an angle α with respect to the x axis, and isopotentials are aligned at an angle α with respect to the y axis.

Uniform Flow: $\alpha = 0$. For the $\alpha = 0$ case (flow from left to right), we substitute in zero for α and find

velocity potential corresponding to a uniform flow at velocity U in the x direction

$$\phi_v = Ux \tag{7.70}$$

and

stream function corresponding to a uniform flow at velocity U in the x direction

$$\psi = Uy \,, \tag{7.71}$$

and for the velocities we find

$$u = U \tag{7.72}$$

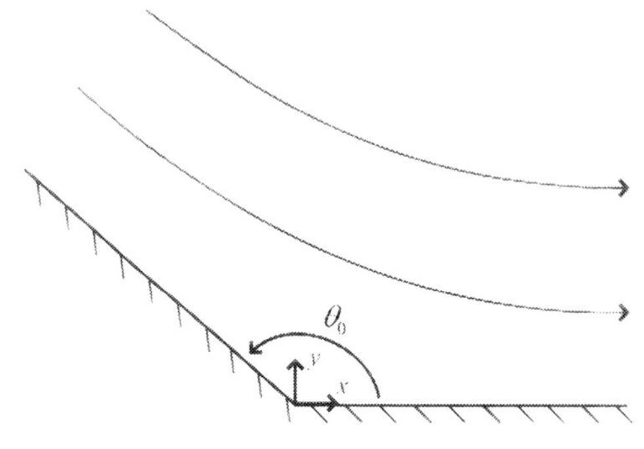

Fig. 7.6 Flow through a corner.

and

$$v = 0 . \tag{7.73}$$

7.3.6 Flow around a corner

Consider a geometry consisting of a wall starting at the origin and proceeding to $\imath \to \infty$ along a line at some angle $\theta = \theta_0$, as well as a second wall starting at the origin and proceeding to $\imath \to \infty$ along the x axis ($\theta = 0$). Consider flow that enters along the $\theta = \theta_0$ wall, runs along the surface, and exits along the x axis (see Fig. 7.6).

This flow is closely related to the uniform flow from the previous section; in fact, the previous section gives the solution for the case in which $\theta_0 = \pi$. In the domain for which $0 < \Delta\theta < \theta_0$, the solution for this flow is given by

$$\underset{\sim}{\phi_v} = A\underset{\sim}{z}^{\pi/\theta_0} , \tag{7.74}$$

where A is a constant.

The corner in this flow can be located arbitrarily (by measuring $\underset{\sim}{z}$ with respect to any point); further, this flow can also be rotated so that it applies to flow entering along a wall at $\Delta\theta = \theta_0 + \alpha$ and exiting along a wall at $\Delta\theta = \alpha$. In this general case, the complex velocity potential is given by

complex velocity potential corresponding to a flow over a corner of angle θ_0

$$\underset{\sim}{\phi_v} = A\left[\underset{\sim}{z} \exp(-j\alpha)\right]^{\pi/\theta_0} . \tag{7.75}$$

Here $\underset{\sim}{z}$ is measured relative to the location of the corner.

The velocity potential for this flow is given by

velocity potential corresponding to a flow over a corner of angle θ_0

$$\phi_v = A\Delta\imath^n \cos\left[n\left(\Delta\theta - \alpha\right)\right] , \tag{7.76}$$

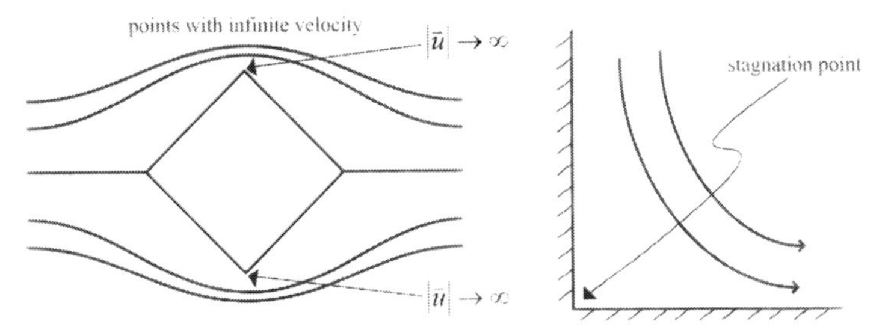

Fig. 7.7 Infinite velocities at sharp convex corners (left) and stagnation at sharp concave corners (right).

where $n = \pi/\theta_0$, and the stream function for this flow is given by

stream function corresponding to a flow over a corner of angle θ_0

$$\psi = A\Delta r^n \sin\left[n\left(\Delta\theta - \alpha\right)\right] . \tag{7.77}$$

The corner flow is commonly observed in any microdevice with a sharp corner. The Laplace equation solutions predict stagnation (zero velocity) at any sharp concave corner and infinite velocity at any sharp convex corner, as shown in Fig. 7.7. In any real system, of course, corners have finite curvature. Thus the velocities at concave corners of finite radius of curvature are low, but finite, and the velocities of convex corners of finite radius of curvature are high, but finite. If these corners had radii of curvature smaller than the Debye length, the use of the Laplace equation to describe the flow outside the boundary layer would be invalid.

7.3.7 Flow over a circular cylinder

A doublet with strength κ and a uniform flow with speed U, both aligned in the same direction, can be superposed to describe the potential flow over a circular cylinder with radius $a = \sqrt{\frac{\kappa}{2\pi U}}$:

$$\underset{\sim}{\phi_{\rm v}} = U\underset{\sim}{z}\exp[-j\alpha] + \frac{\kappa}{2\pi\underset{\sim}{z}}\exp[j\alpha] . \tag{7.78}$$

This flow is observed in shallow microdevices with circular obstacles.

7.3.8 Conformal mapping

For plane-symmetric potential flow, certain mapping functions make it easy to transform problems spatially in a manner that still satisfies the kinematic and mass conservation relations. Because all that is required for a plane-symmetric potential flow to satisfy the Laplace equations is that $\underset{\sim}{\phi_{\rm v}}$ be a function of $\underset{\sim}{z}$ only, any of a variety of transforms can be useful. Rotation in the plane (Subsection 7.3.1) is the simplest transform, and it orients flows with respect to the coordinate axes. Perhaps the most well known of these mapping functions is the Joukowski transform (see Section G.5), which maps a family of conic sections. It is particularly useful when mapping results for circular objects to predict flow over elliptical or linear obstacles. Finally, the Schwarz–Christoffel transform maps the upper half of the complex plane onto arbitrary polygonal shapes and thus facilitates calculation of plane-symmetric potential flow through arbitrary polygonal channels. Given the relative ease of numerical solutions of the Laplace equation, the use of these more advanced transforms by practicing engineers has become relatively uncommon; when the transforms are applicable, though, they can greatly simplify analysis.

7.4 POTENTIAL FLOW IN AXISYMMETRIC SYSTEMS IN SPHERICAL COORDINATES

Axisymmetric systems in spherical coordinates are distinct from plane potential flows in that (a) isopotential contours and streamlines are not generally orthogonal and (b) the use of complex mathematics does not facilitate calculations. For axisymmetric flows, axisymmetric multipolar solutions (described in Appendix F) can be used.

Axisymmetric flows are described by use of the *Stokes stream function*; it is defined by

$$u_r = \frac{1}{r^2 \sin\vartheta} \frac{\partial \psi}{\partial \vartheta} \tag{7.79}$$

and

$$u_\vartheta = -\frac{1}{r \sin\vartheta} \frac{\partial \psi}{\partial r} . \tag{7.80}$$

The velocity potential is given by

$$u_r = \frac{\partial \phi_v}{\partial r} \tag{7.81}$$

and

$$u_\vartheta = \frac{1}{r} \frac{\partial \phi_v}{\partial \vartheta} . \tag{7.82}$$

Axisymmetric potential flows include uniform flows as well as multipolar solutions such as sinks, sources, and doublets. Combinations of these describe flow over *Rankine solids*, such as spheres and ellipsoids. For example, the potential flow over a sphere is given by the sum of a uniform flow (which has a velocity potential of $\phi_v = Ur\cos\vartheta$) with a doublet of strength $\frac{1}{2}Ua^3$ (which has a velocity potential of $\frac{1}{2}\frac{Ua^3}{r^2}\cos\vartheta$) to obtain

$$\phi_v = Ur\left(1 + \frac{1}{2}\frac{a^3}{r^3}\right)\cos\vartheta . \tag{7.83}$$

One immediate conclusion from this solution is that the peak velocity of potential flow around a sphere is $\frac{3}{2}$ that of the freestream. The same, of course, can be said of the magnitude of the electric field around a sphere. Equation (7.83) is important when we are considering the relative motion (electrophoresis) of charged spheres.

For complicated geometries, we typically solve the Laplace equation numerically. Fortunately, the Laplace equation is a well-behaved, elliptical equation, and these numerical simulations are straightforward.

Three-dimensional potential flow is physically identical to 2D potential flow, and the general nature of the flow behavior is identical. Although the solution is still a solution of the Laplace equation, the 3D nature of the solution changes the relevant solution techniques as well as the interpretation of streamlines and potential isocontours.

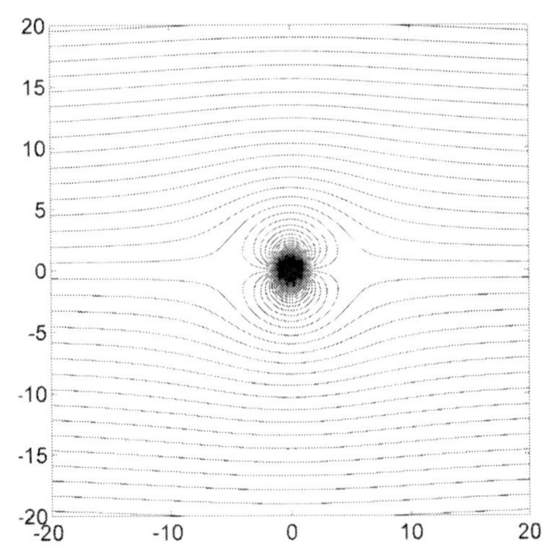

Fig. 7.8 Streamlines for potential flow around a circular cylinder.

EXAMPLE PROBLEM 7.5

Consider 2D potential flow. In the domain $-20 < x < 20$, $-20 < y < 20$, plot the streamlines for a uniform flow in the x direction flowing over a cylinder of radius 5 located at the origin.

SOLUTION: See Fig. 7.8.

7.5 SUMMARY

This chapter discusses the physical rationale for using irrotational flow techniques to describe electroosmotic flows and describes analytical tools for superposing the Green's function solutions to the Laplace equation,

$$\nabla^2 \phi_v = 0\,, \tag{7.84}$$

to identify irrotational flow solutions. Because the vorticity in a purely electroosmotic flow is found only in the EDL, irrotational flow techniques apply for all regions outside the EDL. These solutions, such as the point source, vortex, doublet, and uniform flow, are found by separation of variables on the Laplace equation, as described in Appendix F. Simple analytical solutions for important cases (for example, the potential flow around a sphere) inform study of complex systems of interacting particles.

7.6 SUPPLEMENTARY READING

Potential flow in fluids is a classical subject, and general fluid mechanics texts such as Panton [21], Batchelor [24], Kundu and Cohen [23], and Currie [84] have succinct descriptions of analytical techniques for potential flow. Potential flow in fluids has been discussed in great detail in the aerodynamics literature, because aerodynamic lift on airfoils for attached flow is well predicted by 2D potential flow descriptions combined with

Bernoulli's equation for the pressure. To this end, aerodynamics texts such as Kuethe, Schetzer, and Chow [85] and Anderson [86] cover potential flow in detail, including detailed discussion of distributed multipolar solutions (though they use terminology different from this text). A key difference between the aerodynamic treatment and the potential flow observed in electroosmotic flows is the presence of circulation in aerodynamic flows. Whereas aerodynamic analysis often focuses on circulation owing to its relation to lift forces, purely electroosmotic flows with uniform ζ have no circulation.

Although most practicing engineers solve modest Laplace equation systems with commercial differential equation solvers, large simulations often require more specialized approaches. Numerical solution of the Laplace equation is covered in many texts, including Kundu and Cohen [23] and Chapra and Canale [87]. The transform techniques discussed in this chapter map established solutions, e.g., the flow over a circle, to solve other systems, e.g., flow over an ellipse. For those interested in more advanced transform techniques, including Schwarz–Christoffel transforms, see [23].

7.7 EXERCISES

7.1 Evaluate the velocity components of a plane-symmetric source by examining the velocity potential (real) and stream function (imaginary) components of $\underset{\sim}{\phi}_{\rm v} = \frac{\Lambda}{2\pi} \ln \underset{\sim}{z}$.

7.2 Determine the velocity field induced by a plane-symmetric vortex by differentiation of $\underset{\sim}{\phi}_{\rm v}$.

7.3 Calculate the velocity induced by a plane-symmetric doublet by differentiation of $\underset{\sim}{\phi}_{\rm v}$.

7.4 Evaluate the velocity induced by a doublet of strength κ by evaluating the stream function and velocity potential directly from $\underset{\sim}{\phi}_{\rm v}$ and differentiating the velocity potential or stream function.

7.5 Using complex differentiation of $\underset{\sim}{\phi}_{\rm v} = U\underset{\sim}{z} \exp[-j\alpha]$, evaluate the velocity induced by a uniform flow of velocity U inclined at an angle of α.

7.6 Evaluate the flow induced by a complex potential $\underset{\sim}{\phi}_{\rm v} = U\underset{\sim}{z} \exp[-j\alpha]$ by evaluating the stream function and velocity potential and then differentiating those to obtain the velocity.

7.7 What does the magnitude of $\vec{\vec{\varepsilon}}$ tell you about the applicability of potential flow equations for solving fluid flow problems? How about $\vec{\vec{\omega}}$?

7.8 Define a point in space and assume that a fluid velocity exists at that point. Draw one triangle that uses u and v to decompose the velocity and another that uses $u_{\imath}$ and u_θ. Use this diagram and trigonometric relations to show that

$$u + jv = (u_{\imath} + ju_\theta) \exp j\theta \,. \tag{7.85}$$

7.9 Show that rotation of a vortex flow field by an angle α changes the resulting complex potential but does not affect the resulting velocity field.

7.10 Show that rotation of a source flow field by an angle α changes the resulting complex potential but does not affect the resulting velocity field.

7.11 Consider 2D potential flow. In the domain $-20 < x < 20$, $-20 < y < 20$, plot the streamlines for a uniform flow along the x axis.

7.12 Consider 2D potential flow. In the domain $-20 < x < 20$, $-20 < y < 20$, plot the streamlines for a uniform flow rotated an angle 0.2 rad with respect to the x axis.

7.13 Consider 2D potential flow. In the domain $-20 < x < 20$, $-20 < y < 20$, plot the streamlines for a uniform flow rotated an angle $\pi/5$ rad with respect to the x axis.

7.14 Consider 2D potential flow. In the domain $-20 < x < 20$, $-20 < y < 20$, plot the streamlines for a doublet located at (5, 2) with strength 32π and dipole moment rotated an angle $\pi/5$ rad with respect to the x axis.

7.15 Consider 2D potential flow. In the domain $-20 < x < 20$, $-20 < y < 20$, plot the streamlines for a uniform flow rotated $\pi/5$ with respect to the x direction flowing over a cylinder of radius 4 located at (5, 2).

7.16 Consider 2D potential flow. In the domain $-20 < x < 20$, $-20 < y < 20$, plot the streamlines for a vortex of strength $\Gamma = -16\pi$ located at (5, 2).

7.17 Consider 2D potential flow. In the domain $-20 < x < 20$, $-20 < y < 20$, plot the streamlines for a source of strength $\Lambda = 4\pi$ located at $(-4, -4)$.

7.18 Consider 2D potential flow. In the domain $-20 < x < 20$, $-20 < y < 20$, plot the streamlines for a uniform flow rotated an angle 0.2 rad with respect to the x axis superposed with a source of strength $\Lambda = 4\pi$ located at $(-4, -4)$.

7.19 Consider 2D potential flow. In the domain $-20 < x < 20$, $-20 < y < 20$, plot the streamlines for flow around a semi-infinite flat plate angled at $\pi/4$ with respect to the x axis and terminating at the origin. (*Hint:* This is flow around a sharp corner of angle 2π.)

7.20 Consider 2D potential flow. In the domain $-20 < x < 20$, $-20 < y < 20$, plot the streamlines for flow left to right along a wall aligned with the x axis that turns sharply by an angle $\pi/5$ at $(3, -4)$. (*Hint:* This is flow around a sharp corner of angle $4\pi/5$, which is also rotated by $\pi/5$.)

7.21 Show that the solution given in Eqs. (7.60) and (7.61) satisfies the conservation of mass equations by plugging the solutions into the cylindrical form of $\nabla \cdot \vec{u} = 0$.

7.22 Show that

$$\underset{\sim}{u}^* = u - jv = \frac{\partial \underset{\sim}{\phi}_{\mathrm{v}}}{\partial \underset{\sim}{z}} . \tag{7.86}$$

7.23 Define $\underset{\sim}{z} = \Delta x + j\Delta y$ and $\underset{\sim}{E} = E_x + jE_y$, where $\underset{\sim}{E}$ is a complex representation of a steady 2D electric field in the complex plane, E_x is the x component of the electric field, and E_y is the y component of the electric field. Now define the complex electric potential $\underset{\sim}{\phi} = \phi + j\psi_{\mathrm{e}}$, where ψ_{e} is the electric stream function defined such that

$$E_x = \frac{\partial \psi_{\mathrm{e}}}{\partial y} \tag{7.87}$$

and

$$E_y = -\frac{\partial \psi_{\mathrm{e}}}{\partial x} . \tag{7.88}$$

Show that

$$\underset{\sim}{E} = -\frac{\partial \underset{\sim}{\phi}}{\partial \underset{\sim}{z}} . \tag{7.89}$$

This equation has a different form than Eq. (7.22). Why?

7.24 Define $\underset{\sim}{\phi}$ for a positive line charge with charge per length q'.

7.25 Define $\underset{\sim}{\phi}$ for a positive line doublet aligned in the x direction with dipole moment per unit length p'.

7.26 Consider a body embedded in a potential electroosmotic flow. Integrate Faraday's law over the area of the body and use velocity-current similitude to determine the circulation around the body. If time-varying fields are assumed zero, what must the circulation around the body be?

7.27 Start with the Navier–Stokes equations:

$$\rho \frac{\partial \vec{u}}{\partial t} + \rho \vec{u} \cdot \nabla \vec{u} = -\nabla p + \eta \nabla^2 \vec{u} . \tag{7.90}$$

Take the curl of this equation, and show that the result is given by

$$\rho \frac{\partial \vec{\omega}}{\partial t} + \rho \vec{u} \cdot \nabla \vec{\omega} = \rho \vec{\omega} \cdot \nabla \vec{u} + \eta \nabla^2 \vec{\omega} . \tag{7.91}$$

In so doing, note that the pressure term has been eliminated and is replaced with the *vortex-stretching term* $\vec{\omega} \cdot \nabla \vec{u}$.

7.28 Take the curl of the incompressible, uniform-viscosity Navier–Stokes equations (assume no body force terms) and derive the 2-D velocity–vorticity form of the Navier–Stokes equations:

$$\rho \frac{\partial \vec{\omega}}{\partial t} + \rho \vec{u} \cdot \nabla \vec{\omega} = \eta \nabla^2 \vec{\omega} , \tag{7.92}$$

where $\vec{\omega} = \nabla \times \vec{u}$. Comment on the presence or absence of vorticity source terms in this equation. How does the presence or absence of vorticity source terms relate to irrotationality of electroosmotic flow? Read Ref. [74] and explain the role of the vorticity source term in the context of flow–current similitude.

7.29 In Appendix F, we derive the linear 2D cylindrical multipole expansion, which allows for potentials proportional to $\ln \imath$ as well as solutions proportional to $\imath^{-k}$, where k must be an *integer*. In Subsection 7.3.6, however, we allow complex potentials proportional to $\underset{\sim}{z}^{\pi/\theta_0}$, and we *do not* require that π/θ_0 be an integer. Explain this inconsistency.

7.30 Is it possible for a potential flow solution to lead to recirculatory flow? Why or why not? How does this affect your interpretation of the flow pattern that occurs in an electrokinetic pump at its maximum pressure point?

7.31 Consider purely electroosmotic flow through a microfabricated channel of uniform depth d with a sharp 90° turn whose inside corner is located at the origin, depicted in Fig. 7.9. Focus on the outer flow, i.e., the flow outside the EDL. Assume that the flow far from the corner is uniform across the width of the channel, and assume that the channel can be approximated as being wide, so that the flow near the inside corner can be approximated by the flow around the inside corner in an infinite medium. Informed by Eq. (7.74), write an equation for the velocity distribution *near the inside corner*. Do the same for the velocity distribution *near the outside corner*. In both cases, your solution will have a free constant that is proportional to the inlet and outlet flow velocities, which are left unspecified.

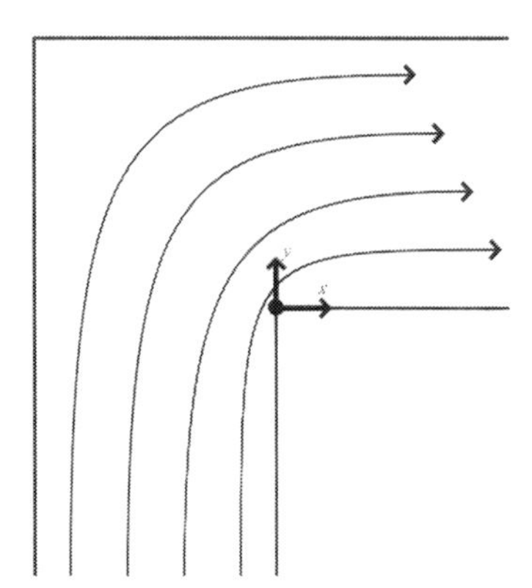

Fig. 7.9 Geometry of flow through a microfluidic device with a 90° turn.

7.32 Consider uniform flow in the x direction with velocity U over an ellipse centered at the origin with a minor axis of length 1 aligned in the y direction and a major axis of length 2 aligned in the x direction. Define the parameters of a Joukowski transform required for transforming this solution to that of a uniform flow over a circle, and write the expression for the velocity potential of this flow.

8 Stokes Flow

The Navier–Stokes equations have not been solved analytically in the general case, and the only available analytical solutions arise from simple geometries (for example, the 1D flow geometries discussed in Chapter 2). Because of this, our analytical approach for solving fluid flow problems is often to solve a simpler equation that applies in a specific limit. Some examples of these simplified equations include the Stokes equations (applicable when the Reynolds number is low, as is usually the case in microfluidic devices) and the Laplace equation (applicable when the flow has no vorticity, as is the case for purely electrokinetic flows in certain limits). These simplified equations guide engineering analysis of fluid systems.

In this chapter, we discuss Stokes flow (equivalently termed *creeping flow*), in which case the Reynolds number is so low that viscous forces dominate over inertial forces. The approximation that leads from the Navier–Stokes equations to the Stokes equations is shown, and analytical results are discussed. The Stokes flow equations provide useful solutions to describe the fluid forces on small particles in micro- and nanofluidic systems, because these particles are often well approximated by simple geometries (for example, spheres) for which the Stokes flow equations can be solved analytically. The Stokes flow equations also lead to simple solutions (Hele-Shaw flows) for wide, shallow microchannels of uniform depths.

8.1 STOKES FLOW EQUATION

We derive the Stokes flow equations by neglecting the unsteady and convective terms when the Reynolds number is low. The Navier–Stokes equations are

$$\rho \frac{\partial \vec{\boldsymbol{u}}}{\partial t} + \rho \vec{\boldsymbol{u}} \cdot \nabla \vec{\boldsymbol{u}} = -\nabla p + \eta \nabla^2 \vec{\boldsymbol{u}}, \tag{8.1}$$

and, rewritten in nondimensional form in which the pressure is normalized by $\eta U/\ell$, are given by

$$Re\, \frac{\partial \vec{\boldsymbol{u}}^*}{\partial t^*} + Re\, \vec{\boldsymbol{u}}^* \cdot \nabla^* \vec{\boldsymbol{u}}^* = -\nabla^* p^* + \nabla^{*2} \vec{\boldsymbol{u}}^*, \tag{8.2}$$

where Re is the Reynolds number, defined by $Re = \rho U \ell/\eta$, η and ρ are fluid properties in the governing equations, and U and ℓ come from the boundary conditions. This nondimensionalization is shown in more detail in Appendix E, Subsection E.2.1. If $Re \to 0$, the unsteady and convective terms can be neglected.

The Stokes flow approximation is valid for $Re \ll 1$, in which case we neglect the unsteady and convective terms, leaving

Stokes equation for low-Re flow

$$\nabla p = \eta \nabla^2 \vec{u} . \tag{8.3}$$

These are the *Stokes flow equations.* These equations are *linear* in both the velocity and the pressure, and they are much easier to solve than the Navier–Stokes equations. Because the unsteady term can be neglected, Stokes flows have the properties of *instantaneity* (no dependence on time except through time-dependent boundary conditions) and *time reversibility* (a time-reversed Stokes flow solves the Stokes equations). Time reversibility also implies that Stokes flow around a symmetric body exhibits fore–aft symmetry. Because the nonlinear term can be neglected, Stokes flows also have the property of *superposability* both for the pressure and for the velocity.

All real flows have a finite Reynolds number, and so the Stokes flow equations are only an approximation of the real flows. Whereas the Reynolds number cutoff depends on the flow, a good rule of thumb is that the Stokes flow solution is a good approximation when $Re < 0.1$.

8.1.1 Different forms of the Stokes flow equations

Equation (8.3) is a function of both the pressure and the velocity, which is sometimes inconvenient. Fortunately, we can convert the Stokes flow equations to two other forms, which are each a function of the pressure field or the velocity field exclusively. Taking the divergence of the Stokes flow equations results in

$$\nabla \cdot \nabla p = \nabla \cdot \eta \nabla^2 \vec{u} . \tag{8.4}$$

From Eq. (C.80), the divergence of the Laplacian of $\vec{u}$ is equal to the Laplacian of the divergence of $\vec{u}$, so

$$\nabla^2 p = \eta \nabla^2 \left(\nabla \cdot \vec{u} \right) , \tag{8.5}$$

and, because the flow is incompressible, $\nabla \cdot \vec{u} = 0$, and thus

Stokes equation for pressure

$$\nabla^2 p = 0 . \tag{8.6}$$

For Stokes flow, the pressure satisfies Laplace's equation. This form of the Stokes flow equations is useful if the boundary conditions are specified exclusively in terms of pressure. Alternatively, we can take the curl of the Stokes flow equations, giving

$$\nabla \times \nabla p = \nabla \times \eta \nabla^2 \vec{u} , \tag{8.7}$$

which eliminates the left hand side because the curl of the gradient of a vector field is zero. From Eq. (C.81), the curl of the Laplacian is equal to the Laplacian of the curl, so

Stokes equation for vorticity

$$\nabla^2 (\nabla \times \vec{u}) = \nabla^2 \vec{\omega} = 0. \tag{8.8}$$

This version of the Stokes flow equations is most useful if the boundary conditions are expressed exclusively in terms of velocities.

The Stokes flow equations can also be written in terms of the stream function. For a 2D flow with plane symmetry, the Stokes equations take the form

Stokes equation for stream function in Cartesian coordinates

$$\nabla^4 \psi = \nabla^2 (\nabla^2 \psi) = 0, \tag{8.9}$$

where ψ is defined by Eqs. (1.3)–(1.4). This is termed the *biharmonic equation,* and ∇^4 is the *biharmonic operator*. For axisymmetric flow, the equation for the Stokes stream function, which is defined in Eqs. (1.9) and (1.10), can be written as

Stokes equation for Stokes stream function in spherical coordinates

$$E^4 \psi_S = E^2 (E^2 \psi_S) = 0, \tag{8.10}$$

where E^2 here is a symbol for the second-order differential operator $\frac{\partial^2}{\partial r^2} + \frac{\sin\vartheta}{r}\frac{\partial}{\partial\vartheta}\frac{1}{\sin\vartheta}\frac{\partial}{\partial\vartheta}$. This operator plays a role similar to the Laplacian operator, but is slightly different owing to the use of curvilinear coordinates.

8.1.2 Analytical versus numerical solutions of the Stokes flow equations

The Stokes flow equations are linear and simple to solve numerically. Thus, when studying low-*Re* flow for a complicated geometry, practicing engineers typically solve the Stokes equations numerically by using any of a variety of numerical codes. We can also solve the Stokes equations analytically for several model problems in a way that is immediately useful for study of flows in microdevices. Sections 8.2 and 8.3 consider analytical solutions for flow in shallow channels and over spheres.

8.2 BOUNDED STOKES FLOWS

For bounded flows (i.e., flows within finite domains), the Reynolds number that determines the applicability of the Stokes approximation uses a mean flow rate through a channel as the characteristic velocity and a channel diameter or depth as the characteristic length. In this section, we focus primarily on the Hele-Shaw solution, an analytical solution that applies to wide, shallow channels typical of many microfabrication processes.

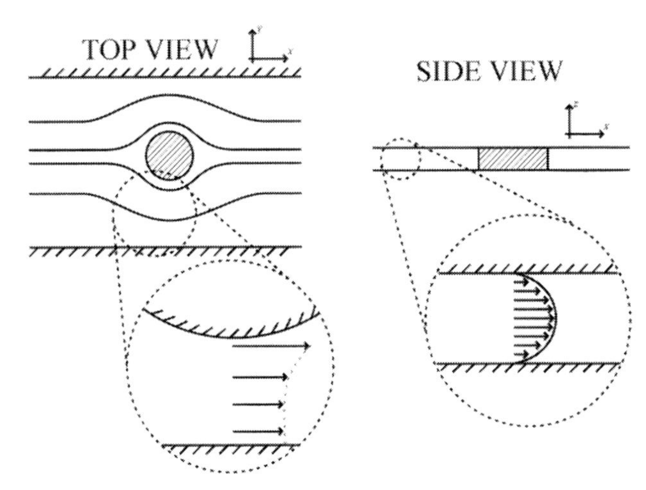

Fig. 8.1 Hele-Shaw flow over a circular spacer. The velocity distribution, when viewed from the top, appears as a potential flow for all regions in which the distance from the side walls is large relative to the depth.

8.2.1 Hele-Shaw flows

One flow configuration that is analytically tractable and relevant to microfluidic devices is the Hele-Shaw cell. A Hele-Shaw cell consists of a domain with a small and uniform depth d in the z direction and much larger dimensions in the x and y directions. Historically, these devices have been fabricated by bringing two flat plates close together, with spacers that serve as obstacles for the flow. In micro- or nanofluidic devices, this geometry occurs naturally, because most microfabrication techniques involve etching a relatively shallow channel into a flat substrate and affixing a flat lid, resulting in a device with x and y dimensions ranging from 5 μm to 1 cm and a uniform z dimension somewhere between 10 nm and 100 μm.

If the depth of the device is uniform and much smaller than the other dimensions, then it is reasonable to assume that $p = p(x, y)$, in which case we can separate variables in $\vec{u}$, writing it as the product of a z function and an xy function. For a channel of depth d with walls at $z = 0$ and $z = d$, this leads to the solution

velocity distribution in a Hele-Shaw cell

$$\vec{u} = -\frac{1}{2\eta} z(d - z) \nabla p . \tag{8.11}$$

At any given z, the velocity solution in the xy plane corresponds to a 2D potential flow in which the velocity potential is proportional to p. The same can be said for the z-averaged velocity. Hele-Shaw flows have no z vorticity, and the streamlines in a Hele-Shaw flow are identical to the streamlines in a potential flow with the same geometry. However, the pressure distribution is different in the two cases. Along the z axis, the flow has the parabolic distribution typical of pressure-driven flow between flat plates.

Computation of a Hele-Shaw flow requires that a 2D Laplace equation be solved for the pressure, at which point Eq. (8.11) can be used to calculate the velocity field. An example of the streamlines for Hele-Shaw flow over an obstacle are shown in Fig. 8.1.

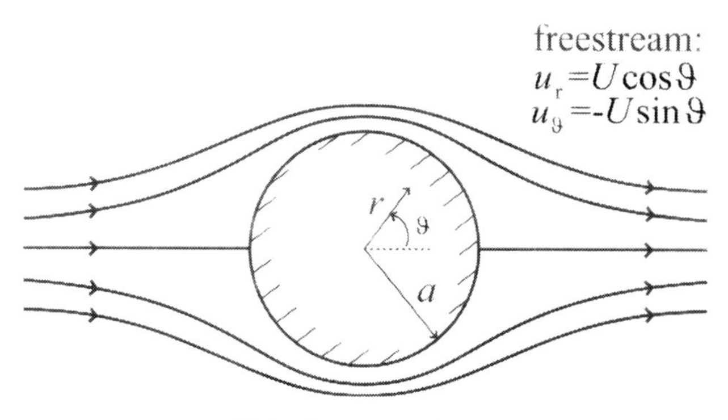

Fig. 8.2 Stokes flow over a sphere.

8.2.2 Numerical solution of general bounded Stokes flow problems

For bounded flows, i.e., flows of fluid within solid walls, the solution approach used by engineers to predict the flow inside a microfluidic device is largely unchanged. Unless the solution is geometrically quite simple (as was the case in Chapter 2) such that direct integration provides a solution, Stokes flows in microchannels are typically solved with numerical techniques.

8.3 UNBOUNDED STOKES FLOWS

Unbounded flows (i.e., flows of fluid around immersed solid objects) are also characterized by the Reynolds number, but in this case the characteristic velocity U and length scale ℓ now come from the velocity and size of the object. Unlike bounded flows, for which numerical approaches are the norm, unbounded flows are commonly treated analytically, and the solution of model problems leads to results that are of immense value in common systems. For example, a microfluidic device may be used to process blood or a cellular suspension. In this case, calculating all details of the flow would be difficult and largely unnecessary, because we can encapsulate the effects of particles in a simple way that can be described analytically. The analytical solutions for flow over a sphere or for the flow response to a point force explain in great part the dynamics (e.g., diffusivity) of macromolecules and particles. Given a flow over a body, the surface stress, integrated over the body surface, is a measure of the drag force on the object as it moves in the fluid. This drag force is used to calculate particle settling times, electrophoretic and dielectrophoretic particle velocities, and the accuracy of particle-image velocimetry (PIV) measurements in microsystems. We start with a discussion of Stokes flow over a sphere in an infinite domain, which can be solved directly.

8.3.1 Stokes flow over a sphere in an infinite domain

Consider axisymmetric flow at velocity U over a sphere of radius a at low Re (Fig. 8.2). Here, the relevant Reynolds number is customarily defined as $Re = \rho U d/\eta$, where $d = 2a$ is the sphere diameter. The governing equations are the Stokes equations, and the boundary conditions are that the velocity is zero at $r = a$ and the velocity is equal to U as $r \to \infty$.

The solution for this flow can be obtained in a number of ways, the simplest of which is to assume that the result can be written in terms of a power series in a/r.

The velocity solution is

radial velocity for flow over a sphere in Stokes flow

$$u_r = U \cos\vartheta \left(1 - \frac{3}{2}\frac{a}{r} + \frac{1}{2}\frac{a^3}{r^3}\right) \tag{8.12}$$

and

colatitudinal velocity for flow over a sphere in Stokes flow

$$u_\vartheta = -U \sin\vartheta \left(1 - \frac{3}{4}\frac{a}{r} - \frac{1}{4}\frac{a^3}{r^3}\right) . \tag{8.13}$$

The variation of the pressure from the freestream value (Δp) is given by

pressure variation for flow over a sphere in Stokes flow

$$\Delta p = -\eta U \frac{3}{2}\frac{a}{r^2} \cos\vartheta . \tag{8.14}$$

The surface forces on the sphere can be integrated to get the total drag force on the sphere:

Stokes drag on a sphere

$$F_{\text{drag}} = 6\pi\eta U a . \tag{8.15}$$

These steady results are applicable for time-varying U owing to the instantaneity of the Stokes equations. For a particle of finite size and Reynolds number, we can confirm this approximation by evaluating the characteristic time for Stokes particles to equilibrate with their fluid surroundings compared with the experimental time scales. Thus we can assume that, for time scales that are long relative to the particle lag time $\tau_p = \frac{2a^2\rho_p}{9\eta}$, the system can be assumed to be in quasi-steady state, i.e., the magnitude of the unsteady term in the equation is small, even when the flow itself is unsteady.

EXAMPLE PROBLEM 8.1

Consider a dense particle (particle density large compared with that of the fluid) moving at a velocity U through a quiescent viscous fluid (presumably owing to some force, perhaps gravity). At time $t = 0$, this force is removed, and the moving particle comes to rest. Assume the Reynolds number is small and the system can be described by Stokes flow. Assume that the steady Stokes flow solution for a sphere applies at all time. Assume that, for time $t > 0$, there is no force on the particle other than the surface force. Given these assumptions, determine the time history of the velocity of the particle as it comes to rest, and show that $\tau_p = \frac{2a^2\rho_p}{9\eta}$ is a characteristic equilibration time for this system. Calculate this quantity for a 1-μm particle suspended in water with a density five times that of the water.

SOLUTION: From the Stokes flow solution, the force on the particle is given by

$$F = 6\pi u \eta a \,, \tag{8.16}$$

where this force tends to *decelerate* the particle. The mass of the particle m is given by

$$m = \frac{4}{3}\pi a^3 \rho_p \,. \tag{8.17}$$

Virtual or added-mass considerations can be ignored because the density of the particle is assumed large relative to that of the fluid. The equation of motion for the particle is simply that the acceleration is given by the force divided by the mass,

$$\frac{\partial u}{\partial t} = -\frac{18\eta}{4\rho_p a^2} u \,, \tag{8.18}$$

which leads to an exponential solution:

$$u(t) = U \exp\left(\frac{t}{\tau_p}\right) \,, \tag{8.19}$$

where

$$\tau_p = \frac{2a^2 \rho_p}{9\eta} = 1.1\ \mu s \,. \tag{8.20}$$

EXAMPLE PROBLEM 8.2

The net force of a fluid on a solid object is given by the integral of the surface stress over the surface:

$$\vec{F} = \int_S \tau dA \,. \tag{8.21}$$

Here, the normal component of τ is the pressure, and the tangent component of τ is the wall viscous stress ($\eta \frac{\partial u_\vartheta}{\partial r}$). Given the Stokes flow solution for flow around a sphere in an infinite domain, specified in Eqs. (8.12) and (8.13), integrate to determine the total drag.

SOLUTION: The drag force per unit area at any ϑ is given by

$$F''_{\text{drag}} = -\cos\vartheta(\Delta p) - \sin\vartheta\left(\eta \frac{\partial u_\vartheta}{\partial r}\right) \,, \tag{8.22}$$

where F''_{drag} is the local drag per unit area (this is also equal to the dot product of the surface stress with the unit vector in the direction of flow). The angle terms come from the dot product of the inward normal with the unit vector in the direction of flow and the dot product of $\hat{\boldsymbol{\vartheta}}$ with the unit vector in the direction of flow, respectively. The pressure difference from ambient (Δp) can be used instead of p because the integral of a uniform normal force around any enclosed surface is zero. Taking the derivative of u_ϑ and plugging in Δp, we get

$$F''_{\text{drag}} = -\cos\vartheta\left(-\eta U\cos\vartheta\frac{3}{2}\frac{a}{a^2}\right) - \sin\vartheta\left(-\eta U\sin\vartheta\frac{3}{2}\frac{a}{a^2}\right), \tag{8.23}$$

which simplifies to

$$F''_{\text{drag}} = \frac{3}{2}\frac{\eta U}{a}. \tag{8.24}$$

The drag per unit area is uniform across the sphere. We can therefore calculate the drag simply by multiplying this relation by the surface area ($4\pi a^2$) to get

$$F_{\text{drag}} = 6\pi\eta U a. \tag{8.25}$$

The Stokes flow solution for flow over a sphere has three terms, which relate to the multipolar solutions discussed in Appendix F. The constant term refers to the uniform freestream velocity – this is the flow that would be observed if the particle were absent. The term proportional to a/r is the *Stokeslet* term – it corresponds to the response of the flow caused by a point force of $6\pi\eta Ua$ applied to the fluid at the center of the sphere. This component of the flow field is shown in Fig. 8.3. The Stokeslet term describes the viscous response of the fluid to the no-slip condition at the particle surface, and this term contains all of the vorticity caused by the viscous action of the particle. The term proportional to a^3/r^3 is the *stresslet* term – it is identical to Eq. (F.6). This term contributes an irrotational flow – it is unrelated to the viscous force of the sphere and is caused by the finite size of the particle. This component of the flow field is shown in Fig. 8.4. Because the stresslet term decays proportionally to r^{-3} whereas the Stokeslet term decays proportionally to r^{-1}, the primary long-range effect of the particle is induced by the Stokeslet. Thus the *net force on the fluid induced by the sphere* is required for prescribing the flow far from a sphere, rather than the particle size or velocity alone. Far from a sphere moving in a Stokes flow, the flow does not distinguish between the effects of one particle that has velocity U and radius $2a$ and another that has velocity $2U$ and radius a, as these two spheres induce the same drag force. Close to these spheres, of course, the two flows are different, as distinguished by the different stresslet terms.

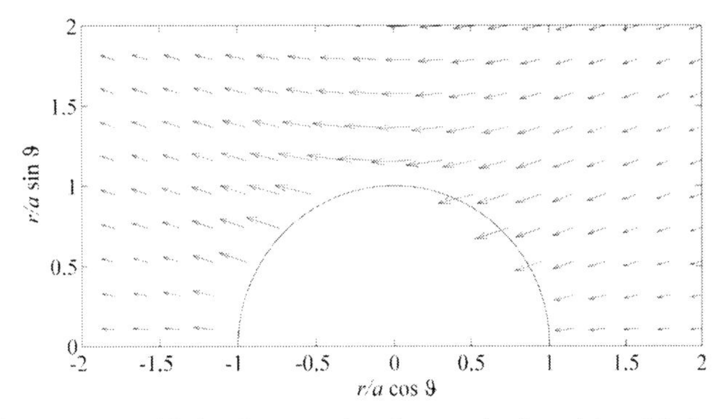

Fig. 8.3 The Stokeslet component of Stokes flow around a sphere moving from right to left along the x axis.

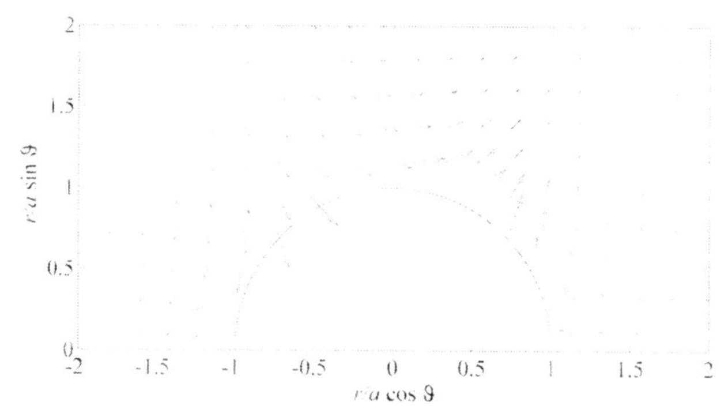

Fig. 8.4 The stresslet component of Stokes flow around a sphere moving from right to left along the x axis. Note, in comparison with Fig. 8.3, how quickly the velocities decay as the distance from the surface increases.

It is customary to define a drag coefficient, which normalizes the drag in Eq. (8.15) by the dynamic pressure of the freestream fluid (in a coordinate reference frame where the particle is motionless) multiplied by the cross-sectional area of the sphere:

$$C_{\mathrm{D}} = \frac{F_{\mathrm{drag}}}{\frac{1}{2}\rho u^2 A_p} = \frac{24}{Re_d}, \tag{8.26}$$

where $A_p = \pi a^2$ is the cross-sectional area of the sphere, and $Re_d = \frac{\rho u d}{\eta}$ is the Reynolds number based on the particle *diameter*.[1]

To predict particle dynamics in microsystems, we require the drag force reported in Eq. (8.15) but not other details of the flow. Because microparticles achieve equilibrium quickly, a force applied to a spherical microparticle induces particle motion at the velocity such that the drag force and the motive force are equal and opposite. Thus the steady-state velocity of a microparticle with radius a in Stokes flow with an applied force $\vec{F}$ is given by

$$\vec{u} = \frac{\vec{F}}{6\pi\eta a}. \tag{8.27}$$

Equation (8.27) relates *force* to *velocity* rather than relating force to *acceleration*. This is typical of steady-state flow relations.

Given a system with small but finite Reynolds number, we can quantify the instantaneity of the particle response by calculating the Stokes number Sk, which is the ratio of the particle lag time to the characteristic time over which the flow changes. The characteristic flow time can come from the characteristic time of an unsteady boundary condition or from the ratio a/U of the characteristic length scale and velocity from a steady boundary condition in a nonuniform flow. Choosing the latter, we have

$$Sk = \frac{U\tau_{\mathrm{p}}}{a} = \frac{2a\rho_{\mathrm{p}}U}{9\eta a}. \tag{8.28}$$

Particles with Stokes number $Sk \ll 1$ can be assumed to be always in steady state with a local velocity field given by the idealized solution derived earlier.

[1] This definition of C_{D} is consistent with the tradition in fluid mechanics, but has the potential to be misleading for Stokes flow, because we have asserted that the flow is independent of the Reynolds number, but then subsequently presented a result in which C_{D} is proportional to Re^{-1}. This apparent contradiction comes from using $\frac{1}{2}\rho U^2$ to normalize this force. If we had used $\eta U/\ell$ instead, C_{D} would have been a constant.

8.3.2 General solution for Stokes flow over a sphere in an infinite domain

The preceding solution applies for a stationary sphere with a uniform flow of velocity U. In microsystems, it is typically more appropriate to consider the fluid as being nominally quiescent and the particle to be in motion. Further, particle motion is rarely aligned with the coordinate system in use, or the presence of multiple particles may make it impossible to define a coordinate system in line with the motion of all particles. To this end, we benefit from writing a general solution. We write this with a hydrodynamic interaction tensor $\vec{\vec{G}}$ that, when dotted against the force applied to the particle, gives the velocity field:

$$\vec{u} = \vec{\vec{G}} \cdot \vec{F} . \tag{8.29}$$

As we have shown earlier, the force $\vec{F}$ applied to a particle induces a particle velocity of $\vec{F}/6\pi\eta a$. The same force induces a fluid velocity field of $\vec{\vec{G}} \cdot \vec{F}$. For a sphere, the hydrodynamic interaction tensor $\vec{\vec{G}}$ is given by

$$\vec{\vec{G}} = \frac{1}{8\pi\eta\Delta r}\left[\left(\vec{\vec{\delta}} + \frac{\vec{\Delta r}\vec{\Delta r}}{\Delta r^2}\right) + \frac{a^2}{\Delta r^2}\left(\frac{1}{3}\vec{\vec{\delta}} - \frac{\vec{\Delta r}\vec{\Delta r}}{\Delta r^2}\right)\right] . \tag{8.30}$$

In this equation, $\vec{\Delta r}$ denotes the distance vector from the sphere center to the fluid location and Δr is the magnitude of that distance. The symbol $\vec{\vec{\delta}}$ denotes the identity tensor. The $\vec{\Delta r}\vec{\Delta r}$ terms lead to dyads, and thus $\vec{\vec{G}}$ is a dyadic tensor. We can find the general solution for the velocity by evaluating the dot product and noting that $\vec{\Delta r}\vec{\Delta r} \cdot \vec{F} = (\vec{\Delta r} \cdot \vec{F})\vec{\Delta r}$:

$$\vec{u} = \vec{\vec{G}} \cdot \vec{F} = \frac{1}{8\pi\eta\Delta r}\left[\left(\vec{F} + \frac{(\vec{\Delta r} \cdot \vec{F})\vec{\Delta r}}{\Delta r^2}\right) + \frac{a^2}{\Delta r^2}\left(\frac{1}{3}\vec{F} - \frac{(\vec{\Delta r} \cdot \vec{F})\vec{\Delta r}}{\Delta r^2}\right)\right] . \tag{8.31}$$

The first term in this equation is the Stokeslet term, and the second term is the stresslet term. The stresslet term becomes unimportant as Δr becomes large, and thus the hydrodynamic interaction tensor is often simplified to the *Oseen–Burgers tensor*:[2]

$$\vec{\vec{G}}_0 = \frac{1}{8\pi\eta\Delta r}\left(\vec{\vec{\delta}} + \frac{\vec{\Delta r}\vec{\Delta r}}{\Delta r^2}\right) . \tag{8.32}$$

By neglecting the stresslet term, the Oseen–Burgers tensor no longer describes the flow caused by a sphere, but rather the flow caused by a *point force* applied at $\Delta r = 0$. At large $\Delta r/a$, the two tensors give the same result.

Now we consider the same rigid spherical particle of radius a moving because of a force $\vec{F}$, and we consider the pressure change caused by the particle motion and attendant flow. This pressure change Δp can be written as

$$\Delta p = \vec{P} \cdot \vec{F} , \tag{8.33}$$

where $\vec{P}$ is the (rank 1) pressure interaction tensor, given by

$$\vec{P} = \frac{1}{4\pi\Delta r^3}\vec{\Delta r} . \tag{8.34}$$

[2] Here, the terminology varies as does the use of constants. Many authors refer to Eq. (8.32) as the Oseen–Burgers tensor, and use *Oseen tensor* to refer to the tensor without the premultiplier $1/8\pi\eta$. In contrast, some authors call Eq. (8.32) the Oseen tensor.

The resulting pressure change owing to a particle of radius a moving with velocity $\vec{\boldsymbol{u}}$ is

$$\Delta p = \frac{3}{2}\eta\frac{a}{\Delta r}\frac{\vec{\boldsymbol{u}}\cdot\vec{\boldsymbol{\Delta r}}}{\Delta r^2}\,. \tag{8.35}$$

The following subsections report how the drag, the drag coefficient, or the particle velocity changes when the shape of the particle changes or it is in proximity to other objects.

EXAMPLE PROBLEM 8.3

Show that $\vec{\vec{\boldsymbol{G}}}$ at the surface of a spherical particle is equal to $\frac{1}{6\pi\eta a}\vec{\vec{\boldsymbol{\delta}}}$.

SOLUTION: We start with $\vec{\vec{\boldsymbol{G}}}$:

$$\vec{\vec{\boldsymbol{G}}} = \frac{1}{8\pi\eta\Delta r}\left[\left(\vec{\vec{\boldsymbol{\delta}}} + \frac{\vec{\boldsymbol{\Delta r}}\vec{\boldsymbol{\Delta r}}}{\Delta r^2}\right) + \frac{a^2}{\Delta r^2}\left(\frac{1}{3}\vec{\vec{\boldsymbol{\delta}}} - \frac{\vec{\boldsymbol{\Delta r}}\vec{\boldsymbol{\Delta r}}}{\Delta r^2}\right)\right], \tag{8.36}$$

and set $\Delta r = a$:

$$\vec{\vec{\boldsymbol{G}}} = \frac{1}{8\pi\eta a}\left[\left(\vec{\vec{\boldsymbol{\delta}}} + \frac{\vec{\boldsymbol{\Delta r}}\vec{\boldsymbol{\Delta r}}}{a^2}\right) + \left(\frac{1}{3}\vec{\vec{\boldsymbol{\delta}}} - \frac{\vec{\boldsymbol{\Delta r}}\vec{\boldsymbol{\Delta r}}}{a^2}\right)\right], \tag{8.37}$$

leading to

$$\vec{\vec{\boldsymbol{G}}} = \frac{1}{8\pi\eta a}\left(\frac{4}{3}\vec{\vec{\boldsymbol{\delta}}}\right), \tag{8.38}$$

and finally

$$\vec{\vec{\boldsymbol{G}}} = \frac{1}{6\pi\eta a}\vec{\vec{\boldsymbol{\delta}}}\,. \tag{8.39}$$

8.3.3 Flow over prolate ellipsoids

For ellipsoidal particles, the flow is a function of a particle's orientation and axis lengths, and thus the drag force is a function of these parameters as well. The force dependence is expressed through changes in both A_p, the cross-sectional area perpendicular to the flow, and C_D, the drag coefficient based on an *effective* particle diameter. For an ellipsoid with axes a_1, a_2, and a_3 ($a_1 > a_2 > a_3$), we use as the effective particle diameter the effective diameter of the cross section (i.e., $2\sqrt{a_2 a_3}$) for flow along the long axis.

For the special case of a prolate ellipsoid ($a_1 > a_2 = a_3$), which is useful because of its similarity to rod-shaped particles, flows along the long axis induce a drag given by

$$C_{\mathrm{D,ellipse}} \simeq \frac{24}{Re}\frac{8}{3}\frac{1}{\sqrt{1-e^2}}\frac{e}{(1+e^2)\ln\frac{1+e}{1-e} - 2e}, \tag{8.40}$$

where the eccentricity[3] is given by $e = \sqrt{1 - \frac{a_2^2}{a_1^2}}$.

[3] The eccentricity e here is distinguished from the elementary charge e used for the electrodynamic portion of this text.

8.3.4 Stokes flow over particles in finite domains

As a moving particle approaches a wall, the fluid velocity field resulting from the moving particle is retarded because of the no-slip boundary condition at the wall. For a given force, the presence of the wall therefore reduces the particle velocity as it approaches the wall. For a sphere of radius a located a distance d from the wall, the force–velocity relation normal to the wall can be approximated by

$$\frac{1}{6\pi\eta a}\frac{\vec{F}\cdot\hat{n}}{\vec{u}\cdot\hat{n}} = 1 + \frac{9}{8}\frac{a}{d}, \tag{8.41}$$

and the force–velocity relation tangent to the wall can be approximated by

$$\frac{1}{6\pi\eta a}\frac{\vec{F}\cdot\hat{t}}{\vec{u}\cdot\hat{t}} = 1 + \frac{9}{16}\frac{a}{d}. \tag{8.42}$$

In both cases, the effects of the wall are small at $d = 10a$.

8.3.5 Stokes flow over multiple particles

If particles are close to each other, the drag force on the particles and therefore the velocity of particles is impacted by particle-particle interactions. As was the case with flow near walls, the isolated sphere relation is accurate as long as particle–particle separations exceed 10 times the particle diameter. The hydrodynamic interaction tensor can be used to evaluate the forces of particles on each other and thus predict the forces and particle velocities.

8.4 MICRO-PIV

PIV is often used in microscale systems to visualize fluid flow. Because it typically involves flow of a suspension of spheres at low *Re*, it is an important example of Stokes flow and the applicability of Stokes flow analysis. PIV operates as follows: First, a fluid flow is seeded with particulate fluid tracers. In microsystems, the most common particles are fluorescent polystyrene latex beads. Then two images of the fluid tracers are recorded in rapid succession. For measurements with high temporal resolution, this usually involves mating a dual-pulse laser to a microscope and recording the fluorescence from the beads with a charge-coupled device (CCD) camera. Figure 8.5 shows an example of a micro-PIV setup. Finally, the two images are correlated. If the two images are separate, this is called a cross-correlation. If the two images are recorded on the same camera image, this is called an autocorrelation. If image one is treated as a $q \times p$ array of brightness values $f(i, j)$ and the second image as $g(i, j)$, then the cross-correlation Φ is given by:

$$\Phi(m, n) = \sum_{i=1}^{q}\sum_{j=1}^{p} f(i, j)g(i + m, j + n), \tag{8.43}$$

and the autocorrelation of a single image f is given by

$$\Phi(m, n) = \sum_{i=1}^{q}\sum_{j=1}^{p} f(i, j)f(i + m, j + n). \tag{8.44}$$

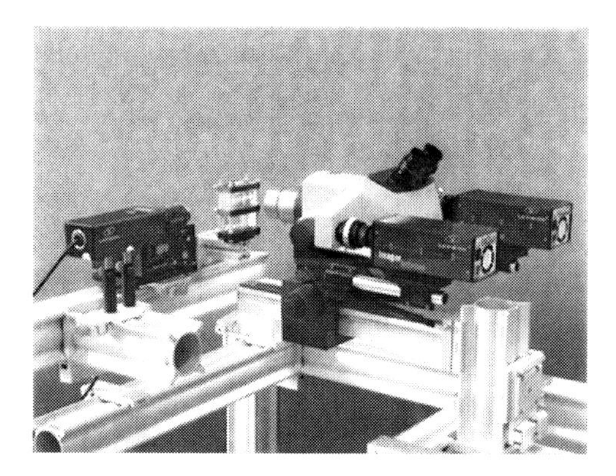

Fig. 8.5 Schematic of a micro-PIV system. (Courtesy LaVision: http://www.lavision.de.) A pulsed laser creates two laser pulses that are focused by an epifluorescent microscope with a high numerical aperture onto a microfluidic device. A microscope lens collects the fluorescent signal from fluorescent particles and focuses the image onto a dual-frame CCD camera. The z axis resolution comes from the narrow focal depth of the collection objective.

Here, m and n indicate pixel offsets in the i and j directions. The cross-correlation thus gives – as a function of the offsets – a measure of how well two images match each other. Figure 8.6 shows this cross-correlation algorithm. The cross-correlation Φ is large when the offsets m and n cause the two images to overlap as perfectly as possible, and small when the offsets cause the two images to misalign. The most likely distance traveled by the fluid between images corresponds to the pixel offset with the maximum value of Φ. The process of finding the maximum of the cross-correlation function is analogous to printing one image on a sheet of paper and printing the second image on a transparent sheet. By moving these two sheets with respect to each other, one can find an offset that best matches the two images to each other. To get a complete flow field, small *interrogation regions* throughout the image are each analyzed to give the velocity in that region, and the flow field is the product of systematically evaluating velocities in interrogation regions throughout the image. Such spatial correlations are straightforward with use of fast Fourier transforms.

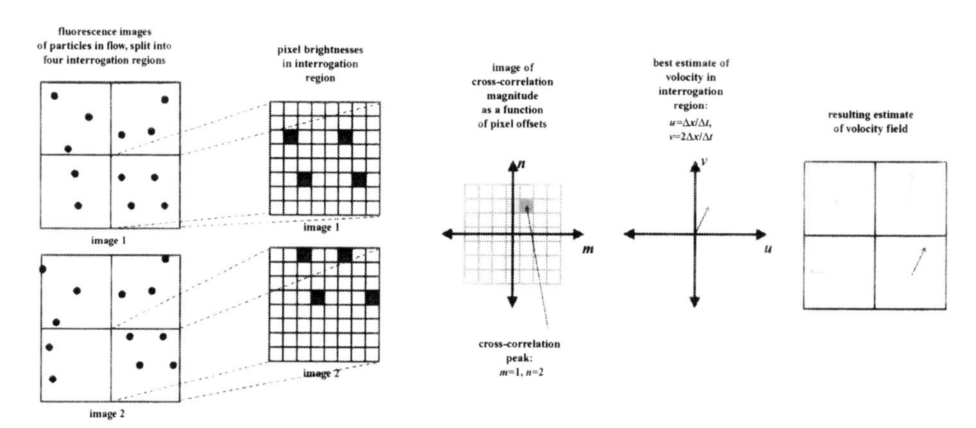

Fig. 8.6 A cross-correlation algorithm for PIV.

8.4.1 Deterministic particle lag

PIV measures the *particle* velocity field. If this is to be used to infer the fluid velocity field, then the relation of the particle velocity field to the fluid velocity field must be established by use of Stokes flow arguments, and two key error sources must be avoided. First, body forces on the particles, e.g., gravity or electric fields, will induce a particle velocity relative to the flow. Second, if the particles are too large or too dense, or the velocity gradients are too high, the particles lag the flow because of their finite inertia. The former issue means that micro-PIV works best in the absence of electric fields and furthermore benefits if neutrally buoyant particles are used. The second issue can be addressed by evaluating the Stokes number of the particle. If the Stokes number is small, then the particle follows the flow and its velocity can be used to infer the fluid velocity – these particles are called *Lagrangian flow tracers*.

8.4.2 Brownian motion

Brownian motion of microparticles is the random motion that is due to the statistical nature of fluid forces on particles. Brownian motion becomes more important as particles become small and the importance of individual particle–fluid collisions becomes larger. The root-mean-square PIV measurement error because of Brownian motion can be related to the particle diffusivity D:

PIV velocity error owing to Brownian motion

$$\langle \Delta u \rangle^2 = \frac{2D}{\Delta t}, \tag{8.45}$$

where Δt is the time between exposures, and the diffusion coefficient for small particles in Stokes flow is given by the Stokes–Einstein relation:

Stokes–Einstein relation for diffusivity of a sphere in the Stokes limit

$$D = \frac{k_B T}{6\pi\eta a}, \tag{8.46}$$

where a is the particle radius, k_B is Boltzmann's constant, and T is temperature. Equation (8.45) assumes that exposures are infinitely fast and determines the errors owing to the random velocity fluctuations *between* exposures. If the exposure time is long, the image can be blurred and there can be errors in the inferred velocity field that are due to the Brownian motion *during* each exposure.

8.5 SUMMARY

The Stokes equations are the limit of the Navier–Stokes equations as the Reynolds number approaches zero, in which case the unsteady and convective terms can be ignored. The Stokes equations are linear, and their solutions exhibit *instantaneity*, *time reversibility*, *linearity*, and *superposability*. The Stokes equations are commonly written in any of three forms:

$$\nabla p = \eta \nabla^2 \vec{\boldsymbol{u}} , \tag{8.47}$$

$$\nabla^2 p = 0 , \tag{8.48}$$

or

$$\nabla^2 (\nabla \times \vec{\boldsymbol{u}}) = \nabla^2 \vec{\boldsymbol{\omega}} = 0 . \tag{8.49}$$

Bounded Stokes flows are solved numerically for most geometries but can be solved with separation of variables and the Laplace equation for Hele-Shaw flows. Unbounded Stokes flows can be treated analytically with the multipolar Stokeslet solutions, described in Appendix F. In particular, the flow around a sphere can be written analytically in terms of a Stokeslet, which describes the response of the fluid to the force applied to the particle, and a stresslet, which describes the irrotational response that the fluid would experience if there was a sphere with a full-slip condition at the surface. A hydrodynamic interaction tensor is used to relate the velocity to the applied force:

$$\vec{\vec{\boldsymbol{G}}} = \frac{1}{8\pi\eta\Delta r}\left[\left(\vec{\vec{\delta}} + \frac{\vec{\boldsymbol{\Delta r}}\vec{\boldsymbol{\Delta r}}}{\Delta r^2}\right) + \frac{a^2}{\Delta r^2}\left(\frac{1}{3}\vec{\vec{\delta}} - \frac{\vec{\boldsymbol{\Delta r}}\vec{\boldsymbol{\Delta r}}}{\Delta r^2}\right)\right] . \tag{8.50}$$

For large spacings of particles, the stresslet term is often ignored and the hydrodynamic interaction tensor is written as the Oseen–Burgers tensor:

$$\vec{\vec{\boldsymbol{G}}}_0 = \frac{1}{8\pi\eta\Delta r}\left(\vec{\vec{\delta}} + \frac{\vec{\boldsymbol{\Delta r}}\vec{\boldsymbol{\Delta r}}}{\Delta r^2}\right) , \tag{8.51}$$

which provides the fundamental means for describing the hydrodynamic interaction of a collection of small particles with each other and with surfaces. Detailed modifications to these relations can be made for special cases, for example, for ellipsoidal particles or for particles in proximity to walls.

8.6 SUPPLEMENTARY READING

Some further concepts are covered in related texts. Low-*Re* number flows from the standpoint of perturbation expansions are covered in [24, 34, 88], including discussions of Stokes' and Whitehead's paradoxes and the Oseen linearization of flow around a circular cylinder. Faxén laws are covered in [30, 32, 88, 89]. A discussion of the forms of the

Stokes flow equations in terms of stream functions and the role of multiple differential operators (i.e., ∇^4 and E^4) is found in [30]. The Lorentz reciprocal theorem is covered in [89]. A variety of theorems relating velocity and stress properties of the Stokes flow fields are discussed in [24, 32, 89]. A stochastic description of Brownian motion of particles can be found in [32], and a similar description with a focus on macromolecules can be found in [90]. Descriptions of droplet motion with surfactant effects can be found in [30].

Stokes flow can be viewed from a thermodynamic standpoint as the fluid-mechanical regime in which the system departs only slightly from equilibrium. Because of this, the thermodynamics of irreversible processes with slight deviations from equilibrium [91, 92] is applicable, and reciprocal relations apply when we link fluid flow with ion flow, as is important in Chapters 9, 10, 13, and 15.

Specialized results relevant to discussion of particle transport in this chapter can be found in [32, 93, 94, 95, 96, 97], and specialized results related to PIV are discussed for macroscale flows in [98] and for microscale flows in [7].

8.7 EXERCISES

8.1 Stokes flow is a low-Reynolds number limit, and the resulting solution in Eqs. (8.12) and (8.13) is independent of the Reynolds number. Why, then, in Eq. (8.26), is the drag coefficient dependent on the Reynolds number? How could the drag coefficient be defined differently such that the drag coefficient is not a function of the Reynolds number?

8.2 Given a particle of radius a in an infinite domain and a freestream velocity in the x direction with magnitude U, solve for the velocity around the particle. Assume that the solution is axisymmetric and write the solution in terms of axisymmetric velocity components u_r and u_ϑ, as is done in Eqs. (8.12) and (8.13).

(a) Write the continuity equation in spherical coordinates, delete terms involving the azimuthal velocity u_φ or derivatives with respect to φ, and show that the continuity equation can be written as

$$\frac{\partial u_r}{\partial r} + \frac{2u_r}{r} + \frac{1}{r}\frac{\partial u_\vartheta}{\partial \vartheta} + \frac{u_\vartheta \cot\vartheta}{r} = 0\,. \tag{8.52}$$

(b) Write the radial momentum equation in spherical coordinates following Eq. (D.16), eliminate the left-hand side owing to the Stokes approximation, delete terms involving the azimuthal velocity u_φ or derivatives with respect to φ, and show that the radial momentum equation for Stokes flow can be written as

$$0 = -\frac{\partial p}{\partial r} + \eta\left[\frac{\partial^2 u_r}{\partial r^2} + \frac{2}{r}\frac{\partial u_r}{\partial r} - \frac{2u_r}{r^2} + \frac{1}{r^2}\frac{\partial^2 u_r}{\partial \vartheta^2} + \frac{\cot\vartheta}{r^2}\frac{\partial u_r}{\partial \vartheta} - \frac{2}{r^2}\frac{\partial u_\vartheta}{\partial \vartheta} - \frac{2u_\vartheta\cot\vartheta}{r^2}\right]. \tag{8.53}$$

(c) Write the colatitudinal momentum equation in spherical coordinates following Eq. (D.16), eliminate the left-hand side owing to the Stokes approximation, delete terms involving the azimuthal velocity u_φ or derivatives with respect to φ, and show that the colatitudinal momentum equation for Stokes flow can be written as

$$0 = -\frac{1}{r}\frac{\partial p}{\partial \vartheta} + \eta\left[\frac{\partial^2 u_\vartheta}{\partial r^2} + \frac{2}{r}\frac{\partial u_\vartheta}{\partial r} - \frac{u_\vartheta}{r^2\sin^2\vartheta} + \frac{1}{r^2}\frac{\partial^2 u_\vartheta}{\partial \vartheta^2} + \frac{\cot\vartheta}{r^2}\frac{\partial u_\vartheta}{\partial \vartheta} + \frac{2}{r^2}\frac{\partial u_r}{\partial \vartheta}\right]. \tag{8.54}$$

(d) Assume that the solutions for the radial and colatitudinal velocities have the following forms:

$$u_r = U\cos\vartheta\left[1 + \frac{C_1}{r} + \frac{C_2}{r^2} + \frac{C_3}{r^3}\right]. \tag{8.55}$$

$$u_\vartheta = -U\sin\vartheta\left[1 + \frac{C_4}{r} + \frac{C_5}{r^2} + \frac{C_6}{r^3}\right]. \tag{8.56}$$

These six constants can be fixed by examination of (a) the boundary conditions at $r = a$, which gives two constraints (one for each component of velocity at the wall); (b) the continuity equation, which gives three constraints (one for each term in the assumed form for velocity); and (c) the Navier–Stokes equation, which gives one new constraint (two total, but one is redundant with one from the continuity). With these constants fixed, the solution is complete.

8.3 Given the flow solution for a sphere moving to the left along the x axis in quiescent fluid at low Reynolds number,

$$u_r = U\cos\vartheta\left[-\frac{3}{2}\frac{a}{r} + \frac{1}{2}\frac{a^3}{r^3}\right] \tag{8.57}$$

and

$$u_\vartheta = -U\sin\vartheta\left[-\frac{3}{4}\frac{a}{r} - \frac{1}{4}\frac{a^3}{r^3}\right], \tag{8.58}$$

and the result for the drag force on the sphere:

$$F_{\text{drag}} = 6\pi\eta U a\,, \tag{8.59}$$

take the limit as $a \to 0$ at constant drag force and determine the flow response to a point source of force.

8.4 Show that Eq. (8.40) simplifies to $C_D = 24/Re$ as $e \to 0$.

8.5 Given a solution of solid microparticles with uniform radius whose volume fraction of solids is ξ, determine the ratio of the average particle spacing to the particle radius. If we wish for this ratio to be larger than 10, at which point particle–particle interactions can largely be ignored, how small must ξ be?

8.6 Given particles of radius a and density ρ_p and a fluid with viscosity $\eta = 1 \times 10^{-3}$ Pa s and density $\rho_w = 1$ kg/m^3, derive a relation for the terminal velocity of the particles when subjected to Earth's gravitational accleration (9.8 m/s^2). If a 20-μm-diameter microchannel has an axis aligned normal to the gravitational axis and is filled with a suspension of these particles, derive a relation for the time it would take for these particles to settle to the bottom of the channel. Calculate the velocity and settling time for

(a) 10-nm-diameter polystyrene beads ($\rho_p = 1300$ kg/m^3),

(b) 1-μm-diameter polystyrene beads ($\rho_p = 1300$ kg/m^3),

(c) 2.5-μm-diameter polystyrene beads ($\rho_p = 1300$ kg/m^3),

(d) human leukocytes (12 μm in diameter; $\rho_p = 1100$ kg/m^3).

8.7 Consider flow through a shallow microfabricated channel of uniform depth d with a sharp 90° turn whose inside corner is located at the origin, depicted in Fig. 8.7. Assume that the flow far from the corner is uniform across the width of the channel, and assume that the channel can be approximated as being wide, so that the flow near the inside corner can be approximated by the flow around the inside corner in an infinite medium. Informed by Eq. (7.74), write an equation for the pressure distribution and, from this, write an equation for the velocity distribution. In both cases, your solution will have a free constant that is proportional to the inlet and outlet flow velocities, which are left unspecified.

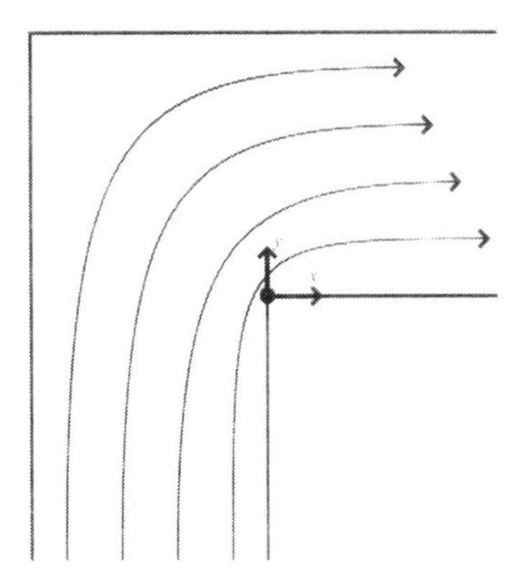

Fig. 8.7 Geometry of flow through a microfluidic device with a 90° turn.

8.8 Problems 7.31 and 8.7 consider the same geometry and result in closely related solutions that nonetheless have several key differences. Compare and contrast these two solutions.

8.9 Consider a microdevice used to capture cells from patient blood, depicted in Fig. 8.8. Assume that the microchannels are 1 cm wide and 100 μm deep and contain circular obstacles of diameter 150 μm. Assume that 1 mL of blood is cycled through the device over a period of 30 min. Approximate the blood as a suspension of rigid 8-μm spheres. Calculate the Stokes number for the spheres in this flow.

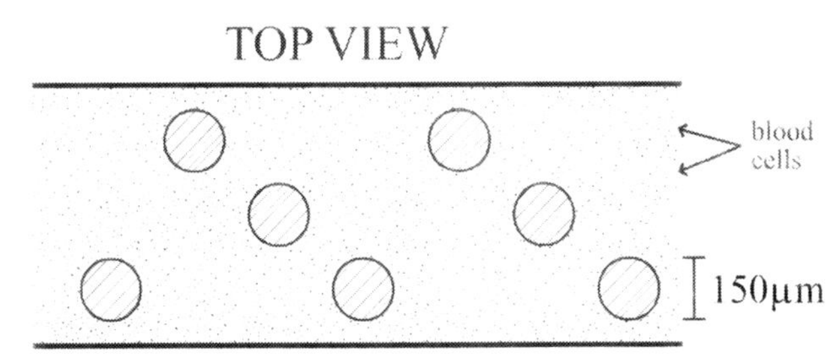

Fig. 8.8 Geometry of a microfluidic device with circular obstacles of diameter 150 μm. The channel is filled with blood. Obstacles are not to scale compared with the total device width.

8.10 Consider a sphere of radius a moving at velocity U from right to left along the x axis in Stokes flow. Consider the instant during which the sphere is at the origin, and thus $\vec{\Delta r} = \vec{r}$ and $\Delta r = r$. Model the effect of the sphere using the Oseen–Burgers hydrodynamic interaction tensor and write an expression for the vector velocity field. Take the dot product of this velocity field with $\vec{r}/r$ to determine the u_r field and compare the result with Eq. (8.12).

8.11 Use the general description of the pressure field in response to Stokes particle motion given in Eq. (8.35) and consider a rigid spherical particle moving in the negative x direction. Consider the instant during which the sphere is at the origin, and thus $\vec{\Delta r} = \vec{r}$ and

$\Delta r = r$. Calculate the pressure change caused by the particle motion and show that it is equal to that given by Eq. (8.14).

8.12 Consider Stokes flow over a rigid sphere located at the origin. Normalized by the radius of the particle, how far must we be from the particle surface for the contribution of the stresslet term to be less than 10% of the Stokeslet term? How far for it to be less than 1%?

8.13 Microdevices used to capture rare cells from blood often involve detailed optimization of the shear stress felt by a cell when the cell comes in contact with a wall whose chemical functionalization is designed to capture the cell.

(a) Consider a spherical cell with radius a sliding along a wall in a Hele-Shaw flow. Calculate the shear stress at the cell center as a function of the Hele-Shaw cell depth d, the radius a, and the local gradient $\frac{dp}{dx}$.

(b) Consider a Hele-Shaw device designed to allow flow from left to right in the x direction. Design the xy geometry of this device such that the shear stress along the x axis varies linearly from the inlet to the outlet of the device. For simplicity, you may make a quasi-1D assumption, i.e., assume that the flow is exclusively in the x direction. Although this assumption is certainly not completely correct, it does lead to a reasonable approximation of the flow rates in the x direction and therefore a reasonable approximation of the shear stress on the centerline.

8.14 Consider the flow visualized in Fig. 8.9, which consists of $a \simeq 50$ μm droplets of water immersed in a uniform 1.09-mm/s mean flow of oil in a 10-μm-deep channel.

Fig. 8.9 Flow of water droplets in an oil medium. Scale bar is 100 μm. From [99].

(a) Model the oil flow as a Hele-Shaw flow. Assume that the oil flow is fast relative to the velocity of the droplets and assume that the oil flow effectively sees a full-slip boundary condition at the oil–water interface. Show that the 2D velocity field is a potential velocity field.

(b) Show that the 2D velocity potential field caused by a single droplet is the superposition of the uniform flow with a point dipole:

$$\phi_v = Ux + R^2 (U - u) \frac{\vec{r}}{r^2} . \tag{8.60}$$

(c) Show that the force of the flow generated by one droplet on another droplet is given by

$$F = \frac{8\pi\eta R^2}{h} \nabla\phi_v(\vec{r}_1 - \vec{r}_2) . \tag{8.61}$$

(d) Given that a droplet in isolation is observed experimentally to travel at 295 μm/s, derive a relation for the velocity of the droplet in an infinitely long series of droplets whose centers are separated by a distance d.

(e) Given this velocity dependence, predict what will happen if a single droplet in an infinitely long series of droplets is perturbed and moves closer to the droplet behind it.

(f) Given a series of droplets, predict what will happen if a single droplet in an infinitely long series of droplets is perturbed and moves transverse to the droplet stream.

8.15 Consider a sphere rotating with angular velocity ω in a simple shear flow, i.e., a flow with velocity given by $u = U + \dot{\gamma} y$. Assume $Re \to 0$. Determine the lift force on the sphere.

8.16 It has been observed experimentally [100, 101] that small spheres embedded in a Poiseuille flow migrate to a single equilibrium radial position. Explain this migration using Stokes flow arguments, or, if this is not possible, explain why Stokes flow arguments cannot explain this phenomenon. Consider the role of the linearity of the Stokes equations when determining your answer.

8.17 Consider a spherical drop of radius a moving at velocity U in an otherwise quiescent fluid. Assume that the droplet remains spherical. Solve for the stream functions for the flow inside and outside of the droplet, and show that they are given, in turn, by

$$\psi_S = \frac{Ua^2 \sin^2\vartheta}{4} \frac{\left(\frac{r}{a}\right)^2 - \left(\frac{r}{a}\right)^4}{\frac{\eta_2}{\eta_1} + 1}, \tag{8.62}$$

and

$$\psi_S = \frac{Ua^2 \sin^2\vartheta}{2} \left[-\frac{r^2}{a^2} + \frac{r}{a} \frac{3\frac{\eta_2}{\eta_1} + 2}{2\left(\frac{\eta_2}{\eta_1} + 1\right)} - \frac{\frac{\eta_2}{\eta_1}}{2\left(\frac{\eta_2}{\eta_1} + 1\right)} \frac{a}{r} \right], \tag{8.63}$$

where the subscript 2 denotes properties inside the droplet and subscript 1 denotes properties in the ambient fluid.

8.18 Show that the flow produced by a sphere rotating with angular frequency ω is given by

$$\vec{u} = \omega \times \vec{r} \frac{a^3}{r^3}. \tag{8.64}$$

8.19 Compute the particle relaxation time and velocity uncertainties for polystyrene beads of diameters 10 nm, 100 nm, and 1 μm used to measure fluid velocities in room-temperature water by using a video rate camera (image separation time $t = 33$ ms).

8.20 Consider a PIV measurement of pressure-driven flow of 1-μm polystyrene beads moving at approximately 100 μm/s in a microchannel filled with room-temperature water. The field being imaged is a turn of radius 500 μm. Two images are taken and correlated to get displacements.

(a) How long must the time lag between exposures be to ensure that the error owing to Brownian motion is less than 3%?

(b) How short must the time lag between exposures be to ensure that the velocity measured by the PIV technique is a good approximation of the real velocity? In other words, how short must the time be to ensure that the particle's path during the time between exposures is approximately linear?

(c) Is it possible to satisfy both of the preceding constraints in this experiment?

8.21 Consider a spherical particle of radius a and density ρ_p in an infinite, quiescent fluid medium with viscosity η. The particle is actuated with a force in the x direction given by $F \cos \omega t$. Assume that the fluid flow is described by the Stokes equations, but the particle's finite inertia must be accounted for. Show that the magnitude of the particle velocity fluctuation Δx is given by

$$\Delta x = \frac{F}{6\pi a\eta} \frac{1}{\sqrt{1 + (2\omega\tau_p)^2}}, \tag{8.65}$$

where

$$\tau_p = \frac{2a^2\rho_p}{9\eta}. \tag{8.66}$$

9 The Diffuse Structure of the Electrical Double Layer

When considering electroosmotic flows, Chapter 6 focuses on *outer solutions*, namely solutions for flow far from boundaries. In that limit, we describe electroosmosis by using an effective slip boundary condition $\vec{u}_{\text{wall}} = \mu_{\text{EO}}\vec{E}$. A 1D integral model of the surface shows that, if the fluid properties are assumed uniform and the electrical potential at the wall is different from the bulk by a value of φ_0, then μ_{EO} is given by $\mu_{\text{EO}} = -\varepsilon\varphi_0/\eta$. The inner distribution of velocity and electrical potential need not be determined.

In this chapter, we address the electrical double layer or EDL (also called the Debye layer) near a charged wall and evaluate the spatial variation of charge and potential in this double layer. This determines the *equilibrium* structure of the fluid boundary layer near a surface in an electroosmotically driven system and describes the spatial variation of velocity near the wall. In the process, we relate the Coulomb force (related to the total wall charge density) and the distribution of the Coulomb force (related to the Debye length) to the velocity distribution. The flows that result are the *inner solutions* of electroosmotic flow problems. In total, solving for ion and potential distributions in the equilibrium EDL leads to predictions of fluid flow and current in electrically driven micro- and nanofluidic systems that do not perturb this equilibrium. As this chapter involves detailed discussion of ion concentrations, a review of the terminology and parameters found in Appendix B is recommended.

9.1 THE GOUY–CHAPMAN EDL

At equilibrium, solid surfaces have a net surface charge density q'' [C/m^2] because of ionization and adsorption processes. The immobile charge on the surface is balanced at equilibrium by a mobile, diffuse volumetric charge density [C/m^3]. Equivalently, ions with like charge to the wall (coions) are repelled from the region near the wall, whereas ions with opposite charge to the wall (counterions) are attracted to the region near the wall. We describe the distribution of ions and electrical potential by using equilibrium between electrostatic forces and Brownian thermal motion. By describing these distributions, we refine the integral result from Chapter 6 (i.e., the bulk electroosmotic velocity) to give the distribution of velocity near the wall of a microdevice.

This model, the Gouy–Chapman model (a schematic of which is shown in Fig. 9.1), accounts for the diffuse nature of the counterion distribution, uses bulk fluid properties throughout, treats the ions and solvent as ideal, and takes the interface properties as given. We build on this later in the chapter by describing modifications to this theory. We start by deriving the ion and potential distributions in the Gouy–Chapman EDL.

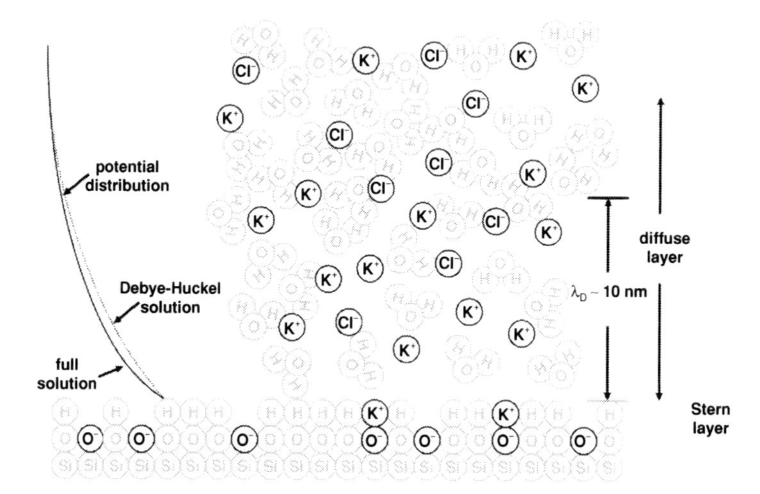

Fig. 9.1 The EDL consists of a region near an interface in which the net charge density is nonzero. Compared with the bulk solution, the counterions (ions with charge opposite that of the wall) are present at higher concentration, whereas the coions (ions with charge of same sign as the wall) are present at lower concentration.

9.1.1 Boltzmann statistics for ideal solutions of ions

Statistical thermodynamics predicts[1] that the likelihood of a system at temperature T being in a specific state with average energy per ion e_1 is proportional to

$$\exp(-e_1/k_B T) . \tag{9.1}$$

Here, k_B is the Boltzmann constant (1.38×10^{-23}J/K).[2] For an ion, the "system" is the ion, and the "state" is its location. For an *ideal solution*, in which ions are treated as non-interacting point charges suspended in a continuum, mean-field solvent, Eq. (9.1) means that the likelihood of an ion being in a specific location is a function of the electrostatic potential energy at that point.

The same relation can be written on a per-mole basis as

$$\exp(-\hat{e}_1/RT) , \tag{9.2}$$

where $\hat{e}_1 = e_1 N_A$ is the energy per mole.[3] R is the universal gas constant [J/mol K], given by

$$R = k_B N_A , \tag{9.3}$$

where N_A is Avogadro's number (6.022×10^{23}). We typically use molar values in this text.

1 The Boltzmann relation used in this section is valid only if the point-charge approximation can be made. For ions of finite size in close proximity, this relation breaks down because (a) the maximum ion concentration is limited by the size of the ion and (b) higher-order ion–ion interactions are possible, especially for multiply valent ions.

2 The reciprocal energy $\beta = 1/k_B T$ is commonly defined in the literature and used to replace $k_B T$ in the equations. We do not use this notation, but it is common and often left undefined in the literature owing to its ubiquity.

3 The symbol e_1 here is a singlet energy and is used to mean the energy of an ion based on its position. This notation becomes more meaningful later, when we distinguish between *singlet energies* e_1 and *pair potentials* e_2. This distinction is important in liquid-state theory or molecular dynamics. This notation also leads to clear distinction between e_1 (the ion energy) and e, the magnitude of the charge of an electron. We use the $\hat{e}_1$ symbol rarely, instead writing energy out as $zF\phi$.

The distribution of ions is thus governed by the potential energy of the ions:

$$\hat{e}_1 = zF\phi, \tag{9.4}$$

where z is the ion valence and F is Faraday's constant, given by $F = eN_A = 96485$ C/mol, e is the magnitude of the charge of an electron ($e = 1.6 \times 10^{-19}$ C), and ϕ denotes the local electrical potential. The preceding relations apply for an *indifferent ion*, which is an ion that interacts with a wall only through its electrostatic interaction, with no chemical interaction.

9.1.2 Ion distributions and potential: Boltzmann relation

Consider first the bulk solution far from any wall. Aqueous solutions contain dissolved ionic species, which we denote with subscript i. Each ion has a bulk concentration c_i and a valence or charge number z_i. For example, if we have a 1-mM solution of NaCl, $c_{Na^+} = 1$ mM and $c_{Cl^-} = 1$ mM and $z_{Na^+} = 1$ and $z_{Cl^-} = -1$. We use the subscript "bulk" or ∞ to define properties in the bulk, far from walls. Further, we define a double-layer potential

$$\varphi = \phi - \phi_{\text{bulk}}, \tag{9.5}$$

and thus φ is zero in the bulk. The value of φ thus specifies how the electrical potential at a point differs from that in the bulk far from walls. The use of φ is possible only when the interface can be assumed smooth, i.e., when the gradients in the bulk electrical potential can be assumed locally parallel to the surface, but this will be the case in most systems. From the arguments of Boltzmann statistics,[4] we can write in general that

$$c_i = c_{i,\infty} \exp\left(-\frac{z_i F \varphi}{RT}\right). \tag{9.6}$$

In the bulk, $\varphi = 0$, and Eq. (9.6) reduces to $c_i = c_{i,\infty}$. We can then write an expression for the local net charge density ρ_E as a function of the local potential:

$$\rho_E = \sum_i c_i z_i F, \tag{9.7}$$

or, inserting Eq. (9.6), we obtain

charge density inside the EDL

$$\rho_E = \sum_i c_{i,\infty} z_i F \exp\left(-\frac{z_i F \varphi}{RT}\right). \tag{9.8}$$

The ideal solution Boltzmann statistics result given in Eq. (9.8) can be combined with the Poisson equation to write a governing equation for the potential (or the ion distributions) in the double layer. We typically write and solve the equation for potential and use that solution and Eq. (9.8) to calculate species concentrations.

[4] Equation (9.6) is equivalent to the statement that, at equilibrium, the chemical potential in solution $\overline{g_i} = g_i^\circ + RT \ln \frac{c_i}{c_i^\circ} + z_i$ is uniform.

9.1.3 Ion distributions and potential: Poisson–Boltzmann equation

The Poisson equation (recall Subsection 5.1.5), which links potential to the local net charge density, is written for uniform ε as

$$\nabla^2 \phi = -\frac{\rho_E}{\varepsilon} . \tag{9.9}$$

Equations (9.8) and (9.9) can be combined to derive the Poisson–Boltzmann equation, which describes double layers if the solvent can be modeled as a mean field and the ions can be approximated as point charges. When we combine these relations, we find

Poisson–Boltzmann equation, uniform fluid properties

$$\nabla^2 \varphi = -\frac{F}{\varepsilon} \sum_i c_{i,\infty} z_i \exp\left(-\frac{z_i F \varphi}{RT}\right) . \tag{9.10}$$

Equation (9.10) is the general formulation of the *Poisson–Boltzmann equation*. This equation can be simplified by normalization, as detailed in Subsection E.2.3. To do this, we normalize the concentrations by the ionic strength in the bulk and normalize the potentials by the thermal voltage:

$$\varphi^* = \frac{F\varphi}{RT} , \tag{9.11}$$

where the *thermal voltage* RT/F is a measure of the voltage (approximately 25 mV at room temperature) that induces a potential energy on an elementary charge of the order of the thermal energy. We normalize the lengths by the *Debye length* λ_D, given by

definition of Debye length

$$\lambda_D = \sqrt{\frac{\varepsilon RT}{2F^2 I_c}}\Bigg|_{\text{bulk}} . \tag{9.12}$$

The Debye length plays an important role in all of our discussions of the EDL. The Debye length, which is a property of the electrolyte solution, gives a rough measure of the characteristic length over which the overpotential at a wall decays into the bulk. The Debye length is calculated with the ionic strength of the *bulk* and is a parameter of the bulk fluid. The nondimensionalized form of the Poisson–Boltzmann equation is given by

nondimensional Poisson–Boltzmann equation

$$\nabla^{*2} \varphi^* = -\frac{1}{2} \sum_i c^*_{i,\infty} z_i \exp\left(-z_i \varphi^*\right) . \tag{9.13}$$

EXAMPLE PROBLEM 9.1

Calculate λ_D for the following conditions. In each case, assume that the system is well approximated by a symmetric electrolyte with the valence z and concentration c specified in parentheses.

1. An electrolyte solution with $z = 1, c = 1 \times 10^{-7}$ M. This might represent deionized water.
2. An electrolyte solution with $z = 1, c = 2 \times 10^{-6}$ M. This might represent water at equilibrium with air – in this case there is dissolved CO_2 and equilibrium HCO_3^-.
3. An electrolyte solution with $z = 2, c = 1 \times 10^{-3}$ M. This might represent 1-mM Epsom salts.
4. An electrolyte solution with $z = 1, c = 0.18$ M. This might approximate phosphate-buffered saline (PBS) – this is often used for biochemical analysis, wound cleaning, or temporary cell storage.

SOLUTION: Recall the definition of λ_D:

$$\lambda_D = \sqrt{\frac{\varepsilon RT}{2F^2 I_c}} . \tag{9.14}$$

Note that I_c must be in SI units, i.e., mM.

At 20°C, we find for DI water $\lambda_D = 960$ nm, for water at equilibrium with air $\lambda_D =$ 215 nm, for Epsom salts $\lambda_D = 4.8$ nm, and for PBS $\lambda_D = 0.72$ nm.

9.1.4 Simplified forms of the nonlinear Poisson–Boltzmann equation

The nonlinear Poisson–Boltzmann equation is difficult to solve analytically owing to the summation of the charge density terms as well as their strongly nonlinear character. General implementation requires numerical solution. However, we must develop a physical intuition for these solutions, and simplified forms of the equations provide straightforward analytical solutions that illustrate key physical concepts.

SIMPLIFIED FORMS OF THE NONLINEAR POISSON-BOLTZMANN EQUATION: 1D

If the curvature of the wall is low (i.e., if the local radius of curvature of the wall is large relative to λ_D), we can simplify Eq. (9.13) by considering a 1D form. Assuming an infinite wall aligned perpendicular to the y axis, we obtain

1D Poisson–Boltzmann equation

$$\frac{\partial^2 \varphi^*}{\partial y^{*2}} = -\frac{1}{2} \sum_i c_{i,\infty}^* z_i \exp(-z_i \varphi^*) . \tag{9.15}$$

SIMPLIFIED FORMS OF THE NONLINEAR POISSON-BOLTZMANN EQUATION: 1D, SYMMETRIC ELECTROLYTE

If the solution is composed of one symmetric electrolyte, we can simplify Eq. (9.15) by noting that $c_1 = c_2 = c_\infty$ and defining $|z_1| = |z_2| = z$ (see Exercise 9.2), resulting in

1D Poisson–Boltzmann equation, symmetric electrolyte

$$\frac{\partial^2 \varphi^*}{\partial y^{*2}} = \frac{1}{z} \sinh(z\varphi^*) . \tag{9.16}$$

SIMPLIFIED FORMS OF THE NONLINEAR POISSON–BOLTZMANN EQUATION: 1D LINEAR POISSON–BOLTZMANN EQUATION

If $z_i \varphi^*$ is small relative to unity, we can replace the exponential term in Eq. (9.13) or (9.15) with a first-order Taylor series expansion by setting $\exp(x) = 1 + x$. Starting with Eq. (9.15), we find

$$\frac{\partial^2 \varphi^*}{\partial y^{*2}} = \frac{1}{2} \sum_i c_{i,\infty}^* {z_i}^2 \varphi^* , \tag{9.17}$$

which, from the definition of the ionic strength and the normalization of the species concentrations in Eq. (E.15), becomes

1D Poisson–Boltzmann equation, Debye–Hückel approximation

$$\frac{\partial^2 \varphi^*}{\partial y^{*2}} = \varphi^* . \tag{9.18}$$

See Exercise 9.3. This approximation is referred to as the *Debye–Hückel approximation*, and Eq. (9.18) is the 1D form of the *linearized Poisson–Boltzmann equation* (the Poisson–Boltzmann equation is sometimes termed the *nonlinear Poisson–Boltzmann equation* to make it clear that the full nonlinear term has been retained). The linearized Poisson–Boltzmann equation is valid for all regions in which φ^* is small. For cases in which φ_0^* is small relative to unity, this relation is valid throughout the flow field, and this linearization simplifies the analysis. Because of this, we often analyze systems in the Debye–Hückel limit in this text to provide qualitative insight. A discussion of the accuracy of the Debye–Hückel approximation for various double-layer parameters is presented in Section 9.4.

9.1.5 Solutions of the Poisson–Boltzmann equation

Solutions of the Poisson–Boltzmann equation are most easily studied starting from the simplified versions (which are amenable to analytical solution) and ending with the more complete solutions (which require numerical solution). We must always check our results to ensure that they are consistent with the approximations used to obtain the results. Two examples, which illustrate regions in which the linearized solution is alternately accurate or inaccurate, are shown in Figs. 9.2 and 9.3.

LINEARIZED 1D POISSON–BOLTZMANN SOLUTION – SEMI-INFINITE DOMAIN

The semi-infinite domain approximation can be used when double layers are thin, i.e., when λ_D is small compared with the depth or width of the channels under examination. In this case, the bulk fluid can effectively be treated as being at infinity. The linearized Poisson–Boltzmann equation (9.18) then leads to an exponential solution. Defining the

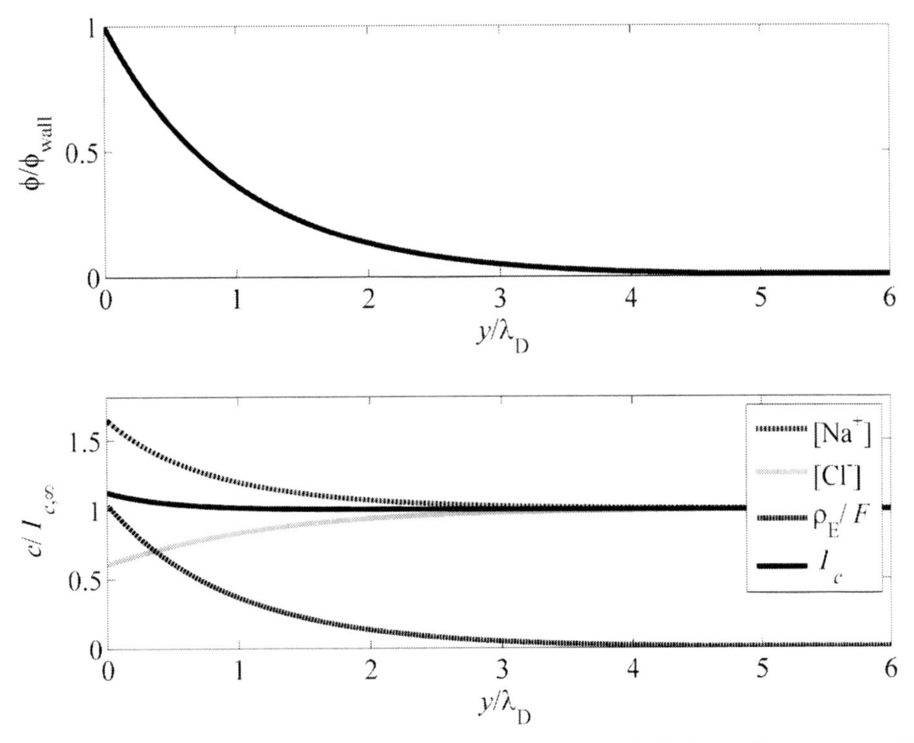

Fig. 9.2 Potential, species concentrations, ionic strength, and charge density in a double layer with surface potential of −12 mV and a (dilute) bulk solution composed of 10-mM NaCl. Species concentrations, ρ_E/F, and I_c are plotted normalized to I_c in the bulk. Here, the wall voltage is below the thermal voltage, and the species concentrations deviate only slightly from those predicted with the Debye–Hückel approximation. The local ionic strength, which is equal to the bulk in the Debye–Hückel limit, increases slightly near the wall.

origin *at the wall* and applying the boundary conditions $\varphi^* = 0$ at $y^* = \infty$ and $\varphi^* = \varphi_0^*$ at $y^* = 0$, we find

$$\varphi^* = \varphi_0^* \exp(-y^*) , \tag{9.19}$$

or, returning to dimensional form, we have

1D Poisson–Boltzmann equation for a semi-infinite domain, Debye–Hückel approximation

$$\varphi = \varphi_0 \exp(-y/\lambda_D) . \tag{9.20}$$

This is one of the solutions shown in Fig. 9.4. Because we have solved for φ^*, we can also write the solution for the concentration of the species:

species concentration distribution for semi-infinite domain around a surface with potential φ_0

$$c_i = c_{i,\infty} \exp[-z_i \varphi_0^* \exp(-y^*)] . \tag{9.21}$$

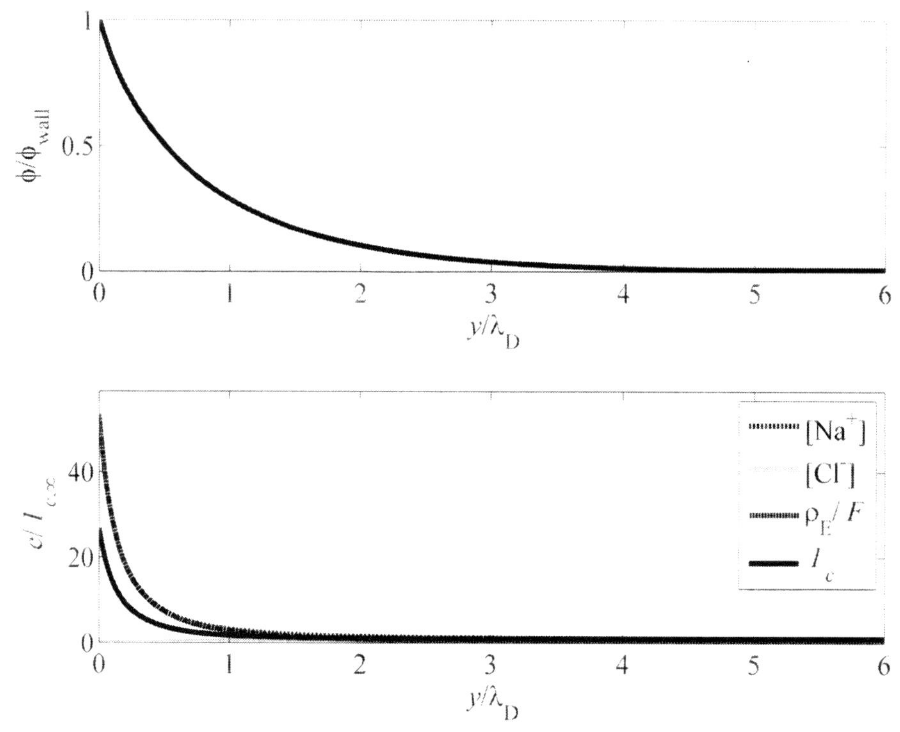

Fig. 9.3 Potential, species concentrations, ionic strength, and charge density in a double layer with surface potential of -100 mV and a (dilute) bulk solution composed of 10-mM NaCl, as predicted by the Poisson–Boltzmann equation. Species concentrations, ρ_E/F, and I_c are plotted normalized to I_c in the bulk. The strong nonlinearity of the species distribution is apparent – populations of the sodium counterion increase by a factor of 50 or so.

If the charge density q'' at the wall is known rather than the wall potential, we can show (Exercise 9.16) that, in the Debye–Hückel limit,

potential solution for semi-infinite domain around a surface with surface charge density q''

$$\varphi^* = -q''^* \exp(-y^*) , \tag{9.22}$$

where $q''^* = q''/\sqrt{2\varepsilon RTI_c}$.

Overall, the linearized solution, despite failing quantitatively for large wall potentials, gives a good qualitative picture of the EDL surrounding most insulating walls.

LINEARIZED 1D POISSON–BOLTZMANN SOLUTION – TWO PARALLEL PLATES

The semi-infinite domain solution is applicable for all cases for which the double layer is thin ($\lambda_D \ll d$, where d is a characteristic dimension of the channel). In that case, the solution near each wall is independent of any other walls. If double layers are not thin, the solution must account for the presence of the other walls. This can be done straightforwardly only for certain geometries.

Given two parallel plates separated by a distance $2d$ with the origin ($y = 0$) set at the midpoint, the hyperbolic cosine solutions of the linearized Poisson–Boltzmann equation (9.18) satisfy the boundary conditions. Applying the boundary conditions $\varphi^* = \varphi_0^*$ at

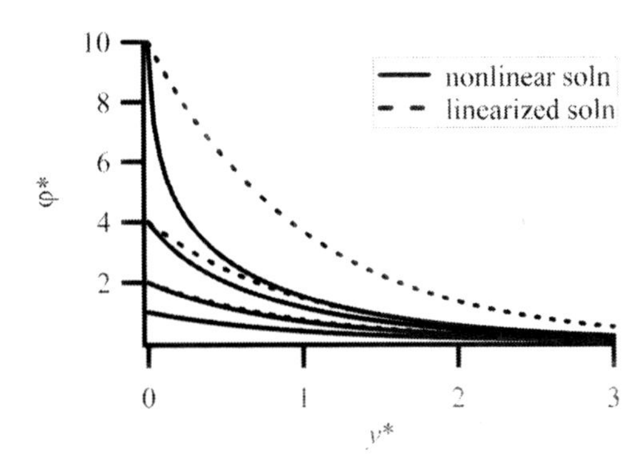

Fig. 9.4 Potential distributions in the EDL for the semi-infinite, symmetric electrolyte case, modeled both with the nonlinear and the linearized Poisson–Boltzmann relations and using $z = 1$. As $\varphi_0^* \to \infty$, the potential distribution far from the wall approaches the linearized solution for $z\varphi_0 = 4$, as derived in Exercise 9.10. The Poisson and Boltzmann relations can become inaccurate as φ^* becomes large.

$y^* = -d^*$ and $y^* = d^*$ and invoking symmetry at $y^* = 0$, we find

$$\varphi^* = \varphi_0^* \frac{\cosh(y^*)}{\cosh(d^*)}, \tag{9.23}$$

where $d^* = d/\lambda_D$. Returning to dimensional form, we get

potential distribution between two infinite parallel plates with potential φ^* at $y = \pm d$, Debye–Hückel approximation

$$\varphi = \varphi_0 \frac{\cosh(y/\lambda_D)}{\cosh(d/\lambda_D)}. \tag{9.24}$$

This solution highlights the effect of proximity of surfaces. If the surface charge density is known rather than the potential, we apply the boundary conditions $q''^* = q_0''^*$ at $y^* = -d^*$ and $y^* = d^*$. Invoking symmetry at $y^* = 0$, we find

$$\varphi^* = \frac{q_0''^*}{\tanh(d^*)} \frac{\cosh(y^*)}{\cosh(d^*)}, \tag{9.25}$$

where $d^* = d/\lambda_D$. Returning to dimensional form, we get

potential distribution between two infinite parallel plates with surface charge density q'' at $y = \pm d$, Debye–Hückel approximation

$$\varphi = \frac{q_0''\lambda_D}{\varepsilon} \frac{\cosh(y/\lambda_D)}{\sinh(d/\lambda_D)}, \tag{9.26}$$

which, evaluated at the wall, gives

relation between surface charge density and surface potential for parallel plates separated by $2d$, Debye–Hückel approximation

$$\varphi_0 = \frac{q_0''\lambda_D}{\varepsilon} \coth(d/\lambda_D). \tag{9.27}$$

Equation (9.27) shows that, when two parallel plates are separated by a distance that is small relative to λ_D, the surface potential must be large to induce a charge density that balances out the surface charge over a short distance. This relation properly predicts that the surface potential must be large, but is quantitatively incorrect because the Debye–Hückel approximation is invalid for large surface potentials.

1D, SYMMETRIC ELECTROLYTE POISSON–BOLTZMANN SOLUTION – SEMI-INFINITE DOMAIN

The 1D, symmetric electrolyte Poisson–Boltzmann equation (9.16) leads to a solution with a sharper-than-exponential dependence. Applying the boundary conditions $\varphi^* = 0$ at $y^* = \infty$ and $\varphi^* = \varphi_0^*$ at $y^* = 0$, we find (see Exercise 9.7)

1D potential distribution, semi-infinite domain, around surface with potential φ_0

$$\frac{\tanh(z\varphi^*/4)}{\tanh(z\varphi_0^*/4)} = \exp[-y^*]\,. \tag{9.28}$$

Given the identity $\tanh^{-1}(x) = \frac{1}{2}\ln\frac{1+x}{1-x}$, this can also be written as

1D potential distribution, semi-infinite domain, around surface with potential φ_0

$$\varphi^* = 2\ln\left(\frac{1+\tanh(z\varphi_0^*/4)\exp[-y^*]}{1-\tanh(z\varphi_0^*/4)\exp[-y^*]}\right)\,. \tag{9.29}$$

Equation (9.28) is reminiscent of the linearized solution in Eq. (9.19) except for the hyperbolic tangent, and the two are equal if $\frac{z\varphi_0^*}{4} \ll 1$. This solution is quantitatively useful and is applicable because many electrolyte systems are symmetric or well approximated by symmetric electrolytes.

Several differences are observed in the nonlinear solution compared with the linearized solution: (a) the potential decays from the wall more sharply than exponentially – this difference is large when φ^* is large relative to unity; (b) λ_D still describes the characteristic decay length in the farfield, where the local φ^* becomes small, but does not in general describe the characteristic decay length near the wall; and (c) the gradients in coions and counterions near the wall are larger than for the linearized solution. Figure 9.5 illustrates on a logarithmic axis how the solution to the Poisson–Boltzmann equation departs from purely exponential behavior at high surface potential.

GENERAL 1D POISSON–BOLTZMANN SOLUTION

For a general electrolyte, the Poisson–Boltzmann equation as written in Eq. (9.10) or (9.13) has no analytical solution and must be solved numerically.

9.2 FLUID FLOW IN THE GOUY–CHAPMAN EDL

Recall from Chapter 6 that the inner solution for purely electroosmotic flow is given by

inner solution for electroosmotic flow for wall with potential φ_0

$$u_{\text{inner}} = -\frac{\varepsilon E_{\text{ext,wall}}}{\eta}(\varphi_0 - \varphi)\,. \tag{9.30}$$

This inner solution assumes that $\vec{E}_{\text{ext}}$ is uniform on the length scales corresponding to the double-layer thickness, and thus is valid only near the wall.

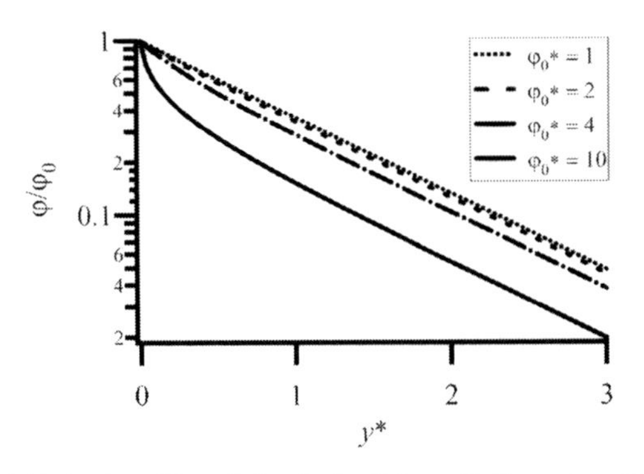

Fig. 9.5 Departure from exponential behavior of the solution to the Poisson–Boltzmann equation as the surface potential increases. Pure exponential behavior (linear behavior on these semilogarithmic axes) is observed far from the wall in all cases. These solutions are for $z = 1$. The Poisson and Boltzmann relations can become inaccurate as φ^* becomes large.

Now that we have determined the spatial dependence of φ in a variety of situations, we can write the spatial variation of the fluid velocity as well. For example, in the Debye–Hückel limit, the inner velocity solution is given by

inner velocity solution for electroosmosis, Debye–Hückel approximation

$$u_{\text{inner}} = -\frac{\varepsilon E_{\text{ext,wall}}}{\eta} \varphi_0 \left[1 - \exp(-y^*)\right] , \tag{9.31}$$

and analogous relations can be written for other flow conditions. For example, for flow between two plates located at $y = \pm d$, in the Debye–Hückel limit,

velocity solution for electroosmosis between plates at $y = \pm d$, Debye–Hückel approximation

$$u_{\text{inner}} = -\frac{\varepsilon E_{\text{ext,wall}}}{\eta} \varphi_0 \left(1 - \frac{\cosh(y/\lambda_D)}{\cosh(d/\lambda_D)}\right) . \tag{9.32}$$

This solution, although valid only for the linearized system, gives a good qualitative picture of electroosmotic flow between two parallel plates. Some example solutions are shown in Fig. 9.6. With the inner solutions calculated from the Gouy–Chapman model of the EDL, the description of electroosmotic flow in a microchannel can now be obtained through a composite asymptotic solution (see Exercise 9.15).

EXAMPLE PROBLEM 9.2

Assuming that the thin-EDL approximation requires that the characteristic depth d of a channel be large compared with λ_D (say, $d > 100\lambda_D$), how deep must a channel be for the thin-EDL approximation to be good if the solution is 1-mM NaCl? 0.01 mM? 100 mM?

SOLUTION:

For 1 mM, λ_D is 9.6 nm and d must be 0.96 μm. For 100 mM, λ_D is 0.96 nm and d must be 96 nm. For 0.01 mM, λ_D is 96 nm and d must be 9.6 μm.

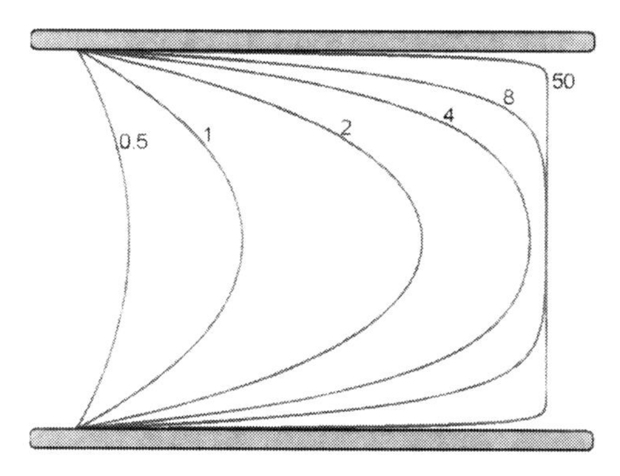

Fig. 9.6 Normalized electroosmotic flow distribution, i.e., $u(y)/\mu_{EO}E$ between two parallel plates located at $y = \pm d^*$, as predicted by the linearized Poisson–Boltzmann equation (and thus representative of the flow if the wall potential is small relative to RT/zF). Velocity distributions are shown for several d^* values. The distribution for $d^* = 50$ is a representative point at which the thin-EDL approximation becomes accurate for the flow distribution.

9.3 CONVECTIVE SURFACE CONDUCTIVITY

Beyond the native bulk conductivity of the electrolyte solution, the perturbed ion distribution in the double layer leads to two additional sources of electrical current when an extrinsic electric field is applied normal to the surface. The excess current beyond that which would exist for an uncharged surface is called *surface current*, and the property of the surface itself is written as a surface conductivity or surface conductance. Fluid flow combined with net charge density leads to a net convective current, and the change in the ion concentrations changes the ohmic conductivity of the fluid in the double layer. In this section, we concentrate on the former. The convective surface current is present in the Debye–Hückel limit, whereas the conductive surface current is zero in the Debye–Hückel limit. Both play a role when the surface potential is large.

When an extrinsic electric field is applied tangent to a charged surface, the electroosmotic flow, combined with the presence of a net charge density in the EDL, causes a net electrical current. In general, convective current (per length parallel to the wall and normal to the electric field) is given by

$$I'_{\text{conv}} = \int_{y=0}^{\infty} u\rho_E \, dy \,, \tag{9.33}$$

where I' denotes current per length. For a channel with perimeter $\mathcal{P}$, the convective surface current is given by $\mathcal{P}I'$. In the Debye–Hückel limit and assuming thin double layers, we can show by evaluating the integral in Eq. (9.33) that the convective surface current per length I' is given by

convective surface current, Debye–Hückel approximation, thin-EDL approximation, simple interface, uniform fluid properties

$$I' = \frac{\varepsilon^2 \varphi_0^2}{2\lambda_D \eta} E \,. \tag{9.34}$$

We can quantify the relative change in current owing to this phenomenon by defining the surface conductivity σ_s, which is the ratio of the convective surface current (normalized by the cross-sectional area of the channel) to the applied electric field:

$$\sigma_s = \frac{I/A}{E} \,. \tag{9.35}$$

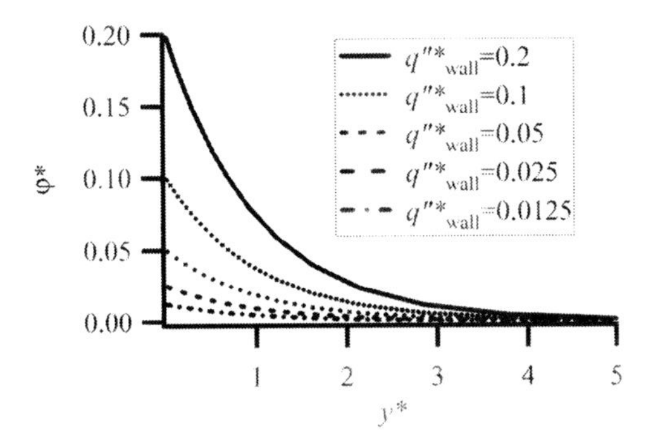

Fig. 9.7 Dependence of the potential distribution on the surface charge density at low surface charge density. Note that the normalized wall potential and normalized wall surface charge density are the same in the low surface charge density limit.

recalling that the hydraulic radius $r_h = 2A/\mathcal{P}$, the convective surface conductivity in the thin-EDL, Debye–Hückel limit is given by

$$\sigma_s = \frac{\varepsilon^2 \varphi_0^2}{\lambda_D \eta r_h} . \tag{9.36}$$

From this, we can see that the convective surface density scales with the surface potential squared, and it is inversely proportional to the hydraulic radius and the Debye length. Thus we expect convective surface conductivity to be significant when the surface potential is large and the channel is small.

Recalling the electrokinetic coupling matrix, first presented in Chapter 3:

$$\begin{bmatrix} Q/A \\ I/A \end{bmatrix} = \chi \begin{bmatrix} -\frac{dp}{dx} \\ E \end{bmatrix} , \tag{9.37}$$

we see that convective surface conductivity is accounted for in the thin-EDL, Debye–Hückel limit by setting

$$\chi_{22} = \sigma + \frac{\varepsilon^2 \varphi_0^2}{\lambda_D \eta r_h} . \tag{9.38}$$

The first term is the contribution of the bulk ohmic conductivity, and the second is the contribution of the convective surface conductivity.

9.4 ACCURACY OF THE IDEAL-SOLUTION AND DEBYE–HÜCKEL APPROXIMATIONS

The two key approximations used in this chapter have been the Debye–Hückel approximation (wall potential small compared with the thermal voltage) and the ideal-solution approximation (point charge, mean field). The accuracy of these approximations is discussed in some detail in this section.

EXAMPLE PROBLEM 9.3

Calculate and plot the dependence of the potential distribution φ^* on the surface charge density for two sets of surface charge densities: first, $q''^*_{\rm wall} =$ 0.0125, 0.025, 0.05, 0.1, 0.2, which corresponds to the low surface charge density limit; and $q''^*_{\rm wall} = 0.5, 1, 2, 4, 8$, which corresponds to high surface charge densities.

SOLUTION: See Figs. 9.7 and 9.8.

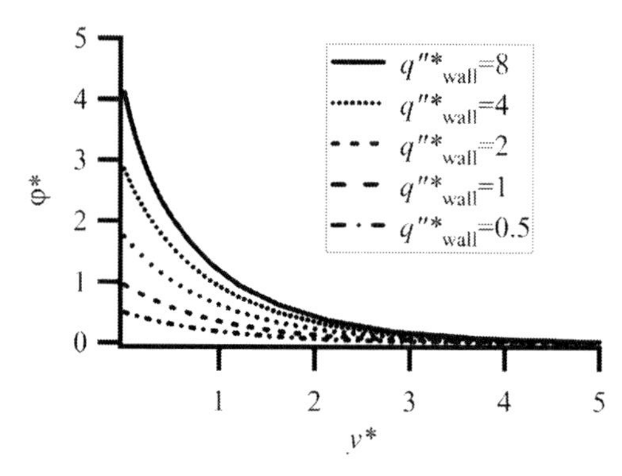

Fig. 9.8 Dependence of the potential distribution on the surface charge density at high surface charge density. Note that increases in the wall charge density lead to less and less increase in the wall potential.

9.4.1 Debye–Hückel approximation

The applicability of the Debye–Hückel approximation is subtle, owing to the nonlinearity of the system and the wide variety of double-layer parameters under consideration. Although it is not mathematically rigorous to do so, the Debye–Hückel approximation is often used in situations in which $z\varphi^*$ is *not* small compared with unity, and we must understand when this approximation is acceptable. Depending on the situation or application, we might be concerned with equations relating the wall potential φ_0, $c(y)$, ρ_E, the double-layer capacitance, or the normal electric field at the wall. The validity of the Debye–Hückel approximation is not equivalent for these different parameters. Table 9.1 lists some parameters and qualitatively indicates the magnitude of the errors caused by the Debye–Hückel approximation. This table is strictly qualitative, as the errors caused by the Debye–Hückel approximation are a strong function of the magnitude of φ_0^*. However, this table gives a rough idea of when the Debye–Hückel approximation is defensible. The conclusions from this table are as follows: Given a specified φ_0^*,

Table 9.1 Predictions resulting from applying the Debye–Hückel approximation to a Poisson–Boltzmann system with uniform permittivity and viscosity, depending on whether the interface potential φ_0 or interface charge density q_0'' is known.

Parameter	Given φ_0	Given q_0''
$\lvert\varphi_0\rvert$	Correctly predicted	Grossly overpredicted
$\lvert u_{\text{bulk}}\rvert$	Correctly predicted	Grossly overpredicted
$\lvert u(y)\rvert$	Slightly underpredicted	Grossly overpredicted
$\lvert\varphi(y)\rvert$	Slightly overpredicted	Grossly overpredicted
Streaming potential magnitude	Slightly underpredicted	Moderately overpredicted
$\lvert\rho_E(y)\rvert$	Grossly underpredicted	Depends on y
Double-layer capacitance	Grossly underpredicted	Correctly predicted
$\left\lvert\frac{\partial u}{\partial y}\right\rvert_{\text{wall}}$	Grossly underpredicted	Correctly predicted
Electric field magnitude normal to wall	Grossly underpredicted	Correctly predicted
Navier slip velocity magnitude at wall	Grossly underpredicted	Correctly predicted
$\lvert q_0''\rvert$	Grossly underpredicted	Correctly predicted

Note: Integral properties are more closely tied to φ_0, whereas differential properties are more closely tied to q_0'', especially near the wall. The absolute magnitude of any error is a strong function of how large φ_0^* or $q_0''^*$ is; this table only estimates which predictions are more or less accurate in a relative sense.

the integrated effect of the double layer on the bulk electroosmotic velocity is independent of the model we use for the variation of potential in the double layer – in fact, the expression for the bulk velocity is derived in Chapter 6 in the absence of a double-layer model. The variations of the velocity and the electrical potential are only weakly affected by the Debye–Hückel approximation (see Fig. 9.4), as is streaming potential (see Subsection 10.6.3), especially because the errors in net charge density caused by the Debye–Hückel approximation are in regions where the flow velocity is low. However, the charge density, normal derivatives of the potential and the velocity at the wall, the related wall velocity predicted by the Navier slip model (Subsection 1.6.4), wall charge density, and double-layer capacitance (Subsection 16.2.1) are predicted poorly by the Debye–Hückel approximation – in fact, these last parameters can be off by a factor of 100 or so for a glass surface at neutral pH in a dilute buffer.

Given φ_0, the results of the Debye–Hückel approximation are reasonably accurate if we are considering integrated values (u_{bulk}) but can be quite inaccurate if we are considering differentiated values ($\frac{\partial u}{\partial y}|_{\text{wall}}$). The opposite is true if q''_{wall} is given – differentiated values are reasonably accurate, but integrated values are not.

9.4.2 Limitations of the ideal solution approximation

Given a known potential at an interface, the Gouy–Chapman model of the EDL describes the equilibrium potential distribution near that wall, as well as the distribution of idealized point electrolytes interacting *only* with that bulk potential field. As such, the Gouy–Chapman model is an equilibrium, mean-field, point-charge formulation that fails to account for solvent structure, ion size, ion–ion correlations, or ion dynamics.

Here we address the mean-field, point-charge aspects of the Gouy–Chapman double layer and highlight both the limits of its applicability and modifications that can be used to extend its applicability. As was the case for the Debye–Hückel approximation, the errors caused by the Boltzmann approximation are generally most dramatic for differential values and most moderate for integrated values.

9.5 MODIFIED POISSON–BOLTZMANN EQUATIONS

Modified Poisson–Boltzmann equation is a catchall term for formulations that use the Poisson–Boltzmann equation with some sort of modification to overcome its limitations. One type of modified Poisson–Boltzmann equation is presented here – one with steric corrections, i.e., corrections that account for the finite space taken up by ions.

9.5.1 Steric correction to ideal solution statistics

The use of ideal solution statistics to describe ion populations requires that the ions be modeled as point charges with an infinitesimal radius. The concentration given by Eq. (9.6) has no upper limit, and the Poisson–Boltzmann equation predicts infinitely high concentration as surface potential goes to infinity. This is inconsistent with the finite size of the ions, which leads to crowding at a surface with a large potential (Fig. 9.9).

If we model the ions as rigid spheres, we can apply a *hard-sphere correction*, and it accounts for the fact that ions have a finite size. This size is often hard to define quantitatively, but we can approximate it by using the diameter of an ion (≈ 1 Å), or the diameter of an ion with a hydration shell (≈ 3 Å). We define an *effective* hard-sphere packing length λ_{HS} such that the maximum ion number density is $n_{\max} = \lambda_{\text{HS}}^{-3}$ and the

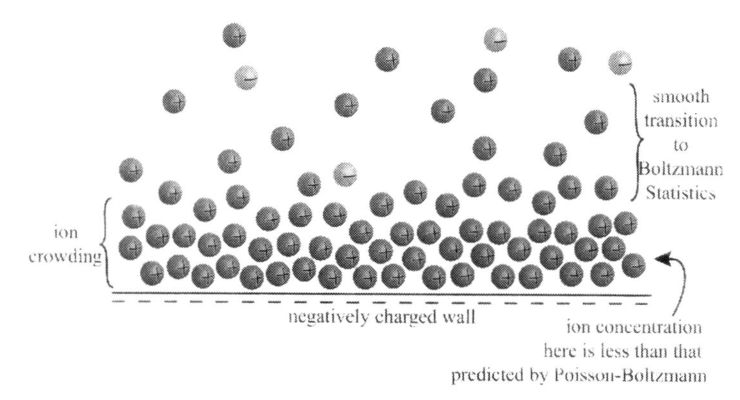

Fig. 9.9 Ion crowding at an interface with large potential.

maximum ion concentration possible is given by

$$c_{\max} = \frac{1}{\lambda_{\text{HS}}^3 N_{\text{A}}}. \tag{9.39}$$

Although the magnitude of this quantity is phenomenological, it can nonetheless be used to modify the Poisson–Boltzmann relation and provide results that better match experimental data.

First, recall the result from ideal solution Boltzmann statistics:

$$c_i = c_{i,\infty} \exp\left(-\frac{z_i F \varphi}{RT}\right). \tag{9.40}$$

We adjust Eq. (9.40) by adding a correction factor in the denominator to account for the space taken up by ions:

steric modification to Boltzmann relation for species concentration dependence on potential

$$c_i = \frac{c_{i,\infty} \exp\left(-\frac{z_i F \varphi}{RT}\right)}{1 + \sum_i \frac{c_{i,\infty}}{c_{i,\max}} \left[\exp\left(-\frac{z_i F \varphi}{RT}\right) - 1\right]}. \tag{9.41}$$

Here, the exp() − 1 term denotes the normalized deviation from the bulk. Multiplied by $c_{i,\infty}$ and summed over i, this is the difference in concentration from the bulk. Divided by $c_{i,\max}$, this is the additional fractional volume taken up by the ions owing to the change in concentration. The denominator is unity plus a term that denotes how much more volume is taken up by the ions than is taken up in the bulk.

This relation can be considered in a number of limits:

- $\varphi \to 0$: here all exponentials are unity, and $c_i = c_{i,\infty}$.
- $c_{\max} \to \infty$ or $\lambda_{\text{HS}} \to 0$: this is the point-charge limit, in which case the denominator equals unity, and the ideal solution Boltzmann statistics result is recovered regardless of the potential.
- *Symmetric electrolyte.* The symmetric electrolyte is convenient if we further assume that $c_{\max}$ is the same for both species. This leads to analytical simplicity. For the

symmetric electrolyte we find

$$c_i = \frac{c_\infty \exp\left(-\frac{z_i F\varphi}{RT}\right)}{1+\xi\left[\cosh\left(-\frac{zF\varphi}{RT}\right)-1\right]} = \frac{c_\infty \exp\left(-\frac{z_i F\varphi}{RT}\right)}{1+2\xi\sinh^2\left(-\frac{zF\varphi}{2RT}\right)}. \tag{9.42}$$

Here we define (for symmetric electrolytes with matched λ_{HS}) a packing parameter $\xi = 2c_\infty/c_{\text{max}} = 2c_\infty N_{\text{A}}\lambda_{\text{HS}}^3$, which is the volume fraction of ions in the bulk. For small ions at millimolar concentrations, ξ is approximately 1×10^{-5}, but the denominator can become large for large ions or high surface potentials, which lead to high concentrations.

- *Symmetric electrolyte,* $\varphi \to \infty$. Here $\cosh(x) \to \frac{1}{2}\exp(x)$ and the constant terms drop out, so for the coions

$$c = 0, \tag{9.43}$$

and for the counterions

$$c = \frac{c_\infty \exp\left(-\frac{z_i F\varphi}{RT}\right)}{\frac{\xi}{2}\exp\left(-\frac{z_i F\varphi}{RT}\right)} = \frac{2c_\infty}{\xi} = \frac{1}{N_{\text{A}}\lambda_{\text{HS}}^3} = c_{\text{max}}, \tag{9.44}$$

which is precisely the maximum packing fraction defined earlier. Thus, at infinite potential, we get a sterically limited concentration rather than an infinite concentration.

9.5.2 Modified Poisson–Boltzmann equation

With the concentrations of individual species specified by Eq. (9.41), we can write a new general modified Poisson–Boltzmann equation:

$$\nabla^2\varphi = -\frac{\rho_{\text{E}}}{\varepsilon} = -\frac{\sum_i z_i c_i F}{\varepsilon}, \tag{9.45}$$

which becomes

sterically modified Poisson–Boltzmann equation

$$\nabla^2\varphi = -\frac{F}{\varepsilon}\frac{\sum_i z_i c_{i,\infty}\exp\left(-\frac{z_i F\varphi}{RT}\right)}{1+\sum_i \frac{c_{i,\infty}}{c_{i,\text{max}}}\left[\exp\left(-\frac{z_i F\varphi}{RT}\right)-1\right]}. \tag{9.46}$$

Nondimensionalizing and making 1D, we have

$$\frac{\partial^2\varphi^*}{\partial y^{*2}} = \frac{\sum_i z_i c_{i,\infty}^*\exp(-z_i\varphi^*)}{1+\sum_i \frac{c_{i,\infty}^*}{c_{i,\text{max}}}\left[\exp(-z_i\varphi^*)-1\right]}. \tag{9.47}$$

The previous equations can be written more simply for symmetric electrolytes with matched λ_{HS}:

$$\nabla^2\varphi = \frac{zFc_\infty}{\varepsilon}\frac{2\sinh\left(\frac{zF\varphi}{RT}\right)}{1+2\xi\sinh^2\left(\frac{zF\varphi}{2RT}\right)}. \tag{9.48}$$

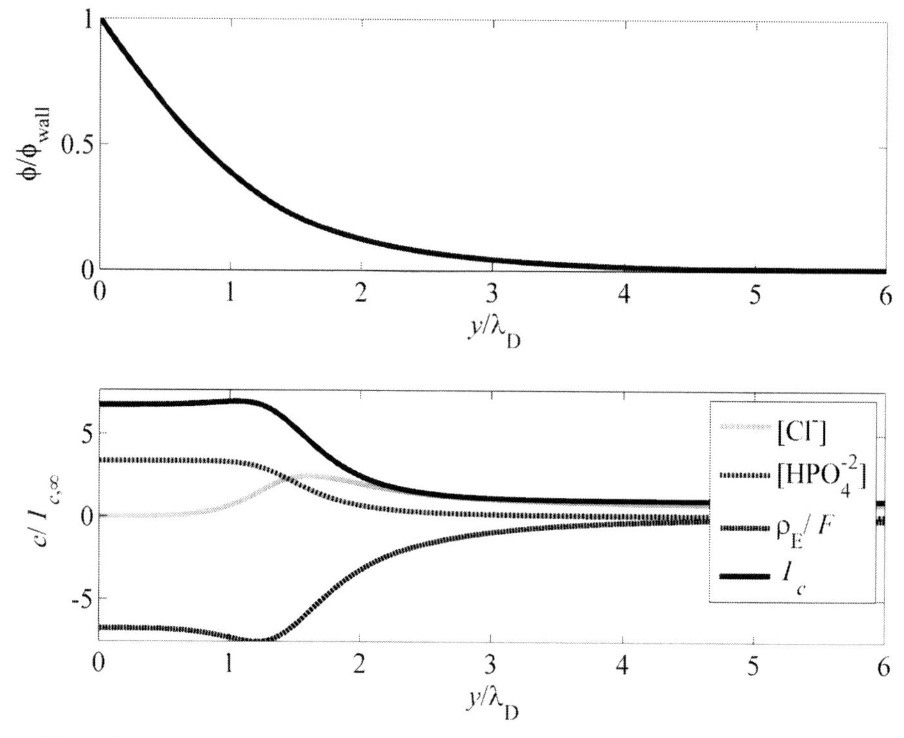

Fig. 9.10 Potential, species concentrations, ionic strength, and charge density in a double layer with surface potential of 250 mV and a (dilute) bulk solution composed of 100-mM NaCl and 10 mM $MgHPO_4$, as predicted by a modified Poisson–Boltzmann equation. Species concentrations, ρ_E/F, and I_c are plotted normalized to I_c in the bulk. At this voltage, which is large relative to the thermal voltage, the modified Poisson–Boltzmann equation predicts a large region of uniform ion concentration, where the most highly valent counterion – in this case, HPO_4^{-2}– is packed to its steric limit. Note also that the concentration of one of the counterions *does not* increase monotonically. This is because steric hindrance "crowds" out counterions of lower valence.

Nondimensionalizing and simplifying to 1D as we did for Eq. (9.16), we find

1D nondimensional sterically modified Poisson–Boltzmann equation, symmetric electrolyte

$$\frac{\partial^2 \varphi^*}{\partial y^{*2}} = \frac{\frac{1}{z}\sinh(z\varphi^*)}{1 + 2\xi \sinh^2(z\varphi^*/2)}. \tag{9.49}$$

An example of a solution to the modified Poisson–Boltzmann equation (9.47) can be seen in Fig. 9.10.

9.5.3 Importance and limitations of Poisson–Boltzmann modifications

The modified Poisson–Boltzmann equation leads to well-behaved solutions in the high-φ limit. Without some accounting for the finite limits of electrolyte concentration, the Poisson–Boltzmann equation predicts a variety of nonphysical results in the large-φ limit, including unbounded electric field at the wall and unbounded wall capacitance (see Chapter 16 for a discussion of EDL capacitance). The correction shown here (which uses only steric hindrance arguments) is a simple fix that makes the predicted solution bounded; however, its accuracy is limited by the limitations of the steric model itself, the accuracy with which c_{max} is known, and other considerations. It also fails to predict the detailed structure of the distribution functions as an integral theory would (see Section H.2). The modification to the Poisson–Boltzmann formulation is most prominent when the dynamic behavior of the EDL is examined, rather than its equilibrium.

9.6 STERN LAYER

The Gouy–Chapman theory is commonly modified by splitting the double layer into a diffuse (Gouy–Chapman) region and a condensed (Stern) region near the wall. In doing so, the condensed region near the wall plays little role directly in terms of fluid mechanics, but it does affect the double-layer capacitance as well as the relation between the chemistry at the surface and the wall potential φ_0. To an extent, the Stern model is simply the limit of the modified Poisson–Boltzmann models we discuss in this chapter.

9.7 SUMMARY

The Poisson and Boltzmann relations predict potential and ion distributions in the equilibrium EDL for ions modeled as point charges and solvents modeled as continua. The combination of these two relations leads to the Poisson–Boltzmann equation, which can be written as

$$\nabla^2\varphi = -\frac{F}{\varepsilon}\sum_i c_{i,\infty} z_i \exp\left(-\frac{z_i F\varphi}{RT}\right). \tag{9.50}$$

Because the Debye length is often small compared with the curvature of microdevice surfaces, this equation is typically solved in 1D form. Although a general solution is impossible, linearizing the equation and assuming that the electrolyte is symmetric both lead to analytically tractable forms of the equation.

The Debye–Hückel approximation is particularly important owing to its ubiquity in double-layer analysis. This linearization leads to satisfactory solutions if the wall potential is known and integral double-layer results are desired, or if the surface charge is known and differential double-layer results are desired. Otherwise, this approximation fails badly.

With the results from Chapter 6, the potential distribution in the EDL also gives the solution for fluid flow near the wall. This solution is the inner solution for electroosmosis problems with equilibrium EDLs.

Because the Gouy–Chapman EDL ignores finite ion sizes and ion condensation at the wall, the Poisson–Boltzmann equation leads to diverging results for large wall potentials. Many modifications to the Poisson–Boltzmann equation have been proposed. In this chapter, a steric hindrance modification is shown that eliminates this divergence. With this approach, the modified Poisson–Boltzmann relation is

$$\nabla^2\varphi = -\frac{F}{\varepsilon}\sum_i z_i \frac{c_{i,\infty}\exp\left(-\frac{z_i F\varphi}{RT}\right)}{1+\sum_i \frac{c_{i,\infty}}{c_{i,\max}}\left[\exp\left(-\frac{z_i F\varphi}{RT}\right)-1\right]}. \tag{9.51}$$

Although the EDL is described in much more detail, we have still assumed that the boundary condition (i.e., φ_0 or q''_0) is known.

9.8 SUPPLEMENTARY READING

This chapter is focused on the Gouy–Chapman description of the diffuse EDLs. The double-layer theory presented here, combined with the presentation of electroosmosis in Chapter 6, constitutes our first foray into the description of fluid flow past a surface in chemical equilibrium with one-way coupling, i.e., in which the EDL induces flow, but the flow leaves the EDL unaffected. This description is our first presentation of the classical theory of linear electrokinetics and has been presented by a number of authors in several useful resources. The Gouy–Chapman double-layer theory is named after Gouy [102] and Chapman [103], who presented early versions of the theory. The EDL is discussed in the context of fluid flow in electrolyte solutions in Probstein's physicochemical hydrodynamics text [29] and Hunter's texts [69, 70]. Bard and Faulkner [59] present the EDL in the context of electrochemistry and EDLs at electrodes. Israelachvili [104] presents the EDL in the generalized context of surface forces, which provides a complementary point of view of the same physics. Colloid science texts, such as those of Russel et al. [32], Hunter [70], and Lyklema [71], provide detailed double-layer information with a view toward surface–surface forces, colloidal stability, and the experimental techniques used to measure interfacial parameters to inform colloidal analysis. A review by Anderson [105] focuses on the motions of particles owing to interfacial phenomena.

We have assumed that the properties of the solution, particularly the electrical permittivity and the viscosity, are uniform throughout. Comments on the limitations of this assumption and proposed models for accounting for this (for example, viscoelectric modeling) can be found in papers by Overbeek [106] and Lyklema [107] as well as Hunter's text [69]. Electrochemical texts such as [58] use differential capacitance measurements to address permittivity variations in the EDL. More detail on the structure and dynamics of these EDLs is presented in Chapters 10 and 16.

Stern [108] introduces the concept of the Stern layer and an early equation resulting from modified Poisson–Boltzmann theory. Steric modifications to Poisson–Boltzmann equations using lattice approaches are generally attributed to Bikerman [109] and have been described more recently by Borukhov et al. [110]. Kilic et al. have implemented this in great detail with a view toward managing the fluid flow in nonequilibrium EDLs developed at electrodes or metallic surfaces [111, 112]. Rather than the symmetric lattice approach, theories from the state theory of liquids (the Carnahan–Starling equation of state) have also been applied to dense ion packing [113, 114, 115, 116]. Other modified Poisson–Boltzmann models include short-range forces between ions [117, 118] and walls [118, 119, 120, 121]. Modifications to Poisson–Boltzmann theories are numerous, and many reviews have been presented [122, 123, 124, 125].

9.9 EXERCISES

9.1 Graph the normalized velocity as a function y^* for a Debye–Hückel double layer.

9.2 Derive Eq. (9.16) from Eq. (9.15).

9.3 Derive Eq. (9.18) from Eq. (9.13).

9.4 Derive Eq. (9.19) from Eq. (9.18).

9.5 Derive Eq. (9.23) from Eq. (9.18).

9.6 Derive Eq. (9.25).

9.7 Derive Eq. (9.28) from Eq. (9.16). This takes several steps:

(a) Multiply both sides by $2(\frac{d\varphi^*}{dy^*})$.

(b) Using dummy variables φ' and y', integrate from a point in the bulk ($\varphi' = 0$; $y' = \infty$) to a point in the double layer ($\varphi' = \varphi^*$; $y' = y^*$) to get an equation in terms of $(\frac{d\varphi^*}{dy^*})^2$.

(c) Note the identity $-1 + \cosh x = 2\sinh^2(x/2)$ and solve for $\frac{d\varphi^*}{dy^*}$.

(d) Note that $\sinh x = 2\sinh(x/2)\cosh(x/2)$, and rearrange to write the equation in terms of $\tanh(\frac{z\varphi^*}{4})$ and the y^* derivative of $\tanh(\frac{z\varphi^*}{4})$.

(e) Solve the ODE to get a relation for $\tanh(\frac{z\varphi^*}{4})$ and thus derive Eq. (9.16).

9.8 Consider a 1-mM NaCl solution of infinite extent at room temperature near an infinite plate aligned along the xz plane. Plot the following relations. For each, consider φ_0 equal to -150 mV, -100 mV, -50 mV, -25 mV, and -5 mV. Include the curves for all five φ_0 values on each plot.

(a) $\frac{\varphi^*}{\varphi_0^*}$ vs. y^*.

(b) φ^* vs. y^*.

(c) φ vs. y.

(d) c_{Na^+} vs. y (use log scale).

(e) c_{Cl^-} vs. y (use log scale).

(f) ρ_E vs. y.

9.9 For a symmetric electrolyte *not* in the Debye–Hückel limit and a given normalized wall charge density $q_0''^*$, evaluate the potential at the wall. Start with the Poisson–Boltzmann equation (9.16) and first derive a relation between φ^* and its derivative at the wall:

(a) Multiply both sides by $2(\frac{d}{dy}\varphi)$.

(b) Integrate both sides, rearranging the derivatives so that on one side you have the derivative of $(\frac{d}{dy}\varphi)^2$ and on the other side you have $d\varphi$.

(c) To evaluate $\frac{d}{dy}\varphi$, integrate over a dummy variable (y' or φ', depending on which side of the equation) from the bulk ($y' = \infty$; $\varphi' = 0$) to some point in the double layer ($y' = y$; $\varphi = \varphi'$).

(d) Derive a relation between φ^* and its derivative at the wall.

Given this result, use the boundary condition relation between the charge density and the gradient of the potential to write the normalized wall potential φ_0^* in terms of the normalized wall charge density $q_0''^*$. Use the identity $\sinh^{-1}x = \ln(x + \sqrt{x^2+1})$. Compare your result with the result obtained in the Debye–Hückel limit. For a given $q_0''^*$, does the Debye–Hückel limit underpredict or overpredict φ_0^*?

9.10 Consider a 1D semi-infinite domain and a symmetric electrolyte. Show that, if $z\varphi_0^* \to \infty$, the solution for φ^* far from the wall approaches $\varphi^* = \frac{4}{z}\exp(-y^*)$.

9.11 Derive a relation for the potential distribution (in terms of φ_0^*, r, and λ_D) around a sphere of radius a and normalized surface potential φ_0^*. Assume $\varphi_0^* \ll 1$ and assume that the electrolyte is symmetric.

Note that you will find the analysis easier if you start by solving for $\varphi^* r^*$.

9.12 Derive a relation for the potential distribution (in terms of φ_0^*, $\imath$, and λ_D) around an infinite cylinder of radius a and normalized surface potential φ_0^*. Assume $\varphi_0^* \ll 1$ and assume that the electrolyte is symmetric.

Note that the solution involves manipulating the governing equation to obtain a modified Bessel's equation.

9.13 Given a 10 mM solution of KCl in contact with an infinite flat wall with a surface charge of -40 mV and an extrinsically applied field of 50 V/cm, calculate the equilibrium velocity distribution $u(y)$.

9.14 Consider two infinite plates located at $y = \pm 40$ nm, separated by a solution with Debye length $\lambda_D = 10$ nm. If the top surface has a potential given by $\varphi_0 = 4$ mV and the bottom surface has a potential given by $\varphi_0 = -2$mV, solve for the equilibrium velocity distribution $u(y)$.

9.15 Consider a plane-symmetric electroosmotic flow induced by a fixed and motionless glass cylinder with radius $a = 1$ cm centered at the origin when an electric field of $E_\infty = 100$ V/cm in the x direction is applied. Assume that the electrokinetic potential $\zeta = \varphi_0 = -10$ mV and the electrolyte solution is 1 mM NaCl in water. You will not need these values for your calculations, but they will allow you to identify the right simplifying approximations.

Note that this problem will use both Cartesian coordinates (x, y) and cylindrical coordinates $(\imath, \theta)$. Note $x = \imath \cos\theta$ and $y = \imath \sin\theta$.

We will use a relatively simple matched asymptotic technique to calculate a composite asymptotic expression for the x velocity at $x = 0$. This will involve finding an *outer* solution for which the charge density is assumed zero and the electric field varies spatially, and an *inner* solution for which the charge density is nonzero but the electric field is assumed uniform.

We can calculate the electrical potential field around an insulating cylinder by superposing the electrical potential field for a uniform field (see Subsection 7.3.5),

$$\phi = -E_\infty x \,, \tag{9.52}$$

with the electrical potential field for a line dipole (see Subsection 7.3.4),

$$\phi = -E_\infty x \frac{a^2}{\imath^2} \,, \tag{9.53}$$

to obtain the total potential field (see Subsection 7.3.7):

$$\phi = -E_\infty x \left(1 + \frac{a^2}{\imath^2}\right) . \tag{9.54}$$

(a) Recall that the *outer* solution for the electroosmotic velocity is given by,

$$\vec{u}_{\text{outer}} = -\frac{\varepsilon\zeta}{\eta} \vec{E} \,, \tag{9.55}$$

where the electrokinetic potential $\zeta = \varphi_0$. Derive a relation for $\phi_v(x, y)$ for the outer solution for the electroosmotic flow.

(b) For simplicity, consider only the line corresponding to $x = 0$. Define an *outer* variable $y^*_{\text{outer}} = \frac{y-a}{a}$ and a normalized velocity $\vec{u}^* = -\vec{u}\eta/\varepsilon\zeta E_\infty$. Plot u^* (the x component of the normalized velocity) as a function of y^*_{outer} from 0 to 10. Does the outer solution give the correct answer at $y = a$? At $y = \infty$?

(c) Recall that the *inner* solution for the electroosmotic velocity is given by

$$u_{\text{inner}} = \frac{\varepsilon E_{\text{wall}}}{\eta}(\varphi - \zeta), \tag{9.56}$$

where again $\zeta = \varphi_0$. Here E_{wall} is the (assumed uniform) magnitude of the tangent electric field at the wall, which we obtained by evaluating the electric field given by Eq. (9.54) at $r = a$. Consider again the line corresponding to $x = 0$, and define an inner variable $y^*_{\text{inner}} = \frac{y-a}{\lambda_D}$. Plot u^* as a function of y^*_{inner} from 0 to 10. Does the inner solution give the correct answer at $y = a$? At $y = \infty$?

(d) Now construct a composite asymptotic solution using the additive formula

$$u_{\text{composite}}(y) = u_{\text{inner}}(y) + u_{\text{outer}}(y) - u_{\text{inner}}(y = \infty). \tag{9.57}$$

This additive relation does not *rigorously* satisfy all of the governing equations, but it is a mathematical tool for creating an analytical solution that transitions smoothly from the inner solution (which is rigorously correct to the extent that the local electric field tangent to the wall is uniform) to the outer solution (which is rigorously correct to the extent that the charge density is zero). Plot $u^*_{\text{composite}}$, u^*_{inner}, and u^*_{outer}. For your independent variable, use either y_{outer} ranging from 1 to 10 on a linear axis or y_{inner} ranging from 1×10^{-3} to 1×10^{9} on a log axis (or both). Does the composite solution give the correct answer at $y = a$? At $y = \infty$?

9.16 Consider a wall with charge density q'' and recall the boundary condition at the wall from Eq. (5.31). Define a nondimensional charge density q''^* given by

$$q''^* = \frac{q''/\varepsilon}{RT/F\lambda_D} = \frac{q''}{\sqrt{2\varepsilon RTI_c}}, \tag{9.58}$$

and show that, in the Debye–Hückel limit, $\varphi^* = -q''^* \exp -y^*$. Give a physical description of the significance of $\sqrt{2\varepsilon RTI_c}$.

9.17 Consider steady-state, purely electroosmotic flow along a charged wall oriented parallel to the x axis. As a function of the fluid viscosity and permittivity, $E_{\text{ext,wall}}$, and q''_{wall}, evaluate (a) the viscous shear stress at the wall, corresponding to $\eta\frac{\partial u}{\partial y}$; and (b) the net electrostatic force on the charge density on the wall. How are these quantities related to each other?

9.18 Show that, in a quiescent EDL at equilibrium, the pressure at a point can be written as

$$p_{\text{edl}} = p_{\text{bulk}} + RT(C - C_\infty), \tag{9.59}$$

where $C = \sum_i c_i$. The product RTC is called the *osmotic pressure*. Is the pressure in an EDL lower or higher than in the bulk? Can a Debye–Hückel approximation of an EDL predict generation of an osmotic pressure?

9.19 Derive Eq. (9.34) by integrating Eq. (9.33). Use the Debye–Hückel approximation and assume that the domain is semi-infinite.

9.20 Derive a relation for the Bjerrum length and evaluate this distance for water at room temperature. The *Bjerrum length* λ_B is the fundamental length scale for ion–ion correlations and is the distance l for which the potential energy of the Coulomb interaction is equal to the thermal energy.

Proceed as follows: Consider two elementary charges (e.g., two protons) separated by a distance l. Use Coulomb's law to calculate the electrostatic attraction force between them as a function of l, and define the electrostatic potential energy such that the attraction force is given by the derivative of the potential energy with respect to l. Set the constant of integration so that the potential energy for $l \to \infty$ is zero, and find the distance l for which the potential energy of the Coulomb interaction is equal to the thermal energy $k_B T$.

9.21 The use of Boltzmann statistics to describe ion populations requires that the ions be interacting with a mean field rather than with other ions, because the energy used in the Boltzmann formula contains only the interaction of the ionic charge with the mean-field potential.

At ion spacings much larger than the Bjerrum length λ_B, ions can be assumed to be interacting primarily with the water around them rather than with other ions, and the mean-field approximation is reasonable. At ion spacings of the order of λ_B or lower, the ions are interacting directly with each other, the Boltzmann relation is in question, and it becomes useful to incorporate ion–ion correlations into an analysis.

For ions present at a total number density n, the mean spacing of ions ℓ_0 is given by $\ell_0 = 1/\sqrt[3]{n}$. The total number density relates to the concentration (of a 1:1 electrolyte) through the relation $c = \frac{n}{2N_A}$. Derive a relation for the c at which the mean ion spacing is equal to the Bjerrum length. In water at room temperature, determine this concentration.

9.22 Consider a 10-mM solution of NaCl in contact with a glass wall with a surface potential given by −170 mV. Assume that λ_{HS} for sodium is well approximated by 1 Å. Evaluate the concentration of sodium ions at the glass surface predicted by Boltzmann statistics, and determine if the hard-sphere packing limit is important in this case.

9.23 Consider a solution with ionic strength 1 M consisting of a 100:10:1 ratio of sodium chloride, barium phosphate, and aluminum citrate. Assume all salts are completely dissociated. Assume that the valences and hard sphere diameters for these six ions are as follows:

Ion	z	λ_{HS}
Na^+	+1	1 Å
Cl^-	−1	2 Å
Ba^{+2}	+2	4 Å
HPO_4^{-2}	−2	2 Å
Al^{+3}	+3	7 Å
$C_3H_5O(COO)_3^{-3}$	−3	2.5 Å

Solve Eq. (9.47) numerically for the following conditions:

(a) $\varphi_0 = 1$ V.

(b) $\varphi_0 = 0.25$ V.

(c) $\varphi_0 = -0.25$ V.

Answer the following questions:

(a) Compare the last two cases to each other. Why are they different?

(b) In the first case, what would we predict for the concentration of $C_3H_5O(COO)_3^{-3}$ at the wall if we used the Poisson–Boltzmann equation?

(c) The Poisson–Boltzmann equation predicts that counterion concentrations will rise monotonically as the wall is approached and coion concentrations will drop monotonically as the wall is approached. Is this the case for your predictions here? What property of the steric modification to the Poisson–Boltzmann equation leads to non-monotonic distributions for some ions?

(d) Compare the potential distribution for these high-voltage cases with a low-voltage case. Are the potential distributions different? Are modified Poisson–Boltzmann equations more important for predicting the potential or for predicting species concentrations?

9.24 Transform the Poisson–Boltzmann equation by setting $c_i = \exp \gamma_i$ and rewriting the equation in terms of γ_i. Why might the transformed equation be more easily manipulated in a numerical solution of the equations?

9.25 Derive a general solution for the Poisson–Boltzmann equation for a $1{:}z$ or $z{:}1$ electrolyte.

9.26 In the Debye–Hückel limit, calculate the y variation of the velocity, shear stress, vorticity, and electrostatic body force for flow along an infinite flat plate.

9.27 Consider the integral formulation of the bulk fluid velocity, performed in Chapter 6 for uniform η, ε, and E_{ext}. Rederive the relation for the bulk fluid velocity, assuming that these parameters are *not* uniform. Show that a general result for the bulk velocity is

$$u_{\text{bulk}} = \int_{\varphi_0}^{0} \frac{\varepsilon E_{\text{ext}}}{\eta} \, d\varphi \,. \tag{9.60}$$

9.28 For a general (not symmetric) electrolyte *not* in the Debye–Hückel limit, evaluate the relation between the surface charge and the surface potential by evaluating the derivative of the potential at the wall:

(a) Start with the Poisson–Boltzmann equation and multiply both sides by $2(\frac{d}{dy}\varphi)$.

(b) Integrate both sides, rearranging the derivatives so that on one side you have the derivative of $(\frac{d}{dy}\varphi)^2$ and on the other side you have $d\varphi$.

(c) To evaluate $\frac{d}{dy}\varphi$, integrate over a dummy variable (y' or φ', depending on which side of the equation) from the bulk ($y' = \infty$; $\varphi' = 0$) to some point in the double layer ($y' = y$; $\varphi = \varphi'$).

(d) Evaluate $-\varepsilon \frac{\partial}{\partial y}\varphi$ at the wall.

Show that the surface charge density is given by the Grahame equation:

$$q'' = \text{sgn}(\varphi_0) \sqrt{2\varepsilon RT I_c \sum_i c_{i,\infty}^* \left[\exp(-\frac{z_i F \varphi_0}{RT}) - 1 \right]} \tag{9.61}$$

or, equivalently,

$$q''^{*} = \text{sgn}(\varphi_0) \sqrt{\sum_i c_{i,\infty}^* \left[\exp(-z_i \varphi_0^*) - 1 \right]} \,. \tag{9.62}$$

9.29 Consider a 100-mM solution of NaCl in water. Model the sodium ions as having $\lambda_{HS} =$ 1.6 Å and the chloride ions as having $\lambda_{HS} = 3.6$ Å. Assume each ion has six water molecules immediately surrounding it. Given the local permittivity changes owing to the electric field induced by the wall and by the individual ions, calculate the permittivity of the solution within an EDL as a function of distance from the surface if the interfacial potential is 1 V.

9.30 Consider a solution of NaCl in water. Model the sodium ions as having $\lambda_{HS} = 1.6$ Å and the chloride ions as having $\lambda_{HS} = 3.6$ Å. Assume each ion has six water molecules immediately surrounding it. Calculate the permittivity of the water molecules in the hydration shell owing to the electric field of the ion. Estimate the dielectric increment of NaCl by calculating the fractional volumes taken up by the ions, the water in the hydration shells, and free water, and volume averaging the permittivities in each volume.

10 Zeta Potential in Microchannels

Previous chapters assert that a potential drop occurs over an EDL, consistent with the fact that chemical reactions occur at the surface to induce ionization of wall species. We now return to this subject in greater detail. Our goal is to be able to predict the equilibrium surface potential at microfluidic device interfaces as a function of the device material and solution conditions. This chapter frames the problem, describes associated parameters, and lists several models that can be used to attack this problem and interpret experimental data. We start by clarifying notation and terminology. We then discuss the chemical origins of surface charge for both Nernstian and non-Nernstian surfaces, discuss techniques for measuring and modifying electrokinetic potentials, and summarize observed zeta potentials for microfluidic substrates. Finally, we discuss how EDL theory is related to interpretation of zeta potential data and the relation between ζ and φ_0.

10.1 DEFINITIONS AND NOTATION

Here we must define the distinct meanings of several terms, namely the zeta potential, the electrokinetic potential, the interfacial potential, the double-layer potential, and the surface potential. These terms have different meanings and are used differently by various authors. Further, some of these terms become equivalent if specific models are used to describe the interface, but have different meanings if other models are used.

Surface potential (or, equivalently, interfacial potential or double-layer potential) typically implies the difference between the potential in a bulk, electroneutral solution and the potential at the wall. The wall here is defined as the point at which the material in the solid phase ends. We write the surface potential in this text as φ_0, but the reader should be aware that this potential is often denoted in other sources as ϕ_0, ψ, ψ_0, or ζ. The surface potential is well defined if the EDL is thin, because interfacial effects decay on a shorter length scale than any spatial variation in the bulk electric field. In this case, we define a "bulk" location that simultaneously is (a) close enough to the wall that it experiences the same extrinsic electric field as the wall; and (b) is far enough from the wall that the net charge density is zero and interfacial effects on the potential can be neglected. The surface potential is also well defined if the bulk electric field is uniform, as is the case for flow over an infinite flat plate or over a small particle. In this case, the bulk is straightforward to define even when double layers are thick. When double layers are thick *and* the extrinsic field is nonuniform, the definition of a surface potential becomes much more difficult.

Electrokinetic potential is specific to a phenomenon (e.g., electroosmosis or electrophoresis). For electroosmosis, the electrokinetic potential is given by $-\mu_{EO}\eta_{bulk}/\varepsilon_{bulk}$, which is the potential that need be inserted into a Helmholtz–Smoluchowski-type equation such as Eq. (6.14) to explain observed electroosmotic flow. We denote the electrokinetic potential as ζ [V] in this text. The electrokinetic potential has units of voltage by definition but need not correspond to a physical potential or potential difference.

The distinction between the electrokinetic potential and the surface potential is critical – the surface potential φ_0 is a measure of the interfacial potential whereas ζ is a measure of fluid or particle flow – but this distinction is confounded by widely varying use of terminology and symbology as well as the fact that the two quantities are equal in magnitude if the uniform-property, integral analysis of Chapter 6 is used. The terms *electrokinetic potential*, *zeta potential*, *double-layer potential*, *surface potential*, and *interfacial potential* are often used interchangeably, and the meaning implied by specific use of these terms is often ambiguous. We use the terms *surface potential*, *interfacial potential*, or *double-layer potential* and the symbol φ_0 to denote the difference in potential between the bulk solution and the interface between the wall and the solution. We use the terms *zeta potential* or *electrokinetic potential* and the symbol ζ to denote $\zeta = -\mu_{EO}\eta_{bulk}/\varepsilon_{bulk}$ for electroosmosis or $\mu_{EP}\eta_{bulk}/\varepsilon_{bulk}$ for electrophoresis.

Recall from Chapter 6 that the bulk velocity for electroosmosis over a flat plate for a system with uniform fluid properties and ion distributions described by Boltzmann statistics is given by $\mu_{EO} = -\varepsilon\varphi_0/\eta$, which implies that

equality of surface potential and electrokinetic potential, simple interface, uniform fluid properties

$$\zeta = \varphi_0 \tag{10.1}$$

for such a system. However, ion crowding or adhesion limits the application of Boltzmann statistics, and the high electric fields and high ion densities give the potential for breakdown of the uniform-property assumption used in Chapter 6.

10.2 CHEMICAL AND PHYSICAL ORIGINS OF EQUILIBRIUM INTERFACIAL CHARGE

The simplest origin of interfacial charge occurs simply if we place two electrodes in contact with a fluid and apply a potential across the two electrodes to create surface charge. In this case, a potential is applied across the system, and the potential is dropped in part across EDLs at the electrodes and in part across the bulk solution. However, the conductivity of the surface prohibits the steady-state transverse extrinsic field one would use to drive steady-state electroosmosis, and thus the double layer at a conductor is irrelevant for *equilibrium* descriptions of *steady* fluid flow.

On insulating surfaces at equilibrium, interfacial charge typically comes from (a) adhesion of charged solutes at the interface or (b) chemical reaction at the surface. This interfacial charge can be described with equilibrium relations. In particular, the electrochemical potential defines equilibrium in a convenient fashion, and that leads to discussion of potential-determining ions, which help frame the parameters that must be examined when interfacial potential is measured or analyzed.

10.2.1 Electrochemical potentials

The molar chemical potential (or partial molar Gibbs free energy) of a species g_i [J/mol] is defined in the ideal solution limit as

definition of chemical potential, ideal solution

$$g_i = g_i^\circ + RT \ln \frac{c_i}{c_i^\circ}, \tag{10.2}$$

where g_i° is the chemical potential at concentration c_i°. Here the term *species* implies both chemical and physical state – thus species in solid, liquid, aqueous, or surface-bound states are all treated differently.

In systems that involve both electrical potential drops and chemical reaction, equilibrium is defined in the ideal solution limit by use of the electrochemical potential $\overline{g_i}$:

definition of electrochemical potential, ideal solution

$$\overline{g_i} = g_i + z_i F \varphi = g_i^\circ + RT \ln \frac{c_i}{c_i^\circ} + z_i F \varphi \,. \tag{10.3}$$

The system is at equilibrium when the electrochemical potentials of all species are uniform – if the electrochemical potential were nonuniform, there would be a driving force to initiate a chemical change. The electrochemical potential relates chemical properties to electrical potential, and thus relates chemistry to the electrostatic boundary condition used to solve our differential equations (e.g., the Poisson–Boltzmann equation) to obtain parameters (e.g., electroosmotic mobility) that directly impact microscale fluid mechanics.

The electrochemical potential formulation in the ideal solution limit is consistent with our earlier derivation of the Poisson–Boltzmann equation. The Boltzmann prediction for chemical species in a double layer, given by Eq. (9.6), can be derived from Eq. (10.3) by setting $\overline{g_i^\circ}$ to be a constant. The electrochemical potential description has the added benefit that it also can be used to treat binding, chemical reaction, and other mechanisms that are the sources of interfacial charge. We can write the electrochemical potential in the mean-field limit but without making the ideal solution approximation:

definition of electrochemical potential for dense solutions

$$\overline{g_i} = g_i + z_i F \varphi = g_i^\circ + RT \ln \frac{a_i}{a_i^\circ} + z_i F \varphi \,, \tag{10.4}$$

where a_i and a_i° are the activities (i.e., effective concentrations) of species i at the current and reference concentrations. Equation (10.4) applies for solutions at all concentrations, whereas Eq. (10.3) applies only at low concentrations. The use of the mean-field electrical potential here ignores the structuring present at high concentration or in the monolayer next to a wall; at high concentration or within an atomic radii or two of a surface, the potential of mean force (see Subsection H.2.1) must be used rather than the mean-field electrical potential.

10.2.2 Potential-determining ions

Describing the equilibrium chemistry at a liquid–solid interface, in general, involves writing an equality for electrochemical potential in the bulk and at the surface,

$$\overline{g_i^\circ}_{\text{bulk}} = \overline{g_i^\circ}_{\text{wall}} \,, \tag{10.5}$$

for every species i. Equation (10.5) thus implies that differences in the reference state chemical potential of a species g_i and the electrical potential φ control the equilibrium activity a_i and thus the concentration c_i.

In determining the charge state at an insulating interface, Eq. (10.5) shows that, if the reference state chemical potential is different for the species bound to a surface compared with the species in solution, the species either preferentially adsorbs or desorbs from the surface. If the species in question carries an electric charge, this process controls the charge density at the surface. Similarly, if a chemical reaction at the surface leads to a change in the charge state of the surface, the reaction equilibrium dictates the charge state at the surface.

Consider, as an example, a glass microchannel in contact with an aqueous solution of sodium chloride and barium chloride. Electrochemical potential arguments can be used to describe the equilibrium of the sodium, barium, chloride, hydronium, and hydroxyl ions in this system. Sodium and chloride are generally (but not always) treated as *indifferent electrolytes* when in contact with a glass surface, meaning that

electrochemical potential relation for indifferent electrolytes

$$g^{\circ}_{i\,\text{bulk}} = g^{\circ}_{i\,\text{wall}} \tag{10.6}$$

for both sodium and chloride ions. Thus the indifferent electrolyte approximation assumes that the interaction between sodium and chloride ions and a glass surface is strictly a bulk electrostatic interaction, without chemical affinity. When this is the case, the equilibrium statement for the sodium and chloride ions is given by

electrochemical potential relation for specifically adsorbing electrolytes

$$RT \ln \frac{a_i}{a_i^{\circ}} + z_i F \varphi = \text{constant}\,, \tag{10.7}$$

which, in the ideal solution limit, is

$$RT \ln \frac{c_i}{c_i^{\circ}} + z_i F \varphi = \text{constant}\,. \tag{10.8}$$

Equation (10.8) can then be rearranged to give the Boltzmann distribution from Eq. (9.6), if c_i° is defined as $c_{i,\infty}$ and the constant is set to zero.

In contrast, barium at a glass surface is an example of a *specifically adsorbing ion*, implying that

$$g^{\circ}_{i\,\text{bulk}} \neq g^{\circ}_{i\,\text{wall}}\,, \tag{10.9}$$

which means that barium has a chemical affinity for a glass surface that is different from its affinity for the water molecules that serve as the solvent. In this case, the Boltzmann statistics of Eq. (9.6) can describe the barium concentration in the bulk but *not* at the wall – the concentration at the wall is then dictated by the energy of adsorption, often described by use of an equilibrium constant:

$$\text{Ba}^{+2}_{\text{bulk}} \overset{K_{\text{eq}}}{\longleftrightarrow} \text{Ba}^{+2}_{\text{wall}}\,. \tag{10.10}$$

In fact, all species in general have a different reference state chemical potential at an interface compared with the bulk – this phenomenon is seen in many aspects of fluid mechanics; for example, the variation of Gibbs free energy at an interface is discussed as the foundation of surface tension in Chapter 1. Thus all species act as surfactants to some extent. Indifferent electrolytes are those for which this energy variation and the attendant effects are small enough to be ignored.

In contrast to the sodium and barium cations, the protons (or hydronium ions) in solution can react chemically with a glass surface. The surface of silicates in contact with water has a large concentration of silanol (SiOH) groups, and these chemical groups undergo acid dissociation as described in Appendix B:

$$\mathrm{SiOH} \overset{K_a}{\longleftrightarrow} \mathrm{SiO}^- + \mathrm{H}^+ , \tag{10.11}$$

where K_a is the acid dissociation constant, often described using the $\mathrm{p}K_a$, defined as $\mathrm{p}K_a = -\log K_a$. Reaction (10.11) thus links the H^+ concentration at the wall to the relative surface density of SiOH and SiO^-. Because the concentrations of OH^- and H^+ are linked by water dissociation, the SiO^-, H^+, and OH^- concentrations are all linked by chemical equilibrium, as described by the Henderson–Hasselbach equation. In the ideal solution limit, the $\mathrm{p}K_a$ is evaluated by use of species concentrations and is a single value. For real systems at high concentrations, the $\mathrm{p}K_a$ must be evaluated by use of species activities, and the $\mathrm{p}K_a$ varies statistically based on the details of the spatial distribution of sites.

If the surface charge density is governed by a single acid dissociation reaction as described by reaction (10.11), the surface charge density q'' is predicted in the ideal solution limit as

$$q'' = -e\Gamma \frac{10^{\mathrm{pH}-\mathrm{p}K_a}}{1 + 10^{\mathrm{pH}-\mathrm{p}K_a}} , \tag{10.12}$$

where Γ is the summed density of SiO^- and SiOH sites $[\mathrm{m}^{-2}]$, e is the elementary charge, and $\mathrm{p}K_a$ is the acid dissociation constant of the site. For typical surfaces, the surface sites exhibit a range of $\mathrm{p}K_a$'s, and the surface charge density is more accurately exhibited with the relation

$$q'' = -e\Gamma \frac{10^{\alpha(\mathrm{pH}-\mathrm{p}K_a)}}{1 + 10^{\alpha(\mathrm{pH}-\mathrm{p}K_a)}} , \tag{10.13}$$

where α is a value between 0 and 1 that describes the spread of $\mathrm{p}K_a$'s; 1 denotes a single $\mathrm{p}K_a$ and lower values correspond to a spread of $\mathrm{p}K_a$'s. Values between 0.3 and 0.7 are common.

Most systems are controlled by a small set of ions, called *charge-determining ions* or *potential-determining ions*. In the preceding example, the charge at the surface is dictated primarily by protonation and deprotonation of silanol groups, and thus the charge is determined by H^+ and OH^- ions. Thus H^+ and OH^- are the primary charge-determining ions for glass, and the electrokinetic potential for a glass surface is a function of pH. Indifferent electrolytes can be added to an electrolyte solution without changing the charge density at the surface – indifferent electrolytes serve only to change the Debye length (which changes the surface potential) and, if present at high enough concentrations, to change fluid properties (which changes the electrokinetic potential). Specific adsorbers or species that react with the surface change the charge density at the surface directly. The process of defining charge-determining ions tends to greatly simplify the process of defining the chemical system that determines the double-layer potential.

10.2.3 Nernstian and non-Nernstian surfaces

A general equilibrium relation for an ion in a solution in contact with a wall is given by

$$\left[g_i^\circ + RT\ln\frac{a_i}{a_i^\circ} + z_i F\varphi\right]_{\text{wall}} = \left[g_i^\circ + RT\ln\frac{a_i}{a_i^\circ} + z_i F\varphi\right]_{\text{bulk}}, \tag{10.14}$$

where a_i is the activity and a_i° is the activity at a reference condition. In the ideal solution limit, $a_i = c_i$. This equation specifies the equilibrium condition but is rarely used in this form, because we rarely know g_i° or a_i at the wall for the ions in question. However, this equation can be manipulated into a form that is useful if some experimental parameters are known. Consider the state for which φ at the wall is zero:

$$\left[g_i^\circ + RT\ln\frac{a_i(\text{pzc})}{a_i^\circ}\right]_{\text{wall}} = \left[g_i^\circ + RT\ln\frac{a_i(\text{pzc})}{a_i^\circ}\right]_{\text{bulk}}, \tag{10.15}$$

where the abbreviation pzc stands for *point of zero charge* and denotes the solution concentrations at which the interfacial potential is zero. We can subtract the equation at zero interfacial potential, i.e., Eq. (10.15), from the general relation, i.e., Eq. (10.14), and solve for φ_0 to get

$$\varphi_0 = \frac{RT}{z_i F}\left[\ln\left.\frac{a_i}{a_i(\text{pzc})}\right|_{\text{bulk}} - \ln\left.\frac{a_i}{a_i(\text{pzc})}\right|_{\text{wall}}\right]. \tag{10.16}$$

This expression relates the surface potential to the activities of the ions in the bulk and on the surface, both at the pzc and at the given experimental condition.

Equation (10.16), in itself, does not identify $a_i(\text{pzc})$ and does not give a_i at the surface at any condition. However, this equation frames the parameters we want to know (φ_0) in terms of parameters we can directly change ($a_{i,\text{bulk}}$, by controlling the concentration of the ions in solution), parameters we can easily measure ($a_{i,\text{bulk}}$ at the pzc, by changing ion concentrations until the surface charge is zero), and, unfortunately, one typically unknown parameter ($a_{i,\text{wall}}$).

Nernstian surfaces lead to a simplification of Eq. (10.16) because, for Nernstian surfaces, the one problematic and unknown parameter is constant across experiments. A Nernstian surface is a surface for which $a_{i,\text{wall}}$ is independent of $a_{i,\text{bulk}}$, meaning that we can add ions to the bulk solution without changing the surface ion activity, i.e., the effective surface ion concentration. This implies that changes in the surface ion concentration, although they are important when we calculate the surface charge for electrokinetic purposes, *do not* affect the role of those ions in surface chemical reaction. An analogous assumption, used when deriving acid dissociation reactions such as the Henderson–Hasselbach equation, is that the concentration of water is unchanged by acid dissociation. This assumption is made for acid dissociation because $[H_2O]$ is enormous relative to $[H^+]$ and changes in $[H^+]$ have little effect on $[H_2O]$. This assumption is sound for surfaces when the concentration of ions at the surface is large relative to the concentration of ions in the solution, and thus typical Nernstian surfaces are weakly soluble ionic crystals such as AgI. Their surface activity is independent of the bulk activity because so many ions cover the surface that the small number of ions that might adsorb or desorb from the surface owing to changes in the bulk activity are only a small perturbation to the effective surface ion concentration. In this case, $a_{i,\text{wall}} = a_{i,\text{wall}}(\text{pzc})$, and thus

$$\varphi_0 = \frac{RT}{z_i F}\left[\ln\left.\frac{a_i}{a_i(\text{pzc})}\right|_{\text{bulk}}\right], \tag{10.17}$$

or, rewriting in terms of base 10 logarithms of the activities, we have

Nernst equation for surface potential of a Nernstian surface

$$\varphi_0 = \frac{RT \ln 10}{z_i F} \left[\log \frac{a_i}{a_i(\text{pzc})} \bigg|_{\text{bulk}} \right], \quad (10.18)$$

where log denotes a base 10 logarithm. In this case, φ_0 can be predicted as a function of a_i as long as an experimental observation of $a_i(\text{pzc})$ is made. For example, the pzc for AgI at room temperature is at $\text{pAg} = -\log[\text{Ag}^+] = 5.5$. We can thus write Eq. (10.18) for AgI in the ideal solution limit at room temperature as

$$\varphi_0 = \frac{RT \ln 10}{z_i F} [\text{pAg} - 5.5] . \quad (10.19)$$

Equation (10.18) is known as the *Nernst equation*. It sets an upper limit for how strongly a surface can be charged owing to electrochemical equilibrium. The quantity $RT \ln 10/z_i F$ is approximately 57 mV at room temperature for monovalent ions, and thus the maximum interfacial potential that a surface can achieve is approximately 57 mV per decade of concentration change from its pzc.

EXAMPLE PROBLEM 10.1

Consider a microfluidic device lined with silver iodide (this would most likely be made by creating a glass device, coating the surface with silver, and then reacting the silver surface with potassium iodide). If the channel were filled with a 1 mM solution of silver nitrate, what will the electroosmotic mobility of the surface be if the integral model for electroosmosis from Chapter 6 is used? How will this be affected by addition of moderate amounts of acid? Sodium chloride?

SOLUTION: Silver iodide is Nernstian, and the surface potential is given by

$$\varphi_0 = \frac{RT \ln 10}{z_i F} [\text{pAg} - 5.5] = \frac{RT \ln 10}{z_i F} [3 - 5.5] = 145 \text{ mV} . \quad (10.20)$$

The electroosmotic mobility is given by

$$\mu_{\text{EO}} = -\frac{\varepsilon \varphi_0}{\eta} = -10.3 \times 10^{-8} \text{ m}^2/\text{Vs} . \quad (10.21)$$

Because the charge-determining ions for AgI are Ag^+ and I^-, the addition of acid should not affect this unless the acid corrodes the surface. Sodium chloride will change λ_D but not φ_0.

Systems that are well described by Eq. (10.18) (because of the independence of the wall ion activity to bulk ion concentration) are called Nernstian systems, whereas systems that are not well described by Eq. (10.18) (because wall ion activity is a function of wall ion concentration and thus a function of bulk ion concentration) are called non-Nernstian systems. Nernstian surfaces (e.g., AgCl, AgBr, AgI, $CaPO_4$) are relatively straightforward to analyze, because the only parameter of the surface that must be known is the bulk concentration of potential-determining ions at the pzc, and the surface potential at a different concentration can be predicted from this parameter if the solution concentrations are known. This is particularly convenient, because the potential at the surface is closely related to the bulk electroosmotic or electrophoretic velocity observed in microsystems with thin double layers. For Nernstian surfaces, the Nernst equation dictates the surface potential and, for thin double layers, the bulk fluid velocity – the surface charge need not be calculated. If it needs to be calculated, the surface charge can be predicted by combining the surface potential with an EDL model such as those discussed in Chapter 9.

Unfortunately, most of the surfaces present in microfabricated systems (e.g., oxides and polymers) are not Nernstian, because the adhesion of ions to the wall or reaction of the ions with the wall has a significant change on the effective concentration or activity of surface species. We thus cannot skip calculation of the surface charge density if our desired final result is the velocity of the fluid or a particle. For non-Nernstian surfaces, we typically must use chemical equilibrium to predict the surface charge density at the interface and use an EDL model to relate this surface charge density to a surface potential or directly to the fluid flow. Further, because of the effect of reaction or adsorption on the surface activity, we observe surface potentials on non-Nernstian surfaces that vary less per decade of concentration change of the potential-determining ions than that of Nernstian surfaces. The complexity of this process and the uncertainty with regards to the Gibbs energy of surface reactions or adsorption require that the surface potentials on non-Nernstian surfaces be measured directly. These observations are described in some detail later in the chapter.

The non-Nernstian nature of most microdevice substrates can also be understood in terms of the density of charge sites. The charge site density for glass at high pH is usually reported as approximately $6/\mathrm{nm}^2$. Such a site density is depicted pictorially in Fig. 10.1, and this schematic indicates that the density of sites is relatively low and the potential at the surface is, in fact, nonuniform within the plane of the surface. Care must thus be used when 1D models are used to describe the EDL.

10.3 EXPRESSIONS RELATING THE SURFACE CHARGE DENSITY, SURFACE POTENTIAL, AND ZETA POTENTIAL

If the Gouy–Chapman model of the EDL is used, the Poisson–Boltzmann equation is assumed to apply for all regions up to the solid wall; the no-slip condition is assumed; and fluid properties are assumed uniform; and the expressions relating between q'', φ_0, and ζ are straightforward. The relation between q'' and φ_0 is determined by the electrostatic boundary condition,

$$q''_{\mathrm{wall}} = -\varepsilon \left.\frac{\partial \varphi}{\partial n}\right|_{\mathrm{wall}} , \tag{10.22}$$

combined with the Poisson–Boltzmann relation for the potential distribution. The result can conveniently be written in general by use of the Grahame equation, which we obtain by transforming the Poisson–Boltzmann equation and solving it for q'' at the wall (see Exercise 9.28):

$$q'' = \mathrm{sgn}(\varphi_0)\sqrt{2\varepsilon RT I_c \sum_i c^*_{i,\infty}\left[\exp(-\frac{z_i F \varphi_0}{RT}) - 1\right]} , \tag{10.23}$$

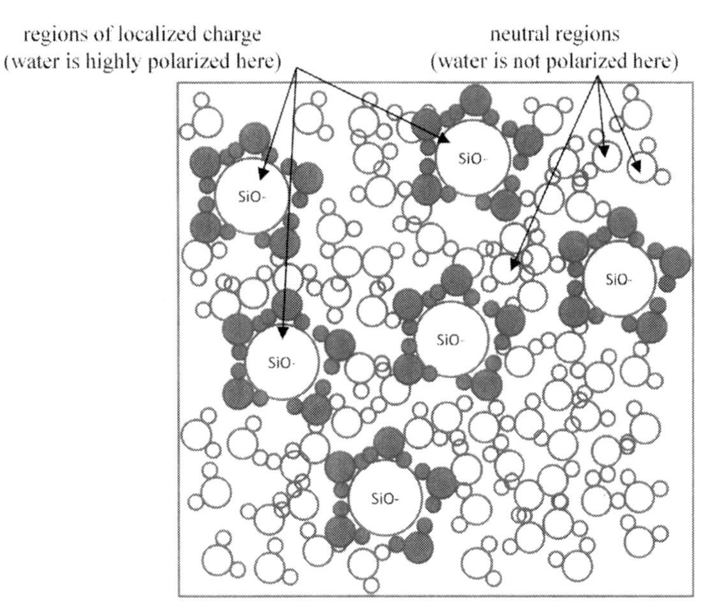

Fig. 10.1 Schematic of charge sites at a $6/\text{nm}^2$ surface, viewed from top. Charge sites are denoted as SiO^-, as would be typical of a glass surface. Water molecules are shown filled for those molecules strongly polarized by interaction with the surface, open for those molecules not strongly polarized by the surface. The inhomogeneity highlights the 3D nature of the EDL, which is ignored in most models.

where the sign function $\text{sgn}(x) = x/|x|$ and I_c is the ionic strength. The Grahame equation can be written for symmetric electrolytes as

$$q''_{\text{wall}} = \frac{\varepsilon\varphi_0}{\lambda_D}\left[\frac{2}{z\varphi_0^*}\sinh\left(\frac{z\varphi_0^*}{2}\right)\right], \tag{10.24}$$

and both forms simplify in the Debye–Hückel limit to $q''_{\text{wall}} = \frac{\varepsilon\varphi_0}{\lambda_D}$. If the no-slip condition applies and fluid properties are uniform and equal to the bulk value, then $\zeta = \varphi_0$.

Despite the material presented so far, prediction of the surface potential or charge density of microchannel substrates has largely eluded researchers. The preceding arguments give the impression that prediction of q'', φ_0, and ζ is rather straightforward – the Poisson–Boltzmann diffuse double-layer theory is well established as a correct description of EDLs if the surface potential or charge density is specified and ion concentration is small. Further, the uniform fluid property analysis from Chapter 6 is a straightforward and intuitive description of the fluid flow induced in an EDL. Unfortunately, if direct experimental measurements of surface charge (e.g., charge titration, described later in this chapter) are combined with the analysis in Chapter 6, the predictions, though qualitatively sound, overpredict electrokinetic velocities and are of relatively poor quantitative accuracy.

Physical models that go beyond the Poisson–Boltzmann relation, uniform fluid properties, and the no-slip condition have been introduced to explain the links among the values of q'', φ_0, and ζ observed experimentally. In particular, these models describe departures from the assumptions that fluid properties are uniform or that all electrolytes are indifferent to the surface. This section explains these physical models. Whereas Chapter 9 discusses the diffuse components of the EDL, this section focuses on adsorbed and condensed regions.

10.3.1 Extended interface models: Modifications to φ_0

The fluid velocity in an electrokinetic system is related to the surface potential and, in turn, the surface charge density. However, experimentally measured surface charge densities predict electroosmotic velocities much higher than observed. Extended interface models seek to reconcile this mismatch by postulating an additional structure to the interface that leads to lower velocities.

The most ubiquitous model is the *Stern layer model*, which postulates that counterions near the surface are so tightly bound that the fluid cannot move with respect to the interface. This layer is termed the *Stern layer*. This model defines an *inner Helmholtz plane* that corresponds to the interface between the wall and the Stern layer, and an *outer Helmholtz plane* that corresponds to the interface between the Stern layer and the diffuse EDL. The Stern model typically defines the fluid boundary condition as $\vec{u} = 0$ at the outer Helmholtz plane and solves the chemical equilibrium equations at the inner Helmholtz plane. In this sense, the effect of a Stern layer on the fluid-mechanical prediction is simply to reduce the "effective" surface charge, thus reconciling the mismatch between charge titration data and electrokinetic data. Similar ideas are used to explain the suppression of ζ by many surface-active polymers.

10.3.2 Fluid inhomogeneity models: Relation between φ_0 and ζ

Various models have been used to describe fluid properties in EDLs, with the key properties being the fluid electrical permittivity and viscosity. Some of these models seek to replace extended interface models such as the Stern layer model, whereas some are designed to be part of an extended interface description.

We start by reconsidering the integral analysis from Chapter 6. Consider a surface with potential φ_0 and an extrinsic electric field applied transverse to the surface. We write the Navier–Stokes equations (with nonuniform viscosity) for a thin region near the wall and include the electrostatic body force term $\rho_E \vec{E}_{ext}$:

Navier–Stokes equation with electrostatic body force term

$$\rho \frac{\partial \vec{u}}{\partial t} + \rho \vec{u} \cdot \nabla \vec{u} = -\nabla p + \nabla \cdot \eta \nabla \vec{u} + \rho_E \vec{E}_{ext} . \tag{10.25}$$

As before, we assume that $\vec{E}_{ext}$ is caused by an external power supply and is uniform within the EDL, with a value equal to $\vec{E}_{ext,wall}$. The intrinsic electric field $E_{int} = -\frac{\partial \varphi}{\partial y}$ is nonuniform. For steady isobaric flow strictly along a wall aligned in the x direction, with velocity and potential gradients *only* in the y direction, Eq. (10.25) reduces to a simple conservation of x-momentum equation:

$$0 = \frac{\partial}{\partial y} \eta \frac{\partial u}{\partial y} + \rho_E E_{ext,wall} , \tag{10.26}$$

and, inserting the Poisson equation with nonuniform permittivity,

$$-\nabla \cdot \varepsilon \nabla \phi = \rho_E , \tag{10.27}$$

using Eq. (10.27) to substitute for ρ_E, and retaining only y gradients, we find

$$0 = \frac{\partial}{\partial y} \eta \frac{\partial u}{\partial y} - \frac{\partial}{\partial y} \varepsilon \frac{\partial \varphi}{\partial y} E_{ext,wall} . \tag{10.28}$$

This can be integrated to obtain

$$\int_{y=0}^{y=\infty} \frac{\partial u}{\partial y} dy = \int_{y=0}^{y=\infty} \frac{C_1}{\eta} dy + E_{\text{ext,wall}} \int_{y=0}^{y=\infty} \frac{\varepsilon}{\eta} \frac{\partial \varphi}{\partial y} dy + C_2 . \tag{10.29}$$

The no-slip condition and boundedness requires that C_1 and C_2 be zero, and rearrangement leads to

outer solution for electroosmosis in general case for nonuniform ε and η

$$u_{\text{bulk}} = E_{\text{ext,wall}} \int_{\varphi=\varphi_0}^{\varphi=0} \frac{\varepsilon}{\eta} d\varphi . \tag{10.30}$$

Equation (10.30) thus describes the bulk velocity observed in the general case in which ε and η vary in the double layer. Equation (10.30) reduces to $u_{\text{bulk}} = -\varepsilon_{\text{bulk}} \varphi_0 E_{\text{ext,wall}}/\eta_{\text{bulk}}$ if ε and η are assumed to retain their bulk values throughout the double layer. Thus the bulk electroosmotic velocity can be determined by numerical integration if the permittivity and viscosity can be predicted as a function of the local potential.

Several models have been proposed to describe water properties under extreme conditions relevant to flow in EDLs and near walls, including effects of electric fields and high ion densities. The *viscoelectric model* postulates that the viscosity of water is a function of the electric field. In particular, because the large intrinsic electric fields in the double layer are aligned normal to the surface whereas the flow induced in a double layer is tangent to the surface, the most relevant component of viscosity is the component normal to the intrinsic electric field. The viscoelectric model postulates that the viscosity normal to the electric field is given by

$$\eta = \eta_{\text{bulk}} \left(1 + k_{\text{ve}} E^2\right) , \tag{10.31}$$

where k_{ve} is the viscoelectric coefficient, postulated by some investigators to be of the order of 1×10^{-15} m^2/V^2 [107]. This model thus predicts that a large electric field (of the order of $k_{\text{ve}}^{-\frac{1}{2}}$) increases the viscosity of fluid moving normal to the electric field.

If the Debye–Hückel approximation is used, double layers are assumed thin, permittivity is assumed uniform, and Eq. (10.31) is used to define the viscosity, the integral for the electroosmotic velocity at a surface can be evaluated analytically, leading to

$$\zeta = \varphi_0 \frac{\tan^{-1}\left(\frac{\sqrt{k_{\text{ve}}}\varphi_0}{\lambda_D}\right)}{\frac{\sqrt{k_{\text{ve}}}\varphi_0}{\lambda_D}} . \tag{10.32}$$

Equation (10.32) shows that the electrokinetic potential ζ is equal to the surface potential φ_0, except for a correction factor that is a function of $\sqrt{k_{\text{ve}}}\varphi_0/\lambda_D$, ranging from unity at low surface potential to zero as the surface potential approaches infinity. This relation provides a quick qualitative estimate of the effects one might expect from a viscoelectric description of the fluid properties in an EDL; however, for the ion concentrations typically used in microfluidic experiments, the correction factor deviates from unity only in a range of surface potential for which the Debye–Hückel approximation is invalid, and the integral analysis of flow in the EDL must be performed numerically.

The viscoelectric model seeks to include electric field effects on viscosity, but disregards the role of ion concentration and retains a dilute solution model for the purposes of modeling the viscosity. Few experimental or numerical results substantiate this theory, but it can explain the electrokinetic potentials observed with dilute electrolytes and modest interfacial potential. The most compelling concern with the viscoelectric models asserted to date is that the magnitude of the viscoelectric coefficient postulated for water (1×10^{-15} m^2/V^2) implies that water's viscosity changes significantly at electric fields of the order of 3×10^7 V/m, at which the water is weakly oriented; pE/kT is approximately 0.03 for water at $E = 3 \times 10^7$ V/m, and thus the ensemble-averaged dipole moment of water is of the order of 0.03 at these voltages. The physical foundation for such a large viscosity change based on such minor orientation is yet to be justified.

Alternatively, models have been proposed inspired by the jamming transition of granular materials, which seek to address complications arising from concentrated electrolytes and large interfacial potentials. The jamming transition in granular materials corresponds to the transition from liquidlike behavior to solidlike behavior as a function of the energy of the system and the particle density. Some everyday examples of these transitions can be seen in the behavior of coffee grounds, sand, and snow. Coffee grounds behave much like a liquid when loosely packed, and can be poured from a bag as if they were liquid. However, if one purchases a vacuum-packed bag of coffee grounds, the bag behaves like a solid. Similarly, a pile of sand dumped onto ground flows like a liquid at first (when the gravitational energy is high) until it settles into a stationary cone shape corresponding to the point where the gravitational energy is no longer enough to jostle the sand grains out of their jammed state. Avalanches correspond to a transition from jammed to flowing state occurring because of a sudden addition of energy to the system.

At high ion concentrations, the viscosity of electrolyte solutions can be described by the relation

$$\frac{1}{\eta} = \frac{1}{\eta_{\text{bulk}}} \left[1 - \left(\frac{c_i}{c_{i,\max}} \right)^{\alpha} \right]^{\beta} , \tag{10.33}$$

where c_i is the concentration of the dominant counterion and $c_{i,\max}$ is the sterically limited concentration of that counterion. Equation (10.33) is inspired by rheological characterization of dense suspensions near the glass transition, and the parameters α and β are typically of the order of unity in rheological characterizations. Physically, this relation implies that, if ions are tightly packed by electrostatic forces near the point where the electron orbitals of the ions are the only repulsion keeping them from collapsing further, those tightly packed ions behave rheologically like a vacuum-packed bag of coffee grounds. Given the absence of data on the rheology of aqueous solutions with highly packed ions, little can be said for what α and β should be in this formulation, and $c_{i,\max}$ is furthermore known only approximately. The significance of this model is that it predicts that viscosity approaches infinity as the electrolyte reaches some maximum concentration.

The permittivity of water in the double layer is another area of interest, both in terms of its electric field dependence and its dependence on ion concentration. The permittivity of water drops precipitously near $E = 1 \times 10^9$ V/m, and drops as small ion concentration increases. See Subsection 5.1.4 for more information on water permittivity.

10.3.3 Slip and multiphase interface models: Hydrophobic surfaces

As discussed in Chapter 1, the tangential stress boundary condition at an interface is usually implemented by use of the no-slip condition. This no-slip condition is central to the integral analysis presented in Chapter 6 and is relevant for hydrophilic surfaces. However, hydrophobic surfaces also exhibit electrokinetic phenomena, in which case the integral analysis must be reexamined. If the velocity boundary condition is replaced with the Navier slip condition and a slip length b, a Debye–Hückel analysis leads to the result

$$\zeta = \varphi_0 \left(1 + \frac{b}{\lambda_D}\right), \tag{10.34}$$

whereas a Poisson–Boltzmann solution with a symmetric electrolyte leads to

$$\zeta = \varphi_0 \left[1 + \frac{b}{\lambda_D}\left(\frac{2}{z\varphi_0^*} \sinh \frac{z\varphi_0^*}{2}\right)\right]. \tag{10.35}$$

10.4 OBSERVED ELECTROKINETIC POTENTIALS ON MICROFLUIDIC SUBSTRATES

The previous sections give some perspective on how electrokinetic potentials can be related to other parameters, but unfortunately, the fundamental interfacial parameters required for calculating interfacial potentials are for the most part unknown. Because of this, electrokinetic potentials are generally measured directly. Electrokinetic potential is a function of both the solid surface and the liquid, including the wall material, impurities in the wall material, chemical reactions performed to the wall, pH, electrolyte concentration, valence, and size, solubility of the wall material in the solvent, surfactants, and even the time history of the surface. This section describes the dependence of electrokinetic potentials on indifferent electrolyte concentration and pH for various microfluidic substrates.

10.4.1 Electrolyte concentration

The role of indifferent electrolytes is to dictate the Debye length and change the relation between the surface charge density and the surface potential. For symmetric electrolytes, the Poisson–Boltzmann relation predicts that the relation between the charge density and the surface potential is given by

$$q'' = \frac{\varepsilon\varphi_0}{\lambda_D} \frac{2}{z\varphi_0^*} \sinh \frac{z\varphi_0^*}{2}. \tag{10.36}$$

If the surface charge density is specified by chemical equilibrium between the charge-determining ions and the wall, addition of indifferent electrolytes reduces φ_0 by decreasing λ_D. In the Debye–Hückel limit, Eq. (10.36) predicts that the surface potential is proportional to the Debye length, and therefore proportional to $1/\sqrt{c_\infty}$. At low concentrations, the nonlinearity of Eq. (10.36) reduces this dependence.

Experimentally, we observe that the zeta potential magnitude is reduced as counterion concentration is increased (Fig. 10.2). Further, this variation is linear if the zeta potential is plotted versus pC, the negative logarithm of the summed counterion concentration, and linear fits are consistent with an intercept at the origin. Although the intercept at the origin has no physical significance (in fact, this intercept is dependent on the choice of units for concentration), this fact is convenient. Because of the linear variation of zeta with pC, we can instead plot ζ/pC, and investigate other parameters while eliminating variations owing to counterion concentration. For example, Fig. 10.3 illustrates the pH dependence of the zeta potential on glass surfaces and uses normalization to collapse data at several concentrations onto one graph.

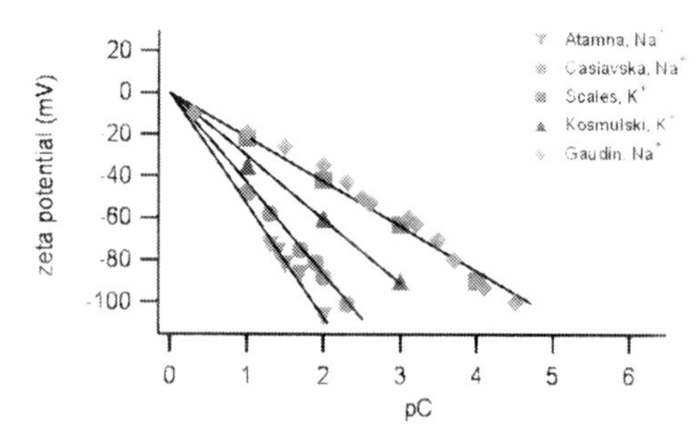

Fig. 10.2 Observed zeta potentials as a function of indifferent electrolyte concentration. pC is defined as the negative logarithm of the summed counterion concentration. (After [126], using data from [127, 128, 129, 130, 131].)

10.4.2 pH dependence

The charge-determining ions of microfluidic substrates are $H^+$and OH^-, and the dependence of ζ on pH is well described by a titration curve – that is, the zeta potential becomes more positive as pH is lowered, and becomes more negative as pH is raised, consistent with Eq. (10.13). Similar curves are seen both for microchannel substrates made of ionizable, acidic surfaces (glass) as well as those for nonionizable surfaces (Teflon), though in the latter case the chemical source of charge is less clear. An example (for Teflon) is shown in Fig. 10.4.

10.5 MODIFYING THE ZETA POTENTIAL

Many techniques can be used to modify the electrokinetic potential in a microchannel system, including changing salt concentrations, adding surfactants, and chemically functionalizing the surface.

10.5.1 Indifferent electrolyte concentrations

As was shown earlier, increasing salt concentrations on most oxide and polymer surfaces tends to reduce the electrokinetic potential. For monovalent ions, this is simply increased shielding of and nonspecific adsorption onto a charged surface. For multivalent ions, this can be caused by specific adsorption or more complicated ion–ion correlation effects.

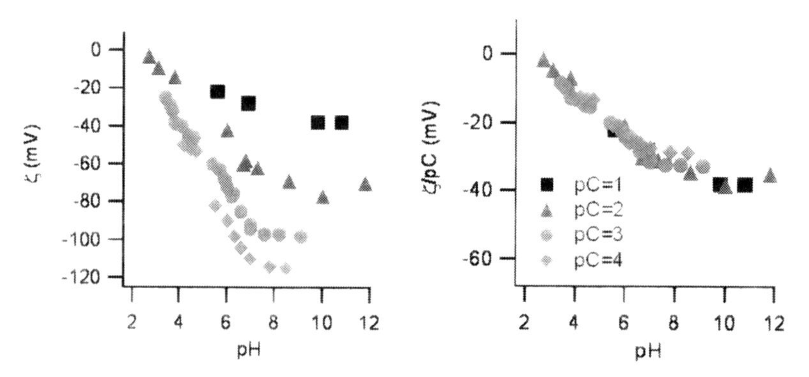

Fig. 10.3 Normalization of observed zeta potentials measured at a variety of electrolyte concentration. (After [126], using data from [129].)

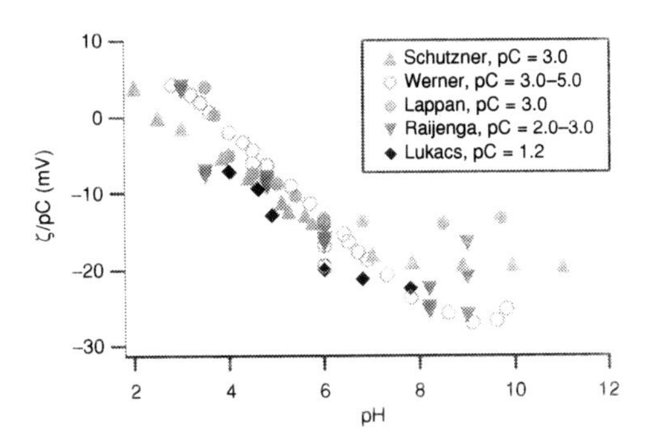

Fig. 10.4 Normalized zeta potentials for Teflon as a function of pH. (After [132].)

10.5.2 Surface-active agents

Surface-active agent (surfactant) is a catchall term for molecules or ions that act at surfaces to change their properties. Many surfactants are known to change the electrokinetic potential on oxides and polymers, including dodecyl sulfate (SDS), cellulose ethers, and polyelectrolytes.

IONIC SURFACTANTS: SDS

Ionic surfactants are usually amphiphilic molecules (i.e., molecules with both hydrophobic and hydrophilic regions) that consist of a hydrophobic alkyl region and a hydrophilic charged group. The hydrophobic end of these molecules often adsorbs to hydrophobic surfaces, extending the charged hydrophilic end of the molecule. The most commonly used ionic surfactant is sodium dodecyl sulfate (i.e., SDS; also called sodium lauryl sulfate), which consists of a $C_{12}H_{25}$ straight-chain alkyl group attached to a charged OSO_4^- group. The solid form of this surfactant is a salt with Na^+. When used on hydrophobic surfaces, SDS often coats the surface, leading to a strongly negative surface charge. Figure 10.5 shows a negative wall potential generated by adhesion of sodium dodecyl sulfate to a hydrophobic surface.

CELLULOSE ETHERS

Cellulose ethers, e.g., hydroxypropylmethylcellulose, cellulose, and methylcellulose, reduce the observed electrokinetic potential of microfluidic substrates by at least an order

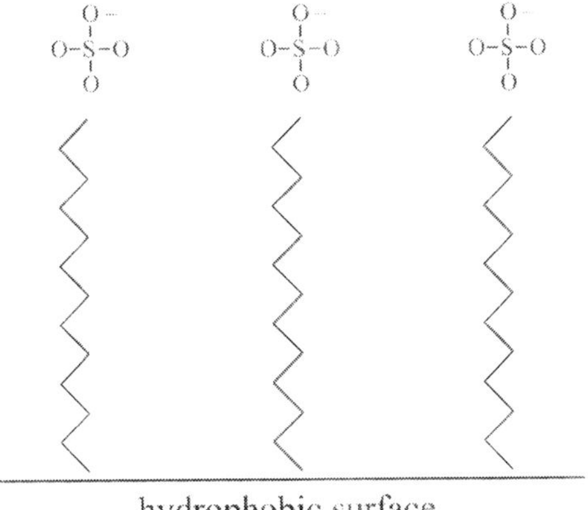

Fig. 10.5 A negative wall potential generated by adhesion of sodium dodecyl sulfate to a hydrophobic surface.

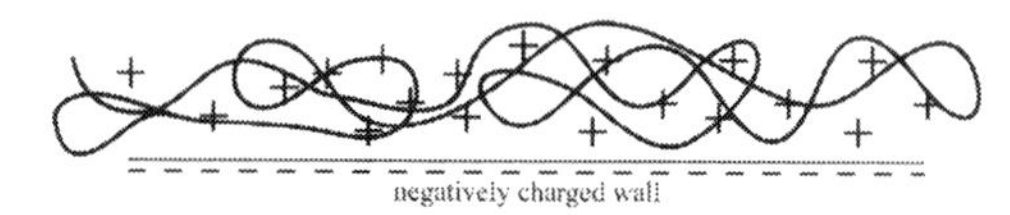

Fig. 10.6 Reversal of the charge of a negative surface by incubation with a positive polyelectrolyte such as poly(styrene sulfonate).

of magnitude and are commonly used for this purpose. These molecules bind to the surface, limiting the mobility of ions in the double layer.

POLYELECTROLYTES

Polyelectrolytes, e.g., poly(styrene sulfonate), are typically polyanions or polycations that adhere electrostatically to charged surfaces. These can be used to reverse the charge of a surface; however, their lifetime is finite. Figure 10.6 shows an example in which a negative surface is rendered positive by incubation with a positive polyelectrolyte.

10.5.3 Chemical functionalizations

Surfaces can be chemically functionalized by use of a variety of techniques. Plasmas and UV sources are often used to oxidize surfaces, usually increasing the negative charges on surfaces. Chemicals with alkoxysilyl or chlorosilyl functional groups can be used to chemically bond to glass or silicon surfaces, releasing alcohols or hydrochloric acid, respectively. These chemicals covalently bond chemical groups to the wall by using a siloxane bond. For example, trimethoxysilylpropylamine reacts to glass, bonding the propylamine group to the wall using a siloxane bond. When these techniques are used, a wide variety of functional groups can be bonded to a glass or silicon surface. Often, polymers are grafted to surfaces by combining self-assembled monolayer processes (e.g., bonding of trimethoxysilylpropyl acrylate) with free-radical chain polymerization or atom-transfer polymerization.

10.6 CHEMICAL AND FLUID-MECHANICAL TECHNIQUES FOR MEASURING INTERFACIAL PROPERTIES

In addition to spectroscopic techniques for characterizing surfaces, chemical and microfluidic techniques for measuring ζ are numerous and widespread. The relation between the observed electrokinetic potential ζ and the surface potential φ_0 depends on the physics at the wall and inside the double layer.

10.6.1 Charge titration

For systems in which H^+ and OH^- are the potential-determining ions, the surface charge as a function of pH can be estimated in a high area–volume ratio system by characterizing the variation of fluid pH as H^+ or OH^- is added to the system. For a system following Eq. (10.13), the pH can be recorded before and after x moles of H^+ is added to the system. The quantity of protons adsorbed per volume is given by

$$A\Gamma\left[\frac{10^{\alpha(\mathrm{p}K_{\mathrm{a}}-\mathrm{pH}_{\mathrm{final}})}}{10^{\alpha(\mathrm{p}K_{\mathrm{a}}-\mathrm{pH}_{\mathrm{final}})}+1}-\frac{10^{\alpha(\mathrm{p}K_{\mathrm{a}}-\mathrm{pH}_{\mathrm{initial}})}}{10^{\alpha(\mathrm{p}K_{\mathrm{a}}-\mathrm{pH}_{\mathrm{initial}})}+1}\right]=N_{\mathrm{A}}\left(10^{-\mathrm{pH}_{\mathrm{initial}}}V-10^{-\mathrm{pH}_{\mathrm{final}}}V+x\right), \tag{10.37}$$

where A is the (known) area of the surface and Γ is the site density on the surface. This measurement essentially measures the buffer capacity of the weak acid sites at the surface, and the pH values as a function of x give Γ, α, and $\mathrm{p}K_{\mathrm{a}}$.

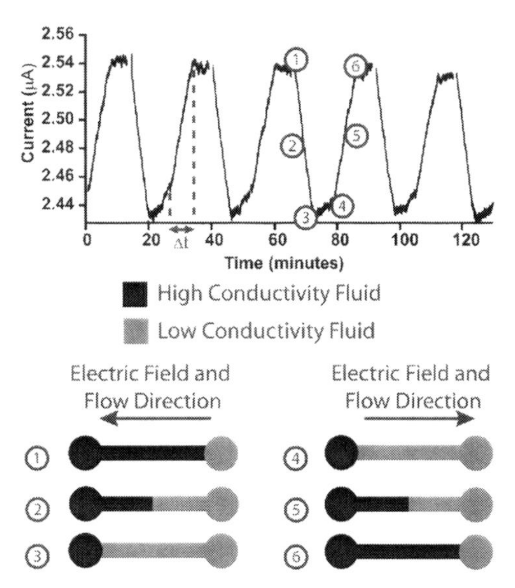

Fig. 10.7 Current monitoring experiment. Current trace is at top, and the corresponding fluid profile is shown at bottom.

10.6.2 Electroosmotic flow

By applying an electric field to a microchannel and observing the flow of fluid in the thin-EDL limit, the electroosmotic mobility of the liquid–solid interface may be measured and, using uniform fluid properties as described previously, we can use the electroosmotic mobility to calculate the electrokinetic potential ζ. In this section, we discuss techniques for measuring a spatially averaged electroosmotic mobility.

CURRENT MONITORING

Current monitoring is a popular method for measuring the electroosmotic mobility of a liquid–solid interface, which can then, in turn, be used to calculate the electrokinetic potential. Current monitoring involves connecting two reservoirs (assumed infinitely large with zero electrical resistance) by a microchannel. The two reservoirs hold fluids of slightly differing conductivity. As an electric field is applied, fluid from one reservoir displaces the fluid in the microchannel, and the resistance of the system is dominated by the conductivity of the fluid in that reservoir (see Fig. 10.7). The direction of the electric field can then be changed, filling the microchannel with fluid from the second reservoir and thus changing the system resistance. Thus fields of alternating direction are applied and the system current goes up or down, and the time required for the current to reach steady state specifies the mean fluid velocity. Current monitoring is a useful technique because the measurement is relatively simple and inexpensive, requiring only a voltage supply and an ammeter, and measurements are readily automated. Its limitations include diffusion, uncertainty owing to pressure gradients and Joule heating, and poor performance when the electroosmotic mobility approaches zero.

NEUTRAL MARKER ELUTION

Another way to monitor flow velocity is to measure the elution time of an uncharged tracer in a capillary electrophoresis system (Chapter 12 describes capillary electrophoresis separations). This is typically used when the research is focused on capillary electrophoresis separations anyway, and the measurement becomes straightforward owing to the immediate availability of the detection system.

10.6.3 Streaming current and potential

A pressure gradient $\frac{dp}{dx}$ applied to a charged microchannel of length L and cross-sectional area A leads to fluid flow, and, in the presence of an EDL that has a net charge density, this flow induces a net current. In the thin-EDL, simple-interface, uniform-property limit, the steady-state current density in the absence of an electric field is given by

$$\frac{I_{\text{str}}}{A} = -\frac{\varepsilon\varphi_0}{\eta}\left(-\frac{dp}{dx}\right). \tag{10.38}$$

This phenomenon is termed *streaming current* and is analogous to electroosmosis, which in the thin-EDL limit can be written as

$$u = -\frac{\varepsilon\varphi_0}{\eta}E \tag{10.39}$$

or, to highlight the analogous structure of the two equations,

$$\frac{Q}{A} = -\frac{\varepsilon\varphi_0}{\eta}\left(-\frac{\partial\phi}{\partial x}\right). \tag{10.40}$$

These phenomena are reported by use of an *electrokinetic coupling matrix*, first presented in Chapter 3. In the absence of EDLs, flow and current are uncoupled. The response of an electrolyte solution to electric fields or pressure gradients in the absence of a charged surface or EDL can be written as

electrokinetic coupling equations for an uncharged wall

$$\begin{bmatrix} Q/A \\ I/A \end{bmatrix} = \begin{bmatrix} r_{\text{h}}^2/8\eta & 0 \\ 0 & \sigma \end{bmatrix}\begin{bmatrix} -\frac{dp}{dx} \\ E \end{bmatrix}. \tag{10.41}$$

In the presence of EDLs, pressure gradients also generate current (by streaming current) and electric fields also generate flow (by electroosmosis), and the matrix equation can be written as

electrokinetic coupling matrix for simple interfaces, thin-EDL approximation, uniform fluid properties

$$\begin{bmatrix} Q/A \\ I/A \end{bmatrix} = \begin{bmatrix} r_{\text{h}}^2/8\eta & -\varepsilon\varphi_0/\eta \\ -\varepsilon\varphi_0/\eta & \sigma + \frac{\varepsilon^2\varphi_0^2}{\lambda_{\text{D}}\eta r_{\text{h}}} \end{bmatrix}\begin{bmatrix} -\frac{dp}{dx} \\ E \end{bmatrix}. \tag{10.42}$$

The structure of the 2×2 coupling matrix exhibits a fundamental property of this system. The coupling matrix is symmetric, and because the coupling matrix is symmetric, the system is said to exhibit *Onsager reciprocity*, a property of many physical systems close to equilibrium in which the coupling between two phenomena is equal in both directions.

The electrokinetic potential ζ of a streaming potential experiment is given by

ζ potential of a streaming current measurement, Debye–Hückel approximation, uniform fluid properties

$$\zeta = \frac{\eta L I_{\text{str}}}{\varepsilon\Delta p}, \tag{10.43}$$

and ζ can thus be measured by measuring current with a low-impedance ammeter. Unfortunately, a measurement of this form is accurate only if both the ammeter and the electrode exhibit low impedance – this implies that the detection electrodes must have a

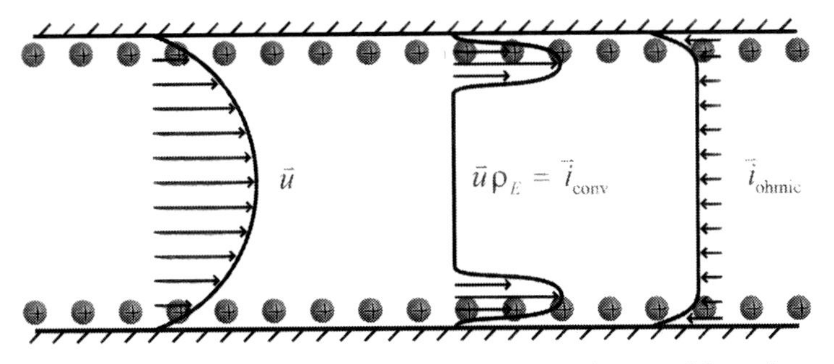

Fig. 10.8 Equilibrium velocity and current distributions during a streaming potential experiment.

catalyst such that the j_0 in Eq. (5.55) is huge and the effective resistance of the electrode is zero. This is typically difficult to achieve.

Experimentally, voltage measurement with a high-impedance voltmeter is more straightforward. If we apply a pressure gradient to a system whose reservoirs are electrically isolated, $-\frac{dp}{dx}$ is specified and $I = 0$. In this case, we can solve the electrokinetic coupling matrix equation to find

$$E = -\frac{\chi_{21}}{\chi_{22}}\left(-\frac{dp}{dx}\right), \tag{10.44}$$

which relates the observed electric field to the applied pressure field. Figure 10.8 shows the velocity and current distributions at equilibrium during a streaming potential experiment. Experimentally, we typically measure the voltage drop and the pressure drop, rather than their gradients. Thus the *streaming potential* ΔV is given by evaluating Eq. (10.44) in terms of these experimental observables:

streaming potential, general description

$$\Delta V = -\frac{\chi_{21}}{\chi_{22}}\Delta p. \tag{10.45}$$

In the thin-EDL, Debye–Hückel limit, with negligible surface conductivity, the electrokinetic potential of a streaming potential measurement is given by

ζ potential for a streaming potential experiment, thin-EDL approximation, Debye–Hückel approximation, negligible surface conductivity

$$\zeta = \frac{\Delta V}{\Delta p}\frac{\sigma\eta}{\varepsilon}. \tag{10.46}$$

This measurement requires that the system be in steady state, but it is independent of the geometry of the microchannel and can be implemented with common electrodes used in electrochemistry, such as platinized platinum electrodes or silver–silver chloride electrodes. Typical values of ΔV might be of the order of volts – for example, for 1-mM KCl on glass, we might observe $\sigma = 180\ \mu$S/cm and $\zeta = -60$ mV, which gives $\Delta V = -1.2$V for $\Delta p = 5$atm. Streaming potential is an isothermal measurement that is independent of the microchannel dimensions in the thin-EDL limit.

Streaming potential is limited to moderate electrolyte concentrations in which the equilibration is suitably fast and the signal levels are high – typically in the 0.1 mM–100 mM range. If double layers are not thin, the conductance of the channel must be measured experimentally.

EXAMPLE PROBLEM 10.2

Consider two infinite parallel plates separated by a distance $2d$. The fluid between the plates is water ($\varepsilon = 80\varepsilon_0$; $\eta = 1$ mPa s) with a symmetric, monovalent electrolyte with bulk conductivity $\sigma_{\text{bulk}} = \sum_i c_i \Lambda_i$. Both the cation and the anion of the electrolyte have the same molar conductivity Λ. Assume d^* large and φ_0^* small such that the thin-EDL and Debye–Hückel approximations can be made. Derive a relation for the net current (i.e., charge flux) $I = \int_S u\rho_E dA$ generated by flow-induced convection of charge if a pressure gradient $\frac{dp}{dx}$ is applied to this system. Presuming the reservoirs at both ends of this system are connected to an open circuit and thus no net current flows in the system, calculate the resulting steady-state potential gradient.

SOLUTION: We write $I = I'h$, where h is the depth of the channel and I' is the current per depth. We determine I' that is due to flow:

$$I' = 2\int_0^d u\rho_E \, dy = 2\lambda_D \int_0^{d^*} u\rho_E \, dy^* , \tag{10.47}$$

$$u(y) = \frac{-1}{2\eta}\frac{dp}{dx}\left(d^2 - y^2\right) = \frac{-1}{2\eta}\frac{dp}{dx}\left(d^{*2} - y^{*2}\right)\lambda_D^2 , \tag{10.48}$$

$$\rho_E = \sum_i c_i z_i F = Fc_\infty \left[\exp(-\varphi^*) - \exp(\varphi^*)\right] . \tag{10.49}$$

Making the Debye–Hückel approximation, $\rho_E = -2Fc_\infty \sinh(\varphi^*) = -2Fc_\infty\varphi^*$, leading to $I' = 2\frac{Fc_\infty\lambda_D^3}{\eta}\frac{dp}{dx}\int_0^{d^*} \varphi^*(d^{*2} - y^{*2})\, dy^*$ and $\varphi^* = \varphi_0^* \frac{\cosh y^*}{\cosh d^*}$. The current per depth is given by the integral:

$$I' = 2\frac{Fc_\infty\lambda_D^3\varphi_0^*}{\eta\cosh d^*}\frac{dp}{dx}\int_0^{d^*} (d^{*2}\cosh y^* - y^{*2}\cosh y^*)\, dy^* . \tag{10.50}$$

Now we integrate by parts, noting that $\int x^2 \cosh x = \left(x^2 + 2\right)\sinh x - 2x\cosh x$:

$$I' = 2\frac{Fc_\infty\lambda_D^3\varphi_0^*}{\eta\cosh d^*}\frac{dp}{dx}\left(2d^*\cosh d^* - 2\sinh d^*\right) . \tag{10.51}$$

Because (1) $\lambda_D^2 = \varepsilon RT/2F^2c_\infty$ for monovalent symmetric electrolytes and (2) $\tanh d^* \ll d^*$, we find

$$I = \frac{2h\varepsilon\varphi_0 d}{\eta}\frac{dp}{dx} . \tag{10.52}$$

Zero net current specifies $I' = 2\sigma Ed = -2\sigma\lambda_D d^* \frac{dV}{dx}$, and setting the sum of the flow-induced current and the ohmic current to zero leads to

$$\frac{dV}{dx} = \frac{dp}{dx}\frac{\varepsilon\varphi_0}{\sigma\eta} . \tag{10.53}$$

10.7 SUMMARY

This chapter describes techniques for predicting the equilibrium surface potential at microfluidic device interfaces as a function of the device material and solution conditions, as well as approaches for relating the surface charge to the electrokinetic potential. In this chapter, distinctions are made among various terms, particularly between interfacial potential (a measure of the potential difference between an interface and the bulk fluid) and electrokinetic potential (a measure of an observed electrokinetic phenomenon, such as fluid velocity, particle velocity, streaming current, or streaming potential). This chapter introduces the electrochemical potential $\overline{g_i}$,

$$\overline{g_i} = g_i + z_i F \varphi = g_i^\circ + RT \ln \frac{c_i}{c_i^\circ} + z_i F \varphi , \qquad (10.54)$$

and discusses indifferent electrolytes, specifically adsorbed electrolytes, and charge-determining electrolytes in the context of the electrochemical potential at the surface and the importance of the electrolytes for controlling surface charge. For Nernstian surfaces, we can describe the surface potential directly by using the Nernst equation:

$$\varphi_0 = \frac{RT \ln 10}{z_i F} \left[\log \left. \frac{a_i}{a_i(\text{pzc})} \right|_{\text{bulk}} \right] , \qquad (10.55)$$

which relates the surface potential to the difference in bulk concentration from the pzc. For non-Nernstian surfaces, surface charge must be described by use of chemical equilibria (which are largely unknown), and the surface potential can then be related to the surface charge with the Grahame equation:

$$q'' = \text{sgn}(\varphi_0) \sqrt{2\varepsilon RT I_c \sum_i c_{i,\infty}^* \left[\exp(-\frac{z_i F \varphi_0}{RT}) - 1 \right]} . \qquad (10.56)$$

Inconsistencies between measured surface charge density and electrokinetic potential are addressed by use of extended interface models that describe a condensed region of ions (Stern models), changes in fluid properties (viscoelectric models), or transverse interfacial velocity (slip models), although none of these models match broad sets of experimental data with satisfying accuracy. Because of this, electrokinetic potentials are still determined primarily by direct measurement and the links between electrokinetic potentials and surface potentials are in dispute. This chapter therefore discusses techniques for measuring electrokinetic potentials and summarizes observed electrokinetic potentials for microfluidic substrates. Because streaming potential is commonly used to measure zeta potential, we discuss streaming current, the Onsager reciprocal of electroosmosis, and write the electrokinetic coupling equation for the simple-interface, thin-EDL limit:

$$\begin{bmatrix} Q/A \\ I/A \end{bmatrix} = \begin{bmatrix} r_h^2/8\eta & -\varepsilon\varphi_0/\eta \\ -\varepsilon\varphi_0/\eta & \sigma \end{bmatrix} \begin{bmatrix} -\frac{dp}{dx} \\ E \end{bmatrix} . \quad (10.57)$$

This relation illustrates the reciprocal relation between electroosmosis and streaming current and highlights how streaming potential measurements combined with conductivity measurements determine the zeta potential.

Because we often want to control the zeta potential, approaches for modifying the zeta potential are also discussed, including addition of electrolytes and surfactants as well as chemical modification of surfaces.

10.8 SUPPLEMENTARY READING

Colloid science texts, such as those of Russel et al. [32], Hunter [69, 70], and Lyklema [71], serve as an introduction to zeta potential and its measurement, with [69] being most focused on the zeta potential. None of these sources focus on microdevice surfaces specifically, being more focused on classical colloids. The primary overlap is for silica, which is relevant both as a colloid and as a microdevice substrate. Some work more focused on microdevice substrates can be found in [126, 132, 133, 134], and review articles on zeta potential include [135]. Work on zeta potential modification can be found in [136].

This chapter focuses on fluid-mechanical concerns in equilibrium double layers, but the physical models have other, related applications. For example, the Stern layer, discussed here in the context of the electrokinetic mobilities of insulating surfaces, also has been proposed as a means for reconciling the inconsistencies between Poisson–Boltzmann descriptions of EDL capacitance and observed double-layer capacitances. This is discussed in Chapter 16 and many analytical electrochemistry sources, for example [59]. Electrokinetic coupling from a general flow–force standpoint is presented in [137], and [91, 92] discuss the thermodynamical underpinnings of the reciprocal hypothesis for coupled thermodynamic systems only slightly out of equilibrium.

Unfortunately, no model of the EDL satisfactorily matches the existing data, and all of the models listed in this chapter fail in some key way. Because the easily observed data (e.g., electroosmotic velocity and streaming potential) are integrated properties, many different models, including those with dubious physical origins, can fit subsets of the existing data. Further, the parametric regions that challenge the models most effectively coincide with those experimental conditions for which experiments are difficult. Fluid property models for the EDL are, in general, poorly established. Viscoelectric models have been described as a success in [107] and a failure in [69]; the central challenge of these models is the absence, to date, of a well-defined set of experiments that generate data for comparison to these models without including multiple confounding variables.

Unfortunately, water is a difficult molecule to model in detail, given finite computational resources and a limited ability to make measurements with the spatiotemporal resolution required for defining all relevant water properties. Because of this, the properties of water at interfaces and in the presence of high electric fields are under current debate. Appendix H gives some perspective on this in terms of the variety of models used to predict water behavior.

10.9 EXERCISES

10.1 The electrokinetic potential of a microchannel is measured as a function of $MgSO_4$ concentration and pH. At low $MgSO_4$ concentration, the electrokinetic potential varies as a function of pH, becoming positive at low pH and negative at high pH. At low pH, addition of large amounts of $MgSO_4$ causes the electrokinetic potential to decrease in magnitude, although it stays positive. At high pH, the addition of large amounts of $MgSO_4$ causes the electrokinetic potential to change signs and become positive. Of the four ions (H^+, OH^-, Mg^{+2}, SO_4^{-2}), which are potential determining? Which are indifferent? Which are specifically adsorbing?

10.2 Derive Eq. (10.32) by integrating the electroosmotic profile for a system with a viscoelectric response in the Debye–Hückel limit.

10.3 Calculate and plot the dependence of the electrokinetic potential on the surface potential by using a viscoelectric model with $k_{ve} = 1 \times 10^{-14}, 3 \times 10^{-15}, 1 \times 10^{-15}, 3 \times 10^{-16}, 1 \times 10^{-16}, 0$. Use $\varphi_0 = 5\,\text{mV}, 50\,\text{mV}, 500\,\text{mV}$.

10.4 Determine ζ as a function of φ_0, λ_D, and the slip length b if fluid properties are assumed uniform. Do not make the Debye–Hückel approximation.

10.5 Consider the Navier slip condition combined with the viscoelectric model. Is the Navier slip condition, as described in Chapter 1, posed in a way that makes it physically sensible if the viscosity in the liquid varies spatially? Rather than defining a slip length as a fundamental property that describes the interface, what description might be more appropriate?

10.6 Derive Eq. (10.34) by performing an integral analysis of the EDL with a Navier slip condition at the wall.

10.7 Considering the charge density–surface potential relationship listed in Eq. (10.36), predict the dependence of surface potential on the concentration of a monovalent symmetric electrolyte in the region between 0.1 and 0.001 M. Compare this result with the phenomenological relation $\zeta \propto \text{pC}$ and comment.

10.8 Consider a glass surface with a reactive site density of 6×10^{18} sites/m^2. Model the surface as a weak acid with a pK_a of 4.5 and assume the surface site behavior is well modeled by ideal solution chemistry. Predict the surface charge density as a function of pH.

10.9 Consider the data shown at left in Fig. 10.3. Assuming that $\zeta = \varphi_0$ for this system, develop a model to describe the surface potential of this system. This model includes a negative surface site density, nominal pK_a, and a parameter α to describe a spread of pK_a's around the nominal pK_a value.

10.10 Consider a Nernstian surface with surface charge determined by the potential-determining ion A^-. Assume NaCl is also in solution as an indifferent electrolyte, and the concentration of NaCl is much larger than that of other ions. Assume ideal solution theory, and consider the case in which the pzc is at pA = 6. Derive a relation for q''_{wall} as a function of $[A^-]$ and the bulk concentration of NaCl.

10.11 Consider pressure-driven flow through a circular capillary with uniform radius R. Using the Debye–Hückel approximation and ignoring surface conductance, evaluate the streaming potential observed at steady state by evaluating the total net convective charge flux caused by the fluid flow and determining the potential required for canceling this with ohmic current.

10.12 Although the change of streaming potential is straightforward to measure, absolute streaming potential measurements are sensitive to electrochemistry at the electrodes. One way of simplifying streaming potential measurements is to make a phase-sensitive measurement, in which case the pressure drop is made to vary sinusoidally and the measured streaming potential is observed over time. In this case, the Fourier transforms of the pressure signal and the potential signal show peaks at the driving frequency, and the magnitudes of these peaks can be used in the equation for the streaming potential. Because the system must be in equilibrium for the streaming potential relation to be valid, determine the criterion for the cycle rate so that the measurement is correct. If the cycle rate is too fast, derive how the observed measurement is in error, and show how the phase lag can be used to correct this error.

10.13 Consider a current-monitoring experiment with a channel of cross-sectional area A and length L with reservoirs of conductivity σ_1 and σ_2. Assume thin double layers and ignore mixing or diffusion between liquids from the two reservoirs. Approximate ζ as uniform throughout the system. Assume that the microchannel is filled with σ_1 fluid at $t = 0$ and an electric field E is applied at $t = 0$, displacing the σ_1 fluid with σ_2.

(a) Write an equation for the current I as a function of time if σ_2 is only slightly different from σ_1.

(b) Write a general equation for the current if σ_2 cannot be assumed close to σ_1.

10.14 Assume that you are using a device made from polystyrene and you would like to operate at a pH such that the electroosmotic mobility is equal to zero. At what pH would you work? Why? Make engineering approximations regarding zeta potential values, using material from the literature.

10.15 Consider a 7-cm-long glass microchannel with radius 10 μm filled with an ampholyte mixture (total ion concentration = 100 mM) that leads to a linear pH gradient ranging from 3 to 10. The solution has viscosity 1 mPa s. Estimate the pressure distribution along this channel if a field of 100 V/cm is applied. To do this, use a strictly 1D analysis that approximates the flow as being everywhere a superposition of a Couette flow (from electroosmosis) and Poiseuille flow (from the electrokinetically generated pressure). Note that this will satisfy the 1D conservation of mass equation but will not satisfy the 2D Navier–Stokes equations. Make engineering approximations regarding zeta potential values, using material from the literature.

10.16 Describe two ways in which a glass microchannel at pH = 7 can be made to have a near-zero electroosmotic mobility. Find one paper in the literature that uses each technique. Reference these papers, and describe briefly what the researchers for each paper did and why.

10.17 Consider a surface with a surface potential φ_0 and a slip length b in the Debye–Hückel limit.

(a) What is the apparent zeta potential ζ_a as a function of φ_0, b, and other relevant fluid and interfacial parameters if one measures the bulk electroosmotic velocity and uses the relation $u_{EO} = -\frac{\varepsilon\zeta_a}{\eta}\vec{E}$ to infer ζ_a? ε is the electrical permittivity of water, ζ_a is the apparent zeta potential, η is the viscosity, and $\vec{E}$ is the magnitude of the extrinsic electric field. Note that the surface potential φ_0 is the potential drop across the double layer, and the apparent zeta potential is the potential that one infers from an experiment if the relation $u_{EO} = -\frac{\varepsilon\zeta_a}{\eta}\vec{E}$ is used. The surface potential is a real potential, whereas the inferred apparent zeta potential is an approximation that is accurate only if the experiment and model are correct.

(b) What is the apparent zeta potential ζ_a as a function of φ_0, b, and other relevant fluid and interfacial parameters if one measures the the streaming potential and uses the relation $\frac{\Delta V}{\Delta p} = \frac{\varepsilon\zeta_a}{\sigma\eta}$ to infer ζ_a? ΔV is $V_{inlet} - V_{outlet}$, Δp is $p_{outlet} - p_{inlet}$, and σ is the bulk conductivity. You should ignore surface conductance and assume thin double layers. Note that the surface potential is the potential drop across the double layer, whereas the apparent zeta potential is the potential that we infer from an experiment. The surface potential is a real potential, whereas the inferred apparent zeta potential is an approximation that is only accurate if the experiment and model are correct.

(c) Given your relations, can simultaneous measurement of both electroosmotic velocity and streaming potential be used in concert in the thin-EDL, Debye–Hückel limit to measure b and φ_0 independently?

10.18 Assume that a surface has a specified charge density q'' and is in contact with a symmetric electrolyte. The electrolyte is indifferent, i.e., there are no surface adsorption processes and all ions behave as predicted by Gouy–Chapman theory. Predict (quantitatively) how the observed surface potential φ_0 and therefore the electroosmotic mobility should vary as the concentration of the electrolyte is changed, keeping in mind the relation between surface charge and the potential gradient at the wall.

10.19 Consider an atomically smooth silicon surface with an oxidized layer, also atomically smooth. Assume that the surface SiO^- density is $6 \times 10^{-18}/m^2$. Model the site as a negative point charge and, at the specified density, calculate the average distance between sites. Ignore the presence of counterions and approximate the charged surface as a circular region with a radius equal to half the average distance between sites. Model water as a sphere of radius 1 Å and calculate how many water molecules on average are in a monolayer on the surface per charged site. Derive a continuum expression for the permittivity of water in the vicinity of this charged site. Calculate and plot the permittivity of water molecules as a function of their radial distance from the ion. What fraction of the water molecules have a permittivity that is more than 20% less than the bulk permittivity?

11 Species and Charge Transport

This chapter describes a general framework for species and charge transport, which assists us in understanding how electric fields couple to fluid flow in *nonequilibrium* systems. The following sections first describe the basic sources of *species* fluxes. These constitutive relations include the diffusivity, electrophoretic mobility, and viscous mobility. The species fluxes, when applied to a control volume, lead to the basic conservation equations for species, the *Nernst–Planck equations*. We then consider the sources of *charge* fluxes, which lead to constitutive relations for the charge fluxes and definitions of parameters such as the conductivity and molar conductivity. Because charge in an electrolyte solution is carried by ionic species (in contrast to electrons, as is the case for metal conductors), the charge transport and species transport equations are closely related – in fact, the charge transport equation is just a sum of species transport equations weighted by the ion valence and multiplied by the Faraday constant. We show in this chapter that the transport parameters D, μ_{EP}, μ_{i}, σ, and Λ are all closely related, and we write equations such as the Nernst–Einstein relation to link these parameters.

These issues affect microfluidic devices because ion transport couples to and affects fluid flow in microfluidic systems. Further, many microfluidic systems are designed to manipulate and control the distribution of dissolved analytes for concentration, chemical separation, or other purposes.

11.1 MODES OF SPECIES TRANSPORT

Chemical species are transported through fluid systems owing to both diffusion and convection, described in the sections to follow. Chemical reactions are omitted for the purposes of this discussion.

11.1.1 Species diffusion

Species diffusion refers to the net migration of species owing to Brownian motion in the system. The thermal energy in the system leads to species motion in random directions with temporally varying velocities whose magnitudes are proportional to the energy of the system (and therefore proportional to RT). The effect is a net migration of species away from high-concentration regions and toward low-concentration regions.

11.1.2 Convection

In addition to the random fluctuations of chemical species that are due to thermal motion, the deterministic motion of chemical species that is due to fluid flow and electric fields (i.e., convection) also leads to a species flux, as the species are carried along when the fluid moves. Considering chemical species as *passive scalars* (i.e., parameters described with scalar variables that are simply carried along with the fluid) leads to the passive scalar diffusion equation (4.6), which describes chemical species diffusion in the absence of electric fields.

In the presence of electric fields, though, charged ions move in response to the Coulomb force they feel in that electric field – a process termed *electrophoresis*. The force exerted by an electric field on an ion is

$$\vec{F}_E = ze\vec{E}\,, \tag{11.1}$$

where e is the magnitude of the charge of an electron ($e = 1.6 \times 10^{-19}$ C) and z is the charge number or valence of the ion (for example, $z = 1$ for Na^+ and $z = -2$ for SO_4^{-2}). The steady-state response of the ion occurs at equilibrium between two equal and opposite forces: the Coulomb force from the electric field and the "drag" force caused by the solvent molecules. This drag force cannot be predicted precisely without detailed molecular dynamics calculations, but we achieve results within an order of magnitude for most ions if we simply model the ion as a sphere in Stokes flow with a radius commensurate with its size:

$$\vec{F}_{\text{drag}} \simeq -6\pi\vec{u}_i\eta r_i\,, \tag{11.2}$$

where $\vec{u}_i$ is the velocity of the ion i and r_i is the hydrated radius of the ion, i.e., the radius of the ion and any water molecules that are bound to it at equilibrium. Clearly the Stokes flow equations *do not* apply at the length scale of an ion; however, the effects have the same scaling, and the magnitude of the "drag" force is surprisingly close, despite the incorrectness of the model. A more physically accurate analogy is the drag force on an object moving through a granular flow, for example, a sphere moving through sand. In the case of a granular flow, a stress is applied to the object to resist motion because the motion of the object displaces the grains and creates local *jamming*, in which grains locally become rigid when the compressive force locks them in a specific configuration that prevents their motion.

At steady state, which for an ion is essentially instantaneous, the electromigratory velocity of species i caused by the electric field is written as

definition of ion electrophoretic mobility

$$\vec{u}_{\text{EP},i} = \mu_{\text{EP},i}\vec{E}\,, \tag{11.3}$$

where $\mu_{\text{EP},i}$ [m^2/V s] is termed the *electrophoretic mobility* of species i. Example electrophoretic mobilities are listed in Table 11.1. Given the motion of the ion with respect to the solvent, the total velocity of the ion $\vec{u}_i$ is given by

total velocity of ion in the presence of fluid flow and electric fields

$$\vec{u}_i = \vec{u} + \vec{u}_{\text{EP},i}\,, \tag{11.4}$$

where $\vec{u}$ refers strictly to the velocity of the fluid and $\vec{u}_i$ refers to the total velocity of the ion.

Table 11.1 Electrophoretic mobilities for some sample ions at 298 K at infinite dilution in water.

Ion	μ_{EP}(m^2/V s)	Ion	μ_{EP}(m^2/Vs)
H^+	36.3×10^{-8}	OH^-	-20.5×10^{-8}
Mg^{+2}	7.3×10^{-8}		
K^+	5.1×10^{-8}		
Na^+	5.2×10^{-8}	Br^-	-8.1×10^{-8}
Li^+	4.0×10^{-8}	Cl^-	-7.9×10^{-8}
Ca^{+2}	3.1×10^{-8}	NO_3^-	-7.4×10^{-8}
Cu^{+2}	2.8×10^{-8}	HCO_3^-	-4.6×10^{-8}
La^{+3}	2.3×10^{-8}	SO_4^{-2}	-4.1×10^{-8}

Note: Values calculated from [138]. The electrophoretic mobilities of H^+ and OH^- are really *effective* mobilities. The observed effective mobilities are anomalously high because, in water, these ions can propagate by reactive mechanisms such as the Grotthus mechanism. For comparison, electroosmotic mobilities are of similar order, for example μ_{EO} for glass-water interfaces might typically be approximately 4×10^{-8} m^2/V s.

11.1.3 Relating diffusivity and electrophoretic mobility: the viscous mobility

Although they find application in different flux terms, species diffusivity and electrophoretic mobility are closely related phenomena and (for small ions) can be calculated from one another with simple equations. Diffusivity is a measure of a species' ability to move randomly because of random thermal molecular motion and is used to describe diffusive fluxes of species. Electrophoretic mobility is a measure of the species' ability to move in response to an electric field and is a component of convective fluxes of species. The molecular collisions that limit both of these motions are the same. The difference is the driving force – thermal motion (proportional to the thermal energy of a mole of ions, and therefore proportional to RT) versus Coulomb force (proportional to the charge of a mole of ions zF). Consistent with this, the diffusivity and electrophoretic mobility of a small ion are related to a single property and to each other by the *Nernst–Einstein relation*:

$$\mu_i = \frac{D_i}{k_B T} = \frac{\mu_{EP,i}}{z_i e} . \tag{11.5}$$

On a per-mole basis, we can equivalently write

Nernst–Einstein relation

$$\frac{\mu_i}{N_A} = \frac{D_i}{RT} = \frac{\mu_{EP,i}}{z_i F} . \tag{11.6}$$

We call μ_i [m/N s] the *viscous mobility* of species i.[1] The viscous mobility measures the ratio of the velocity at which a species moves to the force driving that motion – it

[1] This topic is treated with widely varying terminology and notation. Many authors call this simply the *mobility*. Many authors also use the reciprocal of this value, terming $1/\mu_i$ the *friction constant* or *viscous friction coefficient*. Also, some authors define the mobility as equal to D/RT or μ_{EP}/zF, which is the same conceptually but different numerically by a factor of Avogadro's number, leading to units of $[\frac{\text{mol m/s}}{\text{N}}]$.

is a viscous property of the ion and the solvent, and $1/\mu_i$ [N s/m] is a molecular-scale analog of the drag coefficient that might be defined for a macroscopic object. In fact, a viscous mobility can be defined for a macroscopic object, for example, for a sphere of radius a in Stokes flow (see Subsection 8.3.1), the viscous mobility is given by $\mu = \frac{1}{6\pi\eta a}$.

The Nernst–Einstein relation applies rigorously for species that can be modeled as being a point charge. The point charge assumption is correct for small ions and is approximately correct for proteins (even though the charge on a protein in general is distributed).[2]

11.2 CONSERVATION OF SPECIES: NERNST–PLANCK EQUATIONS

This section describes the phenomena that lead to flux of species. These fluxes, when applied to a control volume, lead to the Nernst–Planck equations.

11.2.1 Species fluxes and constitutive properties

In the absence of chemical reactions, the two mechanisms that lead to flow of species into or out of a control volume are diffusion and convection.

DIFFUSION

In the dilute solution limit with negligible thermodiffusion effects (which is applicable far from walls for most ionic species in the conditions used in microfluidics), Fick's law defines a flux density of species proportional to the gradient of the species concentration and the diffusivity of the species in the solvent:

Fick's law for species flux density in a concentration gradient

$$\vec{j}_{\mathrm{diff},i} = -D_i \nabla c_i \,, \tag{11.7}$$

where $\vec{j}_{\mathrm{diff},i}$ [mol/s m^2] is the diffusive species flux density (i.e., the amount of species i moving across a surface per unit area due to diffusion), D_i is the diffusivity of species i in the solvent (usually water), and c_i is the concentration of species i. Fick's law is a macroscopic way of representing the summed effect of the random motion of species owing to thermal fluctuations. Fick's law is analogous to the Fourier law for thermal energy flux caused by a temperature gradient and the Newtonian model for momentum flux induced by a velocity gradient, and the species diffusivity D_i is analogous to the thermal diffusivity $\alpha = k/\rho c_p$ and the momentum diffusivity η/ρ.

---

[2] A more subtle point is the dependence of diffusion and electrophoresis on the applied electric field. An applied field can lead to changes in the orientation of a molecule. When this is the case, the electrophoretic mobility is proportional to the diffusivity *in the direction of the field*, which might be different from the isotropic diffusivity observed for a statistical sampling of molecules. This is usually a minor correction, important only for systems that are poorly approximated as point charges.

CONVECTION

In addition to the random ion fluctuations that are due to thermal motion, the deterministic ion motion that is due to fluid flow and electric fields (i.e., convection) also leads to a species flux density:

convective flux density owing to fluid flow

$$\vec{j}_{\mathrm{conv},i} = \vec{u}_i c_i \,, \tag{11.8}$$

where $\vec{j}_{\mathrm{conv},i}$ [mol/s m^2] is the convective species flux density (i.e., the amount of species i moving across a surface per unit area due to convection) and $\vec{u}_i$ is the velocity of species i. As described in Eq. (11.4), the velocity of species i is given by the vector sum of $\vec{u}$ (the velocity of the fluid) and $\vec{u}_{\mathrm{EP},i}$ (the electrophoretic velocity of the ion with respect to the fluid:

$$\vec{u}_i = \vec{u} + \vec{u}_{\mathrm{EP},i} \,. \tag{11.9}$$

Because we often use units of moles per liter for species concentration, the species concentrations must be converted to moles per cubic meter if species fluxes are to be in SI units.

11.2.2 Nernst–Planck equations

In general, the transport of a species i in the absence of chemical reactions can be described by use of the Nernst–Planck equations:

Nernst–Planck equations

$$\frac{\partial c_i}{\partial t} = -\nabla \cdot \left[-D_i \nabla c_i + \vec{u}_i c_i \right] . \tag{11.10}$$

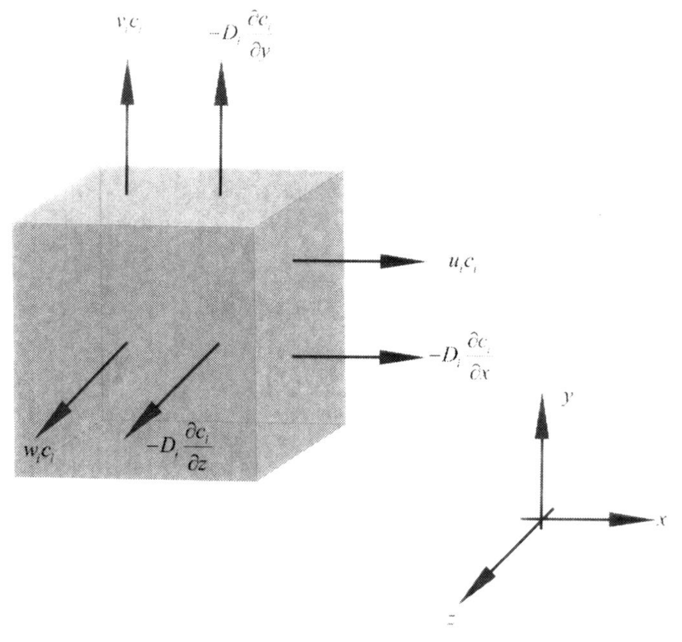

Fig. 11.1 Species fluxes for a Cartesian control volume.

Here, D_i is the diffusivity of species i and $\vec{u}_i$ is the velocity of species i. The symbol $\vec{u}_i$ denotes the vector sum of the fluid velocity $\vec{u}$ and the electrophoretic velocity of the species $\mu_{EP,i}\vec{E}$. The first term in the brackets is the diffusive flux and the second is the convective flux. The Nernst–Planck equations in this form relate the change in the concentration of a species to the divergence of the net species flux density. We can derive this relation by drawing a differential control volume and evaluating the species fluxes. Such a control volume is shown (for a Cartesian system) in Fig. 11.1.

EXAMPLE PROBLEM 11.1

Consider a sodium ion motionless in quiescent water. At time $t = 0$, an electric field of 100 V/cm is applied.

1. Use Eq. (11.1) to calculate the electrophoretic force on the ion.
2. What is the ion acceleration at $t = 0$?
3. Given the known electrophoretic mobility (Table 11.1), what is the terminal velocity of the ion?
4. Approximate the time required to reach terminal velocity as given by the ratio of the terminal velocity to the $t = 0$ acceleration. Calculate this time (technically, this is not the time to reach terminal velocity but rather the exponential decay time of the velocity perturbation).
5. What do you conclude about the process of mobilizing an ion? Can ion motion be assumed instantaneous in response to electric fields, or is there an appreciable inertial lag before an ion reaches its terminal velocity?

SOLUTION: The force is given by

$$F = zeE = 1.6 \times 10^{-15}\,\text{N}\,. \tag{11.11}$$

The sodium ion mass is $23 \times 1.66 \times 10^{-27}$ kg $= 3.8 \times 10^{-26}$ kg. The acceleration is therefore

$$a = \frac{F}{m} = 1.6 \times 10^{-15}/3.8 \times 10^{-26} = 4 \times 10^{10}\,\text{m/s}^2\,. \tag{11.12}$$

The terminal velocity is given by

$$u = \mu_{EP}E = 5.2 \times 10^{-8} \times 1 \times 10^{4} = 5.2 \times 10^{-4}\,\text{m/s}\,, \tag{11.13}$$

and therefore the time to accelerate is

$$t = \frac{u}{a} = 1.3 \times 10^{-14}\,\text{s}\,. \tag{11.14}$$

For practical purposes, ions instantaneously reach their terminal velocity in response to electric fields.

COMPARING THE NERNST-PLANCK, SCALAR CONVECTION-DIFFUSION, AND NAVIER-STOKES EQUATIONS

Equation (11.10) can be reorganized into a form similar to the one we have used for the passive scalar convection–diffusion equation or the Navier–Stokes equations. For example, assuming the diffusivity is uniform and implementing the product rule $\nabla(\vec{u}_i c_i) = \vec{u}_i \nabla c_i + c_i \nabla \cdot \vec{u}_i$, we obtain

$$\frac{\partial c_i}{\partial t} + \vec{u}_i \cdot \nabla c_i + c_i \nabla \cdot \vec{u}_i = D_i \nabla^2 c_i \,. \tag{11.15}$$

Compared with the Navier–Stokes equations for momentum transport in incompressible fluids, the Nernst–Planck equations (shown here without chemical reaction) have neither a source term nor a pressure term. Compared with both the Navier–Stokes equations and the passive scalar convection–diffusion equation, the Nernst–Planck equations have an additional term ($c_i \nabla \cdot \vec{u}_i$) proportional to the divergence of the species velocity. For incompressible fluid flows, the divergence of the fluid velocity $\nabla \cdot \vec{u}$ is zero owing to conservation of mass; however, we cannot in general say that the divergence of the species velocity $\nabla \cdot \vec{u}_i$ is zero, because systems with electric fields and nonuniform conductivity have finite divergence in the species velocity field.

EXAMPLE PROBLEM 11.2

In the case in which diffusivity is uniform, rearrange Eq. (11.10) to obtain Eq. (11.15).

SOLUTION: We start with

$$\frac{\partial c_i}{\partial t} = -\nabla \cdot \left[-D_i \nabla c_i + \vec{u}_i c_i \right] . \tag{11.16}$$

Separating the fluxes and assuming uniform D_i, we get

$$\frac{\partial c_i}{\partial t} = D_i \nabla^2 c_i + \nabla \cdot (\vec{u}_i c_i) \,. \tag{11.17}$$

Using the product rule on the convective term and moving the result to the LHS, we obtain

$$\frac{\partial c_i}{\partial t} + \vec{u}_i \cdot \nabla c_i + c_i \nabla \cdot \vec{u}_i = D_i \nabla^2 c_i \,. \tag{11.18}$$

11.3 CONSERVATION OF CHARGE

In this section, we sum the Nernst–Planck equations for all species i, weighted by $z_i F$, to obtain the *charge conservation equation*. In so doing, we naturally obtain the conductivity σ or the molar conductivity Λ.

11.3.1 Charge conservation equation

The Nernst–Planck equation (11.10) relates the change in species concentration to the divergence of species flux densities. Because the valence of species i is given by z_i, and

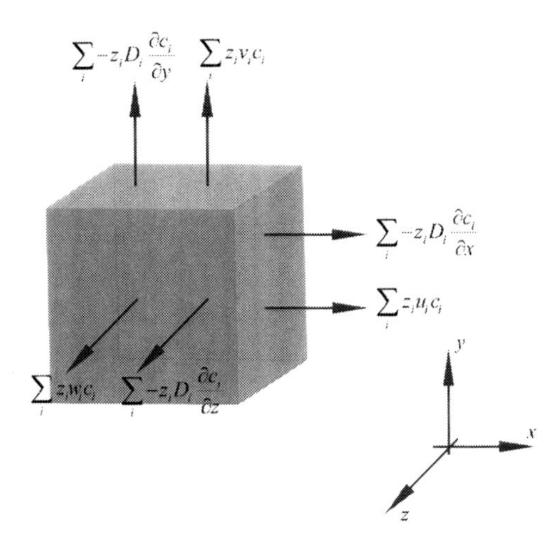

Fig. 11.2 Charge fluxes for a Cartesian control volume.

the charge of a mole of species i is given by $z_i F$, each species conservation equation also describes the charge transport *due to species i*. Summing over all species, we obtain

charge conservation equation

$$\sum_i z_i F \frac{\partial c_i}{\partial t} = -\nabla \cdot \left[-\sum_i z_i F D_i \nabla c_i + \sum_i z_i F \vec{u}_i c_i \right] , \qquad (11.19)$$

The charge fluxes from this equation are shown in Fig. 11.2. Equation (11.19) can be rearranged to give

$$\frac{\partial \rho_E}{\partial t} + \vec{u} \cdot \nabla \rho_E = -\nabla \cdot \left[-\sum_i z_i F D_i \nabla c_i + \sigma \vec{E} \right] , \qquad (11.20)$$

and if the fluid velocity $\vec{u}$ is small compared with the electrophoretic motion of the ions $\mu_{EP,i} \vec{E}$, the $\vec{u} \cdot \nabla \rho_E$ term can be ignored, leaving

$$\frac{\partial \rho_E}{\partial t} = -\nabla \cdot \left[-\sum_i z_i F D_i \nabla c_i + \sigma \vec{E} \right] . \qquad (11.21)$$

If the diffusivities of all species are the same and equal to D, Eq. (11.20) becomes

charge conservation equation, identical species diffusivities

$$\frac{\partial \rho_E}{\partial t} + \vec{u} \cdot \nabla \rho_E = -\nabla \cdot \left[-D \nabla \rho_E + \sigma \vec{E} \right] , \qquad (11.22)$$

and Eq. (11.21) becomes

charge conservation equation, identical species diffusivities, negligible fluid convection

$$\frac{\partial \rho_E}{\partial t} = -\nabla \cdot \left[-D \nabla \rho_E + \sigma \vec{E} \right] . \qquad (11.23)$$

In the preceding equations, σ $[\frac{\mathrm{C}}{\mathrm{V\,m\,s}}]$ is the conductivity and is defined as $\sigma = \sum_i c_i z_i F \mu_{\mathrm{EP},i}$, and the charge density ρ_{E} is defined as $\rho_{\mathrm{E}} = \sum_i c_i z_i F$. The electrical conductivity thus naturally comes from the charge conservation equation and is directly related to $\mu_{\mathrm{EP},i}$. Physically, this is consistent with the notion that charge is conducted in an electrolyte solution owing to ion motion – the higher the electrophoretic mobility of the ionic components, the higher the conductivity of the solution.

11.3.2 Diffusivity, electrophoretic mobility, and molar conductivity

Molar conductivity Λ $[\frac{\mathrm{m^2\,S}}{\mathrm{mol}}]$ is defined such that

definition of molar conductivity

$$\sigma = \sum_i c_i \Lambda_i \,. \tag{11.24}$$

The molar conductivity is convenient if the species transport equation is to be solved simultaneously with the charge conservation equation. The molar conductivity is proportional to the electrophoretic mobility, as we might expect, because ohmic conductivity stems from the ability of charged ions to move in response to an electric field:

relation between molar conductivity and ion electrophoretic mobility

$$\Lambda = zF\mu_{\mathrm{EP}} \,. \tag{11.25}$$

The molar conductivity is always positive, because z and μ_{EP} always have the same sign.

11.4 LOGARITHMIC TRANSFORM OF THE NERNST–PLANCK EQUATIONS

The Nernst–Planck equations,

$$\frac{\partial c_i}{\partial t} = -\nabla \cdot \left[-D_i \nabla c_i + \vec{\boldsymbol{u}}_i c_i\right] , \tag{11.26}$$

can be difficult to solve numerically for microfluidic systems, because the variation in charged species in microfluidic systems (for example, near charged walls) is typically exponential and the derivatives can be difficult to handle. Numerical simulations performed without extreme care often lead to nonphysical solutions, for example, negative concentrations or numerical instability. These problems are commonly observed when the dynamics of EDLs with large applied potentials are predicted, because coion concentrations become extraordinarily small and numerical errors can cause calculated coion concentration to oscillate around zero. These problems stem from the nonlinear response of c_i to applied voltages – for those conditions in which c_i is very small (for example, the coions in an EDL), the physically correct magnitude for c_i (a very small positive magnitude) is numerically close to a physically incorrect magnitude (for example, a negative magnitude). We can address this by making the substitution $c_i = \exp(\gamma_i)$, leading to the equation

logarithmically transformed Nernst–Planck equation

$$\exp(\gamma_i)\frac{\partial \gamma_i}{\partial t} = -\nabla \cdot \left[-D_i \exp(\gamma_i)\nabla\gamma_i + \vec{u}_i \exp(\gamma_i)\right] . \qquad (11.27)$$

This transformed equation has two key advantages. First, the logarithm (γ_i) of the concentration varies linearly when the concentration c_i varies exponentially, and thus γ_i is well suited for solving numerically on regular meshes. Further, numerical errors in γ_i do not lead to nonphysical solutions because c_i is positive for all values of γ_i.

11.5 MICROFLUIDIC APPLICATION: SCALAR-IMAGE VELOCIMETRY

Scalar-image velocimetry (SIV) is a technique used to measure microscale velocity fields; it consists of visualizing a conserved scalar (for example, a dye) at several points in time and using a computational algorithm to infer the velocity field that must have existed to produce such a change. Rigorously, this technique requires resolved 4D information (3D spatial information as well as temporal information). In microsystems, this technique has typically been used with 2D spatial information measured as a function of time and assumes that the z axis has minimal gradients or can be averaged. Most SIV techniques use a molecular-tagging velocimetry approach, meaning that a technique (usually a laser technique) is used to create variations in the conserved scalar, whose motions are then used to infer velocity. Two techniques are described in the next two subsections.

11.5.1 SIV using caged-dye imaging

Caged dyes are dyes that do not fluoresce until they are photodissociated [139]. A UV photodissociation pulse is followed by several visualizations of the dye flow. The strength of this technique is high signal-to-noise ratio and relatively straightforward implementation. The weaknesses are (a) the uncaged dyes are typically charged, and thus this technique is poorly suited for electrokinetic flows; and (b) these dyes degrade quickly over time. Figure 11.3 shows flow visualization obtained using caged-dye imaging.

11.5.2 SIV using photobleaching

Photobleaching implies that an intense pulse of light is used to "bleach" a fluorescent tracer in a specific location. Most dyes, if exposed to light, exhibit decreased fluorescence quantum yield (emitted photons per absorbed photon) because of intersystem crossings, photodissociation, or other quantum effects. Thus a local region in a flow can be bleached, causing the spatial feature whose motion can be used to infer velocity. Photobleaching is typically a low signal-to-noise ratio technique. Figure 11.4 shows results obtained using photobleaching.

11.6 SUMMARY

This chapter describes the basic sources of *species* fluxes, including constitutive relations such as the diffusivity, electrophoretic mobility, and viscous mobility. The species fluxes,

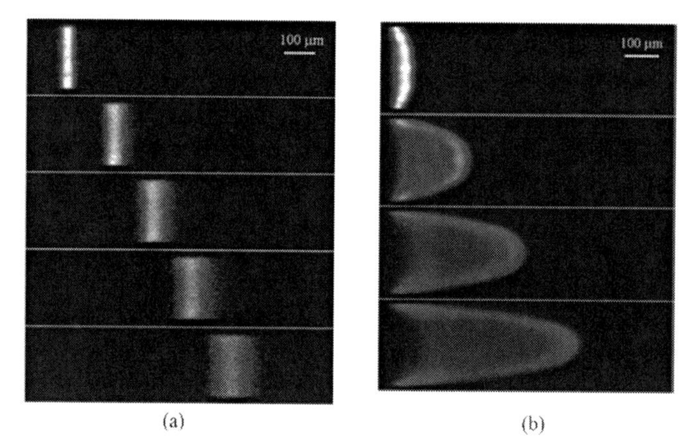

Fig. 11.3 Results for (a) electroosmotic flow and (b) pressure-driven flow in a microchannel, as observed using caged-dye imaging. (Reprinted with permission from [140].)

when applied to a control volume, lead to the basic conservation equations for species, the Nernst–Planck equations:

$$\frac{\partial c_i}{\partial t} = -\nabla \cdot \left[-D_i \nabla c_i + \vec{\boldsymbol{u}}_i c_i \right] . \tag{11.28}$$

This equation can be summed for all species and weighted by the species valence, leading to the charge conservation equation, which is written here for negligible fluid flow and species with identical diffusivities:

$$\frac{\partial \rho_E}{\partial t} = -\nabla \cdot \left[-D \nabla \rho_E + \sigma \vec{\boldsymbol{E}} \right] . \tag{11.29}$$

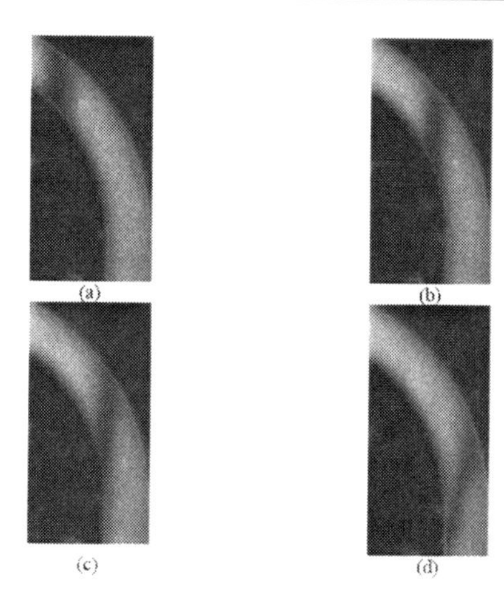

Fig. 11.4 Time history of electroosmotic flow through a turn, visualized using photobleaching of fluorescein [141]. Reprinted with permission.

These equations lead to discussion of transport parameters D, μ_{EP}, μ_i, σ, and Λ, related by the Nernst–Einstein relation,

$$\frac{\mu_i}{N_A} = \frac{D_i}{RT} = \frac{\mu_{EP,i}}{z_i F}, \tag{11.30}$$

and the definitions of conductivity and molar conductivity. These issues are central to nonequilibrium electrokinetic flow and microfluidic manipulation of chemical species.

11.7 SUPPLEMENTARY READING

Probstein [29] offers a thorough treatment of charged species transport and discusses chemical reaction kinetics, which is omitted in this text (except for electrode reactions, discussed briefly in Chapter 5).

Fick's law is described in a simple form that neglects thermodiffusion and assumes dilute solutions. Diffusion becomes more complicated if (a) concentration gradients exist simultaneously with thermal gradients, in which case thermodiffusion or Soret effects occur; or (b) the system is not in the dilute solution limit, in which case the diffusivity is a function of the concentrations of all species components, rather than being a binary property of the species and the solvent. We ignore these phenomena here, as they affect few of the engineering systems on which we are focused.

The Nernst–Einstein relation used in this chapter is correct for point charges. However, it fails when the hydrodynamic effects of diffusion and the electrostatic effects of electrophoresis behave differently owing to the structure of the molecule. It *cannot*, for example, be used to relate the diffusivity of charged macromolecules like DNA to their electrophoretic mobility owing to the difference in the hydrodynamic and electrostatic interactions of the components of the polyelectrolyte. The electrostatic and hydrodynamic interactions of DNA are discussed in Chapter 14 in the context of free-draining and nondraining polymers.

SIV techniques have been described for macroscopic [142, 143] and microscopic [140, 141] flows.

11.8 EXERCISES

11.1 Given the electrophoretic mobilities in Table 11.1 and the approximate relation in Eq. (11.2), estimate the hydrated radius of the following ions:

(a) Na^+,

(b) La^{+3},

(c) Cl^-,

(d) SO_4^{-2}.

11.2 Given the electrophoretic mobilities in Table 11.1 and the Nernst–Einstein relation (11.5), calculate the diffusivity of the following ions in water:

(a) Ca^{+2},

(b) NO_3^-,

(c) HCO_3^-.

11.3 Given the electrophoretic mobilities in Table 11.1 and the definition of molar conductivity in Eq. (11.24), calculate the molar conductivities of the following ions in water:

(a) H^+,

(b) OH^-,

(c) Li^+,

(d) SO_4^{-2}.

11.4 Consider the distribution of an ion of valence z in a 1D potential field $\varphi(y)$. Derive the Einstein relation by

- writing the equilibrium distribution $c(\varphi)$,
- writing the 1D Nernst–Planck equations for ion transport in the y direction, and
- showing that the zero-flux condition at equilibrium requires that the Einstein relation hold.

11.5 Consider the 1D ion flux equation for a chemical species i:

$$j_i = -D_i \frac{\partial c_i}{\partial x} + u_i c_i \,, \tag{11.31}$$

and show that the normalized flux j_i/c_i is proportional to the spatial gradient of the electrochemical potential $\overline{g_i} = g_i^\circ + RT \ln \frac{c_i}{c_i^\circ} + z_i F \varphi$.

11.6 Given the diagram of a Cartesian control volume as shown in Fig. 11.1, derive the Nernst–Planck equations shown in Eq. (11.10).

11.7 Consider the general species conservation equations i.e., the Nernst–Planck equations, as listed in Eq. (11.10). Sum the Nernst–Planck equations over all species i and multiply by F to obtain the charge conservation equation (11.23). Pay particular note to the fluid velocity – what assumption must be made for the fluid velocity to be omitted from the charge conservation equation?

11.8 Consider a solution of several species. If the diffusivities of all species are equal and given by D, show that, in the bulk, $\sigma/D = \varepsilon/\lambda_D^2$.

11.9 We often define the electrical resistance of a microchannel as $R = \frac{L}{\sigma A}$. This representation is correct only if the effect of double layers can be ignored. One way that this relation is corrected to account for double layers is to include a *excess surface conductance* or *surface conductance* G_S, which has units of siemens and represents the additional conductance caused by the excess ion concentration in the EDL.[3] With the surface conductance, the electrical resistance of a microchannel is given by

$$R = \frac{L}{\sigma A + G_S \mathcal{P}} \,. \tag{11.32}$$

Here $\mathcal{P}$ is the perimeter of the microchannel.

[3] Many authors write the surface conductance as σ_S, but in this text we reserve σ for conductivities and use G for conductances.

Consider a Gouy–Chapman model of the EDL, and assume that all ions have the same molar conductivity in the double layer as they do in the bulk solution. Given these approximations, the excess surface conductance ignoring fluid flow is given by

$$G_s = \int_{y=0}^{\infty} \sum_i (c_i - c_{i,\infty})\,\Lambda_i\,dy\,. \tag{11.33}$$

Calculate the surface conductance for a glass surface with a surface potential $\varphi_0 = -50$ mV if the bulk solution is 10-mM KCl.

11.10 The Dukhin number is a dimensionless parameter that indicates the relative importance of surface conductance compared with bulk conductivity. For a microchannel, it is defined as

$$Du = \frac{G_s \mathcal{P}}{\sigma A}\,, \tag{11.34}$$

where $\mathcal{P}$ is the cross-sectional perimeter of the microchannel and A is the cross-sectional area.

For a microparticle, the Dukhin number is typically defined as

$$Du = \frac{G_s}{a\sigma}\,, \tag{11.35}$$

where a is the particle radius. When the Dukhin number is small, surface conductance may be neglected. When it is not small relative to unity, surface conductance is important.

(a) Consider a circular microchannel with radius $R \gg \lambda_D$ and surface potential of φ_0. Using the Debye–Hückel approximation for a Gouy–Chapman double layer, what is the Dukhin number of the microchannel?

(b) Consider a particle of radius $a \gg \lambda_D$ and surface potential of φ_0. Using the Debye–Hückel approximation for a Gouy–Chapman double layer, what is the Dukhin number of the particle?

11.11 Consider the Nernst–Planck equations for a stagnant fluid. Consider a symmetric $z{:}z$ electrolyte and assume that the species diffusivity is uniform and equal to D for all species.

(a) Linearize the equations by replacing c_i with $c_{0,i} + \delta c_i$, where $c_{0,i}$ is the steady-state value and δc_i is a small perturbation from this value. Assume that δc_i is small enough that it can be neglected from the convective term.

(b) Note that the charge density $\rho_E = \sum_i z_i F c_i = \sum_i z_i F \delta c_i$. Subtract the two linearized equations (one for the cation, one for the anion) and derive a transport equation for ρ_E. Use the Nernst–Einstein relation, the Poisson equation, and the definition of the Debye length to show that

$$\frac{1}{D}\frac{\partial \rho_E}{\partial t} = \nabla^2 \rho_E - \frac{\rho_E}{\lambda_D^2}\,. \tag{11.36}$$

This is often called the *Debye–Falkenhagen equation*, and it applies when applied voltages are low and the perturbations to the charge density and concentrations are small.

11.12 Consider an SIV measurement used to measure fluid flow in a system. How do transport properties of the visualized scalar affect velocity measurements for pressure-driven flow with no electric field? For electric-field-driven flow with no pressure gradient?

12 Microchip Chemical Separations

Chemical separations are a critical component of analytical and synthetic chemistry. In microchip applications, a sample comprising multiple chemical species is separated spatially into individual components by inducing the components of a sample to move at differing velocities in a microchannel. This is shown schematically in Fig. 12.1 and a sample experimental result is shown in Fig. 12.2. Separations are achieved by inserting a sample fluid bolus into a microchannel, inducing motion of these species with velocities that differ from species to species, and detecting the concentration of species as a function of time as these species elute (i.e., arrive) at the location of the detector (Fig. 12.1). Many microfluidic separations are modified from capillary or column-based techniques, and draw advantages from more optimal fluid transport, thermal dissipation, or system integration. One example of a chemical separation is an electrophoresis separation, which can be used to separate species that have different electrophoretic mobilities. In this case, species motion is induced by an electric field aligned along the axis of the microchannel, which induces electroosmosis and electrophoresis. Because this technique requires only that electric fields be applied, it integrates easily into microsystem designs, and a large fraction of the microchip analyses developed since 1995 use microchip electrophoresis (see one example in Fig. 12.3). This is true for both protein analysis (Section 12.5) and DNA analysis and sequencing (Chapter 14). This chapter outlines the basic experimental setup and techniques used to realize microchip separations, discussing modes of separation and identifying transport issues related to these separations. In particular, separations motivate discussion of how a discrete bolus of fluid travels through a long straight channel, as well as the diffusive and dispersive effects of the flow on the bolus. Because our focus is on the fluid-mechanical impact on these separations, we dwell on the separations themselves only long enough to motivate the discussion, and use the exercises to encourage implementation of topics described in earlier chapters to these chemical-separation-motivated flows.

12.1 MICROCHIP SEPARATIONS: EXPERIMENTAL REALIZATION

Before delving into the transport issues, we discuss experimental requirements for microchip separations. We realize microchip separations by creating a bolus of sample fluid in a microchannel, actuating that fluid by using pressure or voltage, and detecting the presence of chemical species with some sort of detector. For example, for an electrophoresis separation, a high-voltage power supply would be connected to fluid reservoirs that, in turn, connect to microchannels etched or stamped into a microfluidic chip. Voltage sequences and electroosmosis are used to inject a sample bolus into a separation channel, and the various species i migrate per their net electrokinetic mobility $\mu_{EK,i} = \mu_{EO} + \mu_{EP,i}$. Some sort of detector (electrochemical or laser-induced fluorescence or absorbance) is typically positioned at the end of the separation channel and

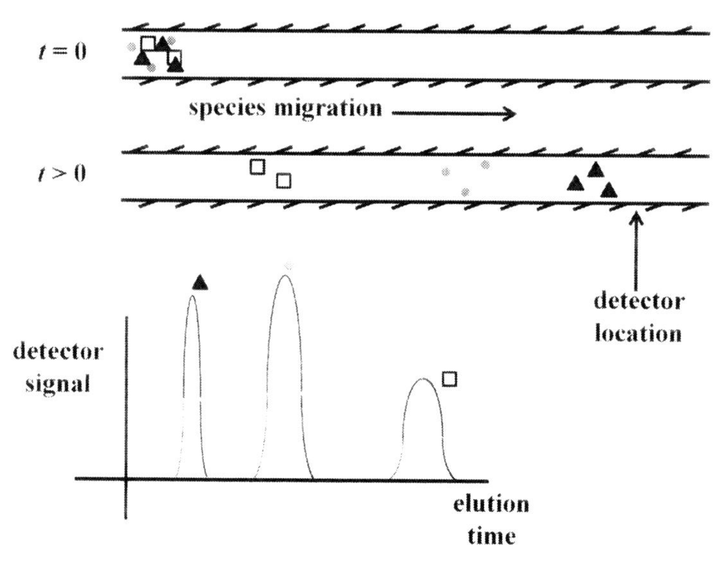

Fig. 12.1 A schematic depiction of a chemical separation.

measures the presence of passing species. An example (which uses conductivity detection) is shown in Fig. 12.4.

12.1.1 Sample injection

Sample injection can be more sophisticated in microfluidic chips compared with capillaries, and efficient sample injection is an important advantage of microdevices for separation. Because of the ability of microdevices to integrate different function, multiple channels can be used and separations can be integrated with other processes such as

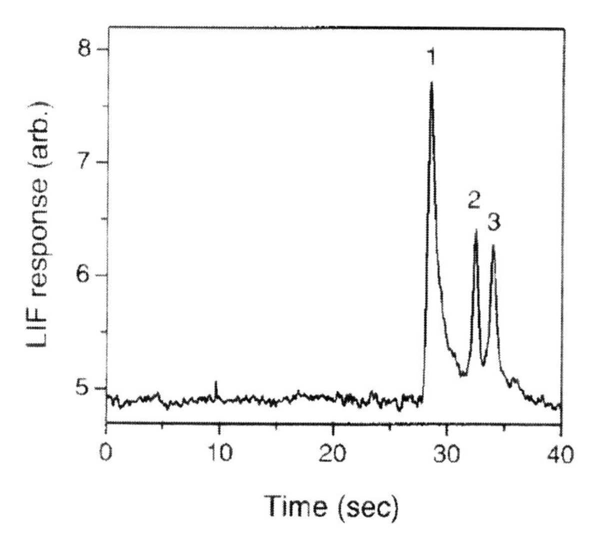

Fig. 12.2 An electrophoretic separation of several proteins, quantified with laser-induced fluorescence. The y axis denotes instantaneous concentration of a species at the detector, which is downstream of the injection. Note that peak 1 corresponds to the species with the highest electrokinetic mobility and peak 3 corresponds to the species with the lowest. (Reproduced with permission from [144].)

Fig. 12.3 The Sandia MicroChemLab chip, designed for capillary zone electrophoresis and capillary gel electrophoresis separations of protein biotoxins.

chemical reaction or incubation. Also, the size of the sample can be exquisitely controlled in a microdevice by controlling the actuation sequence that injects the fluid. The archetypal injection is the pinched electrokinetic injection, shown in Fig. 12.5.

12.1.2 Resolution

Consider a microchannel that extends in the x direction. For the moment, ignore diffusion and fluid flow and treat the problem as 1D. Consider a bolus of fluid of width w_0 centered at $x = 0$ at time $t = 0$ and containing n species. At time $t = 0$, these chemical species are induced to move through the channel with velocities u_i – this may be due to electrophoresis or convection or both. At time t, a bolus corresponding to each species i is centered at the location $x = u_i t$. For any two species whose velocities differ by Δu, the locations of the boluses are separated by Δut, and once $\Delta ut > w_0$, the boluses are spatially separated. We define the *resolution* R as

definition of separation resolution

$$R = \frac{\text{separation of centers}}{\text{bolus width}} = \frac{\Delta ut}{w(t)}, \tag{12.1}$$

where $w(t)$ is the bolus width as a function of time – this width often increases with time owing to diffusion or dispersion. Much of the design of separation systems is focused on maximizing R.

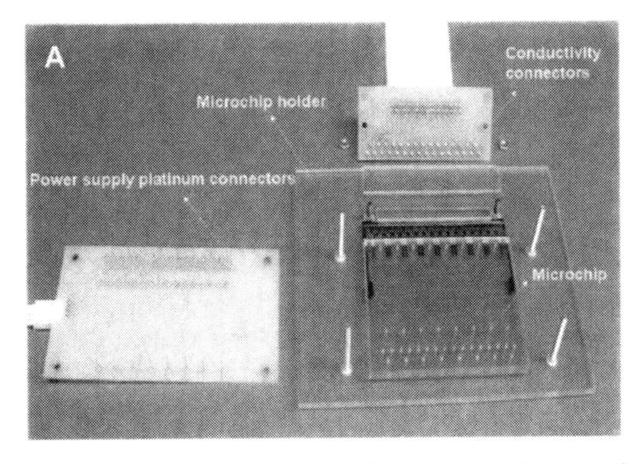

Fig. 12.4 A parallelized device for multiplexed electrophoretic separations and conductivity detection. (Reproduced with permission from [145].)

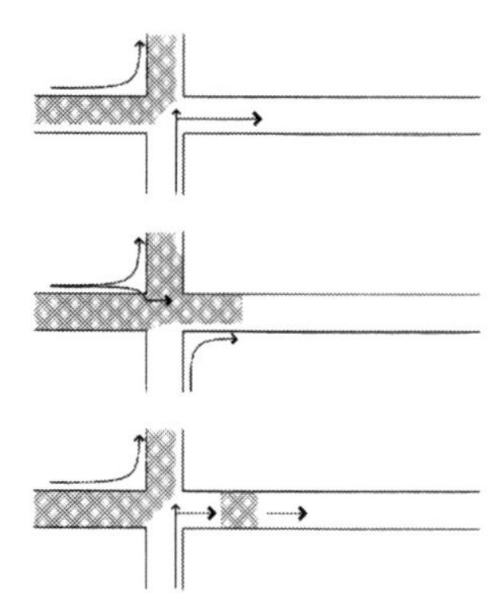

Fig. 12.5 Schematic of loading and injection of a sample in a microfluidic cross chip. Injection is straightforward in microdevices owing to the ease of incorporating multiple channels. A typical injection involves driving fluid from one reservoir to another, briefly switching voltages, then returning to the original voltage configuration.

12.2 1D BAND BROADENING

Many separation modes exist, each of which uses a different approach to give large or unique Δu. Much of the phenomena that predict $w(t)$ are general, and so we begin by describing these *band-broadening* phenomena, which compete with the separation technique and limit the resolution. Because these broadening phenomena involve dispersion and diffusion in the transport of the analyte, we consider these mass transport phenomena. For a model problem, namely a microchip electrophoretic separation, we consider in turn in the following subsections the effects of analyte diffusion, diffusion with electrophoresis, and finally diffusion and electrophoretic separation with dispersive effects such as pressure-driven flow or geometric turns.

12.2.1 Analyte transport: quiescent flow, no electric field

We start by considering quiescent flow to describe how analyte bands diffuse in the absence of separation or dispersion. Consider a 1D distribution of the concentration c of a sample bolus of width w_0 and concentration c_0 centered at $x = 0$ at time $t = 0$. The total number of moles of analyte per unit area Γ is given by $\Gamma = c_0 w_0$.

The governing equation for the evolution of the species concentration is the Nernst–Planck equation:

$$\frac{\partial c_i}{\partial t} = -\nabla \cdot \left[-D_i \nabla c_i + \vec{u}_i c_i \right] . \tag{12.2}$$

Noting that $\vec{u}_i$ is zero given no flow and no electric field, and simplifying to one dimension, we get

$$\frac{\partial c_i}{\partial t} = D_i \frac{\partial^2 c_i}{\partial x^2} , \tag{12.3}$$

which is just a 1D passive scalar diffusion equation. The solution for the distribution of c at large t is given by

solution for diffusion of a point source of species

$$c(x, t) = \frac{\Gamma}{\sqrt{\pi}\sqrt{4Dt}} \exp\left(-\frac{x^2}{4Dt} \right) . \tag{12.4}$$

If we take the limit $w_0 \to 0$ as Γ is held constant, this solution is valid for all t. We obtain this solution by performing a similarity transform on the PDE to obtain an integrable ODE. From this solution, we can conclude that the peak concentration diminishes as $\sqrt{t}$ and the width widens as $\sqrt{t}$. Specifically, the FWHM (full width at half-maximum) w is given by $w = 4\sqrt{\ln 2}\sqrt{Dt}$, and the peak is given by $\frac{\Gamma}{\sqrt{\pi}\sqrt{4Dt}}$. Thus diffusion affects the separation in two ways. First, because of the peak concentration dependence, our ability to experimentally detect the analytes decreases as $\sqrt{t}$. Second, the separation modality must separate species faster than $\sqrt{t}$, or diffusion obscures the separation.

12.2.2 Transport of analytes: electroosmotic flow and electrophoresis

Given the preceding results for pure diffusion, we now add separation to the mix, giving us an idealized result for electrophoretic separations. Here we consider the action of applied electric fields in thin-EDL devices, causing both a separation owing to electrophoresis and uniform velocity owing to electroosmosis.

If electroosmosis in the thin-EDL limit leads to a uniform velocity u_{EO} and the convective transport of an analyte is given by the net electrokinetic velocity $u_{EK} = u_{EO} + u_{EP}$, the solution is the same as Eq. (12.4) except for a spatial transformation:

species transport solution with electrophoresis and electroosmosis, thin-EDL approximation

$$c(x,t) = \frac{\Gamma}{\sqrt{\pi}\sqrt{4Dt}} \exp\left[-\frac{(x - u_{EK}t)^2}{4Dt}\right]. \tag{12.5}$$

The peak decrease and spreading are unaffected by electroosmotic flow because electroosmotic flow is uniform in the thin-EDL limit. Thus electroosmotic flow is *not* dispersive in the thin-EDL limit.

From the preceding relation, we see that applying an electric field in the thin-EDL limit leads to electrokinetic motion of all analytes, leading to separation based on net electrokinetic mobility. Separation of the centers of each bolus is given by $\Delta\mu_{EP}Et$. The width of each bolus expands with time owing to diffusion, and for large t is given by $4\sqrt{\ln 2}\sqrt{Dt}$. Thus the bandwidth scales with $\sqrt{t}$ and the resolution increases with $\sqrt{t}$ as well. This is the ideal result, an example of which is shown in Fig. 12.6.

Effect of pressure-driven flow. As discussed in Section 4.6, pressure-driven flow leads to Taylor–Aris dispersion, and pressure-driven flow with a mean velocity $\overline{u}(x)$ leads to both a net motion of the analyte and an increased effective diffusion. Because these effects often vary with space and time, we write the solution as

species transport solution including Taylor–Aris dispersion

$$\overline{c}(x,t) = \frac{\Gamma}{\sqrt{\pi}\sqrt{4D_{eff}t}} \exp\left[-\frac{\left(x - \int_0^t u_{EK} + \overline{u}\,dt\right)^2}{4D_{eff}t}\right], \tag{12.6}$$

where $D_{eff} = D(1 + \frac{Pe^2}{48})$ per Eq. (4.12). Subsection 12.4.1 discusses sources of pressure-driven flow in microsystems.

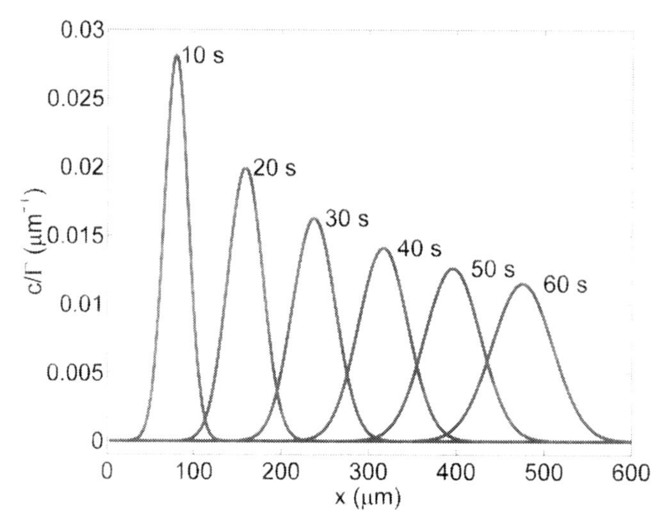

Fig. 12.6 Distribution of a single analyte over time with motion caused by electrophoresis and band broadening owing to diffusion.

12.3 MICROCHIP ELECTROPHORESIS: MOTIVATION AND EXPERIMENTAL ISSUES

Microchips are used for electrophoretic separations for a number of reasons:

1. Small amounts of fluid can be analyzed in small devices.
2. Large electric fields can be used, both because the lengths are small and because microchips dissipate Joule heating better than capillaries.
3. Techniques for injecting a sample bolus to create the initial condition (so far taken for granted) are more sophisticated in microfluidic devices.
4. Long-pathlength separation channels can be compactly folded in microfluidic devices, albeit only with careful geometric design.

Some of these are discussed in the following subsections.

12.3.1 Thermal dissipation

Thermal dissipation in microchips is typically good, simply because the microchip has a larger thermal mass than a capillary. Glass and silicon are relatively good conductors; polymers used for microfluidic devices are poor.

12.3.2 Compact, folded, long-pathlength channels

The separation resolution of an analyte band increases with $\ell^{1/2}$; thus the best separations occur over a long distance. One advantage of microchannels is the ease with which long microchannels can be designed with small footprint. For example, a channel that is 20 μm wide and over 1 m long can be fabricated in a 1 cm $\times$ 1 cm footprints. The fabrication of such a channel takes no more time or money to create than a shorter channel. This is a compelling advantage, but one that can succeed only through the design of low-dispersion turns with special geometries (see Subsection 12.4.2).

12.4 EXPERIMENTAL CHALLENGES

Experimental challenges in microscale separations include (a) pressure-driven flow that is due to hydrostatic head, interface curvature, or electrokinetic potential variations, and (b) analyte dispersion induced by any electroosmotic flow channel that is not 1D.

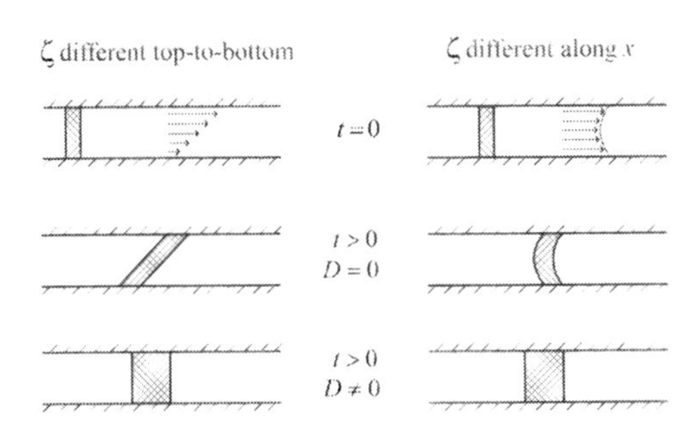

Fig. 12.7 Dispersion caused by nonuniform electrokinetic potential.

12.4.1 Pressure-driven flow

The preceding sections argue that pressure-driven flow leads to Taylor–Aris dispersion, whereas electroosmotic flow does not. This motivates the use of electroosmosis to move analytes (especially large molecules whose diffusivity is low) and electrophoresis to separate them. However, electroosmotic systems often still have pressure-driven flow that degrades their performance.

Ambient pressure gradients. Microchip systems often have pressure gradients stemming from hydrostatic head differences in reservoirs or differences in the radii of curvature of the interface at the top of reservoirs. The effect of this can be limited by including system components with high hydraulic resistance.

Electrokinetically generated pressure gradients. Recall that the electroosmotic flow relations fail to satisfy continuity in an electrokinetic pump owing to the presence of a physical barrier or flow resistance. Similarly, a nonuniform distribution of electrokinetic potential ζ leads to electroosmotic flow solutions with discontinuous mass flow. These flows also induce pressure gradients and Poiseuille-type flows. For microchip electrophoresis separations, uniform zeta potential is critical. Figure 12.7 illustrates how electrokinetic potentials that differ from top to bottom or along the x axis lead to dispersion.

12.4.2 Analyte band dispersion in turns and expansions

Unfortunately, the dispersionless transport enabled by electroosmotic flow in a 1D microchannel does not apply to a 2D channel. Turns, for example, lead to dispersion because analytes on the inside of a turn see both a higher electric field (and higher velocity) and a shorter travel length. This has been addressed through the design of channel turns specifically to minimize dispersion. Figure 12.8 shows this phenomenon, often called the *racetrack effect*, in poorly designed channels, as well as optimized channels that greatly ameliorate these effects. Various solutions to this problem can be found in [82, 83, 146, 147].

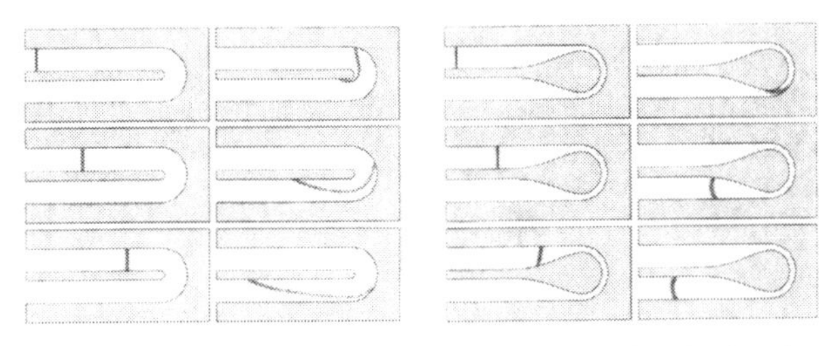

Fig. 12.8 Simulations of the racetrack effect and its elimination in carefully designed microchannel turns. Standard microchannels are shown at left, optimized microchannels are shown at right. (Reproduced with permission from [82].)

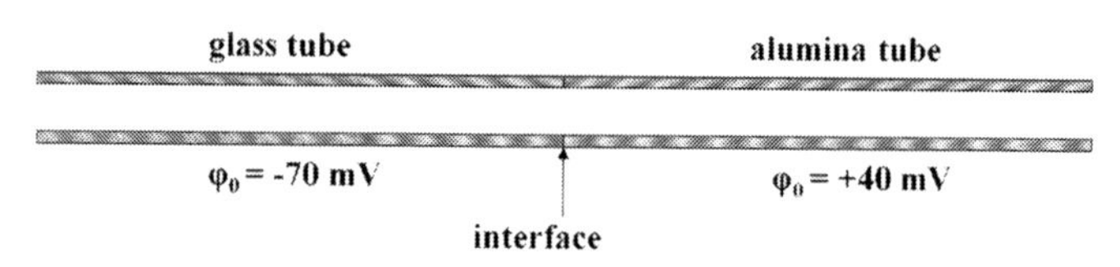

Fig. 12.9 A capillary tube with two regions: glass (left) and alumina (right).

Expansions and other changes in the cross-sectional area also lead to dispersive effects, as does spatial variation in the electroosmotic mobility.

EXAMPLE PROBLEM 12.1

Consider the outer solution for flow of an aqueous solution ($\eta = 1$ mPa s; $\varepsilon/\varepsilon_0 = 78$) through a long, narrow capillary tube with length L and radius R, with $L \gg R$. The tube is made of equal lengths of two different materials: glass ($\varphi_0 = -70$ mV) and alumina ($\varphi_0 = +40$ mV). See Fig. 12.9. An electric field of magnitude 100 V/cm is applied in the x direction. Assume that the EDLs are thin. Derive expressions for the velocity profiles in each of the two sections in terms of r/R. Your expressions need describe the flow only in the regions far from the inlet, outlet, and junction.

SOLUTION: Model the system as a circuit with left port as node 1, right port as node 3, and the interface as node 2. Set $p_1 = p_3$ by definition. Define the flow rate as positive in the x direction. Call the channel in between nodes 1 and 2 channel 1, and the channel in between nodes 2 and 3 channel 2. The given electric field E induces electroosmotic flow in both channels. We assume a simple interface, for which $\zeta = \varphi_0$. In channel 1, we have $Q_{\text{EOF}} = -\frac{\pi R^2 \varepsilon \zeta_1 E}{\eta}$, and in channel 2, we have $Q_{\text{EOF}} = -\frac{\pi R^2 \varepsilon \zeta_2 E}{\eta}$. The mass conservation relation at node 2 is $Q_{\text{EOF},1} + Q_{\text{PDF},1} = Q_{\text{EOF},2} + Q_{\text{PDF},2}$, which can only be satisfied if $p_2 \neq p_1$. Using the hydraulic resistance relation and solving for $(p_2 - p_1)/L$, we find $\frac{p_2 - p_1}{L} = \frac{4\varepsilon E}{R^2}(\zeta_2 - \zeta_1)$, which, for the specified values of the zeta potential, means that the pressure in the center of the channel *increases* owing to the electrokinetic flows. Plugging in this pressure gradient into the Poiseuille flow equation, we find $u_{\text{PDF},1} = -\frac{\varepsilon E}{\eta}(\zeta_2 - \zeta_1)(1 - \frac{r^2}{R^2})$ and $u_{\text{PDF},2} = -\frac{\varepsilon E}{\eta}(\zeta_1 - \zeta_2)(1 - \frac{r^2}{R^2})$. The total flow distributions are thus $u_1 = -\frac{\varepsilon E}{\eta}[(\zeta_2 - \zeta_1)(1 - \frac{r^2}{R^2}) + \zeta_1]$, and $u_2 = -\frac{\varepsilon E}{\eta}[(\zeta_1 - \zeta_2)(1 - \frac{r^2}{R^2}) + \zeta_2]$. With the specified parameters, we find

$$u_1 = -7\ \mu\text{m/mV s}\left[110\ \text{mV}\left(1 - \frac{r^2}{R^2}\right) - 70\ \text{mV}\right], \tag{12.7}$$

$$u_2 = -7\ \mu\text{m/mV s}\left[-110\ \text{mV}\left(1 - \frac{r^2}{R^2}\right) + 40\ \text{mV}\right]. \tag{12.8}$$

The *pressure* generated at the node is proportional to L/R^2, but the *shape of the velocity distribution is independent of L.*

12.5 PROTEIN AND PEPTIDE SEPARATION

Protein measurement, quantification, and separation are central to bioanalysis, because proteins are the primary mechanism by which cells perform tasks. A variety of analyses are important for bioanalytical chemistry, including protein separation, protein concentration, and immunoassays. Here we focus on microfluidic separation of proteins. Also included are separations of peptides, which are sections of proteins, typically obtained through enzymatic digestion.

First, protein properties that affect the transport issues related to separation are discussed, then separation modalities are described, then finally the approaches for combining these modalities to improve separation fidelity are presented.

12.5.1 Protein properties

Proteins have several properties that affect their transport. Proteins are electrically charged. All amino acids have amine groups ($pK_a \simeq 8$) and carboxylic acid groups ($pK_a \simeq 4$); in addition to this, many amino acids (e.g., lysine and arginine) have ionizable side groups. Because of this, proteins in general have measurable properties, such as an electrophoretic mobility that is a function of pH and concentration, as well as an isoelectric point, i.e., a pH at which their electrophoretic mobility is zero. Most water-soluble proteins are roughly spherical and have an inflexible structure in their native state; thus they can be thought of as hard spheres (approximately) with a characteristic hydrated radius of the order of 1–10 nm. Hydrated radius implies the characteristic radius that describes the protein as well as the layer of water molecules that are bound to that protein. Proteins can be denatured for analysis, i.e., their molecular structure can be disrupted, making them behave as long chains rather than as fixed, tight spheroids. This is most commonly done with sodium dodecyl sulfate (SDS). SDS-denatured proteins are linear and highly charged, and behave much like DNA (see Chapter 14). In a denatured state, proteins have a characteristic length rather than a characteristic radius. In denatured form, proteins have measurable properties such as electrophoretic mobility in bulk liquid (which tends to be roughly the same for all proteins) and in gels (which tends to be a function of protein size, with smaller proteins moving faster). The amino acids that make up proteins are of variable hydrophobicity and charge states; thus proteins have different adsorption properties depending on how hydrophobic or hydrophilic they are or how strongly or weakly charged they are. Each of these properties is related to different protein characteristics and motivates different separations, listed in the following subsection.

12.5.2 Protein separation techniques

Many techniques exist (and have been applied on microfluidic substrates) for separating proteins based on subsets of the properties described in the preceding subsection.

CAPILLARY ZONE ELECTROPHORESIS

Capillary electrophoresis (CE) or capillary zone electrophoresis is the technique described in the earlier sections of this chapter and involves injecting a sample bolus of liquid into a channel across which an electric field is applied. Each species i moves with a net electromigratory velocity given by

total ion velocity for purely electric-field-driven flow

$$u_i = u_{EO} + u_{EP,i} \,. \tag{12.9}$$

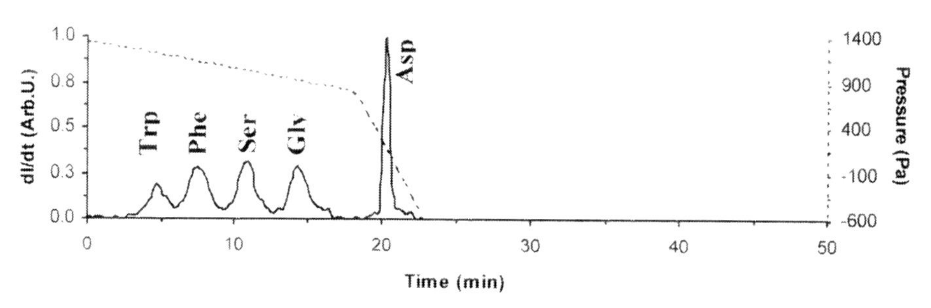

Fig. 12.10 A gradient-enhanced, moving boundary electrophoresis (GEMBE) separation of five amino acids. Whereas pressure often interferes with separation resolution, especially for separations of large molecules, separations of small molecules can be achieved in the presence of pressure gradients because the Peclet number is small and the Taylor–Aris dispersion is relatively minor. GEMBE separations are thus well suited for separations of small molecules. (Reproduced with permission from [148].)

Because the velocities of different species are different, the elution time t for each species is different. On microchips, samples are typically injected by use of pinched electrokinetic injections and detected by laser-induced fluorescence or electrochemistry.

Microchip CE has generally resulted in higher performance than macroscale CE for two reasons. First, CE separations work most rapidly when high electric fields are applied; microchips can be used at high electric fields (a) because their lengths are typically short and even modest high-voltage power supplies can produce high electric fields, and (b) microchips are much better heat sinks than capillaries and high fields can be applied without heating samples excessively. Second, pinched electrokinetic injections can be used to inject small samples ($\sim$ 100 pL), which can then be separated rapidly at high resolution. Many different implementations of electrophoresis have been proposed (see, for example, Fig. 12.10).

LIQUID CHROMATOGRAPHY

Liquid chromatography involves injecting a sample bolus of liquid into a channel and moving this bolus through the channel by using pressure. The surface of the channel and the mobile phase (i.e., the fluid used to carry the protein sample) are both chosen such that proteins occasionally stick to the surface. Proteins elute (i.e., exit the end of the separation column) at a different time depending on their chemical affinity for the surface. For example, a channel with a hydrophobic wall coating causes more hydrophobic proteins to elute more slowly than hydrophilic.

Typically, chromatography is carried out in channels filled with porous media and thus a high ratio of surface area to volume, by use of macroscopic HPLC systems.These porous materials require high pressure gradients, and these techniques are referred to as high-performance liquid chromatography or high-pressure liquid chromatography. Both terms are interchangeable and are abbreviated as HPLC. HPLC has been difficult to integrate into microchip systems, because filling microchips with surface-functionalized porous media can be difficult, high-pressure connections between macroscale devices (e.g., a capillary) and microchips are difficult, as is flow control and injection at high pressures. Despite these challenges, HPLC separations on microchips (Figs. 12.11 and 12.12) have become common.

ISOELECTRIC FOCUSING

All proteins have an *isoelectric point* or pI, namely the pH at which the protein is uncharged on average.[1] Isoelectric focusing or IEF (See Fig. 12.13) concentrates and

[1] In another context, the notation pI can also be used to indicate the negative logarithm of the concentration of iodine, which is unrelated. This is important when we discuss Nernstian surfaces.

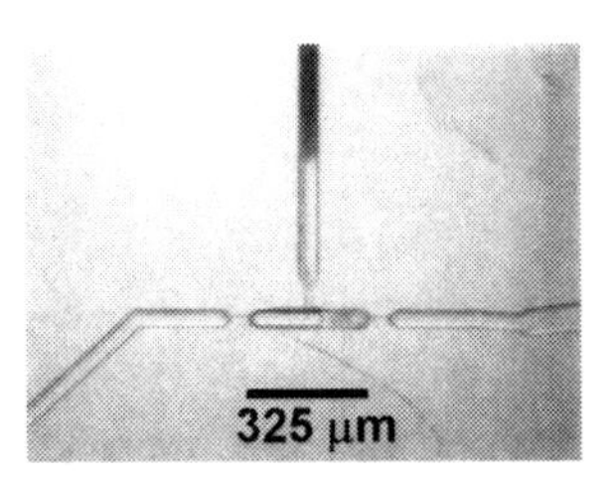

Fig. 12.11 A high-pressure picoliter HPLC injector mated with a photopolymerized reversed-phase HPLC separation column. (Reproduced with permission from [149].)

separates proteins by exposing proteins simultaneously to a pH gradient and also an electric field. Proteins migrate based on their charge. Positively charged proteins migrate toward the cathode, become exposed to higher-pH environments, and become more negatively charged. Negatively charged proteins migrate toward the anode, become exposed to lower-pH environments, and become more positively charged. The steady-state solution has separate bands of proteins, each immobilized at its pI and containing all of the molecules of that protein from that sample. Establishing a pH gradient requires use of carrier *ampholytes*, which are mixtures of polyelectrolytes that help to establish the pH gradient.

12.6 MULTIDIMENSIONAL SEPARATIONS

A chemical separation can be thought of as a technique that transforms a physical property into a spatial position. This spatial position can then be thought of as a position along an axis that indicates that property. For example, if we separate species electrophoretically and record the species with a point detector, the resulting electropherogram could be considered as a plot of concentration versus electrophoretic mobility. Thus the components are separated out on the μ_{EP} axis.

As you might imagine, we can better resolve complex systems if we separate components on multiple axes. *Multidimensional separations* involve performing two or more separations to improve the degree to which the separation resolves individual species.

The language used to describe separations is akin to that from linear algebra. Two separation techniques are said to be *orthogonal* if they separate based on largely unrelated physical properties (i.e., if analyte migration velocity using one technique is largely uncorrelated with analyte migration velocity using the other technique). Often the

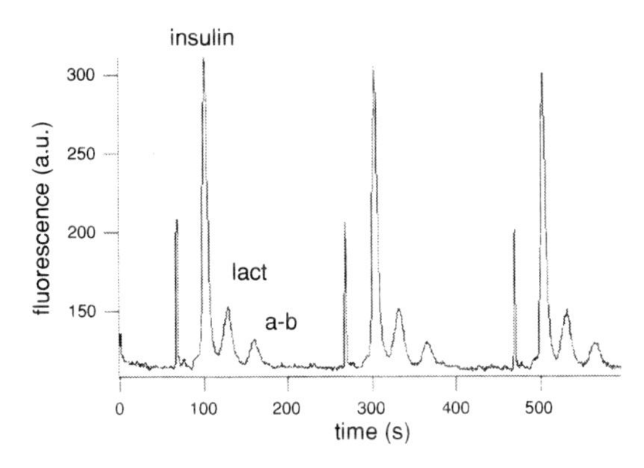

Fig. 12.12 Rapid microchip HPLC separations of a set of three biomolecules (insulin, lactalbumin, and anti-biotin). (Reproduced with permission from [149].)

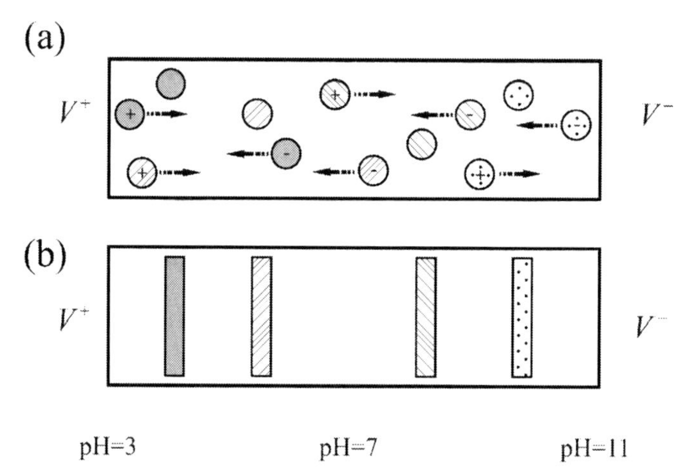

Fig. 12.13 Isoelectric focusing: (a) a mixture of proteins is exposed to a pH gradient as well as an electric field, (b) the steady-state solution, with bands localized at their pI.

degree of orthogonality of the separations is discussed, which refers to the correlation coefficient between the two elution times for a set of analytes.

12.7 SUMMARY

This chapter outlines basic microchip separation techniques and uses these to motivate discussion of related transport issues. In particular, we focus on discussion of how a discrete bolus of fluid travels through a long straight channel. Diffusion and dispersion were seen to reduce the resolution of a chemical separation, and eliminating dispersion motivated a number of specifics of microchip design.

12.8 SUPPLEMENTARY READING

Useful reference texts on separations include [150, 151, 152]. These references contain useful information about important separation techniques that are omitted from this chapter. Some important separations that are omitted included isotachophoresis and field-amplified sample stacking, which separate and concentrate proteins by use of multiple buffer systems; capillary electrochromatography (CEC), which entails using electromigration to move analytes through a channel that is filled with a chromatographic material, and SDS-PAGE (sodium dodecyl sulfate–polyacrylamide gel electrophoresis) which separates proteins based on size in a manner similar to agarose gel separation of DNA. SDS is a surfactant that adheres to and denatures proteins, leaving them in elongated form with large negative charge owing to the sulfate groups. The most common macroscopic 2D separation is an IEF/SDS-PAGE gel that consists of an IEF step, followed by denaturation in SDS and separation in the normal direction by use of PAGE. The protein sample is separated into blobs on the polyacrylamide gel, which can be cut out to extract the protein sample, for example, for mass spectroscopy. For 2D separations on microchips, separation modes include MEKC/CE [144, 153], IEF/CE [154, 155], and CEC/CE [156].

Descriptions of low-dispersion turns in electroosmotic systems include [82, 83, 146, 147]. Microchip HPLC has been described in [149], using high-pressure fixtures presented in [157].

12.9 EXERCISES

12.1 Plot species distributions using Eq. (12.5) for several times ranging from $t = 0$ to $t = 60$ s. Normalize your plots so that they are independent of the total amount of species. Assume there is no electroosmotic flow and the initial width $w = 20\ \mu$m. Assume the species is a protein with a valence of 2 and a diffusivity of $D = 1 \times 10^{-11}$ m^2/s and an electric field of 100 V/cm is applied. The Nernst–Einstein equation (11.30) will be useful.

12.2 Repeat the previous exercise using Eq. (12.6) and considering the case in which the sample bolus is inserted into the middle of a planar (i.e., 2D) microchannel electrokinetic pump with $\mu_{EO} = 1 \times 10^{-4}$ m^2/V s and channel height $d = 10\ \mu$m.

12.3 Consider a long 10-μm radius microchannel with two parts: a region of length ℓ_1 with a zeta potential of ζ_1 and a second region of length ℓ_2 with a zeta potential of ζ_2. Assume double layers are thin compared with the channel radius. The pressures at either end of the channel are equal, the fluid is water with viscosity given by $\eta = 1$ mPa s, and the fluid contains a nucleic acid with diffusivity $D = 1 \times 10^{-11}$ m^2/s.

(a) Derive the 1D flow profile in the fully developed regions of this microchannel (i.e., ignore the points near the interface between the two regions).

(b) For an applied electric field of 100 V/cm, $\zeta_1 = -50$ mV, $\zeta_2 = 0$, $\ell_1 = 1$ cm, and $\ell_2 = 5$ cm, plot the velocity profiles for both region 1 and region 2.

(c) For these parameters, calculate the effective diffusivity of the nucleic acid in each region. What differences, if any, can be seen between the dispersion in the two regions?

12.4 Consider flow through a microchannel of circular cross section and radius R in the presence of a nonuniform electric field but no external pressure gradient. Experimentally, this electric field can be created with a series of electrodes or an ion-permeable membrane. If the electric field varies linearly along the length L ($E = E_0 x/L$), describe the velocity field if the fluid has viscosity η and the surface has electrokinetic potential ζ.

12.5 Electric field gradient focusing uses a gradient of electric field to cause analytes to equilibrate at a location in a microchannel that is a function of the analyte's electrophoretic mobility. Consider flow through a microchannel of circular cross section and radius R in the presence of a nonuniform electric field and an adverse pressure gradient $\frac{dp}{dx} > 0$. Assume the electric field varies linearly along the length L ($E = E_0 x/L$), and complete the following exercises:

(a) Describe the cross-section-averaged velocity field if the fluid has viscosity η and the surface has electrokinetic potential ζ.

(b) Consider only cross-section-averaged velocities, and describe the equilibrium position, if any, of an electrolyte with an electrophoretic mobility μ_{EP}.

(c) Consider the dispersion induced by the full flow field, and describe the width of the distribution of an analyte as a function of its valence.

(d) Now consider that the electric field is changed as a function of time. Assume that two analytes with different electrophoretic velocities (not controlled by the user) are to be separated and held at $x = 0.3L$ and $x = 0.7L$ and the goal of the system is to minimize the width of the distribution of these analytes. Assume the user can observe the location of the analytes at any time. Design an electric field distribution as a function of time that achieves this goal.

12.6 Gradient-elution, moving-boundary electrophoresis causes analytes to be transported through a microchannel with time as a function of the analyte electrophoretic mobility. Consider a 1-mm-long microchannel of circular cross section aligned along the x axis connecting two large reservoirs, the leftmost of which has a set of five cationic electrolytes at 0.1-mM concentration each with electrophoretic mobilities of 0.2, 0.5, 1.0, 1.5, and 2.5×10^{-8} m^2/V s. An adverse ($\frac{dp}{dx} > 0$) pressure gradient is applied, generating flow from the rightmost reservoir to the leftmost reservoir. A positive electric field ($E = 250$ V/cm) is also applied, which generates ion migration in the positive x direction.

(a) Let $\frac{dp}{dx}$ be reduced linearly with time for one minute, expressed by $\frac{dp}{dx} = (60 - t) \times 1 \times 10^3$ Pa. Describe which of the analytes is found in the microchannel as a function of time.

(b) What is the conductivity of the microchannel as a function of time?

12.7 In a variable-temperature system, the conductivity of a buffer system varies owing to two primary phenomena: (1) the reduced viscous friction coefficient of the water at high temperatures allows the mobility (and therefore molar conductivity) of the ions to increase with temperature, and (2) increased temperature causes chemical reactions that increase or decrease the number of ions in solution.

Assume a buffer system is used to control the pH of the system containing a number of proteins that are to be analyzed. Assume that the concentration of the buffer is large relative to the concentration of the protein analytes. Model ion electrophoretic mobilities as increasing by a factor α per degree away from their nominal value at 0°C ($\mu_{EP}/\mu_{EP,0} = 1 + \alpha T$). Assume that the conductivity of the buffer system increases an additional β per degree C owing to increased ion concentrations:

$$\sigma/\sigma_0 = 1 + (\alpha + \beta)T \,. \tag{12.10}$$

Take $\sigma_0 = 2$ S/m, where 1 S = 1/Ω.

Consider a microchannel of circular cross section and radius R connected to electrical heaters such that a uniform temperature gradient is established in the microchannel from $T = T_1$ at $x = 0$ to $T = T_2$ at $x = L$ (assume that this temperature gradient is unperturbed by any fluid flow in the system). A current I is applied from $x = 0$ to $x = L$. Ignore electroosmosis. See Fig. 12.14.

(a) Derive the relation for the x variation of electric field in the microchannel as a function of the experimental parameters.

(b) Derive the relation for the x variation of electrophoretic velocity of an analyte u_{EP} as a function of its $\mu_{EP,0}$ and experimental parameters.

(c) Now consider that a pressure gradient $\frac{dp}{dx}$ is applied to the microchannel. For a given $\frac{dp}{dx}$ and a given x, calculate the $\mu_{EP,0}$ required to lead to a net analyte velocity of zero. Calculate and plot this result for $R = 10\ \mu$m, $\frac{dp}{dx} = 1 \times 10^4$ Pa/m, $\eta = 1$ mPa s, $I = 15\ \mu$A, $\alpha = 0.02$/°C, $\beta = 0.03$/°C, $T_1 = 30$°C, $T_2 = 70$°C, and $L = 5$ cm.

(d) From the previous result, we can assume that analytes are concentrated at an $x = x_0$ such that they stagnate (i.e., their net velocity is zero). Assume that, at equilibrium, the distribution of an analyte around its stagnation point x_0 is given by a Gaussian distribution: $\bar{c}(x) = A \exp[-B(x - x_0)^2]$. Write the 1D transport equation for the cross-sectional average concentration $\bar{c}$, and determine the FWHM width w of the Gaussian distribution at equilibrium. Given the parameters specified previously, calculate and plot $w(x)$.

(e) A detector that measures the local concentration of analytes at any location x is placed at $x = 0$. $\frac{dp}{dx}$ is varied linearly with time (from 0 to 5×10^4 Pa/m) and the signal at the detector is recorded. Qualitatively describe how this signal will vary with time and what the meaning of this signal will be in terms of the analytes in the system.

12.8 You are considering making an electrophoretic separation device out of a glass wafer with etched channels. To complete this device you must make some cover and affix it to your etched glass. You are choosing among glass, alumina, and PMMA. Based strictly on transport phenomena (*not* cost or ease of fabrication or other matters), which of these three cover materials will lead to the best system performance? Explain your answer briefly and qualitatively.

12.9 Model a protein as effectively having two charge sites: an amine, which is neutral at high pH but can gain a proton in a reaction with $pK_a = 8$; and a carboxylic acid, which is neutral at low pH but can lose a proton in a reaction with $pK_a = 4$. Assume $T = 25°C$.

(a) Using the Henderson–Hasselbach equation, calculate the average total charge of the protein as a function of pH. Plot this relation and determine the isoelectric point, i.e., the pH at which the total charge is zero. Although a molecule cannot have an instantaneous partial charge, molecules will gain and lose charge quickly enough that, from the standpoint of time-averaged molecular motion, they will act as if they had a partial charge.

(b) Assume that this protein is inserted into a microchannel used for an IEF separation. Assume the channel is filled with an ampholyte mixture such that the pH ranges linearly in space from pH 3 to pH 10, where pH = 3 is at $x = 0$ and pH = 10 is at $x = 1$ cm. An electric field is applied with a magnitude of 100 V/cm. What must the sign of the electric field be to ensure that the protein will concentrate at a specific location in this microchannel? Where will the protein stabilize if the proper sign of electric field is applied?

(c) Consider the 1D Nernst–Planck equations for the distribution of the concentration of the protein. Linearize the electrophoretic mobility of the protein around the isoelectric point to simplify the math. Given this approximation, write the equation that the concentration profile must satisfy in steady state.

(d) Your equation for the steady-state concentration profile will be satisfied by a distribution with Gaussian form. What is the half-width at half-maximum for the concentration distribution? That is, at what distance from the isoelectric point has the concentration dropped to one half of the concentration at the isoelectric point?

(e) How would the half-width at half-maximum change if the electric field were doubled?

(f) How would the half-width at half-maximum change if the pK_a's of the two reactions were 6.5 and 5.5 instead of 8 and 4?

12.10 Consider an electrophoretic separation. For dispersionless transport, how does the separation resolution depend on the length of the separation channel and the voltage applied across the channel?

12.11 Consider pressure-driven flow through a porous material for an HPLC separation. Design the geometry of a channel required to fabricate a 1-m pathlength in a 1 cm × 1 cm footprint device.

12.12 Consider flow through a long channel of uniform circular cross section with radius $R = 10\ \mu\text{m}$. The mean velocity of the fluid is measured by use of photobleaching with a focused laser beam to define a fluid bolus of thickness with a Gaussian distribution and a FWHM of $w = 10\ \mu\text{m}$ at $t = 0$ and $x = 0$, and imaging the concentration of bleached fluid as a function of time. The velocity is inferred by comparing the x locations of the peaks of the $\imath$-averaged bleached fluid distributions at $t = 0$ and $t = 2$ s. This experiment constitutes a 1D SIV measurement of the fluid flow and the resulting images will be reminiscent of the images in Fig. 11.4.

The ability to resolve the position of a distribution of a scalar is a function of the signal-to-noise ratio of the experiment. Assume that the signal-to-noise ratio of the experiment is such that the location of the peak of the fluid distribution can be measured with an accuracy of 30% of the FWHM of the Gaussian distribution. What is the accuracy of the velocity measurement when this technique is used?

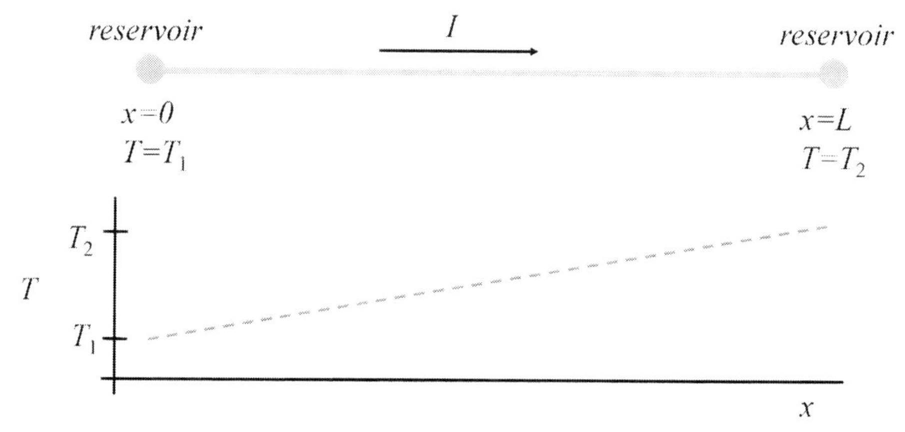

Fig. 12.14 Schematic of a channel with a temperature gradient and an applied current.

13 Particle Electrophoresis

Electrophoresis is the motion of a charged body proportional to an electric field. In contrast to Chapter 11, in which electrophoresis of *molecules* is described; this chapter discusses the motion of charged *particles*. Particle electrophoresis is a straightforward way to manipulate particles in microfluidic devices, both for positioning and for separation; it is ubiquitous if electric fields are used for any purpose.

Particles and molecules differ in that small molecules can be treated as point charges from the standpoint of how they perturb the surrounding electric field, and small molecules cannot support enough counterions to create a continuum electrical double layer. Particles, in contrast, have a large-enough charge that they are surrounded by a continuum EDL. For large particles, electrophoresis is described by use of analyses similar to electroosmosis, albeit with different boundary conditions. For smaller particles, we account for the breakdown of the thin-EDL assumption, leading to size-dependent electrophoretic velocity. Throughout, we discuss both the velocity distribution of the fluid and the velocity of the particle with respect to the bulk fluid.

13.1 INTRODUCTION TO ELECTROPHORESIS: ELECTROOSMOSIS WITH A MOVING BOUNDARY AND QUIESCENT BULK FLUID

Electrophoresis of particles and electroosmosis of fluid are both caused by the same physical phenomena – electrostatic forces on the wall and double layer counteracted at steady state by viscous forces in the fluid or elastic forces in the solid. The key difference between the two is simply the assumed boundary conditions – we use *electroosmosis* to describe fluid motion in a reference frame in which elastic forces in the solid hold the wall motionless, and we use *electrophoresis* to describe particle or droplet motion in a reference frame in which viscous forces in the liquid hold the fluid motionless at infinity. In this section, electroosmosis is described with a moving boundary in preparation for discussion of electrophoresis of an object.

Recall our description of electroosmosis for a surface with a finite bound surface charge density q'' surrounded by an EDL with an integrated mobile charge density equal to $-q''$. Upon the application of an electric field, the wall and the double layer feel equal and opposite electrostatic stresses given by $q'' E_{\mathrm{ext}}$ and $-q'' E_{\mathrm{ext}}$, respectively. However, if we assume that the wall is motionless, we implicitly assume that a mechanical force holds the wall in place; this force has magnitude $-q'' E_{\mathrm{ext}}$. Thus the net force on the charged wall and a quiescent EDL is equal to $-q'' E_{\mathrm{ext}}$, and the equilibrium electroosmotic solution involves fluid motion in the EDL such that the viscous stress at the wall is equal to $q'' E_{\mathrm{ext}}$, canceling the Coulomb force. An electrical field applied to a net neutral system thus induces fluid motion when the electrical force on the wall is counteracted by an external mechanical force.

The same equations can be solved with boundary conditions that specify that the wall moves rather than the bulk fluid. In this case, an electrical field applied to a system with a mobile solid surface but fluid held motionless at infinity leads to motion of the *solid surface*. Fluid motion is induced only within the EDL.

Consider a mobile, infinite flat plate at $y = 0$ with a surface charge density given by q'' and an interfacial potential given by φ_0. The associated EDL consists of diffuse charge with total charge per unit area given by $-q''$. If this flat plate is suspended in an initially quiescent fluid, the fluid and surface will both be mobilized owing to the Coulomb forces.

If we consider, as we do in Chapter 6, steady isobaric flow strictly along a wall aligned in the x direction, with velocity and potential gradients *only* in the y direction, the Navier–Stokes equations reduce to a simple conservation of x-momentum equation:

$$0 = \eta \frac{\partial^2 u}{\partial y^2} + \rho_E E_{\text{ext,wall}} \,. \tag{13.1}$$

Combining this with the uniform-permittivity Poisson equation and retaining only y gradients, we find

$$0 = \eta \frac{\partial^2 u}{\partial y^2} - \varepsilon \frac{\partial^2 \varphi}{\partial y^2} E_{\text{ext,wall}} \,. \tag{13.2}$$

Now we allow the wall to move but assume $u = 0$ far from the wall. Rearranging and integrating from the wall ($y = 0$; where the fluid velocity and plate velocity are the same) to a point outside the double layer ($y \gg \lambda_D$, at which $\varphi = 0$ by definition and $u = 0$ because we have prescribed that the bulk fluid is motionless), we obtain

$$u = \frac{\varepsilon E_{\text{ext,wall}}}{\eta} \varphi \,. \tag{13.3}$$

For large y, $\varphi = 0$ and the fluid is motionless. At the wall, $\varphi = \varphi_0$, and thus the velocity of the plate is given by

$$u_{\text{wall}} = \frac{\varepsilon E_{\text{ext,wall}}}{\eta} \varphi_0 \,. \tag{13.4}$$

If we assume a form for the potential distribution in the EDL, for example, if we make the Debye–Hückel assumption and assume that $\varphi = \varphi_0 \exp[-y/\lambda_D]$, the fluid velocity distribution is given by

$$u = \frac{\varepsilon E_{\text{ext}}}{\eta} \varphi_0 \exp[-y/\lambda_D] = \frac{q'' \lambda_D E_{\text{ext}}}{\eta} \exp[-y/\lambda_D] \,, \tag{13.5}$$

where we have also used the Debye–Hückel result that $\varphi_0 = q''\lambda_D/\varepsilon$. Compared with the result found for a stationary wall and a mobile bulk fluid, this steady-state flow is different only by a velocity offset – the two results are offset by the velocity $\varepsilon E_{\text{ext}} \varphi_0/\eta$. Further, the steady-state plate motion relative to the quiescent fluid in this case ($\varepsilon E_{\text{ext}} \varphi_0/\eta$) is equal to but opposite in sign from the steady-state bulk fluid motion relative to a stationary wall ($-\varepsilon E_{\text{ext}} \varphi_0/\eta$), which is derived in Chapter 6. Whereas the steady-state solution for electrophoresis differs from the steady-state solution for electroosmosis by only a coordinate transform, the startup of electrophoresis, shown for comparison in Fig. 13.1, exhibits a flow profile different from that of the startup of electroosmosis.

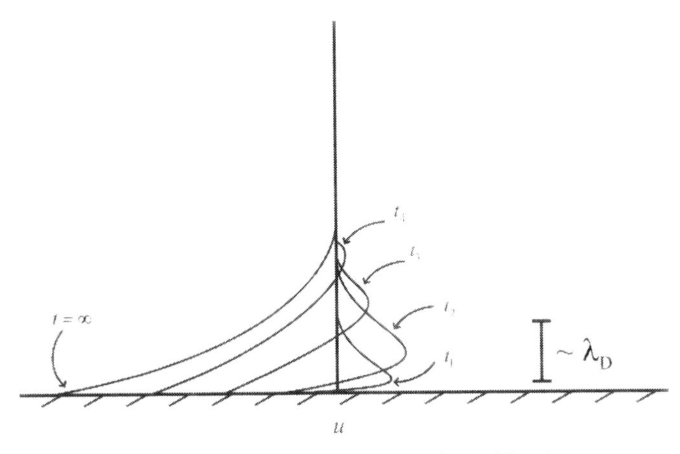

Fig. 13.1 Startup of electrophoresis for a mobile charged flat plate.

As was the case when we considered motionless walls, the relation between the local electric field at the wall and the local velocity distribution can be treated using the 1D analysis if the EDL is thin relative to the radius of curvature of the particle surface.

13.2 ELECTROPHORESIS OF PARTICLES

The 1D unidirectional analysis predicts that a mobile, infinite flat plate suspended in a quiescent fluid with an electric field $E_{\text{ext,wall}}$ applied tangent to the wall will move, at steady state, with a velocity $u = \varepsilon\varphi_0 E_{\text{ext,wall}}/\eta$. More generally, we can show that, if the EDL is thin, a Stokes particle of any shape inserted into a fluid with an otherwise uniform externally applied electric field will move with a velocity given by $u = \varepsilon\varphi_0 E_0/\eta$, where E_0 is the electric field that would have existed if the particle were absent. This follows from the similitude between velocity and electric field for electroosmosis.

Consider first a stationary Stokes particle ($Re \ll 1$) under the appropriate conditions (quasisteady, i.e., changes in the applied electric field or the chemical properties of the surface are slow compared with the characteristic frequency $\eta/\rho a^2$, uniform fluid properties, uniform electroosmotic mobility, thin EDLs, uniform stagnation pressure at infinity, velocity at infinity given by $\vec{u} = \mu_{\text{EO}}\vec{E}_0$). In this case, the flow field is given by $\vec{u} = \mu_{\text{EO}}\vec{E}$. If we consider the same system but transform the boundary conditions at infinity (by requiring that $\vec{u} = 0$ at infinity), the flow field is still irrotational and the unique steady-state solution is

$$\vec{u} = \mu_{\text{EO}}\vec{E} - \mu_{\text{EO}}\vec{E}_0 \,. \tag{13.6}$$

Everywhere on the surface of the particle, this solution has a nonzero velocity normal to the surface given by $\vec{u} = -\mu_{\text{EO}}\vec{E}_0 \cdot \hat{n}$, where $\hat{n}$ is the unit normal to the surface, and thus this flow can occur only when the particle is moving with velocity $\vec{u} = -\mu_{\text{EO}}\vec{E}_0$.

For nonquiescent fluids, the linearity and superposability of the Stokes equations allow us to consider the flow caused by the particle alone and superpose this with the flow that would exist in the absence of the particle. If the distance between the particle and any walls is large compared with λ_D, then the ambient flow can be calculated as if the particle were absent. The additional flow caused by the electric field in the presence of the particle is given by Eq. (13.6); this solution can be added to the ambient solution.

EXAMPLE PROBLEM 13.1

Consider a microchannel of circular cross section with radius $R = 20\ \mu$m and $\mu_{\mathrm{EO}} = 4 \times 10^{-8}\ \mathrm{m}^2/\mathrm{V\,s}$ aligned in the x direction whose centerline passes through the origin. A mobile sphere with radius $a = 1\ \mu$m and $\mu_{\mathrm{EO}} = 2 \times 10^{-8}\ \mathrm{m}^2/\mathrm{V\,s}$ is located on the centerline of the channel. An electric field of 100 V/cm is applied. The particle motion is at steady state. At the instant when the particle is located at the origin, what is the velocity distribution in the channel? Assume that the double-layer thickness can be assumed small relative to the particle radius.

SOLUTION: Because we can model this as Stokes flow, the solution is a superposition of the flow that would exist without the particle plus the flow caused by the particle. In the absence of the particle, the flow solution is uniform:

$$\vec{\boldsymbol{u}} = \mu_{\mathrm{EO}} \vec{\boldsymbol{E}} = 400\hat{\boldsymbol{x}}\ \mu\mathrm{m/s}\,. \tag{13.7}$$

Written in spherical coordinates, this gives

$$\vec{\boldsymbol{u}} = 400\ \mu\mathrm{m/s}\left(\hat{\boldsymbol{r}} \cos\vartheta - \hat{\boldsymbol{\vartheta}} \sin\vartheta\right)\,. \tag{13.8}$$

Because the particle radius is small relative to the radius of the channel, we can treat the flow around the particle as being well approximated by the flow around the particle in an infinite medium. The flow caused by the moving particle is given by

$$\vec{\boldsymbol{u}} = \mu_{\mathrm{EO}}(\vec{\boldsymbol{E}} - \vec{\boldsymbol{E}}_0)\,. \tag{13.9}$$

The electrical potential around the particle can be approximated for $a \ll R$ by the superposition of a uniform field with an axisymmetric dipole with $B_1 = -\vec{\boldsymbol{E}}_0 a^3/2$, leading to an electric field given by

$$\vec{\boldsymbol{E}} = E_0\left[-\cos\vartheta\left(1 - \frac{a^3}{r^3}\right)\hat{\boldsymbol{r}} + \sin\vartheta\left(1 + \frac{1}{2}\frac{a^3}{r^3}\right)\hat{\boldsymbol{\vartheta}}\right]\,. \tag{13.10}$$

The total velocity is thus approximated by

$$\vec{\boldsymbol{u}} \simeq 200\frac{\mu\mathrm{m}}{\mathrm{s}}\left[-\cos\vartheta\left(\frac{a^3}{r^3}\right)\hat{\boldsymbol{r}} + \sin\vartheta\left(\frac{1}{2}\frac{a^3}{r^3}\right)\hat{\boldsymbol{\vartheta}}\right] + 400\frac{\mu\mathrm{m}}{\mathrm{s}}\left(\hat{\boldsymbol{r}}\cos\vartheta - \hat{\boldsymbol{\vartheta}}\sin\vartheta\right)\,. \tag{13.11}$$

This result *does not* strictly satisfy the no-penetration condition at the wall of the channel; however, it is very close to satisfying this condition. The (small) error comes from using an infinite-domain solution for the electric field around the sphere.

We define the *electrophoretic mobility* μ_{EP} of a particle such that

$$\vec{\boldsymbol{u}}_{\mathrm{particle}} = \mu_{\mathrm{EP}} \vec{\boldsymbol{E}}_0\,. \tag{13.12}$$

For particles with thin double layers, the preceding analysis shows that

$$\mu_{\mathrm{EP}} = -\mu_{\mathrm{EO}} = \frac{\varepsilon \varphi_0}{\eta}\,. \tag{13.13}$$

The distinction between μ_{EO} and μ_{EP} is important, as is the distinction between E_0 and $E_{ext,wall}$. The electroosmotic mobility μ_{EO} relates the *outer fluid velocity* near a wall to the *local* electric field $E_{ext,wall}$, whereas μ_{EP} relates the *particle velocity* to the *applied electric field* E_0 that would have existed at the location of the particle center if the particle were absent.

For particles, the thin-EDL approximation implies that the particle radius a is much bigger than λ_D. For $a \gg \lambda_D$, the equilibrium particle motion relative to the fluid is given by

$$\vec{u} = \frac{\varepsilon \vec{E}_0}{\eta} \varphi_0 \,, \tag{13.14}$$

where $\vec{E}_0$ implies the electric field at the location of the particle, or, more precisely, the electric field that would have been at the location of the particle center if the particle were absent. Interestingly, this result indicates that, if the EDL is thin, the velocity of a particle in an electric field is independent of its size or shape, but is dependent only its interfacial potential φ_0. In a manner analogous to electroosmosis, we define an electroosmotic mobility μ_{EP} such that $\vec{u}_{EP} = \mu_{EP} \vec{E}_0$. Based on the preceding relation, μ_{EP} is given by

$$\mu_{EP} = \frac{\varepsilon \varphi_0}{\eta} \,. \tag{13.15}$$

EXAMPLE PROBLEM 13.2

Consider electrophoresis of an infinitely thin flat plate with surface charge density q'' that leads to surface potentials in the Debye–Hückel limit. Assume that the fluid velocity at infinity is zero. Derive the fluid velocity distribution and the velocity of the plate.

SOLUTION: We assume symmetry and consider only the upper surface of the plate at $y \geq 0$. The Coulomb force on the plate is $q''E$. The shear at the wall is $\eta \frac{\partial u}{\partial y}\big|_{wall}$. Setting the sum of these equal to zero gives the boundary condition at equilibrium:

$$\left. \frac{\partial u}{\partial y} \right|_{wall} = -\frac{q''E}{\eta} \,. \tag{13.16}$$

Analyzing the governing equation using the same approach as we did for electroosmosis, we can show that

$$u = \frac{\varepsilon E}{\eta} \varphi + C_1 y + C_2 \,, \tag{13.17}$$

where $C_1 = 0$ to make the velocity bounded and $C_2 = 0$ because at infinity where $\varphi = 0$ the velocity is zero. Thus

$$u = \frac{\varepsilon E}{\eta} \varphi \,. \tag{13.18}$$

For a Debye–Hückel double layer, we know that $\varphi = \varphi_0 \exp[-y/\lambda_D]$. From the boundary condition at the wall, this gives $\varphi_0 = \frac{q''\lambda_D}{\varepsilon}$. Thus the final solution is

$$u = \frac{q'' E \lambda_D}{\eta} \exp[-y/\lambda_D] \,. \tag{13.19}$$

The velocity of the plate is thus given by

$$u_{\text{wall}} = \frac{q'' E \lambda_D}{\eta}. \tag{13.20}$$

13.3 ELECTROPHORETIC VELOCITY DEPENDENCE ON PARTICLE SIZE

In the previous section, the analysis is simple because the flows discussed in that section are unidirectional (and thus $\vec{E}_{\text{ext}}$ is uniform) or the EDLs are thin (and thus $\vec{E}_{\text{ext}}$ is uniform within the double layer). In both of these cases, the convective term $\vec{u} \cdot \nabla \rho_E$ can be ignored because $\nabla \rho_E$ is normal to $\vec{u}$. For the particles we manipulate in microfluidic devices (usually spheres or particles of roughly spherical shape), the flow around the particle is not unidirectional, and we routinely work with particles with a small radius, as low as 5 nm. In these cases, the local fluid flow is not always orthogonal to the gradients in ρ_E, and $\vec{u} \cdot \rho_E$ is nonzero. The electrophoresis of small particles is thus the simplest and most common example of an electrokinetic flow that exhibits *two-way coupling*, in that the flow (via the $\vec{u} \cdot \nabla \rho_E$ term) perturbs the ion distribution in the EDL.

In the general case, in which the particle radius cannot be assumed large relative to the EDL thickness, $\vec{E}_{\text{ext}}$ cannot be assumed uniform throughout the EDL. This leads to a considerable increase in the complexity of the analysis. As previously discussed, electrophoretic velocity in the large-particle, simple-interface, thin-EDL limit (Fig. 13.2) is analogous to the result for electroosmosis:

electrophoretic particle velocity, thin-EDL approximation, simple interface

$$\vec{u}_{\text{EP}} = \frac{\varepsilon \varphi_0}{\eta} \vec{E}. \tag{13.21}$$

However, to describe particles whose radius is not necessarily much greater than λ_D (Fig. 13.3), we rewrite Eq. (13.21) with a correction factor:

electrophoretic particle velocity with correction factor, simple interface

$$\vec{u}_{\text{EP}} = f \frac{\varepsilon \varphi_0}{\eta} \vec{E}, \tag{13.22}$$

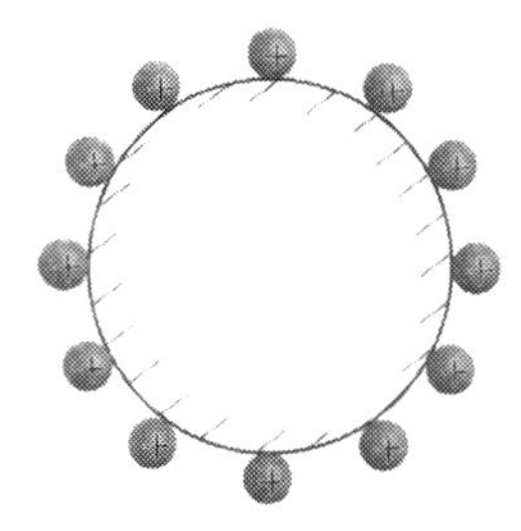

Fig. 13.2 A particle with a thin EDL ($a \gg \lambda_D$).

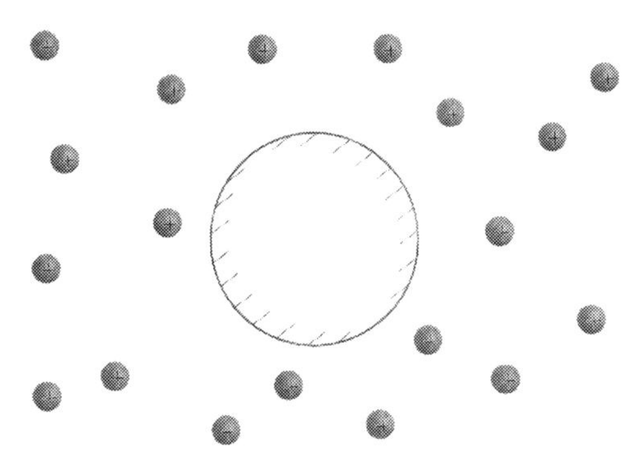

Fig. 13.3 A particle with a thick EDL ($\lambda_D > a$).

where f is a multiplicative factor of the order of 1. This factor accounts for the variation of the local electric field throughout the EDL. The following subsections discuss these results in detail.

13.3.1 Smoluchowski velocity: large particles, small surface potential

Recall that, for electroosmosis, the effect of the double layer can be modeled as a slip velocity as long as the local radius of curvature of the surface is large compared with the double-layer thickness, in which case the surface can be treated as locally flat. Similar arguments can be used to model the flow around a particle. If we consider a motionless particle in an infinite fluid, electroosmosis generates fluid flow relative to the particle with the electroosmotic velocity. As previously discussed, making a coordinate transform to make the fluid motionless and and the particle mobile, we can show that the velocity of the particle is given by

Smoluchowski equation for electrophoretic velocity of large particles

$$\vec{u}_{EP} = \frac{\varepsilon\varphi_0}{\eta}\vec{E}. \tag{13.23}$$

This is the *Smoluchowski equation*. It requires two assumptions, namely that (1) the surface radius of curvature is much larger than the Debye length, and (2) that the double-layer potential φ_0 is small, i.e., that ion distributions (and therefore the local conductivity and therefore the local electric field solution) are only slightly perturbed. In the next subsection, we discuss what happens when these assumptions are relaxed.

13.3.2 Henry's function: effect of finite double layers for small φ_0

Smoluchowski's equation assumes that the double layer is thin and the electric field seen by the ions in the double layer is predicted by the outer solution for the electric field at the surface of the particle. When this is not the case, the electrophoretic mobility changes.

Now assume that the double layer has a finite thickness, but assume that φ_0 is small. Because φ_0 is small, we can neglect any changes in conductivity in the double layer – thus the electric field solution around the particle is unaffected by the double layer. The key *difference* is that the excess ions in the double layer now experience a variety of local electric fields. This means we cannot treat E_{ext} as uniform along the normal coordinate of the EDL. The net effect of this is a retardation of the particle velocity because the component of $\rho_E \vec{E}$ in the direction of the applied field is lower as one moves away from the particle surface along the normal coordinate. This system can be solved with a Stokes flow analysis combined with an *equilibrium* description of a double layer unperturbed by the flow – this approximate solution is appropriate for small wall potentials, for which (a) the ion distribution is only mildly perturbed from the bulk value and (b) ion electromigration is slow compared with diffusion. This approximation leads to a solution written as

definition of Henry function, Debye–Hückel approximation

$$\vec{u}_{EP} = f_0 \frac{\varepsilon \varphi_0}{\eta} \vec{E}, \tag{13.24}$$

where f_0 is the symbol used to denote *Henry's function*. For a spherical particle of radius a with a^* defined as a/λ_D,[1] Henry's function is given by

Henry function for a spherical particle as a function of nondimensional radius

$$f_0 = 1 - \exp(a^*)\left[5E_7(a^*) - 2E_5(a^*)\right], \tag{13.25}$$

where E_n is the exponential integral of order n.[2] This is reasonably well approximated by the relation

$$f_0 \approx \frac{2}{3}\left[1 + \frac{\frac{1}{2}}{\left(1 + \frac{2.5}{a^*}\right)^3}\right], \tag{13.27}$$

which is more straightforward to implement.

In the small-particle limit ($a^* \rightarrow 0$), $f_0 = \frac{2}{3}$, and the resulting electrophoretic mobility relation is termed *Hückel's equation* :

Hückel's equation for electrophoretic particle velocity, thick-EDL approximation

$$\vec{u}_{EP} = \frac{2}{3} \frac{\varepsilon \varphi_0}{\eta} \vec{E}. \tag{13.28}$$

We can write similar relations for an infinite cylinder aligned perpendicularly to the electric field – this is essentially the 2D version of Eq. (13.25). The exact equation is

[1] Many sources define the Debye screening parameter κ such that $\kappa = 1/\lambda_D$; our nondimensionalized radius a^* is thus often written elsewhere as κa.

[2] The exponential integral of order n is defined by

$$E_n(x) = \int_1^\infty \frac{\exp(-xt)dt}{t^n}. \tag{13.26}$$

rather complicated,[3] but an approximate relation is again available[4]:

$$f \approx \frac{1}{2}\left[1 + \frac{1}{\left(1 + \frac{2.5}{a^*}\right)^2}\right]. \qquad (13.30)$$

EXAMPLE PROBLEM 13.3

Consider a silica particle with radius 9.6 nm and surface potential $\varphi_0 = -5$ mV in a pH = 3, room-temperature aqueous solution with 1-mM KCl. What is the particle electrophoretic mobility?

SOLUTION:

$$\mu_{\text{EP}} = f_0 \frac{\varepsilon \varphi_0}{\eta}. \qquad (13.31)$$

We know that $\varepsilon = 80 \times 8.85 \times 10^{-12}$ C/V m. η for water is given by 1×10^{-3} Pa s. The problem specifies that $\varphi_0 = -5 \times 10^{-3}$ V, so the Debye–Hückel approximation is acceptable. What is left is to determine f_0. For 1-mM monovalent ions, the Debye length is 9.6 nm at room temperature. Thus $a^* = 1$. Inserting this into

$$f_0 \approx \frac{2}{3}\left[1 + \frac{1/2}{\left(1 + \frac{2.5}{a^*}\right)^3}\right], \qquad (13.32)$$

we find that $f_0 = 1.01 \times \frac{2}{3}$. Thus, at $a^* = 1$, we are essentially in the thick-EDL limit. The resulting electrophoretic mobility is

$$\mu_{\text{EP}} = 0.24 \times 10^{-8}\ \text{m}^2/\text{V s}. \qquad (13.33)$$

13.3.3 Large surface potential – effect of counterion distribution

The previous relations hold for small zeta potentials, in which case it can be assumed that the EDL exhibits one-way coupling – the electric field applies a force to the double layer, but this force does not perturb the equilibrium ion distribution in the double layer nor does the the presence of double layer perturb the bulk electric field solution. For large surface potentials, this is no longer the case, as convective ion migration can no longer be neglected compared with diffusion, and the f_0 from Henry's equation (13.25) does not accurately describe particle electrophoresis. In particular, if the double-layer thickness

[3] The exact relation is given by

$$f_0 = 1 - \frac{4a^{*4}}{K_0(a^*)}\int_{a^*}^{\infty}\frac{K_0(a')}{a'^5}\,da' + \frac{a^{*2}}{K_0(a^*)}\int_{a^*}^{\infty}\frac{K_0(a')}{a'^3}\,da', \qquad (13.29)$$

where K_0 is the zeroth-order modified Bessel function of the second kind and a' is a dummy integration variable. This is quite similar to the spherical relation (13.25), but with the Bessel functions replacing the exponentials.

[4] The relation is a bit more accurate if 2.55 is used instead of 2.5. Also, both spherical and cylindrical relations can be made more accurate with additional terms. Such details are omitted here.

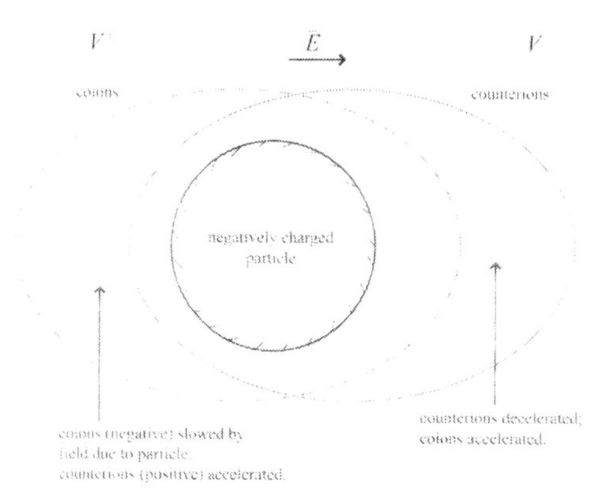

Fig. 13.4 Schematic of double-layer asymmetry caused by surface potential. Here, a negatively charged particle moves from right to left in an electric field. The effect of the surface potential is to accelerate the counterions on the leading surface of the particle and decelerate the counterions on the trailing surface. The net effect is that the counterion concentration behind the trailing edge is larger than that in front of the leading edge. The opposite is the case for the (less numerous) coions. This ion distribution leads to a secondary electric field that cancels out part of the applied electric field.

λ_D is of the same order as the particle radius, the motion of ions through the double layer leads to a distorted double layer that counteracts much of the applied electric field and retards the electrophoretic migration.[5] A schematic of double-layer asymmetry caused by relaxation is shown in Fig. 13.4. The effect of this asymmetric counterion cloud is to make the correction factor f lower in the region near $a^* = 1$, as shown in Fig. 13.5.

In general, a value for Henry's function for large φ_0 can only be solved numerically. Sample results from these numerical calculations are shown in Fig. 13.6. For those wanting to avoid solutions of differential equations, an analytical approximation exists for symmetric electrolytes of valence z and $a^* \gg 10$, derived in [160][6]:

$$f = \left\{1 - \frac{2FH}{\hat{\zeta}(1+F)} + \frac{1}{\hat{\zeta}a^*}\left[-12K\left(t + \frac{t^3}{9}\right) + \frac{10F}{1+F}\left(t + \frac{7t^2}{20} + \frac{t^3}{9}\right)\right.\right.$$
$$-4G\left(1 + \frac{3}{\zeta_{\mathrm{EP}}^{\mathrm{co}}}\right)\left[1 - \exp\left(-\hat{\zeta}/2\right)\right] + \frac{8FH}{(1+F)^2} + \frac{6\hat{\zeta}}{1+F}\left(\frac{G}{\zeta_{\mathrm{EP}}^{\mathrm{co}}} + \frac{H}{\zeta_{\mathrm{EP}}^{\mathrm{ctr}}}\right)$$
$$\left.\left.-\frac{24F}{1+F}\left(\frac{G^2}{\zeta_{\mathrm{EP}}^{\mathrm{co}}} + \frac{H^2}{(1+F)\zeta_{\mathrm{EP}}^{\mathrm{ctr}}}\right)\right]\right\}, \tag{13.34}$$

where $\hat{\zeta}$ is the absolute value of the normalized zeta potential, incorporating ion valence as well:

$$\hat{\zeta} = \frac{zF\,|\zeta|}{RT}, \tag{13.35}$$

[5] Some authors refer to this phenomenon as an *electroviscous* effect.

[6] This relation is simple to evaluate despite its complicated appearance. However, this result stems from an assumed Poisson–Boltzmann distribution, and its predictions at high ζ are inaccurate owing to the nonphysical assumed ion distributions near the wall. Although the original reference claims that this relation is accurate for $a^* > 10$, it actually performs poorly until $a^* \gg 10$.

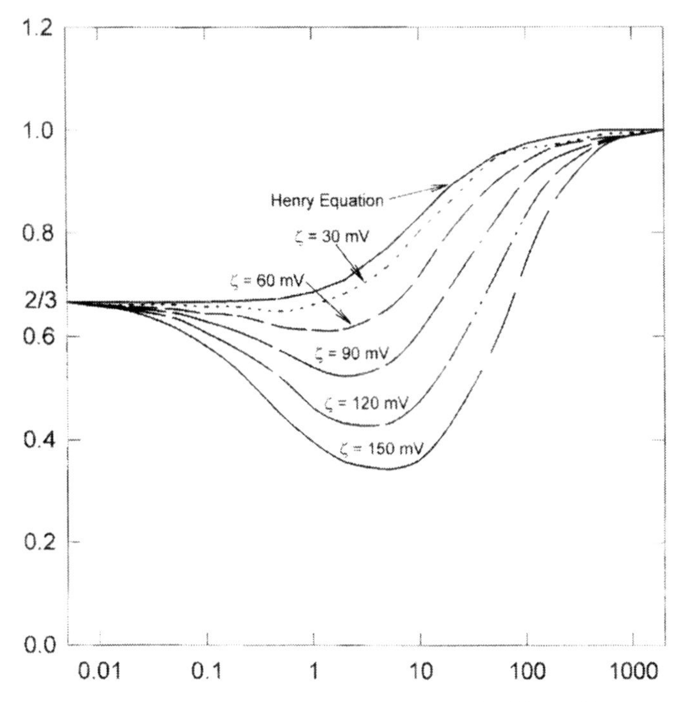

Fig. 13.5 Henry function (ordinate) as a function of a^* (abscissa) and zeta potential. Effects of large zeta potential can be seen even at $a^* = 100$. (Modified from [158].)

$\zeta_{\mathrm{EP}}^{\mathrm{co}}$ and $\zeta_{\mathrm{EP}}^{\mathrm{ctr}}$ are nondimensional ion mobilities of the coions and counterions, respectively[7]:

$$\zeta_{\mathrm{EP},i} = \frac{3\eta F z^2 \left|\mu_{\mathrm{EP},i}\right|}{2\varepsilon RT}, \tag{13.36}$$

and t, F, G, H, and K are abbreviations to simplify Eq. (13.34):

$$t = \tanh\left(\frac{\hat{\zeta}}{4}\right), \tag{13.37}$$

$$F = \frac{2}{a^*}\left(1 + \frac{3}{\zeta_{\mathrm{EP}}^{\mathrm{ctr}}}\right)\left(\exp\left(\hat{\zeta}/2\right) - 1\right), \tag{13.38}$$

$$G = \ln\left(\frac{1 + \exp\left(-\hat{\zeta}/2\right)}{2}\right), \tag{13.39}$$

$$H = \ln\left(\frac{1 + \exp\left(\hat{\zeta}/2\right)}{2}\right), \tag{13.40}$$

$$K = 1 - \frac{25}{3\left(a^* + 10\right)} \exp\left[-\frac{a^*\hat{\zeta}}{6\left(a^* + 6\right)}\right]. \tag{13.41}$$

[7] If the Hückel relation $\mu_{\mathrm{EP}} = 2\varepsilon\zeta/3\eta$ is used to rewrite the observed electrophoretic mobility of an ion in terms of an effective electrophoretic zeta potential, these nondimensional ion mobilities can be written as the effective zeta potential multiplied by z^2 and normalized by RT/F. Alternatively, these values can be described as nondimensionalized diffusivities, with $\zeta_{\mathrm{EP},i} = \frac{D_i}{(\frac{RT}{F})^2|z^3|\frac{2\varepsilon}{3\eta}}$.

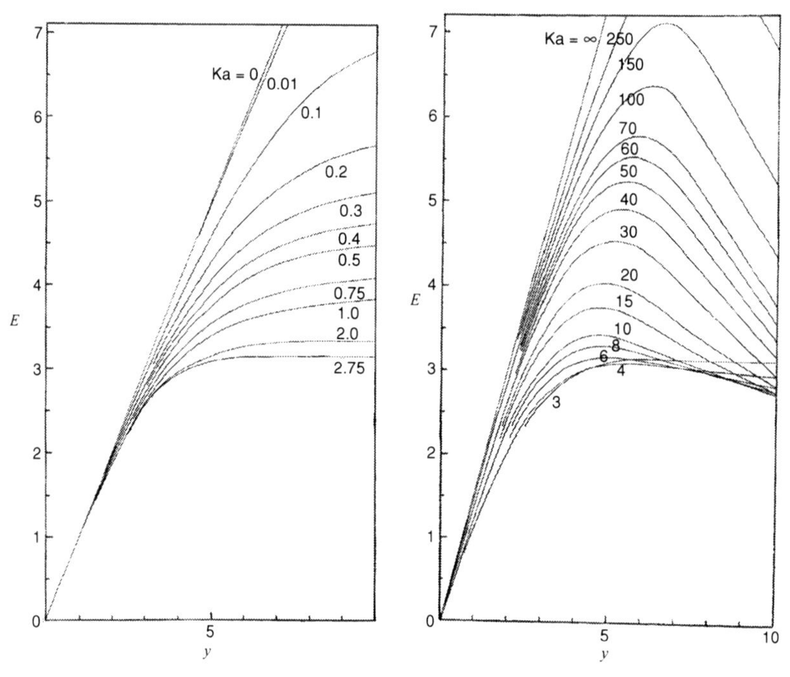

Fig. 13.6 Key results from [159]. In the notation of [159], normalized electrophoretic mobility E is plotted in terms of normalized zeta potential y. In our notation, E corresponds to $3f\hat{\zeta}/2$ and y corresponds to $\hat{\zeta}$. Plots are shown in terms of κa, the ratio of the particle radius to the Debye length (a^* in our notation). The value E/y in the notation of [159] corresponds to $3f/2$ from Eq. (13.24). At small ζ, Henry's equation holds and E is linear with y. At higher ζ, the nonuniformity in the electric field induced by the particle and its motion leads to a nonlinear effect.

Using this analytical approximation, the electrophoretic mobility as a function of zeta potential is shown in Fig. 13.7 for different values of a^*. Similarly, the correction factor as a function of zeta potential is shown in Fig. 13.8 for different values of a^*. Finally, the correction factor as a function of a^* is shown in Fig. 13.9 for different values of $\hat{\zeta}$.

13.4 SUMMARY

This chapter has highlighted that particle electrophoresis comprises the same physics as electroosmosis, with the key differences being (1) that the coordinate system is typically one in which the fluid is motionless, rather than the particle, (2) the curvature of the interface cannot be ignored if double layers are not thin, and (3) the EDL exhibits two-way coupling with the fluid flow if the surface potential is large. Particle electrophoresis transitions analytically between our early study of equilibrium electrokinetics with one-way coupling (electroosmosis with thin double layers) and our later study of electrokinetics with dynamic electric fields (e.g., induced-charge phenomena; Chapter 16) or with two-way coupling (e.g., nanofluidic devices with varying channel cross sections; Chapter 15).

For small surface potentials and thin double layers, the electrophoretic mobility of a particle of any shape is the same as the electroosmotic mobility of a wall made of the

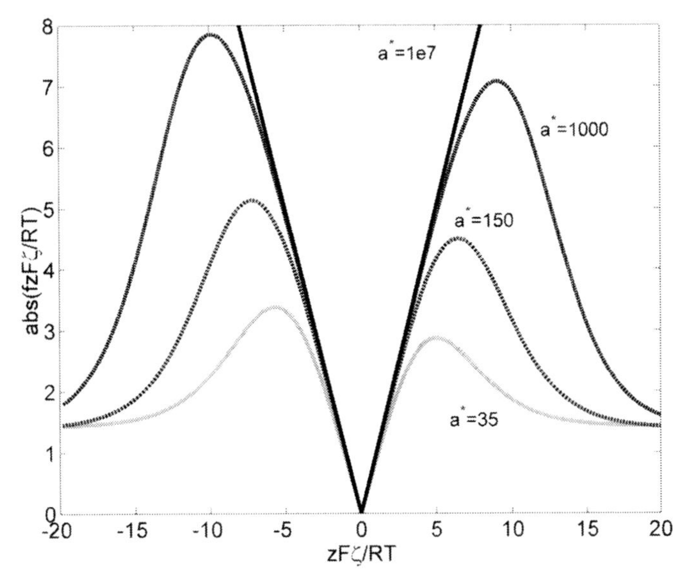

Fig. 13.7 Nondimensional electrophoretic mobility ($f\hat{\zeta}$) as a function of $\hat{\zeta}$ for several values of a^*. Note relaxation effects are more important with slower-moving ions. This example is for H^+ ions and HCO_3^- ions, with almost a factor of 10 difference in ion mobilities. (Cf. Fig. 13.6).

same material – at steady state, the two processes are identical except for a coordinate transformation:

$$\vec{u}_{EP} = \frac{\varepsilon \varphi_0}{\eta} \vec{E}. \tag{13.42}$$

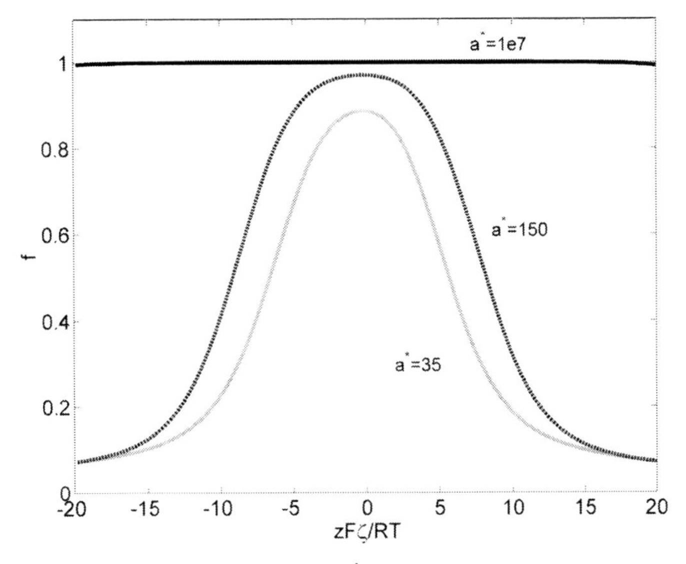

Fig. 13.8 Correction factor as a function of $\hat{\zeta}$. Same parameters as Fig. 13.7.

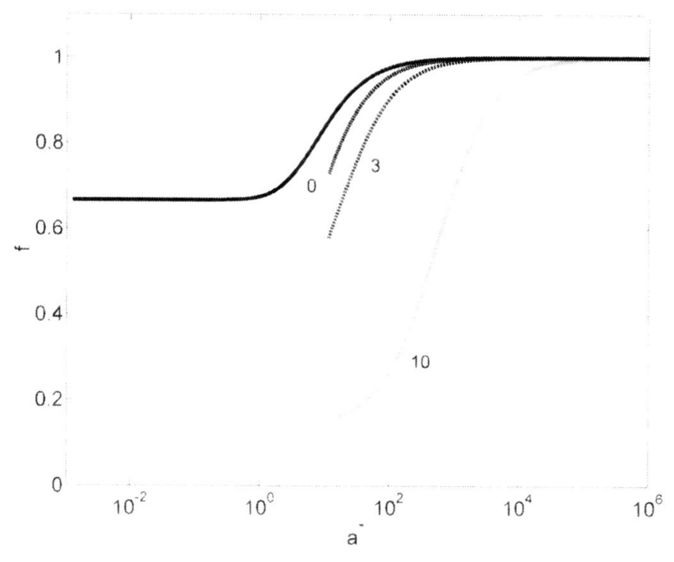

Fig. 13.9 Correction factor as a function of a^*, calculated using the analytical expression for large a^*. Compare with Fig. 13.5.

For particles with small surface potentials but finite double layers, the extension of the double layer causes a reduction in electrophoretic velocity because the net charge density in the double layer experiences lower local electric fields than in the thin-EDL case. For spheres with small surface potentials but finite double layers, we found that Henry's function,

$$f_0 = 1 - \exp(a^*)\left[5E_7(a^*) - 2E_5(a^*)\right], \tag{13.43}$$

describes the correction factor, which reduces the electrophoretic velocity to two thirds of the thin-EDL result as the double layers become thick. For particles with large surface potentials, the system exhibits two-way coupling – the flow perturbs the ion distribution in the EDL and the ion distribution perturbs the bulk electric field. When this happens, the motion of counterions and coions suppresses the local electric field, further reducing the electrophoretic mobility. For spheres, this has been numerically calculated in detail and approximate relations were presented. For more complicated shapes, the electrophoretic mobility can be calculated only with a full analysis of the ion transport in the system using the Poisson, Nernst–Planck, and Navier–Stokes equations.

13.5 SUPPLEMENTARY READING

Electrophoresis of particles is central to colloid science, and material in this chapter is also discussed in physicochemical hydrodynamics and colloid science texts [29, 68, 69, 70, 71]. A review by Anderson [105] focuses on the motions of particles owing to interfacial phenomena and expands to include thermophoresis and other effects. Early papers describing the evolution of Henry's function and corrections to Henry's function are described in [161, 162, 163, 164].

One can find detailed numerical calculations of particle electrophoretic mobility in [159, 165], measurements in [158], and approximate formulas in [160, 166]. These are all fundamentally equilibrium treatments. A review by Dukhin [167] focuses on the equilibrium assumptions made in this chapter and how departure from equilibrium affects colloidal motion and characterization of surfaces.

The preceding analysis is for isolated particles in infinite domains. Corrections must be added for particles in finite domains (e.g., microchannels) or for particles at finite density, such that particle–particle interactions can be prevalent. Suspensions of particles are discussed in [30, 32, 88, 89].

This chapter focuses on electrophoresis of solid particles, but droplets can be analyzed with these techniques as well. Relevant results are presented in [24, 30, 168].

We repeatedly observe that electrokinetic phenomena are most complicated when the double-layer thickness λ_D is of the order of the particle or channel size. This observation, made in this chapter for particles, forms the motivation for most of the analysis in Chapter 15.

13.6 EXERCISES

13.1 The *Smoluchowski limit* (thin-EDL limit) for electrophoresis of a cylinder aligned perpendicularly to the electric field with thin double layers is

$$\vec{u}_{EP} = \frac{\varepsilon\varphi_0}{\eta}\vec{E}, \tag{13.44}$$

whereas the *Hückel limit* (thick-EDL limit) for electrophoresis of a cylinder aligned perpendicularly to the electric field with thick double layers is

$$\vec{u}_{EP} = \frac{1}{2}\frac{\varepsilon\varphi_0}{\eta}\vec{E}. \tag{13.45}$$

Consider the solution for the electric field around an insulating cylinder (which is the superposition of a uniform field and a dipole), and evaluate the maximum electric field on the surface in this case relative to the bulk electric field. In the case of thin double layers, this maximum electric field is applied to the ions in the double layer and gives rise to the particle motion relative to the fluid. In the case of thick double layers, the field applied to the ions in the double layer is for the most part the bulk electric field. Draw a schematic of these two cases and explain why the electrophoretic velocities in these two cases differ by 1/2.

13.2 Plot the exact and approximate relations for Henry's functions for spheres and cylinders. Plot a^* on a log scale from 1×10^{-2} to 1×10^{3}. Estimate the maximum error of the approximate functions.

13.3 An approximate relation for Henry's function for an infinite cylinder aligned perpendicularly to the applied electric field is given above in relation (13.30). If the cylinder is aligned with the field, Henry's function is equal to one because there is no distortion of the field. It can be shown (though we do not do so here) that the time-averaged electrophoretic mobility of an infinite cylinder aligned at randomly varying orientation is equal to the average of the electrophoretic mobilities of the cylinder aligned in each of three orthogonal directions. Given this, derive a relation for the time-averaged Henry's function of an infinitely long cylindrical particle oriented randomly. Plot approximate results for the Henry function for the following, all on one graph, as a function of a^*:

(a) a cylinder aligned perpendicular to the E field,

(b) a cylinder aligned parallel to the E field,

(c) a cylinder aligned randomly with respect to the E field,

(d) a spherical particle, i.e., relation (13.27).

Plot a^* on a log scale from 1×10^{-2} to 1×10^3.

13.4 Plot the electric field lines around an infinite cylinder aligned perpendicular to the field (i.e., field lines around a circle in two dimensions). Recall from the analytical solution that the electric field at the edge of the particle is double the bulk electric field. Draw circles at $r = 1.1a$, $r = 2a$, and $r = 10a$, which nominally denote the outer edge of the double layer for $a^* = 10$, 1, and 0.1, respectively. Comment qualitatively on the fraction of the double layer that sees the high-E region around the circumference of the sphere in these three cases, and relate your observations qualitatively to the approximate Henry function for these three a^* values.

13.5 Here we use a matched asymptotic technique to estimate Henry's function for a cylinder in electrophoresis.

Consider a 2D electroosmotic flow induced by a fixed and motionless glass cylinder with radius a centered at the origin when an electric field of E_∞ in the x direction is applied. Assume that the electrokinetic potential ζ is negative and small in magnitude and the electrolyte solution leads to a Debye length of λ_D.

We can calculate the electrical potential field around an insulating cylinder by superposing the electrical potential field for a uniform field,

$$\phi = -E_\infty x\,, \tag{13.46}$$

with the electrical potential field for a line dipole,

$$\phi = -E_\infty x \frac{a^2}{r^2}\,, \tag{13.47}$$

to obtain the total potential field:

$$\phi = -E_\infty x\left(1 + \frac{a^2}{r^2}\right)\,. \tag{13.48}$$

(a) Recall that the *outer* solution for the electroosmotic velocity for uniform properties and simple interfaces is given by

$$\vec{u}_{\text{outer}} = -\frac{\varepsilon\varphi_0}{\eta}\vec{E}\,. \tag{13.49}$$

Derive a relation for $\phi_v(x, y)$ for the outer solution for the electroosmotic flow.

(b) For simplicity, consider only the line corresponding to $x = 0$. Define an *outer* variable $y^*_{\text{outer}} = \frac{y-a}{a}$ and a normalized velocity $\vec{u}^* = -\vec{u}\eta/\varepsilon\zeta E_\infty$.

(c) Recall that the *inner* solution for the electroosmotic velocity is given by

$$u_{\text{inner}} = \frac{\varepsilon E_{\text{wall}}}{\eta}(\varphi - \varphi_0)\,. \tag{13.50}$$

Here E_{wall} is the (assumed uniform) magnitude of the tangent electric field at the wall, obtained by evaluation of the electric field given by Eq. (13.48) at $r = a$. Consider again the line corresponding to $x = 0$ and define an inner variable $y^*_{\text{inner}} = \frac{y-a}{\lambda_D}$. Write the expression for $u^*_{\text{inner}}(y^*_{\text{inner}})$.

(d) Construct a composite asymptotic solution using the multiplicative formula

$$u_{\text{composite}}(y) = u_{\text{inner}}(y) * u_{\text{outer}}(y)/u_{\text{outer}}(y = a) . \tag{13.51}$$

This multiplicative relation does not *rigorously* satisfy all of the governing equations, but it is a mathematical tool for creating an analytical solution that transitions smoothly from the inner solution (which is rigorously correct to the extent that the local electric field tangent to the wall is uniform) to the outer solution (which is rigorously correct to the extent that the charge density is zero). Whereas an additive composite asymptotic will give good answers *only* for thin double layers, the multiplicative composite expansion will give good answers in all cases.

(e) Consider $a^* = 0.1$, $a^* = 10$, and $a^* = 1000$, where $a^* = a/\lambda_D$. For each value of a^*, make a plot of $u^*_{\text{composite}}$, u^*_{inner}, and u^*_{outer}. For your independent variable, use y_{inner} ranging from 1×10^{-3} to 1×10^9 on a log axis.

(f) For several values of a^* ranging from $a^* = 0.1$ to $a^* = 1000$, numerically evaluate the maximum value that the composite expansion reaches. Normalize your results for the maximum velocity by the value you obtain for $a^* \to \infty$. Plot these on a semilog axis over the range from $a^* = 0.1$ to $a^* = 1000$ and compare your result (on the same graph) with the values for f given by relation (13.30).

(g) Use your results to explain Henry's function physically. How do the size of the double layer and the size of the cylinder interact to make the electrophoretic velocity a function of the particle size? What electric field do the ions in the double layer see in the case of (a) thin double layers, (b) thick double layers, or (c) $a = \lambda_D$?

13.6 Survey the literature and identify the approximate range of observed zeta potentials for (a) mammalian cells, (b) bacterial cells, and (c) virions.

13.7 Show that the electrophoretic velocity of a long cylinder moving along its axis of symmetry is independent of the Debye length. Explain whether such a cylinder will have a preferred orientation if an electric field is applied in a random direction.

14 DNA Transport and Analysis

Microdevices for analyzing deoxyribonucleic acid (DNA) are ubiquitous in biological analysis, and techniques for analyzing DNA in microchips pervade the analytical chemistry literature. Use of nanochannels to study polymer physics has also become common. Owing to DNA's huge biological importance, its chemical properties have been thoroughly studied, and the experimental tools available for chemical analysis of DNA are numerous. The ubiquity and convenience of DNA has also led to extensive study of its physical properties. DNA is therefore an excellent example of how microscale systems facilitate analysis, as well as a model system for examining the effect of nanostructured devices on molecular transport of linear polyelectrolytes. Because the chemistry for fluorescently labeling DNA is relatively inexpensive and available commercially, fluorescence microscopy of DNA is a widely used means for visualizing DNA. It is quite routine to fluorescently label and observe the gross morphology of a single DNA molecule with 1-μm resolution, and thus straightforward experiments can be brought to bear on questions of molecular configuration.

DNA (and other idealized linear polymers) behave physically somewhere in between small molecules (which behave like idealized points) and particles (which behave like rigid continuous solid phases). The behavior observed (and the models that describe this behavior) incorporates aspects of point and particle behavior, and these behaviors are different depending on the type of transport. When we consider DNA behavior within domains (e.g., microchannels or nanochannels) that are small compared with the characteristic size of the DNA molecule, these models must be augmented with explicit consideration of surface interactions. Thus the interaction between the molecule and the confining boundaries requires some sort of physical model of the system that goes beyond typical bulk properties.

This chapter first describes the physicochemical structure of DNA, with particular attention given to the mathematical description of the backbone of linear polyelectrolytes. This treatment of linear polyelectrolytes applies to DNA, but also to a wide variety of other molecules whose backbones have no branched structure. Experimental observations of the bulk properties of DNA are then presented and interpreted in the context of physical models for DNA dynamics. These models then lead to discussion of the behavior of DNA in confining domains. The behavior of DNA translates directly to applications in micro- and nanofluidic devices – bulk diffusion affects the performance of DNA hybridization microarrays, gel electrophoretic mobility affects DNA length separations in microchannels, and DNA behavior upon confinement affects nanofluidic devices for DNA separation and manipulation.

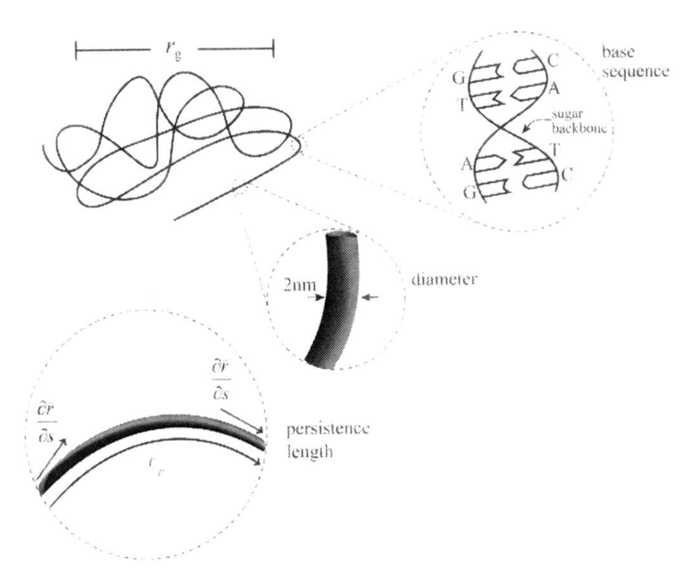

Fig. 14.1 The structure of a dsDNA molecule at several length scales.

14.1 PHYSICOCHEMICAL STRUCTURE OF DNA

DNA exists in a number of double-stranded forms, including *A*-DNA, *B*-DNA, and *Z*-DNA, as well as in single-stranded form. In this chapter and throughout this text, we focus on the (most common) *B*-DNA double-stranded form, with only brief mention of single-stranded DNA. Throughout this chapter, then, "DNA" and "dsDNA" imply double-stranded *B*-DNA unless otherwise specified.

Double stranded *B*-DNA's physical properties are well studied and are prototypical of linear (unbranched) macromolecules. It is thus a model system for exploring introductory polymer physics. Figure 14.1 highlights features of DNA at several length scales. Macroscopically, DNA in aqueous solution appears as a spaghetti-like linear polymer with a characteristic radius $\langle r_g \rangle$. If the local orientation of the polymer chain is viewed in more detail, say on length scales ranging from 10–100 nm, the relative rigidity of the polymer chain is measured with a persistence length ℓ_p. Below 10 nm, the polymer has a diameter of approximately 2 nm, with a molecular structure characterized by a double helix with a negatively charged sugar structure on the exterior and hydrogen bonding between complementary base pairs on the interior. Mathematical definitions of ℓ_p and $\langle r_g \rangle$ are presented in Subsection 14.1.2.

14.1.1 Chemical structure of DNA

A DNA molecule has a hydrophilic sugar (deoxyribose) backbone with negatively charged phosphate groups and a sequence of nitrogenous bases. This sequence consists of the bases adenine (A), guanine (G), cytosine (C), and thymine (T). Biologically, this sequence of bases codes for protein sequences in organisms. Hydrogen bonding between A and T bases and between C and G bases causes two strands of DNA with complementary sequences to spontaneously bind and form a double-helix structure, in which the nitrogenous bases are hydrogen bonded to each other in the interior of the structure and the phosphate groups are pendant on the outside structure. The stability of this structure in aqueous solutions confers relatively high chemical stability to double-stranded DNA in biological systems.

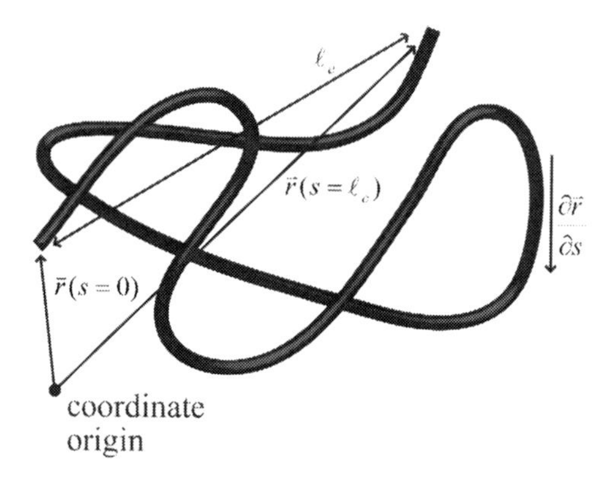

Fig. 14.2 Coordinate definitions for a linear polymer.

The hydrogen bonds that cause two strands of DNA to come together can be overcome by thermal energy, a process called *melting* or *denaturing* the DNA. When this happens, the two strands of DNA fall apart and become single-stranded DNA (ssDNA), a form that is significantly less chemically stable. The temperature at which melting occurs is referred to as the *melting temperature* of the DNA molecule, which varies depending on sequence and solution conditions, but is typically 45–60°C.

Melting is reversible – when the hydrogen bonding overcomes thermal energy, two complementary single-stranded molecules bind with each other to form a double-stranded molecule, a process called *annealing*. When the annealing process is observed between single-stranded molecules from separate sources, we refer to it as *hybridization.* Measuring the degree to which an unknown ssDNA or ribonucleic acid (RNA) sample hybridizes with a known ssDNA molecule is a useful way to identify the genetic makeup of an unknown sample. These hybridization assays, including Southern blotting, northern blotting, and DNA microarrays, are powerful analytical tools.

14.1.2 Physical properties of dsDNA

Double-stranded DNA is a prototypical linear chain polymer, and it has a number of physical properties and characteristic lengths, which are described in the following text, illustrated in Fig. 14.2, and listed in Table 14.1. We consider DNA as a dilute component in a good solvent, and thus we consider DNA's interactions with the solvent only. Water and aqueous solutions are "good" solvents for DNA, and thus this chapter implicitly treats the dynamics of linear polymers in good solvents. The behavior of DNA in a bad solvent, e.g., a nonpolar organic solvent, would be quite different. Of note, the variation between bases (A, T, C, and G) has little effect on the physical properties of DNA.[1] For most DNA molecules, the polymer properties can be approximated as independent of the base sequence. The *physical* properties of DNA are, for the most part, a function of the *length* of the DNA strand only.

This section starts by defining physical and mathematical terminology for flexible linear (unbranched) polyelectrolytes, for which DNA is an excellent prototype. These tools are useful when we develop models to relate intrinsic polymer properties (sequence, contour length) to observed transport properties (diffusivity, electrophoretic mobility).

[1] Differences between A/T-rich regions of DNA compared with G/C-rich regions *can* be measured, but these differences are small.

Table 14.1 Modes of observation and calculation of physical properties and length scales of DNA.

Symbol	Property	When/how observed	When/how calculated
d	Diameter	Observed using x-ray crystallography	From molecular models of DNA structure
ℓ_c	Contour length	Not directly observed	Known from number of base pairs
ℓ_K	Kuhn length (see modeling section)	Not a physical length	Chosen to make idealized models match physical observables
ℓ_e	End-to-end length	Microscopy of fluorophores attached to ends	Fluctuates; only $\langle \ell_e \rangle$ predicted by models
$\langle \ell_e \rangle$	Ensemble-averaged end-to-end length	Microscopy of fluorophores attached to ends	Predicted by models
ℓ_p	Persistence length	Not directly observed	Predicted by some models
$\langle r_g \rangle$	Radius of gyration	Experimentally observed with light scattering or fluorescence microscopy [169]	Predicted by models
D	Diffusivity	Observed directly using fluorescence microscopy of diffusing molecules [170, 171]	Related to contour length using $\langle r_g \rangle$ and Zimm dynamics
μ_{EP}	Electrophoretic mobility	Observed directly with EP velocity if EOF is well-characterized and subtracted	Related to surface charge density by double-layer theory and Rouse dynamics

DEFINITION OF LINEAR (UNBRANCHED) POLYMER PROPERTIES AND LENGTH SCALES

We start by modeling a linear polymer such as DNA mathematically as an idealized contour through space, and we describe the properties and length scales of this contour. Intuitively, we can think of the DNA molecule as behaving somewhat like a microscopic piece of spaghetti suspended in water. In this approximation, we ignore all details of the chemical structure other than the contour of the chemical backbone, and subsume all chemical properties into contour parameters such as the persistence length and radius of gyration, which are subsequently discussed. The geometry of the DNA molecule (as defined by the backbone contour) fluctuates with time owing to the thermal fluctuations in the system. Because of this, we define the molecular state by using both instantaneous properties of the contour (which fluctuate) as well as time or ensemble averages (which do not).

Key length scales and definitions for DNA include the diameter, contour length, persistence length, end-to-end length, and radius of gyration. dsDNA has a uniform diameter of approximately 2 nm. Because DNA has a significant negative charge, the EDL surrounding the DNA leads to electrostatic repulsion between components of the polymer, and the effective diameter can be much larger.

The *contour length* ℓ_c of a DNA molecule is the arc length of the backbone contour, i.e., the distance we would travel if we moved *along the curved backbone* from one end of the molecule to the other. In dsDNA, the base pair spacing is approximately 0.34 nm, and thus the contour length of a DNA molecule with N_{bp} base pairs is $\ell_c \simeq 0.34$ nm $\times$ N_{bp}. DNA molecules can range in length from just a few base pairs (these molecules are typically referred to as oligomers) to hundreds of thousands or millions of base pairs (see Table 14.2). The contour length and the base sequence are the only parameters intrinsic to the DNA molecule, and because the sequence has little effect on the physical

Table 14.2 **Some approximate genome sizes for common species.**

Organism	Genome size (base pairs)	Total DNA length
Virus (λ bacteriophage)	50 kbp	17 μm
Enteric bacterium (*Escherichia coli*)	4 Mbp	1.4 mm
Yeast (*Saccharomyces cerevisiae*)	20 Mbp	6.8 mm
Insect (*Drosophila melanogaster*)	130 Mbp	44 mm
Mammal (*Homo sapiens*)	3.2 Gbp	1 m

Note: Genome sizes vary widely, and are only loosely connected to the complexity of the organism.

properties, the contour length is the only molecular parameter that significantly affects DNA physical properties in aqueous solutions.

We define the coordinate system and notation as follows: We define the scalar s as the arc length (i.e., the distance along the polymer backbone, where $s = 0$ corresponds to one end and $s = \ell_c$ corresponds to the other end. ℓ_c is the *contour length* previously described. We also use s_1 and s_2 to denote two specific points along the polymer, and use Δs defined as $|s_1 - s_2|$ to denote the arc length between these two points. Because the polymer contour is in general curved, the arc length is *not* equal to the linear distance between the points. We also define $\vec{r}(s)$ as the position vector of a point on the backbone with respect to the coordinate system origin. The unit vector tangent to the polymer backbone is $\frac{\partial \vec{r}}{\partial s}$, and the vector quantifying the magnitude and direction of the local curvature of the backbone is proportional to $\frac{\partial^2 \vec{r}}{\partial s^2}$.

The *persistence length* ℓ_p is a measure of the rigidity of a linear polymer, and we evaluate it by determining the distance two points of the DNA polymer need to be from each other for their orientation to become statistically uncorrelated. We use a statistical measure because the position and orientation of a DNA molecule in an aqueous solution is always fluctuating with time owing to thermal perturbations. The persistence length is a measure of the rigidity of the polymer backbone. If the backbone is stiff (imagine uncooked spaghetti), then the components of the backbone for the most part point in the same direction. If the backbone is flexible (now imagine cooked spaghetti), then the parts of the polymer backbone point in random directions.

Persistence length has a precise mathematical definition, namely

$$\ell_p = \frac{-\Delta s}{\ln \left\langle \left. \frac{\partial \vec{r}}{\partial s} \right|_{s_1} \cdot \left. \frac{\partial \vec{r}}{\partial s} \right|_{s_2} \right\rangle}, \tag{14.1}$$

where angle brackets denote the time average of a fluctuating property. The persistence length ℓ_p is meaningful only if it is independent of Δs. The persistence length can equivalently be defined using

$$\left\langle \left. \frac{\partial \vec{r}}{\partial s} \right|_{s_1} \cdot \left. \frac{\partial \vec{r}}{\partial s} \right|_{s_2} \right\rangle = \exp\left(-\frac{\Delta s}{\ell_p} \right). \tag{14.2}$$

This relation compares two points along the polymer backbone that are separated by a distance Δs along the backbone. Over time, it takes the unit tangent vector to the polymer backbone at each of these two points, compares them by taking the dot product of the two vectors (effectively evaluating the cosine of the angle between the two tangents), and averages the result. This relation asserts that the time-averaged cosine of the angle decays exponentially as the arc length between the two points increases. Two proximal points are perfectly correlated, the angle between their tangents is zero, and the dot product between their unit tangent vectors is one. Two points separated by a large arc length are uncorrelated, and the cosine of the angle between them varies randomly between -1 and 1, eventually averaging to zero. In between, the correlation of tangent angles decays exponentially as the arc length separating the points increases. The persistence length is the characteristic length of this exponential decay. dsDNA, for example, has a persistence length approximately equal to 50 nm in concentrated electrolyte solutions at room temperature. Structural biological molecules are stiffer and have a higher persistence length (for example, F-actin's persistence length has been measured to be 17 μm [172]).

The persistence length is important when developing equations to relate DNA transport properties to intrinsic polymer properties and solution conditions. The persistence length provides a means for understanding DNA's configuration, i.e., the shape of the DNA molecule as a function of contour length and external environment. Whereas linear polymers such as DNA typically have a well-defined ℓ_p, many models that describe these polymers do not.

The *end-to-end length* ℓ_e of a DNA molecule is a scalar measure of the *linear* distance (*not* the arc length) between the two endpoints of the molecule at any instant. This can thus be written as

$$\ell_e = \left|\vec{r}(s = \ell_c) - \vec{r}(s = 0)\right| , \tag{14.3}$$

where the vertical bars denote the magnitude of the vector. Often, we are actually more concerned with $\langle \ell_e \rangle$, the time-averaged value of this property, given by

$$\langle \ell_e \rangle = \left\langle \left|\vec{r}(s = \ell_c) - \vec{r}(s = 0)\right| \right\rangle . \tag{14.4}$$

The *radius of gyration* $\langle r_g \rangle$ of a DNA molecule is a statistical measure of the linear distances between different points on the DNA backbone, and $\langle r_g \rangle^3$ is thus an approximate measure of the volume that encloses the DNA molecule. Because the light scattered off of a DNA molecule in solution in certain limits is proportional to $\langle r_g \rangle$, the radius of gyration is usually the most easily measured DNA property in solution. The radius of gyration $\langle r_g \rangle$ is defined as the time average of the root-mean-square of the linear distances between the elements of the contour and the polymer centroid:

$$\langle r_g \rangle^2 = \frac{1}{\ell_c{}^2} \int_{s_2=0}^{\ell_c} \int_{s_1=0}^{s_1=s_2} \left\langle \left|\vec{r}(s_1) - \vec{r}(s_2)\right|^2 \right\rangle ds_1\, ds_2 . \tag{14.5}$$

The size of a microfluidic domain, when compared with $\langle r_g \rangle$, tells us whether the configuration of a DNA molecule in aqueous solution is affected by the fluid boundaries.

14.2 DNA TRANSPORT

DNA transport properties in bulk, in gels, and in nanostructured channels are required to predict the performance of fluid mechanical devices for DNA analysis. This section summarizes the existing experimental measurements of these properties.

14.2.1 DNA transport in bulk aqueous solution

We first consider DNA transport (including diffusion and electromigration) in bulk aqueous domains far from walls. DNA molecules can be large, so DNA's diffusivity is relatively low compared with that of smaller molecules, and the diffusivity is dependent on polymer length, as subsequently described. DNA's electrophoretic mobility is quite high compared with that of most macromolecules, owing to its highly charged sugar backbone, and DNA electrophoretic mobility is largely independent of polymer length.

DNA diffusivity in bulk. Diffusion is the macroscopic description of Brownian motion caused by thermal energy, discussed in Chapter 4 for small molecules. For a macromolecule, we can describe several types of diffusion, including both translational diffusion and rotational diffusion. We focus in this chapter strictly on translational diffusion, i.e., diffusion of the center of mass of the DNA molecule. The translational diffusion is proportional to the thermal energy and thus proportional to $k_B T$, as well as the effective viscous mobility μ. Hydrodynamic diffusion of DNA in bulk aqueous solutions is reasonably well characterized by treating DNA as a *nondraining polymer* that obeys *Zimm dynamics*. The nondraining polymer assumption describes the motion of water near the polymer – the nondraining assumption entails assuming that the motion of water molecules in the region of the polymer is largely suppressed by the presence of the DNA molecule.

The Zimm dynamics approximation is closely related but relates to the motion of various parts of the polymer – Zimm dynamics assumes that the motion of various parts of the polymer are tightly coupled to each other because of the viscous coupling. In the Zimm dynamics approximation (or in the nondraining approximation), the viscous coupling makes the DNA and surrounding water diffuse hydrodynamically as if it were a solid object.[2] Experimentally, we observe that the Zimm model is accurate enough that diffusion of DNA is well approximated by modeling it as a rigid sphere with a radius equal to about $\langle r_g \rangle/3$.

Recall from Chapter 8, for comparison, that the mobility μ for a macroscopic particle of radius a in a liquid of viscosity η is given by the Stokes flow relation:

$$\mu = \frac{1}{6\pi\eta a}, \tag{14.6}$$

leading to the Stokes–Einstein relation for particle diffusivity:

$$D = \frac{k_B T}{6\pi\eta a}, \tag{14.7}$$

and the viscous mobility for an ion with hydrated radius a is approximately given by

$$\mu \simeq \frac{1}{6\pi\eta a}, \tag{14.8}$$

[2] This can be described with much more mathematical rigor with a mobility matrix, in which the polymer is discretized into many elements and the motion of each element is calculated from the set of forces on each element by multiplication with this matrix. For a set of dilute, isolated particles, the mobility matrix has elements along only the diagonal, indicating that the motion of each particle is dependent on only the external force on that particle. For a *free-draining polymer* obeying *Rouse dynamics*, the mobility matrix is the same as for isolated particles – the assumption is that each part of the molecule is unaffected by the presence of other parts of the molecule. For a *nondraining polymer* obeying *Zimm dynamics*, the mobility matrix contains off-diagonal elements that are inversely proportional to the distance between polymer components, typically of a form similar to Eq. (8.32). If the off-diagonal elements of the mobility matrix are zero, the velocities of different parts of the polymer are uncoupled. If the off-diagonal elements of the mobility matrix are large (as assumed for a nondraining polymer), the velocities of different parts of the polymer are strongly coupled. We could think of a solid object as the limit in which the off-diagonal elements linking the motion of various parts of the solid approach infinity.

leading to an approximate relation between ion hydrated radius and diffusivity:

$$D \simeq \frac{k_B T}{6\pi\eta a}. \tag{14.9}$$

Using a Zimm dynamics model and assuming (as inferred from measurements) that DNA's effective solid-particle radius can be approximated by $\langle r_g \rangle/3$, we can approximate DNA diffusivity as

Stokes approximation for DNA diffusivity

$$D \simeq \frac{k_B T}{2\pi\eta\langle r_g \rangle}. \tag{14.10}$$

The relation $a \simeq \langle r_g \rangle/3$ is an approximate value that corrects roughly for the assumptions embedded in the application of a macroscopic relation for flow over a rigid sphere to the transport of a macromolecule. The most important result of the analogy between DNA diffusion and solid particle diffusion is the conclusion that the diffusivity of DNA scales with $\langle r_g \rangle$ in the same manner that the diffusion of solid spherical particles scales with radius a. Zimm dynamics accounts for the fact that the parts of the DNA molecule are all hydrodynamically linked by the water, making the molecule diffuse as one cohesive body. Thus DNA *in bulk solution* acts diffusively (approximately, at least) as if it were a solid particle with a radius approximately equal to $\langle r_g \rangle/3$. The diffusivity D in free solution is independent of both the DNA sequence and the presence or absence of an applied electric field. Solution conditions need be considered only to the (minor) extent that solution conditions affect the radius of gyration or viscosity.

Various models (see later sections in this chapter) predict $\langle r_g \rangle$ as a function of ℓ_c and other properties. Depending on the complexity of the model and the role of the solvent, these models typically predict that $\langle r_g \rangle$ is proportional to $\ell_c^{\frac{1}{2}}$ or $\ell_c^{\frac{3}{5}}$, implying that the bulk diffusivity should be proportional to $\ell_c^{-\frac{1}{2}}$ or $\ell_c^{-\frac{3}{5}}$ and largely independent of other properties. For DNA in water, theoretical treatments lead to the conclusion that $\langle r_g \rangle$ should be proportional to $\ell_c^{-\frac{3}{5}}$.[3] Experiments usually observe an exponent approximately equal to –0.57 (e.g., Fig. 14.3), leading to approximate relations such as

experimental correlation for bulk DNA diffusivity as a function of contour length

$$D \simeq 3 \times 10^{-12} \left(\frac{\ell_c}{1\,\mu\text{m}}\right)^{-0.57} \simeq 3 \times 10^{-10} N_{bp}{}^{-0.57} \text{m}^2/\text{s}, \tag{14.11}$$

[3] Idealized models, as we discuss later in the chapter, ignore the interaction of a polymer backbone with itself and predict that $\langle r_g \rangle$ is proportional to $\ell_c^{\frac{1}{2}}$. We consider idealized models, owing to their simplicity, though for dilute DNA molecules in water they are only qualitatively accurate. These idealized models are accurate for polymer melts, in which case there is no solvent, or for what are commonly termed *theta solvents* – in both cases, the energetics of the polymer interacting with itself are identical to that of the polymer interacting with its environment, and the idealized models (which ignore these energetics) end up giving the correct result. More detailed models account for the interaction of the polymer backbone with itself, and these models explain why different results are observed depending on what the polymer is dissolved in. For dilute polymers in good solvents (like a dilute DNA solution in water), these models predict that $\langle r_g \rangle \propto \ell_c^{\frac{3}{5}}$.

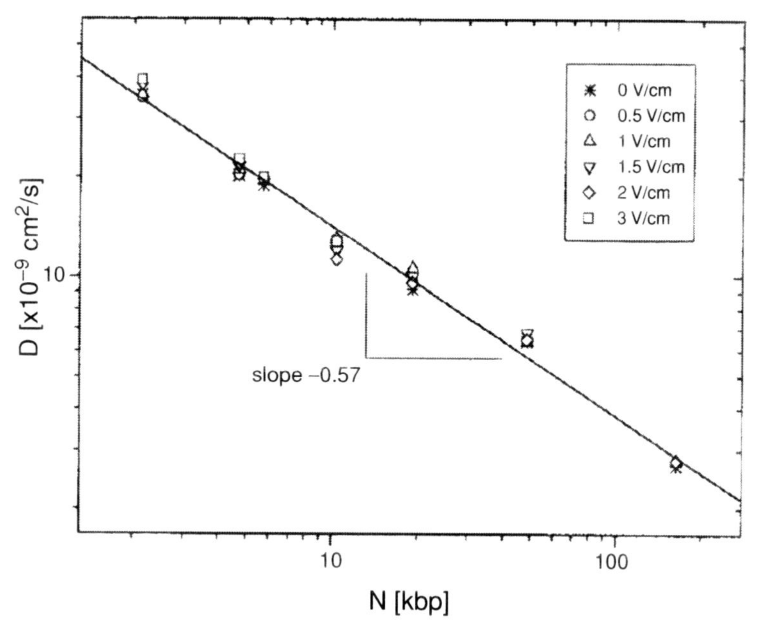

Fig. 14.3 The bulk solution diffusivity of dsDNA as a function of polymer length in 1X TAPS buffer. The different symbols denote different electric fields, and the single curve fit demonstrates that the diffusivity D is independent of the electric field in bulk solution. Reproduced with permission from [173].

which describes the data in [173] for DNA in free solution in the buffer studied. Results for other buffers range from $D \simeq 2\text{–}5 \times 10^{-12}\ \mathrm{m^2/s}(\frac{\ell_c}{1\ \mu\mathrm{m}})^{-0.57}$.

DNA electrophoretic mobility in bulk. Electrophoresis, as discussed in Chapters 11 and 13, is the net electromigration of a molecule induced by Coulomb forces on a charged molecule or particle and, if present, its EDL. Unlike hydrodynamic diffusion of DNA, electrophoresis of long DNA molecules in bulk aqueous solutions of at least modest electrolyte concentration is reasonably well characterized by treating DNA as a *free-draining polymer* that obeys *Rouse dynamics*. The free-draining polymer assumption means that we assume that the motion of water molecules in the region of the polymer is unaffected by the presence of the DNA molecule. This implies that the distance over which the fluid velocity gradients decay (λ_D) is small compared with the spacing of different components of the polymer, leading to free motion of the water with respect to the polymer. Rouse dynamics assumes that the coupling between polymer elements is minor. This phenomenon makes the DNA electrophorese hydrodynamically as if all polymer components were electrophoresing independently.[4]

Despite the complexity of physics that governs DNA electrophoresis, we observe a relatively simple dependence experimentally. The electrophoretic mobility of long DNA in bulk electrolyte solutions is typically in the range $\mu_{EP} \simeq 2\text{–}5 \times 10^{-8}\ \mathrm{m^2/V\,s}$, where the

[4] See the diffusion section for a discussion of the mathematical formulation of the mobility matrix for Rouse and Zimm models. For electrophoresis, the off-diagonal mobility matrix components are proportional to $\exp(-r/\lambda_D)$ rather than $1/r$. Because the spacing between polymer components is typically large relative to λ_D, these off-diagonal components are approximately zero; hence DNA undergoing electrophoresis obeys Rouse dynamics and is free draining.

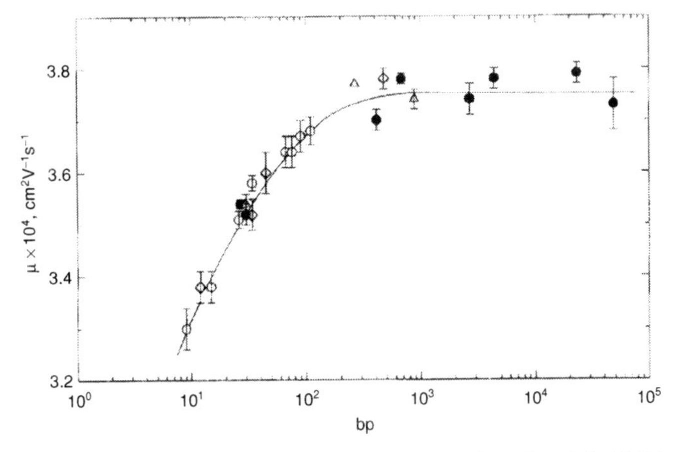

Fig. 14.4 The free-solution electrophoretic mobility of dsDNA as a function of polymer length, in 1X Tris-acetate-EDTA buffer. Different symbols denote different DNA sources. These data illustrate that DNA electrophoretic mobility is roughly constant for DNA strands longer than 100 base pairs. Even below 100 base pairs, the electrophoretic mobility variation is minor – the electrophoretic mobility at 20 base pairs is only 10% lower than that of 2000 base pairs. (Reproduced with permission from [174].)

magnitude is a function of the electrolyte concentration and valence but, for $N_{bp} > 400$, is independent of contour length (Fig. 14.4). This simple relation breaks down for (a) low salt concentrations, in which λ_D becomes large, (b) confined geometries, in which spacing between polymer regions can be reduced by the geometries enforced by confinement, and (c) DNA oligomers, for which many of the geometric approximations of these models are invalid.

Increasing electrolyte concentration reduces the DNA electrophoretic mobility (Fig. 14.5), as does the presence of multivalent cations, akin to that observed for the electrophoretic mobility of particles or the electroosmotic mobility of microdevice walls.

Failure of Nernst–Einstein relation for polyelectrolytes. The distinction between the size dependence of the electrophoretic mobility and the size dependence of the diffusivity is important. This contradicts the Nernst–Einstein relation (11.30), which links the diffusivity of a point charge to its electrophoretic mobility.

For an ion modeled as a point charge, the force that moves the ion (either a Coulomb force or random fluctuations caused by the thermal motion of the solvent) is balanced by the "drag" that the ion feels when it moves through the solvent. This drag is the same regardless of the source of the motion. Thus the Nernst–Einstein relation illustrates that the electrophoretic mobility (when normalized by zF, a measure of the charge of a mole of ions) is equal to the diffusivity (when normalized by RT, related to the thermal energy of a mole of ions). The mean-field, point charge assumption means that each ion behaves independently from all other ions.

For DNA, we can think of the parts of the DNA molecule as a bunch of particles or rods. The hydrodynamic motion of all of the parts of a DNA molecule is akin to the motion of a collection of Stokes particles in close proximity – the surrounding water to a large extent moves along with the particles. The electrophoretic motion of all of the parts of a DNA molecule is akin to the electrophoresis of a collection of particles in close proximity – the surrounding water does not move along with the particles, because the Coulomb forces on the ions in the EDL cause the hydrodynamic perturbations to

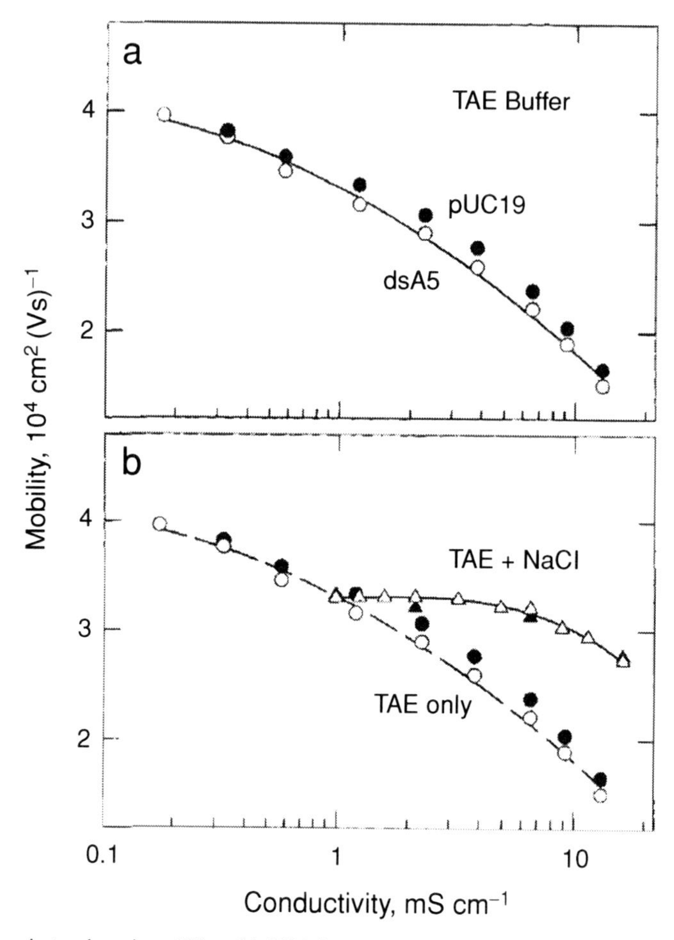

Fig. 14.5 The free-solution electrophoretic mobility of dsDNA for two polymer lengths (dsA5: $N_{bp} = 20$; pUC19: $N_{bp} = 2686$) as a function of buffer and salt concentration (reported using conductivity). Buffer is Tris-acetate-EDTA (ethylene diamine tetraacetic acid). The data show that DNA strands of varying length show similar dependence on ionic strength. The slight dependence of electrophoretic mobility on DNA strand length (seen earlier in Fig. 14.4) is seen in the slightly higher electrophoretic mobility of the filled circles (2686 bp) compared with that of the open circles (20 bp). (Reproduced with permission from [175].)

decay with a characteristic length λ_D. So the different charge components of the backbone are to a great extent unaffected by each other. All feel the same electrophoretic force in the same direction, with minimal interference. These differences make DNA's behavior in diffusion much different than electrophoresis. These differences also create many possibilities for novel DNA separation techniques, as nanofilters (see Chapter 15) take advantage of the contour length dependence of diffusion and the contour length independence of electrophoresis.

14.3 IDEAL CHAIN MODELS FOR BULK DNA PHYSICAL PROPERTIES

Now that we have discussed experimental observation of bulk DNA properties and listed key parameters that describe the idealized polymer backbone, we can discuss some simple models that relate polymer backbone properties to each other and to the bulk transport properties that are important for DNA analysis.

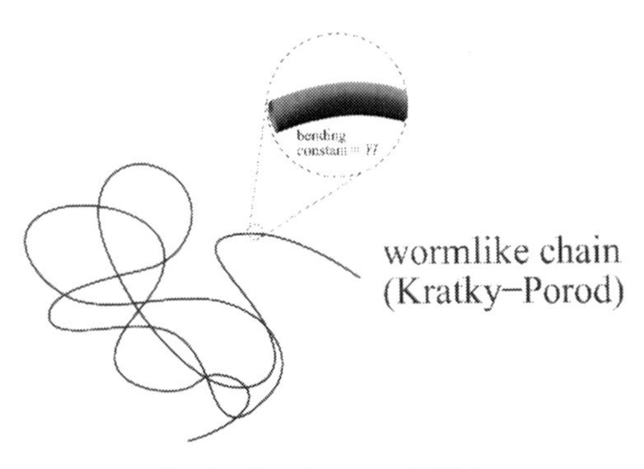

Fig. 14.6 Kratky–Porod model of DNA.

14.3.1 Idealized models for bulk DNA properties

We start by discussing several idealized polymer models that ignore the electrostatic repulsion that various parts of a DNA molecule feel in the vicinity of other parts. We then discuss extensions of the models to account for this repulsion, the "self-avoiding" nature of macromolecules that prevents two parts of the polymer backbone from coexisting at one point. These models are related back to the experimental observations for translational diffusion and electrophoretic mobility discussed earlier.

We seek to develop models that have as fundamental input one intrinsic molecular property, the contour length ℓ_c, and one measurable property of the molecule's configuration in bulk solution, the radius of gyration $\langle r_g \rangle$. The models we develop have certain properties chosen to ensure that $\langle r_g \rangle$ is properly predicted from ℓ_c. We hope also to predict the persistence length ℓ_p, the probability distribution and mean value of the end-to-end length ℓ_e, and other parameters such as the entropy of confinement and the forces induced by confinement. Once we know how microscopic model parameters can reconcile known ℓ_c and $\langle r_g \rangle$ values, these model parameters can then be implemented to predict, for example, how $\langle r_g \rangle$ varies with ℓ_c, and thus how D varies with ℓ_c. Further, these models can then predict how transport properties change upon confinement in nanochannels. These models enable approximate predictions of polymer properties and behavior, but complete description of DNA's physical properties is beyond the idealized models presented here.

Kratky–Porod model of DNA (wormlike chain). The *Kratky–Porod model*, also called the *wormlike chain model*, treats the polymer as if it were an idealized, macroscopic, circularly symmetric beam element with a flexural rigidity YI [N m^2], where Y [Pa] is Young's modulus and I [m^4] is the second moment of area of the beam.[5] This structure is depicted in Figure 14.6. This description is physically reminiscent of the physics of the double helix itself, in which covalent and hydrogen bonds limit the ability of the helix to locally deform. In this model, the atomic-scale forces are greatly oversimplified by subsuming them into the continuum descriptions given by Young's modulus and moment of inertia. Further, the electrostatic repulsion that the sugar backbone feels when folded onto itself is also ignored in this model. Despite these limitations, the properties of long DNA molecules (with a contour length much greater than persistence length) are reasonably

[5] Young's modulus is denoted by E in most solid mechanics texts, but we use Y here to avoid confusion with the electric field $\vec{E}$. The second moment of area I we imply here is given by $I = \int_S r^2 \, dA$, with units of [m^4], and is different from I used elsewhere to denote the electrical current.

well described by treating a DNA molecule as if it were a beam with a flexural rigidity given by

relation between effective flexural rigidity and radius of gyration of a DNA molecule, $\ell_c \gg \ell_p$

$$YI = \frac{3\langle r_g \rangle^2 k_B T}{\ell_c}. \qquad (14.12)$$

At room temperature, the configuration of DNA molecules of *any length* can be well described by modeling the polymer backbone as if it had a flexural rigidity of

effective flexural rigidity of a DNA molecule, concentrated aqueous solution

$$YI \simeq 2 \times 10^{-28}\ \mathrm{N\,m}^2. \qquad (14.13)$$

DNA properties can be predicted from ℓ_c and YI alone. The general expression linking the observed $\langle r_g \rangle$ and the inferred YI is of a complicated form.

The Kratky–Porod model leads to predictions of the end-to-end length, radius of gyration, and persistence length of DNA by noting that a bent beam has an internal energy of bending. The local bending energy per unit length is given by $\frac{1}{2} YI \left| \frac{\partial^2 \vec{r}}{\partial s^2} \right|^2$, and thus the total energy of bending $\mathcal{U}_{\mathrm{bend}}$[J] is given by

$$\mathcal{U}_{\mathrm{bend}} = \frac{1}{2} YI \int_{s=0}^{s=\ell_c} \left| \frac{\partial^2 \vec{r}}{\partial s^2} \right|^2 ds. \qquad (14.14)$$

The total bending energy of the system is thus proportional to the integral of the squared curvature $\left| \frac{\partial^2 \vec{r}}{\partial s^2} \right|^2$ of the polymer backbone. Boltzmann statistics then dictate that the likelihood of a particular conformation is proportional to $\exp\left[-\frac{\mathcal{U}_{\mathrm{bend}}}{k_B T}\right]$. From the Boltzmann analysis (not presented here), we can extract a number of conclusions:

- *Persistence length.* The persistence length for a Kratky–Porod polymer can be shown to be

persistence length for a wormlike chain polymer

$$\ell_p = -\frac{\Delta s}{\ln\left\langle \left. \frac{\partial \vec{r}}{\partial s} \right|_{s_1} \cdot \left. \frac{\partial \vec{r}}{\partial s} \right|_{s_2} \right\rangle} = \frac{YI}{k_B T}, \qquad (14.15)$$

and thus the persistence length is simply proportional to the effective flexural rigidity of the polymer.

- *End-to-end length.* For the Kratky–Porod polymer, we can derive (see Exercise 14.3) that the two ends are separated by a mean linear distance of

mean end-to-end linear distance for a wormlike chain polymer

$$\langle \ell_e \rangle = \sqrt{2\frac{YI}{k_B T}\ell_c - 2\left(\frac{YI}{k_B T}\right)^2\left[1 - \exp\left(-\frac{\ell_c k_B T}{YI}\right)\right]} \tag{14.16}$$

or

mean end-to-end linear distance for a wormlike chain polymer

$$\langle \ell_e \rangle = \sqrt{2\ell_p\ell_c - 2\ell_p{}^2\left[1 - \exp\left(-\frac{\ell_c}{\ell_p}\right)\right]}. \tag{14.17}$$

Equation (14.17) has two simple limits. For $\ell_p \ll \ell_c$ (a flexible cooked-spaghetti-like polymer), this becomes

$$\langle \ell_e \rangle = \sqrt{2\ell_c\ell_p} = \sqrt{2\ell_c\frac{YI}{k_B T}} = \sqrt{6}\langle r_g \rangle . \tag{14.18}$$

We use the long-DNA limit result from Eq. (14.18) to write Eq. (14.12), because the relatively complex form of Eq. (14.17) prevents us from writing a general relationship for YI in terms of $\langle r_g \rangle$. For $\ell_p \gg \ell_c$ (essentially a rigid rod, like a metal beam), Eq. (14.17) becomes

$$\langle \ell_e \rangle = \ell_c . \tag{14.19}$$

Given the experimentally observed persistence length of approximately 50 nm, we can see that the flexible limit applies for DNA molecules with $\ell_c \gg 50$ nm or $N_{bp} \gg 150$. The rigid rod limit applies only for oligomers.

EXAMPLE PROBLEM 14.1

A DNA molecule with a known contour length of 500 μm has a measured $\langle r_g \rangle$ of 2.8 μm in an aqueous solution of unknown nature. Modeling DNA as a wormlike chain, predict what $\langle \ell_e \rangle$ a DNA molecule with contour length $\ell_c = 100$ nm would have in this solution.

SOLUTION: We do not know the solution and thus we do not know the persistence length of DNA in this solution. However, because we are given the two basic observables for a DNA molecule, i.e., the contour length and the radius of gyration, we can determine the persistence length and then determine the mean end-to-end length.

We know that, for a wormlike chain in the limit $\ell_p \ll \ell_c$, the radius of gyration is related to the contour and persistence lengths by the relation $\sqrt{6}\langle r_g \rangle = \sqrt{2\ell_c\ell_p}$. Thus

$$\ell_p = \frac{3\langle r_g \rangle^2}{\ell_c} = 40 \text{ nm} . \tag{14.20}$$

Once the persistence length is known, we can predict the $\langle \ell_e \rangle$ for the molecule by using

$$\langle \ell_e \rangle = \sqrt{2\ell_p \ell_c - 2\ell_p^2 \left[1 - \exp\left(-\frac{\ell_c}{\ell_p}\right)\right]}, \tag{14.21}$$

which gives

$$\langle \ell_e \rangle = 69 \text{ nm}. \tag{14.22}$$

This result illustrates the relative rigidity of a DNA molecule of contour length 100 nm – the end-to-end length $\langle \ell_e \rangle$ is almost as large as ℓ_c.

The Kratky–Porod model describes the structure of DNA well as long as the length scales of interest are 10 nm or greater. Below 10 nm, the Kratky–Porod model does not capture the detailed structure of the polymer. The Kratky–Porod model immediately gives a persistence length and relates that persistence length directly to an intuitive measure of the rigidity of the polymer. Because the Kratky–Porod model is based physically on polymer angles and curvature, it does a good job of predicting orientation (through the persistence length) but can be inconvenient analytically when treating linear distances between parts of the polymer. Other models handle distances well (the Gaussian bead–spring model particularly) but handle angles and persistence lengths poorly. The Kratky–Porod model matches the ℓ_c dependence of observed $\langle r_g \rangle$ and D data reasonably well, although the $\langle r_g \rangle \propto \ell_c^{\frac{1}{2}}$ and $D \propto \ell_c^{-\frac{1}{2}}$ dependences predicted by the model are a bit less than the $\langle r_g \rangle \propto \ell_c^{0.58}$ and $D \propto \ell_c^{-0.57}$ dependences found experimentally. An example value of YI is given in relation (14.13), but the value of YI that matches a Kratky–Porod model to data is dependent on electrolyte concentration and temperature. Most notably, when the electrolyte concentration is low, the large EDL makes the DNA molecule appear more rigid, and larger YI values will match the data better at low electrolyte concentrations.

Idealized freely jointed chain. The freely jointed chain model generates a microscopic molecular model that predicts the DNA configuration and bulk diffusivity, given an experimental measurement of the radius of gyration $\langle r_g \rangle$. Like the Kratky–Porod model, it translates experimental observations of $\langle r_g \rangle$ into a model parameter (ℓ_K) that predicts how DNA properties scale with ℓ_c.

The freely jointed polymer model treats the polymer as a series of freely jointed linkages joined by rigid rods of length given by

Kuhn length for a freely jointed polymer

$$\ell_K = \frac{6\langle r_g \rangle^2}{\ell_c}, \tag{14.23}$$

where ℓ_K is called the *Kuhn length*.[6] The Kuhn length is distinct from the other lengths described in this chapter in that it is *not* a physical length, but rather a parameter

[6] The Kuhn length is often symbolized in other texts as a or b, and the number of Kuhn lengths is often used instead of the contour length ℓ_c.

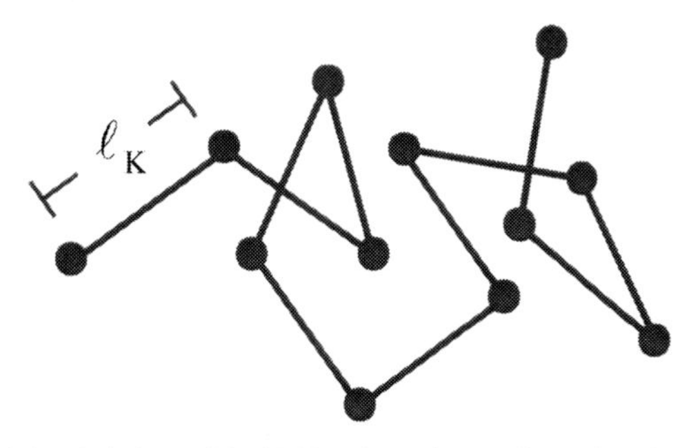

Fig. 14.7 Freely jointed chain model of DNA, drawn in two dimensions for clarity.

used in models that are designed to describe DNA properties. The models that use a Kuhn length *do not* describe DNA properties properly on the scale of the Kuhn length; however, these models *can* describe DNA's macroscopic behavior. Although the Kuhn length is not a physical length, the Kuhn length *can* be related to physical lengths such as the radius of gyration or the end-to-end length.[7]

The structure of a freely jointed chain is depicted in Fig. 14.7. These linkages are free, akin to nanoscopic ball joints. This model, in contrast to the Kratky–Porod model, tells us nothing about the orientation or curvature of individual elements – in fact, the model precludes any correlation of orientations at any distance longer than ℓ_K. The microscopic structure of this model is nonphysical. However, the model does provide a mathematical framework for predicting the relative linear distances between different components of the polymer. This property makes this model (and other jointed models) well suited for handling the physics of macromolecular confinement.

The freely jointed model leads to statistical predictions of polymer geometry by a mathematical analysis of random walk processes. The predictions of this model are as follows:

- *Persistence length.* The persistence length is not defined for the freely jointed chain model. Because the beads are freely jointed, the tangents to the backbone are uncorrelated at distances above ℓ_K.
- *End-to-end length.* The mean end-to-end length $\langle \ell_e \rangle$ can be shown to be

mean linear end-to-end length for a freely jointed chain polymer

$$\langle \ell_e \rangle = \sqrt{6} \langle r_g \rangle = \sqrt{\ell_c \ell_K}\,, \qquad (14.24)$$

and the probability distribution function of ℓ_e is given by

probability density function for end-to-end length for a freely jointed chain polymer

$$\mathcal{P}(\ell_e) = 4\pi \ell_e^2 \left[\frac{1}{4\pi \langle r_g \rangle^2} \right]^{\frac{3}{2}} \exp\left(-\frac{\ell_e{}^2}{4 \langle r_g \rangle^2} \right). \qquad (14.25)$$

[7] For some polymers, such as alkanes, the freely jointed chain model and the freely rotating chain model are closer to physical accuracy on nanometer length scales, and attempts to model those polymers were the genesis of these models. For DNA, the chain models still work reasonably well to predict properties such as the radius of gyration or diffusivity of the polymer, but they lose their nanometer-scale physical relevance.

This relation is valid only for $\ell_e \ll \ell_c$ and $\ell_K \ll \ell_c$. As written here and previously defined, ℓ_e is a *scalar* distance. We use the scalar because, for this model and for most bulk DNA models, the probability of any two parts of the polymer being separated by any particular vector $\vec{\Delta \mathbf{r}}$ is a function of the distance Δr only. Because we write the probability distribution in terms of the scalar end-to-end distance, two terms contribute to this distribution. The Gaussian term in Eq. (14.25) highlights how likely a vector end-to-end distance is – the most likely end-to-end vectors are those with the smallest distance. The $4\pi\ell_e^2$ term is the surface area of a sphere i.e., the number of vector distances that correspond to the scalar distance ℓ_e. Thus, although the most likely vector distance is the zero vector, the most likely end-to-end scalar distance is nonzero.

- *Linear distance between two parts of the polymer.* Given two specified locations on the polymer separated by an arc length $\Delta s \gg \ell_K$,[8] the mean linear distance $\langle \Delta r \rangle$ between the two locations is given by

mean linear distance between chain elements for a freely jointed chain polymer

$$\langle \Delta r \rangle = \sqrt{\frac{6\Delta s}{\ell_c}} \langle r_g \rangle , \tag{14.26}$$

and the probability distribution function of Δr is given by

probability density function for linear distance between chain elements for a freely jointed chain polymer

$$\mathcal{P}(\Delta r) = 4\pi \Delta r^2 \left[\frac{\ell_c}{4\pi \Delta s \langle r_g \rangle^2} \right]^{\frac{3}{2}} \exp\left(-\frac{\ell_c \Delta r^2}{4\Delta s \langle r_g \rangle^2} \right) . \tag{14.27}$$

Again, this relation is valid only for $\ell_e \ll \ell_c$ and $\ell_K \ll \ell_c$. Substituting ℓ_c in for Δs in Eqs. (14.26) and (14.27) gives the mean and PDF for the polymer end-to-end length.

- *Relation of ℓ_K to other models.* Equation (14.24) shows us that the ℓ_K used in the freely jointed chain model is equal to double the ℓ_p observed for the Kratky–Porod model. Thus DNA in concentrated solutions at room temperature is well approximated by a freely jointed model with $\ell_K \simeq 100$ nm.

EXAMPLE PROBLEM 14.2

Given $\langle r_g \rangle$, what is the most likely scalar end-to-end distance of a linear polymer described by the freely jointed chain model?

SOLUTION: To find the most likely distance, we set the derivative of the probability density function with respect to Δr equal to zero. Rather than keep all the terms for the moment, we write $\mathcal{P}(\Delta r)$ in the form $\mathcal{P}(\Delta r) = A\Delta r^2 \exp B\Delta r^2$. Taking the derivative of this with respect to Δr and setting the result equal to zero, we obtain

$$\Delta r = \sqrt{-\frac{1}{B}} . \tag{14.28}$$

[8] By definition, Δs is an integral multiple of ℓ_K for the freely jointed chain.

Because, from Eq. (14.25), B for a freely jointed polymer is given by $-1/4\langle r_g\rangle^2$, the most likely distance is given by

$$\Delta r = 2\langle r_g\rangle . \tag{14.29}$$

Despite using an approach that is distinct from the Kratky–Porod model, the freely jointed chain model matches the ℓ_c dependence of observed $\langle r_g\rangle$ and D data with essentially the same accuracy. As was the case for the Kratky–Porod model, when the electrolyte concentration is low, the large EDL makes the DNA molecule appear more stiff, and ℓ_K values must be adjusted to match the data at low electrolyte concentrations.

The freely jointed chain model does not describe the microstructure of DNA well. This model fails to provide a well-defined persistence length. The freely jointed chain model gives a Gaussian distribution of distances between points in a polymer *only* if the distance along the backbone is relatively large.

Idealized freely rotating chain. As with the freely jointed polymer, the freely rotating chain generates a microscopic model that predicts properties given the contour length ℓ_c and a bulk measurement of the radius of gyration $\langle r_g\rangle$.

The freely rotating chain model treats the polymer as a series of jointed linkages joined by rigid rods. Here, the linkages are at a specific colatitudinal angle ϑ, but these linkages are free to rotate in the azimuthal direction. This model is inspired by the structure of alkane polymers and has been used to predict their properties. The linkages have a length (the Kuhn length ℓ_K) given by

Kuhn length for a freely rotating chain polymer

$$\ell_K = \frac{6\langle r_g\rangle^2}{\ell_c}\frac{1-\cos\vartheta}{1+\cos\vartheta} . \tag{14.30}$$

This structure is depicted in Fig. 14.8. This model leads to predictions similar to those of the freely jointed chain model:

- *Persistence length.* Unlike other jointed chain models, the freely jointed chain model does give a persistence length once Δs is large compared with ℓ_K:

$$\ell_p = \frac{1}{2}\ell_K\frac{1+\cos\vartheta}{1-\cos\vartheta} . \tag{14.31}$$

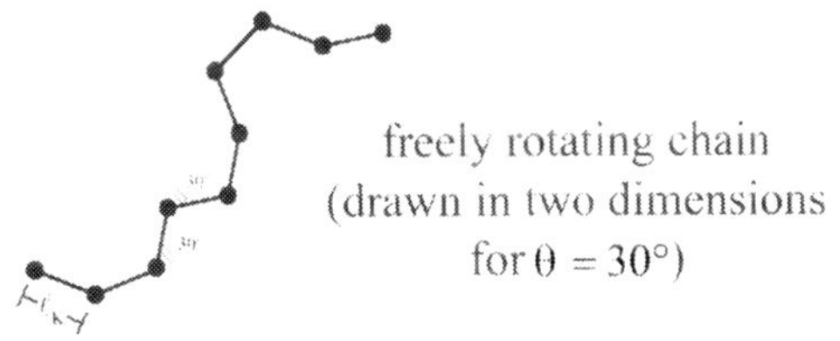

freely rotating chain
(drawn in two dimensions
for $\theta = 30°$)

Fig. 14.8 Freely rotating chain model of DNA.

- *End-to-end length.* The mean end-to-end length $\langle \ell_e \rangle$ is given for a freely rotating chain by

mean linear end-to-end distance for a freely rotating chain polymer

$$\langle \ell_e \rangle = \sqrt{6}\langle r_g \rangle = \sqrt{\frac{1+\cos\vartheta}{1-\cos\vartheta}\ell_c \ell_K}\,, \tag{14.32}$$

and the probability distribution function of ℓ_e is given (for $\ell_e \ll \ell_c$ and $\ell_K \ll \ell_c$) by

probability density function for linear end-to-end distance for a freely rotating chain polymer

$$\mathcal{P}(\ell_e) = 4\pi {\ell_e}^2 \left[\frac{1}{4\pi\langle r_g\rangle^2}\right]^{\frac{3}{2}} \exp\left(-\frac{{\ell_e}^2}{4\langle r_g\rangle^2}\right). \tag{14.33}$$

- *Linear distance between two parts of the polymer.* Given two specified locations on the polymer separated by a distance $\Delta s \gg \ell_K$ along the backbone,[9] the mean linear distance $\langle \Delta r \rangle$ between the two locations is given by

mean linear distance between segments of a freely rotating chain polymer

$$\langle \Delta r \rangle = \sqrt{\frac{6\Delta s}{\ell_c}}\langle r_g \rangle = \sqrt{\frac{1+\cos\vartheta}{1-\cos\vartheta}\Delta s \ell_K}\,, \tag{14.34}$$

and the probability distribution function of Δr is given by

probability density function for linear distance between segments of a freely rotating chain polymer

$$\mathcal{P}(\Delta r) = 4\pi \Delta r^2 \left[\frac{\ell_c}{4\pi\Delta s\langle r_g\rangle^2}\right]^{\frac{3}{2}} \exp\left(-\frac{\ell_c \Delta r^2}{4\Delta s\langle r_g\rangle^2}\right). \tag{14.35}$$

Substituting ℓ_c in for Δs in Eqs. (14.34) and (14.35) gives the mean and PDF for the polymer end-to-end length.
- *Relation of ℓ_K to other models.* Equation (14.31) shows us that the ℓ_K used in the freely jointed chain model is related to the ℓ_p observed for the Kratky–Porod model by a factor that is dependent on ϑ. As $\vartheta \to 0$, the ℓ_p for a given ℓ_K approaches infinity. For $\vartheta \to 90°$, the ℓ_p–ℓ_K relation becomes equal to that of the freely jointed chain.

Here, we see that the freely rotating chain is quite similar to the freely jointed chain. The presence of ϑ in many of the equations and the dependence of ℓ_K on ϑ illustrates that this and the other rigid linkage models are quite arbitrary in their microscopic detail. The microscopic details are arbitrary, because the mathematics of many random-walk processes turns out to be the same as the mathematics of the energy of bending of many bonds in the presence of thermal fluctuations. Thus, if the polymer is long, the details of the microscopic model often have little effect on the $\langle r_g \rangle$ and $\langle \ell_e \rangle$ predictions of the

[9] By definition, Δs is an integral multiple of ℓ_K.

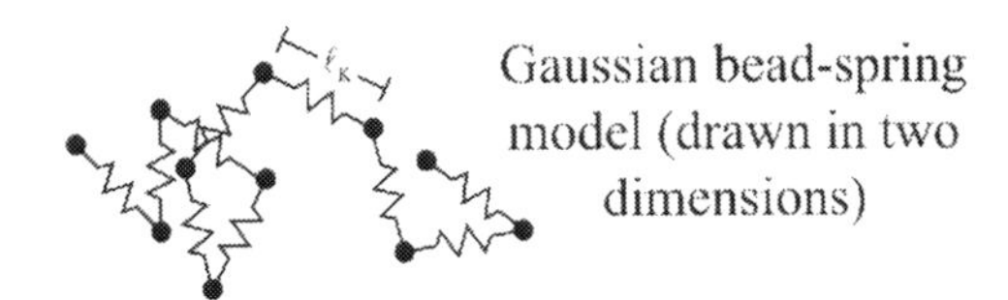

Fig. 14.9 Gaussian bead–spring model of DNA.

model. Because of this, we typically adopt an idealized bead–spring model (described in the next subsection), which is the least physically accurate of the freely jointed models at the atomic level but is mathematically convenient at molecular levels.

Idealized Gaussian or "bead–spring" chain. The bead–spring chain model treats the polymer as a series of freely jointed linkages (the "beads") separated by springs with average length ℓ_K given by

Kuhn length for bead–spring chain

$$\ell_K = \frac{6\langle r_g \rangle^2}{\ell_c}, \tag{14.36}$$

and furthermore assumes that these linkages, rather than being rigid, are springs with spring constant k given by

bead–bead spring constant for bead–spring chain

$$k = \frac{3k_B T}{\ell_K{}^2} = \frac{k_B T \ell_c{}^2}{12\langle r_g \rangle^4}. \tag{14.37}$$

This structure is depicted in Fig. 14.9. It differs from the freely jointed chain model in that the presence of the springs make the PDF of bead–bead distances Gaussian for *all* Δs, not just large Δs. This model leads to identical predictions compared with the freely jointed chain, except that Δs never need be assumed large:

- *Persistence length.* The persistence length is not defined for the bead–spring model. Because the beads are freely jointed, the tangents to the backbone are uncorrelated at distances above ℓ_K.
- *End-to-end length.* The mean end-to-end length $\langle \ell_e \rangle$ is, by definition, given for a linear polymer by (see Exercise 14.5)

mean end-to-end linear distance for bead–spring chain

$$\langle \ell_e \rangle = \sqrt{6}\langle r_g \rangle = \sqrt{\ell_c \ell_K}, \tag{14.38}$$

and the probability distribution function of ℓ_e is given by

probability density function for end-to-end linear distance for bead–spring chain

$$\mathcal{P}(\ell_e) = 4\pi\ell_e^2 \left[\frac{1}{4\pi\langle r_g \rangle^2}\right]^{\frac{3}{2}} \exp\left(-\frac{\ell_e{}^2}{4\langle r_g \rangle^2}\right). \tag{14.39}$$

- *Linear distance between two parts of the polymer.* Given two specified locations on the polymer separated by a distance Δs along the backbone,[10] the mean linear distance $\langle \Delta r \rangle$ between the two locations is given by (see Exercise 14.6)

mean linear distance between segments of bead–spring chain

$$\langle \Delta r \rangle = \sqrt{\frac{6 \Delta s}{\ell_c}} \langle r_g \rangle , \tag{14.40}$$

and the probability distribution function of Δr is given by

probability density function for linear distance between segments of bead–spring chain

$$\mathcal{P}(\Delta r) = 4\pi \Delta r^2 \left[\frac{\ell_c}{4\pi \Delta s \langle r_g \rangle^2} \right]^{\frac{3}{2}} \exp\left(-\frac{\ell_c \Delta r^2}{4 \Delta s \langle r_g \rangle^2} \right) . \tag{14.41}$$

Substituting ℓ_c in for Δs in Eqs. (14.40) and (14.41) gives the mean and PDF for the polymer end-to-end length.
- *Relation of ℓ_K to other models.* Equation (14.38) shows us that the ℓ_K used in the bead–spring model is equal to double the ℓ_p observed for the Kratky–Porod model.

The bead–spring model makes no attempt to describe the microstructure of DNA well. For example, DNA has a well-defined persistence length, but the bead–spring model does not, and the microscopic spring constant of stretched DNA cannot be approximated by Eq. (14.37). However, this model is analytically simple because the distances between *any* two parts of the polymer have a Gaussian probability distribution function given by Eq. (14.41). This model thus describes *linear distances* easily, though it gives no information about angle. Although we might hope that the Kuhn length required to properly describe the radius of gyration, as specified by Eq. (14.36), would be universal and thus independent of the length of the DNA molecule, the required ℓ_K is a function of the length of the DNA molecule.[11]

SUMMARY OF IDEAL MODELS

The Kratky–Porod model treats DNA as having two properties: ℓ_c and an effective flexural rigidity YI, which is approximately given by $YI = 2 \times 10^{-28}$ N m^2, with some dependence on solution conditions. This leads to the most complete description of the mean

---

[10] By definition, Δs is an integral multiple of ℓ_K.

[11] In this chapter, we take $\langle r_g \rangle$ as given and ask what properties a model must have to match it. The bead–spring chain model can predict $\langle r_g \rangle$ properly; however, to do so requires that ℓ_K be chosen differently depending on ℓ_c.

Most polymer physics texts (e.g., Refs. [90, 176] start with the model parameters (ℓ_K and ℓ_c) and determine what $\langle r_g \rangle$ the model predicts. When one takes this approach, the bead–spring model predicts that $\langle r_g \rangle \propto (\frac{\ell_c}{\ell_K})^{\frac{1}{2}}$. Experiments show that real polymers (like DNA) have a radius of gyration that scales like $\langle r_g \rangle \propto \ell_c^{0.57}$. If the bead–spring model is to account for this, ℓ_K must be chosen differently depending on ℓ_c. Physically, this is because the bead–spring model does not account for the self-avoidance of the polymer, i.e., the fact that the polymer backbone is electrostatically repelled from itself.

end-to-end length $\langle \ell_e \rangle$, which applies for any contour length. The Kratky–Porod model also gives ℓ_p, and ℓ_p helps to highlight the rigid and flexible limits of a DNA molecule. The Kratky–Porod model gives orientations easily but gives distances only by integration of the orientations.

Freely jointed models treat DNA as having two properties: ℓ_c and a Kuhn length ℓ_K. For the Gaussian bead–spring model, ℓ_K is approximately equal to 100 nm for concentrated solutions at room temperature, with some dependence on solution conditions. Freely jointed models work best for long DNA molecules, for which the nonphysical aspects of the model do not interfere with its predictions. These models are nonphysical at the atomic level, but accurately describe polymer behavior on a molecular scale. The fact that the freely jointed chain model and the freely rotating chain model can both be made to give the same answers is evidence of this. The Gaussian bead–spring model is the mathematically most convenient form and the one most commonly used. The Gaussian chain easily gives linear distances between polymer segments but does not provide orientation information. If one is to use any of these models to *predict* $\langle r_g \rangle$ rather than to use $\langle r_g \rangle$ as an input, some model for YI or ℓ_K must be developed. Usually, measurements at one condition are used to extrapolate to another condition with some simple argument, for example that YI or ℓ_K is a constant.

EXAMPLE PROBLEM 14.3

Assume that you have a DNA molecule with contour length 60 μm whose radius of gyration in an unknown solution is 1.1 μm. Using the bead–spring model, infer the appropriate Kuhn length required to model this DNA molecule in this solution.

1. If a DNA molecule with contour length 240 μm is suspended in this solution, what will its radius of gyration be?
2. From the experimentally observed diffusion data presented earlier in the chapter, one can infer that the radius of gyration is more accurately given by the relation $\langle r_g \rangle \propto \ell_c{}^{0.57}$. Using this relation, recalculate a second prediction of what the radius of gyration of the 240-μm DNA molecule would be.

SOLUTION: For a linear polymer with $\ell_c \gg \ell_K$,

$$\sqrt{6}\langle r_g \rangle = \sqrt{\ell_c \ell_K}\,. \tag{14.42}$$

For the specified values,

$$\ell_K = \frac{6\langle r_g \rangle^2}{\ell_c} = 121 \text{ nm}\,. \tag{14.43}$$

The radius of gyration of an $\ell_c = 240$ μm polymer is

$$\langle r_g \rangle = \sqrt{\frac{\ell_c \ell_K}{6}} = 2.2\ \mu\text{m}\,. \tag{14.44}$$

Using the relation $\langle r_g \rangle \propto \ell_c^{0.57}$, we can predict that

$$\langle r_g \rangle = 1.1\ \mu\text{m} \times \left(\frac{240\ \mu\text{m}}{60\ \mu\text{m}} \right)^{0.57}, \qquad (14.45)$$

which leads to

$$\langle r_g \rangle = 2.42\ \mu\text{m}. \qquad (14.46)$$

14.3.2 Dependence of transport properties on contour length

As previously discussed, DNA is free draining during electrophoresis, and μ_{EP} is not a function of the model used to describe the DNA configuration. However, because DNA is nondraining during diffusion, the models predict D by predicting $\langle r_g \rangle$. The preceding models all add a second DNA property (YI or ℓ_K) that combines with the model to predict DNA properties as a function of ℓ_c. If YI and ℓ_K are assumed independent of ℓ_c, the preceding models all predict that, for reasonably long strands of DNA, the radius of gyration $\langle r_g \rangle$ is proportional to $\sqrt{\ell_c}$, and thus the diffusivity of a DNA molecule is predicted by these models to be proportional to $\ell_c^{-\frac{1}{2}}$ or, equivalently, $N_{bp}^{-\frac{1}{2}}$. Thus we predict the diffusivity of DNA in bulk solution rather easily – given knowledge of ℓ_c and a reasonable approximation of $\langle r_g \rangle$ or ℓ_p or YI or ℓ_K, we can use the approximate Stokes–Einstein-type relation (14.10) to approximate the DNA diffusivity.

The previous scaling result can be compared with the experimental observation in relation (14.11), which shows a dependence proportional to $\ell_c^{-0.57}$ or $N_{bp}^{-0.57}$ and thus points out a limitation in the model prediction's accuracy. The key physical failure of the idealized models for DNA is that they do not account for the self-avoidance of the polymer owing to the electrostatic repulsion between parts of the polymer once they come within approximately λ_D of each other. This self-avoidance causes $\langle r_g \rangle$ to be slightly more strongly dependent on ℓ_c than it would be otherwise. Thus the models that assume YI or ℓ_K independent of ℓ_c capture most, but not all, of the structural features of the molecule.

The properties of idealized models are summarized in Table 14.3.

14.4 REAL POLYMER MODELS

The idealized models described in the previous section lead to a simple statistical description of the chain conformation, and a strictly *entropic* description of the response of the polymer to external forces. For engineering purposes, these idealized models are quite reasonable for predicting bulk DNA behavior. To be sure, the experimentally observed data indicate that DNA diffusivity is proportional to $\ell_c^{-0.57}$ whereas the idealized models predict that diffusivity is proportional to $\ell_c^{-\frac{1}{2}}$, but the idealized models nonetheless lead quickly to reasonable estimates of DNA behavior.

Unfortunately, idealized polymer models are much less successful at predicting the behavior of DNA confined to nanoscale channels. The idealization that different parts of the molecule can coexist in the same location ignores the *energetics* of DNA molecule conformation. Although this does not lead to enormous errors in bulk, this idealization leads to quite incorrect predictions in confining geometries.

Table 14.3 **Physical properties and length scales of DNA: relation to models.**

Model	Wormlike chain (Kratky–Porod)	Freely jointed chain (FJC)
Physics and input parameters	Rod with flexural rigidity YI	Rigid rods of length ℓ_K joined by linkages free to rotate around both axes
Strengths	Easy to calculate ℓ_p; predicts $\langle \ell_e \rangle$ reasonably for all ℓ_c, ℓ_p limits	Easy to calculate link–link distances if $\Delta s \gg \ell_K$
Weaknesses	Spacing between polymer components is awkward to compute; no self-avoidance	Does not predict ℓ_p correctly; no self-avoidance
ℓ_K	Not used	$\ell_K = \frac{6\langle r_g \rangle^2}{\ell_c}$
$\langle \ell_e \rangle$ in terms of $\langle r_g \rangle$	$\langle \ell_e \rangle = \frac{\langle r_g \rangle}{\sqrt{6}}$ if $\ell_c \gg \ell_p$; otherwise difficult to compute	$\langle \ell_e \rangle = \frac{\langle r_g \rangle}{\sqrt{6}}$ if $\ell_c \gg \ell_K$; otherwise difficult to compute
$\langle \ell_e \rangle$ in terms of model parameters	$\langle \ell_e \rangle = \sqrt{2\ell_p \ell_c - 2\ell_p^2 \left[1 - \exp\left(-\frac{\ell_c}{\ell_p}\right)\right]}$	$\langle \ell_e \rangle = \sqrt{\ell_c \ell_K}$ for $\ell_c \gg \ell_K$
$\langle r_g \rangle$ in terms of model parameters	$\langle r_g \rangle = \sqrt{\frac{\ell_c YI}{3k_B T}}$ for $\ell_c \gg \ell_p$	$\langle r_g \rangle = \sqrt{\ell_c \ell_K / 6}$ for $\ell_c \gg \ell_K$
ℓ_p	$\ell_p = YI/k_B T$	ℓ_p not defined
Model	**Freely rotating chain**	**Gaussian (bead–spring)**
Physics and input parameters	Rigid rods of length ℓ_K joined by linkages at subtended angle ϑ free to rotate around one axis	Rods with spring constant k and length ℓ_K joined by linkages free to rotate around both axes
Strengths	Same as FJC; slightly more physical	Same as FJC but statistics hold for all ℓ_c and all Δs; analytically the most tractable
Weaknesses	Does not predict ℓ_p correctly; no self-avoidance	Does not predict ℓ_p correctly; no self-avoidance
ℓ_K	$\ell_K = \frac{6\langle r_g \rangle^2}{\ell_c} \frac{1-\cos\vartheta}{1+\cos\vartheta}$	$\ell_K = \frac{6\langle r_g \rangle^2}{\ell_c}$
$\langle \ell_e \rangle$ in terms of $\langle r_g \rangle$	$\langle \ell_e \rangle = \frac{\langle r_g \rangle}{\sqrt{6}}$ if $\ell_c \gg \ell_K$; otherwise difficult to compute	$\langle \ell_e \rangle = \frac{\langle r_g \rangle}{\sqrt{6}}$ for all ℓ_c
$\langle \ell_e \rangle$ in terms of model parameters	$\langle \ell_e \rangle = \sqrt{\ell_c \ell_K \frac{1+\cos\vartheta}{1-\cos\vartheta}}$ for $\ell_c \gg \ell_K$	$\langle \ell_e \rangle = \sqrt{\ell_c \ell_K}$
$\langle r_g \rangle$ in terms of model parameters	$\langle r_g \rangle = \sqrt{\frac{\ell_c \ell_K}{6} \frac{1+\cos\vartheta}{1-\cos\vartheta}}$ for $\ell_c \gg \ell_K$	$\langle r_g \rangle = \sqrt{\ell_c \ell_K / 6}$
ℓ_p	$\ell_p = \frac{1}{2}\ell_K \frac{1+\cos\vartheta}{1-\cos\vartheta}$ for $\Delta s \gg \ell_K$	ℓ_p not defined

EFFECT OF SELF-AVOIDANCE

Self-avoidance is the property of a polymer of never having two parts of the polymer existing in exactly the same place at the same time. Although this property is self-evident, the effects of this property are not.

In the preceding modeling sections, we saw that an idealized polymer could be modeled in terms of bending energy or in terms of a random walk of nanoscopic ball joints. Either way, the modeling did not preclude the polymer folding back onto itself, and $\langle r_g \rangle$ was proportional to $\sqrt{\ell_c}$.

As already discussed, the power-law scaling of $\langle r_g \rangle$ with ℓ_c does not quite match experiments. We use the previous idealized models because they are mathematically simple and provide an intuitive understanding of DNA behavior, particularly bulk DNA behavior. If we are to match observation more closely, especially observation in confining geometries, we must address self-avoidance. We do this most often by modifying the picture of the jointed chain models previously described by using the mathematics of *self-avoiding walks*. We note several critical conclusions:

- $\mathcal{P}(\ell_e)$. The probability distribution for a self-avoiding polymer is particularly low for short end-to-end vectors – one end of a self-avoiding polymer is less likely to be close to the other end than an ideal polymer. This likelihood is proportional to $\ell_e^{7/3}$ for self-avoiding polymers instead of $\ell_e{}^2$. Further, the exponential term decays for large ℓ_e proportional to $\exp(-\ell_e^{5/2})$ instead of $\exp(-\ell_e^2)$ [176].
- $\langle r_g \rangle(\ell_c)$. The radius of gyration $\langle r_g \rangle$ and mean end-to-end length $\langle \ell_e \rangle$ of a self-avoiding polymer can be shown to be theoretically proportional to $\ell_c^{\frac{3}{5}}$ by assuming that, at equilibrium, the net volumetric electrostatic repulsion of the polymer from itself balances the compressive force from entropy (subsequently described) [176, 177]. Although more accurate than the idealized calculation, this model still has rather severe approximations.

EXAMPLE PROBLEM 14.4

Consider a Kratky–Porod molecule that has a radius of gyration of 4 μm when suspended in a 1-μM solution of monovalent salts. What will the molecule's radius of gyration be if salt is added to make the solution 100-mM NaCl?

SOLUTION: The Debye length of a 1-μM solution of monovalent salts at standard temperature is 304 nm. Using this as an approximation for ℓ_p for DNA in this solution, we can infer the contour length from Eq. (14.18) as

$$\ell_c = \frac{3\langle r_g \rangle^2}{\ell_p} = 158\ \mu\text{m}\,. \tag{14.47}$$

For a 100-mM solution, the Debye length is approximately 1 nm, and thus the persistence length is approximately 50 nm. The radius of gyration can then be given by

$$\langle r_g \rangle = \sqrt{\frac{\ell_p \ell_c}{3}}\,, \tag{14.48}$$

leading to

$$\boxed{\langle r_g \rangle = 1.6\ \mu\text{m}\,.} \tag{14.49}$$

EFFECT OF CHARGE AND THE EDL ON POLYMER PROPERTIES

The charge of the polymer backbone can influence the conformation of DNA, as observed physically in $\langle r_g \rangle$ and ℓ_c and as modeled theoretically by the model parameters ℓ_K and YI. The Debye length λ_D gives the characteristic length over which the potential induced by the charge decays, and thus an approximation of the spatial extent of electrostatic effects.

The flexibility of the DNA polymer is a function of λ_D. If λ_D is small, the electrostatic effects are shielded over small distances, and the ℓ_p of DNA is dictated by the conformation of the covalent and hydrogen bonds that make the double helix. In this case, the ℓ_p is approximately the 50 nm quoted as the "native" persistence length of DNA.

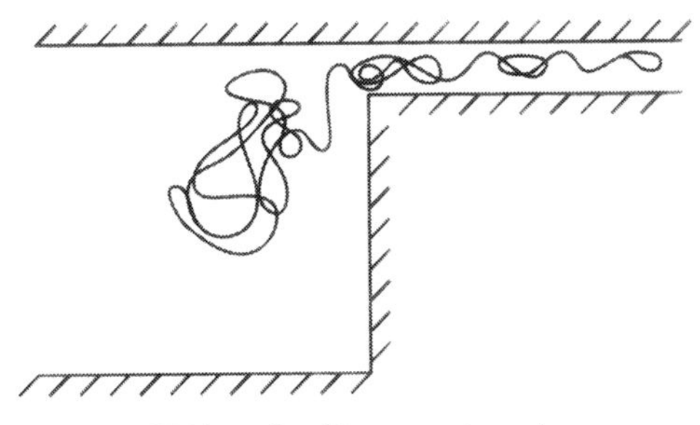

Fig. 14.10 DNA confined in a nanochannel.

However, if λ_D is large relative to the native ℓ_p (i.e., if, for DNA, $\lambda_D > 50$ nm), then the electrostatic repulsion dominates the flexibility rather than the covalent/hydrogen bonds, and the ℓ_p becomes approximately equal to λ_D. The resulting $\langle r_g \rangle$ and $\langle \ell_e \rangle$ are larger. If $\lambda_D \gg \ell_c$, then $\ell_p \gg \ell_c$ and $\ell_e \simeq \ell_c$, and the DNA is effectively a rigid rod.

The EDL becomes particularly important when excluded volume effects play a strong role. The diameter of the DNA molecule is approximately 2 nm, but the effective diameter is approximately equal to 2 nm plus $2\lambda_D$. As the electrolyte concentration goes down and the effective diameter increases, the volume taken up by the DNA tends to increase. This is even more important when the DNA molecule is confined to a nanoscale domain.

14.5 dsDNA IN CONFINING GEOMETRIES

By fabricating devices with micro- and nanoscale channels, we can confine molecules such as DNA in a controlled fashion. This is of critical importance because this enables observation of individual molecules (rather than groups of molecules) with well-defined boundary conditions. These detailed measurements thus complement measurements on ensembles of molecules made with more conventional techniques.

The models previously described for DNA in a bulk solution work well as long as the size of the channel is large relative to the radius of gyration of the molecule. When the channel size is smaller than the radius of gyration of the molecule, or when the ends of the DNA are controlled by attaching the ends of the DNA molecule to a wall or a particle whose location is controlled, the conformation of the molecule is affected by these external forces. Figure 14.10 shows DNA confined in a nanochannel, with the attendant change in the polymer configuration caused by the confining geometry.

The importance of DNA dynamics in confining geometries is threefold: first, geometries that are small relative to the molecule radius of gyration offer potential improvements in our ability to manipulate, sort, and separate DNA; second, studying a well-defined and well-known polymer in these geometries enables fundamental study of polymer physics; and finally, DNA is often in a confining geometry in living cells, and these geometries may have important implications for DNA's biological function.

14.5.1 Energy and entropy of controlled polymer extension

The ends of DNA molecules can be attached to surfaces or microbeads, and the end-to-end length can be controlled by use of external forces applied to these beads. Such an experiment highlights DNA's entropic spring constant. This spring constant is *not* equal to the spring constant of the individual molecular bonds nor to the spring constant used in a bead–spring model. Rather, this spring constant illustrates the entropic forces

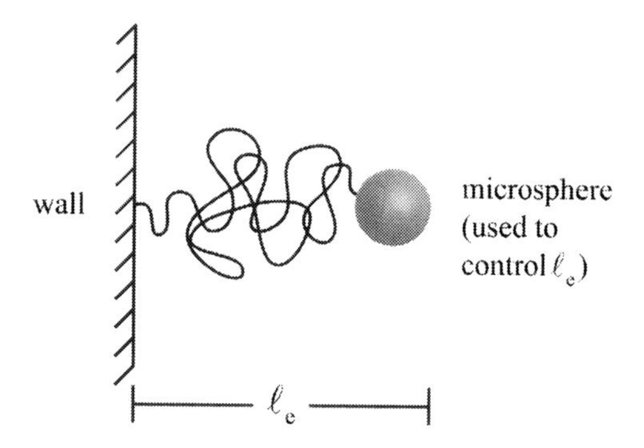

Fig. 14.11 A DNA molecule whose ends are controlled by attachment to a wall and manipulation of a microbead.

on the DNA molecule as a whole. Figure 14.11 shows a DNA molecule whose ends are controlled by attachment to a wall and manipulation of a microbead.

Consider the Gaussian bead–spring chain. From the distribution of ℓ_e, we can show by using the Boltzmann entropy relation that the instantaneous entropy of the DNA molecule is given by

$$\mathcal{S}(\ell_e) = \mathcal{S}(0) - \frac{3k_B \ell_e^2}{2\ell_c \ell_K}, \tag{14.50}$$

where $\mathcal{S}$ is the instantaneous entropy of the molecule. This relation shows how entropy is related to the end-to-end length of the molecule. If we pull a DNA molecule tight, so that $\ell_e = \ell_c$, then ℓ_e^2 is much greater than $\ell_c \ell_K$, and the second (negative) term is large in magnitude, making the entropy low. A DNA molecule pulled tight has only one degree of freedom, i.e., its alignment with respect to the coordinate axes, and thus it has low entropy. In contrast, a system with $\ell_e = 0$ has enormous freedom and thus high entropy. The high entropy state is also the most likely.

We can cast the entropy result in terms of force by calculating the conformational free energy and equating the force to the derivative of the free energy. First we write the conformational Helmholtz free energy, using $\Delta \mathcal{A} = \Delta \mathcal{U} - T\Delta \mathcal{S}$:

$$\mathcal{A}(\ell_e) = \mathcal{A}(0) - T\left[\mathcal{S} - \mathcal{S}(0)\right] = \mathcal{A}(0) + \frac{3k_B T \ell_e^2}{2\ell_c \ell_K}, \tag{14.51}$$

where the molecule internal energy $\mathcal{U}$ is independent of the conformation. Then we take the derivative of Eq. (14.51) with respect to ℓ_e to get the instantaneous restoring force F felt on the ends of the molecule:

entropic spring force for a bead–spring chain

$$F = -\frac{\partial \mathcal{A}}{\partial \ell_e} = -\frac{3k_B T \ell_e}{\ell_c \ell_K}. \tag{14.52}$$

The entropic force tends to pull the ends of the molecule together. For an ideal polymer, this conformational spring constant is strictly entropic and does not consider any electrostatic (i.e., energetic) interactions between parts of the molecule.

From this, we can treat the ends of the molecule as if they were connected by a spring with a restoring force $F = -k\ell_e$ defined by a spring constant k, which has a value

entropic spring constant for a bead–spring chain

$$k = \frac{3k_B T}{\ell_c \ell_K} . \tag{14.53}$$

The Kuhn length ℓ_K itself is a function of temperature and solution conditions. Equation (14.53) can also be written in general for any contour distance Δs as $k = 3k_B T/\Delta s \ell_K$, indicating that entropy tends to act as a spring pulling the components of the molecule together. This balances thermal fluctuations and self-avoidance, which disperse the parts of the polymer.

EXAMPLE PROBLEM 14.5

One end of an $\ell_c = 250$ μm DNA molecule is tethered to the wall of a microdevice and the other end is tethered to a 5-μm-diameter polystyrene bead with density $\rho_p = 1500$ kg/m^3. The system is at 300 K. A focused laser beam (an optical tweezer) is used to pull the polystyrene bead to a distance of 200 μm from the wall. The laser is turned off, and the system returns to equilibrium. Modeling the polymer as a Gaussian bead–spring system with Kuhn length $\ell_K = 100$ nm and ignoring Brownian motion of the bead, write the equations of motion for the bead.

SOLUTION: Defining x as the distance from the wall, the entropic spring force for the DNA molecule is given by $F = -\frac{3k_B T}{\ell_c \ell_K} x$, which, for the specified values, gives

$$F = -5.0 \times 10^{-10} \text{ kg/s}^2 \times x . \tag{14.54}$$

There is a hydrodynamic drag on the polymer, but it is safely assumed smaller than that on the particle, especially when the polymer is so fully extended. Thus the key hydrodynamic force is the Stokes drag on the particle, which can be written with the bulk Stokes flow relation because the distance from the wall is more than ten times the radius: $F = -6\pi\eta u a$. Note that the negative sign is used because u is the velocity of the particle and this relation is the drag force *on the particle* caused by the fluid. For water and a 2.5 μm radius particle, the result is given by

$$F = -4.7 \times 10^{-8} \text{ kg/s} \times \frac{dx}{dt} . \tag{14.55}$$

The acceleration of the bead is given by the summed force divided by the particle mass, which is

$$m = \frac{4}{3}\pi a^2 \rho_p = 9.8 \times 10^{-14} \text{ kg} . \tag{14.56}$$

With x and dx/dt written in SI units, the equation of motion for the bead is thus

$$\frac{d^2x}{dt^2} = -4.8 \times 10^5 \frac{dx}{dt} - 5.1 \times 10^3 x . \tag{14.57}$$

Note that the expression for the spring constant is valid for a Gaussian bead–spring chain of $\ell_c = 250$ μm, but this same expression would *not* be valid for a freely jointed chain. These equations of motion predict that the bead will accelerate to a velocity near -2 μm/s and move toward the wall. As the particle moves toward the wall, the velocity will decrease proportionally to the distance. As the particle gets close to the wall, the Stokes relation will break down. In the physical system, Brownian motion would also add random motion to the deterministic action of the polymer.

14.5.2 Energy and entropy of confinement for ideal polymers

The bead–spring molecule can describe the effects when a DNA molecule is confined. This is of particular interest when we are considering the motion of DNA in nanochannels, whose dimensions can be small compared with $\langle r_g \rangle$. We might confine DNA in nanochannels for a variety of reasons, including (1) validating DNA polymer dynamic models by observing DNA behavior in a controlled environment with well-known geometry; (2) physically separating DNA molecules based on contour-length-dependent transport variations observed at this length scale; or (3) extending the DNA molecule to sequence it or otherwise characterize it chemically. These relations are not derived directly here, but the results are presented. The properties of a bead–spring chain that is confined in an idealized rectangular box with dimensions L_x, L_y, and L_z follow from the change in the partition function and therefore entropy that a bead–spring chain has when confined.

For a bead–spring molecule confined in this box, the partition function in a dimension i (where $i = x, y, z$) is given by

$$Z_i = \frac{8}{\pi^2} L_i \sum_{p=0}^{\infty} \frac{1}{(2p+1)^2} \exp\left[-\frac{\pi^2 \langle r_g \rangle^2 (2p+1)^2}{L_i{}^2} \right], \tag{14.58}$$

where $\langle r_g \rangle$ is the mean radius of gyration observed for the polymer *in the bulk*. The partition function of the system is given by

$$Z = Z_x Z_y Z_z, \tag{14.59}$$

and the total entropy is given by

$$S = k_B \ln Z = k_B \ln Z_x + k_B \ln Z_y + k_B \ln Z_z. \tag{14.60}$$

We evaluate the partition function (and therefore entropy) in two limits. First, for the case in which $\langle r_g \rangle \ll L_i$, i.e., the case in which the polymer is not confined by its surroundings, we get $Z_i = L_i$. For the case in which $\langle r_g \rangle \gg L_i$, i.e., the case in which the polymer is greatly confined, we get

$$Z_i = \frac{8}{\pi^2} L_i \exp\left[-\frac{\pi^2 \langle r_g \rangle^2}{L_i{}^2} \right]. \tag{14.61}$$

From this, we can show that the difference $\Delta \mathcal{S}$ in entropy caused by finite $\langle r_g \rangle$ effects in one dimension is negative and given by

entropy decrease owing to confinement in one dimension, ideal polymer

$$\Delta \mathcal{S} = -k_B \frac{\pi^2 \langle r_g \rangle^2}{L_i^2} . \quad (14.62)$$

The change $\Delta \mathcal{A}$ in Helmholtz free energy caused by finite $\langle r_g \rangle$ effects is positive and given by

excess Helmholtz free energy owing to confinement in one dimension, ideal polymer

$$\Delta \mathcal{A} = k_B T \frac{\pi^2 \langle r_g \rangle^2}{L_i^2} , \quad (14.63)$$

It has become the norm in the polymer physics literature to refer to systems with only one degree of confinement (i.e., channels with nanoscale depth but large width) as *nanoslits*, whereas systems with two degrees of confinement (i.e., nanoscale depth *and* width) are referred to as *nanochannels*. Many other communities do not make this distinction.

In nanofluidic systems, we might typically confine DNA in a long, narrow nanochannel. If we assume that the channel length is much larger than $\langle r_g \rangle$ but the channel cross section is square with width and depth d, the DNA is confined in two dimensions and the free energy change is

excess Helmholtz free energy owing to confinement in two dimensions, ideal polymer

$$\Delta \mathcal{A} = 2 k_B T \frac{\pi^2 \langle r_g \rangle^2}{d^2} . \quad (14.64)$$

Thus the force required to confine DNA in the long narrow channel can be determined from free-energy considerations. This force causes DNA to pool in large reservoirs rather than remain in a confined area and is central to a number of separation techniques, e.g., [178], which are discussed in Chapter 15.

The preceding relations focus strictly on idealized bead–spring polymer chains. The work presented here can be extended to address self-avoidance and volume exclusion, with improved accuracy. In particular, idealized models do not predict the experimental observation that confinement in one direction causes expansion in the other directions, but self-avoiding models do.

14.5.3 DNA transport in confined geometries

Confinement changes the configuration of DNA, as previously described, leading to changes also in DNA's transport properties. Most notably, DNA confined to geometries smaller than the radius of gyration is no longer hydrodynamically coupled in a roughly spherical ball of a diameter given by the radius of gyration – rather, the molecule is extended into a nonspherical conformation. In this case, the viscous mobility of this extended configuration is inversely proportional to ℓ_c. In this case, the diffusivity of the DNA is inversely proportional to the contour length ($D \propto \ell_c^{-1}$) [170, 171]. The dependence of the diffusivity on the height h of a nanoslit is currently under dispute – different theories (e.g., blob models, reflecting rod models) give different results. Experiments observe dependences ranging from $D \propto h^{\frac{1}{2}}$ to $D \propto h^{\frac{2}{3}}$. Because DNA is free draining during electrophoresis, the role of confining geometry on DNA electrophoretic mobility is related to EDL overlap and can be described by similar equations.

14.6 DNA ANALYSIS TECHNIQUES

As mentioned earlier, DNA is of enormous biological importance, and its analysis is central to many biological studies. These analytical techniques are often implemented in microfluidic chips, so we briefly discuss a number of DNA analysis techniques in the following subsections.

14.6.1 DNA amplification

DNA has the unique property that it can be amplified by any of several amplification techniques, most notably the polymerase chain reaction (PCR). PCR amplification requires (1) adding *primers*, i.e., DNA strands of known sequence that are designed to initiate replication of a specific DNA sequence; (2) adding other chemicals, such as polymerase and oligonucleotides; and (3) thermally cycling the DNA sample. When this is done properly, a single DNA strand can be amplified into millions of DNA strands, making detection rather simple. From a microfluidic standpoint, PCR requires that microchips meter reagents and cycle temperature in a controlled manner.

14.6.2 DNA separation

Macroscopically, DNA is typically separated by size by use of electrophoresis in agarose or polyacrylamide slab gels. This involves inserting a DNA sample (as well as a set of DNA of known lengths) onto the gel, applying an electric field, and staining the resulting separated DNA bands. Microscopically, the separation function of these gels is to force the DNA to follow a tortuous path that creates molecular elongation and causes the longer DNA to travel more slowly (see [179, 180]). Because the electrophoretic mobility of DNA is insensitive to contour length in bulk solution owing to the free-draining nature of DNA in bulk, the gel facilitates the separation by causing the elution time to be dependent on molecular elongation and confinement. In microfluidic devices, separations are performed on DNA in gels much as they are in bulk for any other set of molecules with distinguishable electrophoretic velocities. An example of a microdevice for PCR amplification and DNA separation is shown in Fig. 14.12. Nanofluidic devices [178, 181, 182] have been used as well. Sequencing requires different chemical reactions, but still involves electrophoretic separations on reaction products (see [183]).

SANGER SEQUENCING

Sanger sequencing of DNA uses (1) DNA replication with DNA polymerase and fluorescently labeled nucleotides combined with (2) electrophoretic separation to infer the base sequence of the DNA. Many copies of the DNA to be sequenced are incubated with an oligomeric primer, DNA polymerase, and free nucleotides, both fluorescently labeled (a small fraction) and unlabeled (the majority of the nucleotides). Four colors are used,

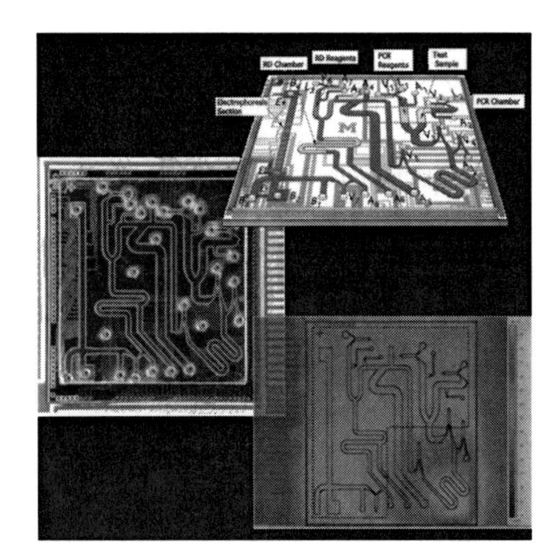

Fig. 14.12 A microchip for PCR amplification and DNA separation. (From Mark Burns's lab, University of Michigan.)

one each for each type of nucleotide. The fluorescently labeled nucleotides, owing to their dideoxy configuration, terminate the polymerization process. Incubation, in this case, creates copies of DNA with a variety of lengths and one of four fluorophores – for each molecule, polymerization occurs until the fluorophore is attached, at which point polymerization ceases. Each fluorophore denotes what was the terminal base on the copy. For example, if a DNA oligomer of sequence ACTGATT is sequenced in the presence of unlabeled nucleotides A, C, T, G, and fluorescently labeled nucleotides A*, C§, T†, and G‡, the possible oligomers are A*, AC§, ACT†, ACTG‡, ACTGA*, ACTGAT†, and ACTGATT†. These seven oligomers can then be separated electrophoretically in a gel, with the shortest oligomer eluting earliest and the longest eluting latest. By monitoring the colors of the eluted peaks, we can infer the sequence of the DNA. Obviously, the speed and resolution with which the DNA molecules can be separated are closely related to the ability of Sanger sequencing to sequence DNA molecules. Figure 14.13 shows the steps in Sanger sequencing schematically.

14.6.3 DNA microarrays

DNA microarrays are essentially flat surfaces that have been functionalized with a large array (> 1000) of complementary DNA (cDNA) strands that correspond to different sections of DNA or RNA from an organism. By incubating a microarray with a solution from a biological system and an intercalating dye, fluorescent signal on each spot indicates how much DNA or RNA was present in the sample. Figure 14.14 shows a schematic of the process. From a fluid mechanical standpoint, DNA microarrays are a challenge because the diffusion time scale for transport and hybridization in a microarray can be quite large. Advanced low-Re chaotic mixing systems have been applied to increase hybridization efficiency [47, 48, 49, 50, 51].

14.7 SUMMARY

This chapter describes the physicochemical structure of DNA, its transport properties, and its physical models, all with specific application to aqueous solutions of DNA in

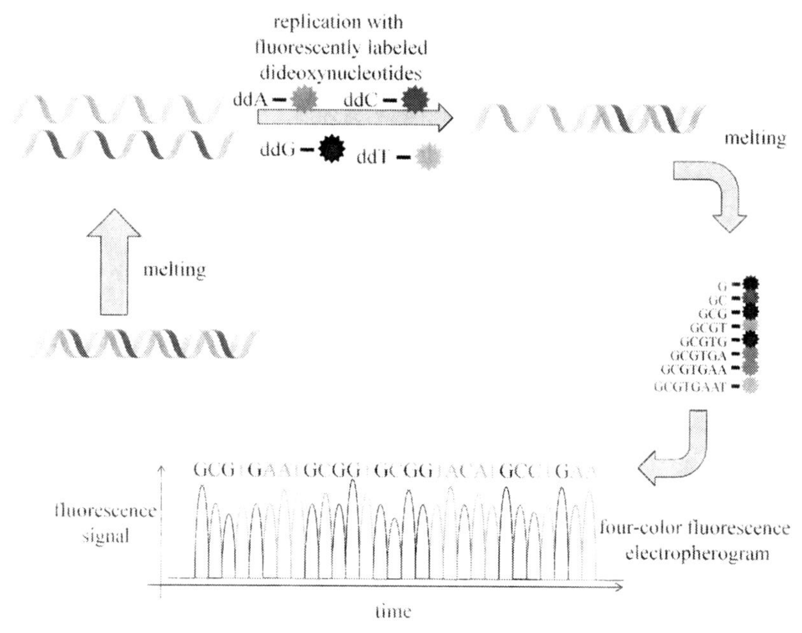

Fig. 14.13 Sanger sequencing. Following melting of DNA, PCR replication is performed with a small fraction of fluorescent dideoxynucleotides, which terminate the polymerization process. Following replication, the strands, each of which terminates with a fluorophore, are separated and the readout gives the sequence.

microfluidic devices. A variety of models are shown to be able to explain DNA conformation and provide predictive scaling for both the conformation and the resulting transport properties, especially diffusion. Much of the DNA analysis is performed in gels, but analysis of DNA transport through gels is omitted in this chapter. However, DNA

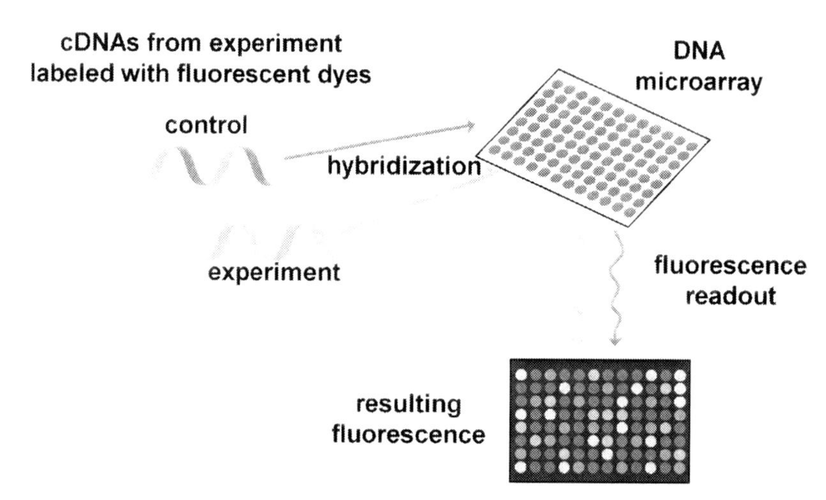

Fig. 14.14 A DNA microarray used for a comparative study of gene expression. Complementary DNA from two sources (typically a control source and an unknown source) are labeled with two different dyes, hybridized to DNA strands immobilized in spots on the microarray, and quantified by reading out the fluorescence from the spots. A key issue for operation of these microarrays is the time required for all DNA from the sample to explore all of the spots on the microarray, an issue discussed in Chapter 4.

properties in bulk solution are discussed, leading to empirical relations for the diffusivity of DNA,

$$D \simeq 3 \times 10^{-12}\ \mathrm{m^2/s} \left(\frac{\ell_c}{1\ \mu\mathrm{m}} \right)^{-0.57}, \tag{14.65}$$

and for the electrophoretic mobility,

$$\mu_{EP} \simeq 4 \times 10^{-8}\ \mathrm{m^2/V\,s}. \tag{14.66}$$

Ideal polymer models combined with the nondraining or Zimm dynamics assumption lead to the approximation that D should be proportional to $\ell_c^{-\frac{1}{2}}$, whereas self-avoiding polymer models lead to the approximation that D should be proportional to $\ell_c^{-\frac{3}{5}}$. The free-draining or Rouse dynamics assumption lead to the approximation that μ_{EP} should be independent of ℓ_c. The effects of confinement of ideal polymers in nanochannels is discussed, both in terms of thermodynamic predictions of DNA response to confinement as well as reference to recent research on DNA dynamics in nanochannels and nanoslits.

14.8 SUPPLEMENTARY READING

Several excellent texts on polymer dynamics, including those by Flory [177], Doi [184], Doi and Edwards [90], and deGennes [176], provide a useful background. Doi and Edwards [90], in particular, provide relations for polymer mobility matrices and partition functions, facilitating more detailed analysis of several of the concepts presented in this chapter. The classical models for linear polymers are treated in all of these texts, albeit with a different approach compared with this chapter; readers will benefit from experiencing both approaches. Rubinstein and Colby [36] provide an accessible introduction with derivations of many relations and descriptions of experimental techniques for measuring $\langle r_g \rangle$. The physics of random-walk processes is largely ignored in this chapter but is fundamental to most descriptions of polymer properties and dynamics; Refs. [33, 36] discuss these processes.

Because countless bioanalytical techniques (including Sanger sequencing, as described in Subsection 14.6.2) use DNA separations, separations require DNA transport properties that are dependent on intrinsic DNA properties. DNA diffusivity varies with DNA contour length, but that is generally a difficult property to use for separation. Electrophoretic mobility, on the other hand, is relatively straightforward to use for separations (see Chapter 12) but varies with DNA length only for small DNA molecules. Consequently, bulk electrophoretic separations for DNA are rare.

Because of the relative insensitivity of *bulk* electrophoretic mobility to contour length, DNA separations on molecules with more than approximately 200 base pairs are instead performed in gels made of agarose or polyacrylamide. The molecular structure of the gel forces the DNA to travel through the interstitial spaces between gel molecules, resulting in an electrophoretic mobility that is dependent on DNA size.

Gel electrophoresis separation of DNA is a workhorse technique that has been widely successful experimentally. Analytically, the process of predicting DNA mobility in gels has been a difficult one, owing to the confluence of (a) the irregular structure inherent in gels, (b) complex and interdependent physical phenomena, and (c) the dearth of experimental data resolving DNA conformation and transport with base pair resolution. A thorough review of the work and modeling in this area is provided in [179]. Because the dependence of the electrophoretic mobility on DNA contour length is complicated, most experiments are performed with simultaneous calibration samples with known contour lengths.

The recent growth of nanofluidic devices and their impact on the study of DNA physics has led to an explosion of work on the physics of DNA, a small subset of which can be found at [171, 185, 186, 187, 188, 189].

DNA separation in micro- and nanofluidic structures and gels has been summarized in a number of reviews, for example from a polymer dynamics point of view in [179] and from a system integration standpoint in [180]. A number of nanofluidic DNA separation and analysis applications are discussed in Edel and de Mello's text [190], and some examples of note include [178, 181, 182].

DNA's biochemical properties are discussed in [191].

14.9 EXERCISES

14.1 Draw cartoons of the shape of the following:

(a) a 3-kbp strand of dsDNA in water.

(b) a 30-bp strand of dsDNA in water.

14.2 Predict approximately the relative change in $\langle r_g \rangle$ and D of a DNA molecule in bulk solution if the monovalent electrolyte solution is changed from 1 M to 1×10^{-5} M.

14.3 Presume that you have a strand of DNA with a well-defined persistence length ℓ_p. Consider the mean end-to-end length $\langle \ell_e \rangle$, which can be written as

$$\langle \ell_e \rangle^2 = \left\langle \left| \vec{r}(s = \ell_c) - \vec{r}(s = 0) \right|^2 \right\rangle . \tag{14.67}$$

Thus $\langle \ell_e \rangle$ can be found by integrating the angles:

$$\langle \ell_e \rangle^2 = \int_{s_1=0}^{\ell_c} ds_1 \int_{s_2=0}^{\ell_c} ds_2 \left\langle \frac{\partial \vec{r}(s_1)}{\partial s} \cdot \frac{\partial \vec{r}(s_2)}{\partial s} \right\rangle . \tag{14.68}$$

Evaluate the mean end-to-end length $\langle \ell_e \rangle$ by evaluating this integral for a DNA polymer with a well-defined ℓ_p. Simplify your result for $\langle \ell_e \rangle$ in the limit where $\ell_p \ll \ell_c$ (a long flexible polymer) and also in the limit where $\ell_p \gg \ell_c$ (a short, rigid rod). What value of ℓ_K must be used in an ideal polymer model to make the end-to-end length predicted by the ideal polymer model match your result for $\ell_p \ll \ell_c$?

14.4 Plot $\langle \ell_e \rangle$ as a function of N_{bp} for $YI = 2 \times 10^{-28}$ N m^2. Use the range $N_{bp} = 100$ to $N_{bp} = 10{,}000$ and plot on a log–log scale. Compare this result with the equivalent plot for a Gaussian chain polymer with $\ell_K = 100$ nm.

14.5 In the bead–spring model of DNA, the probability density function for ℓ_e is given by Eq. (14.39):

$$\mathcal{P}(\ell_e) = 4\pi {\ell_e}^2 \left[\frac{1}{4\pi\langle r_g\rangle^2}\right]^{\frac{3}{2}} \exp\left(-\frac{{\ell_e}^2}{4\langle r_g\rangle^2}\right). \tag{14.69}$$

Evaluate the mean value of ℓ_e by integrating $\mathcal{P}(\ell_e)\ell_e$ from $\ell_e = 0$ to $\ell_e = \infty$. Show that the result corresponds to Eq. (14.38):

$$\langle \ell_e \rangle = \sqrt{6}\langle r_g\rangle. \tag{14.70}$$

14.6 In the bead–spring model of DNA, the probability density function for Δr is given by Eq. (14.41):

$$\mathcal{P}(\Delta r) = 4\pi \Delta r^2 \left[\frac{\ell_c}{4\pi\Delta s\langle r_g\rangle^2}\right]^{\frac{3}{2}} \exp\left(-\frac{\ell_c \Delta r^2}{4\Delta s\langle r_g\rangle^2}\right). \tag{14.71}$$

Evaluate the mean value of Δr by integrating $\mathcal{P}(\Delta r)\Delta r$ from $\Delta r = 0$ to $\Delta r = \infty$. Show that the result corresponds to Eq. (14.40):

$$\langle \Delta r \rangle = \sqrt{\frac{6\Delta s}{\ell_c}}\langle r_g\rangle. \tag{14.72}$$

14.7 Confirm the correctness of the result for the mean value of Δr for a Gaussian bead–spring chain by showing that the mean-square displacement of polymer segments is equal to $\langle r_g\rangle$. That is, show that

$$\frac{1}{{\ell_c}^2}\int_{s_2=0}^{s_2=\ell_c}\int_{s_1=0}^{s_1=s_2}\left\langle \left|\vec{r}(s_2) - \vec{r}(s_1)\right|^2\right\rangle ds_1\, ds_2 = \langle r_g\rangle. \tag{14.73}$$

A general conclusion stemming from this sort of analysis is that $\langle r_g\rangle = \frac{\langle \ell_e\rangle}{\sqrt{6}}$ for any linear polymer whose probability density function for Δr is Gaussian. This relationship breaks down, though, for short DNA molecules that are not Gaussian but rather well approximated by a rigid rod.

14.8 Evaluate $\langle r_g\rangle$ for a polymer that is well approximated by an infinitely rigid linear rod of length ℓ_c (i.e.,$\langle \ell_e\rangle = \ell_c$), and show that $\langle r_g\rangle \neq \frac{\langle \ell_e\rangle}{\sqrt{6}}$ in this case. Given the same $\langle \ell_e\rangle$, does a rigid rod have a larger or smaller $\langle r_g\rangle$ compared with a freely jointed chain? How about for the same ℓ_c?

14.9 Given that the partition function in a dimension i (where $i = x, y, z$) is given by

$$Z_i = \frac{8}{\pi^2}L_i \sum_{p=0}^{\infty} \frac{1}{(2p+1)^2} \exp\left[-\frac{\pi^2\langle r_g\rangle^2(2p+1)^2}{{L_i}^2}\right], \tag{14.74}$$

derive the partition function in the limit where $\langle r_g\rangle \ll L_i$.

14.10 Given that the partition function in a dimension i (where $i = x, y, z$) is given by

$$Z_i = \frac{8}{\pi^2}L_i \sum_{p=0}^{\infty} \frac{1}{(2p+1)^2} \exp\left[-\frac{\pi^2\langle r_g\rangle^2(2p+1)^2}{{L_i}^2}\right], \tag{14.75}$$

derive the partition function in the limit where $\langle r_g\rangle \gg L_i$.

14.11 Assume that a linear polymer is confined to and uniformly fills a spherical domain of radius R. Calculate $\langle r_g\rangle$ for this polymer configuration.

14.12 Consider a Gaussian bead–spring model. Consider the point on the polymer that is a fraction α of the distance along the polymer backbone. Show that the mean-square distance between this point and the center of mass of the polymer is given by

$$\frac{\ell_c \ell_K{}^2}{3}\left[1 - 3\alpha(1-\alpha)\right] . \tag{14.76}$$

14.13 Consider a Gaussian bead–spring model of a linear polymer and assume that the two ends are fixed at a distance ℓ_e, but the polymer is otherwise free to equilibrate. Consider the component of the mean-square radius of gyration in the direction along the line connecting the two ends of the polymer. Show that this component of the radius of gyration is given by

$$\frac{1}{36}\ell_c \ell_K \left(1 + 3\frac{\ell_e^2}{\ell_c \ell_K}\right) . \tag{14.77}$$

14.14 Derive the persistence length of a freely rotating chain.

14.15 Consider a Gaussian bead–spring chain model of a linear polymer in water. Assume that the chain has a positive charge $+q$ on one end and a negative charge $-q$ at the other end. Assume that an electric field of 100 V/cm is applied to the polymer at room temperature. If the polymer has $\ell_c = 5\ \mu$m and $\ell_K = 20$ nm, what will $\langle \ell_e \rangle$ be for this polymer in this field? At this distance, can the Coulomb interaction between the two charges be ignored?

14.16 Consider a 50-kbp DNA molecule with a contour length of 22 μm and a radius of gyration (at room temperature) of 0.75 μm that is forced into a long channel with square cross section. Model this DNA molecule as an ideal bead–spring chain. How small must the channel depth/width d be for the free energy of confinement to be 10 $k_B T$?

14.17 The expression for the probability density function for ℓ_e for the freely jointed chain and the freely rotating chain both allow for the possibility that ℓ_e will exceed ℓ_c. Explain why this apparently contradictory situation exists.

14.18 Consider an ideal freely jointed chain. If the ends of the chain are separated from each other with a force F, the resulting mean end-to-end length $\langle \ell_e \rangle$ is given by

$$\langle \ell_e \rangle = \frac{F \ell_c \ell_K}{3 k_B T} , \tag{14.78}$$

if $\langle \ell_e \rangle$ is small relative to ℓ_c. If $\langle \ell_e \rangle$ cannot be assumed small relative to ℓ_c, a more precise relation is

$$\langle \ell_e \rangle = \ell_c \left[\coth\left(\frac{F \ell_K}{k_B T}\right) - \frac{k_B T}{F \ell_K}\right] , \tag{14.79}$$

where $\coth(x) - \frac{1}{x}$ is referred to as the *Langevin function*. Show that $\lim_{\frac{F\ell_K}{k_B T} \to 0} \langle \ell_e \rangle = \frac{F \ell_c \ell_K}{3 k_B T}$. Plot both relations.

14.19 Derive the entropy of a Gaussian chain polymer as a function of ℓ_e.

14.20 Develop a Rouse (i.e., free-draining) model for DNA diffusion. Model the DNA as an ideal Gaussian chain of beads linked by springs. Assume that the viscous force on the springs is zero, but assume that each bead feels a force described by the Stokes flow relation for drag on an isolated sphere in an infinite medium.

(a) Assume all components of the polymer are moving with velocity U and write the force on the DNA molecule as a function of ℓ_c, ℓ_K, and the bead radius a.

(b) Note that the viscous mobility μ relates F and U by $U = \mu F$. Write μ for the DNA molecule.

(c) Using Einstein's relation $D = \mu k_B T$, write D for DNA using the Rouse model.

(d) Comment on the dependence of D on ℓ_c. How does this fail to match experimental observations for DNA in aqueous solution? What is the physical inaccuracy of the Rouse model for bulk diffusion of DNA?

15 Nanofluidics: Fluid and Current Flow in Molecular-Scale and Thick-EDL Systems

To this point, we have considered flow in channels whose dimension was large relative to the Debye length or the size of any molecules or particles suspended in the flow. When we use channels with shallow (e.g., nanoscale) depths d, we cannot separate the EDL from the bulk fluid by using boundary-layer theory; instead, we must account for the presence of net charge density in the bulk flow field. Even if the double layers remain thin, the perturbative effects of double layers (for example, surface conductance) are more significant as the channel becomes small. Because these phenomena typically coincide with transport through nanoscale channels, the term *nanofluidics* is often used to refer to flows with small d^* or flows with molecules or particles comparable to the size of the channel, though the scale need not be nanoscopic for these phenomena to be important, and these phenomena are unimportant in some nanoscale flows. Despite this, some authors use the term *nanofluidics* to refer specifically to flows in nanoscale channels with no reference to molecular size or λ_D. Because our interest is the interplay of electrokinetic effects with channels and molecular-scale confinement, our focus is on channels with *molecular scale* or of a *size comparable to* λ_D, and we pay only cursory attention to the absolute dimension of the channel.

For unidirectional flow in infinitely long, uniform-cross-section channels, thick-EDL effects are observed primarily through changes in the elements of the electrokinetic coupling matrix. In this case, the absence of gradients in the direction of flow keeps the system analytically straightforward, but system properties normal to the wall must be integrated to determine a cross-section-averaged channel property. These systems exhibit only one-way coupling (the double layer affects the flow but not the other way around) because the $\vec{u} \cdot \nabla c_i$ terms are zero. Real engineering systems have added complexity because they include cross-sectional variations and/or interfaces between regions with different dimensions. In these cases, the flow and the double layer exhibit two-way coupling, meaning that the EDL affects the bulk fluid flow through the $\rho_E \vec{E}$ source term and the bulk fluid flow affects the EDL through nonzero $\vec{u} \cdot \nabla c_i$ terms. This leads to phenomena such as concentration polarization and ion-current rectification. Interfaces also provide obstacles for macromolecules such as DNA and proteins, and the transport properties of these macromolecules are dependent on the geometry.

15.1 UNIDIRECTIONAL TRANSPORT IN INFINITELY LONG NANOCHANNELS

For channels whose dimension is not large relative to λ_D, the electroneutral "bulk" is no longer clearly defined. This phenomenon is commonly referred to as *double-layer overlap*. However, if we consider unidirectional transport in infinitely long channels, the EDL is still described by equilibrium relations, because the gradients are normal to the flow direction.

15.1.1 Fluid transport

In the unidirectional case, the Poisson–Boltzmann description,

Poisson–Boltzmann equation, dilute solution limit

$$\nabla^2 \varphi = -\frac{F}{\varepsilon} \sum_i c_{i,\infty} z_i \exp\left(-\frac{z_i F \varphi}{RT}\right), \tag{15.1}$$

applies, and, for uniform properties and a simple interface, the velocity is given by

$$u = \frac{\varepsilon E_{\text{ext,wall}}}{\eta}(\varphi - \varphi_0). \tag{15.2}$$

We use these solutions in the detailed descriptions in the following subsections. For channels that are small relative to λ_D, equations such as Eq. (15.2) still apply, but the definition of φ becomes more complicated. Since the channel is small relative to λ_D, there is no longer a well-defined bulk where $\varphi = 0$. In these cases, φ is measured relative to a hypothetical location where the fluid is electroneutral and the potential energy of an ion is independent of its valence.

The electroosmotic flow for narrow wall spacings can be determined by applying the boundary conditions that $\varphi^* = \varphi_0^*$ at the wall and $\frac{\partial \varphi}{\partial y} = 0$ at the axis of symmetry. This leads to the flow distributions shown in Fig. 9.6 and represented in Eq. (9.32). Compared with the thin-EDL limit, for which the flow is approximately uniform, electroosmotic flow when $\lambda_D \simeq d$ is nonuniform and therefore dispersive – because the velocity is a function of y throughout the flow field, a sample bolus will be dispersed by the flow. Further, the total flow rate is reduced because the potential difference between the wall and the bulk fluid is reduced.

15.1.2 Electrokinetic coupling matrix for thick-EDL transport

Unidirectional flow through long, narrow channels can be analyzed with a cross-sectional-area-averaged solution of the transport equations, which motivates the use of the electrokinetic coupling matrix. This analysis is valid when the nondimensional quantity $Pe\frac{d}{L}$ is small for ions, where the velocity U used in the Peclet number is the characteristic velocity of ion motion through the channel, d is a measure of the channel depth, and L is a measure of the channel length. This criterion ensures that the ion distribution normal to the direction of flow is governed by electrochemical equilibrium. Because d/L is routinely as small as 1×10^{-5} in nanochannels, this criterion is satisfied for essentially all ions in nanochannels. The simplification created by the use of cross-sectional-area-averaged properties is enormous, and the resulting errors are negligible for channels of uniform cross section. With this approach, we can compare electrokinetic coupling phenomena in small channels both in terms of the absolute channel size (i.e., in terms of d) and in terms of the relative thickness of the channel with respect to the double layer (i.e., in terms of $d^* = d/\lambda_D$).

Recall that we can write the electrokinetic coupling matrix as

$$\chi = \begin{bmatrix} \chi_{11} & \chi_{12} \\ \chi_{21} & \chi_{22} \end{bmatrix}, \tag{15.3}$$

such that the electrokinetic coupling equation becomes

$$\begin{bmatrix} Q/A \\ I/A \end{bmatrix} = \chi \begin{bmatrix} -\frac{dp}{dx} \\ E \end{bmatrix} . \tag{15.4}$$

For *vanishingly thin* double layers and small interfacial potentials, the electrokinetic coupling matrix was given by

$$\chi = \begin{bmatrix} r_{\mathrm{h}}^2/8\eta & -\varepsilon\varphi_0/\eta \\ -\varepsilon\varphi_0/\eta & \sigma \end{bmatrix} , \tag{15.5}$$

where $r_{\mathrm{h}} = R$ for a circular channel of radius R and $r_{\mathrm{h}} = 2d$ for infinite parallel plates separated by $2d$. For vanishingly thin λ_{D}, we assume $d^* \to \infty$, the system is independent of λ_{D}, and the only size dependence is on d through the hydraulic radius dependence on χ_{11}. For λ_{D} small relative to d but not vanishingly small, we show in Eq. (9.36) that the convective surface conductivity in the Debye–Hückel limit is given by the first term in the perturbation expansion for small λ_{D}/d:

$$\chi = \begin{bmatrix} r_{\mathrm{h}}^2/8\eta & -\varepsilon\varphi_0/\eta \\ -\varepsilon\varphi_0/\eta & \sigma + \frac{\varepsilon^2\varphi_0^2/\lambda_{\mathrm{D}}^2}{\eta r_{\mathrm{h}}/\lambda_{\mathrm{D}}} \end{bmatrix} . \tag{15.6}$$

Here, $\varepsilon^2\varphi_0^2/\lambda_{\mathrm{D}}^2$ is equal to the square of the surface charge density, and the surface conductivity is thus proportional to the squared surface charge density normalized by the viscosity and the ratio of the hydraulic radius to the Debye length. Taking the limit as $\lambda_{\mathrm{D}} \to 0$ at constant $\varepsilon\varphi_0/\lambda_{\mathrm{D}}$ recovers Eq. (15.5).

Equation (15.6) is derived using the assumption that the double layer is given by the semi-infinite solution and can be integrated to infinity. As the channel size d is reduced in comparison to λ_{D} and d^* can no longer be assumed large, the integral over finite domain must be employed, and several of the elements of this matrix take on a different functional form. This matrix analysis highlights the effects of double-layer overlap on fluid and ion transport. We can parameterize these changes in the electrokinetic coupling matrix by writing it as

$$\chi = \begin{bmatrix} c_{11} r_{\mathrm{h}}^2/8\eta & -c_{12}\varepsilon\varphi_0/\eta \\ -c_{21}\varepsilon\varphi_0/\eta & c_{22}\sigma \end{bmatrix} . \tag{15.7}$$

In this form, the coefficients c_{11}, c_{12}, c_{21}, and c_{22} illustrate the *ratio* of system response in a thick-EDL system to that in a system with vanishingly small λ_{D}.

ELECTROKINETIC COUPLING COEFFICIENTS FOR THICK DOUBLE LAYERS

The elements of the electrokinetic coupling matrix vary as a function of the characteristic size of a channel. The hydraulic coupling coefficient (or hydraulic conductivity) $\chi_{11} = r_{\mathrm{h}}^2/8\eta$ retains its functional dependence on r_{h} as long as viscosity is assumed independent of the local ion concentration or electric field, and thus χ_{11} is independent of λ_{D}. Because r_{h} is linearly proportional to the channel size, χ_{11} decreases as channel size is decreased. The electrical conductivity χ_{22} increases with increasing φ_0^* and decreasing d^*, owing both to increased convective surface current and increased ohmic conductivity in the double layer. The effective electroosmotic mobility χ_{12} increases with increasing φ_0, but it decreases with decreasing d^*, as double-layer overlap leads to a reduction in the total flow. The effective streaming current coefficient χ_{21} increases as d^* decreases, because more of the net charge density is in regions of relatively fast fluid flow. When the wall potential is large *and* double layers are thick, the values of χ_{12} and χ_{21} are unequal; thus large-potential, thick-EDL systems do not exhibit Onsager reciprocity.

In the dilute solution limit, we can calculate the electroosmotic correction coefficient c_{12} by solving the 1D nonlinear Poisson–Boltzmann equation and the 1D Navier–Stokes equation and integrating:

$$c_{12} = -\frac{\eta}{\varepsilon \varphi_0} \frac{1}{A} \int_S \frac{u}{E} \, dA. \tag{15.8}$$

We require approximate algebraic relations for c_{12} for common systems, for example, for an electrolyte between two plates at $y^* = \pm d^*$, with uniform fluid properties. In the Debye–Hückel limit, this can be determined analytically:

electroosmotic correction coefficient, Debye–Hückel approximation, uniform fluid properties, simple interface

$$c_{12} = 1 - \frac{\tanh d^*}{d^*}. \tag{15.9}$$

For wall potentials that cannot be assumed small, we can solve the equation numerically and extract phenomenological relations to approximate c_{12}. The c_{12} expression in Eq. (15.9) is reminiscent of a logistic curve when plotted on a logarithmic axis, and for large wall potentials this curve shifts to lower d^* and spreads out. This is because, at high wall potentials, the characteristic decay of the potential distribution occurs over a length shorter than λ_D. Because of this behavior, we can approximate c_{12} phenomenologically for a symmetric monovalent electrolyte with a logistic curve, where the parameters depend on φ_0:

electroosmotic correction coefficient, symmetric monovalent electrolyte, $\varphi_0 < 6$, uniform fluid properties, simple interface

$$c_{12} \simeq \frac{1}{1 + \exp\left[-\alpha(\log d^* - \log d_0^*)\right]}, \tag{15.10}$$

where

$$\alpha \simeq -0.15 \left|\varphi_0^*\right| + 3.5 \tag{15.11}$$

and

$$d_0^* \simeq -0.01\varphi_0^{*2} - 0.1 \left|\varphi_0^*\right| + 2. \tag{15.12}$$

The numerical values for the φ_0^* dependence of α and d_0^* come from a fit to full numerical solutions and are accurate to within 10% for $\varphi_0^* < 6$. Results for $\varphi_0^* = 0, 3, 6$ are shown in Fig. 15.1.

In the dilute solution limit, we can also calculate the streaming current correction coefficient c_{21} by solving the 1D nonlinear Poisson–Boltzmann equation and the 1D Navier–Stokes equation and integrating:

$$c_{21} = -\frac{\eta}{\varepsilon \varphi_0} \frac{1}{A} \int \frac{u \rho_E}{-\frac{dp}{dx}} \, dA. \tag{15.13}$$

Here, u is the velocity profile generated by a pressure gradient. For example, for flow between plates at $y^* = \pm d^*$ with uniform fluid properties,

$$u = -\frac{1}{2\eta} \frac{dp}{dx} \left(d^2 - y^2\right) = -\frac{1}{2\eta} \frac{dp}{dx} \left(d^{*2} - y^{*2}\right) \lambda_D^2. \tag{15.14}$$

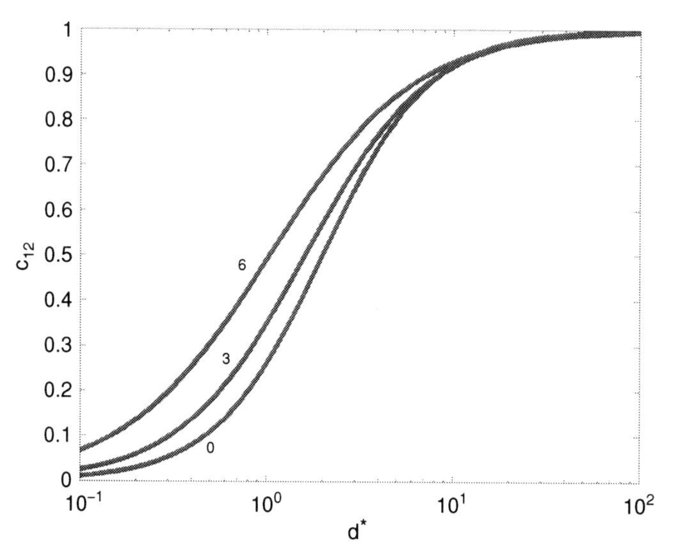

Fig. 15.1 Correction factor c_{12} for nanoslit geometries for symmetric monovalent electrolytes.

In the Debye–Hückel limit, c_{21} can be determined analytically:

streaming current correction coefficient, Debye–Hückel approximation, simple interface, uniform fluid properties

$$c_{21} = 1 - \frac{\tanh d^*}{d^*}, \qquad (15.15)$$

which echoes the result in Eq. (15.9) – Onsager reciprocity holds for thick double layers if the wall potential is low, because the deviation of ion distributions from the bulk concentrations is minor. For wall potentials that cannot be assumed small, no closed-form exact solution exists. Although we do not do so here, this can also be solved numerically to extract approximate relations for c_{21}.

The thick-EDL modification for c_{22} has two components, stemming both from ohmic and convective contributions to the current. Recall that the convective surface conductivity has already been presented in the thin-EDL, Debye–Hückel limit in Eq. (9.36). The average current density $\bar{i} = I/A$ in any 1D system is given by

$$\bar{i} = \frac{1}{d}\int_0^d \sum_i c_i u_i z_i F \, dy. \qquad (15.16)$$

Because the velocity magnitude of a species u_i can be separated into $u_i = u + \mu_{\mathrm{EP},i} E$, we can split this into a convective component and an electrophoretic component:

$$\bar{i} = \frac{1}{d}\int_0^d \sum_i c_i u z_i F dy + \int_0^d \sum_i c_i \mu_{\mathrm{EP},i} E z_i F \, dy. \qquad (15.17)$$

No diffusion term appears in the previous equation because we have assumed that the system has no gradients in the x direction. Simplifying the preceding relation, we get

$$\bar{i} = \frac{1}{d}\int_0^d u\rho_{\mathrm{E}} \, dy + \frac{E}{d}\int_0^d \sigma dy, \qquad (15.18)$$

where the net charge density $\rho_E = \sum_i c_i z_i F$ and the conductivity $\sigma = \sum_i c_i \mu_{EP,i} z_i F$. For a system in which the double layer is vanishingly thin, the current density $\vec{i}$ is uniform and given by $\vec{i} = \sigma \vec{E}$. When the double layer is finite, the first term is nonzero owing to nonzero ρ_E in the EDL. When the wall potential is large, the second term is modified because σ is a function of y and increases in the EDL.

Equation (15.18) can be evaluated by numerical integration for general species with species-dependent μ_{EP}; however, we find it useful to approximate the μ_{EP} of all species as equal, resulting in a general relation for the enhancement of the effective conductivity of the solution as a function of the species valences and wall potential alone. Assuming, for simplicity, that $|\mu_{EP}|$ is the same for all species, we can normalize the current density by $\sigma_{bulk} E$, resulting in

$$\frac{\bar{i}}{\sigma_{bulk} E} = \left|\frac{\mu_{EO}}{\mu_{EP}}\right| \frac{\frac{1}{d}\int_0^d \left|\frac{\varphi - \varphi_0}{\varphi_0}\right| \sum_i c_i z_i}{\sum_i c_{i,\infty} |z_i|} dy + \frac{1}{d}\int_0^d \sum_i \frac{c_i}{c_{i,\infty}} dy\,, \tag{15.19}$$

where μ_{EP} is the electrophoretic velocity of the ions and $\mu_{EO} = -\varepsilon\varphi_0/\eta$ is the electroosmotic mobility of the surface in the thin-EDL limit. From the form of the first term, we can see that the relative contribution of the electroosmotic fluid flow to the current is large if the surface electroosmotic mobility is large compared with the ion electrophoretic mobility. The relative contribution of the electroosmotic fluid flow to the current is also large when there is considerable overlap between the electroosmotic fluid velocity and the charge density, which occurs when the double layers are thick. The second term accounts for the ohmic surface conductance, i.e., the increase in conductivity from the increased ion concentration in EDLs.

From the preceding relations, c_{22} can be written as

conductivity correction coefficient, uniform fluid properties, simple interface

$$c_{22} = \left|\frac{\mu_{EO}}{\mu_{EP}}\right| \frac{\frac{1}{d}\int_0^d \left|\frac{\varphi - \varphi_0}{\varphi_0}\right| \sum_i c_i z_i}{\sum_i c_{i,\infty} |z_i|} dy + \frac{1}{d}\int_0^d \sum_i \frac{c_i}{c_{i,\infty}} dy\,. \tag{15.20}$$

The ohmic contribution is highest for large φ_0 and small d^*. The convective contribution is highest for large φ_0 and d^* near 1, where the velocity magnitude is high and substantial overlap between the charge density and the velocity leads to large net charge flux.

OBSERVED SYSTEM DEPENDENCE ON d AND d^* WHEN FORCING FUNCTIONS ARE CONTROLLED

We can consider the pressure and electric field to be the forcing functions of the electrokinetic coupling equation and consider flow and current to be the outcomes. If we control the forcing functions and observe the outcomes, the results are each a function of only one element of the electrokinetic coupling matrix. If we apply an electric field but fix the pressure gradient at zero, our measured current density I/A is given by

$$\frac{I}{A} = \chi_{22} E\,, \tag{15.21}$$

and the cross-section-averaged electroosmotic velocity Q/A is given by

$$\frac{Q}{A} = \chi_{12} E\,. \tag{15.22}$$

This and the correction factors described earlier indicate that thick-EDL systems (typically found in nanochannels) exhibit increased conductivity but decreased electroosmosis relative to thin-EDL systems (typically found in microchannels).

If we apply a pressure gradient $\left(-\frac{dp}{dx}\right)$ and fix the electric field at zero, we find that the mean flow rate Q/A is given by

$$\frac{Q}{A} = \chi_{11}\left(-\frac{dp}{dx}\right) \quad (15.23)$$

and the streaming current I/A is given by

$$\frac{I}{A} = \chi_{21}\left(-\frac{dp}{dx}\right). \quad (15.24)$$

The streaming current magnitude increases as d^* decreases (because χ_{21} increases). Thus thick-EDL systems have increased streaming current, and small-diameter systems have reduced pressure-driven flow.

OBSERVED SYSTEM DEPENDENCE ON d AND d^* WHEN ONE FORCING FUNCTION AND ONE OUTCOME ARE CONTROLLED

If we consider control of one forcing function and one outcome, we find that the results are a function of two or four of the electrokinetic coupling coefficients.

If we apply an electric field in a mechanically closed system, E is specified and $Q = 0$. In this case, we can solve the matrix equation to find

$$-\frac{dp}{dx} = -\frac{\chi_{12}}{\chi_{11}}E \quad (15.25)$$

and

$$\frac{I}{A} = \left[\chi_{22} - \frac{\chi_{12}\chi_{21}}{\chi_{11}}\right]E. \quad (15.26)$$

Equation (15.25) shows that the pressure gradient generated by an electrokinetic pump is increased as d decreases owing to the d dependence of χ_{11} until double layers overlap and performance is degraded by the decrease in χ_{12}. Equation (15.26) shows that the current required to drive an electrokinetic pump is affected by double-layer overlap in three ways: The conductivity is increased because the ohmic contribution goes up and the convective ion transport goes up, peaking at d^* near 1, but the net current is reduced by the adverse streaming current caused by the adverse pressure-driven flow (which is, in turn, caused by the electroosmosis). The ohmic current is typically the dominant term, but the d dependence of χ_{11} makes the adverse convective current important when d is small. The current density required to drive an electrokinetic pump increases as d^* decreases, but this effect is attenuated if d becomes so small that adverse streaming current plays a prominent role.

If we apply a pressure gradient to a system whose reservoirs are electrically isolated, $-\frac{dp}{dx}$ is specified and $I = 0$. In this case, we can solve the matrix equation to find

$$E = -\frac{\chi_{21}}{\chi_{22}}\left(-\frac{dp}{dx}\right) \quad (15.27)$$

and

$$\frac{Q}{A} = \left[\chi_{11} - \frac{\chi_{12}\chi_{21}}{\chi_{22}}\right]\left(-\frac{dp}{dx}\right). \quad (15.28)$$

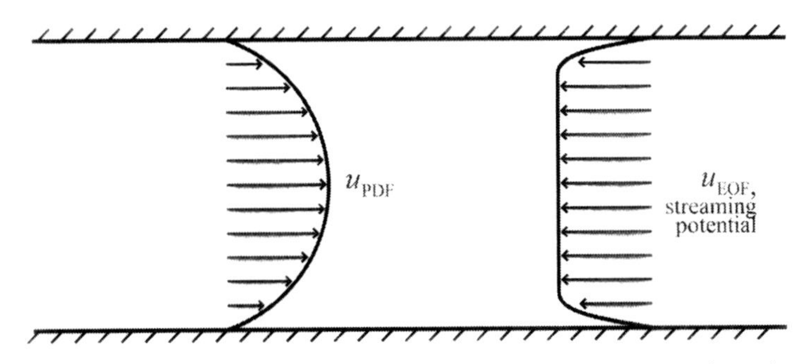

Fig. 15.2 Adverse electroosmosis induced by streaming potential – the cause of apparent electroviscosity.

Equation (15.27) shows that streaming potential is dependent on only the ratio of the size of the channel to λ_D – the streaming potential is reduced when double layers overlap, because χ_{22} increases more than χ_{12} when the ion distribution of finite double layers is accounted for. Equation (15.28) shows that the mean fluid velocity through an electrically isolated channel is primarily a function of size, owing to the dependence of χ_{11} on d, but this flow is furthermore reduced at low d^* (electroviscosity) because the streaming current induced by the pressure also creates an adverse electroosmotic flow. Because the first term is proportional to d^2 and the second term is proportional to λ_D^2, the relative importance of the two terms in the brackets in Eq. (15.28) is dictated by d^{*2}.

ELECTROVISCOSITY IN THICK-EDL SYSTEMS

Electroviscosity is the term commonly used for the decrease in flow rate owing to adverse electroosmosis induced by streaming potential. Figure 15.2 illustrates the adverse electroosmosis induced by streaming potential in a system in which a pressure is applied and the net current is zero. The term *electroviscosity* has the potential to be misleading, because no physical change in viscosity is implied – rather, the flow is attenuated owing to electrokinetic effects. For nanochannels with dimensions of the same order as those of the double layer, the flow reduction can be significant – approximately a factor of two in an extreme case. If electrokinetic effects are ignored when flow rates are interpreted, this makes the fluid appear to be more viscous or the channel to be of smaller diameter than it really is. Electroviscosity is unrelated to the *viscoelectric effect*, described by Eq. (10.31) and different in that the viscoelectric effect is a postulated change in the viscosity of water owing to the local electric field.

As discussed earlier, the flow induced by pressure for a system whose reservoirs are otherwise isolated electrically is given by

$$\frac{Q}{A} = \left[\chi_{11} - \frac{\chi_{12}\chi_{21}}{\chi_{22}}\right]\left(-\frac{dp}{dx}\right). \tag{15.29}$$

Although many authors commonly define an "effective viscosity" for this system by writing

$$\eta_{\text{eff}} = \frac{\frac{r_h^2}{8}\left(-\frac{dp}{dx}\right)}{Q/A} \tag{15.30}$$

or

$$\eta_{\text{eff}} = \frac{\eta}{\left[1 - \frac{\chi_{12}\chi_{21}}{\chi_{11}\chi_{22}}\right]} \tag{15.31}$$

we avoid this construction in this text in favor of a direct measure of the attenuation of the flow:

$$\frac{\frac{Q}{A}}{-\frac{dp}{dx}} = \left[1 - \frac{\chi_{12}\chi_{21}}{\chi_{11}\chi_{22}}\right] . \tag{15.32}$$

We evaluate the flow attenuation in the vanishingly thin-EDL limit – the result in this limit overestimates electroviscous effects if applied to a thick-EDL system, but it nonetheless provides an upper estimate of the magnitude of this flow attenuation and identifies situations for which electroviscosity can be ignored. Substituting the vanishingly thin-EDL relations for the components of χ, and noting that this approach sets an upper bound for the flow attenuation, we find

first-order estimate of the electroviscous effect

$$1 > \frac{\frac{Q}{A}}{-\frac{dp}{dx}} > 1 - 8\frac{\varepsilon^2\varphi_0^2}{\eta\sigma r_h^2} . \tag{15.33}$$

This relation is reminiscent of the thin-EDL, Debye–Hückel relation for convective surface conductivity, though it is a second-order perturbation in r_h/λ_D and is less often significant. Although the rightmost expression in relation (15.33) overestimates the flow attenuation significantly, its results are useful for determining those cases for which electroviscosity can be ignored. For example, the correction term $8\frac{\varepsilon^2\varphi_0^2}{\eta\sigma r_h^2}$ is less than 8% if the hydraulic radius is larger than

$$r_h = \frac{10\varepsilon\varphi_0}{\sqrt{\eta\sigma}} . \tag{15.34}$$

If the hydraulic radius is smaller than this value, a full calculation to estimate the electroviscosity is warranted.

VALENCE-DEPENDENT ION TRANSPORT IN THICK-EDL SYSTEMS

Small ion transport in systems with d^* near unity exhibits valence-dependent velocity even in channels of uniform cross section. Because double-layer overlap leads to a velocity distribution that is nonuniform over the entire channel depth, variations in species distributions normal to the channel surfaces lead to a difference in the net cross-sectional average of the species velocity. In systems with d^* near 1, the distribution of small ions (i.e., ions that can be modeled as point charges) is affected by the valence of the ion. Thus the convective velocity of monovalent ions is larger than the convective velocity of divalent or trivalent ions.

15.1.3 Circuit models for nanoscale channels

Recall from Chapter 3 that 1D fluid flow systems can be predicted by use of equivalent circuits and the Hagen–Poiseuille law $Q = \Delta p/R_h$. This is used to create sets of algebraic equations that can be solved in matrix form for the pressure at nodes and the volumetric flow rate in channels. Further, in Chapter 5 we use an identical approach to solve electrical circuit equations in matrix form for the voltage at nodes and the current in channels. These two systems of equations are uncoupled if walls are uncharged, but they are coupled if walls are charged. For a single channel, this is illustrated in Eqs. (10.41) and (10.42) in Chapter 10 by showing that the resulting 2×2 matrix has off-diagonal elements that are zero if the walls are uncharged and are nonzero if the walls are charged. For a system with N nodes and M channels, the system for uncharged walls can be written as two separate $N + M \times N + M$ matrix systems, whereas the system for charged walls must be written as one coupled $2N + 2M \times 2N + 2M$ matrix system. For channels whose cross section is uniform and for which gradients are normal to the flow, the only difference between thick-EDL and thin-EDL systems is in the electrokinetic coupling coefficients, as described earlier in this section.

Equivalent circuit models are effective only when effects of interfaces and nonuniform cross sections are negligible. Equivalent circuits can use algebraic relations to describe flow in long, narrow channels because of several key approximations – for example, in thin-EDL systems we pay little attention to the details of the geometry of the interface between several channels, presuming that the electrical and hydraulic resistance in a long, straight channel is much larger than that of the intersection itself. This approach inherently stems from a thermodynamic argument that local equilibrium can be applied to the system, even though global nonequilibrium holds. This is analogous to the Stokes approximation, which implies local equilibrium of fluid momentum by neglecting the unsteady and convective terms. The effect on the flows is seen in Chapter 8 through the observation of instantaneity (i.e., local equilibrium) – Stokes flows can vary in time only through time-dependent boundary conditions (i.e., global nonequilibrium). From an ion transport standpoint, the use of an equilibrium distribution (Poisson–Boltzmann) for ion distributions normal to a surface implies equilibrium in ion distributions normal to surfaces (i.e., local equilibrium), which vary only if there is time-dependent surface potential (i.e., global nonequilibrium).

For systems with d^* near unity, equivalent circuit models fail at interfaces and in regions of varying cross section, because there is no longer a clear separation between local phenomena that are assumed to be in equilibrium and global phenomena that are assumed to be out of equilibrium, and the length scale for double-layer equilibration is similar to the characteristic scales of the global system. If φ_0^* cannot be assumed small (and thus the perturbation to ion distributions owing to the wall is significant), then significant global effects evince the absence of local equilibrium. This occurs also during electrophoresis of particles with a^* near one (see Fig. 13.5). In the case of particles with radius near λ_D, the flow distortion of the double layer leads to an adverse electrical field that attenuates the particle flow. The electric field and flow are not everywhere normal to ion gradients, and the resulting $\vec{u} \cdot \nabla c_i$ terms are large enough to lead to significant effects. In nanochannel systems, the effects of note occur when thick-EDL channels have cross-section changes or at the location where these channels intersect. At these intersections or for systems with variable cross section, we can no longer use local equilibrium to separate the ion distribution solution normal to the wall from the flow parallel to the wall. The net effect of this is concentration polarization, in which ion concentrations are increased at one end of a nanochannel and decreased at the other. Concentration polarization at the interfaces between channels can often dominate the transport properties of a nanofluidic network.

15.2 TRANSPORT THROUGH NANOSTRUCTURES WITH INTERFACES OR NONUNIFORM CROSS-SECTIONAL AREA

The double-layer description used in earlier sections of this text considers variation of species normal to nanochannel walls but assumes that there is no variation in the direction of flow. With that approximation, all $\vec{u} \cdot \nabla c_i$ terms are zero and the equilibrium distribution of ions normal to the surface is independent of the flow or applied electric field. This is appropriate for infinitely long channels with constant cross section. We now consider systems with finite extent or nonuniform cross section, which exhibit gradients in the direction of flow.

If gradients in the direction of flow are present, we can no longer ignore convective fluxes on the equilibrium of ion distributions – the Poisson equation must then be coupled with the Nernst–Planck equations to determine the dynamics or equilibrium of ion distributions. In general, this system involves both the Nernst–Planck equations for each species i,

$$\frac{\partial c_i}{\partial t} = -\nabla \cdot \left[-D_i \nabla c_i + u_i c_i\right] , \quad (15.35)$$

coupled with the Poisson equation,

$$-\nabla \cdot \varepsilon \nabla \phi = \sum_i c_i z_i F . \quad (15.36)$$

Unlike for unidirectional flow in uniform-cross-section channels, the electric field used in the Poisson equation cannot be separated from the electric field used in the Nernst–Planck equations. Equations (15.35) and (15.36), combined with the Navier–Stokes equations,

$$\rho \frac{\partial \vec{u}}{\partial t} + \rho \vec{u} \cdot \nabla \vec{u} = -\nabla p + \eta \nabla^2 \vec{u} , \quad (15.37)$$

and conservation of mass,

$$\nabla \cdot \vec{u} = 0 , \quad (15.38)$$

comprise $n + 5$ coupled equations for p, ϕ, the three components of $\vec{u}$, and the n species distributions c_i. The coupled Poisson and Nernst–Planck equations are often termed the Poisson–Nernst–Planck equations, to contrast with the Poisson–Boltzmann formulation described in Chapter 9.

We do not attempt analytical solution of this set of $n + 5$ coupled equations; however, we instead discuss the results obtained in the 1D equilibrium limit. This 1D equilibrium limit can capture the qualitative aspects of the *two-way coupling* of these systems, in which fluid flow and ion transport change the ion distribution in the double layer.

15.2.1 1D equilibrium model

Cross-sectional ion distributions can be approximated at varying levels of detail. The simplest is to implement a 1D equilibrium model, appropriate for long, narrow channels for which $L/r_h > Pe$, where L is the channel length. In this model, we average properties across the cross section when writing the 1D transport equations, and assume that equilibrium transverse to the flow is always maintained. Thus convective transport affects the total ion concentrations at each streamwise location but does not perturb the equilibrium relations between coions and counterions. This hybrid approach allows nonequilibrium in the axial direction of a channel (where the fluid velocity leads to net convective flux) but maintains equilibrium in the transverse direction (where variations do not contribute to convective flux and are thus less important). This approach is the simplest one that captures convective ion flux and therefore perturbation of the ion distributions by the flow.

If transverse equilibrium is assumed, the ion distributions transverse to the flow direction satisfy the Poisson–Boltzmann equation, and the ion distributions can be solved if the charge density of potential at the wall is specified. These distributions can be averaged to give cross-section-averaged ion concentrations of each species. We can also write

$$\int_S \rho_E \, dA + \int_C q''_{\text{wall}} \, ds = 0 \,, \tag{15.39}$$

where C denotes the contour around the perimeter of the cross section of the channel and S denotes the surface of the cross section. This relation implies that the net charge density in the double layer cancels out the charge density at the wall, making the channel electroneutral when averaged over the entire cross section. Assuming q''_{wall} is uniform on C, Eq. (15.39) leads to the averaged relation

$$\overline{\rho}_E = \frac{-2q''_{\text{wall}}}{r_h} \,. \tag{15.40}$$

For a symmetric electrolyte, the system is even simpler, and the Poisson–Boltzmann equation is not necessary, allowing us to focus on the effects of convective ion flux. We consider a symmetric electrolyte with valence z, for which the difference in concentrations between the positive and negative electrolyte can be written as

$$\overline{c}_+ - \overline{c}_- = \frac{-2q''_{\text{wall}}}{zFr_h} \,, \tag{15.41}$$

and we start by writing the 1D Nernst–Planck equation for species i,

$$\frac{\partial c_i}{\partial t} = -\frac{\partial}{\partial x} j_i = -\frac{\partial}{\partial x}\left[-D_i \frac{\partial c_i}{\partial x} + u_i c_i\right] , \tag{15.42}$$

where j_i is the flux density of species i and the total current density $\overline{i} = I/A$ is given by $\sum_i j_i z_i F$. Evaluating the total current density for a symmetric electrolyte with equal diffusivities and electrophoretic mobility magnitudes gives

$$\overline{i} = -DzF \frac{\partial}{\partial x}(\overline{c}_+ - \overline{c}_-) + \overline{u}zF(\overline{c}_+ - \overline{c}_-) + EzF\,|\mu_{EP}|\,(\overline{c}_+ + \overline{c}_-) \,, \tag{15.43}$$

or

$$\overline{i} = -D \frac{\partial}{\partial x}\left(\frac{-2q''_{\text{wall}}}{r_h}\right) + \overline{u}\left(\frac{-2q''_{\text{wall}}}{r_h}\right) + EzF\,|\mu_{EP}|\,C \,, \tag{15.44}$$

where $\overline{u}$ is the mean velocity of the *fluid* and $C = \sum_i c_i$. Equilibrium requires that $\overline{i}$ be uniform:

1D Nernst–Planck equation for a nanochannel, symmetric electrolyte, transverse equilibrium

$$0 = -D \frac{\partial^2}{\partial x^2}\left(\frac{-2q''_{\text{wall}}}{r_h}\right) + \frac{\partial}{\partial x}\left(\frac{-2q''_{\text{wall}}}{r_h}\overline{u}\right) + \Lambda \frac{\partial}{\partial x}(EC) \,, \tag{15.45}$$

where $\Lambda = zF|\mu_{EP}|$. Solving for the equilibrium concentration of ions thus requires solution of Eq. (15.45) simultaneous with charge conservation i.e., $\frac{\partial}{\partial x}(AEC) = 0$, where A is the channel cross-sectional area.

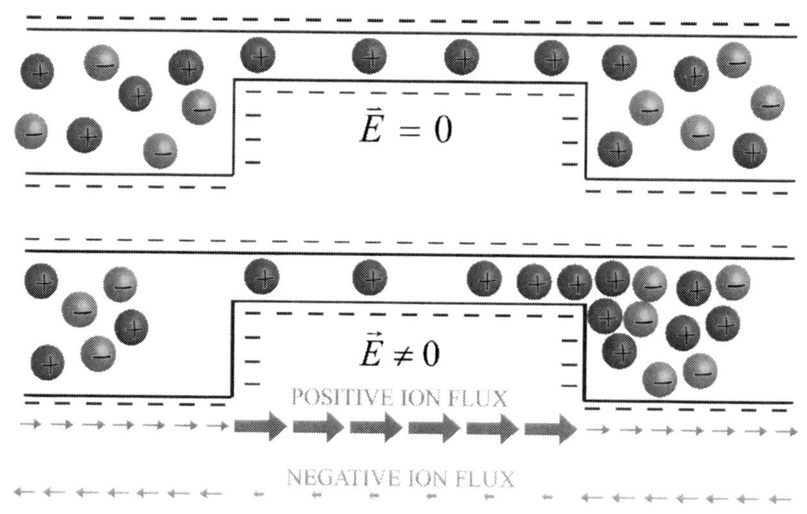

Fig. 15.3 Concentration polarization caused by an extrinsic electric field. In this system, intrinsic electric fields are present normal to the walls everywhere; intrinsic electric fields tangent to the walls exist at the entrance and exit to the shallow channel, where the hydraulic radius varies spatially.

Equation (15.45) shows that variations in channel cross-section or surface charge density create intrinsic electric fields at steady state. With no fluid motion, we can see that

$$0 = -D\frac{\partial^2}{\partial x^2}\left(\frac{-2q''_{\text{wall}}}{r_{\text{h}}}\right) + \Lambda\frac{\partial}{\partial x}(EC) , \tag{15.46}$$

and any variations in the surface charge or the channel hydraulic radius must be balanced by intrinsic electric fields in the axial direction. This also corresponds to the low-E limit, in which diffusion of ions dominates convection of ions.

For finite E, solution of Eq. (15.45) describes the balance between ion diffusion, ion convection, and ion electrophoresis along the channel caused by an extrinsic electric field or pressure gradient. For the device with negatively charged walls depicted in Fig. 15.3, a shallow region between deep regions leads to a higher positive charge density in the shallow region. At each interface, diffusion leads to a flux of positive ions toward the deep channel and a flux of negative ions toward the shallow channel. These are balanced by an electric field at each interface pointing toward the shallow channel, which causes a flux of positive ions toward the shallow channel and a flux of negative ions toward the deep channel.

CONCENTRATION POLARIZATION AND CURRENT RECTIFICATION

When electric fields are applied to a system with nonuniform $q''_{\text{wall}}/r_{\text{h}}$, *concentration polarization* occurs. A schematic of this phenomenon for a system with nonuniform r_{h} is shown in Fig. 15.3. The presence of an extrinsic electric field changes the balance of ion concentrations – wherever $q''_{\text{wall}}/r_{\text{h}}$ varies spatially, the extrinsic electric field will lead to an increase or decrease in C and the concentration of both positive and negative ions in that location. For example, for a shallow channel between deep channels, as depicted in Fig. 15.3, we find that $-q''/r_{\text{h}}$ has a smaller positive value in the deep channels and a larger positive value in the shallow channels. Thus equilibrium requires that C be high at the right interface and low at the left interface. For systems with asymmetric geometries, this phenomenon can lead to a voltage–current response that is asymmetric, and systems with asymmetric geometry can rectify current.

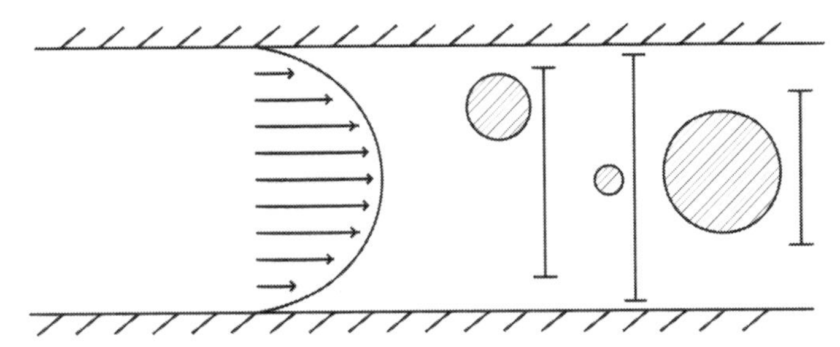

Fig. 15.4 Hydrodynamic particle separation. Larger particles sample only the central, high-velocity regions of the flow. Smaller particles sample the entire flow and are thus slower on average.

15.2.2 Large molecule and particle transport

Nanochannels with geometric complexity can manipulate the transport of large molecules, with the potential for separation of different materials. In fact, even unidirectional flow through channels of constant cross section leads to differential transport of particles and molecules if the particle or molecule size is of the order of the channel depth. We discuss these phenomena in the following subsections.

HYDRODYNAMIC DISPERSION IN PRESSURE-DRIVEN FLOW

Hydrodynamic dispersion generally refers to phenomena in which the speed of an object in a dispersive flow is dependent on a property that affects the analyte's transverse position. This is closely related to Taylor–Aris dispersion, but focuses on the time-averaged velocity rather than the effective diffusivity. A simple example of this is convection of a rigid particle in a channel with pressure-driven flow. Point particles sample all transverse positions with equal frequency and migrate with a time-averaged velocity equal to the cross-section-averaged fluid velocity. However, a rigid particle with finite radius a cannot have its center less than a away from either surface, and thus the particle does not sample the slowest region of the flow adjacent to the wall (Fig. 15.4). In such a flow, particles with larger radii move with a higher time-averaged velocity. For flow between two infinite parallel plates separated by a distance $2d$, a rigid particle with radius a and no lift force samples a space whose fluid moves faster than the mean fluid by a factor of

$$\frac{d^3 - \frac{3}{2}a^2 d + \frac{1}{2}a^3}{d^2\,(d-a)}. \tag{15.47}$$

This factor is a first-order approximation of the increased speed of the particle; however, particles that are large relative to the channel size cannot be assumed to move with the geometry of the unperturbed mean flow at the particle center. Finite particles modify the flow and feel a hydrodynamic lift that moves the particles away from the channel surface. This lift is absent at $Re = 0$ but is present and can be significant for finite but small Re.

OGSTON SIEVING

Ogston sieving refers to the retardation of flow of a molecule during transport into and out of pores that are not much larger than the molecule. For molecules that can be treated as semirigid particles (for example, globular proteins or DNA molecules with $\ell_c \simeq \ell_p$), motion through a channel that is of the same order of but larger than the molecule size is dictated by the entropic energy barrier posed by the channel. This process is particularly important for electrophoretic transport of macromolecules through gels and nanochannels.

As an example, we show in Eq. (14.63) that a DNA molecule with radius of gyration $\langle r_g \rangle$ confined in a channel of dimension d has a potential energy of confinement given approximately by

$$\Delta \mathcal{A} \simeq k_B T \frac{\pi^2 \langle r_g \rangle^2}{d^2}, \tag{15.48}$$

where we have grossly approximated the excess energy described by the partition function in Eq. (14.58). The absolute value of this energy has no effect on the transport. However, if the path the macromolecule takes through the device goes back and forth from confined to unconfined geometries (consider, for example, a periodic array of deep and shallow channels), then the macromolecule must travel through regions of varying potential energy. This variation of potential energy can change the electrophoretic mobility of the molecule. If there is no confinement, an applied field makes the DNA move toward the positive electrode, and the velocity dependence of the molecule is a function of only the bulk viscous mobility of the molecules (in Chapter 14, we show that the viscous mobility of large DNA polymers undergoing electrophoresis is independent of molecule size). However, if there is confinement, large molecules experience large potential energy barriers whereas smaller molecules experience smaller potential energy barriers. If these potential energy barriers are significant compared with the potential energy generated by the electric field, they slow down the electromigration of the molecule. In the Ogston sieving limit, DNA electrophoresing through nanochannels of varying cross section or through materials with varying confinement (for example, agarose or polyacrylamide gels) move faster if the molecule is small and slower if the molecule is large.

We can approximate this process by arguing that the electrophoretic mobility of a macromolecule varies with the channel constriction according to the extra time spent overcoming the energy barrier. Large fields reduce the delay time, and large molecule size and large barrier height retard the molecule electromigration.

ENTROPIC TRAPPING

As discussed in the previous subsection, rigid molecules are governed by the entropic energy barrier caused by a gradient of the degree of confinement, leading to a size dependence of electrophoretic mobility when molecules experience these gradients. For large, flexible molecules, e.g., DNA with $\ell_c \gg \ell_p$, the molecule has enough degrees of freedom that its entropic forces vary along the contour of the polymer backbone (see, for example, Fig. 14.10, in which some of the polymer is confined and some is not). Thus regions of confinement produce an entropic force *per unit length* along the polymer backbone. Consider a DNA molecule threading from a large channel to a small channel. As the polymer enters the small channel, its total entropy decrease and its Helmholtz free energy increase are proportional to the length of DNA in the small channel. The force (i.e., gradient of free energy) of this process is independent of length. Because of this, the μ_{EP} of long DNA through variations in channel size does not decrease as ℓ_c increases – rather, μ_{EP} increases as ℓ_c increases – because the limiting process is actually getting a part of the DNA started into the small channel. The starting process is more efficient if ℓ_c is large, because each region of the polymer has an equal chance of being the one to start the threading process.

15.3 SUPPLEMENTARY READING

Nanofluidic transport has been the subject of several recent books and reviews, for example [52, 190, 192]. The membrane literature, e.g., [193] and parts of [29], is an excellent source for information about nanofluidic issues, although nanoporous membranes typically do not afford the flexibility of geometry and experimental technique required for exploring the physics completely.

Nanoscale transport in long straight channels is described in [194, 195, 196], including discussion of valence-dependent electroosmosis. Relevant discussion of the motion of particles in small channels is in [100, 101]. A treatment of current rectification using 1D equilibrium is found in [197], and a number of models are discussed in [52]. Nanochannels also motivate considering EDLs outside the dilute solution limit, because nanoconfinement can change the role of ion size on ion distributions. Some recent work on this includes implementation of modified Poisson–Boltzmann equations stemming from [198, 199, 200] by Liu et al. [201]. Other modified Poisson–Boltzmann and molecular dynamics simulations include [118, 120].

Some descriptions of macromolecular transport in nanochannels can be found in [181, 178]. The matrix formulation of nanochannels in this chapter is reminiscent of the area-averaged treatment of porous and gel materials [137].

15.4 EXERCISES

15.1 Define a geometry-dependent effective electroosmotic mobility $\mu_{EO,eff}$ such that the flow rate per unit length Q' of an electroosmotic flow between two infinite plates separated by a distance $2d$ is given by

$$Q' = 2d\mu_{EO,eff}\vec{E}, \tag{15.49}$$

where $d^* = d/\lambda_D$. This effective electroosmotic mobility thus gives the spatially averaged flow rate in this system. $\mu_{EO,eff}$ is a function of d^*.

(a) Graph the velocity distribution for several values of d^* ranging geometrically from 0.2 to 50.

(b) Evaluate $\mu_{EO,eff}$ as a function of d^* and plot this relation from $d^* = 0.2$ to $d^* = 50$.

15.2 Consider two infinite parallel plates separated by a distance $2d$. Assume the fluid between the plates is water ($\varepsilon = 80\varepsilon_0$; $\eta = 1\,\text{mPa s}$) with a symmetric, monovalent electrolyte with bulk conductivity $\sigma_{bulk} = \sum_i c_i \Lambda_i$. Assume that both the cation and the anion of the electrolyte have the same molar conductivity Λ. Assume the normalized potential at the wall is given by $\varphi_0^* = -0.2$.

(a) Calculate and plot the velocity profile between the two plates for the cases in which λ_D is equal to 5, 0.2, and 0.02 times d. Use $E = 100\,\text{V/cm}$.

(b) Define ζ_{eff} as the effective electrokinetic potential such that the cross-sectional-area-averaged velocity $\bar{u}$ is given by $-\frac{\varepsilon\zeta_{eff}}{\eta}E$. Calculate and plot ζ_{eff}/φ_0 vs. d^* for $0.1 < d^* < 100$.

(c) Define σ_{eff} as the effective conductivity of the channel such that the channel electrical resistance is given by $R = L/\sigma_{eff}A$. Calculate and plot $\sigma_{eff}/\sigma_{bulk}$ vs. d^* for $0.1 < d^* < 100$.

15.3 Consider two infinite parallel plates separated by a distance $2d$. Assume the fluid between the plates is water ($\varepsilon = 80\varepsilon_0$; $\eta = 1\,\text{mPa s}$) with a symmetric, monovalent electrolyte with bulk conductivity $\sigma_{bulk} = \sum_i c_i \Lambda_i$. Assume that both the cation and the anion of the electrolyte have the same molar conductivity Λ. Assume the normalized potential at the wall is given by $\varphi_0^* = 0.3$. Assume d^* is not large relative to unity.

(a) Derive a relation for the net current (i.e., charge flux) $I = \int_A u\rho_E dA$ generated by flow-induced convection of charge if a pressure gradient $\frac{dp}{dx}$ is applied to this system.

(b) Presume the reservoirs at both ends of this system are connected to an open circuit (i.e., presume there can be no net current in the system). In this case, there is (at equilibrium) a potential gradient generated in the system. Calculate this potential gradient.

15.4 Consider a nanochannel with equal concentrations of NaCl and $MgSO_4$ and a depth $2d = 2\lambda_D$. For a wall potential of $\varphi_0 = -3RT/F$, estimate the net electromigration of Na^+, Cl^-, Mg^{+2}, and SO_4^{-2} averaged over the cross section when an electric field E is applied.

15.5 Consider a nanochannel of depth 30 nm, width 40 μm, and length 120 μm aligned along the x axis. Assume that the leftmost 60 μm of the channel has a surface charge of 3 mC/m and the rightmost 60 μm of the channel has a surface charge of zero. A voltage is applied at infinite reservoirs at the ends of the nanochannel.

Model the electrolyte as symmetric and nonreacting with mobility equal to 7.8×10^{-8} m^2/V s, and for simplicity assume that the electrolyte concentration is uniform across the cross section and of a magnitude such that the net charge density at the wall is canceled by the net charge density in the fluid.

(a) Write the 1D Nernst–Planck equations for transport of the positive and negative electrolytes.

(b) Given a surface charge density that is nonuniform in x, use equilibrium electroneutrality to define the difference between the cation and anion concentrations.

(c) Integrate the resulting equations to predict the ion distributions in the channel as a function of applied voltage and electrolyte bulk concentration.

(d) Compare your results with the experimental data in Fig. 15.5, taken from [197]. Does your model accurately match the experimental results? What does this tell you about the rectification of current in a nanofluidic channel with a surface charge density discontinuity?

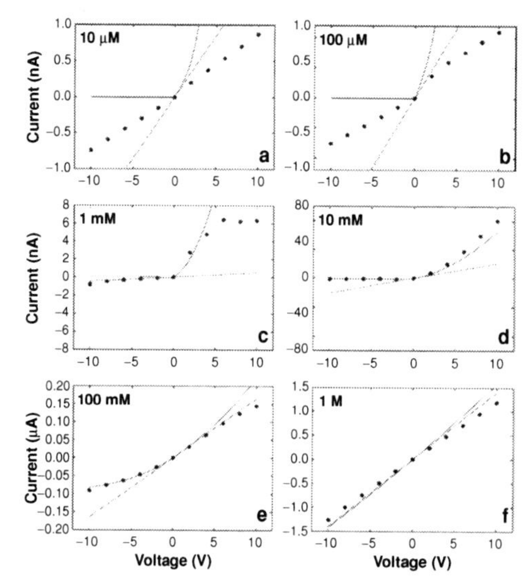

Fig. 15.5 Current–voltage response of a nanofluidic channel with two regions of differing net surface charge density. The positive electrode for calculation of the voltage is the rightmost reservoir. Symbols denote experimental results; lines denote model predictions generated by the study's authors. (From [197]. Used with permission.)

15.6 Set $q'' = 0$ in Eq. (15.44) and show that the result is $\bar{i} = \sigma E$.

15.7 Assume q''/r_h is uniform in Eq. (15.44) and show that, in the Debye–Hückel limit, the result is $\bar{i} = \sigma E + 2\frac{\varepsilon^2 \varphi_0^2}{\lambda_D \eta r_h}$. Explain why the convective surface current differs from that predicted in Eq. (9.36) by a factor of two.

15.8 Consider flow through an infinite channel of circular cross section and radius R with a dilute suspension of rigid particles of radius a. Assume that the particles are evenly distributed across the channel, except for the steric repulsion that prevents the particle center from approaching any closer than a to the wall. Approximate the particle velocity as the velocity of the fluid at the particle center location if the particle were absent, and calculate the factor by which the particles move faster than the mean fluid velocity.

15.9 Consider a fluid channel of width 100 μm. The channel has a shallow region of depth 500 nm and length 4 mm surrounded by regions of depth 2 μm and length 4 mm. A suspension of 50% by volume, 150-nm diameter particles is introduced at one end of the device. Assume the particles are evenly distributed other than their steric repulsion from the wall. What is the volume fraction of the particles in the 500-nm channel? The *Fahraeus effect* in blood is the observation that blood hematocrit (i.e., blood cell volume fraction) decreases if a blood sample goes through a narrow tube. Can the Fahraeus effect be explained, in part, by your calculations?

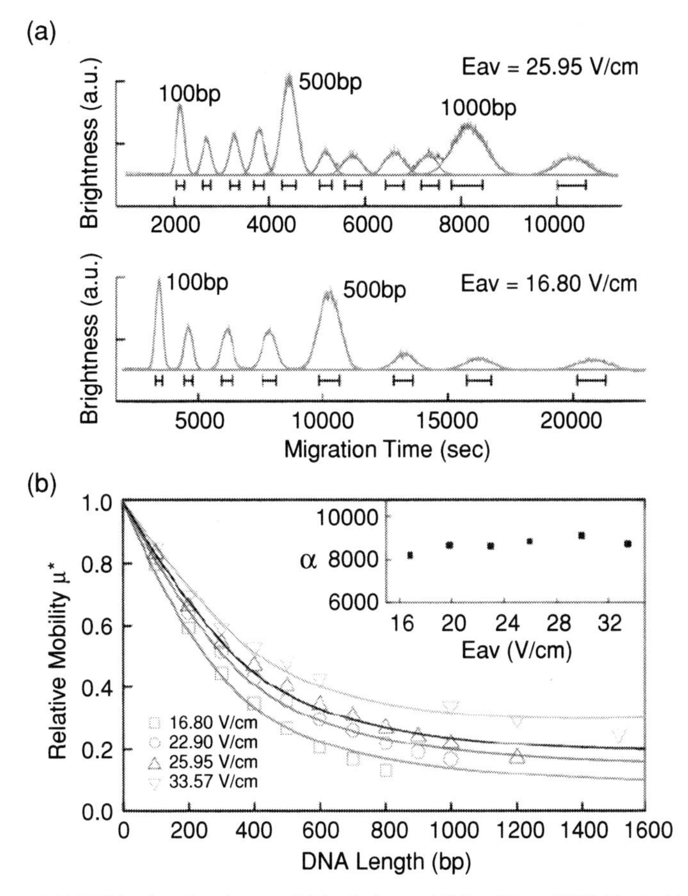

Fig. 15.6 Measurements of (a) DNA migration times and (b) relative mobilities. (From [181]. Used with permission.)

15.10 Consider a mixture of 50-nm and 100-nm particles moving through a wide channel of depth 200 nm at mean velocity 10 μm/s. Calculate the different mean velocities of these particles, and determine if a hydrodynamic separation of these particles is possible in this configuration.

15.11 Consider the data in Fig. 15.6. These data involve measurements of migration time of DNA of varying sizes in nanochannels with periodically varying depths. Given the data, fit the result by using

$$\mu_{\mathrm{EP}} = \frac{\mu_{\mathrm{EPbulk}}}{1 + \frac{\alpha \ell_c}{E^2 \exp\left(-\frac{\langle r_g \rangle^2}{d^2}\right)} \exp\left(-\frac{\Delta \mathcal{A}}{k_B T}\right)}, \tag{15.50}$$

where α is a fit parameter. Evaluate the efficacy of this model in explaining the DNA transport data.

16 AC Electrokinetics and the Dynamics of Diffuse Charge

Equilibrium models of the EDL (Chapter 9) assume that the ion distribution is in equilibrium and use a Boltzmann statistical description to predict ion distributions. The equilibrium assumption is appropriate for the EDL at an electrically insulating surface such as glass or most polymers, because the ion distribution processes are typically fast relative to the phenomena that change the boundary condition φ_0 (e.g., surface adsorption or changes in electrolyte concentration or pH).

In this chapter, we address the *dynamics* of diffuse charge. We focus primarily on the formation of thin double layers at *electrodes* with attention to the dynamics of double-layer formation and equilibration. Unlike for the double layer formed at the surface of an insulator, the double-layer equilibration at an electrode (owing to the potential applied at that electrode) is not necessarily fast compared with the variation of the voltage at the electrode – high-frequency voltage sources can vary rapidly compared with double-layer equilibration. Thus the dynamic aspects of double-layer equilibration are critically pertinent.

A full continuum description of these phenomena comes from the Poisson, Nernst–Planck, and Navier–Stokes equations combined with boundary conditions describing electrode kinetics. Because such analysis is daunting, we approximate the problem as that of predicting surface electroosmosis with time-dependent electrokinetic potentials, and use 1D models of the EDL to form equivalent circuits that can be used to model the temporal response of φ_0. We focus on thin EDLs because AC electrokinetic effects are most prominent in microscale channels.

In addition to the characteristic times of double-layer equilibration, the geometries of interest are different and usually more complicated when we are considering flow induced by the potential at a conducting surface. In particular, the straightforward geometry used in our early discussion of electroosmosis – a potential at an insulating surface inducing an electric field normal to the surface and a largely uncoupled extrinsically applied electric field parallel to the surface – is not possible at a conducting surface because the conducting surface precludes the existence of the parallel electric field. The basic system we use to introduce the material (ion motion normal to two parallel electrodes) can be used to calculate the characteristic equilibration times, but this system does not induce fluid flow. The geometries that do induce fluid flow are typically not 1D and numerical solution is typically required.

By considering a more complicated picture of double-layer dynamics, we can identify a number of interesting effects, which are generally classified as AC electrokinetic phenomena. Two specific types of flows we will consider have been termed AC electroosmosis and induced-charge electroosmosis. We focus on the dynamics of EDLs for a 1D geometry and on equivalent circuit models of this process. By stressing *dynamics*, we imply that our central goal is to identify not just the steady-state nature of the double layer, but also the characteristic time required for the double layer to achieve steady state upon experiencing a change in surface potential. By identifying this characteristic time, we can describe the response of the system to AC signals and predict a variety of AC phenomena.

Diffuse charge is not generated solely near surfaces. If the properties of a fluid are nonuniform, the spatial variations in electrical conductivity and permittivity lead to a perturbation in the electric field and the net charge density field. The net charge density created leads to an electrostatic body force, which also leads to fluid flow. These flows are usually called electrothermal flows, because thermal inhomogeneity is their most common source.

16.1 ELECTROOSMOSIS WITH TEMPORALLY VARYING INTERFACIAL POTENTIAL

We showed in Subsection 6.3.2 that outer solutions for electroosmosis can be predicted (for thin double layers) by replacing the body force applied to the EDL with an effective slip. In this case, the net charge density source terms in the Navier–Stokes and Poisson equations are ignored, and their effects are replaced with an effective electroosmotic slip velocity boundary condition:

effective electroosmotic slip velocity, simple interface, uniform fluid properties

$$\vec{u}_{\text{outer,wall}} = -\frac{\varepsilon \varphi_0}{\eta} \vec{E}_{\text{ext,wall}} , \tag{16.1}$$

where φ_0 can be a function of space and time. If the time dependence of the potential drop across the EDL can be determined, the fluid response to AC voltages can be described using the Navier–Stokes equations with this time-dependent boundary condition, and the only info needed from the structure of the EDL is the characteristic time constant of its response, which is used to calculate $\varphi_0(t)$.

We now consider double-layer potentials induced by voltages applied at electrodes, which requires that we solve the dynamic equations to predict the potential drop across the EDL (φ_0) as a function of the voltage applied at the electrodes.

The time required for a double layer to charge up and reach its equilibrium state is central to predicting the electrokinetic response of systems with unsteady potential boundary conditions. Analytically, our ability to model these systems hinges on the spatial discretization of the system and the accuracy of the models used to describe the system or its elements. As a first step, we begin by identifying characteristic frequencies by using equivalent electrical circuit models, and we use EDL models to predict the differential double-layer capacitance and describe the EDL dynamics in this equivalent-circuit context. The predicted double-layer capacitance (and therefore the predicted system response) is a strong function of the double-layer model – in fact, the study of double-layer capacitance is usually a much more rigorous test of EDL models than, for example, the study of electroosmosis. Thus more detailed models of the double layer lead to better predictions in the equivalent circuit framework.

16.2 EQUIVALENT CIRCUITS

Much can be learned about double-layer dynamics by simply modeling the bulk fluid as a resistor and the EDLs as capacitors. In so doing, we discretize the spatial variations in the system (which are in reality continuous) into simplified and isolated components and use the circuit analysis tools from Chapter 5. Bulk fluid can be modeled as a resistor, because it conducts current but does not store charge. Double layers can be modeled as capacitors because (a) they primarily *store* charge and (b) they conduct current only when the double layer is changing. The capacitor analogy for double layers works best for surfaces that have intrinsic voltages caused by their surface chemistry but transmit no current, or for electrodes that do not lead to electrode reactions (Faradaic reactions) because the voltages applied are low or the electrode material is poor at catalyzing Faradaic reactions. If Faradaic reactions are occurring, modeling becomes more complicated. We restrict our discussion here to surfaces that do not conduct current.

RC circuit theory indicates that the characteristic time for response of a circuit is given by $\tau = RC$; this is the basis for low- and high-pass *RC* circuits, as well as many much more complicated systems. Specifically, for current passed between two electrodes through an electrolyte solution, if the applied voltage is a sinusoid at $\omega = 1/RC$, half of the voltage is dropped across the EDLs and half is dropped across the bulk solution. For $\omega \ll 1/RC$, all voltage is dropped across the EDL, and for $\omega \gg 1/RC$, all voltage is dropped across the bulk fluid. This means that for $\omega \ll 1/RC$, the applied voltage creates charge that can induce electroosmosis, whereas for $\omega \gg 1/RC$ the applied voltage does not have enough time to localize charge at the interface and thus the electroosmotic effects are minimal. If we create an EDL by applying voltage at an electrode, the temporal relation between φ_0 and the applied voltage V is dictated by an *RC* circuit model. If we then wish to predict fluid flow generated by these voltages, we must combine an *RC* model to predict $\varphi_0(t)$ combined with the Navier–Stokes equations. To predict the dynamics of EDLs, then, we need model only the resistance of the fluid in the system (for simple geometries, $R = L/\sigma A$) as well as the capacitance.

16.2.1 The double layer as a capacitor

Given the overall structure just described, we need to predict the electrical circuit properties of the EDL. To so do, we review the definition of a capacitor and evaluate the differential capacitance of an EDL.

A capacitor is a fundamental electrical circuit element that stores charge in response to a voltage applied across it:

definition of capacitance

$$C = \frac{dq}{d\Delta V}, \tag{16.2}$$

where ΔV is the (positive) voltage drop across the capacitor and q is the (positive) difference in charge between the capacitor ends. The capacitance is positive by definition. Ideal capacitors have a linear charge–voltage response, so the total capacitance $\frac{q}{\Delta V}$ is the same as the differential capacitance $C = \frac{dq}{d\Delta V}$ for all ΔV. Most physics and circuits texts define the capacitance by using the total capacitance; however, the differential capacitance is the property that governs the AC performance of a circuit or the dynamics of double-layer charging, so the differential capacitance is what is used here.

A parallel-plate capacitor is a medium of thickness d and permittivity ε separating two conductors of area A, where d^2 is small in comparison with A. The capacitance of a parallel-plate capacitor is

capacitance of a parallel-plate capacitor

$$C = \frac{\varepsilon A}{d}. \tag{16.3}$$

THE DOUBLE LAYER AS A CAPACITOR

We now wish to model an EDL as a capacitor, because the diffuse EDL stores charge in response to a potential drop. If we assume that the charged surface does not pass any current, the EDL is purely *capacitive*. Because double layers are typically thin, the assumptions for parallel-plate capacitors are good, and we typically refer to capacitances per unit area C'':

capacitance per unit area of an EDL

$$C'' = \frac{dq''_{\mathrm{edl}}}{d\Delta V}, \tag{16.4}$$

where the charge per unit area in the EDL is defined as $q''_{\mathrm{edl}} = q_{\mathrm{edl}}/A$. If the double layer is thin, the charge per unit area in the EDL is well defined and is balanced by an equal and opposite charge on the surface:

$$q''_{\mathrm{edl}} = -q''_{\mathrm{wall}}. \tag{16.5}$$

For a surface connected to an infinite reservoir of fluid, the wall is a voltage source, the bulk reservoir is the ground, and the double layer is the capacitor in between.

EXAMPLE PROBLEM 16.1

For a symmetric electrolyte in the Debye–Hückel limit, evaluate the net charge in a thin double layer by evaluating the derivative of the potential at the wall. Show that the net charge in the double layer is given by $q''_{\mathrm{edl}} = -\varepsilon\varphi_0/\lambda_D$. Show that the capacitance per unit area is given by $C'' = \varepsilon/\lambda_D$.

SOLUTION: The slope of the potential distribution at the wall is

$$\left.\frac{\partial \varphi}{\partial y}\right|_{\mathrm{wall}} = -\frac{\varphi_0}{\lambda_D}. \tag{16.6}$$

Now recall

$$\left.\frac{\partial \varphi}{\partial y}\right|_{\mathrm{wall}} = -\frac{q''_{\mathrm{wall}}}{\varepsilon} = \frac{q''_{\mathrm{edl}}}{\varepsilon}. \tag{16.7}$$

Given the Debye–Hückel solution for the potential in an EDL, the charge density on the wall is given by

$$q'' = \frac{\varepsilon}{\lambda_D}\varphi_0, \tag{16.8}$$

and the charge density in the double layer is given by

$$q'' = -\frac{\varepsilon}{\lambda_D}\varphi_0 \,. \tag{16.9}$$

The capacitance is given by

$$C'' = -\frac{q''_{\rm edl}}{\varphi_0} \,, \tag{16.10}$$

giving

$$C'' = \frac{\varepsilon}{\lambda_D} \,. \tag{16.11}$$

DOUBLE-LAYER CHARGE

The total charge in a double layer can be determined from the potential slope at the wall or by integrating the net charge distribution. For a symmetric electrolyte at low wall potentials (Debye–Hückel approximation), we can show that the net charge in the double layer is given by $q''_{\rm edl} = -\varepsilon\varphi_0/\lambda_D$, and the capacitance per unit area is given by $C'' = \varepsilon/\lambda_D$. Thus a Debye–Hückel double layer acts as a parallel-plate capacitor with plate spacing λ_D.

Differential double-layer capacitance: model dependence. Recall that, if we make the Debye–Hückel approximation, we can show that the linearized Poisson–Boltzmann equation predicts that the net charge (per unit area) in the double layer is given by

capacitance per unit area of an EDL, Debye–Hückel approximation

$$q''_{\rm edl} = -\varepsilon\varphi_0/\lambda_D \,. \tag{16.12}$$

The capacitance per unit area is given by the ratio of the difference of the stored charge between the ends to the magnitude of the potential drop. For a wall in isolation, the second surface of the capacitor is assumed to be at infinity. The capacitance per unit area is thus given by

$$C'' = -\frac{q''_{\rm edl}}{\varphi_0} \,, \tag{16.13}$$

where the sign is negative because, if φ_0 is positive, the end that carries the positive charge is at infinity, and thus the differential charge is $-q''_{\rm edl}$. Alternatively, if φ_0 is negative, the differential charge is $q''_{\rm edl}$ but the potential drop is $-\varphi_0$. From Eq. (16.13), the capacitance per unit area for a thin, Debye–Hückel double layer in equilibrium is $C'' = \varepsilon/\lambda_D$. If we do not make the Debye–Hückel assumption, we can show

(Exercise 16.4) that the Poisson–Boltzmann equations predict that, for a symmetric electrolyte, the net charge in the double layer is given by

$$q''_{\mathrm{edl}} = -\frac{\varepsilon\varphi_0}{\lambda_D}\left[\frac{2}{z\varphi_0^*}\sinh\left(\frac{z\varphi_0^*}{2}\right)\right] \tag{16.14}$$

and that the differential capacitance per unit area $dq''/d\varphi_0$ is given by

differential capacitance per unit area, symmetric electrolyte

$$C'' = \frac{\varepsilon}{\lambda_D}\left[\cosh\left(\frac{z\varphi_0^*}{2}\right)\right]. \tag{16.15}$$

In both cases, the nonlinearity is seen through a hyperbolic correction factor in brackets, which approaches unity as $\varphi_0^* \to 0$. Similar but more unwieldy results can be derived in the general case by use of the Grahame equation. The Poisson–Boltzmann model improves on the Debye–Hückel model in that, in addition to correctly predicting the capacitance for small voltages, it correctly describes the differential capacitance for surface potentials that are of the order of RT/F; however, this model predicts that $\lim_{\varphi_0\to\infty} C'' = \infty$, which is not observed in experiments. The inaccuracy of the Poisson–Boltzmann model stems from its point-charge approximation, which leads to unrealistically large ion concentrations at large wall voltages. Thus the Poisson–Boltzmann model erroneously predicts that the charge stored by the EDL goes up and the effective double-layer thickness is unchanged. Clearly a correction is required for handling large voltages.

Several approaches have been implemented to correct the Poisson–Boltzmann result at large voltages. We discuss the Stern and modified Poisson–Boltzmann approaches here. The Stern model postulates that the interface consists both of (1) the diffuse double layer described by Gouy–Chapman theory and (2) a thin layer of condensed ions. By *condensed,* we mean that these ions do not move normal to the surface. This thin layer (the Stern layer) is given a thickness of λ_S and a permittivity of ε_S, leading to a capacitance per unit area of $C'' = \varepsilon_S/\lambda_S$. Using the series relation for a capacitor and modeling the Stern layer as a linear capacitor, whose differential capacitance is not a function of potential drop, we find that the total double-layer capacitance is given by

double-layer capacitance per unit area including Stern layer, symmetric electrolyte

$$\frac{1}{C''} = \frac{1}{\frac{\varepsilon}{\lambda_D}\left[\cosh\left(\frac{z\varphi_0^*}{2}\right)\right]} + \frac{1}{\varepsilon_S/\lambda_S}. \tag{16.16}$$

Figure 16.1 shows the condensed layer used in the Stern EDL model. From Eq. (16.16), we see that the Stern layer model puts an upper limit on the capacitance of the system (the limit as $\varphi_0^* \to \infty$ is $C'' = \varepsilon_S/\lambda_S$), which matches experiments better. For aqueous electrolyte solutions near electrodes, ε_S is usually modeled with a value ranging from $6\varepsilon_0$ to $30\varepsilon_0$ and λ_S is modeled with a value ranging from 1 to 10 Å. In this form, the Stern model adds an ad hoc capacitance limit to the system.

An alternative approach is to incorporate a modified Poisson–Boltzmann model with steric hindrance (for which the ion distribution is described in Chapter 9), which similarly limits the capacitance. Given finite ion size, the region near the wall becomes

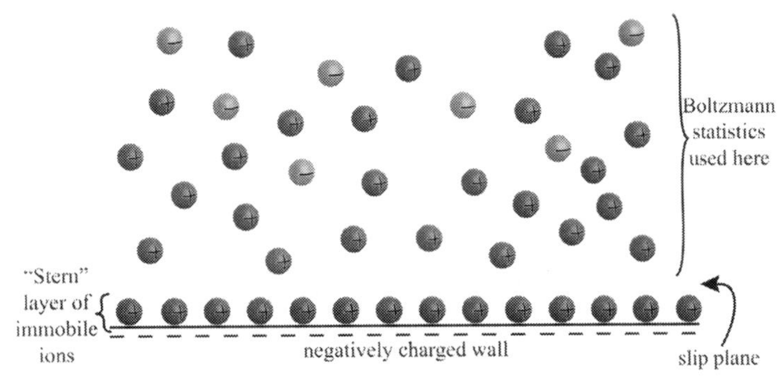

Fig. 16.1 Stern model.

saturated at high φ_0^*. When this happens, the thickness of the dense ion layer expands and the layer becomes less able to store charge. Both of these results correspond to lower differential capacitance. Quantitatively, the modified Poisson–Boltzmann model predicts (again for a symmetric electrolyte) that

$$q''_{\mathrm{edl}} = -\frac{\varepsilon\varphi_0}{\lambda_D}\left\{\frac{-1}{z\left|\varphi_0^*\right|}\sqrt{\frac{2}{\xi}\ln\left[1+2\xi\sinh^2\left(\frac{z\varphi_0^*}{2}\right)\right]}\right\}, \quad (16.17)$$

in which we have defined (for symmetric electrolytes) a packing parameter $\xi = 2c_\infty N_A \lambda_{HS}^3$, which is the volume fraction of ions in the bulk. The modified Poisson–Boltzmann model also predicts that the differential capacitance per unit area $-dq''/d\varphi_0$ is given by

differential capacitance per unit area, symmetric electrolyte, with steric correction

$$C'' = \frac{\varepsilon}{\lambda_D}\left[\frac{\sinh\left(z\varphi_0^*\right)}{\left(1+2\xi\sinh^2\left(\frac{z\varphi_0^*}{2}\right)\right)\sqrt{\frac{2}{\xi}\ln\left(1+2\xi\sinh^2\left(\frac{z\varphi_0^*}{2}\right)\right)}}\right]. \quad (16.18)$$

The effect of steric hindrance is to make the differential capacitance much lower at high wall voltages than that predicted by the Poisson–Boltzmann relation – in fact, the differential capacitance is a nonmonotonic function of the wall voltage.

Even within the applicability of equivalent circuit analysis, both techniques presented have problems. The Stern model discretizes the EDL and its properties in a nonphysical way, and the parameters that go into the model are poorly known and generally treated as phenomenological. The modified Poisson–Boltzmann model removes discretization, but also depends on a poorly known parameter (the ion hard-sphere radius). The modified Poisson–Boltzmann model cannot account for specific ion adsorption to walls (whereas the Stern model can) and ignores the variation of solution permittivity with large electric field (which is handled, albeit discretely, with the Stern layer permittivity). A more thorough equivalent circuit model requires that the equations that predict ion distribution be solved in more detail, to provide better equivalent circuits to add to the model. Beyond that, full treatment of the governing equations for the whole system further improves modeling of the dynamics of the double layer.

Despite the limitations of the capacitor models, they nonetheless frame approximately what the capacitance of the EDL is and thus predict approximately what the potential drop is across the double layer. Per the Helmholtz–Smoluchowski equation ($u_{\text{bulk}} = -\frac{\varepsilon\varphi_0}{\eta} E_{\text{ext}}$), the potential drop across the double layer specifies the effective electroosmotic slip at the interface; this boundary condition combined with the fluid flow equations predicts the fluid flow from voltages applied at conducting surfaces.

EXAMPLE PROBLEM 16.2

Consider two electrodes separated by a distance $2l$ and containing a symmetric electrolyte with all ions having mobilities and diffusivities of equal magnitude. Suppose that two different voltages are applied to the two electrodes. How long does it take for the double layers at the electrodes to form?

Treat the bulk fluid as a resistor. Show that the resistance $R = \frac{l}{\sigma A}$ (where σ is the bulk conductivity and A is the cross-sectional area of the electrolyte linking the two electrodes) can also be written as $\frac{\lambda_D^2 l}{\varepsilon D A}$.

Treat each double layer in the Debye–Hückel limit as a Helmholtz capacitor with thickness equal to λ_D ($C = \frac{\varepsilon A}{\lambda_D}$). Model the system as a capacitor, resistor, and capacitor in series, and show that RC for this model system is equal to $\frac{\lambda_D l}{D}$.

Given this result, what is the characteristic time for the double layer to equilibrate if you have two microelectrodes separated by 40 μm in a 1-mM NaCl solution? Approximate D for these ions as 1.5×10^{-9} m^2/s.

SOLUTION: For the capacitor, we have $C = \frac{\varepsilon A}{\lambda_D}$, and for two capacitors in series, $C = \frac{\varepsilon A}{2\lambda_D}$. For the resistor, we have $R = \frac{2l}{\sigma A}$. To define σ in terms of λ_D, we write

$$\sigma = \sum_i c_i \Lambda_i = 2c\Lambda = 2czF\mu_{\text{EP}}, \tag{16.19}$$

leading to

$$\sigma = 2czF\frac{zFD}{RT} = \varepsilon D\frac{2z^2F^2c}{\varepsilon RT} = \frac{\varepsilon D}{\lambda_D^2}. \tag{16.20}$$

Substituting in for σ, we find

$$R = \frac{2\lambda_D^2 l}{\varepsilon D A}, \tag{16.21}$$

leading to

$$RC = \frac{\lambda_D l}{D}, \tag{16.22}$$

$$\tau = \frac{10 \times 10^{-9}\text{ m} \times 20 \times 10^{-6}\text{ m}}{1.5 \times 10^{-9}\text{ m}^2/\text{s}} = 133\ \mu\text{s}. \tag{16.23}$$

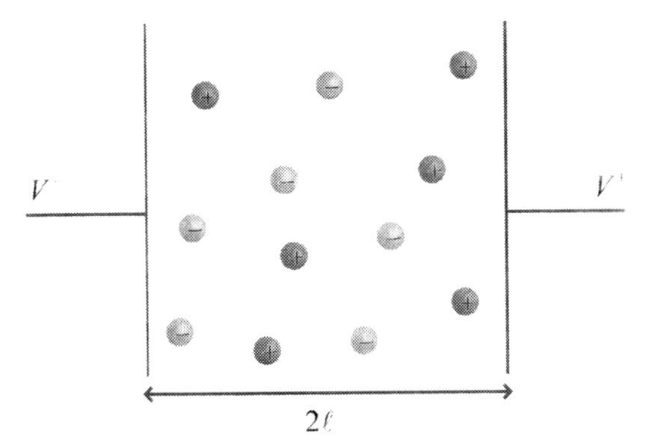

Fig. 16.2 Schematic of a circuit between two parallel-plate electrodes.

16.3 INDUCED-CHARGE FLOW PHENOMENA

Native double layers in equilibrium with a surface charge density or surface potential caused by chemical reactions or adsorption at the wall are described by a straightforward geometric picture: One field, which we term the *intrinsic field*, is maintained by equilibrium chemical or adsorption processes and leads to the EDL. A separate field, which we term the *extrinsic field*, moves the double-layer ions (and thus the fluid) around in a microdevice. Because the intrinsic field is maintained by an insulating surface, it is simple to separate these fields when we are considering native electroosmosis and electrophoresis. In this section, we consider double layers that are caused by applied electric fields rather than being caused by reaction or adsorption. Because the intrinsic and extrinsic fields can no longer be separated, these processes are inherently nonlinear.

16.3.1 Induced-charge double layers

Consider a finite 1D system with two idealized parallel-plate electrodes at which there is neither current nor reaction. If a potential is applied between these electrodes (Fig. 16.2), the electrolyte reorients in response to the electric field, and in so doing, causes the potential drop at equilibrium to occur mostly over the EDLs rather than the bulk fluid. After a time τ (in microfabricated systems this is often approximately 1 ms), the system is at equilibrium, and each electrode has a double layer composed primarily of counterions – the cathode attracts cations and the anode attracts anions. These double layers have been termed *induced-charge* double layers – the modifier *induced-charge* implying that the charge density is induced by the motion of ions in response to an applied field, rather than occurring naturally owing to reaction or absorption. Because larger voltages can be applied with electrodes than are observed in native double layers, this gives the potential for more dramatic double-layer effects, faster fluid flows, and faster particle flows.

Comments in the previous paragraph notwithstanding, the electrode configuration just described does not lead to any fluid flow. Recall that electroosmosis requires both a charge density in the double layer itself, and a transverse electrical field that actuates the net charge density in the double layer. In the case of a 1D parallel-plate system, a double layer is created, but there is no transverse field. In addition, the ideal zero-current electrode (i.e., electrode without current or chemical reactions at the metal–electrolyte interface) can be approximated experimentally only if AC fields are used – in the AC case, reactions (which always happen to some extent) are reversed every

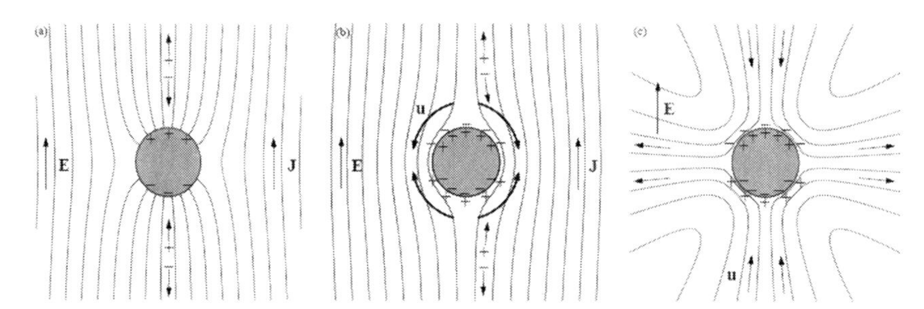

Fig. 16.3 Electric field lines (a) before and (b) after double-layer charging in response to a suddenly applied DC field. (c) the resulting flow streamlines. (Reproduced with permission from [202].)

time the polarity changes and reaction products do not build up. Because of this, electroosmosis owing to induced charges requires at minimum a 2D variation in the electric field.

16.3.2 Flow due to induced-charge double layers – induced-charge electroosmosis

In the previous case, the simplicity of the geometry prevented the field that induced the double layer from simultaneously actuating the net charge density created. However, we need perturb the geometry only slightly to see an effect. For example, consider a circular conductor in an applied 2D electric field. If the cylinder is treated as an ideal conductor with no Faradaic reactions at the surface, then a suddenly applied electric field causes one side of the cylinder to acquire a net negatively charged double layer, whereas the other side acquires a net positively charged double layer. The resulting space charge alters the electric field so that it becomes similar to that for the electric field around an insulator. Unlike the 1D case, this field actuates the double layers, leading to a quadrupolar flow. This can be seen in Fig. 16.3. This flow is approximately independent of the sign of the applied electric field. Because of this, an AC field can drive this DC flow. The frequency f of the field must be such that $f\tau_D$ is not $\gg 1$.

The full solution of induced-charge electroosmosis problems requires simultaneous solution of the Poisson, Nernst–Planck, and Navier–Stokes equations. However, we can obtain an approximate solution for thin-EDL systems by calculating the double-layer solutions in the $t = 0$ and $t \to \infty$ limits, and then assuming that the time constant of the transition between these two limits is given by the RC time constant of the equivalent resistor–capacitor circuit. With this estimate, the spatiotemporal dependence of the double-layer potential is known, and the bulk electric field and electroosmosis can thus be calculated.

16.3.3 Flow due to induced-charge double layers – AC electroosmosis

Another geometry in which electric fields simultaneously create and actuate double layers is on alternating positive and negative electrodes actuated with AC signals. This configuration is similar to the interdigitated arrays often used for electrochemical detection or for dielectrophoretic trapping (geometries for dielectrophoresis are described in Chapter 17). In this geometry, double layers are created at both electrodes, first in the inner regions and later in the outer regions. No fluid flow is generated at short times because no double layer yet exists, and no fluid flow is generated at long times because the field is screened by the double layer. However, when the double layers have been formed in the inner regions of the electrodes but not at the outer regions, the inner

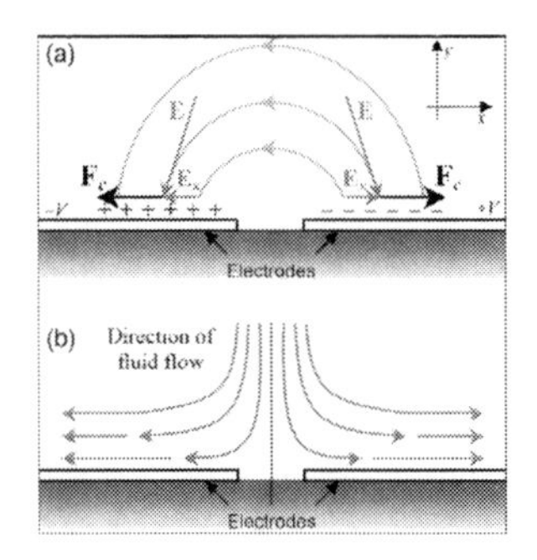

Fig. 16.4 (a) field lines and (b) streamlines for AC electroosmosis. Reproduced with permission from [203].

regions provide the double layer and the outer regions provide unshielded electrodes and therefore a bulk field. Thus DC fluid pumping (see Fig. 16.4 for a schematic of field lines and induced flow in these geometries) occurs if the frequency of the applied AC field corresponds to the time required for *part* of the electrodes to form double layers but not all. This flow is visualized in experiments and computation in Figs. 16.5 and 16.6. Figure 16.5 shows particle pathlines for flow induced by AC electroosmosis. Figure 16.6 shows comparisons between pathlines and computed streamlines.

16.4 ELECTROTHERMAL FLUID FLOW

For uniform fluid properties, there is no net accumulation of bound charge, and the Coulomb body force is the only electrostatic body force:

$$\vec{f} = \rho_E \vec{E}. \tag{16.24}$$

However, if the fluid permittivity varies (for example, because of temperature variations), we must also include a dielectric force:

$$\vec{f} = \rho_E \vec{E} - \frac{1}{2}\vec{E} \cdot \vec{E}\nabla\varepsilon. \tag{16.25}$$

The $\rho_E \vec{E}$ term describes the force on the free charge, whereas the $-\frac{1}{2}\vec{E} \cdot \vec{E}\nabla\varepsilon$ term describes the force on accumulated bound charge. The combined Coulomb and dielectric forces drive flows in bulk fluid most often when temperature gradients exist in a microsystem; these flows are termed *electrothermal* flows. Unlike electroosmosis, for which the net charge density is an equilibrium phenomenon caused by a charged wall, electrothermal flows are driven by dynamic fluctuations in net charge density caused by spatial inhomogeneities in the fluid properties. Recall that, for uniform fluid properties far from walls, the charge conservation equation and Poisson's equation both reduce to Laplace's equation. However, if a localized temperature increase exists, the local conductivity is increased and the permittivity is decreased. Equilibrium conservation of charge indicates that the local electric field must be reduced, whereas an instantaneous Gauss's law formulation indicates that the local electric field can be reduced only if a local charge density exists. Thus temperature fluctuations lead to dynamic fluctuations in net charge density, which in turn lead to electrostatic body forces.

Using Gauss's law to write ρ_E in terms of ε and $\vec{E}$, substituting this into the charge conservation equation, and neglecting convective charge transport leads to

$$\nabla\sigma \cdot \vec{E} + \sigma\nabla \cdot \vec{E} + \frac{\partial}{\partial t}\left[\nabla\varepsilon \cdot \vec{E}\right] = 0. \tag{16.26}$$

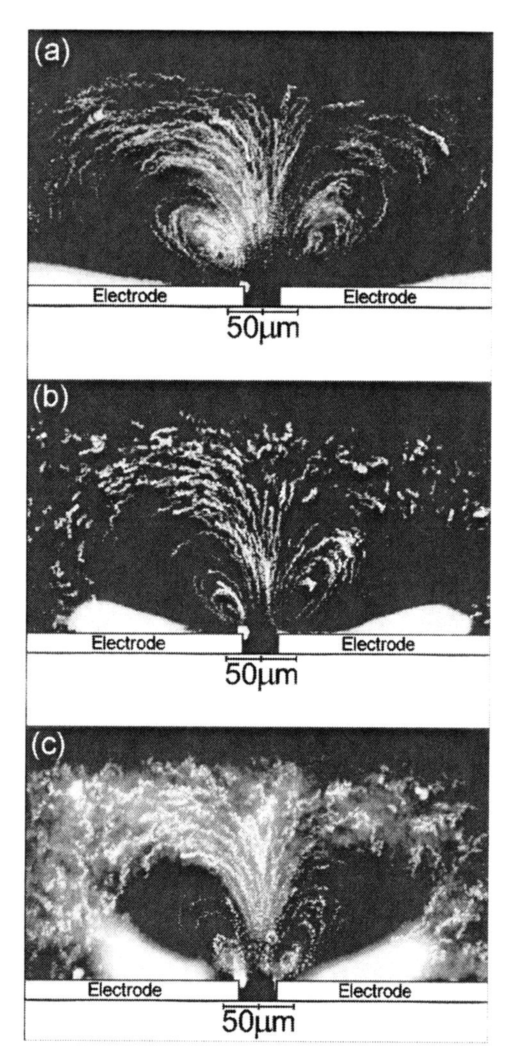

Fig. 16.5 Particle streaklines for flow induced by 2-V AC signals at (a) 100 Hz, (b) 300 Hz, and (c) 1000 Hz. Electrolyte is 2.1 mS/m KCl. (Reproduced with permission from [203].)

Simplifying this relation with a perturbation expansion and substituting it into Eq. (16.25) leads to a solution for the body force (see Exercise 16.12). If we assume that the applied field is sinusoidal, we can use the analytic representation of the field to calculate a time-averaged force. Further, this force can be written in terms of the field that would have existed but for the temperature perturbations. Writing the analytic representation of this applied field as $\underset{\sim}{\vec{E}} = \underset{\sim}{\vec{E}_0} \exp j\omega t$, then we can write the time-averaged body force as

$$\left\langle \vec{f} \right\rangle = \frac{1}{2}\mathrm{Re}\left[\left(\frac{(\sigma \nabla \varepsilon - \varepsilon \nabla \sigma) \cdot \underset{\sim}{\vec{E}_0}}{\sigma + j\omega\varepsilon} \right) \underset{\sim}{\vec{E}_0}^* - \frac{1}{2} \underset{\sim}{\vec{E}_0} \cdot \underset{\sim}{\vec{E}_0}^* \nabla \varepsilon \right], \tag{16.27}$$

where σ and ε have been assumed nonuniform but constant. For fields with multiple frequencies, the time-averaged forces $\langle \vec{f} \rangle$ from each frequency can be calculated independently and summed to determine the total force, because the time-averaged force from cross terms are zero owing to orthogonality of the sinusoidal functions.

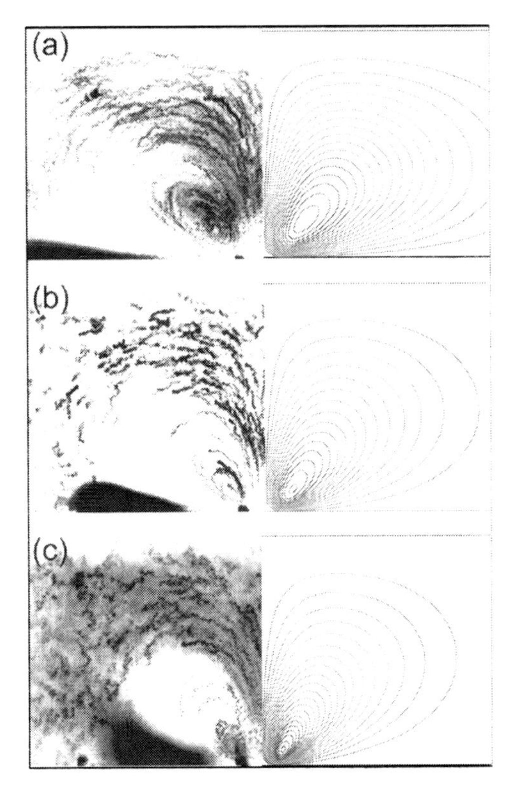

Fig. 16.6 Comparison between particle pathlines (left) and computed streamlines (right). (Reproduced with permission from [203].)

16.5 SUMMARY

In this chapter, we have addressed the dynamics of diffuse charge in EDLs near non-Faradaic electrodes and identified the characteristic times for these double layers as functions of the double-layer capacitance and the bulk solution resistance. A key result is that, for symmetric electrolytes, the characteristic time of the equilibration of Debye–Hückel EDLs on electrodes spaced by L is given by

$$\tau = RC = \frac{\lambda_D L}{D}, \qquad (16.28)$$

indicating that the characteristic time is a function both of double-layer thickness and electrode separation. In general, the differential capacitance of the EDL can be modeled with varying degrees of accuracy by the Debye–Hückel model,

$$C'' = \varepsilon/\lambda_D\,; \qquad (16.29)$$

the Poisson–Boltzmann model,

$$C'' = \frac{\varepsilon}{\lambda_D}\left[\cosh\left(\frac{z\varphi_0^*}{2}\right)\right]; \tag{16.30}$$

the Gouy–Chapman–Stern model,

$$C'' = \frac{\frac{\varepsilon_S \varepsilon}{\lambda_S \lambda_D}\left[\cosh\left(\frac{z\varphi_0^*}{2}\right)\right]}{\frac{\varepsilon_S}{\lambda_S} + \frac{\varepsilon}{\lambda_D}\left[\cosh\left(\frac{z\varphi_0^*}{2}\right)\right]}; \tag{16.31}$$

and the modified Poisson–Boltzmann model,

$$C'' = \frac{\varepsilon}{\lambda_D}\left[\frac{\sinh\left(z\varphi_0^*\right)}{\left[1+2\xi\sinh^2\left(\frac{z\varphi_0^*}{2}\right)\right]\sqrt{\frac{2}{\xi}\ln\left[1+2\xi\sinh^2\left(\frac{z\varphi_0^*}{2}\right)\right]}}\right]. \tag{16.32}$$

With a specified EDL capacitance, *RC* circuit models can predict the double-layer potential φ_0 as a function of the frequency of the applied signal, and $\varphi_0(t)$ can be combined with the Navier–Stokes equations to predict the fluid flow. Two examples of this type of fluid flow are flow around a conducting obstacle for flow induced by AC fields applied in interdigitated electrodes.

In the presence of spatially varying fluid properties (usually owing to thermal variations), applied electric fields lead to diffuse charge throughout the flow field, leading to an electrostatic body force term and electrothermal flows. This electrostatic body force term can be written in time-averaged form for sinusoidal fields:

$$\left\langle \vec{f} \right\rangle = \frac{1}{2}\mathrm{Re}\left[\left(\frac{(\sigma\nabla\varepsilon - \varepsilon\nabla\sigma)\cdot \vec{\underset{\sim}{E}}_0}{\sigma + j\omega\varepsilon}\right)\vec{\underset{\sim}{E}}_0^{\,*} - \frac{1}{2}\vec{\underset{\sim}{E}}_0 \cdot \vec{\underset{\sim}{E}}_0^{\,*}\nabla\varepsilon\right]. \tag{16.33}$$

16.6 SUPPLEMENTARY READING

Chang and Yeo [52] cover AC electrokinetic effects in detail and go well beyond the coverage in this chapter. Early derivations of double-layer capacitances can be found in [108, 204]. Bazant and co-workers have studied EDL dynamics in great detail with a view toward managing the fluid flow in nonequilibrium EDLs developed at electrodes or metallic surfaces [111, 112, 202, 205, 206, 207, 208], and this chapter draws directly from that work. Green et al. [203, 209], Ramos et al. [210], and Gonzales et al. [211] have presented descriptions of AC electroosmosis effects, and [212] shows net pumping by asymmetric electrodes. Ref. [213] shows use of traveling waves to generate DC pumping. di Caprio et al. have presented modified Poisson–Boltzmann theories with attention to capacitance effects [214]. Much current attention is focused on the challenges regarding quantitative predictions of electrokinetic phenomena at electrodes – the fluid velocity magnitudes (and even signs) predicted by analysis described in this chapter often do not match experiments, an issue discussed in detail in [208]. A review by Dukhin [167] focuses on the equilibrium assumptions made in Chapter 13 and how departure from equilibrium affects colloidal motion and characterization of surfaces.

Although the surfaces discussed in this chapter are all conducting, interfacial charge is also created when an electric field is applied normally to any interface with mismatched permittivity or conductivity – so electric fields applied to any insulating particle whose properties are not matched to its suspending medium creates interfacial charge. The resulting dipole on a particle is the source of dielectrophoretic forces, described in Chapter 17. This interfacial charge also induces fluid flow, with characteristic frequencies similar to those of the ACEO and ICEO flows described in this chapter. In fact, these flows are related to the variation in observed dielectrophoretic responses at kilohertz frequencies described in the dielectric spectroscopy literature as the alpha relaxation. Compared with the flows at conducting surfaces, though, the fluid flows induced at insulating objects are smaller in magnitude.

16.7 EXERCISES

16.1 Consider the equilibration of the double layer around a small particle or macromolecule with radius $a \ll \lambda_D$. Explain why the relevant time scale for double-layer equilibration is $\tau = \lambda_D^2/D$.

16.2 Consider the equilibration of the double layer around a conducting particle with radius $a \gg \lambda_D$. Explain why the relevant time scale for double-layer equilibration is $\tau = a\lambda_D/D$.

16.3 For a symmetric electrolyte in the Debye–Hückel limit, evaluate the net charge in the double layer by integrating the charge distribution. Show that the net charge in the double layer is given by $q''_{edl} = -\varepsilon\varphi_0/\lambda_D$. Show that the capacitance per unit area is given by $C'' = \varepsilon/\lambda_D$.

16.4 For a symmetric electrolyte outside the Debye–Hückel limit, evaluate the net charge in the double layer by evaluating the derivative of the potential at the wall:

(a) Start with the Poisson–Boltzmann equation (9.16) and multiply both sides by $2(\frac{d}{dy}\varphi)$.

(b) Integrate both sides, rearranging the derivatives so that on one side you have the derivative of $(\frac{d}{dy}\varphi)^2$ and on the other side you have $d\varphi$.

(c) To evaluate $\frac{d}{dy}\varphi$, integrate over a dummy variable (y' or φ', depending on which side of the equation) from the bulk ($y' = \infty$; $\varphi' = 0$) to some point in the double layer ($y' = y$; $\varphi' = \varphi'$).

(d) Evaluate $\varepsilon\frac{d}{dy}\varphi$ at the wall.

Show that the net charge in the double layer is given by

$$q''_{edl} = -\frac{\varepsilon\varphi_0}{\lambda_D}\left[\frac{2}{z\varphi_0^*}\sinh\left(\frac{z\varphi_0^*}{2}\right)\right]. \tag{16.34}$$

and that the differential capacitance per unit area $dq''/d\varphi_0$ is given by

$$C'' = \frac{\varepsilon}{\lambda_D}\left[\cosh\left(\frac{z\varphi_0^*}{2}\right)\right]. \tag{16.35}$$

16.5 For a symmetric electrolyte, plot the capacitance (normalized by ε/λ_D) versus $z\varphi_0^*$ for (a) the nonlinear Poisson–Boltzmann double layer and (b) a Poisson–Boltzmann diffuse double layer in series with a Stern layer with $\varepsilon_S = \varepsilon/10$ and $\lambda_S = \lambda_D/1000$. Plot the modified Poisson–Boltzmann model for $\xi = 1 \times 10^{-4}$, $\xi = 1 \times 10^{-4.5}$, and $\xi = 1 \times 10^{-5}$. Plot results over the range $0 < z\varphi_0^* < 35$ and $0 < C''\lambda_D/\varepsilon < 650$. Compare the results of the Stern modification with the results of the steric hindrance modification.

16.6 Consider two electrodes separated by a distance $2l$ and containing a symmetric electrolyte with all ions having mobilities and diffusivities of equal magnitude. Suppose that two different voltages are applied to the two electrodes. How long does it take for the double layers at the electrodes to form?

Treat the bulk fluid as a resistor. Show that the resistance $R = \frac{l}{\sigma A}$ (where σ is the bulk conductivity and A is the cross-sectional area of the electrolyte linking the two electrodes) can also be written as $\frac{\lambda_D^2 l}{\varepsilon D A}$.

Treat each double layer in the Debye–Hückel limit as a Helmholtz capacitor with thickness equal to λ_D ($C = \frac{\varepsilon A}{\lambda_D}$). Model the system as a capacitor, resistor, and capacitor in series, and show that RC for this model system is equal to $\frac{\lambda_D l}{D}$.

Given this result, what is the characteristic time for the double layer to equilibrate if you have two microelectrodes separated by 40 μm in a 1-mM NaCl solution? Approximate D for these ions as 1.5×10^{-9} m^2/s.

16.7 Reconsider the previous problem by assuming that the double layer has two parts, which can be considered as capacitors in series (Fig. 16.7). The first part is the diffuse double layer, which can be modeled as being of thickness λ_D with a permittivity of ε. The second is the Stern layer and is modeled as having a capacitance of $C_S = \frac{\varepsilon}{\lambda_S}$, where λ_S is an effective thickness that accounts for both the thickness of the Stern layer and its electrical permittivity (which is typically below the permittivity of the bulk solution). Show that the previous result must be modified by a factor of $(1 + \frac{\lambda_S}{\lambda_D})^{-1}$.

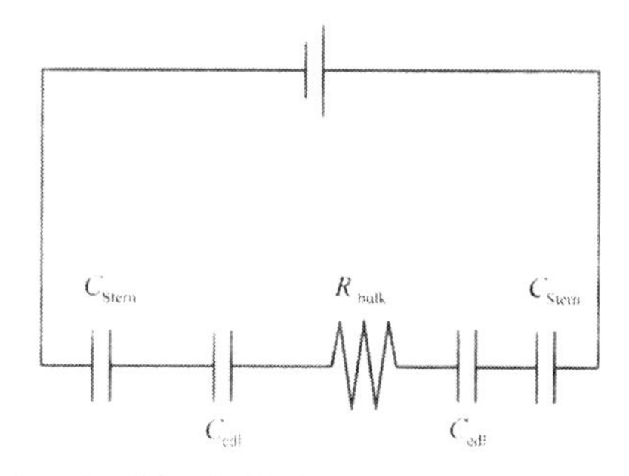

Fig. 16.7 Equivalent circuit for double-layer charging with Stern layers.

16.8 Estimate the velocities and time dependence of velocities at the electrode surface during an ACEO cycle.

16.9 Given what you know about induced-charge electroosmosis, consider an AC field applied to an infinitely thin metal plate in one of two positions:

(a) With the long edge tangent to the applied electric field.

(b) With the long edge normal to the applied electric field.

Assume the metal plate is approximately 10 μm long and infinitely thin. Qualitatively predict the flow induced by an AC field applied to this system. Assume the frequency of the AC field is low enough that the field oscillation is slow compared with the double-layer charging time. Include sketches of the electric field lines at various points during the field cycle.

16.10 Consider a 20-μm-deep insulating microchannel that is infinitely wide and 1 cm long. Assume that, halfway along the length of the channel, a 200-μm patch of conducting metal is patterned along the wall. Estimate the magnitude of the flow resulting from a 100-Hz signal of 150 V applied at one end of the microchannel while the other end is grounded. Use a Schwarz–Christoffel transform or a numerical simulation to calculate the electric field in the channel.

16.11 Consider the previous problem, but now assume that the conducting metal is covered with a layer of glass that is 1 nm thick. How does this change the flow? How is the flow changed by a layer that is 1 μm thick?

16.12 Derive Eq. (16.27) as follows:

(a) Write the Poisson equation as $\rho_E = \nabla \cdot \varepsilon \vec{E}$. Expand $\vec{E}$ in terms of the uniform-property solution $\vec{E}_0$ (which is large in magnitude and divergence free) plus a perturbation owing to the property variations (which is small in magnitude but of finite divergence). Replace ρ_E in the electrostatic body force term with an expression using these two electric fields.

(b) Write the charge conservation equation and assume that convective charge fluxes are small compared with electrophoretic charge fluxes. Defend this assumption. Derive Eq. (16.26).

(c) Assume sinusoidal fields and write the fields as their analytic representations. Solve the resulting equation for the divergence of the perturbation field, and thus derive Eq. (16.27).

16.13 Describe the governing equations and boundary conditions used to solve for the outer solution for AC electroosmosis between two electrodes. The electrodes are each 40 μm wide and separated by 40 μm. The microchannel depth is 1 μm, and the ionic strength of the medium is 10 mM. The electrode at left is connected to electrical ground, and the electrode at right is energized to a voltage given by $V = V_0 \cos \omega t$. In particular, the RC time constant for different regions of the electrodes should be different, and thus the boundary conditions on the outer solution for the Laplace equation will be spatially varying.

16.14 Consider two infinite parallel plates located at $y = \pm h$, $h \gg \lambda_D$, connected to an AC power supply that generates a differential voltage of $V = V_0 \cos \omega t$. The plates are covered with a dielectric thin film of $\varepsilon = 5\varepsilon_0$ and thickness 50 nm, which is treated as an additional capacitor in parallel with the fluid, which is a 1-mM solution of KCl. A small transverse electric field $\vec{E} = \hat{x} E_0 \cos(\omega t + \alpha)$ is applied. What is the time-averaged flow field as a function of h and α?

16.15 Consider a 1D system of length L with three domains, each of length $L/3$. Although the system is water throughout, assume that the domains at left and right are at temperature T_1, with permittivity ε_1 and conductivity σ_1. The domain in the center is at temperature T_2, with permittivity ε_2 and conductivity σ_2. For times $t < 0$, the applied electric field is zero, as is the net charge density and interfacial charge.

(a) At time $t = 0$, voltage V is applied at left and the voltage at right is held at $V = 0$. For a time that is long compared with orientational relaxation of the water but slow compared with motion of free charge in the water, what is the net charge density in the water? What is the charge density at the interfaces between the three regions?

(b) What is the charge density in the water and at the interfaces for $t \to \infty$? What is the pressure in the central region?

16.16 Assume that a focused laser beam maintains an elevated temperature at a specific point in an otherwise uniform flow. Assume that this temperature is maintained regardless of any fluid flow or other processes. Model this point as a sphere of radius R with a conductivity that is $\Delta\sigma$ different than the surrounding fluid and a permittivity that is $-\Delta\varepsilon$ different than the surrounding fluid. Model the force generated by this hot spot and show that the resulting flow can be approximated by a stresslet whose stress vector is aligned with the applied electric field.

17 Particle and Droplet Actuation: Dielectrophoresis, Magnetophoresis, and Digital Microfluidics

Microsystems use several techniques to actuate particles beyond the electrophoresis discussed in Chapter 13. Two physical phenomena are described in this chapter – dielectrophoresis and magnetophoresis – which are commonly used in microdevices to manipulate particles or droplets in suspension. This chapter also discusses digital microfluidics, which is not a physical phenomenon, but rather a system concept for manipulating fluid droplets using AC electric fields.

17.1 DIELECTROPHORESIS

Dielectrophoresis (DEP) is often used in microsystems as a mechanism for manipulating particles. It is appealing because the dielectrophoretic force on a particle scales with the characteristic length scale of the system to the -3 power, and dielectrophoretic forces are quite large when small devices are used. Further, particle response varies based on the frequency and phase of the applied field. Because the user can change particle response by changing a setting on a function generator, DEP measurements afford great flexibility to the user. Because of this, DEP has been used for many applications, with one example shown in Fig. 17.1.

The term dielectrophoresis refers to the Coulomb response of an electrically polarized object in a nonuniform electric field. In contrast to linear electrophoresis, it (a) does not require that the object have a net charge and (b) has a nonzero time-averaged effect even if AC electric fields are used.

Consider, as an example, a spherical, uncharged, uniform, ideal dielectric particle with a finite polarizability, expressed using its electrical permittivity ε_p, suspended in empty space. If a uniform electrical field is applied to this system, the sphere polarizes, and a net positive charge is present at one end of the sphere whereas a net negative charge is present on the other end of the sphere. Given that the electric field is uniform, the Coulomb forces on either end of the sphere are equal and opposite, and the net Coulomb force is zero. If the electric field is nonuniform, however, the side of the sphere with the larger electric field feels a larger attractive force, and the net force moves the particle toward the region of high electric field. Motion toward regions of high electric field is termed *positive dielectrophoresis*.

Microfluidic applications involve particles suspended in a medium (usually an aqueous solution) with electrical permittivity ε_m. For the moment, we assume that the medium is a perfect dielectric as well. In this case, the arguments are similar to the preceding ones, except that the particle and the medium *both* polarize. For a particle suspended in a medium, the net force on the particle is dependent on the *difference* between the polarization of the particle and the polarization of the medium. If the medium polarizes less than the particle, the particle experiences positive DEP and moves toward the regions of high electric field. If the medium polarizes more than the particle,

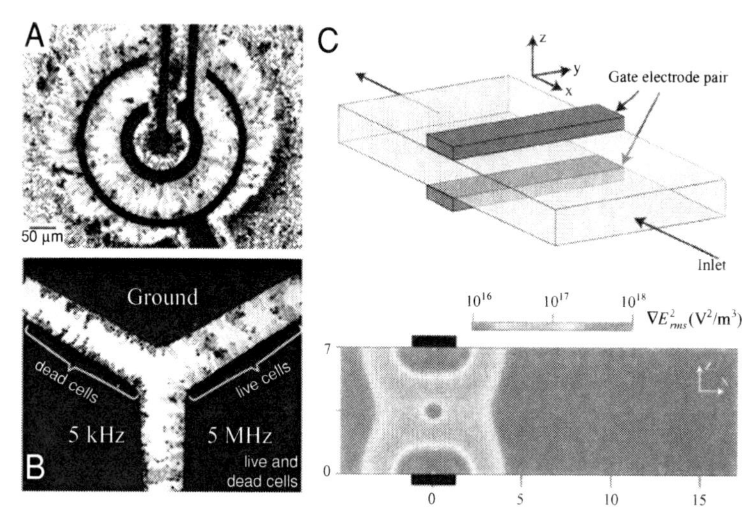

Fig. 17.1 Capture and sorting of cells using DEP. (A), (B) Live and dead cells are captured in distinct regions [215] using multiple electrode configurations and multiple-frequency DEP. (C) A dielectrophoretic "gate" [216] controls passage of selected particles downstream of the gate. Reproduced with permission.

the particle experiences *negative dielectrophoresis* and moves toward the regions of low electric field. In both cases, the direction of motion of the particle is a function of the electric field magnitude, but not its polarity. Thus the dielectrophoretic response of an *uncharged, uniform, ideal dielectric* object in an ideal dielectric medium is independent of whether a DC or AC field is used, or even the frequency of the field, as long as the permittivity is independent of frequency.

The preceding description highlights the basic physics of DEP – by controlling the polarization of a particle with respect to the suspending medium, charge is generated at the interface between the particle and the medium, and this charge leads to a net force if the electric field is nonuniform. This charge is called *Maxwell–Wagner interfacial charge*. The motion of the particle is dictated by the sign and magnitude of this charge. In the upcoming sections, we quantify the response of a uniform, uncharged sphere and extend the analysis to include media and particles with finite conductivity and permittivity; we describe Maxwell equivalent body techniques for describing nonuniform isotropic particles; and we extend the analysis to account for surface charge and the attendant EDL. We also describe the response of nonspherical particles as well as electrorotation and traveling-wave DEP.

Although the analysis in this section is mostly focused on predicting the dielectrophoretic force $\vec{F}_{\mathrm{DEP}}$, the velocity of a particle in a nonuniform field is the end result. In a manner analogous with descriptions of electrophoretic mobility or electroosmotic mobility, the time-averaged dielectrophoretic velocity can be described by use of a dielectrophoretic mobility, μ_{DEP}:

definition of dielectrophoretic mobility

$$\langle \vec{u}_{\mathrm{DEP}} \rangle = \mu_{\mathrm{DEP}} \nabla \left| \vec{E}_{\mathrm{ext},0} \right|^2 , \tag{17.1}$$

which is written in terms of the magnitude of the *applied* electric field, i.e., the field that would have existed if the particle were absent. The subsections to follow derive expressions for this dielectrophoretic mobility.

17.1.1 Inferring the Coulomb force on an enclosed volume from the electric field outside the volume

All dielectrophoretic analysis techniques are enabled by the fundamental concept that the only information required for determining the Coulomb force on an enclosed volume is the material properties and electric field on the outer surface of that enclosed volume. This property follows mathematically from Maxwell's equations – for example, in the absence of magnetic fields, we can show that the force on an enclosed volume is given by

Coulomb force on an enclosed volume, zero magnetic fields

$$\vec{F} = \int_S \left[\varepsilon \vec{E}\vec{E} - \frac{1}{2}\bar{\bar{\delta}} \left(\varepsilon \vec{E} \cdot \vec{E} \right) \right] \cdot \hat{n} \, dA, \tag{17.2}$$

where S is the surface of the volume and $\hat{n}$ is a unit outward normal.

Predicting the force on a particle is greatly simplified if all required information lies outside the particle. Equation (17.2) illustrates that the force on the particle can be determined from only the electric field and permittivity at the surface. Thus two different particles, if they result in the same solution for the electric field in the suspending medium, feel the same electrostatic force, even if they lead to a *different solution* for the electric field inside the particle. Further, given two alternative systems suspended in a uniform medium and an arbitrary volume large enough to enclose either one, if the two systems lead to the same electric field on the outside of that volume, the force on each system is the same.

Multiple analytical tools are enabled by the independence of the force on the field internal to the particle. For example, if we can replace a particle with a simpler structure that creates the same electric field at the surface of the particle, then the force on the simple structure predicts the force on the physical particle. This is the basis for the effective dipole approximation for the dielectrophoretic force on a sphere, or the Maxwellian equivalent body for particles of regular shape but nonuniform material properties. Even if such a simple analytical approach is not possible (for example, for irregularly shaped objects), to find the solution we need only integrate Eq. (17.2), the so-called *Maxwell stress tensor* approach. The electric field inside the particle need not be known.

17.1.2 The force on an uncharged, uniform, isotropic sphere in a linearly varying electric field with uniform, isotropic phase

We solve for the force on a particle in terms of material properties and the applied field $\vec{E}_{ext}(\vec{r})$. By *applied* field, we refer to the electric field that would exist if the particle were absent. The electric field and potential in the presence of the particle are written as $\vec{E}(\vec{r})$ and $\phi(\vec{r})$.

We can derive the dielectrophoretic force on an uncharged, uniform, isotropic, dielectric sphere (Fig. 17.2) of radius a located at the origin in a *steady* linearly varying electric field with uniform, isotropic phase by first solving for the electrical potential ϕ around a sphere in a uniform applied field (here we take $\vec{E}_{ext} = E_0\hat{z}$) to evaluate its effective dipole and then by calculating the force on that effective dipole owing to a linear variation in electric field. Thus we solve Gauss's law for $\vec{E}(\vec{r})$ by using the Laplace equation,

$$\nabla \cdot \vec{D} = -\varepsilon \nabla^2 \phi = 0, \tag{17.3}$$

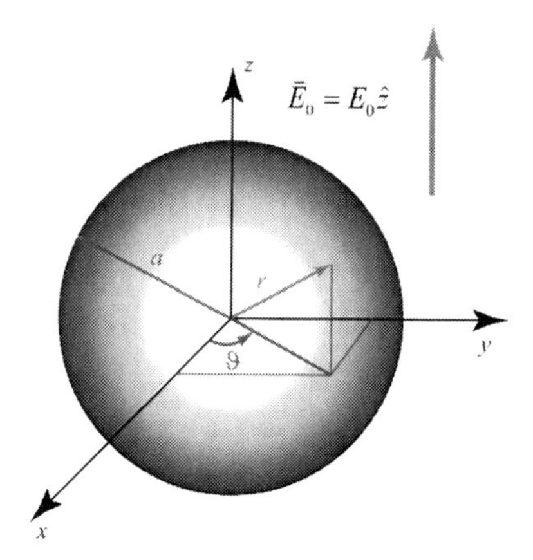

Fig. 17.2 A sphere of radius a under the influence of an electric field aligned with the z axis.

with boundary conditions at $r = \infty$,

$$\phi = -E_0 z, \tag{17.4}$$

and at $r = a$,

$$\varepsilon_{\mathrm{m}} \left.\frac{\partial \phi}{\partial n}\right|_{\mathrm{m}} - \varepsilon_{\mathrm{p}} \left.\frac{\partial \phi}{\partial n}\right|_{\mathrm{p}} = q'' = 0, \tag{17.5}$$

where subscripts m and p refer to the medium and particle, respectively, and the n direction is normal to the surface and directed outward, toward the suspending medium.

Rather than solving this steady problem, however, we immediately generalize it to include AC electric fields. For AC fields, the general approach is the same, but we combine the charge conservation and Poisson equations to create a complex version of the Laplace equation, which we can then solve in terms of the complex permittivity.

To determine the effective dipole induced in a sphere by a uniform applied electric field, we consider an uncharged sphere with uniform, isotropic properties ε_{p} and σ_{p} embedded in a medium with uniform, isotropic properties ε_{m} and σ_{m}. We assume a uniform applied sinusoidal electric field,

$$\vec{\boldsymbol{E}}_{\mathrm{ext}} = \vec{\boldsymbol{E}}_0 \cos \omega t = E_0 \hat{\boldsymbol{z}} \cos \omega t, \tag{17.6}$$

where E_0 is the peak magnitude of the applied electric field. We write the analytic representation of the applied field as

$$\underset{\sim}{\vec{\boldsymbol{E}}_0} = \vec{\boldsymbol{E}}_0 \exp j\omega t = E_0 \hat{\boldsymbol{z}} \exp j\omega t. \tag{17.7}$$

As is always the case when phasors are used in this text, the undertilde denotes a complex quantity and the subscript 0 denotes the phasor. As described in Section 5.3, the charge conservation equation (without convection),

$$\frac{\partial \rho_{\mathrm{E}}}{\partial t} + \nabla \cdot \sigma \vec{\boldsymbol{E}} = 0, \tag{17.8}$$

and the Poisson equation,

$$\nabla \cdot \varepsilon \vec{\boldsymbol{E}} = \rho_{\mathrm{E}}, \tag{17.9}$$

can be combined and written as one equation by using the complex permittivity $\underset{\sim}{\varepsilon} = \varepsilon + \sigma / j\omega$,

$$\nabla \cdot \underset{\sim}{\varepsilon} \underset{\sim}{\vec{\boldsymbol{E}}} = 0. \tag{17.10}$$

Equation (17.10) can also be written as an equation for the electrical potential,

$$\underset{\sim}{\varepsilon}\nabla^2\underset{\sim}{\phi} = 0\,, \tag{17.11}$$

with boundary conditions at $r \to \infty$,

$$\underset{\sim}{\phi} = -E_0 z \exp j\omega t\,, \tag{17.12}$$

and at $r = a$,

$$\underset{\sim}{\varepsilon}_{\mathrm{m}} \left.\frac{\partial\underset{\sim}{\phi}}{\partial n}\right|_{\mathrm{m}} - \underset{\sim}{\varepsilon}_{\mathrm{p}} \left.\frac{\partial\underset{\sim}{\phi}}{\partial n}\right|_{\mathrm{p}} = 0\,. \tag{17.13}$$

Here subscripts m and p indicate medium (fluid) and particle, respectively. The electrostatic condition (Gauss's law) and the electrodynamic condition (charge conservation) are simultaneously specified with this complex equation, and the solution provides the response at all frequencies as long as fluid convective transport of current can be ignored. The solution is written in terms of several complex quantities, which can be directly related to the dynamics of system response. This system has a solution in terms of Legendre polynomials, as described in Appendix F. Inside the sphere ($r < a$), the solution is

$$\underset{\sim}{\phi} = -\frac{3\underset{\sim}{\varepsilon}_{\mathrm{m}}}{\underset{\sim}{\varepsilon}_{\mathrm{p}} + 2\underset{\sim}{\varepsilon}_{\mathrm{m}}} E_0 r \cos\vartheta \exp j\omega t\,, \tag{17.14}$$

whereas the solution outside the sphere ($r > a$) is

$$\underset{\sim}{\phi} = -E_0 r \cos\vartheta \exp j\omega t + \frac{\underset{\sim}{\varepsilon}_{\mathrm{p}} - \underset{\sim}{\varepsilon}_{\mathrm{m}}}{\underset{\sim}{\varepsilon}_{\mathrm{p}} + 2\underset{\sim}{\varepsilon}_{\mathrm{m}}} E_0 \frac{a^3}{r^2} \cos\vartheta \exp j\omega t\,. \tag{17.15}$$

The solution outside the sphere consists of a term corresponding to the applied electric field ($\underset{\sim}{\phi} = -E_0 r \cos\vartheta \exp j\omega t = E_0 z \exp j\omega t$) and a term corresponding to a point dipole (Subsection F.1.1) at the origin. In terms of the Legendre polynomials bounded at infinity, i.e., the multipolar expansion, this dipole has coefficient $B_1 = E_0 \frac{\underset{\sim}{\varepsilon}_{\mathrm{p}} - \underset{\sim}{\varepsilon}_{\mathrm{m}}}{\underset{\sim}{\varepsilon}_{\mathrm{p}} + 2\underset{\sim}{\varepsilon}_{\mathrm{m}}} a^3 \exp j\omega t$, corresponding to an electrical dipole with dipole moment phasor

effective dipole moment of a isotropic, homogeneous, uncharged sphere in a uniform applied electric field

$$\vec{\underset{\sim}{p}}_0 = 4\pi\varepsilon_{\mathrm{m}} E_0 a^3 \frac{\underset{\sim}{\varepsilon}_{\mathrm{p}} - \underset{\sim}{\varepsilon}_{\mathrm{m}}}{\underset{\sim}{\varepsilon}_{\mathrm{p}} + 2\underset{\sim}{\varepsilon}_{\mathrm{m}}} \hat{z}\,. \tag{17.16}$$

The quantity

$$\underset{\sim}{f}_{\mathrm{CM}} = \frac{\underset{\sim}{\varepsilon}_{\mathrm{p}} - \underset{\sim}{\varepsilon}_{\mathrm{m}}}{\underset{\sim}{\varepsilon}_{\mathrm{p}} + 2\underset{\sim}{\varepsilon}_{\mathrm{m}}} \tag{17.17}$$

is called the *Clausius–Mossotti factor*; the Clausius–Mossotti factor is equivalent to the first-order (dipolar) polarization coefficient, often denoted by $\underset{\sim}{K}^{(1)}$.

The electrical potential *outside* $r = a$ is therefore identical to the electrical potential that would exist if the sphere were replaced with a point dipole whose magnitude oscillates at frequency ω with dipole moment phasor

effective dipole moment phasor for a sphere exposed to a sinusoidal electric field

$$\vec{\underset{\sim}{p}}_0 = 4\pi\varepsilon_{\mathrm{m}} E_0 a^3 \underset{\sim}{f}_{\mathrm{CM}} \hat{z}\,. \tag{17.18}$$

Because this effective dipole accurately predicts the field outside the particle, the full solution is typically ignored and the analysis proceeds with this effective dipole solution. Returning to real quantities, this *effective* dipole moment is given by

$$\vec{p} = \mathrm{Re}[\underset{\sim}{\vec{p}}] = 4\pi\varepsilon_m E_0 a^3 \hat{z}\left[\mathrm{Re}(\underset{\sim}{f_{CM}})\cos\omega t + \mathrm{Im}(\underset{\sim}{f_{CM}})\sin\omega t\right], \quad (17.19)$$

or, equivalently,

$$\vec{p} = \mathrm{Re}[\underset{\sim}{\vec{p}}] = 4\pi\varepsilon_m E_0 a^3 \hat{z}\left[\left|\underset{\sim}{f_{CM}}\right|\cos(\omega t + \angle \underset{\sim}{f_{CM}})\right], \quad (17.20)$$

where the magnitude of the Clausius–Mossotti factor is

$$\left|\underset{\sim}{f_{CM}}\right| = \sqrt{\mathrm{Re}(\underset{\sim}{f_{CM}})^2 + \mathrm{Im}(\underset{\sim}{f_{CM}})^2}, \quad (17.21)$$

and its angle is

$$\angle \underset{\sim}{f_{CM}} = \mathrm{atan2}\left[\mathrm{Im}(\underset{\sim}{f_{CM}}), \mathrm{Re}(\underset{\sim}{f_{CM}})\right]. \quad (17.22)$$

The magnitude of the dipole moment is proportional to $|\underset{\sim}{f_{CM}}|$, the magnitude of $\underset{\sim}{f_{CM}}$, and is proportional to the magnitude of the net charge buildup at either end of the sphere; the phase lag between the applied electric field and the effective dipole moment is $\angle \underset{\sim}{f_{CM}}$, the angle of $\underset{\sim}{f_{CM}}$. Equivalently, $\mathrm{Re}(\underset{\sim}{f_{CM}})$ is the portion of the effective dipole moment in phase with the applied electric field, and $\mathrm{Im}(\underset{\sim}{f_{CM}})$ is the portion of the effective dipole moment that is 90° out of phase with the applied electric field.

Because a point dipole at the origin can be used to create the same electric field outside the particle, and given our earlier observations in Subsection 17.1.1, the force on the volume enclosing the particle is the same as the force on any volume enclosing a point dipole at the origin with dipole moment specified by Eq. (17.16). Because the instantaneous force on a dipole is given by $\vec{F} = \vec{p}\cdot\nabla\vec{E}_{ext}$, the force on a particle is given by the dot product of the applied electric field gradient and the effective dipole moment. This relation is valid if the applied electric field varies linearly, and this linear variation is small enough over the length of the sphere that the effective induced dipole calculated for a *uniform* field is still accurate. The force on the sphere is thus given by the force on a dipole in a linearly varying electric field ($\vec{F} = \vec{p}\cdot\nabla\vec{E}_{ext}$ for the steady-state case).

The analytic representations of $\vec{p}$ and $\vec{E}$ are suitable as long as linear equations are used; however, $\vec{p}\cdot\nabla\vec{E}$ is a nonlinear expression and thus evaluating $\vec{p}\cdot\nabla\vec{E}$ using phasors must be executed carefully. We therefore use Eq. (G.33) to write the real dot product of $\vec{p}$ and $\nabla\vec{E}$ in terms of the phasors $\underset{\sim}{\vec{p}_0}$ and $\underset{\sim}{\vec{E}_0}$:

$$\vec{F} = \frac{1}{2}\left\{\mathrm{Re}\left(\underset{\sim}{\vec{p}_0}\cdot\nabla\underset{\sim}{\vec{E}_0}^*\right) + \cos\left[2\omega t + \angle\mathrm{Re}\left(\underset{\sim}{\vec{p}_0}\cdot\nabla\underset{\sim}{\vec{E}_0}\right)\right]\right\}. \quad (17.23)$$

The resulting instantaneous force is given by

$$\vec{F} = 2\pi\varepsilon_m a^3 \vec{E}_0\cdot\nabla\vec{E}_0\left(\mathrm{Re}(\underset{\sim}{f_{CM}}) + \mathrm{Re}(\underset{\sim}{f_{CM}})\cos 2\omega t + \mathrm{Im}(\underset{\sim}{f_{CM}})\sin 2\omega t\right), \quad (17.24)$$

or, applying the inverse chain rule for curl-free fields,

$$\vec{F} = \pi\varepsilon_m a^3 \nabla(\vec{E}_0\cdot\vec{E}_0)\left(\mathrm{Re}(\underset{\sim}{f_{CM}}) + \mathrm{Re}(\underset{\sim}{f_{CM}})\cos 2\omega t + \mathrm{Im}(\underset{\sim}{f_{CM}})\sin 2\omega t\right). \quad (17.25)$$

Thus the force on the particle has a DC component proportional to $\mathrm{Re}(\underset{\sim}{f_{CM}})$ and a component at 2ω with magnitude proportional to $|\underset{\sim}{f_{CM}}|$.[1] Although the *force* at 2ω is significant, the observed *velocity* at 2ω is damped by the viscous drag of the fluid surrounding the particle – for the same force magnitude, the velocity magnitude at 2ω is smaller than the DC motion by the factor $\sqrt{1+(2\omega\tau_p)^2}$, where $\tau_p = 2a^2\rho_p/9\eta$ is the characteristic time of particle equilibration. Thus at high frequencies, the oscillatory particle motion is damped. Because of this, we focus on the time-averaged force and velocity. The time-averaged force is

time-averaged DEP response, spherical, uniform, isotropic, uncharged particle, AC field

$$\left\langle \vec{F}_{\mathrm{DEP}} \right\rangle = \pi \varepsilon_m a^3 \mathrm{Re}(\underset{\sim}{f_{CM}}) \nabla (\vec{E}_0 \cdot \vec{E}_0), \tag{17.26}$$

where the angle brackets denote time averaging. For a DC applied field $\vec{E}_{\mathrm{ext}}$, the result has a similar form, but the expression has a different constant multiplier,

time-averaged DEP response, spherical, uniform, isotropic, uncharged particle, DC field

$$\left\langle \vec{F}_{\mathrm{DEP}} \right\rangle = 2\pi \varepsilon_m a^3 \mathrm{Re}(\underset{\sim}{f_{CM}}) \nabla (\vec{E}_{\mathrm{ext}} \cdot \vec{E}_{\mathrm{ext}}), \tag{17.27}$$

because the average value of $E_0{}^2$ is $E_0{}^2$, but the average value of $(E_0 \cos \omega t)^2$ is $\frac{1}{2} E_0{}^2$.

Different components of Eq. (17.26) illustrate the role of different physical phenomena. The real part of the Clausius–Mossotti factor, $\mathrm{Re}(\underset{\sim}{f_{CM}})$, ranges from 1 (for $|\underset{\sim}{\varepsilon}_p| \gg |\underset{\sim}{\varepsilon}_m|$) to $-\frac{1}{2}$ (for $|\underset{\sim}{\varepsilon}_m| \gg |\underset{\sim}{\varepsilon}_p|$) and captures the phase relation between the effective dipole and the applied electric field. The sign of $\mathrm{Re}(\underset{\sim}{f_{CM}})$ determines whether particles are attracted to, or repelled from, regions of high electric field magnitude. In the high-frequency limit, $\underset{\sim}{\varepsilon}$ can be replaced by ε in Eq. (17.17), and DEP effects are strictly due to the polarization of the medium and particle as expressed by their permittivities. In the low-frequency limit, $\underset{\sim}{\varepsilon}$ can be replaced by σ in Eq. (17.17), and DEP effects are strictly a function of the conductivities of the medium and particle. Because the force is proportional to a^3, the force on a sphere is proportional to the sphere volume. DEP is a second-order electrokinetic effect that scales with applied voltage squared, and if the separation between electrodes is changed at constant voltage, dielectrophoretic effects scale with electrode spacing to the -3 power.

From Eq. (17.26), we can immediately write the dielectrophoretic mobility of a sphere. For a spherical particle of radius a, for which the drag force is $\vec{F} = 6\pi\vec{u}\eta a$, the dielectrophoretic mobility μ_{DEP} is given by

DEP mobility of a uniform, homogeneous sphere; no surface charge; linearly varying electric field

$$\mu_{\mathrm{DEP}} = \frac{a^2 \varepsilon_m \mathrm{Re}(\underset{\sim}{f_{CM}})}{6\eta}. \tag{17.28}$$

[1] Equation (17.25) is often written with the RMS value of the applied electric field, in which case $\vec{E}_0$ is replaced with $\vec{E}_{\mathrm{RMS}}$ and the premultiplicative factor changes from $\pi\varepsilon_m a^3$ to $2\pi\varepsilon_m a^3$.

Because the effective dipole $\vec{p}$ and the applied electric field $\vec{E}_{\text{ext}}$ are collinear, the torque $\vec{p} \times \vec{E}_{\text{ext}}$ on a uniform sphere in a field with isotropic phase is zero.

17.1.3 Maxwellian equivalent body for inhomogeneous, spherically symmetric particles

Although the result for an uncharged homogeneous sphere, listed in Eq. (17.27), is straightforward to evaluate, many objects are heterogeneous. In particular, biological cells have a complicated internal structure, a gross model of which might include a lipid bilayer encapsulating a conductive cytosol with a defined nucleus. If the body can be approximated as spherically symmetric with several discrete layers, however, an effective dipole can again be defined.

Consider an uncharged sphere with a radius a_2 and a core of radius a_1 suspended in a uniform medium. Let the properties of the core be ε_1 and σ_1, the properties of the shell be ε_2 and σ_2, and the properties of the medium be ε_m and σ_m. Given a uniform harmonic applied field, this system consists of the complex equation

$$\underset{\sim}{\varepsilon} \nabla^2 \underset{\sim}{\phi} = 0 , \tag{17.29}$$

with boundary conditions at $r \to \infty$,

$$\underset{\sim}{\phi} = -E_0 z , \tag{17.30}$$

at $r = a_1$,

$$\underset{\sim}{\varepsilon_2} \left. \frac{\partial \underset{\sim}{\phi}}{\partial n} \right|_2 - \underset{\sim}{\varepsilon_1} \left. \frac{\partial \underset{\sim}{\phi}}{\partial n} \right|_1 = 0 , \tag{17.31}$$

and at $r = a_2$,

$$\underset{\sim}{\varepsilon_m} \left. \frac{\partial \underset{\sim}{\phi}}{\partial n} \right|_m - \underset{\sim}{\varepsilon_2} \left. \frac{\partial \underset{\sim}{\phi}}{\partial n} \right|_2 = 0 . \tag{17.32}$$

This system, although algebraically more tedious than the one for a homogeneous sphere, can also be solved by use of Legendre polynomials, and the solution outside the particle can again be written in terms of the uniform applied field plus the response of a dipole at the origin. The solution for the effective dipole phasor can be written in a form identical to Eq. (17.16),

$$\underset{\sim}{\vec{p}_0} = 4\pi\varepsilon_m E_0 a^3 \frac{\underset{\sim}{\varepsilon_p} - \underset{\sim}{\varepsilon_m}}{\underset{\sim}{\varepsilon_p} + 2\underset{\sim}{\varepsilon_m}} \hat{z} , \tag{17.33}$$

if we set the radius in Eq. (17.33) equal to the outer radius of the composite particle ($a = a_2$) and write the effective complex permittivity $\underset{\sim}{\varepsilon_p}$ of the particle as a function of the properties and thicknesses of the two particle components:

effective complex permittivity of a sphere with core of radius a_1 and total radius of a_2

$$\underset{\sim}{\varepsilon_p} = \underset{\sim}{\varepsilon_2} \left[\frac{\frac{a_2^3}{a_1^3} + 2\frac{\underset{\sim}{\varepsilon_1} - \underset{\sim}{\varepsilon_2}}{\underset{\sim}{\varepsilon_1} + 2\underset{\sim}{\varepsilon_2}}}{\frac{a_2^3}{a_1^3} - \frac{\underset{\sim}{\varepsilon_1} - \underset{\sim}{\varepsilon_2}}{\underset{\sim}{\varepsilon_1} + 2\underset{\sim}{\varepsilon_2}}} \right] . \tag{17.34}$$

As was the case for a homogeneous uncharged sphere, we can describe the force on the uncharged core–shell sphere with an effective dipole because this effective dipole induces an electric field outside the particle that is identical to that generated by the core–shell sphere itself. For particles with more than one shell, the effective permittivity of the particle is determined by repeated application of Eq. (17.34), starting from the core, calculating the effective permittivity of the core plus one shell, then calculating the effective permittivity of that plus the second shell, and so on, working out through all of the shells.

MAXWELLIAN EQUIVALENT BODY FOR THIN OUTER SHELLS

When the outer shell is vanishingly thin relative to the core, Eq. (17.34) takes on a simplified form. Consider the solution from Eq. (17.34), but let $\Delta a = a_2 - a_1$ with $\Delta a \ll a_2$. We can write a linear expansion ($a_2^3/a_1^3 = 1 + 3\Delta a/a_2$) and show that the effective permittivity is given by

effective complex permittivity of a sphere with total radius of a_2 and thin outer shell of thickness Δa

$$\underset{\sim}{\varepsilon}_p = \underset{\sim}{\varepsilon}_2 \frac{\underset{\sim}{\varepsilon}_1 + \frac{\Delta a}{a_2}\left(\underset{\sim}{\varepsilon}_1 + 2\underset{\sim}{\varepsilon}_2\right)}{\underset{\sim}{\varepsilon}_2 + \frac{\Delta a}{a_2}\left(\underset{\sim}{\varepsilon}_1 + 2\underset{\sim}{\varepsilon}_2\right)} . \tag{17.35}$$

If $\sigma_2 \gg \sigma_1$ and $\varepsilon_2 \gg \varepsilon_1$, as might be the case for an electrolyte film at the surface of a polystyrene bead, this simplifies to

effective complex permittivity of a sphere with total radius of a_2 and thin, highly permittive/conductive outer shell of thickness Δa

$$\underset{\sim}{\varepsilon}_p = \underset{\sim}{\varepsilon}_1 + 2\underset{\sim}{\varepsilon}_2 \frac{\Delta a}{a_2} , \tag{17.36}$$

which is reminiscent of the parallel relation for capacitors, because, in this limit, the core is the primary resistance to current, current goes primarily through the shell layer, and thus the field lines are almost parallel to the shell surface. If $\sigma_2 \ll \sigma_1$ and $\varepsilon_2 \ll \varepsilon_1$, as might be observed in a biological cell with a low-conductivity, low-permittivity lipid bilayer at its surface, this simplifies to

effective complex permittivity of a sphere with total radius of a_2 and thin, negligibly permittive/conductive outer shell of thickness Δa

$$\underset{\sim}{\varepsilon}_p = \frac{\underset{\sim}{\varepsilon}_1 \underset{\sim}{\varepsilon}_2}{\frac{\Delta a}{a_2}\underset{\sim}{\varepsilon}_1 + \underset{\sim}{\varepsilon}_2} , \tag{17.37}$$

which is reminiscent of the series relation for capacitors, because in this limit the shell is the primary resistance to current, and thus the field lines are normal to the shell surface.

When inferring particle properties from experimental measurements, it is often difficult to independently measure Δa and $\underset{\sim}{\varepsilon}_2$, because experimental data are a function of their product, as in Eq. (17.36), or their ratio, as in Eq. (17.37). Thus experiments on cells usually measure a complex cell membrane capacitance per unit area $\underset{\sim}{C}'' = \frac{\underset{\sim}{\varepsilon}_2}{\Delta a}$, whereas experiments on particles with thin conductive shells usually measure a complex shell conductance $\underset{\sim}{G} = j\omega\underset{\sim}{\varepsilon}_2 \Delta a$.

17.1.4 Dielectrophoresis of charged spheres

The analysis for spheres with no surface charge often gives a good estimate for the performance of relatively large spheres in the thin-EDL limit, especially when the particle itself is relatively conductive. However, if the particle is not conductive or if the double layer is not thin, the presence of charge on the surface, and particularly the double layer, affects the net dipole on the particle. In fact, the response of the double layer routinely dominates the dielectrophoretic response of nanoparticles.

Response owing to fixed surface charge. In a system with no free charge (e.g., a solid particle suspended in space), the presence of uniform fixed surface charge on the surface of a particle does not directly lead to a change in the dielectrophoretic response of the particle. For example, consider a particle with radius a and a surface charge density q''. If this system is exposed to a uniform DC or AC electric field, the solution for the electric field outside the particle is unchanged except for an additional steady term given by

$$\phi = \frac{a^2 q''}{\varepsilon r}, \tag{17.38}$$

which is just the electric potential induced by a charge $4\pi a^2 q''$, the total charge on the particle, at the origin. Thus the presence of fixed charge leads to a net Coulomb force on the particle, but it does not change the induced dipole caused by the electric field, or the dielectrophoretic response.

Response owing to property variation in the EDL. Although a fixed charge density on the surface of an object does not lead to a change in the induced dipole, in aqueous systems, the surface charge density is balanced by the charge in the EDL, and the ion distribution in the EDL *does* change the dielectrophoretic response. If the double layer is spherically symmetric and the electromigration of ions can be approximated with an effective ohmic conductivity, the effect of the ion concentration in the double layer can be incorporated into a general multishell model for the effective dipole response of the particle, or, if the double layer is thin, the effect of the increased conductivity and reduced permittivity can be handled with a thin-shell model, as discussed in Subsection 17.1.3. This approach neglects the *fluid convection* and accounts for all ion electromigration (including convective transport) through the use of an effective ohmic double-layer conductivity. If convective processes distort the double layer and cause fluid properties in the double layer to vary from spherical symmetry (cf. Fig. 13.4), then spherically symmetric shell models cannot predict the effective dipole of the particle – no simple model exists for describing dielectrophoretic response when this occurs. Double-layer asymmetry occurs at low frequencies (the so-called alpha relaxation), when the cycle is long enough that an appreciable amount of charge can build up in the double layer on either end of a particle and convection can carry this charge beyond the boundaries of the static double layer. We also discuss double-layer asymmetry in Subsection 13.3.3 as a reason for nonlinearity in the electrophoretic response of particles – there, electrophoretic response is nonlinear when the double layer becomes distorted and generates an adverse electric field.

17.1.5 Dielectrophoresis of nonspherical objects or objects in nonlinearly varying fields

DEP of spheres is analytically convenient because the effective dipole and the dielectrophoretic force can be calculated straightforwardly. For nonspherical objects, the calculations are more complex. This section describes two general techniques, namely the Maxwell stress tensor approach and the multipole expansion approach, as well as specific results for DEP of ellipsoids.

MAXWELL STRESS TENSOR APPROACH

We can determine the dielectrophoretic force on an uncharged particle with irregular shape by solving the Laplace equation for the potential distribution inside and outside the particle, and integrating the Maxwell stress tensor $\vec{\vec{T}}$, given in the absence of magnetic fields by

definition of Maxwell stress tensor, zero magnetic fields

$$\vec{\vec{T}} = \varepsilon \vec{E}\vec{E} - \frac{1}{2}\vec{\vec{\delta}}\left(\varepsilon \vec{E} \cdot \vec{E}\right), \tag{17.39}$$

to give the force,

force on an enclosed volume in terms of the Maxwell stress tensor at the surface

$$\vec{F} = \int_{S} (\vec{\vec{T}} \cdot \hat{n})\, dA, \tag{17.40}$$

where S denotes the surface of the particle and $\hat{n}$ is an outward-pointing unit normal.

For a harmonic electric field specified by $\vec{\underset{\sim}{E}}_{ext} = \vec{\underset{\sim}{E}}_0 \exp j\omega t$, the time-averaged form of the stress tensor is

$$\left\langle \vec{\vec{T}} \right\rangle = \frac{1}{4}\mathrm{Re}(\underset{\sim}{\varepsilon})\left[(\vec{\underset{\sim}{E}}_0 \vec{\underset{\sim}{E}}_0^{\,*} + \vec{\underset{\sim}{E}}_0^{\,*} \vec{\underset{\sim}{E}}_0) - |\vec{\underset{\sim}{E}}_0|^2 \vec{\vec{\delta}} \right], \tag{17.41}$$

where $\vec{\vec{\delta}}$ is the unit or identity tensor, $\langle \cdots \rangle$ denotes time averaging, and the asterisk denotes complex conjugation. The symbol $\vec{\underset{\sim}{E}}_0$ denotes the phasor of the electric field.

The Maxwell stress tensor technique is the most general, but because of its computational demands, it is used only when other analytical techniques are inapplicable to the particle under study.

MULTIPOLE EXPANSION

A multipolar expansion, in general, is an expansion of the electric field in powers of $1/r$ corresponding to solutions of the Laplace equation obtained by separation of variables (described in Appendix F). Approximate solutions may be obtained by truncating the expansion, resulting in solutions that are exact for large r but that have finite error when r is finite. For some simple geometries, an exact result can be obtained with a small number of terms. For example, for a sphere in a linearly varying electric field, the exact result requires only a uniform field term and an effective dipole term. Multipole expansions are always implemented with a finite number of terms, and in that sense are always approximations of the (exact) Maxwell stress tensor description. This approach is useful in cases in which the particle geometry is simple, but either (a) variations in the electric field are on the same length scale as the particle, or (b) the dipole moment (the $n = 1$

term of the general multipole expansion) is zero, such as locations where the local field is zero but the gradients are nonzero.

Although a particle of general shape must be described by a general multipole expansion, axisymmetric particles in an axisymmetric field can be described using *linear* multipoles generated by use of separation of variables applied to the axisymmetric Laplace equation (see Subsection F.1.1). Generalizing the previous result for a sphere in a linearly varying field to obtain the result for a sphere in a general axisymmetric AC field aligned with the z axis ($\vec{\underset{\sim}{E}} = E_0 \hat{z} \exp j\omega t$), we find that the effective multipole moment phasors are

$$\vec{\underset{\sim}{p}}_0^{(k)} = \frac{4\pi\varepsilon_m a^{2k+1}}{(k-1)!} \underset{\sim}{K}^{(k)} \left(\frac{\partial^{k-1}}{\partial z^{k-1}} E_0 \right)^2 \hat{z} , \tag{17.42}$$

where $\underset{\sim}{K}^{(k)}$ is

$$\underset{\sim}{K}^{(k)} = \frac{\underset{\sim}{\varepsilon}_p - \underset{\sim}{\varepsilon}_m}{k\underset{\sim}{\varepsilon}_p + (k+1)\underset{\sim}{\varepsilon}_m} . \tag{17.43}$$

For the preceding results in this chapter for a linearly varying field, the only k with a nonzero multipole moment phasor is the $k = 1$ (dipolar) term. The time-averaged force on the linear multipole is given by

$$\left\langle \vec{F} \right\rangle = \sum_{k=1}^{N} \left\langle \vec{F}_k \right\rangle = \sum_{k=1}^{N} \frac{2\pi\varepsilon_m a^{2k+1}}{k!(k-1)!} \mathrm{Re}\left(\underset{\sim}{K}^{(k)} \right) \left(\frac{\partial^{k-1}}{\partial z^{k-1}} E_0 \right)^2 \hat{z} , \tag{17.44}$$

The accuracy of this approximation is good if the electric field is close to axisymmetric on the length scale of the particle.

DEP OF ELLIPSOIDS

In general, the shape of a polarized particle influences the electric field it creates. When a particle's shape deviates from spherical, the applicability of the equivalent dipole representation based on a spherical particle rapidly decreases. Consider an ellipsoidal particle with a long axis, a_1, and minor axes a_2 and a_3. The effective dipole moment has three vector components:

$$\underset{\sim}{p}_{0,i} = 4\pi a_1 a_2 a_3 \varepsilon_m \underset{\sim}{f}_{\mathrm{CM},i} E_{0,i} , \tag{17.45}$$

where each axis has its own Clausius–Mossotti factor,

$$\underset{\sim}{f}_{\mathrm{CM},i} = \frac{\underset{\sim}{\varepsilon}_p - \underset{\sim}{\varepsilon}_m}{3\left[\underset{\sim}{\varepsilon}_m + (\underset{\sim}{\varepsilon}_p - \underset{\sim}{\varepsilon}_m) L_i\right]} , \tag{17.46}$$

defined with *depolarization factors* L_i,

$$L_i = \frac{a_1 a_2 a_3}{2} \int_0^\infty \frac{1}{(\ell + a_i{}^2)\sqrt{(\ell + a_1{}^2)(\ell + a_2{}^2)(\ell + a_3{}^2)}} d\ell , \tag{17.47}$$

where ℓ is a dummy integration variable.

This analysis leads to three dipole moments, one along each axis of the ellipsoid. If the field gradient is not aligned with the principal axis, the effective dipole will lead to a torque that rotates the particle's principal axis to be in line with the applied field. If the electric field gradient is aligned with the principal axis, there is no net torque on the particle, and the dielectrophoretic force can be evaluated in terms of the dipole moment along the principal axis.

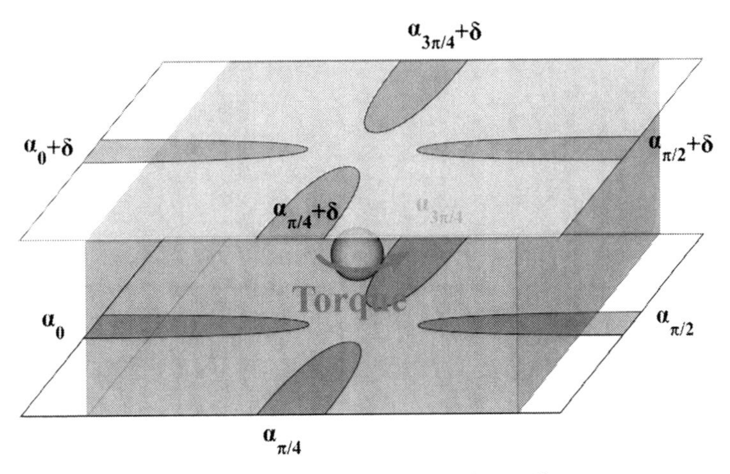

Fig. 17.3 An octode cage used for electrorotation studies.

For prolate ellipsoids, for which $a_2 = a_3$, we can use a simpler relation for the depolarization factor along the long axis,

$$L_1 = \frac{a_2{}^2}{2a_1{}^2e^3}\left[\ln\left(\frac{1+e}{1-e}\right) - 2e\right], \tag{17.48}$$

where the eccentricity e is given by

$$e = \sqrt{1 - \frac{a_2{}^2}{a_1{}^2}}. \tag{17.49}$$

Prolate ellipsoids are commonly used to model rod-shaped bacteria.

We find the dielectrophoretic mobility for a prolate ellipsoid with its long axis parallel to the direction of the applied electric field by combining Eqs. (5.39), (8.40), and (17.45):

dielectrophoretic mobility for a prolate ellipsoid with long axis parallel to the applied electric field

$$\mu_{\text{DEP}} = \frac{a_2{}^2\varepsilon_{\text{m}}}{16\eta e^3}\left[(1+e^2)\ln\frac{1+e}{1-e} - 2e\right]\text{Re}\left[\frac{\underset{\sim}{\varepsilon}_{\text{p}} - \underset{\sim}{\varepsilon}_{\text{m}}}{\underset{\sim}{\varepsilon}_{\text{m}} + (\underset{\sim}{\varepsilon}_{\text{p}} - \underset{\sim}{\varepsilon}_{\text{m}})L_1}\right]. \tag{17.50}$$

17.1.6 Nonuniform and anisotropic phase effects

A harmonic electric field with uniform and isotropic phase (as assumed in Subsection 17.1.2) leads to a particle force proportional to $\text{Re}(\underset{\sim}{f_{\text{CM}}})$, and the spatial dependence comes from the $\nabla(\vec{E}_0 \cdot \vec{E}_0)$ term. The torque on a particle with an applied field with uniform, isotropic phase is zero. However, a particle feels a torque if the phase is anisotropic. This is normally referred to as *electrorotation* (ROT). When the phase is anisotropic, i.e., when the phases of the different vector components of the electric field are unequal, we refer to the field as a circularly polarized or *rotating electric field*. Figure 17.3 shows an electrode setup used for electrorotation.

If the phase is nonuniform, the particle experiences an additional force and torque. The *force* resulting from this nonuniform phase is the basis of traveling-wave DEP techniques. Figure 17.4 illustrates an interdigitated electrode array used for traveling-wave DEP.

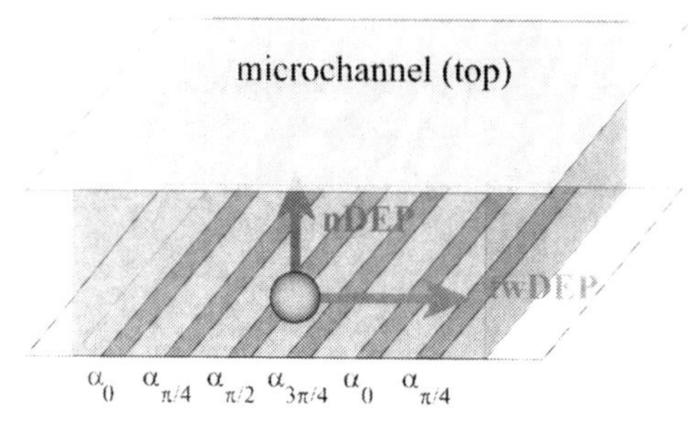

Fig. 17.4 Configuration and forces in a traveling-wave DEP electrode array.

Both electrorotation and traveling-wave DEP are analyzed by allowing the phase of the applied signal to vary as a function of both the position and the electric field component. Thus each component $i = 1, 2, 3$ of the applied electric field is written as $E_{\text{ext},i}(\vec{r}) = E_{0,i}(\vec{r}) \cos\left[\omega t + \alpha_i(\vec{r})\right]$. The analytic representation of each component is $\underset{\sim}{E}_{\text{ext},i} = E_{0,i} \exp j\left[\omega t + \alpha_i(\vec{r})\right]$. Writing each component $\underset{\sim}{E}_{0,i} = E_{0,i} \exp j\alpha_i$ defines a complex vector quantity $\underset{\sim}{\vec{E}}_0(\vec{r})$ that captures the nonuniformity and anisotropy. The applied electric field can then be written as

$$\underset{\sim}{\vec{E}}_{\text{ext}} = \underset{\sim}{\vec{E}}_0 \exp j\omega t\,. \tag{17.51}$$

Comparing this with Eq. (17.7), we find that the only difference is that the real vector $\vec{E}_0$ has been replaced with the phasor $\underset{\sim}{\vec{E}}_0$. The effects of nonuniform phase are described in detail in the following subsections.

ELECTROROTATION

Electrorotation refers to the asynchronous rotation of an object in a rotating electric field owing to the torque on the induced dipole when the induced dipole lags the (anisotropic) phase of the electric field. The instantaneous torque on an induced dipole is given by $\vec{T} = \vec{p} \times \vec{E}_{\text{ext}}$, and the time-averaged torque is given by $\langle \vec{p} \times \vec{E}_{\text{ext}} \rangle$. The analytical representation of the effective dipole for a spherical particle in a uniform, harmonic field is

$$\underset{\sim}{\vec{p}} = 4\pi\varepsilon_{\text{m}} \underset{\sim}{\vec{E}}_0 \underset{\sim}{f}_{\text{CM}} a^3\,. \tag{17.52}$$

Evaluating the time-averaged cross product $\langle \vec{p} \times \vec{E}_{\text{ext}} \rangle$ by using the guidelines in Section G.3, we find that the torque $\vec{T}$ is given by

time-averaged torque on a spherical particle in a uniform, rotating electric field

$$\left\langle \vec{T} \right\rangle = -4\pi\varepsilon_{\text{m}} a^3 \text{Im}(\underset{\sim}{f}_{\text{CM}}) \left[\text{Re}(\underset{\sim}{\vec{E}}_0) \times \text{Im}(\underset{\sim}{\vec{E}}_0)\right]. \tag{17.53}$$

The torque is proportional to $\text{Re}(\underset{\sim}{\vec{E}}_0) \times \text{Im}(\underset{\sim}{\vec{E}}_0)$ and porportional to $\text{Im}(\underset{\sim}{f}_{\text{CM}})$. The vector product $\text{Re}(\underset{\sim}{\vec{E}}_0) \times \text{Im}(\underset{\sim}{\vec{E}}_0)$ has three components that quantify how much the applied electric field is rotating around the three coordinate axes. If the phase is isotropic, then $\text{Re}\left(\underset{\sim}{\vec{E}}_0\right)$ and $\text{Im}\left(\underset{\sim}{\vec{E}}_0\right)$ point in the same direction, and the torque is zero. The imaginary component $\text{Im}(\underset{\sim}{f}_{\text{CM}})$ is a measure of the dipole component that is 90° out of phase with the applied field. The dipole moment in phase with the field induces no torque, because it points in the same direction as the electric field, but the dipole moment 90° out of phase with the field is oriented normal to the electric field and induces a torque. Thus the cross product quantifies how much the field is rotating, and the imaginary part of the Clausius–Mossotti factor quantifies what fraction of the effective dipole lags behind to induce the torque.

Electrorotation is often quantified by measuring particle rotation rates as a function of electric field rotation frequency – this measurement is often termed an *ROT spectrum* and the process is termed *electrorotation spectroscopy*. Because the real and imaginary parts of the Clausius–Mossotti factor are related by the Kramers–Krönig relation,[2] ROT spectra inform dielectrophoretic experiments and vice versa, and electrorotation magnitudes are maximum when DEP forces are zero.

Rotating electric fields can be created by forming a quadrupolar configuration and driving each electrode at a different phase (usually, $\alpha = 0, \pi/2, \pi$, and $3\pi/2$) or by using an octode cage consisting of two planar, quadrupolar electrode arrays assembled facing one another (Fig. 17.3). In many experimental systems, one quadrupolar array is offset by a few degrees with respect to the other by adding a phase increment between the signals driving the quadrupolar arrays.

TRAVELING-WAVE DEP

Unlike electrorotation, which depends on phase anisotropy alone and leads to particle rotation, traveling-wave dielectrophoresis (twDEP) depends on phase anisotropy *and* nonuniformity and leads to particle motion. Consider an applied field specified by $\underset{\sim}{\vec{E}}_{\mathrm{ext}} = \underset{\sim}{\vec{E}}_0 \exp j\omega t$, where $\vec{E}_0$ is the vector made up of $E_{0,i}$ terms and $\underset{\sim}{\vec{E}}_0$ is the vector made up of $E_{0,i} \exp j\alpha_i$ terms. For this extrinsically applied field, a time-averaged analysis of the DEP force on a sphere leads to

time-averaged dielectrophoretic force for nonuniform and anisotropic field phase

$$\left\langle \vec{F}_{\mathrm{DEP}} \right\rangle = \pi a^3 \varepsilon_{\mathrm{m}} \mathrm{Re}(\underset{\sim}{f}_{\mathrm{CM}}) \nabla (\vec{E}_0 \cdot \vec{E}_0) - 2\pi a^3 \varepsilon_{\mathrm{m}} \mathrm{Im}(\underset{\sim}{f}_{\mathrm{CM}}) \nabla \times \mathrm{Re}(\underset{\sim}{\vec{E}}_0) \times \mathrm{Im}(\underset{\sim}{\vec{E}}_0) . \tag{17.56}$$

The first term in this equation is equal to the time-averaged dielectrophoretic force with uniform phase, indicating that the regular DEP force is insensitive to phase anisotropy. The second term arises only in the presence of a nonuniform electric field phase. Unlike the uniform-phase effects, which are maximized when the dipole moment and field are *in phase* and the gradient of the field squared is high, the effects of spatially varying phase are maximized when the dipole moment and field are 90° out of phase and when the curl of the real and imaginary parts of the electric field are large.

[2] For the Clausius–Mossotti factor, the Kramers–Krönig relationships are

$$\mathrm{Re}[\underset{\sim}{f}_{\mathrm{CM}}(\omega)] = \frac{2}{\pi} \int_0^{\infty} \frac{\varpi \, \mathrm{Im}[\underset{\sim}{f}_{\mathrm{CM}}(\varpi)]}{\varpi^2 - \omega^2} \, d\varpi + \mathrm{Re}(\underset{\sim}{f}_{CM,\infty}) \tag{17.54}$$

and

$$\mathrm{Im}[\underset{\sim}{f}_{\mathrm{CM}}(\omega)] = \frac{2}{\pi} \int_0^{\infty} \frac{\mathrm{Re}[\underset{\sim}{f}_{\mathrm{CM}}(\varpi)] - \mathrm{Re}(\underset{\sim}{f}_{CM,\infty})}{\varpi^2 - \omega^2} \, d\varpi + \mathrm{Re}(\underset{\sim}{f}_{CM,\infty}) . \tag{17.55}$$

where $\underset{\sim}{f}_{CM,\infty}$ is the limit of the complex Clausius–Mossotti factor as $\omega \to \infty$.

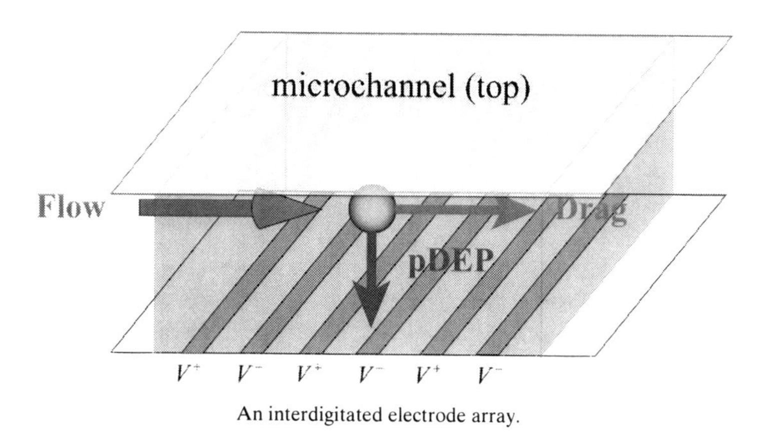

Fig. 17.5 An interdigitated electrode array.

twDEP is typically implemented experimentally by use of an array of electrodes patterned in a microfluidic channel. The electrode array is composed of alternating, independently driven electrodes with different phases (e.g., $\alpha = 0, \pi/4, \pi/2, 3\pi/4, 0, \ldots$). The electrode array is typically aligned at an angle to the direction of flow. These signals – irrespective of phase – levitate particles against gravity within the flow field owing to irrotational, negative DEP, and the varying phase drives particles transverse to the direction of flow according to the imaginary component of the Clausius–Mossotti factor.

ELECTRODE CONFIGURATIONS

Microfabricated electrodes are generally the most practical and straightforward method for creating nonuniform electric fields. They benefit from a long history of fabrication techniques and technologies and are quite flexible in terms of implementation. In practical applications, their potential major limitations are fouling and electrolysis at low electric field frequencies. Some device designs may also require complex multilevel microfabrication, increasing the cost of devices as well as the time required for creating them.

As previously discussed, changing the shape and orientation of electrodes, in addition to modulating the frequency and phase applied, can give rise to dielectrophoretic particle trapping, dielectrophoretic sorting, electrorotation, and twDEP effects. We now explore specific geometries and their key experimental parameters.

Interdigitated Electrode Array. The interdigitated electrode array is a common electrode configuration used in DEP studies. The electrode array consists of two sets of electrodes, grounded and energized, that alternate spatially. This creates a nonuniform field in the region of the electrode array that traps particles against a flow (Fig. 17.5).

Castellated Electrodes. Castellated electrodes are similar to interdigitated electrodes. Rather than straight electrodes, however, the castellated electrode array consists of square-wave-shaped electrodes. These patterns are usually placed parallel to each other and create alternating regions of low and high electric field. Symmetric and offset configurations of castellated electrodes are shown in Fig. 17.6.

Traps. Electrode-based DEP traps generally consist of geometries that trap single particles. The goal is usually to create a system of addressable particle traps to observe individual particle responses to a stimulus or to study biological particle interactions as

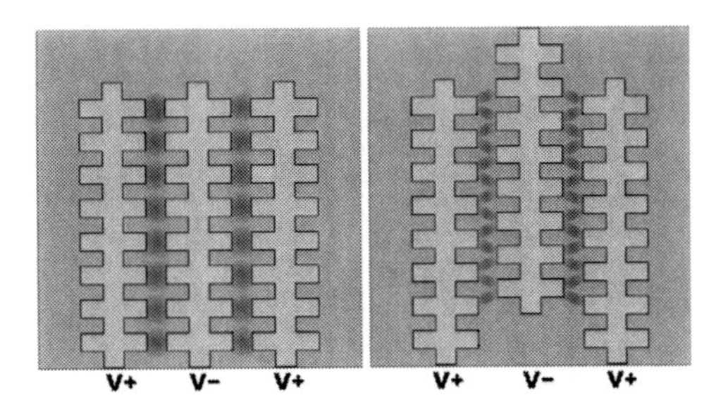

Fig. 17.6 A castellated, interdigitated electrode array.

a function of distance. Methods used to achieve addressable trapping include point-lid, quadrupole cages, ring-dot, and "DEP microwell" geometries (Fig. 17.7) [217].

Other Geometries. Although the most common way to shape electric fields has been patterned electrodes, patterned insulators can also be used to shape the electric field and effect dielectrophoretic particle motion (Fig. 17.8). This technique is termed *insulative dielectrophoresis* or *electrodeless dielectrophoresis*.

17.2 PARTICLE MAGNETOPHORESIS

Magnetophoresis is analogous to DEP, but leads to vastly different experimental realizations owing to the enormous variation in magnetic permeability observed between different materials, the nonlinearity of magnetic response of most materials to common experimental magnetic fields, and the different experimental tools available to generate magnetic fields. Magnetic fields are an excellent way to apply forces to a controlled subset of the particles or cells or analytes in a system, because only specialized materials are affected by magnetic fields enough to be successfully actuated in most microsystems.

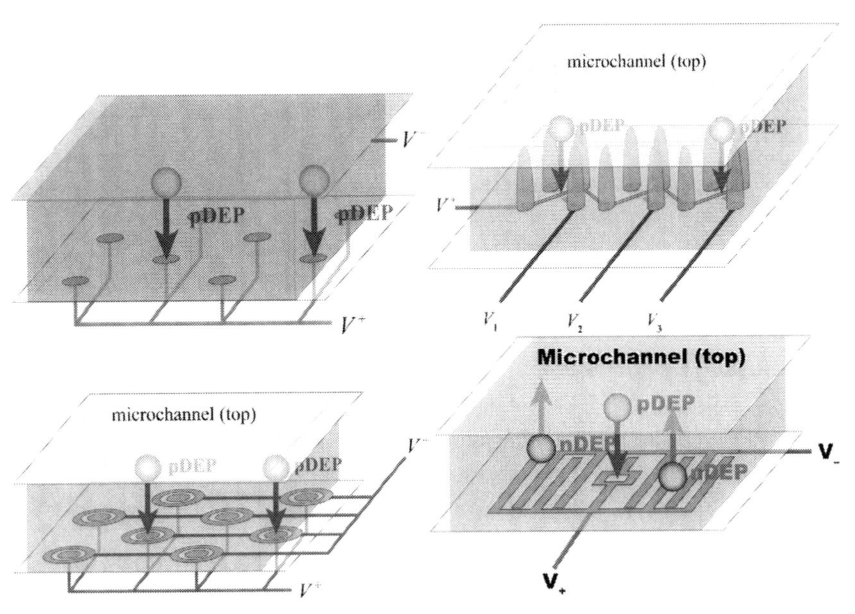

Fig. 17.7 Various electrode-based DEP trap geometries.

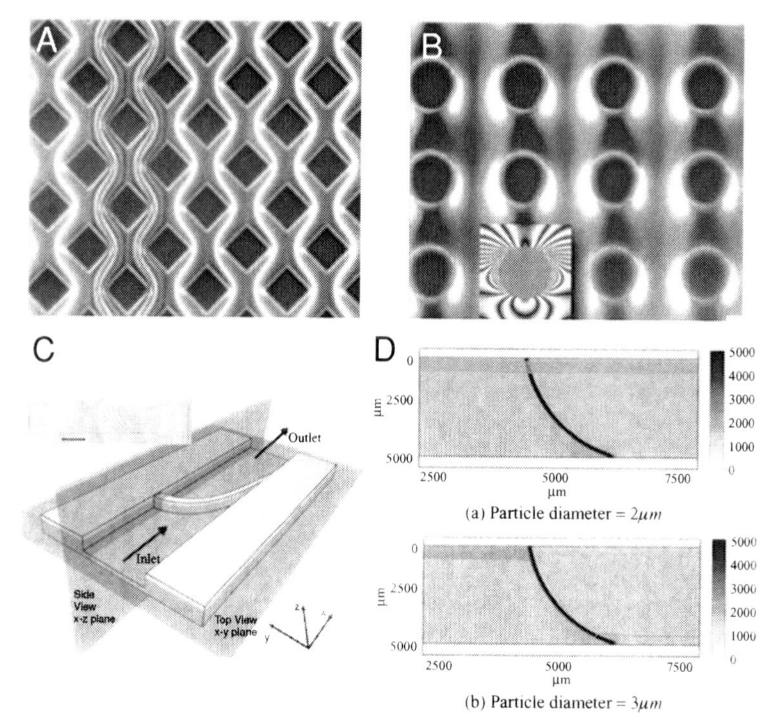

Fig. 17.8 (A) A post array using insulative DEP separates particles into streams or (B) captures them at the posts [218]. (C, D) A curved constriction and insulative DEP [219] is used to separate 2-μm microspheres from 3-μm microspheres. Reproduced with permission.

17.2.1 Origin of magnetic fields in materials

Classically, magnetic fields in materials can be thought of as being induced by orbital rotation of electrons, analogous to the magnetic field generated by current flowing through a wire loop. This classical representation can qualitatively explain differences between different material classes (e.g., diamagnetic, paramagnetic, ferromagnetic) but cannot predict how these properties stem from molecular structure – that requires a quantum-mechanical description.

Diamagnetism. Diamagnetic materials are those whose electrons are all paired. Diamagnetic effects stem from changes in electron orbital motion induced by a magnetic field. It leads to a dipole aligned against the magnetic field, i.e., the diamagnetic component of magnetic permeability is negative. Diamagnetic effects are small enough to be neglected for our purposes.

Paramagnetism. Paramagnetic effects occur in materials whose electrons are unpaired. In this case, the spins of the unpaired electrons (and the resulting magnetic moment) align with the external magnetic field. Paramagnetism is exhibited when thermal fluctuations prevent the magnetic dipoles from *locking* in orientation aligned with the field. In this case, the paramagnetic component of magnetic permeability is positive but small.

Ferromagnetism. Ferromagnetism is exhibited in materials with unpaired electrons when thermal fluctuations are small compared with the forces that lead magnetic dipoles to *lock* in orientation aligned with the field. In this case, the ferromagnetic component

Table 17.1 **Properties of magnetic materials.**

Material Class	Examples	Typical χ_m	*B-H* relationship	Comments
Diamagnetic	Water	-1×10^{-5}	Linear (constant χ_m)	No hysteresis
Paramagnetic	Aluminum	2×10^{-5}	Linear (constant χ_m)	No hysteresis; becomes ferromagnetic below Curie temp
Ferromagnetic	Iron	3×10^{3}	Nonlinear; χ_m is $f(B)$	Shows hysteresis
Ferrimagnetic	$MnZn(Fe_2O_4)_2$	2.5×10^{-3}	Nonlinear; χ_m is $f(B)$	Shows hysteresis

of magnetic permeability is positive and large. The only difference between a magnet and the material to which magnets are attracted (e.g., iron) is the size of the magnetic domains. All ferromagnetic materials have magnetic domains in which the magnetic dipoles are aligned, but in materials such as iron, the magnetic domains are small and randomly oriented, leading to no net orientation on a large scale. A magnet, in contrast, has large, permanently oriented domains.

17.2.2 Attributes of magnetism

Although magnetism is closely related to electricity, and the equations are similar, key differences between the two make the engineering application of magnetic effects different from application of electrical effects. Electricity is controlled primarily by monopoles (point charges, for example, an ion) and dipoles (equal and opposite charges that are separated by a distance, for example, a nucleus and an electron cloud that has been polarized by an electric field). Magnetism has no monopoles and is thus controlled primarily by dipole or polarization effects. Thus electricity is characterized by both conductivity and electrical permittivity, whereas magnetism is described only by magnetic permeability. The magnetic permeabilities of most materials are essentially the same as free space, but the magnetic permeability of a few select materials (such as iron, nickel, ferrite, etc.) can be 3–6 orders of magnitude higher. This can be seen quantitatively in Table 17.1. In contrast, electric permittivities vary continuously among materials and by not more than a factor of 80 or so from the lowest (air; 1) to the highest (water; 80). The impact of this is that most materials do not experience magnetic forces that are significant, and the materials that do experience magnetic forces are specialized and relatively easy to engineer. In a microsystem, magnetic fields can manipulate magnetic beads without affecting other attributes of the flow. This, in general, cannot be said for electrical effects like electrophoresis or DEP. Because of this, magnetic beads can be used to isolate forces in specific areas or to specific particles.

The magnetic permeability μ_{mag} expresses the magnetic behavior of a material:

$$\vec{\boldsymbol{B}} = \mu_{mag,0}(\vec{\boldsymbol{M}} + \vec{\boldsymbol{H}}) = \mu_{mag,0}(1 + \chi_m)\vec{\boldsymbol{H}} = \mu_{mag}\vec{\boldsymbol{H}}, \quad (17.57)$$

where $\vec{\boldsymbol{B}}$ is the applied magnetic field and $\vec{\boldsymbol{H}}$ is the induced magnetic field. $\vec{\boldsymbol{M}}$ is the resulting magnetization of the material. The magnetic permeability of free space $\mu_{mag,0}$ is $4\pi \times 10^{-7}$ H/m.[3] The magnetic permeabilities of other materials are given by $\mu_{mag} = \mu_{mag,0}(1 + \chi_m)$, where χ_m is the magnetic susceptibility of the material. Most materials have values of χ_m that are small as compared to unity, and thus magnetic effects are tiny. Only ferromagnetic, antiferromagnetic, and ferrimagnetic materials experience significant magnetic effects in systems of interest.

[3] The unit of inductance H is termed a henry and is equal to 1 V s/A.

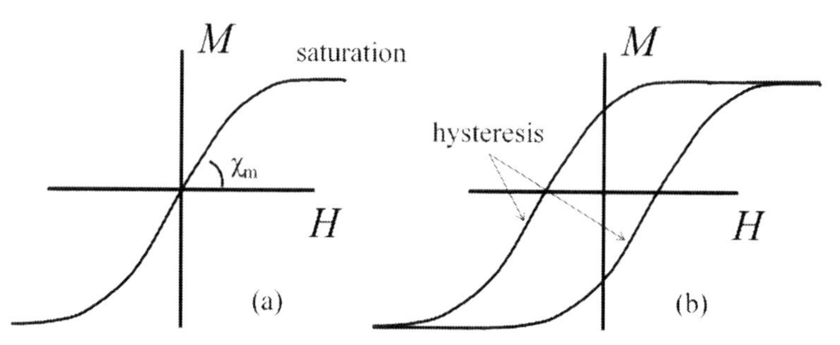

Fig. 17.9 (a) Saturation in a magnetically soft material with no hysteresis. This is representative of superparamagnetism, the property magnetic beads are designed to exhibit. Superparamagnetism is typically observed in small domains of ferromagnetic materials. (b) Saturation in a magnetically hard material with hysteresis, such as iron or magnetite at long length scales. (After [61].)

SATURATION

Saturation is a key property that prevents magnetophoresis from being directly analogous to DEP. Materials have a saturation magnetism (e.g., 2.2T for iron) that corresponds to their spins being fully aligned. Increased magnetic field does not lead to further spin or further magnetic flux density.[4] This leads to a magnetic permeability that is dependent on the applied magnetic field (See Fig. 17.9a). In *magnetically soft* materials, the B-H curve[5] is nonlinear but shows no hysteresis. In *magnetically hard* materials, the B-H curve also shows hysteresis, as seen in Fig. 17.9b, owing to the fact that nonequilibrium states are metastable.

17.2.3 Magnetic properties of superparamagnetic beads

Superparamagnetic beads typically consist of a polystyrene matrix (of roughly 1μm diameter) filled to about 15% mass fraction with roughly 10nm ferromagnetic or ferrimagnetic particles. The 10-nm particles are described as superparamagnetic, which means that they respond strongly to external magnetic fields but do not retain permanent magnetism and do not show hysteresis. Owing to their small domain size, the nanoparticles can orient and equilibrate themselves quickly, so they show strong effects like a ferromagnetic or ferrimagnetic material and avoid hysteresis and permanent magnetism.

17.2.4 Magnetophoretic forces

Magnetophoresis is the motion of objects with respect to a surrounding medium caused by the net interaction of a magnetization with a magnetic field gradient. Magnetophoresis is analogous to dielectrophoresis of a particle except that magnetic materials have a nonlinear induced dipole response when exposed to typical experimental conditions.

The magnetophoretic force on a particle $\vec{F}_{\text{mag}}$ is given by

$$\vec{F}_{\text{mag}} = \left(\mu_{\text{mag,m}} m_{\text{eff}} \cdot \nabla\right) \vec{H} , \tag{17.58}$$

where $\vec{H}$ is the external magnetic field, $\mu_{\text{mag,m}}$ is the magnetic permeability of the medium, and m_{eff} is the magnetic dipole moment (i.e., magnetization) induced by the

[4] This saturation limit is critical for magnetism because experimentally realistic magnetic fields that do not lead to any chemical reaction in magnetic materials will induce saturation. Electrical dipoles oriented by electric fields also eventually go nonlinear, as described in Subsection 5.1.4.

[5] Or, equivalently, the M-H curve.

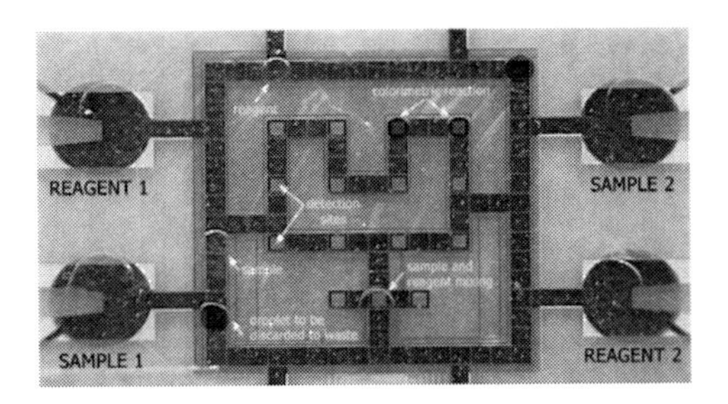

Fig. 17.10 A digital microfluidics lab-on-a-chip device for manipulating physiological fluids and monitoring glucose levels. (Reproduced with permission from [220].)

field. This equation is physically analogous to Eq. (17.26) even though it does not match notationally – the magnetic permeability is included here, whereas no electric permittivity is found in Eq. (17.26), because magnetic moments are defined with different conventions than electric moments.

17.2.5 DC magnetophoresis of spheres – linear limit

If we assume a steady magnetic field and particles that respond linearly to the magnetic field, then magnetophoresis is quite similar to DC DEP. For spheres of diameter a, we can show that the averaged magnetophoretic force $\vec{F}_{\mathrm{mag}}$ is given by

DC magnetophoretic force for magnetically linear materials

$$\vec{F}_{\mathrm{mag}} = 2\pi\mu_{\mathrm{mag,m}}a^3 \left(\frac{\mu_{\mathrm{mag,p}} - \mu_{\mathrm{mag,m}}}{\mu_{\mathrm{mag,p}} + 2\mu_{\mathrm{mag,m}}}\right) \nabla \left|\vec{H}\right|^2 . \tag{17.59}$$

Here, as before, subscripts m and p indicate medium and particle, respectively. Here the Clausius–Mossotti factor uses magnetic permeabilities rather than electrical permittivities, as was the case for DEP. Because we are considering only DC magnetic fields, the magnetic Clausius–Mossotti factor is real. Commercial particles with roughly 10%–15% ferrite or magnetite have effective magnetic susceptibilities χ_{m} of the order of 3 and a roughly linear range up to applied magnetic fields of approximately $H = 5 \times 10^4$ A/m. Owing to the nonlinearities in magnetic systems, calculations using Eq. (17.59) are approximate.

17.3 DIGITAL MICROFLUIDICS

Capillarity and surface tension are important for understanding how channels fill with fluid and how bubbles move in microchannels. For microsystem design purposes, liquid–gas or liquid–liquid multiphase flows give the potential for what has been termed *digital microfluidics*.

Digital microfluidics [220, 221] implies manipulation of liquid droplets, typically on arrays of electrodes. Figure 17.10 shows an example of a digital microfluidic device. Digital microfluidics depends on DEP or electrowetting forces on individual droplets. The physics behind these forces is the focus of this section.

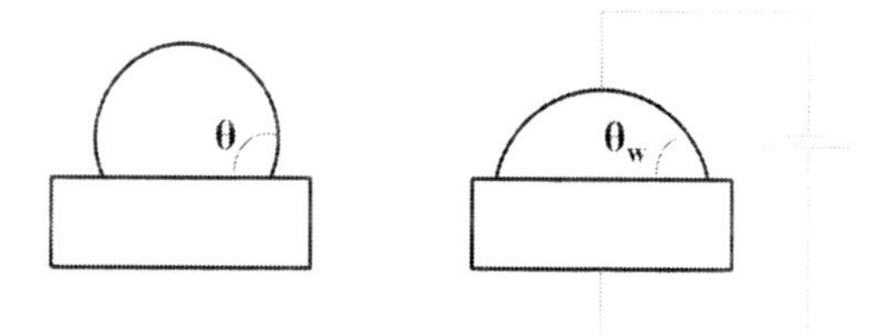

Fig. 17.11 Electrowetting-on-dielectric.

17.3.1 Electrocapillarity and electrowetting

Electrocapillarity refers to the change in the interfacial tension when an electrical potential is applied across an interface, which is attributable to the induced charge in the electrical double layer at the interface. The term stems from the Lippman electrometer, in which a potential applied to mercury reservoirs separated by an electrolyte changes the meniscus position of the mercury-electrolyte interface in a capillary. Electrocapillarity is typically described by the Lippman equation,

$$\gamma_w = \gamma - \frac{1}{2}C''V^2, \tag{17.60}$$

where γ_w is the interfacial surface tension in the presence of the applied field, γ is the surface tension with no applied field, C'' is the interfacial capacitance per unit area, and V is the applied voltage drop across the interface. Equation (17.60) is consistent with the postulate that the change in surface tension per voltage change is equal to the induced surface charge density $C''V$, and can be derived by evaluating the surface stress predicted by the Maxwell stress tensor when a voltage is applied across a thin capacitive film – the existence of an electric field creates a surface charge and generates a Gibbs free energy term associated with this charge. Equation (17.60) assumes that the surface is uniform, with uniform induced charge density, and that all voltage is dropped across the electrical double layer at the interface and not the bulk fluid.

Electrocapillarity and dielectrophoresis are fundamentally the same phenomenon, namely the change in the state of a system associated with induced surface charge attributable to an applied electric field. Electrocapillarity is a natural bookkeeping approach for systems with static but deformable interfaces (e.g., a static droplet) whose shape indicates changes in the surface tension and therefore surface charge, whereas dielectrophoresis is a natural bookkeeping approach for systems with rigid but mobile interfaces, e.g., a suspended particle.

Electrowetting is used to describe the change in contact angle of a droplet on a surface when an electric field is present at an interface. In microfluidic systems, we typically observe the effects of applied fields by creating a thin dielectric layer (often Teflon AF deposited by spinning) on an electrode and applying a voltage across the interface to observe changes in droplet contact angle (Figs. 17.12 and 17.11). A rudimentary description of static electrowetting can be generated by applying the Lippman equation to a droplet interface, and assuming that the errors associated with the uniform-surface-charge assumption are small. Recasting the Lippman equation in terms of contact angles, we can write

electrowetting change in contact angle

$$\cos\theta_w = \cos\theta + \frac{1}{2\gamma_{lg}}C''V^2, \tag{17.61}$$

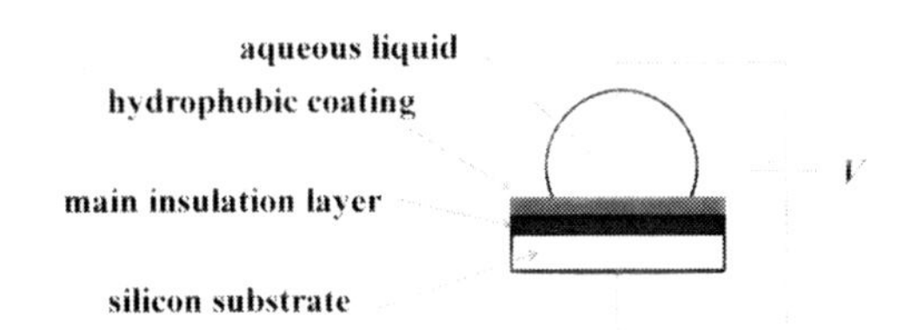

Fig. 17.12 Side view of substrate and coatings for electrowetting on dielectric.

where θ_w is the contact angle and θ is the contact angle without an applied field. Here, the capacitance per unit area C'' and voltage V refer to the properties of the dielectric electrode coating and the primary voltage drop is assumed to be across this coating. Equation (17.61) describes the change in static contact angle of a droplet predicted if a uniform surface charge is generated on a dielectric thin film on which the droplet rests.

Although Eq. (17.61) can be used to quantitatively predict the pressures on a droplet in a digital microfluidic device, we avoid making such predictions here, because this description has serious limitations and the match with experimental data is poor. First, *contact angle hysteresis* is observed in the motion of droplets along surfaces, and the magnitude of the contact angle when the droplet is moving cannot be predicted from static considerations alone. Further, the uniform-charge assumption is quite inaccurate at the triple point. Thus Eq. (17.61) gives a good qualitative description of why the static contact angle changes upon application of voltage but is insufficient to quantitatively predict droplet motion.

Modeling the dielectric layer as if it were a Helmholtz capacitor, we can observe that C'' is given by $C'' = \varepsilon / w$, where w and ε are the thickness and electrical permittivity of the dielectric layer. From this, the contact angle can also be written as

electrowetting change in contact angle in terms of insulating layer properties

$$\cos \theta_w = \cos \theta + \frac{\varepsilon}{2 w \gamma_{lg}} V^2 , \tag{17.62}$$

which highlights the thickness and material dependence of the contact angle change on the insulating layer properties. An array of electrodes can then be used to manipulate droplets by creating a nonuniform distribution of charge on the surface of the droplet, which is observed through changes in the contact angle.

17.4 SUMMARY

This chapter summarizes particle actuation methods, including dielectrophoresis and magnetophoresis, and droplet actuation methods, typically described using a digital microfluidics system approach. Analysis of dielectrophoresis of objects leads to a description of the time-averaged force on an uncharged sphere,

$$\left\langle \vec{F}_{\mathrm{DEP}} \right\rangle = \pi \varepsilon_{\mathrm{m}} a^3 \mathrm{Re}(\underset{\sim}{f_{\mathrm{CM}}}) \nabla (\vec{E}_0 \cdot \vec{E}_0) , \tag{17.63}$$

highlighting the importance of the real part of the Clausius–Mossotti factor in describing the magnitude and sign of the dielectrophoretic force. Central to this derivation is the concept of the Maxwell stress tensor and Maxwell equivalent body, which allows for the sphere to be described using a dipole approximation. For isotropic and spherically symmetric objects, nonuniformity can be handled using algebraic descriptions of the equivalent dipole. Relations are also presented for charged spheres and ellipsoids, in the absence of electrical double layers. Anisotropy and nonuniformity of the phase of the applied electric field is shown to lead to electrorotation and traveling-wave dielectrophoresis, with dependence on the imaginary portion of the Clausius–Mossotti factor.

Similar relations can be derived to describe magnetophoresis; however, the magnetophoretic response is much different owing to the differences in magnetic permeability of materials. In particular, few materials have a significant magnetic response, and those that do typically respond in a nonlinear fashion, saturating at magnetic fields commonly used in experiments. Thus relations assuming a linear material response are limited in their quantitative usefulness.

Digital microfluidics is a systems approach to manipulating droplets of fluid. It typically uses voltages applied to arrays of electrodes and uses nonuniformity of electric field to generate charge at surfaces, leading to a force on the droplet. For droplets, this force is most naturally visualized and conceived in terms of changes in the observed contact angle, and this manipulation is typically described using electrowetting analyses. In the static case, the contact angle of a droplet is changed by the application of a voltage according to the relation

$$\cos \theta_w = \cos \theta + \frac{1}{2\gamma_{lg}} C'' V^2 , \tag{17.64}$$

which describes how applied voltages reduce the contact angle.

17.5 SUPPLEMENTARY READING

Excellent references on the electromechanics of particles include [60, 61, 52, 222]. These sources discuss DEP and electrorotation in detail, including particle–particle interactions, detailed multipolar analysis, and double-layer effects. Jones [61] treats magnetophoresis in detail as well as DEP, and Chang and Yeo [52] discuss double-layer effects on dielectrophoresis in detail. DEP involves induced surface charge, and in that sense has similarities to the AC electrokinetic effects described in Chapter 16; however, in the case of DEP, the surface charge is caused by a discontinuity of material properties in lossy dielectrics rather than a spatial gradient of material properties (e.g., electrothermal effects) or discontinuity between a lossy dielectric and an ideal conductor (e.g., ICEO, ACEO).

Dielectrophoretic analysis often requires decisions among various levels of analysis to use, ranging from the Maxwell stress tensor to the effective dipole. In real systems, analytical solutions are either numerical or approximate. With specific reference to DEP, Wang et al. [223, 224] present an analytical formalism for DEP forces and torques derived from the Maxwell stress tensor expression [224, 225, 226, 227]. Another body of work is devoted to numerical simulations and solutions for the DEP force and particle trajectories [225, 226, 227, 228, 229, 230, 231]. These findings indicate that the Maxwell stress tensor must be integrated to find accurate solutions when the particle size approaches (within an order of magnitude) the characteristic length scale of spatial variations of the electric field. Multipolar descriptions are described in detail in [61, 232]. Several different geometries have been considered in detail [97, 226, 227, 233, 234, 235, 236, 237, 238]. Although we restrict ourselves in this chapter to spherically symmetric and isotropic materials, DEP and electrorotation have been used to gain insight into internal structure or composition [236, 239, 240, 241, 242, 243, 244, 245, 246, 247, 248, 249, 250]. Electrode geometries implemented and studied include interdigitated electrodes [251], castellated electrodes [252, 253], and angled electrodes [254, 255, 2].

DEP applications range from fractionating particles based on their "electrical phenotype" [217] to precise manipulation of single particles for property interrogation to new strategies for the creation of engineered tissues and organs. A number of dielectrophoretic trapping techniques have been used; examples of separation by cell types can be found in [1, 2, 216]. Cell populations have been used to discern physiologic differences such as activation of mitosis [256], cell-cycle phase [257], exposure to drugs [258, 259], induced cell differentiation [223, 260], and cell death [215, 261, 262, 263, 264]. Multiple-frequency techniques are presented in [215, 265, 266]. A subset of the published work uses "electrodeless" techniques, in which the electric field nonuniformity is created through constrictions in the channel geometry [218, 219, 264]. Many of these techniques are suitable for continuous-flow separation, also implemented with electrodes in [267, 268, 269]. These continuous-flow separations have been applied for polystyrene spheres [219, 270], bacteria [271], yeast [272], and mammalian cells [257]. Individual cells have been addressed in DEP electrode arrays to facilitate single-cell capture, analysis, and release, both in solution [273, 274, 275] and in photopolymerized gels [276].

This chapter omits dielectric spectroscopy, which can be used on colloidal suspensions to infer double-layer and particle properties. The dielectric spectroscopy community describes many of the phenomena described in this chapter using different terminology – for example, the terms *alpha* and *beta* relaxations are used to describe the low- and moderate-frequency changes in the dielectric response owing to double-layer asymmetry and fluid flow (alpha dispersion) or Maxwell–Wagner polarization (beta dispersion).

Chang and Yeo [52] discuss EWOD theory as well as its limitations. Some examples of digital microfluidics are presented in [221, 277].

17.6 EXERCISES

17.1 Consider a protein whose permanent dipole moment is 100 D. Model this protein as having one positive charge on one end of the protein and another positive charge on the other end of the protein. How far apart must these charges be to explain the observed permanent dipole moment?

17.2 Consider a hydrogen atom, consisting of a nucleus with charge $q = e$ and an electron cloud of $q = -e$, where e is the magnitude of the electron charge. Treat the nucleus and electron cloud as if they were each point charges. By applying an electric field with magnitude $E = 100$ V/cm in the x direction, the nucleus and electron cloud will be displaced relative to each other until the field caused by one charge on the other that holds them together,

$$E = \frac{q}{4\pi\varepsilon_0} \frac{1}{\Delta x^2}, \tag{17.65}$$

balances out the external electric field. Derive a relation for the displacement Δx in terms of the other given parameters and calculate the magnitude of this displacement. Is this displacement small or large compared with an atomic radius?

17.3 Derive the Maxwell stress tensor using Gauss's law.

17.4 A polystyrene bead of radius $a = 1\ \mu\text{m}$ is suspended in an electrolyte solution. The permittivity of polystyrene is approximately $2\varepsilon_0$ and the conductivity is $2\ \mu\text{S/cm}$. A silica bead of radius $a = 1\ \mu\text{m}$ is also suspended ($\varepsilon = 4\varepsilon_0$; $\sigma = 0.2\ \mu\text{S/cm}$). The solution is deionized water ($\varepsilon = 80\varepsilon_0$; $0.18\ \mu\text{S/cm}$). A spatially varying field proportional to $E_0 \cos \omega t$ is applied. Calculate and plot the real part of the Clausius–Mossotti factor for the polystyrene and silica beads as a function of ω. Is there a range of frequencies at which these two beads will feel forces in opposite directions (i.e., one toward regions of high electric field, one towards regions of low electric field)? What is this range?

17.5 Consider a polystyrene bead of radius $a = 1\ \mu\text{m}$ suspended in an electrolyte solution in a microchannel. The permittivity of polystyrene is approximately $2\varepsilon_0$ and the conductivity is $2\ \mu\text{S/cm}$. The solution is deionized water ($\varepsilon = 80\varepsilon_0$; $0.18\ \mu\text{S/cm}$). In these conditions, the electrophoretic mobility of this particle is $\mu_{\text{EP}} = 1 \times 10^{-8}\ \text{m}^2/\text{V s}$.

Presume that the geometry of the microchannel is designed such that the potential in the channel (which ranges from $x = 0$ to $x = L$) varies according to the following equation: $\phi = Ax^2$. If A is a constant, what value of A will cause the particle to stagnate at $x = L/2$?

17.6 Assume a combined DC plus AC field is applied to a particle:

$$\vec{E} = \text{Re}\left[\vec{E}_{\text{DC}} + \vec{E}_{\text{AC}} \exp(j\omega t)\right] \qquad (17.66)$$

Where Re indicates "real part of". Assume the AC and DC fields are aligned in the same direction, and define $\alpha = \vec{E}_{\text{AC}}/\vec{E}_{\text{DC}}$.

Assume that the Clausius–Mossotti factor for the particle and the medium is given for both DC fields ($\underset{\sim}{f}_{\text{CM,DC}}$) and AC fields ($\underset{\sim}{f}_{\text{CM,AC}}$).

(a) Note that the instantaneous effective dipole moment on a sphere is given by

$$\vec{p}_{\text{eff}} = \text{Re}\left[4\pi\varepsilon_{\text{m}}a^3 \underset{\sim}{f}_{\text{CM}} \vec{E}\right] \qquad (17.67)$$

Given this, derive a relation for the effective dipole moment $\vec{p}_{\text{eff}}$ for the DC-offset AC field in terms of α, Clausius–Mossotti factors, and $\vec{E}_{\text{DC}}$, as well as ε_{m} and a.

(b) Evaluate the *instantaneous* DEP force. Note that, for a purely AC field with magnitude $\vec{E}_{\text{AC}}$, the *time-averaged DEP force* is given by

$$\left\langle \vec{F} \right\rangle = \pi\varepsilon_{\text{m}}a^3 \text{Re}\left[\underset{\sim}{f}_{\text{CM}}\right] \nabla \left|\vec{E}_{\text{AC}}\right|^2 \qquad (17.68)$$

Your instantaneous force relation will be more complicated than the time-averaged relation, because it will involve *two* Clausius–Mossotti factors, as well as time-dependent terms. Rearrange your instantaneous force relation so that (a) there are no nonlinear trigonometric terms – i.e., ensure that there are no $\cos^2$ or sin cos terms, and so that (b) the real and imaginary parts of the Clausius–Mossotti factors are listed separately.

(c) In the case where $\underset{\sim}{f}_{\text{CM,DC}} = \underset{\sim}{f}_{\text{CM,AC}}$, derive the time-averaged force in terms of α, the Clausius–Mossotti factor, and $\vec{E}_{\text{DC}}$, as well as ε_{m} and a.

17.7 Consider a homogeneous solid sphere of radius a located at the origin of a axisymmetric spherical coordinate system and surrounded by a homogeneous medium. Let the sphere have properties ε_p and σ_p and let the medium have properties ε_m and σ_m. Assume a sinusoidal electric field is applied, whose analytic representation is $\underset{\sim}{\vec{E}} = E_0\hat{z}\exp j\omega t$, where $\hat{z}$ is the unit vector in the z axis direction and it is understood that the electric field is given by the real part of $\underset{\sim}{\vec{E}}$ (specifically, $\vec{E} = E_0\hat{z}\cos j\omega t$). Assume that the analytical representation of the electrical potential $\underset{\sim}{\phi}(r, \theta, t) = \phi_0(r, \theta)\exp j\omega t$.

Assume that any net charge density at interfaces in this system is induced by the electric field and is also sinusoidal.

Laplace's equation is the governing equation inside the sphere and in the medium outside the sphere. In axisymmetric spherical coordinates, Laplace's equation is given by

$$\frac{\partial}{\partial r}\left(r^2\frac{\partial}{\partial r}\phi_0\right) + \frac{1}{\sin\vartheta}\frac{\partial}{\partial\vartheta}\left(\sin\vartheta\frac{\partial}{\partial\vartheta}\phi_0\right) = 0 \tag{17.69}$$

The general solution of this equation can be written in terms of Legendre polynomials:[6]

$$\phi_0(r, \vartheta) = \sum_{k=0}^{\infty}\left(A_k r^k + B_k r^{-k-1}\right) P_k(\cos\vartheta), \tag{17.70}$$

where the Legendre polynomials $P_k(x)$ are given by

$$P_k(x) = \frac{1}{2^k k!}\left(\frac{d}{dx}\right)^k \left(x^2 - 1\right)^k. \tag{17.71}$$

In particular, $P_1(x) = x$; you can show that this is the only Legendre polynomial required for this problem.

The boundary conditions for this problem are

- $\phi_{0,p} = \phi_{0,m}$ at $r = a$
- $\underset{\sim}{\varepsilon}_p\frac{\partial\phi_{0,p}}{dn} = \underset{\sim}{\varepsilon}_m\frac{\partial\phi_{0,m}}{dn}$ at $r = a$
- $\phi_0(r = \infty) = -E_0 z = E_0 r\cos\vartheta$
- $\phi_0(r = 0)$ is bounded

(a) Solve for ϕ_0 inside and outside the sphere. Write both solutions in terms of $\underset{\sim}{f_{CM}}$.

(b) Show that the result for the potential outside a sphere can we written as the sum of (i) the sinusoidally varying applied electric field and (ii) a field that is equivalent to that created by a dipole at the origin with a sinusoidally varying dipole moment but a different phase. What is the magnitude of this dipole? What is the phase lag?

(c) Use Gauss's law at the interface between particle and medium to calculate the induced charge density at the interface.

(d) Given the result for the time-varying potential induced by a sphere subjected to a sinusoidal electric field, derive the relation for the DEP force as a function of particle size, Clausius–Mossotti factor, and other parameters.

[6] This is a classic example of solution of a differential equation by separation of variables and infinite sums of orthogonal functions. This is analogous to Fourier and Bessel function solutions in Cartesian and cylindrial coordinates, and the Legendre polynomial solution for a dielectric sphere in an infinite medium is solved in most electrodynamics texts.

17.8 Show that, in a uniform medium with properties ε_m and σ_m and a uniform applied field $\vec{E}_{ext}$, the induced dipole on a particle of radius a_2 consisting of a core of radius a with properties ε_p and σ_p surrounded by a shell with properties ε_m and σ_m is the same as the induced dipole on a particle with radius a with properties ε_p and σ_p.

17.9 Consider pressure-driven flow of a particle suspension through a microchannel that has two electrodes as shown in Fig. 17.13. If an AC signal is passed between the two electrodes (labeled V_1 and GND), what will happen to the particles? What physical phenomena govern this system? Describe *in detail* the parameters that govern the device, fluid, and particles, in this system. What parameters must be known to be able to predict the performance of this system? How could you change parameters to

- make particle trajectories independent of the electric field?
- make different particles react differently to the electric field?
- separate different types of particles?
- trap and store particles?

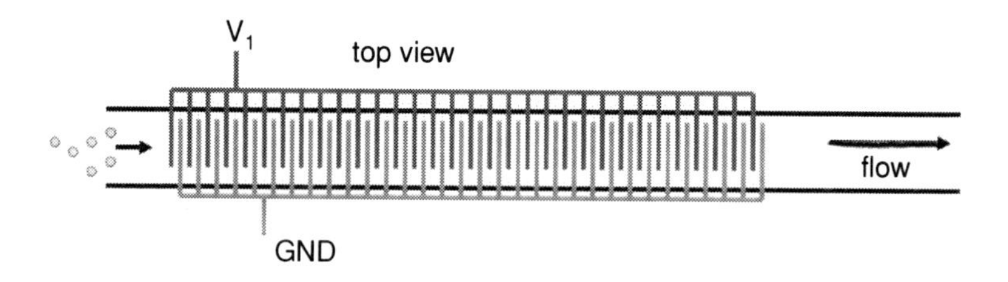

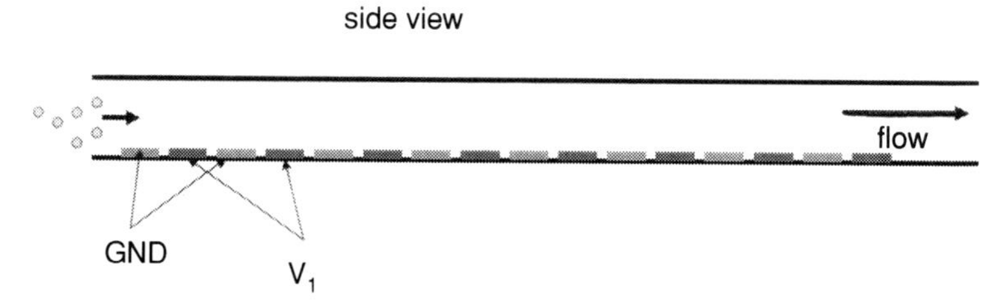

Fig. 17.13 A proposed microfluidic device.

17.10 Calculate the force on a 10-μm aluminum particle suspended in water if it is in a magnetic field aligned in the x direction with magnitude 1×10^4 A/m and field gradient (in the same direction) of 1×10^8 A/m^2.

17.11 Consider an idealized 10-μm magnetic particle with a magnetic susceptibility of 3 and an infinitely large linear range, suspended in a solution whose magnetic susceptibility can be approximated as zero.

(a) Calculate the force on this particle if it is in a magnetic field with magnetic field in the x direction with magnitude 1×10^4 A/m and field gradient (in the same direction) of 1×10^8 A/m^2.

(b) Assume that the force described above is applied in a 20-μm microchannel by an external magnet to induce the particles to become trapped on one wall of the channel so that the system may be flushed. Assume that the field and field gradient can be approximated as being uniform for the purposes of this calculation. Of interest is the settling time, i.e., the time required for all magnetic particles in the system to be pulled to one wall. Assuming the solution is water and the particles can be modeled as Stokes spheres, calculate this settling time.

17.12 Show how Eq. (17.59) follows from Gauss's law of magnetism, listed in Eq. (5.5), in the limit where the magnetic susceptibilities of particle and medium are independent of space, magnetic field, and magnetic field history.

17.13 Consider the limit in which the particle magnetic susceptibility varies with the *applied magnetic field* but can be assumed constant *within the particle* because the magnetic field within the particle is approximately constant. Show how the force on a magnetic particle is affected by saturation.

17.14 Derive Eq. (17.61) using geometric arguments and the Young–Laplace equation.

17.15 If the contact angle of water on a 50-nm-thick layer of Teflon AF ($\varepsilon/\varepsilon_0 = 2$) is 120° and the surface tension of water in contact with air is 73 mN/m, calculate the voltage required to reduce the contact angle to 85°.

17.16 Consider a sphere in a uniform harmonic applied field with frequency ω. Show that, as $\omega \to 0$, the Clausius–Mossotti factor approaches

$$\underset{\sim}{f}_{\mathrm{CM}} = \frac{\sigma_{\mathrm{p}} - \sigma_{\mathrm{m}}}{\sigma_{\mathrm{p}} + 2\sigma_{\mathrm{m}}} . \tag{17.72}$$

17.17 Consider a sphere in a uniform harmonic applied field with frequency ω. Show that, as $\omega \to \infty$, the Clausius–Mossotti factor approaches

$$\underset{\sim}{f}_{\mathrm{CM}} = \frac{\varepsilon_{\mathrm{p}} - \varepsilon_{\mathrm{m}}}{\varepsilon_{\mathrm{p}} + 2\varepsilon_{\mathrm{m}}} . \tag{17.73}$$

17.18 Two electrodes are located at $y = \pm 50\ \mu$m and are driven with a voltage difference of $V_0 \cos \omega t$, with $V_0 = 1$ V. Two electrodes are located at $x = \pm 50\ \mu$m and are driven with a voltage difference of $V_0 \sin \omega t$. Assume that the field at the origin is well approximated by $E_0 \sin \omega t \hat{\boldsymbol{x}} + E_0 \cos \omega t \hat{\boldsymbol{y}}$, where $E_0 = 100$ V/cm. Derive $\underset{\sim}{\vec{E}}_0$ and evaluate $\mathrm{Re}(\underset{\sim}{\vec{E}}_0) \times \mathrm{Im}(\underset{\sim}{\vec{E}}_0)$ for this electric field. What is the direction of the pseudovector defining the rotation of a particle at the origin if $\mathrm{Im}(\underset{\sim}{f}_{\mathrm{CM}}) < 0$?

17.19 Starting with the general relation for the effective permittivity of a spherical particle with a core and one outer shell:

$$\underset{\sim}{\varepsilon}_{\mathrm{p}} = \underset{\sim}{\varepsilon}_2 \left[\frac{\frac{a_2^3}{a_1^3} + 2\frac{\underset{\sim}{\varepsilon}_1 - \underset{\sim}{\varepsilon}_2}{\underset{\sim}{\varepsilon}_1 + 2\underset{\sim}{\varepsilon}_2}}{\frac{a_2^3}{a_1^3} - \frac{\underset{\sim}{\varepsilon}_1 - \underset{\sim}{\varepsilon}_2}{\underset{\sim}{\varepsilon}_1 + 2\underset{\sim}{\varepsilon}_2}} \right] , \tag{17.74}$$

derive a relation for the effective permittivity if the outer shell has a thickness $\Delta a \ll a_2$.

17.20 Consider a sphere of radius a with uniform properties ε_{p} and σ_{p} embedded in a medium with properties ε_{m} and σ_{m}. Show that, if the interface has a charge density q'', the solution for the electrical potential outside the sphere if an applied field $\vec{E}_{\mathrm{ext}} = E_0 \hat{\boldsymbol{z}} \cos \omega t$ has a term given by

$$\phi = \frac{a^2 q''}{\varepsilon r} . \tag{17.75}$$

17.21 Consider a polystyrene microsphere with $\varepsilon_p = 2.5\varepsilon_0$ and $\sigma_p = 0.05\ \mu S/cm$, and an interfacial potential of -50 mV when embedded in a pH = 7, 10 mM KCl solution. A harmonic, uniform electric field is applied.

Model the EDL as a conducting layer with a surface conductance owing to the increased ion concentration in the double layer. Ignore surface conductance from electroosmotic convection, and simplify the math by approximating the potential distribution in the EDL as exponential. How does your prediction of the dielectrophoretic force felt by the sphere change depending on whether surface conductance is included?

17.22 Integrate $\vec{\vec{T}} \cdot \hat{\boldsymbol{n}}$ over the surface of a sphere exposed to a uniform, harmonic applied field, and show that the force felt by the sphere is given by

$$\vec{F} = \pi\varepsilon_m a^3 \nabla(\vec{E}_0 \cdot \vec{E}_0)\left(\mathrm{Re}(\underset{\sim}{f_{CM}}) + \mathrm{Re}(\underset{\sim}{f_{CM}})\cos 2\omega t + \mathrm{Im}(\underset{\sim}{f_{CM}})\sin 2\omega t\right). \tag{17.76}$$

17.23 Consider a spherical particle with radius a_2, with a core with properties ε_1 and σ_1 and a thin shell with thickness Δa and properties ε_2 and σ_2, where $\varepsilon_2 \gg \varepsilon_1$ and $\sigma_2 \gg \sigma_1$. Estimate the effect of this thin shell on the effective properties of the particle by replacing the sphere with a 1D geometry consisting of a cube of side length a_2 with properties corresponding to the core, and two thin layers of thickness Δa on the right and left side with properties corresponding to the shell. Model the electric field as being uniform and applied vertically, and model the system as consisting of three capacitors in parallel: one for the core and one each for the two layers on either side. Using the Helmholtz capacitor relation $C = \varepsilon A/d$, determine the capacitances for the components and the system. Using the total capacitance and the geometry of the entire system, derive the effective permittivity of the system and show that it is given by

$$\underset{\sim}{\varepsilon_p} = \underset{\sim}{\varepsilon_1} + 2\underset{\sim}{\varepsilon_2}\frac{\Delta a}{a_2}. \tag{17.77}$$

This geometry is a suitable approximation for a sphere with a conductive or high-permittivity shell, because the field lines for such a sphere can for the most part be approximated as being parallel to the shell.

17.24 Consider a spherical particle with radius a_2, with a core with properties ε_1 and σ_1 and a thin shell with thickness Δa and properties ε_2 and σ_2, where $\varepsilon_2 \ll \varepsilon_1$ and $\sigma_2 \ll \sigma_1$. Estimate the effect of this thin shell on the effective properties of the particle by replacing the sphere with a 1D geometry consisting of a cube of side length a_2 with properties corresponding to the core, and a thin layer of thickness Δa on the top of the cube with properties corresponding to the shell. Model the electric field as being uniform and applied vertically, and model the system as consisting of two capacitors in series: one for the core and one for the thin layer on top. Using the Helmholtz capacitor relation $C = \varepsilon A/d$, determine the capacitances for the components and the system. Using the total capacitance and the geometry of the entire system, derive the effective permittivity of the system and show that it is given by

$$\underset{\sim}{\varepsilon_p} = \frac{\underset{\sim}{\varepsilon_1}\underset{\sim}{\varepsilon_2}}{\frac{\Delta a}{a_2}\underset{\sim}{\varepsilon_1} + \underset{\sim}{\varepsilon_2}}. \tag{17.78}$$

This geometry is a suitable approximation for a sphere with an insulating or low-permittivity shell because the field lines for such a sphere can for the most part be approximated as being normal to the shell.

17.25 Consider an ideal dielectric sphere of radius a and uniform permittivity ε_p located at the origin and embedded in an ideal dielectric medium of permittivity ε_m. A point charge q is located outside the particle at a distance z_q from the origin along the z axis.

If the distance between the point charge and a point in the medium is defined as Δr, the expression $1/\Delta r$ can be written (for $r < z_q$) in terms of Legendre polynomials as

$$\frac{1}{\Delta r} = \frac{1}{z_q} \sum_{k=0}^{\infty} \left(\frac{r}{z_q} \right)^k P_k(\cos \vartheta) . \tag{17.79}$$

(a) Solve for the electric field inside and outside the particle in terms of Legendre polynomials.

(b) Define the multipole moment $p^{(k)} = 4\pi\varepsilon_m A_k$, where A_k is the Legendre polynomial coefficient from Appendix F, and define $p^{(k)}$ in terms of q, z_q, a, and material properties.

(c) *At the origin*, evaluate the electric field and the successive partial derivatives of the electric field with respect to z. Report these functions in terms of material properties, q, and z_q. Your result at the origin should give, for the electric field,

$$\vec{E} = -\left(\frac{q}{4\pi\varepsilon_m {z_q}^2} \right) \hat{z} , \tag{17.80}$$

and for the derivatives, you should obtain

$$\frac{\partial^k \vec{E}}{\partial z^k} = -\left(\frac{(k+1)!q}{4\pi\varepsilon_m z_q^{k+2}} \right) \hat{z} . \tag{17.81}$$

(d) Write the multipole moment in terms of the material properties, a, and the field and its derivatives at the origin, i.e., show that the multipole moment can be written as

$$p^{(k)} = \frac{4\pi\varepsilon_m K^{(k)} a^{2k+1}}{(k-1)!} \frac{\partial^{k-1} E_z}{\partial z^{k-1}} , \tag{17.82}$$

where the kth-order polarization coefficient $K^{(k)}$ is written as

$$K^{(k)} = \frac{\varepsilon_p - \varepsilon_m}{k\varepsilon_p + (k+1)\varepsilon_m} , \tag{17.83}$$

and E_z is the z component of the electric field.

(e) Given that the problem considers an ideal dielectric sphere in an ideal dielectric medium, evaluate the dipole moment $p^{(1)}$ and show that it is equal to the dipole caused by a uniform electric field; also, evaluate the first-order polarization coefficient $K^{(1)}$ and show that it is identical to the Clausius–Mossotti factor.

(f) By induction, write the multipole moment $\underset{\sim}{p}^{(k)}$ and polarization coefficient $\underset{\sim}{K}^{(k)}$ for a particle with complex permittivity $\underset{\sim}{\varepsilon}_p$ embedded in a medium with complex permittivity $\underset{\sim}{\varepsilon}_m$ that has a fixed charge at $z = z_q$ with an oscillatory magnitude $q = q_0 \cos \omega t$.

17.26 It is illustrative to examine the transient response of a uniform, spherical particle's effective dipole moment to the instantaneous application of a uniform, DC field, that is

$E_{\mathrm{ext}} = 0$ for all $t < 0$ and $E_{\mathrm{ext}} = E_0\hat{\mathbf{z}}$ for all $t \geq 0$. In this case, the effective dipole moment is given by

$$\vec{\boldsymbol{p}}(t) = 4\pi\varepsilon_{\mathrm{m}}a^3 E_0\hat{\mathbf{z}}\left\{\left(\frac{\sigma_{\mathrm{p}}-\sigma_{\mathrm{m}}}{\sigma_{\mathrm{p}}+2\sigma_{\mathrm{m}}}\right)\left[1-\exp\left(-\frac{t}{\tau_{\mathrm{MW}}}\right)\right] + \left(\frac{\varepsilon_{\mathrm{p}}-\varepsilon_{\mathrm{m}}}{\varepsilon_{\mathrm{p}}+2\varepsilon_{\mathrm{m}}}\right)\exp\left(-\frac{t}{\tau_{\mathrm{MW}}}\right)\right\}. \tag{17.84}$$

Evaluate the effective dipole in the two limits $t \ll \tau_{\mathrm{MW}}$ and $t \gg \tau_{\mathrm{MW}}$ where $\tau_{\mathrm{MW}} = \frac{\varepsilon_{\mathrm{p}}+2\varepsilon_{\mathrm{m}}}{\sigma_{\mathrm{p}}+2\sigma_{\mathrm{m}}}$.

APPENDIX A

Units and Fundamental Constants

A.1 UNITS

Key units are summarized in Table A.1.

Table A.1 **Fundamental units.**

Quantity	Units	Symbol	Other unit relations
Number	Mole	mol	
Mass	Kilogram	kg	
Time	Second	s	
Force	Newton	N	$kg\ m/s^2$
Pressure (stress)	Pascal	Pa	N/m^2
Energy (work)	Joule	J	N m; C V
Power	Watt	W	J/s; C A
Temperature	Kelvin	K	Celsius temperature: $T[C] = T[K] + 273.15$
Mass density	Kilogram per cubic meter	kg/m^3	
Species flux density	Millimolar per second per square meter	$mM/s\ m^2$	$mol/s\ m^5$
Electric current (electrical charge flux)	Ampere	A	
Electric charge	Coulomb	C	A s
Electric potential difference	Volt	V	
Electrical resistance	Ohm	Ω	V/A; V s/C
Electrical resistivity	Ohm meter	Ω m	
Electrical conductance	Siemens	S	A/V; C/V s
Electrical conductivity	Siemens per meter	S/m	
Electric field	Volts per meter	V/m	
Electrical current density	Ampere per square meter	A/m^2	
Capacitance	Farad	F	C/V
Inductance	Henry	H	V s/A
Magnetic flux	Weber	Wb	V s
Molar concentration	Millimolar	mM	mol/m^3
Normality	Millinormal	mN	mol/m^3
Osmolarity	Milliosmolar	mOsm	mol/m^3

A.2 FUNDAMENTAL PHYSICAL CONSTANTS

A number of important fundamental physical constants are summarized in Table A.2.

Table A.2 Fundamental constants.

Constant	Value
D, the debye, a unit of dipole moment	3.34×10^{-30} C m
ε_0, electrical permittivity of free space	8.85×10^{-12} C/V m
$F = N_A e$, Faraday constant	9.649×10^4 C/mol
k_B, Boltzmann constant	1.38×10^{-23} J/K
$\mu_{mag,0}$, magnetic permeability of free space	$4\pi \times 10^{-7}$ H/m
N_A, Avogadro's number	6.022×10^{23} mol^{-1}
e, magnitude of the electron charge	1.60×10^{-19} C
$R = N_A k_B$, universal gas constant	8.314 J/mol K

APPENDIX B

Properties of Electrolyte Solutions

Our interest in flows of electrolyte solutions in microdevices requires that we keep track of electrolyte solutions themselves, as well as acid–base chemistry at surfaces and in buffers. The coupling of electric fields and fluid mechanics that is common in microfluidics leads also to coupling between fluid mechanics and chemistry, because acid–base chemistry describes most interfacial charge and the charge state of many common analytes. Acid–base reactions at interfaces dictate the interfacial charge and, in turn, the electroosmotic mobility of the interface. Acid–base reactions, for example, between water and DNA or between water and proteins dictate the electrophoretic mobility of proteins and DNA in solution.

Because of this, this appendix provides a description of the properties of water, electrolyte solutions, and the associated acid–base chemistry. It defines solution terminology, derives the Henderson–Hasselbach equation for dissociation equilibrium, shows how the Henderson–Hasselbach equation relates reaction pK_a to acid dissociation, and shows how the water dissociation equation leads to a simple relation between pH and pOH for water at room temperature. This basic understanding explains, for example, the pH dependence of the electroosmotic mobility.

B.1 FUNDAMENTAL PROPERTIES OF WATER

Water is a unique molecule with several properties that are unusual compared with other liquids. A number of important fundamental properties of water are summarized in Table B.1.

B.2 AQUEOUS SOLUTIONS AND KEY PARAMETERS

We use the term *aqueous solutions* to describe solutions of water with dissolved solutes. We use the term *electrolyte solutions* to define solutions in which some or all of the solutes are ionized, and we use the term *electrolyte* to refer to the ionized solutes. Each species of electrolyte has a number of properties, including molar concentration, normality, and valence. Molar concentration c is defined as the moles of solute per liter of solution, and it is expressed in units of molar or M. For example, 1×10^{-3} moles in 1×10^{-1} liters is 1×10^{-2} M or 10 mM. One mole is equal to N_A or 6.02×10^{23} molecules. Because a liter is not an SI unit ($1\ \mathrm{L} = 1 \times 10^{-3}\ \mathrm{m}^3$), M is *not* an SI unit.[1] The *normality* N of a species is a measure of the number of H^+ or OH^- available to be gained or lost per liter of solution.[2] Thus 1 M of H_2SO_4 is 2 N, because two H^+ ions dissociate per

[1] The SI unit for concentration is moles per cubic meter: $1\,\frac{\mathrm{mol}}{\mathrm{m}^3} = 1$ mM.

[2] For oxidizing and reducing agents, this can also be a measure of electrons available to be lost or gained per liter.

Table B.1 **Properties of liquid water.**

Property	Value	Temperature [°C]
D, self-diffusivity	2.27×10^{-9} m^2/s	25 [278]
α, thermal diffusivity	14.6×10^{-7} m^2/s	25 [279]
η, dynamic viscosity	1.0 mPa s	20 [279]
	0.89 mPa s	25 [279]
p_w, dipole moment	2.95 D	27 [280]
quadrupole moments in D Å(xx, yy, zz)	$\begin{bmatrix} -4.27 \\ -7.99 \\ -5.94 \end{bmatrix}$	25 [281]
octupole moments in D Å$^2(xxz, yyz, zzz)$	$\begin{bmatrix} -1.75 \\ -0.55 \\ -1.981 \end{bmatrix}$	25 [282]
ρ, density	1000 kg/m^3 997 kg/m^3	20 [279] 25 [279]
λ_B, Bjerrum length	0.7 nm	25
$\varepsilon/\varepsilon_0$, dielectric constant	87.9	0 [283]
	78.4	25 [283]
	55.6	100 [283]

Note: Numbers in brackets refer to the references. Axes for the water molecule are defined in Fig. B.1.

H_2SO_4 molecule.[3] The valence or charge number z of an ion is the charge of the ion, normalized by the elementary charge. For example, the valence of Na^+ is 1, and the valence of SO_4^{-2} is −2. A solution in total has an ionic strength, written as I_c and defined by

$$I_c = \frac{1}{2} \sum_i c_i z_i^2 , \tag{B.1}$$

where i refers to an individual chemical species. For example, for 2-M KCl, $I_c = 2$ M, and for 2-M $MgSO_4$, $I_c = 8$ M. Certain solutions are described as being made up of symmetric electrolytes. A *symmetric electrolyte* is an electrolyte system in which the magnitudes of the charges of the anion and cation are the same. Examples include NaCl and $MgSO_4$.

These solutes have a number of transport properties, such as diffusivity, electrophoretic mobility, and molar conductivity, which are described in Chapter 11. They also have dielectric properties such as the dielectric increment, described in Section B.4.

B.3 CHEMICAL REACTIONS, RATE CONSTANTS, AND EQUILIBRIUM

Although this text does not consider reaction rates in detail, we discuss chemical reaction rates to establish equilibrium chemical conditions and to consider species in different states (e.g., bound to a wall or in the bulk solution), as these affect interfacial charge and electrokinetic transport.

[3] Some other units of solute concentration include % w/v, the weight in grams of solute per 100 mL of solution; weight percent (% w/w), the weight in grams of solute per 100 g of solution; *molality*, the moles of solute per kilogram of solvent; mole fraction, the ratio of moles of solute to total moles in the solution, and *osmolarity*, the summed molarity of all solutes, reported in milliosmolar or mOsm.

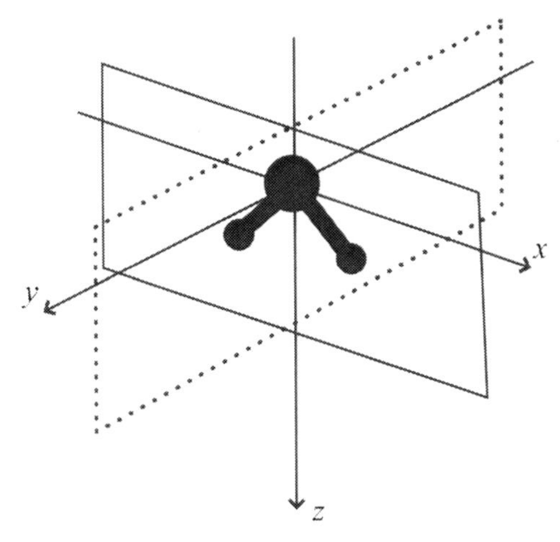

Fig. B.1 Axes for the internal structure of a water molecule.

Chemical reactions have rates that can be defined by rate constants and molar concentration. For reversible reactions, this leads to the definition of an equilibrium constant that describes the equilibrium of the reaction system. Consider the general acid dissociation and recombination reactions:

$$\mathrm{HA} \xrightarrow{k_1} \mathrm{H}^+ + \mathrm{A}^-\,, \tag{B.2}$$

$$\mathrm{H}^+ + \mathrm{A}^- \xrightarrow{k_{-1}} \mathrm{HA}\,, \tag{B.3}$$

where A refers to any species. The rate for reaction (B.2) is $k_1[\mathrm{HA}]$ and the rate for reaction (B.3) is $k_{-1}[\mathrm{H}^+][\mathrm{A}^-]$, where [X] denotes the molar concentration of species X. Because these two rates must be equal when the system is at equilibrium, an equilibrium constant K_{eq} or *acid dissociation constant* K_{a} can be defined for reactions (B.2) and (B.3):

$$K_{\mathrm{eq}} = K_{\mathrm{a}} = \frac{k_1}{k_{-1}} = \left.\frac{[\mathrm{H}^+][\mathrm{A}^-]}{[\mathrm{HA}]}\right|_{\mathrm{equilibrium}}\,. \tag{B.4}$$

Here, K_{eq} symbolizes an equilibrium constant and can be used for any reaction, and K_{a} symbolizes acid dissociation constant, applicable only when the dissociation of an acid is being discussed, as in reactions (B.2) and (B.3). K_{eq} can be derived for any reversible reaction by taking the ratio of forward and backward reaction rate coefficients.

B.3.1 Henderson–Hasselbach equation

Given the existence of equilibrium constants for any reversible reaction, we can create a convenient notation system and methodology for analyzing acid–base reactions. These reactions are central to micro- and nanoscale fluid flow because they determine the charge of walls and macromolecules in the aqueous systems under study.

The Henderson–Hasselbach equation is a useful way to relate solution pH and acid dissociation constant $\mathrm{p}K_{\mathrm{a}}$ to the degree to which acids dissociate and the concentrations of both acid and conjugate base. The Henderson–Hasselbach equation thus provides a simple quantitative framework for predicting the charge of macromolecules and surfaces. For acid dissociation reactions, we can derive the Henderson–Hasselbach equation from the equilibrium equation:

$$K_{\mathrm{a}} = \frac{[\mathrm{H}^+][\mathrm{A}^-]}{[\mathrm{HA}]}\,. \tag{B.5}$$

Table B.2 **List of weak acids and corresponding pK_a's.**

Weak Acid	pK_a
H_3PO_4	2.1
$H_2PO_4^-$	7.2
HPO_4^{-2}	12.3
$Tris^+$	8.3
$HEPES^+$	7.5
$TAPS^+$	8.4
borate	9.24
citrate	3.06
citrate	4.74
citrate	5.40
$ACES^+$	6.9
$PIPES^+$	6.8
Acetic acid	4.7

Taking logarithms and defining pK_a = − log K_a and pH = − log[H^+], we can rearrange this to obtain

Henderson–Hasselbach equation, ideal solution

$$\log\frac{[A^-]}{[HA]} = pH - pK_a\,. \tag{B.6}$$

This is the *Henderson–Hasselbach equation*, which shows how the ionization states of weak acids can be predicted by the pK_a of the reaction combined with the pH of the solution. If the pK_a of the weak acid is known (pK_a values are listed for several weak acids in Table B.2), the ion concentrations can be readily calculated. For example, when the pH of the solution is equal to the reaction pK_a, the weak acid is 50% dissociated. If the pH is more acidic, say one or two pH units below the pK_a, then the fraction of the acid in a dissociated state (A^-) is reduced (to 9% and 0.9%, respectively). If the pH is more basic, say one or two pH units above the pK_a, then the fraction of the acid in a dissociated state is increased (to 91% and 99%, respectively). These results are shown graphically in Fig. B.2. Acids and bases are termed *strong* when they are completely ionized upon dissolution in water, i.e., when they have infinite dissociation constants.

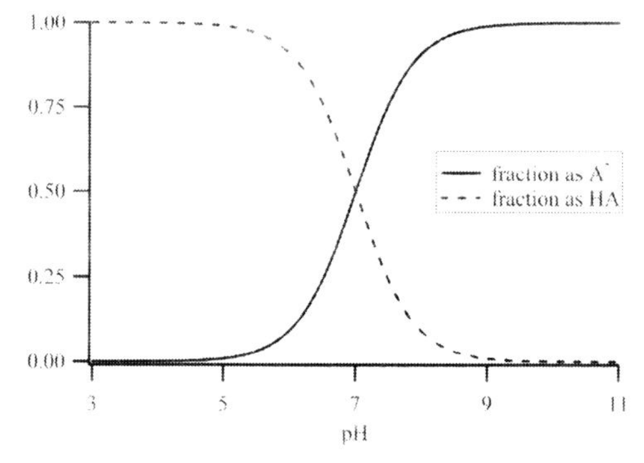

Fig. B.2 Fraction of molecules in the form of weak acid (HA) versus conjugate base (A^-) as a function of pH for a reaction with pK_a = 7.

EXAMPLE PROBLEM B.1

1 mmol of a weak acid, HA, is added to pure water, and the resulting pH of the system is measured to be 5.5. What is the pK_a of the weak acid?

SOLUTION: Because the resulting pH is 5.5, we know $[H^+] = 1 \times 10^{-5.5}$ and therefore $[A^-] = 1 \times 10^{-5.5}$. From this, we know $[HA] = 1 \times 10^{-3} - 1 \times 10^{-5.5}$. We can substitute these into

$$\mathrm{p}K_a = \mathrm{pH} - \log \frac{[A^-]}{[HA]} \tag{B.7}$$

to get

$$\mathrm{p}K_a = 8.00 \,. \tag{B.8}$$

B.3.2 Conjugate acids and bases; buffers

Many definitions of acids and bases are used; for our purposes here, the most useful one is the Brønsted definition, which defines acids as species that donate protons and bases as species that accept protons. When an acid loses a proton, the resulting solute is its *conjugate base*, and when a base gains a proton, the resulting solute is its *conjugate acid*.

Buffers consist of mixtures of weak acids and their conjugate base and are so named because the pH of a solution that contains these components changes less upon addition of H^+ or OH^- than a solution without these components. The reason why the presence of a weak acid buffers the system against perturbation by H^+ or OH^- is that the equilibrium relation between the weak acid and its conjugate base adjusts in response to these perturbations, largely canceling out the change in H^+ or OH^-. Thus, in a buffered system, addition of H^+ or OH^- has little effect on the pH, but has a large effect on the relative concentrations of weak acid and conjugate base. Because pH controls the charge states of ions and surfaces that we care about, whereas the concentrations of weak acid and conjugate base do not, the use of a buffer in an electrolyte solution makes its electrokinetic performance (and thus its fluid flow properties) insensitive to perturbations caused, for example, by passing current through the system. Buffers are thus critical to obtaining reliable flow results in electrokinetically driven microfluidic systems.

B.3.3 Ionization of water

By writing an acid dissociation equation for water and calculating the concentrations of H^+ and OH^-, we can derive a simple relation for pH and pOH. For water, we can write an acid dissociation reaction:

$$H_2O \longleftrightarrow H^+ + OH^- \,. \tag{B.9}$$

The form that protons take in an aqueous solution is H_3O^+, but we use H^+ and H_3O^+ interchangeably in this text. For this reaction, the equilibrium constant is given by

$$K_{eq} = \frac{[H^+][OH^-]}{[H_2O]} \,. \tag{B.10}$$

For this reaction at 25°C, $K_{eq} = 1.8 \times 10^{-16}$ mol/L. If we assume that $[H_2O]$ is constant, then $[H_2O]$ can be calculated from the density of water and its molar mass, leading to

water dissociation relation

$$pH + pOH \simeq 14\,, \tag{B.11}$$

where pOH is the negative logarithm of $[OH^-]$. This relation links the concentration of H^+ and OH^- in aqueous systems at room temperature and defines a neutral pH at pH = 7.

EXAMPLE PROBLEM B.2

Given its molecular weight (41.05 g/mol) and density (782 kg/m^3), calculate the molarity of pure acetonitrile.

SOLUTION:

$$[CH_3CN] = \frac{782\ \text{g}}{\text{L}} \times \frac{\text{mol}}{41.05\ \text{g}} = 19.0\ \frac{\text{mol}}{\text{L}} = 19.0\ \text{M}\,. \tag{B.12}$$

B.3.4 Solubility product of weakly soluble salts

As was the case for dissociation of weak acids, dissociation of weakly soluble salts leads to a simple relation between the concentrations of the dissolved ions. For a salt, we can write a dissociation reaction:

$$BA(s) \longleftrightarrow B^+(aq) + A^-(aq)\,, \tag{B.13}$$

where (s) denotes solid phase and (aq) denotes aqueous i.e., dissolved phase. If we assume that ample solid is available, the concentration of the solid-phase ions is a constant, and the equilibrium can be described in terms of a *solubility product* $K_{sp} = [B^+][A^-]$. This is often written in terms of the negative logarithm of the concentrations:

solubility product relation

$$pA + pB \simeq pK_{sp}\,, \tag{B.14}$$

where pA and pB are the negative logarithms of $[A^-]$ and $[B^+]$, respectively. This relation links the concentration of the ions to one fundamental property of the salt. For example, the pK_{sp} of AgI at room temperature is 16.5.

B.3.5 Ideal solution limit and activity

An *ideal solution* is one in which solutes can be assumed to interact exclusively with the solvent – no solute–solute interactions affect the solution properties. In the ideal solution limit, the properties of ions and molecules are not affected by their concentration. In real solutions, however, finite solute concentrations lead to a certain amount of solute–solute interactions. These interactions affect the chemical interactions of the solute and reduce the apparent molarity of the solute for chemical equilibrium. We thus define the activity a as the effective or apparent molarity of a solute. When studying chemical equilibrium in real solutions, we use the activity rather than the concentration to account for these effects.

B.3.6 Electrochemical potentials

The molar chemical potential (or partial molar Gibbs free energy) of a species [J/mol] is defined as

definition of molar chemical potential

$$g_i = g_i^\circ + RT \ln \frac{a_i}{a_i^\circ}, \tag{B.15}$$

where g_i° is the chemical potential at the reference activity a_i°. For ideal solutions, $a = c$ and the concentration can be used instead of the activity.

Equilibrium between systems that involve electrical potential drops and chemical reaction is defined by means of the electrochemical potential $\overline{g_i}$:

electrochemical potential

$$\overline{g_i} = g_i + z_i F \varphi = g_i^\circ + RT \ln \frac{a_i}{a_i^\circ} + z_i F \varphi. \tag{B.16}$$

Again, $a = c$ for dilute solutions. We use chemical and electrochemical potentials to treat surface adsorption and ion distributions in an electric field.

B.4 EFFECTS OF SOLUTES

In the dilute solution limit, the effects of added solvents or dissolved solutes are assumed to be negligible. However, as the concentrations of these other solvents or solutes increase, the effects of these solutes on the liquid properties cannot be ignored. As a first approximation, these effects can be linearized and treated in terms of differential properties.

B.4.1 Dielectric increments

Solutions of electrolytes have electrical permittivities that are different from that of pure water. This is typically reported in terms of a *dielectric increment* $\frac{\partial \varepsilon}{\partial c}$, which stems from a first-order Taylor series approximation of the electrical permittivity as a function of concentration:

$$\varepsilon(c) = \varepsilon(0) + c\frac{d\varepsilon}{dc}. \tag{B.17}$$

Table B.3 **Dielectric increments of a variety of solutes.**

Solute	Dielectric increment, $\frac{\partial \varepsilon}{\partial c}$ [M^{-1}]
Na^+	−8 [284]
K^+	−8 [284]
H^+	−17 [284]
Mg^{+2}	−24 [284]
Cl^-	−3 [284]
OH^-	−13 [284]
SO_4^{-2}	−7 [284]
Glycine	24 [62]
Lactose	−8.2 [285]
Sucrose	−8.2 [285]
Trehalose	−7.6 [285]
Maltose	−8.9 [285]
Sorbitol	−2.9 [285]
Mannitol	−2.5 [285]
Inositol	−1.9 [285]
Ethanol	−4.6 [286]
Methanol	−2.1 [286]
Glycerol	−1.9 [285]
Gly-Gly	72 [62]
Gly-Gly-Gly	126 [62]
Cyclohexylaminoethane sulfonate, $4.0 < pH < 6.3$	23 [67]
Cyclohexylaminopropane sulfonate, $4 < pH < 7.4$	38 [67]
Cyclohexylaminobutane sulfonate, $4 < pH < 7.7$	68 [67]
Trimethylammoniopropane sulfonate, $4 < pH < 10$	42–52 [66, 67]
Triethylammonioethane sulfonate, $4 < pH < 10$	42 [287]
Triethylammoniopropane sulfonate, $4 < pH < 10$	59 [287]
Triethylammoniobutane sulfonate, $4 < pH < 10$	73 [287]

Note: Numbers in brackets refer to the references. The linear model for the dielectric increment in general is applicable only for finite concentrations, typically up to 3 M.

Table B.3 lists dielectric increments for several solutes. Small ions reduce the dielectric constant of water when in solution. Zwitterionic salts often increase the dielectric constant of water because these molecules have a permanent dipole larger than water when they are ionized. The pH ranges listed for some solutes refer to the pH range in which their positive and negative sites can be assumed fully ionized – outside this range, the dielectric increment is reduced by the fact that the charges are no longer fully ionized. Homologous sets, i.e., groups of molecules that have similar structure, show how greater separation of charge leads to a larger dipole moment – so Gly-Gly-Gly has a larger dipole moment than Gly because three dipoles are lined up in order, and triethylammoniobutane sulfonate has a higher dipole moment than triethylammonioethane sulfonate because its charges (i.e., the quaternary ammonium and the sulfonate) are separated by a C_4 linker rather than a C_2 linker. Larger dipole moments lead to larger permittivities and therefore larger positive dielectric increments. These relationships work for relatively small chains, e.g., 2–6 carbon links. Above that, the molecule has enough flexibility that the positive and negative charges can loop around and attract each other, reducing the dipole moment and thus the dielectric increment.

EXAMPLE PROBLEM B.3

What are the dielectric constants of the following solutions at 25 C?

1. 1-mM NaCl
2. 10-mM NaCl
3. 100-mM NaCl
4. 1-M NaCl

SOLUTION: The dielectric increment of Na^+ is $-8\ M^{-1}$, and the dielectric increment of Cl^- is $-3\ M^{-1}$. Thus the total effect of NaCl is $-11\ M^{-1}$.

Assuming that the 25 C dielectric constant of water is 78.4, the dielectric constants of these solutions are 78.4, 78.3, 77.3, and 67.4.

B.5 SUMMARY

In this appendix, it is asserted that the key reactions required for modeling fluid flow in aqueous electrolytes in microdevices are primarily acid–base reactions. Basic terminology is summarized for describing electrolyte solutions as well as equilibria of acid–base reactions. The key results are the Henderson–Hasselbach relation between acids and conjugate bases,

$$\log \frac{[A^-]}{[HA]} = \mathrm{pH} - \mathrm{p}K_a \,, \tag{B.18}$$

and the relation between the H^+ and OH^- for water at room temperature,

$$\mathrm{pH} + \mathrm{pOH} \simeq 14 \,. \tag{B.19}$$

B.6 SUPPLEMENTARY READING

A number of texts [62, 278, 279, 283, 284] cover water properties in detail and serve as useful background for the values cited here. Segel's text [288] is excellent for solution properties and acid–base chemistry, as it provides many useful definitions, numerous worked exercises, and problems with solutions. It also provides extensive reference pK_a and activity data.

B.7 EXERCISES

Assume dilute solution results throughout.

B.1 Derive the Henderson–Hasselbach equation from the rate equations for acid dissociation and recombination.

B.2 Given the room-temperature K_{eq} for water dissociation, show that relation (B.11) is approximately correct.

B.3 Consider two separate systems: one that is 100 mL of pure water at and one with 100 mL containing 100 mM each of KH_2PO_4 and K_2HPO_4. What is the resulting pH if 1 mL of 1 N HCl is added to each solution? What does this tell you about the buffering action of the phosphate ions?

B.4 Given its molecular weight (32.04 g/mol) and density (791 kg/m^3), calculate the molarity of pure methanol.

B.5 Calculate the pH of the following solutions:

(a) 1-M HCl,

(b) 0.02-M H_2SO_4,

(c) 10-mM KOH.

B.6 0.1 mol of a weak acid, HA, is added to pure water, and the resulting pH of the system is measured to be 2.2. What is the pK_a of the weak acid?

B.7 1 mol of a weak acid, HA, is added to pure water, and the resulting pH of the system is measured to be 4.15. What is the pK_a of the weak acid?

B.8 25 mmol of a weak acid, HA, is added to pure water, and the resulting pH of the system is measured to be 5.9. What is the pK_a of the weak acid?

B.9 Phosphoric acid (H_3PO_4) has three protons that systematically dissociate with pK_a's of 2.1, 7.2, and 12.3 to form $H_2PO_4^-$, HPO_4^{-2}, and PO_4^{-3}. Thus we can write three equilibrium equations:

$$H_3PO_4 \overset{pK_{a1}}{\longleftrightarrow} H^+ + H_2PO_4^-, \tag{B.20}$$

$$H_2PO_4^- \overset{pK_{a2}}{\longleftrightarrow} H^+ + HPO_4^{-2}, \tag{B.21}$$

$$HPO_4^{-2} \overset{pK_{a3}}{\longleftrightarrow} H^+ + PO_4^{-3}. \tag{B.22}$$

If a phosphate ion exists in the form PO_4^{-3}, we would say that three protons have dissociated. If it existed in the form HPO_4^{-2}, we would say that two protons have dissociated, and so on. At any given pH, there is a statistical distribution of the four possible dissociation states of phosphate and thus an average number of protons that have dissociated. So we can define the concentration of dissociated protons as

$$3[PO_4^{-3}] + 2[HPO_4^{-2}] + [H_2PO_4^{-}], \tag{B.23}$$

and the total concentration of phosphate ions (in all forms) is

$$[PO_4^{-3}] + [HPO_4^{-2}] + [H_2PO_4^{-}] + [H_3PO_4]. \tag{B.24}$$

Calculate and plot (as a function of pH) the average number of protons that have dissociated per phosphate ion.

APPENDIX C

Coordinate Systems and Vector Calculus

This appendix briefly summarizes coordinate systems and vector/tensor operations to clarify notation and to serve as a reference for material discussed in the main text. Throughout this text, we use symbolic or Gibbs notation, which uses the nabla symbol (∇) and vector operations such as the dot product ($\cdot$) and cross product ($\times$). We also make a number of tacit, simplifying assumptions throughout. We are interested in using governing equations to describe continuous regions of physical fluid flows, so we typically assume the field properties are differentiable and bounded except at point singularities or regions outside the physical domain.

C.1 COORDINATE SYSTEMS

We use coordinate systems to depict positions in space. Common coordinate systems (Fig. C.1) include Cartesian coordinates (x, y, z), cylindrical coordinates $(\imath, \theta, z)$, and spherical coordinates (r, ϑ, φ), all of which are discussed in the following subsection.

C.1.1 3D coordinate systems

Cartesian coordinates denote a position by the three lengths x, y, and z as well as the three unit *basis vectors* $\hat{\boldsymbol{x}}$, $\hat{\boldsymbol{y}}$, and $\hat{\boldsymbol{z}}$. Cylindrical coordinates denote the z position similarly, but use the coordinate $\imath$ and the angle θ $(0 < \theta < 2\pi)$ to denote the orthogonal projection of a point onto the xy plane. The unit basis vectors for cylindrical coordinates are denoted as $\hat{\imath}$, $\hat{\boldsymbol{\theta}}$, and $\hat{\boldsymbol{z}}$. Unlike for Cartesian coordinates, whose unit vectors are fixed, the unit vectors for cylindrical coordinates are dependent on location – specifically, both $\hat{\imath}$ and $\hat{\boldsymbol{\theta}}$ are functions of θ, i.e.,

$$\hat{\imath} = \hat{\boldsymbol{x}}\frac{x}{\sqrt{x^2+y^2}} + \hat{\boldsymbol{y}}\frac{y}{\sqrt{x^2+y^2}} = \hat{\boldsymbol{x}}\cos\theta + \hat{\boldsymbol{y}}\sin\theta \tag{C.1}$$

and

$$\hat{\boldsymbol{\theta}} = \hat{\boldsymbol{x}}\frac{-y}{\sqrt{x^2+y^2}} + \hat{\boldsymbol{y}}\frac{x}{\sqrt{x^2+y^2}} = -\hat{\boldsymbol{x}}\sin\theta + \hat{\boldsymbol{y}}\cos\theta\,. \tag{C.2}$$

Spherical coordinates are described with respect to a reference axis, usually the positive z axis but occasionally positive x axis in this text. Spherical coordinates use the colatitude angle ϑ $(0 < \vartheta < \pi$; defined as the angle from the reference axis to the point), the azimuthal angle φ $(0 < \varphi < 2\pi$; defined as the angle from the positive x axis to the orthogonal projection of the point onto the xy plane), and the distance r. The unit basis vectors for spherical coordinates are denoted as $\hat{\boldsymbol{r}}$, $\hat{\boldsymbol{\vartheta}}$, and $\hat{\boldsymbol{\varphi}}$.[1] All of these unit vectors

[1] The symbols used for spherical coordinates vary widely from source to source, as do the order of the colatitudinal and azimuthal coordinates.

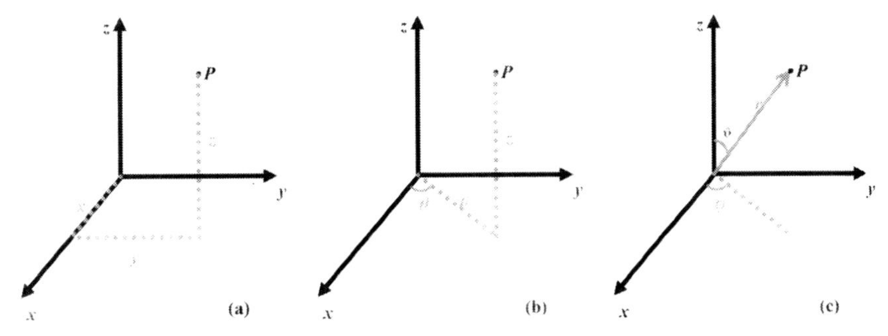

Fig. C.1 The location of a point P denoted with (a) Cartesian, (b) cylindrical, and (c) spherical coordinate systems.

are functions of location:

$$\hat{\boldsymbol{r}} = \hat{\boldsymbol{x}} \sin\vartheta \cos\varphi + \hat{\boldsymbol{y}} \sin\vartheta \sin\varphi + \hat{\boldsymbol{z}} \cos\vartheta\,, \tag{C.3}$$

$$\hat{\boldsymbol{r}} = \hat{\boldsymbol{x}} \frac{x}{\sqrt{x^2+y^2+z^2}} + \hat{\boldsymbol{y}} \frac{y}{\sqrt{x^2+y^2+z^2}} + \hat{\boldsymbol{z}} \frac{z}{\sqrt{x^2+y^2+z^2}}\,, \tag{C.4}$$

$$\hat{\boldsymbol{\vartheta}} = \hat{\boldsymbol{x}} \cos\vartheta \cos\varphi + \hat{\boldsymbol{y}} \cos\vartheta \sin\varphi - \hat{\boldsymbol{z}} \sin\vartheta\,, \tag{C.5}$$

$$\hat{\boldsymbol{\vartheta}} = \hat{\boldsymbol{x}} \frac{xz}{\sqrt{x^2+y^2}\sqrt{x^2+y^2+z^2}} + \hat{\boldsymbol{y}} \frac{yz}{\sqrt{x^2+y^2}\sqrt{x^2+y^2+z^2}} - \hat{\boldsymbol{z}} \frac{\sqrt{x^2+y^2}}{\sqrt{x^2+y^2+z^2}}\,, \tag{C.6}$$

and

$$\hat{\boldsymbol{\varphi}} = -\hat{\boldsymbol{x}} \sin\vartheta \sin\varphi + \hat{\boldsymbol{y}} \sin\vartheta \cos\varphi\,. \tag{C.7}$$

$$\hat{\boldsymbol{\varphi}} = \hat{\boldsymbol{x}} \frac{-y}{\sqrt{x^2+y^2+z^2}} + \hat{\boldsymbol{y}} \frac{x}{\sqrt{x^2+y^2+z^2}}\,. \tag{C.8}$$

In this text, we use Cartesian, 2D cylindrical, and axisymmetric spherical coordinates; because of this, the symbol φ for the azimuthal coordinate is rarely, if ever, needed.[2] The Cartesian coordinate system is a rectangular coordinate system – a special case in which all unit vectors are of the same length and independent of location. This simplifies the math considerably. Cylindrical and spherical coordinate systems are curvilinear coordinate systems, which are more complicated. In curvilinear coordinate systems, each coordinate can have different magnitudes and even units. Unit vectors and velocities all have the same units, but their direction is a function of the location.

C.1.2 2D coordinate systems

For physical situations in which a third dimension exhibits symmetry, we can simplify our analysis by considering only the two dimensions in which parameters vary. In Cartesian coordinate systems, we neglect one of the dimensions (usually the z dimension) and consider only a plane (usually the xy plane). In cylindrical coordinate systems with plane symmetry, we neglect the z dimension and consider the $\imath\theta$ plane, whereas for axial symmetry we ignore θ and consider the $\imath z$ plane. In spherical coordinate systems with axial symmetry, we neglect the azimuthal (φ) dimension and consider the $r\vartheta$ plane.

[2] Because the φ is rarely used, we use φ for the electrical double-layer potential throughout this text.

C.2 VECTOR CALCULUS

Spatial phenomena naturally give rise to vector descriptions, and the calculus of these parameters is central to all of the analysis in this text. This section describes scalars, vectors, and tensors, as well as calculus operations on these quantities.

C.2.1 Scalars, vectors, and tensors

We use scalars, vectors, and tensors to denote a variety of properties. *Scalars* indicate values that have a magnitude but no direction, for example pressure p and temperature T. Scalars are described by $3^0 = 1$ value. *Vectors* indicate values that have both magnitude and direction, for example the velocity $\vec{u}$. Vectors are described by $3^1 = 3$ values, and, although we do not explore such transformations in this text, true vectors can be transformed into different coordinate systems through the use of a direction cosine matrix.[3] In this text, we use arrows to distinguish between vectors (e.g., the velocity $\vec{u}$) and scalars (e.g., the magnitude of the x velocity u). We use hats to denote unit vectors, so $\hat{x}$, for example, denotes a unit vector in the x direction. We distinguish among position vectors, distance vectors, velocity vectors, coordinate vectors, unit vectors, and vector magnitudes. For example, $\vec{r}$ denotes the position vector relative to the origin, $\hat{r}$ denotes a unit vector in that direction, and r denotes the magnitude of $\vec{r}$. These are related by $\hat{r} = \vec{r}/r$. The distance vector between two points is denoted by $\vec{\Delta r}$.

Using unit vectors, we can write a vector (for example, the velocity vector $\vec{u}$) in Cartesian coordinates as

$$\vec{u} = \hat{x}\,u + \hat{y}\,v + \hat{z}\,w\,, \tag{C.9}$$

in cylindrical coordinates as

$$\vec{u} = \hat{\imath}\,u_{\imath} + \hat{\theta}\,u_{\theta} + \hat{z}\,u_z\,, \tag{C.10}$$

and in spherical coordinates as

$$\vec{u} = \hat{r}\,u_r + \hat{\vartheta}\,u_{\vartheta} + \hat{\varphi}\,u_{\varphi}\,. \tag{C.11}$$

The components of a Cartesian vector can be written as a matrix. This matrix notation is convenient and useful for Cartesian vectors, because matrix-like operations can determine the result of a Cartesian vector operation. However, the matrix notation is *not* well suited for cylindrical or spherical coordinate systems, because the vector operations for curvilinear systems do not correspond to simple matrix-like operations. Because of this, we specifically *avoid* matrix representations for curvilinear coordinate systems.

In Cartesian coordinates, we find it useful to write vectors as a matrix, for example,

$$\vec{u} = \begin{bmatrix} u & v & w \end{bmatrix}\,. \tag{C.12}$$

Conversions between coordinate systems. We often find it useful to convert back and forth between coordinate systems. Most often, we convert back and forth between Cartesian coordinates and other coordinate systems. For cylindrical coordinates, we have:

$$x = \imath \cos\theta\,, \tag{C.13}$$

$$y = \imath \sin\theta\,, \tag{C.14}$$

$$z = z\,, \tag{C.15}$$

[3] We use the term *pseudovector* to describe a magnitude and direction described by three values whose transformation into different coordinate systems with a direction cosine matrix sometimes introduces a sign change. Properties that describe rotation, like vorticity, are pseudovectors because their sign depends on the handedness of the coordinate system.

and

$$\imath = \sqrt{x^2 + y^2}\,, \tag{C.16}$$

$$\theta = \text{atan2}\,(y, x)\,, \tag{C.17}$$

$$z = z\,, \tag{C.18}$$

where we use atan2 to denote the two-argument inverse tangent, whose value (in the range 0–2π) depends on the signs of y and x, as well as on the value of y/x:

$$\text{atan2}(y, x) = \begin{cases} \tan^{-1}(y/x) & \text{if} \quad x > 0,\, y > 0 \\ \tan^{-1}(y/x) + \pi & \text{if} \quad x < 0 \\ \tan^{-1}(y/x) + 2\pi & \text{if} \quad x > 0,\, y < 0 \\ \pi/2 & \text{if} \quad x = 0,\, y > 0 \\ 3\pi/2 & \text{if} \quad x = 0,\, y < 0 \end{cases}. \tag{C.19}$$

For spherical coordinates, we have

$$x = r \sin\vartheta \cos\varphi\,, \tag{C.20}$$

$$y = r \sin\vartheta \sin\varphi\,, \tag{C.21}$$

$$z = r \cos\vartheta\,, \tag{C.22}$$

and

$$r = \sqrt{x^2 + y^2 + z^2}\,, \tag{C.23}$$

$$\vartheta = \text{atan2}\left(\sqrt{x^2 + y^2},\, z\right)\,, \tag{C.24}$$

$$\varphi = \text{atan2}\,(y, x)\,. \tag{C.25}$$

EXAMPLE PROBLEM C.1

Convert the following position vectors between coordinate systems as specified.

1. Convert $(\imath, \theta, z) = (2, -1, -1)$ from cylindrical to Cartesian coordinates.
2. Convert $(r, \vartheta, \varphi) = (1, \pi, \pi)$ from spherical to Cartesian coordinates.
3. Convert $(x, y, z) = (1, 1, 1)$ from Cartesian to cylindrical coordinates.
4. Convert $(x, y, z) = (10, -5, 6)$ from Cartesian to spherical coordinates.

SOLUTION:

1. $x = \imath \cos\theta = 1.08$; $y = \imath \sin\theta = 1.68$; $z = -1$.
2. $x = r \sin\vartheta \cos\varphi = 0$; $y = r \sin\vartheta \sin\varphi = 1$; $z = r \cos\vartheta = 0$.
3. $\imath = \sqrt{x^2 + y^2} = \sqrt{2}$, $\theta = \text{atan2}(1, 1) = \pi/4$, $z = 1$.
4. $r = \sqrt{x^2 + y^2 + z^2} = 12.69$, $\vartheta = \text{atan2}(11.2, 6) = 1.08$, $\varphi = \text{atan2}(-5, 10) = 5.82$.

Magnitude of a velocity vector. The magnitude of a velocity vector $\vec{u}$ is denoted in this text with absolute value symbols: $|\vec{u}|$. Because each component of velocity is orthogonal to all other components and has the same units, the expression for the magnitude of a velocity vector is essentially the same in any of the three coordinate systems we typically use. For Cartesian velocity vectors $\vec{u} = \hat{x}\,u + \hat{y}\,v + \hat{z}\,w$, the magnitude is given by the

Euclidean norm,

$$|\vec{u}| = \sqrt{u^2 + v^2 + w^2}\,, \tag{C.26}$$

the magnitude of a cylindrical velocity vector $\vec{u} = \hat{\imath} u_{\imath} + \hat{\boldsymbol{\theta}} u_\theta + \hat{z} u_z$ is given by

$$|\vec{u}| = \sqrt{{u_{\imath}}^2 + {u_\theta}^2 + {u_z}^2}\,, \tag{C.27}$$

and the magnitude of a spherical velocity vector $\vec{u} = \hat{r} u_r + \hat{\boldsymbol{\vartheta}} u_\vartheta + \hat{\varphi} u_\varphi$ is given by

$$|\vec{u}| = \sqrt{{u_r}^2 + {u_\vartheta}^2 + {u_\varphi}^2}\,. \tag{C.28}$$

Length of position vector. The length of a position vector $\vec{r}$ is also denoted in this text with absolute value symbols: $|\vec{r}|$. Unlike velocity vectors, whose components all have units of length per time, position vectors can have components with different units. For example, the coordinates x, y, z, $\imath$, and r have units of length, whereas θ, ϑ, and φ are unitless. Because of this, the relations for the length of a position vector vary between coordinate systems.

For Cartesian position vectors $\vec{r} = (x,\ y,\ z)$, the magnitude is given by the Euclidean norm,

$$|\vec{r}| = \sqrt{x^2 + y^2 + z^2}\,, \tag{C.29}$$

the magnitude of a cylindrical position vector $\vec{r}$ specified by $\imath$, θ, and z is given by

$$|\vec{r}| = \sqrt{\imath^2 + z^2}\,, \tag{C.30}$$

and the magnitude of a spherical position vector $\vec{r}$ specified by r, ϑ, and φ is given by

$$|\vec{r}| = r\,. \tag{C.31}$$

EXAMPLE PROBLEM C.2

Consider two particles located at the specified positions and moving at the specified velocities. Calculate (a) the distance vector between the first particle and the second and (b) the relative velocity of particle 2 with respect to the velocity of particle 1.

1. Cartesian coordinate system. Particle 1: $(x, y, z) = (1, 0, 3)$; $(u, v, w) = (4, \pi, 6)$; Particle 2: $(x, y, z) = (2, \pi, -1)$; $(u, v, w) = (7, -\pi, 1)$.
2. Cylindrical coordinate system. Particle 1: $(\imath, \theta, z) = (1, 0, 3)$; $(u_{\imath}, u_\theta, u_z) = (4, \pi, 6)$; Particle 2: $(\imath, \theta, z) = (2, \pi, -1)$; $(u_{\imath}, u_\theta, u_z) = (7, -\pi, 1)$.

SOLUTION:

1. $(\Delta x, \Delta y, \Delta z) = (1, \pi, -4)$, $(\Delta u, \Delta v, \Delta w) = (3, -2\pi, -5)$.
2. Here we convert to Cartesian coordinates. After conversion, $(x, y, z) = (1, 0, 3)$ for particle 1 and $(x, y, z) = (-2, 0, -1)$ for particle 2. Similarly, $(u, v, w) = (4, \pi, 6)$ for particle 1 and $(u, v, w) = (-7, \pi, 1)$ for particle 2. We find $(\Delta x, \Delta y, \Delta z) = (-3, 0, -4)$ and $(\Delta u, \Delta v, \Delta w) = (-11, 0, -5)$. Converting back to cylindrical coordinates: $(\Delta \imath, \Delta\theta, \Delta z) = (3, \pi, -4)$ and $(\Delta u_{\imath}, \Delta u_\theta, \Delta u_z) = (11, 0, -5)$.

Length of distance vector. The distance vector $\vec{\Delta r}$ from a point denoted by position vector $\vec{r}_1$ to a point denoted by position vector $\vec{r}_2$ is written as $\vec{\Delta r} = \vec{r}_2 - \vec{r}_1$. In Cartesian coordinates, the values of the components of this vector are simply the differences between the components of the position vectors:

$$\vec{\Delta r} = (u_2 - u_1, v_2 - v_1, w_2 - w_1), \tag{C.32}$$

and the length of this vector can be calculated by evaluating the Euclidean norm. In cylindrical and spherical coordinates, the length of the distance vector is usually calculated by first converting to Cartesian coordinates.

Tensor. Generally, an *nth-order tensor* or a *rank n tensor* refers to a set of 3^n values that transform by means of n direction cosine matrices; these values have a magnitude and n directions. A scalar is a zeroth-order tensor (no direction) and a vector is a first-order tensor (one direction). In this text, we do not use any tensors beyond rank 2, so we use the term *tensor* to refer specifically to rank 2 tensors, which have $3^2 = 9$ components. Rank 2 tensors can be thought of as having two directional components, or they can be thought of as a relation between two vectors. For example, the velocity gradient $\nabla\vec{u}$ has values that are a function of two directions: the direction of the velocity component and the direction of its gradient. In this text, we use arrows or double arrows to distinguish among tensors (e.g., $\vec{\vec{\varepsilon}}$, the strain rate tensor), vectors (e.g., $\vec{u}$, the velocity), and scalars (e.g., u, the magnitude of the x velocity). In the text (but not in the figures), vectors and tensors are also set in bold type to highlight their presence.

For Cartesian coordinates, the components of a tensor can usefully be written as a matrix. We typically write Cartesian tensors as 3×3 matrices, for example

$$\nabla\vec{u} = \begin{pmatrix} \frac{\partial u}{\partial x} & \frac{\partial u}{\partial y} & \frac{\partial u}{\partial z} \\ \frac{\partial v}{\partial x} & \frac{\partial v}{\partial y} & \frac{\partial v}{\partial z} \\ \frac{\partial w}{\partial x} & \frac{\partial w}{\partial y} & \frac{\partial w}{\partial z} \end{pmatrix}. \tag{C.33}$$

C.2.2 Vector operations

Key vector operations include the dot product and the cross product. These operations are the vector analogs of scalar multiplication, providing results that are proportional to the products of the magnitudes of the vectors, but with additional information related to the direction or orientation of the vectors with respect to each other, the coordinate system, or both.

DOT PRODUCT

The *dot product* or scalar product of two vectors is a scalar equal to the product of the length of the two vectors and the cosine of the subtended angle (Fig. C.2):

$$\vec{A} \cdot \vec{B} = |\vec{A}||\vec{B}| \cos\alpha, \tag{C.34}$$

where α is the angle between vectors $\vec{A}$ and $\vec{B}$. The dot product commutes, so $\vec{A} \cdot \vec{B} = \vec{B} \cdot \vec{A}$.

For position vectors, the fact that the coordinates have different units for different coordinate systems makes the definition of the dot product specific to the coordinate system. For two Cartesian position vectors $\vec{A} = (x_1, y_1, z_1)$ and $\vec{B} = (x_2, y_2, z_2)$, the dot product is given by

$$\vec{A} \cdot \vec{B} = x_1x_2 + y_1y_2 + z_1z_2, \tag{C.35}$$

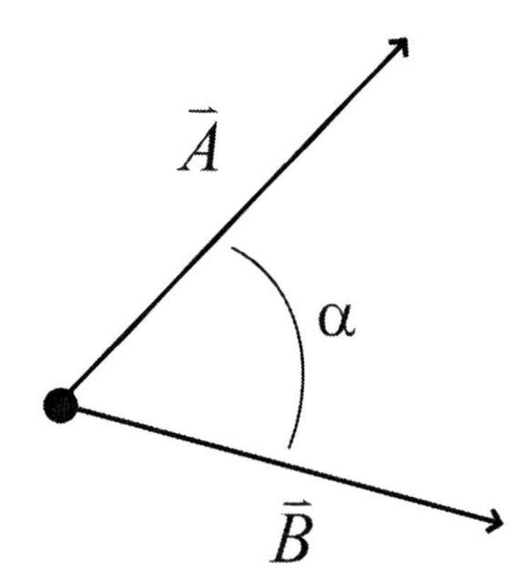

Fig. C.2 Definition of α for the calculation of the dot product.

but for two cylindrical position vectors $\vec{\boldsymbol{A}} = (\mathscr{r}_1, \theta_1, z_1)$ and $\vec{\boldsymbol{B}} = (\mathscr{r}_2, \theta_2, z_2)$, the dot product is given by

$$\vec{\boldsymbol{A}} \cdot \vec{\boldsymbol{B}} = \mathscr{r}_1 \mathscr{r}_2 \cos(\theta_1 - \theta_2) + z_1 z_2\,, \tag{C.36}$$

and for two spherical position vectors $\vec{\boldsymbol{A}} = (r_1, \vartheta_1, \varphi_1)$ and $\vec{\boldsymbol{B}} = (r_2, \vartheta_2, \varphi_2)$, the dot product is given by

$$\vec{\boldsymbol{A}} \cdot \vec{\boldsymbol{B}} = r_1 r_2 \sin\varphi_1 \sin\varphi_2 \cos(\vartheta_1 - \vartheta_2) + \cos\varphi_1 \cos\varphi_2\,. \tag{C.37}$$

The dot product is used in several ways. One important use is to determine what component of a vector is pointing in a specific direction. For example, we often evaluate fluxes across surfaces of control volumes; when we do this we invariably calculate the dot product of a vector with the unit outward normal. Some examples of this implementation are Eqs. (1.19), (1.26), and (5.9).

Matrix representation for Cartesian vectors and tensors. For two Cartesian vectors $\vec{\boldsymbol{A}} = (x_1, y_1, z_1)$ and $\vec{\boldsymbol{B}} = (x_2, y_2, z_2)$, we can write the dot product as:

matrix representation of Cartesian vector

$$\begin{bmatrix} x_1 & y_1 & z_1 \end{bmatrix} \cdot \begin{bmatrix} x_2 & y_2 & z_2 \end{bmatrix} = x_1 x_2 + y_1 y_2 + z_1 z_2\,, \tag{C.38}$$

and the dot product for a Cartesian vector $\vec{\boldsymbol{u}} = (u, v, w)$ and a Cartesian tensor τ can be written as:

matrix representation of dot product of Cartesian vectors

$$\begin{bmatrix} u & v & w \end{bmatrix} \cdot \begin{bmatrix} \tau_{xx} & \tau_{xy} & \tau_{xz} \\ \tau_{yx} & \tau_{yy} & \tau_{yz} \\ \tau_{zx} & \tau_{zy} & \tau_{zz} \end{bmatrix} = \begin{bmatrix} u\tau_{xx} + v\tau_{xy} + w\tau_{xz} \\ u\tau_{yx} + v\tau_{yy} + w\tau_{yz} \\ u\tau_{zx} + v\tau_{zy} + w\tau_{zz} \end{bmatrix}. \tag{C.39}$$

The notation adopted in this text for matrix-like manipulation of Cartesian vectors and tensors is different from standard matrix notation. In the notation of this text, the

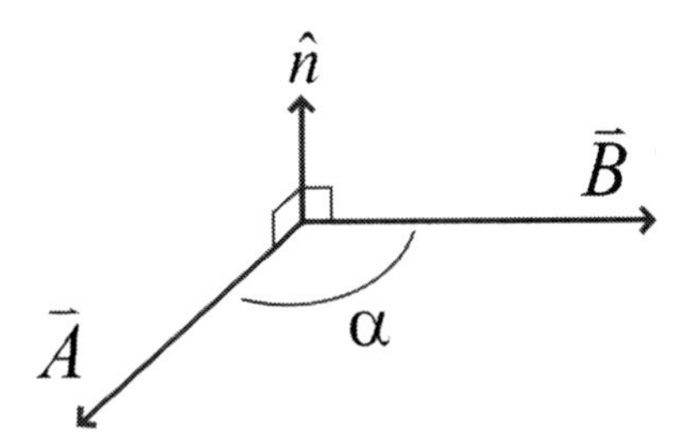

Fig. C.3 Definition of α and $\hat{\boldsymbol{n}}$ for the calculation of the cross product. The direction of $\hat{\boldsymbol{n}}$ is defined by the right-hand rule.

dot product of two vectors is written as

$$\begin{bmatrix} x_1 & y_1 & z_1 \end{bmatrix} \cdot \begin{bmatrix} x_2 & y_2 & z_2 \end{bmatrix} = x_1 x_2 + y_1 y_2 + z_1 z_2 , \qquad \text{(C.40)}$$

in which the elements of the row of the first vector are multiplied in turn by the elements of the row of the second vector term by term. In contrast, standard matrix multiplication would require that a row vector be multiplied by a column vector, and would omit the dot symbol $(\cdot)$:

$$\begin{bmatrix} x_1 & y_1 & z_1 \end{bmatrix} \begin{bmatrix} x_2 \\ y_2 \\ z_2 \end{bmatrix} = x_1 x_2 + y_1 y_2 + z_1 z_2 . \qquad \text{(C.41)}$$

This notation is used to highlight that vector operations and matrix operations are not identical and do not follow the same rules. For example, the dot product of two vectors commutes, but matrix multiplication does not.

For Cartesian vectors and tensors, the vector operations are simplified by matrix-like notation. This approach works *only* for rectangular coordinate systems, and curvilinear coordinate systems are not typically described with matrix operations.

Independent of the coordinate system, the dot product of a vector with a vector results in a scalar, and the dot product of a vector with a second-order tensor results in a vector.

CROSS PRODUCT

The cross product or vector product of two vectors is a pseudovector whose magnitude is equal to the product of the length of the two vectors and the sine of the subtended angle, and whose direction is normal to plane of the two vectors (Fig. C.3):

$$\vec{\boldsymbol{A}} \times \vec{\boldsymbol{B}} = \hat{\boldsymbol{n}} \, |\vec{\boldsymbol{A}}| |\vec{\boldsymbol{B}}| \sin \alpha , \qquad \text{(C.42)}$$

where α is the angle between the two vectors and $\hat{\boldsymbol{n}}$ is a unit normal to the plane of the two vectors $\vec{\boldsymbol{A}}$ and $\vec{\boldsymbol{B}}$, with a direction given by the right-hand rule. The cross product anticommutes, i.e., $\vec{\boldsymbol{A}} \times \vec{\boldsymbol{B}} = -\vec{\boldsymbol{B}} \times \vec{\boldsymbol{A}}$. Cross products and results of related vector operations (such as torque or vorticity or angular momentum) are *pseudovectors* because they have "handedness" and their sign can change if the handedness of the coordinate system changes.

Because the coordinates have different units for different coordinate systems, the definition of the cross product is specific to the coordinate system. For two Cartesian vectors $\vec{\boldsymbol{A}} = (x_1, y_1, z_1)$ and $\vec{\boldsymbol{B}} = (x_2, y_2, z_2)$, the cross product is given by

$$\vec{\boldsymbol{A}} \times \vec{\boldsymbol{B}} = \hat{\boldsymbol{x}} \, (y_1 z_2 - y_2 z_1) + \hat{\boldsymbol{y}} \, (x_2 z_1 - x_1 z_2) + \hat{\boldsymbol{z}} \, (x_1 y_2 - x_2 y_1) . \qquad \text{(C.43)}$$

For cylindrical and spherical coordinates, we convert to Cartesian coordinates to evaluate the cross product.

Matrix representation for Cartesian vectors. For Cartesian tensors, we can write the cross product for two vectors:

matrix representation of cross product for Cartesian vectors

$$\begin{bmatrix} a & b & c \end{bmatrix} \times \begin{bmatrix} d & e & f \end{bmatrix} = \det \begin{bmatrix} \hat{\boldsymbol{x}} & \hat{\boldsymbol{y}} & \hat{\boldsymbol{z}} \\ a & b & c \\ d & e & f \end{bmatrix}, \qquad \text{(C.44)}$$

where det denotes the determinant. The cross product of a vector with a vector results in a pseudovector that is orthogonal to both of the original vectors.

C.2.3 Del or nabla operations

The symbol ∇ is referred to as the *del operator* or *nabla operator*.[4] The del operator is a *vector operator*. It is *not* a vector.[5] Vector operators operate on scalars, vectors, or tensors to give a result, which may be a scalar or vector or tensor, depending on the operation.

We use three different vector operations that are denoted by the symbol ∇: the gradient (∇), divergence ($\nabla\cdot$), and curl ($\nabla\times$) operations. The divergence and curl symbols are reminiscent of the dot and cross product operations, as is their implementation, especially in Cartesian coordinates. The gradient is a way of taking derivatives that gives the direction of the maximum rate of change as well as the magnitude of that rate of change. The divergence of a velocity field tells us how much the flow is expanding or contracting. The curl of a velocity field tells us how much the flow is rotating.

GRADIENT OPERATOR

The gradient (∇) is a 3D spatial derivative. It gives the direction of the maximum rate of change as well as the magnitude of that rate of change.

The gradient of a scalar is a vector pointing in the direction in which the partial derivative of the scalar in that direction is maximum and has a magnitude equal to the spatial derivative with respect to that direction. The gradient of a vector is a second-order tensor that, in Cartesian coordinates, is made up of vectors corresponding to the gradients of the components of the vector.

Gradient Operator – Cartesian Coordinates. In Cartesian coordinates, the gradient of a scalar ϕ is given by

$$\nabla\phi = \hat{\boldsymbol{x}}\frac{\partial\phi}{\partial x} + \hat{\boldsymbol{y}}\frac{\partial\phi}{\partial y} + \hat{\boldsymbol{z}}\frac{\partial\phi}{\partial z}. \qquad \text{(C.45)}$$

[4] Some authors use *del* to refer to the operator and use *nabla* to refer to the symbol.

[5] This is analogous to the partial derivative operator, for example $\frac{\partial}{\partial x}$. Although $\frac{\partial}{\partial x}$ itself has no value, if $\frac{\partial}{\partial x}$ operates on something, say u, we get $\frac{\partial u}{\partial x}$, which has a value. Hence a vector operator has no value by itself and is thus not a vector.

Matrix representation of Cartesian gradients. In Cartesian coordinates, the gradient of a scalar a is given by

gradient of scalar, Cartesian coordinates

$$\nabla a = \begin{bmatrix} \frac{\partial a}{\partial x} & \frac{\partial a}{\partial y} & \frac{\partial a}{\partial z} \end{bmatrix}, \tag{C.46}$$

and the gradient of a vector (u, v, w) is given by

gradient of vector, Cartesian coordinates

$$\nabla \vec{u} = \begin{bmatrix} \frac{\partial u}{\partial x} & \frac{\partial u}{\partial y} & \frac{\partial u}{\partial z} \\ \frac{\partial v}{\partial x} & \frac{\partial v}{\partial y} & \frac{\partial v}{\partial z} \\ \frac{\partial w}{\partial x} & \frac{\partial w}{\partial y} & \frac{\partial w}{\partial z} \end{bmatrix}. \tag{C.47}$$

The gradient of a scalar results in a vector, and the gradient of a vector results in a second-order tensor. The units of the gradient are equal to the units of the scalar or vector operated on, divided by length.

Gradient Operator – Cylindrical Coordinates. Cartesian coordinates are relatively straightforward, because x, y, and z are in the same units and all of the relations in the previous section had a convenient symmetry to them. In cylindrical coordinates, however, θ is an angle rather than a length. Specifically, at a given radius r, if we rotate by an angle $d\theta$, we move a distance $r d\theta$ in total.[6] Thus derivatives with respect to θ differ from derivatives with respect to r by a factor of r. To account for this, the definition of the gradient operator in cylindrical coordinates is different than that for Cartesian coordinates. In cylindrical coordinates, the gradient of a scalar ϕ is given by

$$\nabla \phi = \hat{r}\frac{\partial \phi}{\partial r} + \hat{\boldsymbol{\theta}}\frac{1}{r}\frac{\partial \phi}{\partial \theta} + \hat{z}\frac{\partial \phi}{\partial z}. \tag{C.48}$$

Gradient Operator – Spherical Coordinates. In spherical coordinates, r has units of length, but ϑ and φ are angles. Specifically, at a given radius r, if we rotate by an angle $d\vartheta$, we move a distance $r d\vartheta$, and if we rotate by an angle $d\varphi$, we move a distance $r \sin\vartheta d\varphi$. Thus derivatives with respect to φ differ from derivatives with respect to r by a factor of $r \sin\vartheta$. To account for this, the definition of the gradient operator in spherical coordinates takes a different form than it does in Cartesian or cylindrical coordinates. The gradient of a scalar ϕ in spherical coordinates is given by

$$\nabla \phi = \hat{\boldsymbol{r}}\frac{\partial \phi}{\partial r} + \hat{\boldsymbol{\vartheta}}\frac{1}{r}\frac{\partial \phi}{\partial \vartheta} + \hat{\boldsymbol{\varphi}}\frac{1}{r \sin\vartheta}\frac{\partial \phi}{\partial \varphi}. \tag{C.49}$$

DIVERGENCE OPERATOR

Given a vector that measures the flux density of a property, the divergence $(\nabla \cdot)$ of a vector measures the net flux of that property at a point in space. So the divergence of the velocity vector (which measures volumetric flux density) measures whether there is

[6] For example, for motion through a a complete circle, $d\theta$ is 2π and the total distance moved is equal to the circumference $2\pi r$.

a net flux of volume at a point in space. For incompressible systems, this is equivalent to measuring if mass is created or destroyed. In incompressible systems, conservation of mass implies that the divergence of velocity is zero. Formally, the divergence of a vector $\vec{A}$ is defined as

$$\nabla \cdot \vec{A} = \lim_{\Delta \mathcal{V} \to 0} \frac{\int_{\mathcal{S}} \vec{A} \cdot \hat{n}\, dA}{\Delta \mathcal{V}}, \tag{C.50}$$

where $\Delta \mathcal{V}$ is a volume, $\mathcal{S}$ is its surface, and dA is a differential area element along that surface. The unit vector $\hat{n}$ is directed outward from the surface. This definition naturally leads to the *divergence theorem*, which relates the integral of the flux of a vector through a surface to the volume integral of the divergence:

divergence theorem

$$\int_{\mathcal{V}} \nabla \cdot \vec{A} dv = \int_{\mathcal{S}} \vec{A} \cdot \hat{n}\, dA. \tag{C.51}$$

Divergence Operator – Cartesian Coordinates. In Cartesian coordinates, the divergence of the velocity vector $\vec{u}$ is given by

divergence operator, Cartesian coordinates

$$\nabla \cdot \vec{u} = \frac{\partial u}{\partial x} + \frac{\partial v}{\partial y} + \frac{\partial w}{\partial z}. \tag{C.52}$$

Divergence Operator – Cylindrical Coordinates. As was the case with the gradient operator, the fact that cylindrical coordinates have different units makes the divergence more complicated. As before, the $\frac{\partial}{\partial \theta}$ term must be corrected by a factor of $\imath$. Also, the $\frac{\partial}{\partial \imath}$ term must account for the fact that the geometry changes with $\imath$. Thus the divergence of the velocity vector $\vec{u}$ in cylindrical coordinates is given by

divergence operator, cylindrical coordinates

$$\nabla \cdot \vec{u} = \frac{1}{\imath} \frac{\partial}{\partial \imath} (\imath u_{\imath}) + \frac{1}{\imath} \frac{\partial u_{\theta}}{\partial \theta} + \frac{\partial u_z}{\partial z}. \tag{C.53}$$

Divergence Operator – Spherical Coordinates. As for the cylindrical case, the derivative terms in the divergence operator must account for the fact that the geometry changes with r and ϑ. Thus, the divergence of the velocity vector $\vec{u}$ in spherical coordinates is given by

divergence operator, spherical coordinates

$$\nabla \cdot \vec{u} = \frac{1}{r^2} \frac{\partial}{\partial r} \left(r^2 u_r\right) + \frac{1}{r \sin \vartheta} \frac{\partial (u_{\vartheta} \sin \vartheta)}{\partial \vartheta} + \frac{1}{r \sin \vartheta} \frac{\partial u_{\varphi}}{\partial \varphi}. \tag{C.54}$$

CURL OPERATOR

The curl ($\nabla\times$) of the velocity vector quantifies to what extent the fluid is rotating. Formally, the curl of a vector $\vec{\mathbf{A}}$ is defined as

$$\nabla \times \vec{\mathbf{A}} = \hat{\boldsymbol{n}} \lim_{\Delta \mathcal{S} \to 0} \frac{\int_C \vec{\mathbf{A}} \cdot \hat{\boldsymbol{t}}\, ds}{\Delta \mathcal{S}}, \quad \text{(C.55)}$$

where $\Delta\mathcal{S}$ denotes the area of a surface, C is a contour encapsulating that surface, and ds is a differential element along that contour. The vector $\hat{\boldsymbol{n}}$ is a unit normal specified by the right-hand rule given the path of the contour, and $\hat{\boldsymbol{t}}$ is a unit vector along the contour. This definition naturally leads to *Stokes' theorem*, which relates the integral of the flux of the curl of a vector across a surface to the contour integral of the vector:

Stokes' theorem

$$\int_{\mathcal{S}} \left(\nabla \times \vec{\mathbf{A}}\right) \cdot \hat{\boldsymbol{n}}\, dA = \int_C \vec{\mathbf{A}} \cdot \hat{\boldsymbol{t}}\, ds. \quad \text{(C.56)}$$

Curl Operator – Cartesian Coordinates. The curl $\nabla \times \vec{\boldsymbol{u}}$ of a vector $\vec{\boldsymbol{u}}$ is given, in Cartesian coordinates, as

curl operator, Cartesian coordinates

$$\nabla \times \vec{\boldsymbol{u}} = \hat{\boldsymbol{x}}\left[\frac{\partial w}{\partial y} - \frac{\partial v}{\partial z}\right] + \hat{\boldsymbol{y}}\left[\frac{\partial u}{\partial z} - \frac{\partial w}{\partial x}\right] + \hat{\boldsymbol{z}}\left[\frac{\partial v}{\partial x} - \frac{\partial u}{\partial y}\right]. \quad \text{(C.57)}$$

Matrix representation of curl for Cartesian coordinates. The curl in Cartesian coordinates can also be written as

curl operator, Cartesian coordinates

$$\nabla \times \vec{\boldsymbol{u}} = \det \begin{bmatrix} \hat{\boldsymbol{x}} & \hat{\boldsymbol{y}} & \hat{\boldsymbol{z}} \\ \frac{\partial}{\partial x} & \frac{\partial}{\partial y} & \frac{\partial}{\partial z} \\ u & v & w \end{bmatrix}, \quad \text{(C.58)}$$

where $\frac{\partial}{\partial x}$, $\frac{\partial}{\partial y}$, and $\frac{\partial}{\partial z}$ are used as shorthand to denote derivative operations on the velocity components in the third row of the matrix.

Curl Operator – Cylindrical Coordinates. The curl $\nabla \times \vec{\boldsymbol{u}}$ of a vector $\vec{\boldsymbol{u}}$ is given, in cylindrical coordinates, as

curl operator, cylindrical coordinates

$$\nabla \times \vec{\boldsymbol{u}} = \hat{\boldsymbol{r}}\left[\frac{1}{r}\frac{\partial u_z}{\partial \theta} - \frac{\partial u_\theta}{\partial z}\right] + \hat{\boldsymbol{\theta}}\left[\frac{\partial u_r}{\partial z} - \frac{\partial u_z}{\partial r}\right] + \hat{\boldsymbol{z}}\left[\frac{1}{r}\left(\frac{\partial}{\partial r}(r u_\theta) - \frac{\partial u_r}{\partial \theta}\right)\right]. \quad \text{(C.59)}$$

Curl Operator – Spherical Coordinates. The curl $\nabla \times \vec{\boldsymbol{u}}$ of a vector $\vec{\boldsymbol{u}}$ is given, in spherical coordinates, as

curl operator, spherical coordinates

$$\begin{aligned} \nabla \times \vec{\boldsymbol{u}} = \hat{\boldsymbol{r}}\left[\frac{1}{r\sin\vartheta}\left(\frac{\partial u_\varphi \sin\vartheta}{\partial\vartheta} - \frac{\partial u_\vartheta}{\partial\varphi}\right)\right] \\ + \hat{\boldsymbol{\vartheta}}\left[\frac{1}{r}\left(\frac{1}{\sin\vartheta}\frac{\partial u_r}{\partial\varphi} - \frac{\partial(r u_\varphi)}{\partial r}\right)\right] + \hat{\boldsymbol{\varphi}}\left[\frac{1}{r}\left(\frac{\partial}{\partial r}(r u_\vartheta) - \frac{\partial u_r}{\partial\vartheta}\right)\right] \end{aligned} \quad \text{(C.60)}$$

LAPLACIAN AND E^2 OPERATORS

The Laplacian (∇^2) of a scalar is a scalar that is one of several measures of the concavities of the scalar's distribution in space. Formally, the Laplacian of a scalar ϕ is defined as

$$\nabla^2\phi = \nabla \cdot \nabla\phi. \quad \text{(C.61)}$$

The Laplacian (∇^2) of a vector is a vector quantity and is computed in the same way:

$$\nabla^2\vec{\boldsymbol{u}} = \nabla \cdot \nabla\vec{\boldsymbol{u}}. \quad \text{(C.62)}$$

Laplacian Operator – Cartesian Coordinates. The Laplacian of a scalar in Cartesian coordinates is the sum of the second derivatives of the scalar with respect to the coordinates:

$$\nabla^2\phi = \frac{\partial^2\phi}{\partial x^2} + \frac{\partial^2\phi}{\partial y^2} + \frac{\partial^2\phi}{\partial z^2}. \quad \text{(C.63)}$$

In Cartesian coordinates, the components of the Laplacian $\nabla^2\vec{\boldsymbol{u}}$ of a vector $\vec{\boldsymbol{u}}$ are given by the Laplacians of the vector components:

$$\begin{aligned} \nabla^2\vec{\boldsymbol{u}} = \hat{\boldsymbol{x}}\left(\frac{\partial^2 u}{\partial x^2} + \frac{\partial^2 u}{\partial y^2} + \frac{\partial^2 u}{\partial z^2}\right) \\ + \hat{\boldsymbol{y}}\left(\frac{\partial^2 v}{\partial x^2} + \frac{\partial^2 v}{\partial y^2} + \frac{\partial^2 v}{\partial z^2}\right) \\ + \hat{\boldsymbol{z}}\left(\frac{\partial^2 w}{\partial x^2} + \frac{\partial^2 w}{\partial y^2} + \frac{\partial^2 w}{\partial z^2}\right). \end{aligned} \quad \text{(C.64)}$$

Laplacian Operator – Cylindrical Coordinates. The Laplacian of a scalar in cylindrical coordinates is given by

$$\nabla^2\phi = \frac{1}{\varrho}\frac{\partial}{\partial\varrho}\varrho\frac{\partial\phi}{\partial\varrho} + \frac{1}{\varrho^2}\frac{\partial^2\phi}{\partial\theta^2} + \frac{\partial^2\phi}{\partial z^2}. \quad \text{(C.65)}$$

Unfortunately, the Laplacian of a vector in curvilinear coordinates is not given by the the sum of the Laplacians of the components of the vector. Rather, the Laplacian $\nabla^2\vec{\boldsymbol{u}}$ of a vector $\vec{\boldsymbol{u}}$ in cylindrical coordinates is given by

Laplacian of vector, cylindrical coordinates

$$\begin{aligned} \nabla^2\vec{\boldsymbol{u}} = \hat{\boldsymbol{\varrho}}\left\{\frac{\partial}{\partial\varrho}\left[\frac{1}{\varrho}\frac{\partial}{\partial\varrho}(\varrho u_\varrho)\right] + \frac{1}{\varrho^2}\frac{\partial^2 u_\varrho}{\partial\theta^2} + \frac{\partial^2 u_\varrho}{\partial z^2} - \frac{2}{\varrho^2}\frac{\partial u_\theta}{\partial\theta}\right\} \\ + \hat{\boldsymbol{\theta}}\left\{\frac{\partial}{\partial\varrho}\left[\frac{1}{\varrho}\frac{\partial}{\partial\varrho}(\varrho u_\theta)\right] + \frac{1}{\varrho^2}\frac{\partial^2 u_\theta}{\partial\theta^2} + \frac{\partial^2 u_\theta}{\partial z^2} + \frac{2}{\varrho^2}\frac{\partial u_\varrho}{\partial\theta}\right\} \\ + \hat{\boldsymbol{z}}\left\{\frac{1}{\varrho}\frac{\partial}{\partial\varrho}\left(\varrho\frac{\partial u_z}{\partial\varrho}\right) + \frac{1}{\varrho^2}\frac{\partial^2 u_z}{\partial\theta^2} + \frac{\partial^2 u_z}{\partial z^2}\right\}. \end{aligned} \quad \text{(C.66)}$$

Laplacian Operator – Spherical Coordinates. The Laplacian $\nabla^2\phi$ of a scalar ϕ is given, in spherical coordinates, as

Laplacian of scalar, spherical coordinates

$$\nabla^2\phi = \frac{1}{r^2}\frac{\partial}{\partial r}r^2\frac{\partial\phi}{\partial r} + \frac{1}{r^2\sin\vartheta}\frac{\partial}{\partial\vartheta}\left(\sin\vartheta\frac{\partial\phi}{\partial\vartheta}\right) + \frac{1}{r^2\sin^2\vartheta}\frac{\partial^2\phi}{\partial\varphi^2}. \tag{C.67}$$

The Laplacian of a vector $\vec{u}$ is given in spherical coordinates by

Laplacian of vector, spherical coordinates

$$\begin{aligned}\nabla^2\vec{u} = \hat{r}&\left\{\tfrac{1}{r^2}\tfrac{\partial}{\partial r}\left(r^2\tfrac{\partial u_r}{\partial r}\right) + \tfrac{1}{r^2\sin\vartheta}\tfrac{\partial}{\partial\vartheta}\left(\sin\vartheta\tfrac{\partial u_r}{\partial\vartheta}\right) + \tfrac{1}{r^2\sin^2\vartheta}\tfrac{\partial^2 u_r}{\partial\varphi^2}\right.\\
&\left. - \tfrac{2}{r^2}\left(u_r + \tfrac{\partial u_\vartheta}{\partial\vartheta} + u_\vartheta\cot\vartheta\right) + \tfrac{2}{r^2\sin\vartheta}\tfrac{\partial u_\varphi}{\partial\varphi}\right\}\\
+\hat{\vartheta}&\left\{\tfrac{1}{r^2}\tfrac{\partial}{\partial r}\left(r^2\tfrac{\partial u_\vartheta}{\partial r}\right) + \tfrac{1}{r^2\sin\vartheta}\tfrac{\partial}{\partial\vartheta}\left(\sin\vartheta\tfrac{\partial u_\vartheta}{\partial\vartheta}\right)\right.\\
&\left. + \tfrac{1}{r^2\sin^2\vartheta}\tfrac{\partial^2 u_\vartheta}{\partial\varphi^2} + \tfrac{2}{r^2}\tfrac{\partial u_r}{\partial\vartheta} - \tfrac{1}{r^2\sin^2\vartheta}\left(u_\vartheta + 2\cos\vartheta\tfrac{\partial u_\varphi}{\partial\varphi}\right)\right\}\\
+\hat{\varphi}&\left\{\tfrac{1}{r^2}\tfrac{\partial}{\partial r}\left(r^2\tfrac{\partial u_\varphi}{\partial r}\right) + \tfrac{1}{r^2\sin\vartheta}\tfrac{\partial}{\partial\vartheta}\left(\sin\vartheta\tfrac{\partial u_\varphi}{\partial\vartheta}\right)\right.\\
&\left. + \tfrac{1}{r^2\sin^2\vartheta}\tfrac{\partial^2 u_\varphi}{\partial\varphi^2} + \tfrac{1}{r^2\sin^2\vartheta}\left(2\tfrac{\partial u_r}{\partial\varphi} + 2\cos\vartheta\tfrac{\partial u_\vartheta}{\partial\varphi} - u_\varphi\right)\right\}.\end{aligned} \tag{C.68}$$

E^2 operator – spherical coordinates. The E^2 operator is similar to the Laplacian operator, and in fact is equal to the Laplacian operator when used in plane-symmetric systems. In axisymmetric spherical coordinates, however, the Laplacian and E^2 operators are different. The E^2 operator in axisymmetric spherical coordinates, when applied to a scalar ψ_S, is given by

$$E^2\psi_S = \frac{\partial^2}{\partial r^2}\psi_S + \frac{\sin\vartheta}{r}\frac{\partial}{\partial\vartheta}\frac{1}{\sin\vartheta}\frac{\partial}{\partial\vartheta}\psi_S. \tag{C.69}$$

C.2.4 Biharmonic and E^4 operators

The biharmonic operator ∇^4 of a scalar or vector is the Laplacian of the Laplacian, for example,

$$\nabla^4\psi = \nabla^2\left(\nabla^2\psi\right). \tag{C.70}$$

The E^4 operator, similarly, is the result when the E^2 operator is applied twice, for example,

$$E^4\psi_S = E^2\left(E^2\psi_S\right). \tag{C.71}$$

C.2.5 Vector identities

A list of some common vector identities follows.

Null operator combinations. The curl of the gradient of a scalar field is always zero:

$$\nabla\times\nabla c = 0. \tag{C.72}$$

The divergence of the curl of a vector field is always zero:

$$\nabla \cdot (\nabla \times \vec{\boldsymbol{u}}) = 0\,. \tag{C.73}$$

Product rules. The product rule for the gradient is

$$\nabla(bc) = b\nabla c + c\nabla b\,. \tag{C.74}$$

Vector operations on the product of a scalar and a vector include

$$\nabla \cdot (c\vec{\boldsymbol{u}}) = c(\nabla \cdot \vec{\boldsymbol{u}}) + (\vec{\boldsymbol{u}} \cdot \nabla)c\,, \tag{C.75}$$

$$\nabla \times (c\vec{\boldsymbol{u}}) = c(\nabla \times \vec{\boldsymbol{u}}) + (\nabla c) \times \vec{\boldsymbol{u}}\,. \tag{C.76}$$

The gradient of the vector dot product is given by

$$\nabla(\vec{\boldsymbol{u}} \cdot \vec{\boldsymbol{E}}) = \vec{\boldsymbol{u}} \times (\nabla \times \vec{\boldsymbol{E}}) + \vec{\boldsymbol{E}} \times (\nabla \times \vec{\boldsymbol{u}}) + (\vec{\boldsymbol{u}} \cdot \nabla)\vec{\boldsymbol{E}} + (\vec{\boldsymbol{E}} \cdot \nabla)\vec{\boldsymbol{u}}\,. \tag{C.77}$$

The divergence of the vector cross product is given by

$$\nabla \cdot (\vec{\boldsymbol{u}} \times \vec{\boldsymbol{E}}) = \vec{\boldsymbol{E}} \cdot (\nabla \times \vec{\boldsymbol{u}}) - \vec{\boldsymbol{u}} \cdot (\nabla \times \vec{\boldsymbol{E}})\,. \tag{C.78}$$

The curl of the vector cross product is given by

$$\nabla \times (\vec{\boldsymbol{u}} \times \vec{\boldsymbol{E}}) = \vec{\boldsymbol{u}}(\nabla \cdot \vec{\boldsymbol{E}}) - \vec{\boldsymbol{E}}(\nabla \cdot \vec{\boldsymbol{u}}) + (\vec{\boldsymbol{E}} \cdot \nabla)\vec{\boldsymbol{u}} - (\vec{\boldsymbol{u}} \cdot \nabla)\vec{\boldsymbol{E}}\,. \tag{C.79}$$

Commutations of vector operations with the Laplacian. The cross product, dot product, and gradient all commute with the Laplacian:

$$\nabla \cdot (\nabla^2 \vec{\boldsymbol{u}}) = \nabla^2(\nabla \cdot \vec{\boldsymbol{u}})\,, \tag{C.80}$$

$$\nabla \times (\nabla^2 \vec{\boldsymbol{u}}) = \nabla^2(\nabla \times \vec{\boldsymbol{u}})\,, \tag{C.81}$$

$$\nabla(\nabla^2 \vec{\boldsymbol{u}}) = \nabla^2(\nabla \vec{\boldsymbol{u}})\,. \tag{C.82}$$

EXAMPLE PROBLEM C.3

In Cartesian coordinates, show that the operator $\vec{\boldsymbol{u}} \cdot \nabla$, when operating on the vector $\vec{\boldsymbol{a}} = (a\,, b\,, c)$, gives the result

$$\begin{bmatrix} u\frac{\partial a}{\partial x} + v\frac{\partial a}{\partial y} + w\frac{\partial a}{\partial z} \\ u\frac{\partial b}{\partial x} + v\frac{\partial b}{\partial y} + w\frac{\partial b}{\partial z} \\ u\frac{\partial c}{\partial x} + v\frac{\partial c}{\partial y} + w\frac{\partial c}{\partial z} \end{bmatrix}. \tag{C.83}$$

SOLUTION: This is a matter of using the definitions and multiplying out the terms and operators. Starting with the $\vec{\boldsymbol{u}} \cdot \nabla$ operator:

$$(\vec{\boldsymbol{u}} \cdot \nabla)\,\vec{\boldsymbol{a}}\,, \tag{C.84}$$

this can then be written as

$$\vec{\boldsymbol{u}} \cdot \nabla\vec{\boldsymbol{a}}\,, \tag{C.85}$$

which implies that we first evaluate the gradient of $\vec{a}$ and then evaluate the dot product of $\vec{u}$ with the result. Writing out the terms, we have

$$\begin{bmatrix} u & v & w \end{bmatrix} \cdot \begin{bmatrix} \frac{\partial a}{\partial x} & \frac{\partial a}{\partial y} & \frac{\partial a}{\partial z} \\ \frac{\partial b}{\partial x} & \frac{\partial b}{\partial y} & \frac{\partial b}{\partial z} \\ \frac{\partial c}{\partial x} & \frac{\partial c}{\partial y} & \frac{\partial c}{\partial z} \end{bmatrix} . \tag{C.86}$$

We then evaluate the dot product to obtain

$$\begin{bmatrix} u\frac{\partial a}{\partial x} + v\frac{\partial a}{\partial y} + w\frac{\partial a}{\partial z} \\ u\frac{\partial b}{\partial x} + v\frac{\partial b}{\partial y} + w\frac{\partial b}{\partial z} \\ u\frac{\partial c}{\partial x} + v\frac{\partial c}{\partial y} + w\frac{\partial c}{\partial z} \end{bmatrix} . \tag{C.87}$$

C.2.6 Dyadic operations

The superposition of two vectors (for example, $\vec{A}\vec{B}$) denotes a second-rank dyadic tensor, and the dot product of a dyadic tensor with a vector results in a vector.

$$(\vec{A}\vec{B}) \cdot \vec{C} = (\vec{B} \cdot \vec{C})\vec{A}. \tag{C.88}$$

The dyadic tensor $\vec{A}\vec{B}$, when multiplying by a vector $\vec{C}$, gives a vector that is in the direction of $\vec{A}$ with a magnitude given by the dot product $\vec{B} \cdot \vec{C}$. Unlike the dot product of vectors, the dot product of a dyadic with a vector is not in general equal to the dot product of the vector with the dyadic:

$$\vec{C} \cdot (\vec{A}\vec{B}) = (\vec{A} \cdot \vec{C})\vec{B} \neq (\vec{B} \cdot \vec{C})\vec{A}. \tag{C.89}$$

C.3 SUMMARY

This appendix defines Cartesian, cylindrical, and spherical coordinate systems, their notation, and unit vectors. Vector operators such as the gradient, curl, divergence, and Laplacian are also defined in these coordinate systems, leading to definitions of the divergence and Stokes theorems.

C.4 SUPPLEMENTARY READING

For the introductory vector analysis presented here, Wilson [289], Aris [290], and Greenberg [291] all provide useful treatment.

This appendix lists out vector operations in a number of coordinate systems as a guide to the reader. It omits the more general approach, which is to write vector operations in terms of general orthogonal curvilinear coordinate systems, as discussed in [22, 24]. Reference [28], furthermore, covers semiorthogonal interfacial coordinates in detail.

This text focuses on symbolic or Gibbs notation and omits Cartesian or Einstein notation. Cartesian notation provides a much more compact treatment of vector and tensor operations in rectangular coordinate systems, and is discussed in [21, 23, 30]. A particularly lucid treatment of both Cartesian tensors and their associated notation is given in [292].

C.5 EXERCISES

C.1 Compare the length of the position vectors for these three cases:

(a) $x = 1, y = 1, z = 1$.

(b) $\mathscr{r} = 1, \theta = 1, z = 1$.

(c) $r = 1, \vartheta = 1, \varphi = 1$.

Compare your results with the magnitude of the velocity vectors for these three cases:

(a) $u = 1, v = 1, w = 1$.

(b) $u_{\mathscr{r}} = 1, u_\theta = 1, u_z = 1$.

(c) $u_r = 1, u_\vartheta = 1, u_\varphi = 1$.

C.2 In Cartesian coordinates, show that the operator $\nabla \cdot \nabla$, when operating on the vector $\vec{\boldsymbol{a}} = (a, b, c)$, gives the result

$$\begin{bmatrix} \dfrac{\partial^2 a}{\partial x^2} + \dfrac{\partial^2 a}{\partial y^2} + \dfrac{\partial^2 a}{\partial z^2} \\ \dfrac{\partial^2 b}{\partial x^2} + \dfrac{\partial^2 b}{\partial y^2} + \dfrac{\partial^2 b}{\partial z^2} \\ \dfrac{\partial^2 c}{\partial x^2} + \dfrac{\partial^2 c}{\partial y^2} + \dfrac{\partial^2 c}{\partial z^2} \end{bmatrix} . \tag{C.90}$$

C.3 Show that $\nabla \cdot (-p\vec{\vec{\boldsymbol{\delta}}}) = -\nabla p$.

C.4 By converting both to Cartesian form, show that cylindrical unit vectors can be written in terms of spherical unit vectors as follows:

$$\hat{\mathscr{r}} = \hat{\boldsymbol{r}} \sin\vartheta + \hat{\boldsymbol{\vartheta}} \cos\vartheta , \tag{C.91}$$

$$\hat{\boldsymbol{\theta}} = \hat{\boldsymbol{\varphi}} , \tag{C.92}$$

$$\hat{\boldsymbol{z}} = \hat{\boldsymbol{r}} \cos\vartheta - \hat{\boldsymbol{\vartheta}} \sin\vartheta . \tag{C.93}$$

C.5 By converting both to Cartesian form, show that spherical unit vectors can be written in terms of cylindrical unit vectors as follows:

$$\hat{\boldsymbol{r}} = \hat{\mathscr{r}} \frac{\mathscr{r}}{\sqrt{\mathscr{r}^2 + z^2}} + \hat{\boldsymbol{z}} \frac{z}{\sqrt{\mathscr{r}^2 + z^2}} , \tag{C.94}$$

$$\hat{\boldsymbol{\vartheta}} = \hat{\mathscr{r}} \frac{z}{\sqrt{\mathscr{r}^2 + z^2}} - \hat{\boldsymbol{z}} \frac{\mathscr{r}}{\sqrt{\mathscr{r}^2 + z^2}} , \tag{C.95}$$

$$\hat{\boldsymbol{\varphi}} = \hat{\boldsymbol{\theta}} . \tag{C.96}$$

C.6 By converting to Cartesian coordinates, show that the components of a position vector can be converted from spherical to cylindrical with the following relations:

$$\mathscr{r} = r \sin\vartheta , \tag{C.97}$$

$$\theta = \varphi , \tag{C.98}$$

$$z = r \cos\vartheta . \tag{C.99}$$

C.7 By converting to Cartesian coordinates, show that the components of a position vector can be converted from cylindrical to spherical with the following relations:

$$r = \sqrt{\varrho^2 + z^2}\,, \tag{C.100}$$

$$\vartheta = \operatorname{atan2}(\varrho, z)\,, \tag{C.101}$$

$$\varphi = \theta\,. \tag{C.102}$$

APPENDIX D

Governing Equation Reference

The governing equations in this text are usually reported by use of Gibbs or symbolic notation. Interpreting these governing equations explicitly requires managing the equation in the coordinate system relevant to the problem by using the guidelines in Appendix C. In this appendix, key governing equations are written out explicitly, as a guide and reference for the student.

D.1 SCALAR LAPLACE EQUATION

The Laplace equation for a scalar potential (either electrical or velocity potential) is written as

$$\nabla^2 \phi = 0 . \tag{D.1}$$

In Cartesian coordinates, this is written as

$$\frac{\partial^2 \phi}{\partial x^2} + \frac{\partial^2 \phi}{\partial y^2} + \frac{\partial^2 \phi}{\partial z^2} = 0 . \tag{D.2}$$

In cylindrical coordinates, this is written as

$$\frac{1}{r} \frac{\partial}{\partial r} \left(r \frac{\partial \phi}{\partial r} \right) + \frac{1}{r^2} \frac{\partial^2 \phi}{\partial \theta^2} + \frac{\partial^2 \phi}{\partial z^2} = 0 . \tag{D.3}$$

In spherical coordinates, this is written as

$$\frac{1}{r^2} \frac{\partial}{\partial r} \left(r^2 \frac{\partial \phi}{\partial r} \right) + \frac{1}{r^2 \sin \vartheta} \frac{\partial}{\partial \vartheta} \left(\sin \vartheta \frac{\partial \phi}{\partial \vartheta} \right) + \frac{1}{r^2 \sin^2 \vartheta} \frac{\partial^2 \phi}{\partial \varphi^2} = 0 . \tag{D.4}$$

D.2 POISSON–BOLTZMANN EQUATION

The Poisson–Boltzmann equation for the electrical potential is written as

uniform fluid properties

$$\nabla^2\phi = -\frac{F}{\varepsilon}\sum_i c_{i,\infty} z_i \exp\left(-\frac{z_i F \varphi}{RT}\right). \tag{D.5}$$

In Cartesian coordinates, this is written as

uniform fluid properties

$$\frac{\partial^2\phi}{\partial x^2} + \frac{\partial^2\phi}{\partial y^2} + \frac{\partial^2\phi}{\partial z^2} = -\frac{F}{\varepsilon}\sum_i c_{i,\infty} z_i \exp\left(-\frac{z_i F \varphi}{RT}\right). \tag{D.6}$$

In cylindrical coordinates, this is written as

uniform fluid properties

$$\frac{1}{r}\frac{\partial}{\partial r}\left(r\frac{\partial\phi}{\partial r}\right) + \frac{1}{r^2}\frac{\partial^2\phi}{\partial\theta^2} + \frac{\partial^2\phi}{\partial z^2} = -\frac{F}{\varepsilon}\sum_i c_{i,\infty} z_i \exp\left(-\frac{z_i F \varphi}{RT}\right). \tag{D.7}$$

In spherical coordinates, this is written as

uniform fluid properties

$$\frac{1}{r^2}\frac{\partial}{\partial r}\left(r^2\frac{\partial\phi}{\partial r}\right) + \frac{1}{r^2\sin\vartheta}\frac{\partial}{\partial\vartheta}\left(\sin\vartheta\frac{\partial\phi}{\partial\vartheta}\right) + \frac{1}{r^2\sin^2\vartheta}\frac{\partial^2\phi}{\partial\varphi^2} = -\frac{F}{\varepsilon}\sum_i c_{i,\infty} z_i \exp\left(-\frac{z_i F \varphi}{RT}\right). \tag{D.8}$$

D.3 CONTINUITY EQUATION

The incompressible, uniform-property conservation of mass equation is written in symbolic notation as

$$\nabla \cdot \vec{u} = 0. \tag{D.9}$$

In Cartesian coordinates, this is written as

uniform fluid properties

$$\frac{\partial u}{\partial x} + \frac{\partial v}{\partial y} + \frac{\partial w}{\partial z} = 0. \tag{D.10}$$

In cylindrical coordinates, this is written as

uniform fluid properties

$$\frac{1}{r}\frac{\partial}{\partial r}\left(r u_r\right) + \frac{1}{r}\frac{\partial u_\theta}{\partial \theta} + \frac{\partial u_z}{\partial z} = 0. \tag{D.11}$$

In spherical coordinates, this is written as

uniform fluid properties

$$\frac{1}{r^2}\frac{\partial}{\partial r}\left(r^2 u_r\right) + \frac{1}{r\sin\vartheta}\frac{\partial}{\partial\vartheta}\left(u_\vartheta \sin\vartheta\right) + \frac{1}{r\sin\vartheta}\frac{\partial u_\varphi}{\partial\varphi} = 0. \tag{D.12}$$

D.4 NAVIER–STOKES EQUATIONS

The incompressible, uniform-property, Newtonian Navier–Stokes equations in symbolic or Gibbs notation are given by

uniform fluid properties

$$\rho\frac{\partial \vec{\boldsymbol{u}}}{\partial t} + \rho\vec{\boldsymbol{u}}\cdot\nabla\vec{\boldsymbol{u}} = -\nabla p + \eta\nabla^2\vec{\boldsymbol{u}} + \sum_i \vec{\boldsymbol{f}}_i. \tag{D.13}$$

In Cartesian coordinates, this is written as

uniform fluid properties

$$\begin{aligned}
\rho\tfrac{\partial u}{\partial t} + \rho u\tfrac{\partial u}{\partial x} + \rho v\tfrac{\partial u}{\partial y} + \rho w\tfrac{\partial u}{\partial z} &= -\tfrac{\partial p}{\partial x} + \eta\tfrac{\partial^2 u}{\partial x^2} + \eta\tfrac{\partial^2 u}{\partial y^2} + \eta\tfrac{\partial^2 u}{\partial z^2} + \textstyle\sum_i \vec{\boldsymbol{f}}_{x,i}, \\
\rho\tfrac{\partial v}{\partial t} + \rho u\tfrac{\partial v}{\partial x} + \rho v\tfrac{\partial v}{\partial y} + \rho w\tfrac{\partial v}{\partial z} &= -\tfrac{\partial p}{\partial y} + \eta\tfrac{\partial^2 v}{\partial x^2} + \eta\tfrac{\partial^2 v}{\partial y^2} + \eta\tfrac{\partial^2 v}{\partial z^2} + \textstyle\sum_i \vec{\boldsymbol{f}}_{y,i}, \\
\rho\tfrac{\partial w}{\partial t} + \rho u\tfrac{\partial w}{\partial x} + \rho v\tfrac{\partial w}{\partial y} + \rho w\tfrac{\partial w}{\partial z} &= -\tfrac{\partial p}{\partial z} + \eta\tfrac{\partial^2 w}{\partial x^2} + \eta\tfrac{\partial^2 w}{\partial y^2} + \eta\tfrac{\partial^2 w}{\partial z^2} + \textstyle\sum_i \vec{\boldsymbol{f}}_{z,i}.
\end{aligned} \tag{D.14}$$

Here, $\vec{\boldsymbol{f}}_{x,i}$, $\vec{\boldsymbol{f}}_{y,i}$, and $\vec{\boldsymbol{f}}_{z,i}$ are the x, y, and z components of the body force terms.

In cylindrical coordinates, this is written as

uniform fluid properties

$$\begin{aligned}
&\rho\tfrac{\partial u_r}{\partial t} + \rho u_r\tfrac{\partial u_r}{\partial r} + \rho\tfrac{u_\theta}{r}\tfrac{\partial u_r}{\partial\theta} + \rho u_z\tfrac{\partial u_r}{\partial z} - \rho\tfrac{u_\theta{}^2}{r} \\
&\quad = -\tfrac{\partial p}{\partial r} + \eta\tfrac{1}{r}\tfrac{\partial}{\partial r}\left(r\tfrac{\partial u_r}{\partial r}\right) + \eta\tfrac{1}{r^2}\tfrac{\partial^2 u_r}{\partial\theta^2} + \eta\tfrac{\partial^2 u_r}{\partial z^2} - \eta\tfrac{2}{r^2}\tfrac{\partial u_\theta}{\partial\theta} - \eta\tfrac{u_r}{r^2} + \textstyle\sum_i \vec{\boldsymbol{f}}_{r,i}, \\
&\rho\tfrac{\partial u_\theta}{\partial t} + \rho u_r\tfrac{\partial u_\theta}{\partial r} + \rho\tfrac{u_\theta}{r}\tfrac{\partial u_\theta}{\partial\theta} + \rho u_z\tfrac{\partial u_\theta}{\partial z} + \rho\tfrac{u_r u_\theta}{r} \\
&\quad = -\tfrac{1}{r}\tfrac{\partial p}{\partial\theta} + \eta\tfrac{1}{r}\tfrac{\partial}{\partial r}\left(r\tfrac{\partial u_\theta}{\partial r}\right) + \eta\tfrac{1}{r^2}\tfrac{\partial^2 u_\theta}{\partial\theta^2} + \eta\tfrac{\partial^2 u_\theta}{\partial z^2} + \eta\tfrac{2}{r^2}\tfrac{\partial u_r}{\partial\theta} - \eta\tfrac{u_\theta}{r^2} + \textstyle\sum_i \vec{\boldsymbol{f}}_{\theta,i}, \\
&\rho\tfrac{\partial u_z}{\partial t} + \rho u_r\tfrac{\partial u_z}{\partial r} + \rho\tfrac{u_\theta}{r}\tfrac{\partial u_z}{\partial\theta} + \rho u_z\tfrac{\partial u_z}{\partial z} \\
&\quad = -\tfrac{\partial p}{\partial z} + \eta\tfrac{1}{r}\tfrac{\partial}{\partial r}\left(r\tfrac{\partial u_z}{\partial r}\right) + \eta\tfrac{1}{r^2}\tfrac{\partial^2 u_z}{\partial\theta^2} + \eta\tfrac{\partial^2 u_z}{\partial z^2} + \textstyle\sum_i \vec{\boldsymbol{f}}_{z,i}.
\end{aligned} \tag{D.15}$$

Here, $\vec{f}_{r,i}$, $\vec{f}_{\theta,i}$, and $\vec{f}_{z,i}$ are the r, θ, and z components of the body force terms. In spherical coordinates, this is written as

uniform fluid properties

$$
\begin{aligned}
&\rho\frac{\partial u_r}{\partial t}+\rho u_r\frac{\partial u_r}{\partial r}+\rho\frac{u_\vartheta}{r}\frac{\partial u_r}{\partial\vartheta}+\rho\frac{u_\varphi}{r\sin\vartheta}\frac{\partial u_r}{\partial\varphi}-\rho\frac{u_\vartheta{}^2+u_\varphi{}^2}{r}\\
&\quad=-\frac{\partial p}{\partial r}+\eta\frac{1}{r^2}\frac{\partial}{\partial r}\left(r^2\frac{\partial u_r}{\partial r}\right)+\eta\frac{1}{r^2\sin\vartheta}\frac{\partial}{\partial\vartheta}\left(\sin\vartheta\frac{\partial u_r}{\partial\vartheta}\right)\\
&\quad+\frac{1}{r^2\sin^2\vartheta}\eta\frac{\partial^2 u_r}{\partial\varphi^2}-\eta\frac{2}{r^2}\left(u_r+\frac{\partial u_\vartheta}{\partial\vartheta}+u_\vartheta\cot\vartheta\right)+\eta\frac{2}{r^2\sin\vartheta}\frac{\partial u_\varphi}{\partial\varphi}+\sum_i\vec{f}_{r,i}\,,\\
&\rho\frac{\partial u_\vartheta}{\partial t}+\rho u_r\frac{\partial u_\vartheta}{\partial r}+\rho\frac{u_\vartheta}{r}\frac{\partial u_\vartheta}{\partial\vartheta}+\rho\frac{u_\varphi}{r\sin\vartheta}\frac{\partial u_\vartheta}{\partial\varphi}+\rho\frac{u_r u_\vartheta-u_\varphi{}^2\cot\vartheta}{r}\\
&\quad=-\frac{1}{r}\frac{\partial p}{\partial\vartheta}+\eta\frac{1}{r^2}\frac{\partial}{\partial r}\left(r^2\frac{\partial u_\vartheta}{\partial r}\right)+\eta\frac{1}{r^2\sin\vartheta}\frac{\partial}{\partial\vartheta}\left(\sin\vartheta\frac{\partial u_\vartheta}{\partial\vartheta}\right)\\
&\quad+\frac{1}{r^2\sin^2\vartheta}\eta\frac{\partial^2 u_\vartheta}{\partial\varphi^2}+\eta\frac{2}{r^2}\frac{\partial u_r}{\partial\vartheta}-\eta\frac{1}{r^2\sin^2\vartheta}\left(u_\vartheta+2\cos\vartheta\frac{\partial u_\varphi}{\partial\varphi}\right)+\sum_i\vec{f}_{\vartheta,i}\,,\\
&\rho\frac{\partial u_\varphi}{\partial t}+\rho u_r\frac{\partial u_\varphi}{\partial r}+\rho\frac{u_\vartheta}{r}\frac{\partial u_\varphi}{\partial\vartheta}+\rho\frac{u_\varphi}{r\sin\vartheta}\frac{\partial u_\varphi}{\partial\varphi}+\rho\frac{u_r u_\varphi+u_\vartheta u_\varphi\cot\vartheta}{r}\\
&\quad=-\frac{1}{r\sin\vartheta}\frac{\partial p}{\partial\varphi}+\eta\frac{1}{r^2}\frac{\partial}{\partial r}\left(r^2\frac{\partial u_\varphi}{\partial r}\right)+\eta\frac{1}{r^2\sin\vartheta}\frac{\partial}{\partial\vartheta}\left(\sin\vartheta\frac{\partial u_\varphi}{\partial\vartheta}\right)\\
&\quad+\frac{1}{r^2\sin^2\vartheta}\eta\frac{\partial^2 u_\varphi}{\partial\varphi^2}+\eta\frac{1}{r^2\sin^2\vartheta}\left(2\frac{\partial u_r}{\partial\varphi}+2\cos\vartheta\frac{\partial u_\vartheta}{\partial\varphi}-u_\varphi\right)+\sum_i\vec{f}_{\varphi,i}\,.
\end{aligned}
\tag{D.16}
$$

Here, $\vec{f}_{r,i}$, $\vec{f}_{\vartheta,i}$, and $\vec{f}_{\varphi,i}$ are the r, ϑ, and φ components of the body force terms.

D.5 SUPPLEMENTARY READING

Good sources for explicit rendering of these governing equations are typically textbooks on fluid mechanics or electrodynamics [17, 18, 19, 20, 21, 22, 23, 28, 55, 56, 57].

APPENDIX E

Nondimensionalization and Characteristic Parameters

This appendix outlines the role of several key dimensional and nondimensional parameters in micro- and nanoscale fluid mechanics that come from nondimensionalization of governing equations. A key advantage of nondimensionalization is that it leads to a compact description of flow parameters (i.e., *Re*) and thus leads to generalization. Nondimensionalization can be a powerful tool, but it is useful only if implemented with insight into the physics of the problems. Our stress here is the *process* of nondimensionalization, rather than a listing of nondimensional parameters, and we focus on only a few examples.

E.1 BUCKINGHAM Π THEOREM

The Buckingham Π theorem is a theorem in dimensional analysis that quantifies how many nondimensional parameters are required for specifying a problem. It also provides a process by which these nondimensional parameters can be determined. The Buckingham Π theorem states that a system with n independent physical variables that are a function of m fundamental physical quantities can be written as a function of $n - m$ nondimensional quantities. As an example, the steady Navier–Stokes equations have four parameters: a characteristic length ℓ, a characteristic velocity U, the viscosity η, and the fluid density ρ. These are a function of three fundamental physical quantities: mass, length, and time. Thus the system can be described in terms of $4 - 3 = 1$ nondimensional quantity, and it can be shown that the nondimensional quantity must be proportional to $\rho U \ell / \eta$ to some power.

The Buckingham Π theorem does not define unique nondimensional quantities, nor does it help identify which nondimensional quantities are most physically meaningful or useful; however, it does frame general problems from a dimensional standpoint.

E.2 NONDIMENSIONALIZATION OF GOVERNING EQUATIONS

Many of our nondimensional parameters come from nondimensionalization of governing equations. Nondimensionalizing the governing equations makes the equations simpler and highlights which terms are the most important.

E.2.1 Nondimensionalization of the Navier–Stokes equations: Reynolds number

The Reynolds number *Re* plays several roles and stems from fluid-mechanical considerations in several ways. In this section, we discuss the nondimensionalization of the Navier–Stokes equations and the relation of this nondimensionalization to *Re*.

Consider the incompressible Navier–Stokes equations for uniform-viscosity, Newtonian fluids with no body forces:

$$\rho \frac{\partial \vec{u}}{\partial t} + \rho \vec{u} \cdot \nabla \vec{u} = -\nabla p + \eta \nabla^2 \vec{u}. \tag{E.1}$$

This governing equation has two parameters: ρ and η. In addition, the boundary conditions have a size characterized by a length ℓ and velocities characterized by a velocity U. The characteristic velocity U is a representative fluid velocity in the flow domain, specified perhaps by the velocity at an inlet or at infinity (if available) or by some mean measure of the flow in the flow field (usually employed if the boundary conditions are specified using pressures). The characteristic length ℓ characterizes the lengths over which the velocities change by an amount proportional to U. Furthermore, if the boundary conditions are time dependent, then a characteristic time t_c could denote the time over which the boundary condition changes, perhaps the inverse of the frequency if the boundary conditions are cyclic. Thus five parameters define an unsteady Navier–Stokes problem and four parameters define a steady Navier–Stokes problem. These lead to two (for unsteady) or one (for steady) nondimensional parameter(s). In nondimensionalizing the equations, the structure of the Navier–Stokes equations naturally leads to the definition of the Reynolds number.

We can define nondimensional variables denoted by starred properties, namely,

$$x^* = \frac{x}{\ell}, \tag{E.2}$$

$$y^* = \frac{y}{\ell}, \tag{E.3}$$

and

$$z^* = \frac{z}{\ell}. \tag{E.4}$$

Spatial derivatives naturally follow from the nondimensional coordinates, so

$$\nabla^* = \frac{\nabla}{1/\ell} \tag{E.5}$$

and

$$\nabla^{*2} = \frac{\nabla^2}{1/\ell^2}. \tag{E.6}$$

We normalize the velocity by the characteristic velocity U:

$$\vec{u}^* = \frac{\vec{u}}{U}. \tag{E.7}$$

We must choose a time t_c that is characteristic of the flow, either the natural characteristic time of the flow (ℓ/U) or the characteristic time t_{BC} over which the boundary conditions change (for example, if the boundary conditions are oscillatory, this is the period of oscillation of the boundary condition). We pick the faster of these two to use as t_c. If the boundary conditions change rapidly, i.e., if $t_{BC} < \ell/U$, then we define $t_c = t_{BC}$ and set $t^* = \frac{t}{t_c}$. The *Strouhal number*, which is defined as $St = t_c U/\ell$, is then given by $St = t_{BC} U/\ell$. If the boundary conditions are steady or change slowly (i.e., if $t_{BC} > \ell/U$), then we define $t^* = \frac{t}{t_c}$ by using $t_c = \ell/U$; therefore $St = 1$ by definition.

With these definitions, we can substitute and rearrange the unsteady Navier–Stokes equations into nondimensional form. If we use $p^* = \frac{p}{\eta U/\ell}$, we obtain

nondimensional Navier–Stokes equation using viscous stress to normalize p, uniform fluid properties

$$\frac{Re}{St}\frac{\partial \vec{u}^*}{\partial t^*} + Re\,\vec{u}^* \cdot \nabla^* \vec{u}^* = -\nabla^* p^* + \nabla^{*2}\vec{u}^* \,, \tag{E.8}$$

and if we use $p^* = \frac{p}{\rho U^2}$, we obtain

nondimensional Navier–Stokes equation using dynamic pressure to normalize p, uniform fluid properties

$$\frac{1}{St}\frac{\partial \vec{u}^*}{\partial t^*} + \vec{u}^* \cdot \nabla^* \vec{u}^* = -\nabla^* p^* + \frac{1}{Re}\nabla^{*2}\vec{u}^* \,, \tag{E.9}$$

where the Reynolds number is defined as $Re = \rho U\ell/\eta$ and the Strouhal number[1] is defined as $St = t_c U/\ell$. Because a number of different characteristic lengths are often available, it is customary to use a subscript on Re to clarify what characteristic length is used to calculate the Reynolds number. For example, $Re_x = \rho Ux/\eta$ and $Re_L = \rho UL/\eta$. The difference between Eqs. (E.8) and (E.9) is simply the premultiplier on the pressure term, which varies based on how we normalized the pressure. Mathematically speaking, our choices for how we nondimensionalize our parameters are arbitrary. We can nondimensionalize the equation in any of a variety of forms. We are able to confirm that our nondimensionalization is physically meaningful only when we use the nondimensionalized equation to generate physical insight about the system, perhaps by using the nondimensional parameters to correlate experimental data or to neglect certain terms in the equations. We know that our nondimensionalization is incorrect if it suggests that we neglect terms that are actually important (this happens if we define t^* or p^* incorrectly), or if we are unable to correlate experimental data (this happens if we choose U or ℓ incorrectly).

From the format of Eqs. (E.8) and (E.9), we can see that the Reynolds and Strouhal numbers are measures of the relative magnitude of the different terms in the Navier–Stokes equations. For example, Eq. (E.8) is used in Chapter 8 to illustrate why the terms on the left-hand side of the Navier–Stokes equations can be ignored at low Reynolds numbers. Although not relevant for microscale flows, Eq. (E.9) can be used to derive the Euler equations, valid for high Reynolds numbers. Consider a steady flow (for which the Strouhal number is unity) and consider the $Re \to 0$ and $Re \to \infty$ limits. Clearly, the $Re \to 0$ limit will lead to elimination of the convection and unsteady terms, and the $Re \to \infty$ limit will lead to elimination of the viscous term. However, the role of the pressure term depends on which definition we use for p^*. To clarify this, recall that the system to be solved has four equations (one for mass and three for momentum) and four unknowns (one pressure and three velocity components). If the pressure term is eliminated, we have four equations in three unknowns. Thus *the Navier–Stokes equations in general cannot be solved if the pressure term is neglected,* and the pressure gradients in a fluid system can be neglected only in degenerate cases (for example, Couette

[1] Some authors define the Strouhal number as $St = \ell/t_c U$, in which case the form of the first term of the nondimensional Navier–Stokes equation is different. We do not do so in this text, but the Strouhal number can also be used to characterize flows with steady boundary conditions but an oscillatory flow solution, such as moderate Re flow over a circular cylinder.

or purely electroosmotic flow). The Reynolds number helps us to identify which *velocity* terms to keep, but the value of the Reynolds number never motivates us to eliminate the pressure term. A consequence of this is that the most physically meaningful form of p^* is the one that leads to retention of the pressure term when a limit of the Reynolds number is taken. At low *Re*, we use Eq. (E.8) because physically we know that the pressure gradients are primarily caused by viscous effects, and further because mathematically we know that Eq. (E.8) will retain the pressure term in the $Re \to 0$ limit. At high *Re*, we use Eq. (E.9) because physically we know that the pressure gradients are caused primarily by inertial effects, and further because mathematically we know that Eq. (E.9) will retain the pressure term in the $Re \to \infty$ limit.

The Reynolds number does more than eliminate terms in certain limits. We refer to two flows as being *dynamically similar* if they have the same Reynolds number and if their geometry is similar. The nondimensional solution to the Navier–Stokes equations will be identical for two systems if the geometry and Reynolds number are matched. Thus fluid-mechanical results can be meaningfully compared across many different experimental realizations. In microscale flows, the Reynolds number is small relative to unity and small relative to *St*, and thus we often solve the Stokes flow equations (Chapter 8), for which the unsteady and convective terms are neglected.

EXAMPLE PROBLEM E.1

Consider the following flows. Specify the characteristic velocity U and the characteristic length ℓ. Explain why these velocities and lengths are appropriate.

1. Laminar flow through a tube with circular cross section of diameter d with pressure gradient $\frac{dp}{dx}$.
2. Flow through a curved microchannel with maximum velocity u_{max} with circular cross section whose cross-sectional radius is a and whose radius of curvature is R.
3. Flow at mean velocity U in a microchannel of width and depth d with a dilute suspension of particles of radius a moving at a velocity u with respect to the flow.

SOLUTION:

1. The characteristic velocity U could be the maximum or mean of the resulting flow, calculated with techniques from Chapter 2. The length ℓ is d.
2. The characteristic velocity U is u_{max}. The characteristic length ℓ is a, not R.
3. There are two relevant Reynolds numbers. The Reynolds number for the bulk flow has U and d as its characteristic values, whereas the Reynolds number for the flow around the particle has u and a.

E.2.2 Nondimensionalization of the passive scalar transfer equation: Peclet number

This subsection discusses the nondimensionalization of the passive scalar transfer equation (applicable to passive transport of mass or temperature) and the relation of this nondimensionalization to *Pe*. We ignore active transport mechanisms and source terms, such as electromigration of charged chemical species or chemical reaction.

Consider the mass transfer equations for dilute solutions of species i in the absence of an electric field:

$$\frac{\partial c_i}{\partial t} + \vec{\boldsymbol{u}} \cdot \nabla c_i = D_i \nabla^2 c_i \,, \tag{E.10}$$

where c_i is the molar concentration of species i and D_i is the binary diffusivity of species i in the solvent. If n species are being considered, these equations have n parameters in the governing equations – the n species diffusivities D_i. The boundary conditions have two characteristic parameters (U and ℓ) if the boundary conditions are steady, and a third parameter (t_{BC}) if the boundary conditions are unsteady. This equation has two units (length and time), and thus the Buckingham Π theorem predicts that a system with unsteady boundary conditions will have $n+3$ parameters minus 2 fundamental physical quantities, leading to $n+1$ nondimensional parameters that govern the system. Following a similar approach to that used for the Navier–Stokes equations, we find

nondimensional passive scalar transport equation, uniform fluid properties

$$\frac{1}{St}\frac{\partial c_i^*}{\partial t^*} + \vec{u}^* \cdot \nabla^* c_i^* = \frac{1}{Pe_i}\nabla^{*2} c_i^* , \tag{E.11}$$

where the *mass transfer Peclet number* for each species i is defined as $Pe_i = U\ell/D_i$ and the Strouhal number is given by $St = t_c U/\ell$. For each species, the Peclet number is a measure of the relative magnitude of the diffusion term in the mass transfer equations compared with the convection term. For all species, the Strouhal number gives the relative magnitude of the unsteady term to the convection term, and for steady boundary conditions, the Strouhal number is unity. Compared with the Navier–Stokes equations, the passive scalar transport equation is different because D_i for the species we study in microdevices varies widely from species to species, and is often orders of magnitude smaller than η/ρ for water; thus Pe varies more widely than Re and is small less often than Re is.

EXAMPLE PROBLEM E.2

A system of two fluids is traveling through a microchannel with width 250 μm, depth 10 μm, and length 1 cm. Fluid 1 is in the leftmost 125 μm of the channel, and fluid 2 is in the rightmost 125 μm of the channel. Which of these lengths should be used as the length scale in calculating the Reynolds number? Which should be used as the length scale in calculating the Peclet number?

SOLUTION: The length scale for the Peclet number is the length that characterizes the distance over which the species concentration changes. In this case, this is the width, and thus the relevant length scale is proportional to the width. Two natural lengths might be 125 μm or 250 μm. Although this system will have some species concentration variations over the depth of the channel, the *primary* variation is over the width.

The length scale for the Reynolds number is the length that characterizes the distance over which the velocity changes. In this case, this is the depth, and thus the relevant length scale is proportional to the depth, 10 μm. Some might use the half-depth 5 μm. Although this system will have some velocity concentration variations over the width of the channel, the system is well approximated for the most part as being infinitely wide, and thus the flow solution is close to that between two infinite plates, at least for most of the cross section.

E.2.3 Nondimensionalization of the Poisson–Boltzmann equation: Debye length and thermal voltage

The Poisson–Boltzmann equation can be nondimensionalized using techniques similar to those in the preceding subsections. The Poisson–Boltzmann equation is a bit different in three key ways. First, because the Poisson–Boltzmann equation has many more parameters, there is more flexibility with regard to how the nondimensional groups are formed. Second, because of this flexibility, the Poisson–Boltzmann equation can be nondimensionalized without using any of the parameters from the boundary conditions. Thus the nondimensionalization of the Poisson–Boltzmann equation can be carried out independently of its voltage and length boundary conditions. Third, the standard nomenclature that stems from this nondimensionalization focuses on the characteristic length and voltage that arise from manipulation of the governing equation, rather than the resulting nondimensional groups. The process is largely the same, but the resulting nomenclature focuses on different parameters. Because nondimensionalization of the Navier–Stokes equations is more common, we make a number of comparisons between the process for nondimensionalizing the Poisson–Boltzmann equation with that for nondimensionalizing the Navier–Stokes equations.

We begin with the nonlinear Poisson–Boltzmann equation, written as

$$\nabla^2 \varphi = -\frac{F}{\varepsilon} \sum_i c_{i,\infty} z_i \exp\left(-\frac{z_i F \varphi}{RT}\right) . \tag{E.12}$$

This equation has $2n + 6$ parameters, where n is the number of chemical species. The governing equation has $n + 4$ parameters, namely ε, T, F, R, and the species valences z_i, whereas the boundary conditions have $n + 2$ parameters, namely the surface potential φ_0, a characteristic length scale ℓ, and the n bulk species concentrations $c_{i,\infty}$. The fundamental physical quantities in this equation are fourfold (C, V, m, K), and thus we expect $2n + 2$ nondimensional groups. The Poisson–Boltzmann equation is an equilibrium equation and thus has no characteristic time.

We proceed in an order different from that used for the Navier-Stokes equations. For the Navier–Stokes equations, we presumed we knew how to nondimensionalize the terms, and substituted the nondimensional forms into the equation, resulting in two nondimensional parameters (Re and St). The process for the Poisson–Boltzmann equation can start by simply rearranging the governing equation with no attention to the boundary condition. First, we notice that the argument of the exponential term must be dimensionless, and it already highlights a nondimensional ratio. We thus define a nondimensional potential as follows:

$$\varphi^* = \frac{F\varphi}{RT} . \tag{E.13}$$

This effectively normalizes the potential by the *thermal voltage* RT/F, which is a measure of the voltage (about 25 mV at room temperature) that induces a potential energy on an elementary charge equal to the thermal energy. This leads to

$$\nabla^2 \varphi^* = -\frac{F^2}{\varepsilon RT} \sum_i c_{i,\infty} z_i \exp\left(-z_i \varphi^*\right) . \tag{E.14}$$

This is philosophically different from the steps we used for the Navier–Stokes equation. For that process, we normalized the key parameter we were solving for ($\vec{u}$) by a characteristic value from the boundary conditions. Here, the Poisson–Boltzmann equation provides a parameter (RT/F) that can be used for this nondimensionalization – the boundary condition is not necessary.

Next, we normalize concentrations by the ionic strength of the *bulk* solution:

$$c^*_{i,\infty} = \frac{c_{i,\infty}}{I_{c,\text{bulk}}} , \tag{E.15}$$

leading to

$$\nabla^2 \varphi^* = -\frac{1}{2} \frac{2F^2 I_c}{\varepsilon RT} \sum_i c^*_{i,\infty} z_i \exp\left(-z_i \varphi^*\right) . \tag{E.16}$$

This is more closely akin to our previous approach, in that we are normalizing a property by a characteristic value specified at the boundary (in this case, at infinity).

Next, we note that both right- and left-hand sides of this equation have units of length^{-2}. Thus the premultiplier of the sum on the right-hand side can be interpreted as a characteristic length to the -2 power. We thus define the *Debye length* λ_D as follows,

definition of Debye length

$$\lambda_D = \sqrt{\frac{\varepsilon RT}{2F^2 I_c}}\Bigg|_{\text{bulk}} , \tag{E.17}$$

and normalize all spatial variables x, y, and z by the Debye length:

$$x^* = x/\lambda_D , \tag{E.18}$$

$$y^* = y/\lambda_D , \tag{E.19}$$

$$z^* = z/\lambda_D . \tag{E.20}$$

In so doing, we define a nondimensional del operator:

$$\nabla^* = \left(\frac{\partial}{\partial x^*}, \frac{\partial}{\partial y^*}, \frac{\partial}{\partial z^*}\right) . \tag{E.21}$$

Again, the key difference here compared with nondimensionalization of the Navier–Stokes equations is that the governing equation provides a characteristic length, and the boundary conditions need not be used. Implementing the nondimensional del operator leads to the nondimensional form of the Poisson–Boltzmann equation:

nondimensional Poisson–Boltzmann equation, uniform fluid properties

$$\nabla^{*2} \varphi^* = -\frac{1}{2} \sum_i c^*_{i,\infty} z_i \exp\left(-z_i \varphi^*\right) . \tag{E.22}$$

The $2n+2$ nondimensional parameters that govern this system are the n valences z_i, the n normalized bulk concentrations $c^*_{i,\infty}$, the normalized characteristic length (ℓ/λ_D), and the normalized double-layer potential $\varphi_0^* = F\varphi^*/RT$. The characteristic length (ℓ) and the characteristic voltage drop across the double layer (φ_0) are not in the nondimensional governing equation, but are found in the boundary conditions. We have given names (thermal voltage and Debye length) to the characteristic length and voltage that evolve from the Poisson–Boltzmann equations, rather than to their nondimensional forms as used in the boundary conditions.

The Debye length gives an estimate of the length scale over which an electrostatic perturbation (such as a charged surface) is shielded by rearrangement of ions. The use of this parameter has some analogies with the use of the electrical permittivity – whereas the electrical permittivity describes how the polarization of a medium cancels out much of the field a charge would have caused if it were a vacuum, the Debye length describes the characteristic length scale over which an electrolyte cancels out the remaining electric field by rearrangement of ions. The normalized length ℓ/λ_D is a measure of how large an object is relative to the EDL surrounding it. Chapter 13, for example, uses $a^* = a/\lambda_D$ to characterize the electrophoretic mobility of a particle with radius a.

The thermal voltage nondimensionalizes the surface potential. The nondimensionalized surface potential indicates how much of a perturbation the surface makes on ion concentrations – if φ_0^* is small, the perturbations are small, the conductivity of the medium remains uniform, and the Poisson–Boltzmann equation can be linearized with minimal error. If φ_0^* is large, the perturbations are large, ion distributions are drastically changed, and the system is strongly nonlinear.

E.3 SUMMARY

This appendix summarizes nondimensionalization of the Navier–Stokes, scalar transport, and Poisson–Boltzmann equations. In the case of the Navier–Stokes equations, this process leads naturally to the Strouhal and Reynolds numbers:

$$St = t_c \frac{U}{\ell}, \tag{E.23}$$

$$Re = \frac{\rho U \ell}{\eta}. \tag{E.24}$$

Two choices in the nondimensionalization include the choice of the characteristic time and the choice of the characteristic pressure. The characteristic time t_c is the flow time ℓ/U if the boundary conditions are steady or slowly changing, and t_{BC} if the boundary condition changes rapidly. The characteristic pressure is $\eta U/\ell$ for low-*Re* flows and ρU^2 for high-*Re* flows, owing to the need to retain the pressure term so that the system is not overconstrained. The passive scalar transport equation is similar to the Navier–Stokes equation, but generally simpler. It leads to a Peclet number for each scalar:

$$Pe = \frac{U\ell}{D}. \tag{E.25}$$

Finally, the Poisson–Boltzmann equation is nondimensionalized by defining the thermal voltage RT/F and the Debye length λ_D,

$$\lambda_D = \sqrt{\frac{\varepsilon RT}{2F^2 I_c}}\bigg|_{\text{bulk}}, \qquad \text{(E.26)}$$

with no attention to the boundary conditions. The boundary condition for the potential and the geometric scale of the system can then be normalized by these parameters.

E.4 SUPPLEMENTARY READING

A number of texts contain thorough lists of nondimensional parameters with brief descriptions of their meaning; one example can be found in [33]. Nondimensionalization of the Navier–Stokes equations and dynamic similitude is discussed precisely in [24]. For those interested in a presentation of transport that thoroughly stresses nondimensionalization throughout, Ref. [30] makes nondimensionalization a focus.

This text neglects many nondimensional parameters, for example, parameters related to buoyancy, such as the Bond, Grashof, and Rayleigh numbers; parameters related to chemical reaction, such as the Damköhler numbers; parameters related to macromolecule relaxation, such as the Weissenberg and Deborah numbers; and parameters related to thermal transport, such as the Brinkman, Biot, and Fourier numbers. A thorough microfluidic review with a focus on the role of nondimensionalization (and mention of many parameters omitted here) can be found in [293].

E.5 EXERCISES

E.1 Derive Eq. (E.8) from the dimensional Navier–Stokes equations by substituting in the relations in Eqs. (E.2)–(E.6).

E.2 We wish to study the flow of water through a circular microchannel with radius $R = 4\,\mu$m and mean velocity 100 μm/s. We expect experimental resolution of the velocity profile to 1/100 of the channel. Unfortunately, most experimental techniques for measuring fluid velocity cannot localize the measurement to a region as small as 40 nm. Describe how you might select the channel size and fluid such that you could make a dynamically similar measurement in a larger channel and achieve the desired resolution.

E.3 Consider the Navier–Stokes equations and the nondimensionalization that leads to the Stokes flow approximation. Rather than nondimensionalizing the pressure by $\eta U/\ell$, one could nondimensionalize the pressure by ρU^2. Show that this nondimensionalization, in the limit where Re $\rightarrow 0$, leads to the equation

$$\nabla^2 \vec{u}^* = 0\,. \qquad \text{(E.27)}$$

Is it possible to solve this equation and the continuity equation if the only variables are the components of the velocity vector? Comment on the mathematical form of the governing equations for momentum and mass as they relate to pressure.

E.4 Derive the nondimensional passive scalar mass transfer equation (E.11) from the dimensional passive scalar mass transfer equation (E.10).

E.5 From the standpoint of dimensional analysis, define the physical parameters and the fundamental physical quantities for the Poisson–Boltzmann equation, and explain why two of the nondimensional quantities that govern the equation are the nondimensional voltage and nondimensional length.

APPENDIX F

Multipolar Solutions to the Laplace and Stokes Equations

The Laplace and Stokes equations are both linear in the fluid velocity and are both amenable to solution by superposition of Green's functions; further, each equation has a set of solutions whose superposition is termed the *multipolar expansion*. This appendix details these solutions.

F.1 LAPLACE EQUATION

The Laplace equation governs key electromagnetic solutions and, in the case of purely electroosmotic flow far from walls, the Laplace equation can govern fluid flow as well. Although *numerical* approaches are the norm for solving the governing equations in complex geometries, a great wealth of *analytical* solutions to the Laplace equation can be found by solving by separation of variables, especially for systems with symmetry. These solutions have mathematical importance because they lead to a convenient series expansion (the multipolar expansion) that approximates the correct solutions, and these solutions have physical importance because the building blocks of electrical systems (point charges), magnetic systems (point magnetic dipoles), and fluid systems (sources, sinks, vortexes) correspond to individual terms in this expansion.

This appendix describes solutions to the Laplace equation with both axial and plane symmetry. The solutions for plane-symmetric flows forecast the potential flows discussed in Chapter 7, and the solutions for axial symmetry forecast the use of the multipolar expansion for modeling the dielectrophoretic response of particles (Chapter 17).

F.1.1 Laplace equation solutions for axisymmetric spherical coordinates: separation of variables and multipolar expansions

This subsection presents the separation of variables solution for the spherical axisymmetric Laplace equation and describes the linear axisymmetric multipolar expansion for the Laplace equation. This multipolar expansion comprises a subset of the terms in the separation of variables solution and is designed to be accurate for large r.

THE GENERAL LEGENDRE POLYNOMIAL SOLUTION TO THE AXISYMMETRIC LAPLACE EQUATION

In axisymmetric spherical coordinates, the Laplace equation for a scalar ϕ is given by

axisymmetric spherical Laplace equation

$$\frac{\partial}{\partial r}\left(r^2\frac{\partial \phi}{\partial r}\right)+\frac{1}{\sin\vartheta}\frac{\partial}{\partial \vartheta}\left(\sin\vartheta\frac{\partial \phi}{\partial \vartheta}\right)=0\,. \tag{F.1}$$

Table F.1 **Equations for the first five Legendre polynomials.**

k	$P_k(x)$
0	1
1	x
2	$\frac{1}{2}(3x^2 - 1)$
3	$\frac{1}{2}(5x^3 - 3x)$
4	$\frac{1}{8}(35x^4 - 30x^2 + 3)$

Here, this scalar could be any property that satisfies the Laplace equation – for example, the electrical potential or the velocity potential. This equation can be solved by separation of variables (see Exercise F.2), in which case the general solution of this equation is written in terms of polynomial terms in r and Legendre polynomials in terms of $\cos\vartheta$ (Fig. F.1)

$$\phi(r, \vartheta) = \sum_{k=0}^{\infty} \left(A_k r^k + B_k r^{-k-1}\right) P_k(\cos\vartheta), \tag{F.2}$$

where the Legendre polynomials $P_k(x)$ are listed for a few k in Table F.1 and are given by the general formula

$$P_k(x) = \frac{1}{2^k k!}\left(\frac{d}{dx}\right)^k \left(x^2 - 1\right)^k. \tag{F.3}$$

Using the Legendre polynomials for axisymmetric spherical coordinates is analogous to using Fourier and Bessel function solutions for Cartesian and cylindrical coordinates, respectively.

The coefficients A_k and B_k in Eq. (F.2) are chosen to satisfy the boundary conditions of a given problem. In particular, the B_k terms are unbounded at $r = 0$ and are thus useful to describe ϕ only for $r > 0$, which in microfluidic systems typically means the

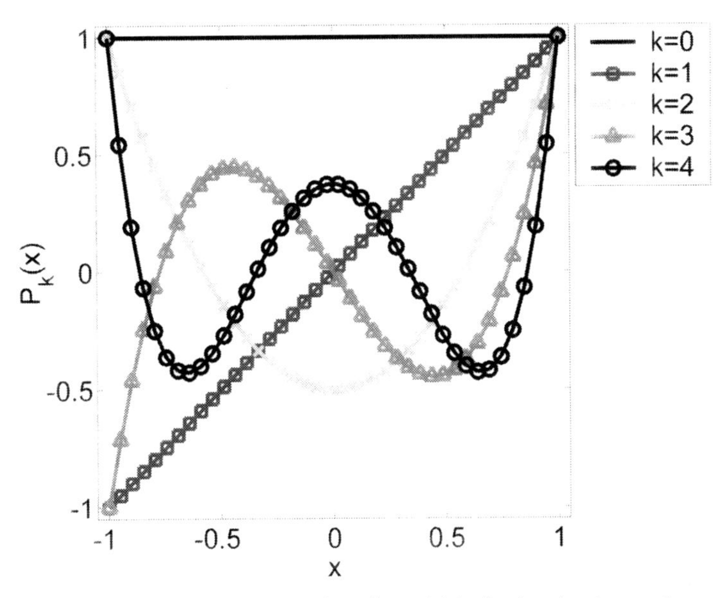

Fig. F.1 The value of the first five Legendre polynomials in the domain $-1 < x < 1$.

region *outside* a particle or the region outside a microdevice feature. The A_k terms (at least for $k > 0$) are unbounded at $r \to \infty$ and are thus useful to describe ϕ only for finite r, which typically means *inside* a particle or microdevice feature.

MULTIPOLE SOLUTIONS - GENERAL FORM AND APPLICABILITY

Solutions in which specific B_k values are nonzero (whereas all A_k values are zero) are referred to as *multipole solutions* or *singular solutions*. The sum of all multipole solutions,

multipole expansion solution of Laplace equation

$$\phi(r, \vartheta) = \sum_{k=0}^{\infty} B_k r^{-k-1} P_k(\cos\vartheta), \tag{F.4}$$

is referred to as the multipole expansion. To be specific, Eq. (F.4) is the *linear axisymmetric multipole expansion for the Laplace equation* and must be distinguished from other multipole expansions for other coordinate symmetries or equations.

The multipole expansion is an expansion in inverse powers of r. Thus a finite number of terms can approximate the correct result for ϕ as long as r is large enough. In addition, some physical objects correspond exactly or approximately to multipoles – for example, a point charge corresponds to an electrical monopole, and an iron atom corresponds to a magnetic dipole.

We distinguish between a *mathematical* or *ideal* multipole and a *physical* multipole. A mathematical multipole corresponds to a term in a series expansion. In contrast, a physical multipole is a physical object that leads to a solution of the Laplace equation that is well approximated by the corresponding mathematical multipole. For example, the O-H bond of a water molecule is asymmetric, with the hydrogen atom containing a partial positive charge and the oxygen atom containing a partial negative charge. These charges are separated by the length of the O-H bond (which is about 1 Å). This asymmetric bond is a physical dipole and is well approximated by, but not identical to, a mathematical dipole.

The multipole expansion is an infinite sum but is typically truncated at a finite number of terms. For certain simple geometries, the solution requires only a single term (an example of this is the electric field or potential flow around a sphere, which is generated by adding a dipole to a uniform field or uniform flow). For more complicated geometries, we might model the system with several terms. The nature of the multipolar expansion is such that the approximation becomes better as r becomes larger, meaning that this sort of approximation improves as the distance from the object improves.

CREATING HIGH-ORDER MULTIPOLE SOLUTIONS FROM LOWER-ORDER MULTIPOLES

We can show that multipole solutions can be made from combinations of lower-order multipoles. Mathematically, this elucidates a fundamental property of the Legendre polynomials. Physically, this is important because the physical objects that generate multipole solutions are often combinations of monopoles or dipoles.

Any multipole of order $k+1$ can be made by combining two multipoles of order k in a limiting process. These two multipoles must have infinitely high (but opposite) Legendre polynomial coefficients, and be separated by an infinitesimal distance δd (i.e., the positive k-order multipole must be at $z = \delta d/2$, and the negative k-order multipole must be at $z = -\delta d/2$. The superposition of these two solutions results in a multipole with a Legendre polynomial coefficient of $B_{k+1} = k B_k \delta d$. Thus two monopoles brought

Table F.2 **Multipliers for constructing multipoles from superposition of lower-order multipoles.**

$z=-2\delta d$	$z=\frac{-3\delta d}{2}$	$z=-\delta d$	$z=\frac{-\delta d}{2}$	$z=0$	$z=\frac{\delta d}{2}$	$z=\delta d$	$z=\frac{3\delta d}{2}$	$z=2\delta d$	
				1					order k
			−1		1				order $k+1$
		1		−2		1			order $k+2$
	−1		3		−3		1		order $k+3$
1		−4		6		−4		1	order $k+4$

together as their strength increases and the distance decreases lead to a dipole. Two dipoles brought together as their strength increases and the distance decreases lead to a quadrupole. And so on.

Using similar arguments, any multipole can be made of lower-order multipoles by superposing positive and negative multipoles of any lower order. These multipoles are added and subtracted in a manner delineated by a form of Pascal's triangle (Table F.2), and the resulting Legendre polynomial coefficient is $B_{k+n} = n!\,B_k \delta d^n$.

The mathematical multipole solutions are always generated in the *limiting case* in which $\delta d \to 0$, $B_k \to \infty$, and B_{k+n} is finite. Physical multipoles correspond to monopoles or dipoles of finite strength displaced a finite distance from each other.

For the case of electrical multipoles, Fig. F.2 shows the first four linear multipolar configurations and how they are constructed from the fundamental physical unit for electrical multipoles (a point charge). Other physical systems may have different fundamental units – for example, magnetic systems do not have monopoles, and the fundamental unit is a dipole.

POTENTIAL SOLUTIONS FROM LINEAR AXISYMMETRIC MULTIPOLE SOLUTIONS

The first four multipole solutions have their own names and are important owing to their utility in describing a multitude of physical systems. These solutions are described in the following subsections. Isocontours of potential for these linear axisymmetric multipole solutions are also shown in Fig. F.3.

MONOPOLE

The monopole solution (zeroth-order multipole) corresponds to $B_0 \neq 0$, all other $B_k = 0$. Some authors add a factor of 4π in the denominator for this and other multipole solutions. Such a constant multiplier does not affect calculations as long as it is applied consistently; it simply redefines the B values:

axisymmetric spherical monopole

$$\phi = \frac{B_0}{r}. \tag{F.5}$$

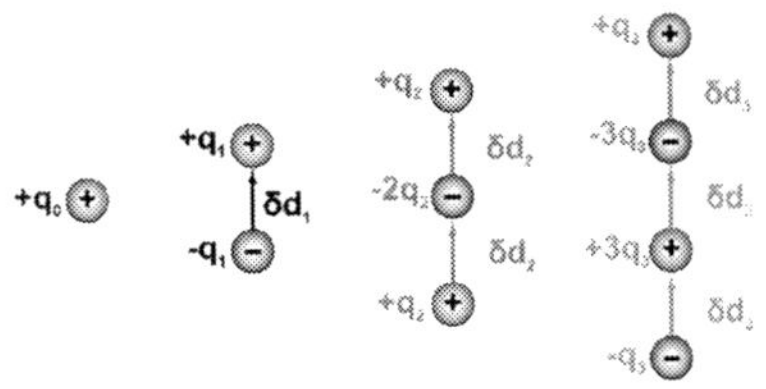

Fig. F.2 Physical configurations of a linear electrical monopole, dipole, quadrupole, and octupole. An electrical monopole is a point charge; all higher-order electrical multipoles correspond physically to *distributions* of electrical charge.

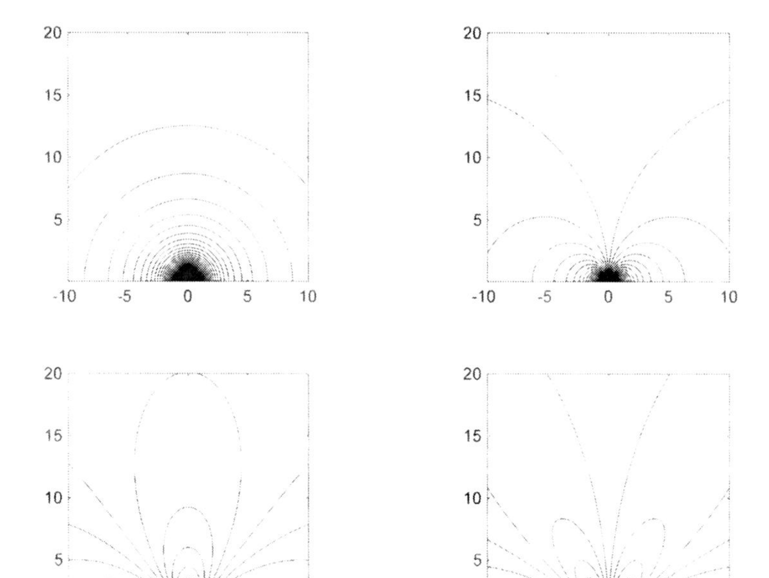

Fig. F.3 Isocontours of potentials stemming from linear axisymmetric multipoles. The r dependences of the dipole, quadrupole, and octupole are purposely distorted here for visualization purposes.

DIPOLE

The dipole solution (first-order multipole) corresponds to $B_1 \neq 0$, all other $B_k = 0$:

axisymmetric spherical dipole

$$\phi = \frac{B_1}{r^2} \cos \vartheta \,. \tag{F.6}$$

EXAMPLE PROBLEM F.1

A water molecule exhibits a dipole moment of approximately 2.9 D. Characterize the electrical potential caused by an individual water molecule suspended in a water solution (treat the solution as a continuum) and report this field in terms of a multipolar coefficient. Assume that the water is aligned such that the molecule plane of symmetry is on the x axis, with the positive charges aligned in the positive x direction and the negative charge aligned in the negative x direction.

SOLUTION: An object with a dipole moment is an electrical dipole, and the electric potential it induces is given by

$$\phi = \frac{B_1}{r^2} \cos \vartheta \,. \tag{F.7}$$

The strength of a monopole is given by $B_0 = \frac{q}{4\pi\varepsilon}$. The strength of a corresponding dipole is given by $B_1 = 1! B_0 \delta d$, which is

$$B_1 = \frac{q\delta d}{4\pi\varepsilon} = \frac{p}{4\pi\varepsilon} \,, \tag{F.8}$$

where p is the magnitude of the dipole moment. Thus the electric field is given by

$$\phi = \frac{2.9\ \text{D}}{4\pi\varepsilon r^2}\cos\vartheta \tag{F.9}$$

or

$$\phi = \frac{1.1 \times 10^{-21}\ \text{V m}^2}{r^2}\cos\vartheta\,. \tag{F.10}$$

QUADRUPOLE

The quadrupole solution (second-order multipole) corresponds to $B_2 \neq 0$, all other $B_k = 0$:

axisymmetric spherical quadrupole

$$\phi = \frac{B_2}{r^3}\frac{3\cos^2\vartheta - 1}{2}\,. \tag{F.11}$$

OCTUPOLE

The octupole solution (third-order multipole) corresponds to $B_3 \neq 0$, all other $B_k = 0$:

axisymmetric spherical octupole

$$\phi = \frac{B_3}{r^4}\frac{5\cos^3\vartheta - 3\cos\vartheta}{2}\,. \tag{F.12}$$

Other multipoles are typically referred to as 16-poles, 32-poles, and so on.

EXAMPLE PROBLEM F.2

Assume 5 Cl^- ions are equispaced between $x = -2$ nm and $x = 2$ nm along the x axis in water. Approximate the electrical potential in the system caused by these ions by an equivalent multipole expansion and write an expression for ϕ. Treat the water as a continuum.

SOLUTION: The net charge at $x = (-2$ nm, -1 nm, 0 nm, 1 nm, 2 nm$)$ is $(-1, -1, -1, -1, -1)$. We need a linear superposition of multipoles to give this charge distribution. This can be satisfied with a 16-pole of strength $-1 \times 4!$ $(-1, 4, -6, 4, -1)$, a quadrupole of strength $-5 \times 2!$ $(0, -5, 10, -5, 0)$, and a monopole of strength $-5 \times 0!$ $(0, 0, -5, 0, 0)$. Thus the multipole solution is given by

$$\phi = \frac{B_0}{r} + \frac{B_2}{r^3}\frac{3\cos^2\vartheta - 1}{2} + \frac{B_4}{r^5}\frac{35\cos^4\vartheta - 30\cos^2\vartheta + 3}{8}\,. \tag{F.13}$$

The multipole coefficients are given by

$$B_0 = -5(0!)\frac{e}{4\pi\varepsilon} = \frac{-5(1.6 \times 10^{-19}\ \text{C})}{4\pi(80)(8.85 \times 10^{-12}\ \text{C/V m}} = -9.0 \times 10^{-11}\ \text{V m}\,, \tag{F.14}$$

$$B_2 = -5(2!)\delta d^2 \frac{e}{4\pi\varepsilon} = \frac{-10(1 \times 10^{-9}\,\mathrm{m})^2(1.6 \times 10^{-19}\,\mathrm{C})}{4\pi(80)(8.85 \times 10^{-12}\,\mathrm{C/V\,m})} = -1.8 \times 10^{-28}\,\mathrm{V\,m^3}\,, \tag{F.15}$$

$$B_4 = -1(4!)\delta d^4 \frac{e}{4\pi\varepsilon} = \frac{-24(1 \times 10^{-9}\,\mathrm{m})^4(1.6 \times 10^{-19}\,\mathrm{C})}{4\pi(80)(8.85 \times 10^{-12}\,\mathrm{C/V\,m})} = -4.3 \times 10^{-46}\,\mathrm{V\,m^5}\,. \tag{F.16}$$

Thus the potential field is

$$\phi = \frac{-9.0 \times 10^{-11}\,\mathrm{V\,m}}{r} + \frac{-1.8 \times 10^{-28}\,\mathrm{V\,m^3}}{r^3}\frac{3\cos^2\vartheta - 1}{2} + \frac{-4.3 \times 10^{-46}\,\mathrm{V\,m^5}}{r^5}\frac{35\cos^4\vartheta - 30\cos^2\vartheta + 3}{8}\,. \tag{F.17}$$

F.1.2 Systems with plane symmetry: 2D cylindrical coordinates

For systems with plane symmetry, multipole solutions can be generated for 2D cylindrical coordinates just as well as axially symmetric spherical coordinates. The multipole expansion is less useful in two dimensions, because the set of *linear* multipoles is not enough to serve as an expansion for any plane-symmetric Laplace solution. However, the cylindrical expansion is useful because it makes the 2D cylindrical potential flow elements discussed in Chapter 7 more transparent.

THE GENERAL HARMONIC SOLUTION TO THE PLANE-SYMMETRIC CYLINDRICAL LAPLACE EQUATION

In 2D cylindrical coordinates, the Laplace equation for a potential ϕ is given by

plane-symmetric cylindrical Laplace equation

$$\frac{1}{\imath}\frac{\partial}{\partial \imath}\left(\imath\frac{\partial\phi}{\partial r}\right) + \frac{1}{\imath^2}\frac{\partial^2\phi}{\partial\theta^2} = 0\,. \tag{F.18}$$

This equation can be solved by separation of variables (see Exercise F. 3), in which case the general solution of this equation is written in terms of harmonic functions:

$$\phi(r,\theta) = A_0 + B_0 \ln \imath + \sum_{k=1}^{\infty}\left(A_k \imath^k + B_k \imath^{-k}\right)\cos(k\theta + \alpha_k)\,. \tag{F.19}$$

This is similar to the axisymmetric spherical solution, except that a pure sinusoid replaces the Legendre polynomial in the ODE for the θ equation, and the polynomial solution for the $\imath$ equation is slightly different. Because the orientation of a multipole in spherical axisymmetric coordinates is by definition along the axis of symmetry, but the orientation of a multipole in 2D cylindrical coordinates can vary arbitrarily, the 2D cylindrical solution has an angle α that denotes how the multipole is rotated with respect to the x axis.

MULTIPOLE SOLUTIONS – GENERAL FORM AND APPLICABILITY

As before, solutions in which specific B_k values are nonzero but all $A_k = 0$ are the multipole solutions. The sum of all multipole solutions,

2D cylindrical multipole expansion

$$\phi(r, \theta) = B_0 \ln \mathscr{r} + \sum_{k=1}^{\infty} B_k \mathscr{r}^{-k} \cos(k\theta + \alpha_k), \tag{F.20}$$

can be referred to as the *2D cylindrical multipole expansion.*

EXAMPLES OF 2D CYLINDRICAL MULTIPOLE SOLUTIONS

2D cylindrical multipole solutions are analogous to spherical axisymmetric multipoles. Examples follow.

2D MONOPOLE

The monopole solution (zeroth-order multipole) corresponds to $B_0 \neq 0$, all other $B_k = 0$:

plane-symmetric cylindrical monopole

$$\phi = B_0 \ln \mathscr{r}. \tag{F.21}$$

In 2D potential flow, the 2D monopole corresponds to a source or sink.

2D DIPOLE

The dipole solution (first-order multipole) corresponds to $B_1 \neq 0$, all other $B_k = 0$:

plane-symmetric cylindrical dipole

$$\phi = \frac{B_1}{\mathscr{r}} \cos(\theta + \alpha_1). \tag{F.22}$$

In 2D potential flow, the 2D dipole corresponds to a doublet.

LINEAR 2D QUADRUPOLE

The linear quadrupole solution (second-order multipole) corresponds to $B_2 \neq 0$, all other $B_k = 0$:

plane-symmetric cylindrical quadrupole

$$\phi = \frac{B_2}{\mathscr{r}^2} \cos(2\theta + \alpha_2). \tag{F.23}$$

LINEAR 2D OCTUPOLE

The linear octupole solution (third-order multipole) corresponds to $B_3 \neq 0$, all other $B_k = 0$:

plane-symmetric cylindrical octupole

$$\phi = \frac{B_3}{\imath^3}\cos(3\theta + \alpha_3)\,. \tag{F.24}$$

F.2 STOKES EQUATIONS

The Stokes equations govern key viscous flows, as discussed in Chapter 8. As was the case for the Laplace equation, numerical approaches are the norm for solving the governing equations in complex geometries, but analytical solutions to the Stokes equation can be found by considering the Green's function solutions to the Stokes equations. These solutions have mathematical importance because they lead to a convenient series expansion (the multipolar expansion) that approximates the correct solutions. Although the multipolar expansion for the Stokes equations is a bit more complicated than the multipolar expansion for the Laplace equations, we nonetheless are able to consider small viscous objects and their hydrodynamic interaction by using the multipolar formulation. These multipolar formulations facilitate calculation of the forces on large numbers of interacting microparticles.

F.2.1 The Green's function for Stokes flow with a point source

The Stokes equations with a point force $\vec{F}$ applied at a point can be written as

$$\nabla p - \eta\nabla^2\vec{u} = \vec{F}\delta(\vec{\Delta r})\,, \tag{F.25}$$

where $\delta(\vec{\Delta r})$ is the Dirac delta function and $\vec{\Delta r}$ is the distance from the point of application of the force. For an incompressible flow, we also have

$$\nabla\cdot\vec{u} = 0\,. \tag{F.26}$$

As discussed in Chapter 8, the solution to these equations can be written as

$$\vec{u} = \vec{\vec{G}}_0\cdot\vec{F}\,, \tag{F.27}$$

where

$$\vec{\vec{G}}_0 = \frac{1}{8\pi\eta\Delta r}\left(\vec{\vec{\delta}} + \frac{\vec{\Delta r}\vec{\Delta r}}{\Delta r^2}\right)\,, \tag{F.28}$$

and

$$\Delta p = \vec{P}_0\cdot\vec{F}\,, \tag{F.29}$$

where

$$\vec{P}_0 = \frac{1}{4\pi\Delta r^2}\frac{\vec{\Delta r}}{\Delta r}\,. \tag{F.30}$$

$\vec{\vec{G}}_0$ and $\vec{P}_0$ are the Green's functions, respectively, for the velocity and pressure caused by a point force in Stokes flow. $\vec{\vec{G}}_0\cdot\vec{F}$ is referred to as a Stokeslet and constitutes the monopole of the multipolar expansion for Stokes flow. The multipole solutions for

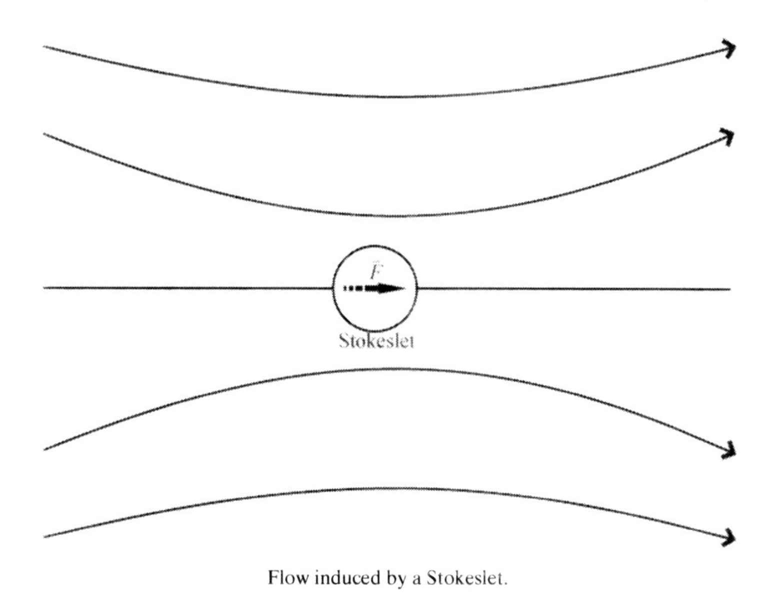

Fig. F.4 Flow induced by a Stokeslet.

Stokes flow are also referred to as singular solutions or fundamental solutions. Figure F.4 shows a Stokeslet and a schematic of the resulting flow pattern.

F.3 STOKES MULTIPOLES: STRESSLET AND ROTLET

The multipolar solutions for Stokes flow are obtained in a manner similar to the multipole solutions for the Laplace equation (Table F.3), with the key difference being that the Stokes flow is not irrotational, and thus the velocity gradient tensor resulting from the Stokes dipole has both a symmetric (strain) and antisymmetric (rotation) component. We thus divide the Stokes dipole into these components (which are degenerate quadrupoles), and refer to them as the stresslet and rotlet, respectively. In a manner analogous to that for creating dipoles for Laplace equation solutions, we obtain the Stokes dipole (shown schematically in Fig. F.5) by superimposing two Stokeslets of infinitely large, equal, and opposite strengths $\vec{F}$, separated by an infinitesimal distance $\delta\vec{d}$, such that the separation is normal to the force direction. The stresslet (Fig. F.6) is given by the addition of two Stokes dipoles oriented at 90° with respect to each other. The velocity induced by a stresslet is given by

$$\vec{u} = \vec{\vec{G}}_S \cdot \vec{F}. \tag{F.31}$$

Table F.3 **Analogies between multipolar solutions for axisymmetric Laplace and axisymmetric Stokes equations.**

Topic	Laplace equation	Stokes equation
Monopole	Green's function solution to Laplace equation	Green's function solution to Stokes equation
Superposition to make multipoles	Monopoles aligned along axis of symmetry	Monopoles aligned normal to axis of symmetry
Mathematical relation between multipoles	Multipoles are related to derivatives of lower multipoles by Rodrigues formula	Multipoles are related to derivatives of lower multipoles

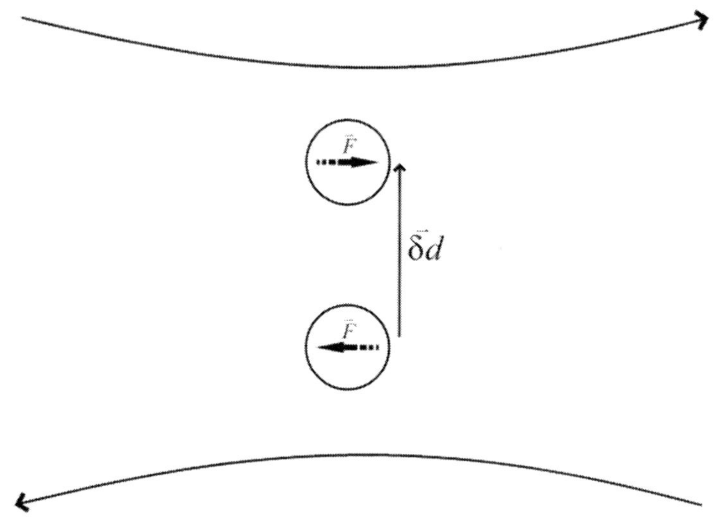

Fig. F.5 Flow induced by a Stokes dipole.

The resulting flow has a magnitude equal to the product of the point forces and the separation distance. The stresslet interaction tensor $\vec{\vec{G}}_s$ is given by

$$\vec{\vec{G}}_s = \frac{\left|\vec{\delta d}\right|}{8\pi\eta\Delta r^3}\left(\vec{\vec{\delta}} - 3\frac{\vec{\Delta r}\vec{\Delta r}}{\Delta r^2}\right). \tag{F.32}$$

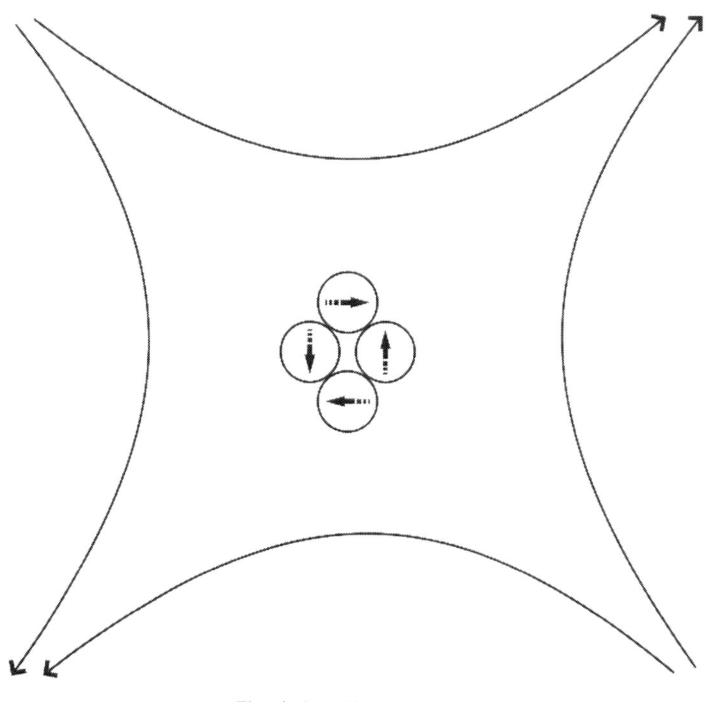

Fig. F.6 Flow induced by a stresslet.

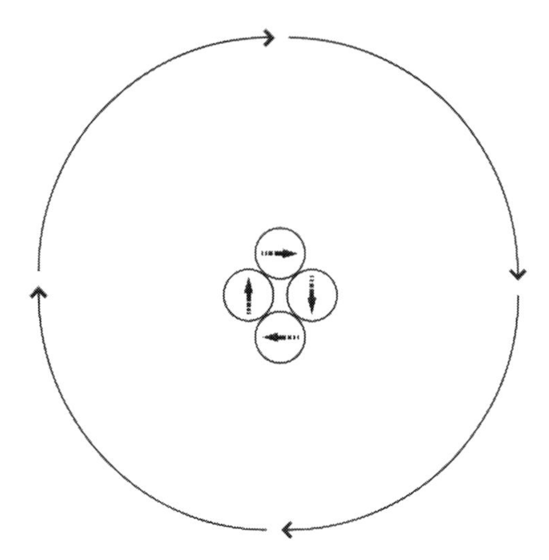

Fig. F.7 Flow induced by a rotlet.

The rotlet (or couplet; Fig. F.7) is given by the difference between two Stokes dipoles oriented at 90° with respect to each other. The velocity induced by a rotlet is given by

$$\vec{\boldsymbol{u}} = \vec{\vec{\boldsymbol{G}}}_r \cdot \vec{\boldsymbol{T}}, \tag{F.33}$$

where the torque pseudovector $\vec{\boldsymbol{T}}$ is given by $\vec{\boldsymbol{\delta d}} \times \vec{\boldsymbol{F}}$, and the rotlet interaction tensor $\vec{\vec{\boldsymbol{G}}}_r$ is given by

$$\vec{\vec{\boldsymbol{G}}}_r = \frac{\vec{\boldsymbol{\Delta r}}}{8\pi\eta\Delta r^3}. \tag{F.34}$$

F.4 SUMMARY

This appendix outlines multipolar solutions for the Laplace equation in both plane-symmetric and axially symmetric geometries. For axially symmetric geometries, the linear multipole solution is given by

$$\phi(r, \vartheta) = \sum_{k=0}^{\infty} B_k r^{-k-1} P_k(\cos\vartheta). \tag{F.35}$$

For plane-symmetric flows, the linear multipole solution is given by

$$\phi(r, \theta) = B_0 \ln \imath + \sum_{k=1}^{\infty} B_k \imath^{-k} \cos(k\theta + \alpha_k). \tag{F.36}$$

For the Laplace equations, these multipoles correspond directly to physical objects, most notably point charges.

We have also outlined multipolar solutions for the Stokes equations with axial symmetry. For the Stokes equations, the multipoles correspond directly to a force or torque applied at a point. For a point force, the velocity is given by

$$\vec{u} = \vec{\vec{G}}_0 \cdot \vec{F} \tag{F.37}$$

and

$$\vec{\vec{G}}_0 = \frac{1}{8\pi\eta\Delta r}\left(\vec{\vec{\delta}} + \frac{\vec{\Delta r}\vec{\Delta r}}{\Delta r^2}\right). \tag{F.38}$$

The impact of these analytical solutions is that they both provide critical insight into the distance dependence of the velocity fields and enable solutions in simplified cases.

F.5 SUPPLEMENTARY READING

Griffiths [55] and Jones [61] are both excellent sources for Laplace equation solutions. Griffiths [55] is expository and provides an accessible introduction. It also covers the separation of variables solution for axisymmetric coordinates. Jones [61] covers the general multipolar theory in considerable detail in an appendix, with specific attention to the solution of electrostatic and electrodynamic problems involving particles in a medium. Jackson [56] provides the most thorough treatment of separation of variables and the Green's function solution techniques.

Derivations of the Green's function solutions for Stokes flow can be found that use Fourier transform techniques [32] or a limiting approach in which a sphere's radius approaches zero in a limiting process. Kim and Karrila [89] present a detailed description of the multipolar solutions for Stokes flow, which focuses on mathematical formulation. Russel et al. [32] include a detailed discussion of this material, including a general formulation for Stokes multipoles. Happel and Brenner [88] also provide a useful resource.

F.6 EXERCISES

F.1 By substitution, show that each of the first three Legendre polynomials, defined by Eq. (F.3), satisfy Legendre's differential equation:

$$\frac{d}{dx}\left[\left(1-x^2\right)\frac{d}{dx}P_k(x)\right] + k(k+1)P_k(x) = 0. \tag{F.39}$$

F.2 Consider the axisymmetric Laplace equation, Eq. (F.1). Find a general solution using separation of variables.

(a) Start by assuming that the solution for ϕ can be written as the product of two functions, one which is a function of r only and one which is a function of ϑ only:

$$\phi = R(r)\Theta(\vartheta). \tag{F.40}$$

(b) Insert this relation into the Laplace equation and rearrange the equation so that all terms involving r are on one side and all terms involving ϑ are on the other.

(c) Because each side of this equation must be a constant, each side can be rewritten as an ordinary differential equation with an arbitrary constant. This arbitrary constant is typically written in a form proportional to $k(k+1)$ in anticipation of the boundary conditions that the physical problem will need to satisfy. The R side of the equation can be rearranged to obtain an Euler differential equation:

$$r^2 \frac{d^2}{dr^2} R + 2r \frac{d}{dr} R = k(k+1) R, \tag{F.41}$$

which is satisfied by polynomial solutions in r. The ϑ side of this equation is most easily handled by rewriting the equation by applying the transform $x = \cos\vartheta$ (the variable x here is not a distance or coordinate but simply a transform variable). With this transform, the equation can be rearranged into Legendre's differential equation:

$$\frac{d}{dx}\left[\left(1 - x^2\right) \frac{d}{dx} P_k(x)\right] + k(k+1) P_k(x) = 0, \tag{F.42}$$

which allows for Legendre polynomial solutions.

(d) Substitute $\cos\vartheta$ back in for x, and write the general solution as a product of the R and Θ functions, and identify the allowable values of k by noting that the solution for $\phi = 0$ must be equal to the solution for $\phi = 2\pi$. With the allowable values of k specified, write the full infinite sum solution as written in Eq. (F.2).

F.3 Consider the 2D cylindrical Laplace equation, Eq. (F.18). Find a general solution using separation of variables.

(a) Start by assuming that the solution for ϕ can be written as the product of two functions, one that is a function of $\mathscr{r}$ only and one that is a function of θ only:

$$\phi = R(\mathscr{r})\Theta(\theta). \tag{F.43}$$

(b) Insert this relation into the Laplace equation and rearrange the equation so that all terms involving $\mathscr{r}$ are on one side and all terms involving θ are on the other.

(c) Because each side of this equation must be a constant, each side can be rewritten as an ordinary differential equation with an arbitrary constant. This arbitrary constant is typically written in a form proportional to k^2 in anticipation of the boundary conditions that the physical problem will need to satisfy. The R side of the equation can be rearranged to obtain an Euler differential equation,

$$\mathscr{r}^2 \frac{\partial^2 R}{\partial \mathscr{r}^2} + \mathscr{r} \frac{\partial R}{\partial \mathscr{r}} = k^2 R, \tag{F.44}$$

which is satisfied by polynomial solutions in $\mathscr{r}$. The θ side of this equation can be rearranged to obtain

$$\frac{\partial \Theta}{\partial \theta} + k^2 \Theta = 0, \tag{F.45}$$

which leads to sine and cosine solutions.

(d) Write the general solution as a product of the R and Θ functions, and identify the allowable values of k by noting that the solution for $\phi = 0$ must be equal to the solution for $\phi = 2\pi$. With the allowable values of k specified, write the full infinite sum solution as written in Eq. (F.19).

F.4 List the positions and strengths of the dipoles required for approximating a linear axisymmetric octupole at the origin aligned along the z axis.

F.5 List the positions and strengths of the quadrupoles required for approximating a linear axisymmetric octupole at the origin aligned along the z axis.

F.6 List the positions and strengths of the monopoles required for approximating a linear axisymmetric quadrupole at the origin aligned along the z axis.

F.7 Consider two linear axisymmetric monopoles of strength $\pm B_0$, located at finite z locations $\pm\frac{1}{2}d$. Along the z axis, compare the resulting values of ϕ calculated in two ways:

(a) By superposing the two monopole solutions;

(b) By approximating these two monopoles as a dipole with strength $B_1 = B_0 d$.

Plot the two results. At what values of z/d is the dipole solution within 5% of the exact (two-monopole) solution?

F.8 Show that Eq. (F.11) is a solution of the axisymmetric spherical Laplace equation (F.1).

F.9 Show that Eq. (F.22) is a solution of the 2D cylindrical Laplace equation (F.18).

F.10 Calculate and plot the velocity induced by a stresslet with $\vec{S} = \frac{\sqrt{2}}{2}\hat{x} + \frac{\sqrt{2}}{2}\hat{z}$.

F.11 Calculate and plot the velocity induced by a rotlet with $\vec{T} = \hat{z}$.

F.12 Show that the velocity fields of a stresslet and a Stokeslet can be combined to give the velocity field around a sphere in Stokes flow.

F.13 Consider a sphere that is rotating, perhaps a cell rotating because of a rotating electric field. Calculate the velocity field induced by this sphere.

F.14 Consider five spherical, 1-μm-diameter particles at (0,0,0), (10,10,10), (0,−10, 0), (5,−5,5), and (−20,0,0), in which all locations are specified in units of μm. Assume uniform flow in the x direction. What is the viscous force on the particle at the origin owing to the flow?

APPENDIX G

Complex Functions

We use complex functions throughout this text to facilitate calculations involving harmonic functions because of the relatively simple mathematics of complex exponentials. Also, when complex variables denote distance in a 2D plane, analytic functions automatically solve the Laplace equation. This appendix summarizes basic properties and manipulation of complex variables.

G.1 COMPLEX NUMBERS AND BASIC OPERATIONS

A complex number $\underset{\sim}{C}$ is written in the form $\underset{\sim}{C} = a + jb$, where a and b are real numbers and $j = \sqrt{-1}$. The *real part* a of the complex number is denoted by $\mathrm{Re}[\underset{\sim}{C}]$, and the *imaginary part* b of the complex number is denoted by the real value $\mathrm{Im}[\underset{\sim}{C}]$. Complex numbers are denoted with undertildes throughout this text.

The *complex conjugate* of $\underset{\sim}{C}$ is denoted by $\underset{\sim}{C}^*$ and is found by reversing the sign of the imaginary part:

$$\underset{\sim}{C}^* = a - jb\,. \tag{G.1}$$

The product of a number and its complex conjugate is purely real.

A complex number $\underset{\sim}{z}$ can be thought of as a position vector in a 2D coordinate system ($\underset{\sim}{z} = x + jy$). Alternatively, cylindrical coordinates (magnitude and angle) can specify position in this plane. The *magnitude* or absolute value of a complex number is equivalent to the radial distance in the complex plane and is denoted by $|\underset{\sim}{z}|$ and given by

$$|\underset{\sim}{z}| = \sqrt{x^2 + y^2}\,, \tag{G.2}$$

whereas the *angle* of a complex number is equivalent to the azimuthal coordinate of the position in the complex plane:

$$\angle(\underset{\sim}{z}) = \mathrm{atan2}\,(y, x)\;, \tag{G.3}$$

where we use atan2 to denote the two-argument, four-quadrant inverse tangent, whose value (in the range 0–2π) depends on the signs of y and x, as well as the value of y/x:

$$\mathrm{atan2}(y, x) = \begin{cases} \tan^{-1}(y/x) & \text{if } \; x > 0,\, y > 0 \\ \tan^{-1}(y/x) + \pi & \text{if } \; x < 0 \\ \tan^{-1}(y/x) + 2\pi & \text{if } \; x > 0,\, y < 0 \\ \pi/2 & \text{if } \; x = 0,\, y > 0 \\ 3\pi/2 & \text{if } \; x = 0,\, y < 0 \end{cases}. \tag{G.4}$$

Given Euler's formula, $\exp\, j\theta = \cos\theta + j\sin\theta$, we can thus equivalently write

$$\underset{\sim}{z} = |\underset{\sim}{z}| \exp\, j\angle(\underset{\sim}{z})\,, \tag{G.5}$$

a notation termed the *polar form* of the complex number.

G.1.1 Arithmetic operations

Complex numbers can be added and subtracted by adding and subtracting the real and imaginary parts separately, so

$$(a + jb) + (c + jd) = (a + c) + j(b + d) \tag{G.6}$$

and

$$(a + jb) - (c + jd) = (a - c) + j(b - d)\,. \tag{G.7}$$

Complex numbers can be multiplied term by term, noting that $j^2 = -1$:

$$(a + jb)(c + jd) = (ac - bd) + j(bc + ad)\,. \tag{G.8}$$

Complex numbers are divided by first multiplying the numerator and denominator by the complex conjugate of the denominator. The resulting real denominator divides both the real and imaginary parts of the numerator:

$$\frac{a + jb}{c + jd} = \frac{(a + jb)(c - jd)}{(c + jd)(c - jd)} = \left(\frac{ac + bd}{c^2 + d^2}\right) + j\left(\frac{bc - ad}{c^2 + d^2}\right)\,. \tag{G.9}$$

G.1.2 Calculus operations

Differentiation of a complex function $f(\underset{\sim}{z})$ with respect to a complex number can be evaluated in a number of ways. If the desired result is a function of $\underset{\sim}{z}$, the derivative is often taken symbolically, as is the case throughout most of Chapter 7. When evaluating the derivative for plotting in the xy plane, though, we often want to write the derivative with respect to x or y. We do this by evaluating the limit in the definition of the derivative along either the real axis or the imaginary axis. This leads to

derivative relation for complex functions in the complex plane

$$\frac{\partial f}{\partial \underset{\sim}{z}} = \frac{\partial f}{\partial x} = -j\frac{\partial f}{\partial y}\,. \tag{G.10}$$

There is a critical distinction here between differentiation of complex numbers compared with real numbers. For differentiability of real functions, the derivative,

$$f'(x) = \lim_{\Delta x \to 0} \frac{f(x + \Delta x) - f(x)}{\Delta x}\,, \tag{G.11}$$

must be the same for Δx approaching zero from both the positive direction and the negative direction. If these two are not the same, we say the function is not differentiable at x. For example, $|x|$ is not differentiable at $x = 0$, because the limit in Eq. (G.11) evaluated with positive Δx is 1, whereas the derivative evaluated with negative Δx is -1. For a derivative of a complex function with respect to a complex variable,

$$f'(\underset{\sim}{z}) = \lim_{\Delta \underset{\sim}{z} \to 0} \frac{f(\underset{\sim}{z} + \Delta \underset{\sim}{z}) - f(\underset{\sim}{z})}{\Delta \underset{\sim}{z}}\,. \tag{G.12}$$

For a complex function, two requirements must be satisfied for differentiability. First, the derivative must be the same regardless of whether $\Delta \underset{\sim}{z}$ is positive or negative. Second, the derivative must be the same regardless of whether $\underset{\sim}{\Delta z}$ is real or imaginary. The second condition is satisfied if Eq. (G.10) holds:

$$\frac{\partial f}{\partial x} = -j\frac{\partial f}{\partial y}. \tag{G.13}$$

This means that differentiability for complex functions has much stricter requirements than differentiability for real functions does. For real functions, we think of a differentiable function as one that is smooth. For complex functions, differentiability implies not just smoothness, but that Eq. (G.13) provides a specific relation between the derivatives of the function in the real and imaginary directions. If we use a complex variable to represent spatial distances in a 2D system, as we do in Chapter 7, then the differentiability of the complex function (a relation between derivatives along the real axis versus the imaginary axis) implies a specific spatial relation between the derivatives in the x and y directions. To show this, consider the complex function $\underset{\sim}{\phi_v} = \phi_v + j\psi$. By taking the derivatives of $\underset{\sim}{\phi_v}$ along the real and complex directions, we can show that

$$\frac{\partial \phi_v}{\partial x} = \frac{\partial \psi}{\partial y} \tag{G.14}$$

and

$$\frac{\partial \psi}{\partial x} = -\frac{\partial \phi_v}{\partial y}. \tag{G.15}$$

These two relations are the *Cauchy–Riemann equations* for complex functions. Combining these two relations (by taking partial derivatives with respect to x or y), we obtain two relations:

Laplace equation for real part of complex function

$$\frac{\partial^2 \phi_v}{\partial x^2} + \frac{\partial^2 \phi_v}{\partial y^2} = 0 \tag{G.16}$$

and

Laplace equation for imaginary part of complex function

$$\frac{\partial^2 \psi}{\partial x^2} + \frac{\partial^2 \psi}{\partial y^2} = 0. \tag{G.17}$$

Functions in the complex plane are therefore differentiable only if their real and imaginary parts each satisfy the Laplace equation. In Chapter 7, we take advantage of this by defining $\underset{\sim}{\phi_v}$ in terms of $\underset{\sim}{z}$ only. These functions, if differentiable, satisfy the Laplace equation. Many of the functions used in Chapter 7 are differentiable except at singularities; in this case, the Laplace equation is satisfied everywhere except at the singularity.

EXAMPLE PROBLEM G.1

Consider the analytic function $\underset{\sim}{\phi_v} = \phi_v + j\psi$. By taking the derivatives of $\underset{\sim}{\phi_v}$ along the real and complex directions, show that

$$\frac{\partial \phi_v}{\partial x} = \frac{\partial \psi}{\partial y} \tag{G.18}$$

and

$$\frac{\partial \psi}{\partial x} = -\frac{\partial \phi_v}{\partial y}. \tag{G.19}$$

SOLUTION: If $\underset{\sim}{\phi_v}$ is analytic, it is differentiable in the complex plane and therefore

$$\frac{\partial \underset{\sim}{\phi_v}}{\partial x} = -j\frac{\partial \underset{\sim}{\phi_v}}{\partial y}. \tag{G.20}$$

Expanding, we find

$$\frac{\partial \phi_v}{\partial x} + j\frac{\partial \psi}{\partial x} = -j\left(\frac{\partial \phi_v}{\partial y} + j\frac{\partial \psi}{\partial y}\right), \tag{G.21}$$

and simplifying, we obtain

$$\frac{\partial \phi_v}{\partial x} + j\frac{\partial \psi}{\partial x} = \frac{\partial \psi}{\partial y} - j\frac{\partial \phi_v}{\partial y}. \tag{G.22}$$

Equating the real and imaginary parts of each side of the equation, we obtain

$$\frac{\partial \phi_v}{\partial x} = \frac{\partial \psi}{\partial y} \tag{G.23}$$

and

$$\frac{\partial \psi}{\partial x} = -\frac{\partial \phi_v}{\partial y}. \tag{G.24}$$

G.2 USING COMPLEX VARIABLES TO COMBINE ORTHOGONAL PARAMETERS

We often use complex variables with linear systems to combine two related (and, in some way, orthogonal) parameters into one parameter. Some examples of this are complex distances, i.e., locations in the complex plane, in which we combine the two directions x and y into one parameter:

$$\underset{\sim}{z} = x + jy. \tag{G.25}$$

The benefits of this approach stem from the fact that x and y are spatially orthogonal to each other, and functions automatically satisfy the Laplace equation if they are dependent on $\underset{\sim}{z}$ only.

Similarly, streamlines and isopotential contours are orthogonal to each other in 2D potential flow, and if we define a complex velocity potential,

$$\underset{\sim}{\phi_v} = \phi_v + j\psi \, , \tag{G.26}$$

and the fluid velocity can be related to derivatives of this function with respect to $\underset{\sim}{z}$. For electric fields, we similarly write

$$\underset{\sim}{\phi} = \phi + j\psi_e \, . \tag{G.27}$$

Parameters can also be temporally orthogonal – in particular, harmonic functions are orthogonal to each other if they have different frequencies or if they have the same frequency but are out of phase by 90°, because the integral of their product over a cycle is zero in either case. For example, ohmic and displacement current in response to a sinusoidal electric field are orthogonal to each other, and we can combine the electric displacement and the ohmic current to define a complex permittivity:

$$\underset{\sim}{\varepsilon} = \varepsilon + \frac{\sigma}{j\omega} \, . \tag{G.28}$$

G.3 ANALYTIC REPRESENTATION OF HARMONIC PARAMETERS

The *complex representation* or *analytic representation* of a real sinusoidal signal consists of the complex exponential whose real part is equal to the sinusoidal signal. So, if $V = V_0 \cos(\omega t + \alpha)$, the voltage's analytic representation $\underset{\sim}{V}$ is given by

analytic representation of a sinusoidal real function

$$\underset{\sim}{V} = \underset{\sim}{V}_0 \exp j\omega t \, . \tag{G.29}$$

The *phasor* $\underset{\sim}{V}_0 = V_0 \exp j\alpha$ is what remains after the analytic representation of the parameter is normalized by $\exp j\omega t$. The phasor is complex, and its angle captures phase differences of the real signal with respect to a reference sinusoid. In the applications in this text, the reference sinusoid is usually an applied electric or pressure field, and the phase lag is related to fluid or charge buildup. Symbolically, the analytic representation of a harmonic parameter is denoted by placing an undertilde under the symbol, so the analytic representation of p is given by $\underset{\sim}{p}$. The phasor is always denoted with a subscript zero and has an undertilde: $\underset{\sim}{p}_0$. The phasor magnitude is real and is denoted with subscript zero but no undertilde. For example, the analytic representation of $p = p_0 \cos(\omega t + \alpha)$ is written as $\underset{\sim}{p}$ and given by $\underset{\sim}{p} = \underset{\sim}{p}_0 \exp j\omega t$, where $\underset{\sim}{p}_0 = p_0 \exp j\alpha$. The phasor is $\underset{\sim}{p}_0$, and the phasor magnitude is p_0. For the pressure signal $p = p_0 \cos \omega t$, there is no phase lag and the phasor is real ($\underset{\sim}{p}_0 = p_0$). The analytic representation is exponential, leading to straightforward mathematical manipulation facilitated by the properties of complex algebra. The analytic representation is particularly effective for handling linear equations, in which case all analysis can be done with the analytic representation, and the real part of the analytic result corresponds to the parameters in the physical system.

G.3.1 Applicability of the analytic representation

The analytic representation exists for any sinusoidal function or any function that can be represented as a sum of sinusoidal functions. The analytical and phasor representations, however, are *useful* only when the relations are linear and all of the important parameters vary at the same frequency.

G.3.2 Mathematical rules for using the analytic representation of harmonic parameters

Generating the analytic representation from a real parameter. Given a real harmonic function f, we find the analytic representation by replacing $\cos(\omega t + \alpha)$ with $\exp[j(\omega t + \alpha)]$ and replacing $\sin(\omega t + \alpha)$ with $\exp[j(\omega t - \pi/2 + \alpha)]$.

Generating the real parameter from its analytic representation. Given an analytic representation $\underset{\sim}{f}$ of a function f, we can determine the function by simply taking the real part of $\underset{\sim}{f}$ or, equivalently, taking the average of $\underset{\sim}{f}$ and its complex conjugate:

real part of analytic function

$$f = \mathrm{Re}[\underset{\sim}{f}] = \frac{1}{2}(\underset{\sim}{f} + \underset{\sim}{f}^*) . \qquad \text{(G.30)}$$

Using analytic representations directly in linear relations. The analytic representation of a harmonic function satisfies a linear relation directly. Examples of important linear relations include $V = IR$ and $\vec{\boldsymbol{D}} = \varepsilon \vec{\boldsymbol{E}}$.

Using analytic representations in nonlinear relations. The analytic representation cannot be used directly with any nonlinear equation or to evaluate a nonlinear parameter, because the analytic representation of the product of two functions is not equal to the product of the analytic representations of the functions:

$$\underset{\sim}{fg} \neq \underset{\sim}{f}\,\underset{\sim}{g} . \qquad \text{(G.31)}$$

For example, consider a harmonic electric field E given by $E = E_0 \cos \omega t$ and let its analytic representation be $\underset{\sim}{E} = E_0 \exp j\omega t$. The parameter E^2 is given by $E^2 = E_0{}^2 \cos^2 \omega t = E_0{}^2 \left(\frac{1}{2} + \frac{1}{2} \cos 2\omega t\right)$. The square of the analytic representation of E is $\underset{\sim}{E}^2 = E_0{}^2 \exp 2j\omega t$. Unfortunately, the analytic representation of E^2 (i.e., $\underset{\sim}{E^2}$) is clearly not equal to the square of the analytic representation of E (i.e., $\underset{\sim}{E}^2$). Thus, although analytic representations considerably simplify linear analysis, they are of little use for nonlinear analysis.

When evaluating a nonlinear relation (for example, the product of two harmonic parameters, say, the power generated in a circuit or the force on an induced dipole), we must evaluate the nonlinear function by use of real quantities, or we must use specific relations designed for this purpose. Here we write the relation for the real product of two harmonic functions in terms of their phasors. We do not report the result for the analytic representation of this product, as it has little use. If $f = f_0 \cos \omega t + \alpha_f$ and

$g = g_0 \cos \omega t + \alpha_g$ and the analytic representations of these functions are $\underset{\sim}{f} = \underset{\sim}{f_0} \exp j\omega t$ and $\underset{\sim}{g} = \underset{\sim}{g_0} \exp j\omega t$, we can show that

$$fg = \frac{1}{2}\left[\mathrm{Re}\left(\underset{\sim}{f_0}\,\underset{\sim}{g_0}^*\right) + \mathrm{Re}\left(\underset{\sim}{f_0}\,\underset{\sim}{g_0}\right)\cos 2\omega t + \mathrm{Im}\left(\underset{\sim}{f_0}\,\underset{\sim}{g_0}\right)\sin 2\omega t\right], \tag{G.32}$$

or, equivalently,

$$fg = \frac{1}{2}\left\{\mathrm{Re}\left(\underset{\sim}{f_0}\,\underset{\sim}{g_0}^*\right) + \cos\left[2\omega t + \angle\mathrm{Re}\left(\underset{\sim}{f_0}\,\underset{\sim}{g_0}\right)\right]\right\}. \tag{G.33}$$

Because $\mathrm{Re}(\underset{\sim}{f_0}\,\underset{\sim}{g_0}^*) = \mathrm{Re}(\underset{\sim}{f_0}^*\,\underset{\sim}{g_0})$, these equations can be written using $\mathrm{Re}(\underset{\sim}{f_0}^*\,\underset{\sim}{g_0})$ as well.

From this, we can show that the time average of the product of two harmonic functions of the same frequency is given by

time average of product of real functions in terms of their analytic representations

$$\langle fg \rangle = \frac{1}{2}\mathrm{Re}\left(\underset{\sim}{f_0}\,\underset{\sim}{g_0}^*\right). \tag{G.34}$$

Owing to orthogonality of the harmonic functions, the time average of the product of two harmonic functions of different frequencies is zero. With Eqs. (G.33) (in general) and (G.34) (for time-averaged values) we can directly calculate real nonlinear functions from the analytic representations without the need to return to the real representation of the harmonic function.

G.4 KRAMERS–KRÖNIG RELATIONS

For an analytic function $\underset{\sim}{\varepsilon}(\omega)$ (i.e., a differentiable complex function) whose value approaches zero as $\omega \to \infty$, we can show that

$$\underset{\sim}{\varepsilon}(\omega) = \frac{1}{j\pi}\int_{-\infty}^{\infty} \frac{\underset{\sim}{\varepsilon}(\varpi)}{\varpi - \omega}\, d\varpi, \tag{G.35}$$

where ϖ is a dummy variable used for the integration. This relation relates the function $\underset{\sim}{\varepsilon}$ at a specific frequency to an integral of that function over all frequencies. Because the integral is premultiplied by $1/j\pi$, this relation effectively relates the real and imaginary parts of an analytic function to each other.

The integrals from 0 to ∞ and from $-\infty$ to 0 can be shown equal to each other if the system under study is real. With this, Eq. (G.35) can be separated into real and imaginary components to obtain

$$\varepsilon'(\omega) = \frac{2}{\pi}\int_{0}^{\infty} \frac{\varpi\varepsilon''(\varpi)}{\varpi^2 - \omega^2}\, d\varpi \tag{G.36}$$

and

$$\varepsilon''(\omega) = -\frac{2}{\pi}\int_{0}^{\infty} \frac{\omega\varepsilon'(\varpi)}{\varpi^2 - \omega^2}\, d\varpi. \tag{G.37}$$

The Kramers–Krönig relations describe the relation between the reactive response of a system and its dissipative response when excited with a forcing function. For example, the Kramers–Krönig relations relate the polarization of a medium to the heating of a medium as an electric field is applied.

G.5 CONFORMAL MAPPING

For systems solved by the Laplace equation, certain mapping functions transform problems spatially and simplify solution of the governing equations.

G.5.1 Joukowski transform

One common mapping function for potential flow or electric field solutions is the Joukowski transform,

$$\mathcal{J}\left[\underset{\sim}{z}\exp(-j\alpha), b\right] = \underset{\sim}{z}\exp(-j\alpha) + \frac{b^2}{\underset{\sim}{z}\exp(-j\alpha)}. \tag{G.38}$$

This transform maps a distance $\underset{\sim}{z}$ to a new distance $\mathcal{J}(\underset{\sim}{z})$. In particular, a line of length $4b$ centered on $\underset{\sim}{z} = 0$ and rotated an angle α with respect to the x axis is transformed into a circle of radius b centered on $\underset{\sim}{z} = 0$. This transform, for example, maps a uniform flow into a flow over a cylinder. If the complex potential for a uniform flow is given by $\underset{\sim}{\phi_v} = U\underset{\sim}{z}\exp(-j\alpha)$, then the transformed flow is given by $\underset{\sim}{\phi_v} = U\mathcal{J}\left[\underset{\sim}{z}\exp(-j\alpha), b\right]$, or

$$\underset{\sim}{\phi_v} = U\left[\underset{\sim}{z}\exp(-j\alpha) + \frac{b^2}{\underset{\sim}{z}\exp(-j\alpha)}\right], \tag{G.39}$$

which is identical to Eq. (7.78) for $a = b$. Because any function of $\mathcal{J}(\underset{\sim}{z})$ only is also a function of $\underset{\sim}{z}$ only, these transformed solutions also satisfy the Laplace equations and thus are solutions of the flow equations, but with a spatially transformed boundary.

The inverse Joukowski transform maps a circle of radius b onto a line. The inverse Joukowski transform, though generally more useful, is harder to implement, because the transform is dual valued in a manner that is not immediately obvious. The inverse Joukowski transform $\mathcal{J}^{-1}$ is given by

$$\begin{aligned}\mathcal{J}^{-1}&\left[\underset{\sim}{z}\exp(-j\alpha), b\right] \\ &= \tfrac{1}{2}\underset{\sim}{z}\exp(-j\alpha) + \operatorname{sgn}\left[\operatorname{Re}(\underset{\sim}{z})\right]\operatorname{sgn}\left(|\underset{\sim}{z}| - b\right)\exp(-j\alpha)\tfrac{1}{2}\sqrt{\left[\underset{\sim}{z}\exp(-j\alpha)\right]^2 - 4b^2},\end{aligned} \tag{G.40}$$

where $\operatorname{sgn}(x) = x/|x|$ is the sign function of x. The sign functions account for the dual-valued inverse, which involves a square root. If the point in question is either (a) outside the circle and in the right half of the complex plane or (b) inside the circle and in the left half of the complex plane, we use the principal square root, and the inverse Joukowski transform is given by

$$\mathcal{J}^{-1}\left[\underset{\sim}{z}\exp(-j\alpha), b\right] = \frac{1}{2}\underset{\sim}{z}\exp(-j\alpha) + \exp(-j\alpha)\frac{1}{2}\sqrt{\left[\underset{\sim}{z}\exp(-j\alpha)\right]^2 - 4b^2}, \tag{G.41}$$

and otherwise, we use the negative square root, and the inverse transform is given by

$$\mathcal{J}^{-1}\left[\underset{\sim}{z}\exp(-j\alpha), b\right] = \frac{1}{2}\underset{\sim}{z}\exp(-j\alpha) - \exp(-j\alpha)\frac{1}{2}\sqrt{\left[\underset{\sim}{z}\exp(-j\alpha)\right]^2 - 4b^2}. \tag{G.42}$$

If we define shapes as loci of points defined by distances $\underset{\sim}{z}$ from some reference point, we can describe how the Joukowski transform and its inverse map these shapes. The inverse Joukowski transform maps a circle of radius b centered on $\underset{\sim}{z}=0$ onto a line. It simultaneously maps circles centered on $\underset{\sim}{z}=0$ with radius greater than a to ellipses. This becomes useful if we want to analytically calculate the flow around an ellipse. For example, the complex potential for flow around a cylinder of radius a is given by

$$\underset{\sim}{\phi}_{\mathrm{v}}\left[\underset{\sim}{z}\exp(-j\alpha)\right]=U\left[\underset{\sim}{z}\exp(-j\alpha)+\frac{a^2}{\underset{\sim}{z}\exp(-j\alpha)}\right], \tag{G.43}$$

and, if transformed by the inverse Joukowski transform:

$$\underset{\sim}{\phi}_{\mathrm{v}}\left\{\mathcal{J}^{-1}\left[\underset{\sim}{z}\exp(-j\alpha)\right]\right\}=U\left\{\mathcal{J}^{-1}\left[\underset{\sim}{z}\exp(-j\alpha)\right]+\frac{a^2}{\mathcal{J}^{-1}\left[\underset{\sim}{z}\exp(-j\alpha)\right]}\right\}, \tag{G.44}$$

the result is flow around an ellipse with semimajor axis $a+b^2/a$ and semiminor axis $a-b^2/a$.

G.5.2 Schwarz–Christoffel transform

Given locations in 2D planes denoted by complex variables $\underset{\sim}{z}=x+jy$, the Schwarz–Christoffel transform maps an arbitrary polygon onto a half-space. This is useful primarily for approximating electric fields in polygonal microchannels, because many microfabrication techniques lead to polygonal microchannels (e.g., rectangular or trapezoidal). The Schwarz–Christoffel formula states that, given specific constraints that are met for physically well-posed problems, there exists a conformal mapping between a half-space in a complex plane and a polygon in an alternate complex plane. Given a location $\underset{\sim}{z}$ in the upper half-plane, i.e., for $\mathrm{Im}(\underset{\sim}{z})>0$, $\mathcal{S}(\underset{\sim}{z})$ is a complex number that gives the location of that point inside a specified polygon. The function that gives this mapping, $\mathcal{S}(\underset{\sim}{z})$, is given by

$$\mathcal{S}(\underset{\sim}{z})=C_2+C_1\int\prod_{i=1}^{n}(\underset{\sim}{z}-\underset{\sim}{z}_i)^{\alpha_i/\pi}\,d\underset{\sim}{z}, \tag{G.45}$$

where the polygon consists of n vertices with locations $\underset{\sim}{z}_i$, each of which has an exterior angle α_i defined in the range $-\pi<\alpha_i<\pi$.

G.6 SUMMARY

This appendix has briefly summarized manipulation of complex variables, their use in providing solutions to the 2D Laplace equation, their use in handling orthogonal harmonic functions, and the Joukowski and Schwarz–Christoffel conformal mapping techniques.

G.7 SUPPLEMENTARY READING

Many readers do not need any complex analysis beyond this appendix for analysis of micro- and nanofluidic systems – the basics of complex number definitions and discussion of differentiability and the Cauchy–Riemann equations are typically enough. A slightly more formal discussion of the calculus of analytic functions in the context of plane-symmetric potential flows is presented in [84], and an accessible but more rigorous mathematical presentation is given in [294]. Conformal mapping is discussed specifically in [295]. Some examples of microfluidic research using conformal mapping can be found in [253, 296, 297].

G.8 EXERCISES

G.1 Derive Eq. (G.16).

G.2 Derive Eq. (G.17).

G.3 Derive Eq. (G.34).

APPENDIX H

Interaction Potentials: Atomistic Modeling of Solvents and Solutes

Most of this text uses continuum, ideal-solution theory to describe transport in micro- and nanoscale systems. Molecule-scale interactions, however, require that the interactions of the solvent, dissolved electrolytes, and macromolecules be treated in more detail. The first example of this in the text is the steric modification of the Poisson–Boltzmann description of the EDL presented in Chapter 9. Other examples of this include the intermolecular potentials used in atomistic simulations, excluded-volume modeling for predictions of DNA conformation in nanochannels, and colloidal simulations. This appendix focuses on the general concepts of interaction potentials and distribution functions, with a focus on atomistic modeling. As is often done in the atomistic/molecular dynamics literature, we use the term *atom* in this appendix in the original sense of an *indivisible unit* rather than the modern chemical sense of a nucleus surrounded by an electron cloud. Thus the term *atom* in this appendix refers in general to any entity that we model based on its potential energy – e.g., solvent molecules or dissolved ions or even boundaries.

H.1 THERMODYNAMICS OF INTERMOLECULAR POTENTIALS

Atoms that interact rarely are straightforward to model thermodynamically. That these atoms rarely interact is the basic premise of the ideal gas law and the ideal solution approximation. For example, the Boltzmann approximation for ions in an electrolyte consists of treating the ions as if they do not interact with each other, interacting only with a continuum electrical field. The solvent molecules (typically water) are ignored. The only effect of the solvent molecules is to dictate the electrical permittivity ε of the mean field. The Boltzmann approximation leads to the simple conclusion that the energy of an atom is given exclusively by the singlet energy $e_1(\vec{r}_i)$ – namely, that

$$e_1(\vec{r}_i) = z_i k_B \phi(\vec{r}_i), \tag{H.1}$$

which was presented earlier as Eq. (9.1), fully describes the potential energy landscape that dictates atom distributions. This is a *singlet potential*, meaning that the potential is a function of position but not a function of the location of or properties of other atoms. The potential energy of a system owing to singlet energies has the mathematical form $E = \sum_i e_1(\vec{r}_i)$, where E is the potential energy, i denotes atom i, e_1 is the singlet energy, and $\vec{r}_i$ is the position vector of atom i. These terms carry the effects of external fields on the atoms, but do not account for atom–atom interactions. For example, if walls are at $x = 0$ and $x = L$, e_1 would be zero far from the walls, but would become large as x approached 0 or L.

Unfortunately, as systems become more dense, we must consider the *interaction* of systems. This leads to *pair potentials,* which are a function of two atoms and their relative location, *triplet potentials,* which are a function of the relative positions of sets of three atoms, and so on. The following sections discuss pair potentials in some detail. Compared with an ideal-solution model, considering these pair potentials in detail leads to vastly different solutions for the distribution of atoms. A dramatic example of this is the counterion distribution function near a wall, because real counterions described atomistically exhibit a nonmonotonic distribution function. We describe these in the coming sections with reference to the multipolar theory from Appendix F.

Pair potentials imply energies that are given by the interactions between pairs of atoms. These energies have the mathematical form $E = \sum_{j>i} \sum_i e_2(\vec{r}_i, \vec{r}_j)$. The sum denoted by $\sum_{j>i}$, in this context, means that the energy from the interactions between molecules i and j is counted only once. The most fundamental challenge and limitation of molecular dynamics is the definition and evaluation of the pair potential e_2. In particular, pair potentials can be precisely defined only if a large amount of data (including fluid properties, quantum modeling, and spectroscopic data) is used. They can be rather precisely defined for simple atoms (e.g., liquid argon), but they are difficult to define for more complex structures and are more difficult to define for water. Even once a pair potential formula is agreed on, evaluating the potential numerically is computationally expensive. If all pair potentials in a system with N molecules is considered, order N^2 evaluations are required. Given that N is typically large, this is onerous, and approximate techniques must be used. When the number of evaluations is reduced with some sort of engineering approximation (say, by only calculating energies for molecules that are reasonably close to each other), the true pair potential no longer conserves momentum and energy, and typically must be replaced with an *effective* pair potential (or some other approximation) that corrects for these limitations and for triplet and higher-order energies.

H.1.1 Monopole pair potentials

The simplest pair potential considers the electrostatic interaction between the charges on atoms. Recall that the electrostatic potential energy in a vacuum between two atoms with point charges q_1 and q_2 is

$$e_2(\vec{r}_1, \vec{r}_2) = \frac{q_1 q_2}{4\pi\varepsilon_0 \Delta r_{12}}, \tag{H.2}$$

where Δr_{12} is the length of the vector between the two locations $\vec{r}_1$ and $\vec{r}_2$. The force is simply the derivative of the potential with respect to Δr_{12}:

$$F(\vec{r}_1, \vec{r}_2) = -\frac{q_1 q_2}{4\pi\varepsilon_0 \Delta r_{12}^2}. \tag{H.3}$$

If the atoms are no longer in a vacuum, we must approximate the interactions of the atoms with all of the other atoms (i.e., the solvent atoms) somehow. Here, we often make a continuum approximation that the medium in which the points exist can be described by a single value (the electric permittivity) that describes how the potential is attenuated by the dielectric screening (see Subsection 5.1.2):

$$e_2(\vec{r}_1, \vec{r}_2) = \frac{q_1 q_2}{4\pi\varepsilon \Delta r_{12}}. \tag{H.4}$$

The only difference is that ε_0 has been replaced with ε, the electrical permittivity of the medium.

H.1.2 Spherically symmetric multipole pair potentials

All atomic interaction potentials that are not described by point charges (i.e., monopolar interactions) correspond to distributions of charge (i.e., multipoles). The forces caused by electrostatic interactions between electron orbitals of molecules are typically described by a time-averaged potential that accounts for all of the multipolar electrostatic interactions. We consider only spherically symmetric potentials, because most models for water involve a spherically symmetric pair potential combined with monopole pair potentials to account for the partial charge on each of the atoms.

HARD-SPHERE POTENTIAL

A strong, short-ranged repulsion prevents multiple atoms from existing in the same location at the same time – this stems from the Pauli exclusion principle as applied to the electron orbitals. A simple way to handle this is to treat the atoms as hard spheres. From the standpoint of pair potentials, this corresponds to

$$e_2(\vec{r}_1, \vec{r}_2) = \begin{cases} \infty, & \Delta r_{12} < d_{12} \\ 0, & \Delta r_{12} > d_{12} \end{cases}, \tag{H.5}$$

where d_{12} is the effective hard-sphere diameter of the two-body system. Hard-sphere systems are simple to study and exhibit some semblance of reality but fail to predict a number of simple phenomena, such as a liquid–gas transition.

LENNARD–JONES POTENTIAL

Real atoms always show long-range Coulomb attraction known as Van der Waals attraction that scales inversely with the sixth power of the distance.[1] The short-range repulsive potential is large but neither infinite nor discontinuous, and it has been deemed convenient to approximate this with a Δr_{12}^{-12} dependence. There is no real physical source for the scaling of this term, but it works well and liquid properties are not a strong function of the functional form used for the repulsion term. More important, it is computationally more efficient to calculate an r^{-12} term because the r^{-12} term is the square of the r^{-6} term. Thus the most common form of pair potential has historically been a Lennard–Jones (LJ) potential (alternatively called an LJ 6-12 potential), with general form

Lennard–Jones potential

$$e_2(\Delta r_{12}) = 4\varepsilon_{LJ}\left[\left(\frac{\Delta r_{12}}{\sigma_{LJ}}\right)^{-12} - \left(\frac{\Delta r_{12}}{\sigma_{LJ}}\right)^{-6}\right], \tag{H.6}$$

where ε_{LJ} is the depth of the potential well and σ_{LJ} is the point at which the pair potential is zero.[2] This potential is straightforward to define mathematically, efficient

[1] Real electron orbitals fluctuate, leading to fluctuating dipoles. Two atoms thus experience an interaction because each has a fluctuating dipole and the two dipoles interact. The net effect is nonzero because the fluctuations correlate with each other and thus the time-average of the interaction is finite. Monopole–monopole potentials scale as Δr_{12}^{-1}, as seen in Eq. (H.4), whereas monopole–dipole potentials scale as Δr_{12}^{-2} and dipole–dipole potentials scale as Δr_{12}^{-3}. The degree to which the dipoles correlate also scales as Δr_{12}^{-3}, so the net potential scales as Δr_{12}^{-6}.

[2] The atomistic literature for the most part denotes the well depth as ε and the zero point of the pair potential as σ. We use the LJ subscript to highlight that these values are unrelated to the conductivity and mean-field permittivity of a medium.

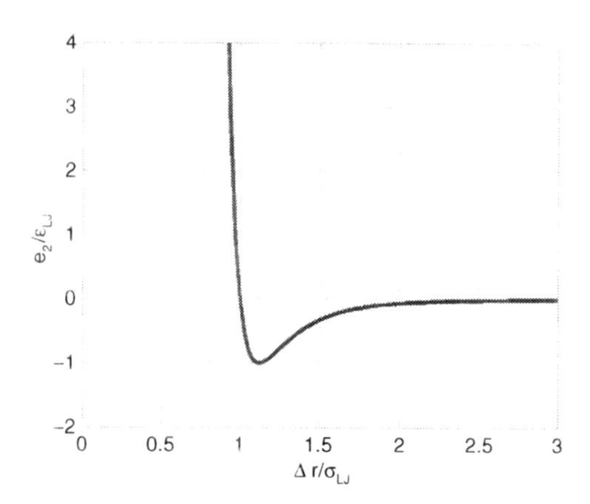

Fig. H.1 The Lennard–Jones potential.

computationally, and roughly approximates real pair potentials for spherically symmetric atoms. For asymmetric atoms, the Lennard–Jones potential is typically used in combination with other terms. An example is shown in Fig. H.1. The value of σ_{LJ} for water atoms is approximately 3.1 Å, and the value of ε_{LJ}/k_B for water atoms is approximately 430 K.

The depth of the potential well naturally leads to a means for normalizing the temperature. We define a reference temperature ε_{LJ}/k_B, which is the temperature at which $k_B T$ is equal to the well depth. Then a nondimensional temperature T^* can be defined as

nondimensionalized temperature

$$T^* = \frac{k_B T}{\varepsilon_{LJ}}. \tag{H.7}$$

At T^* smaller than one, we expect the intermolecular potentials to strongly affect the atom distributions. At T^* much larger than one, we expect the intermolecular potentials to not affect the atom distributions much at all.

H.2 LIQUID-STATE THEORIES

Modified Poisson–Boltzmann equation models account for the finite size of atoms, but do not account for the limitations of mean-field representations of the solvent.

The solvent can be treated reasonably well as a mean field (i.e., atoms can be assumed to be interacting primarily with an electric field rather than with all of the surrounding atoms) as long as the atom–atom spacing is larger than the *Bjerrum length* λ_B (see Exercise 9.20). This length scale defines the atom spacing (and therefore atom concentration) at which atom–atom correlations become important and the mean-field approximation breaks down. For aqueous electrolytes at room temperature, ionic spacing approaches the Bjerrum length only when ion concentrations exceed 1 M.

Liquid-state theories describe the state of liquids in part by describing correlation functions or radial distribution functions that describe the relation of atoms to each other. These theories typically involve integrodifferential equations and are quite unwieldy, even for 1D problems. A common model is the *hypernetted-chain* model, which is a specific simplification to the general Ornstein–Zernike equation. The results of these theories include atom–atom, radial distribution functions, equations of state, and descriptions of atom distributions near solid surfaces.

H.2.1 Integral techniques for concentration profiles

Integral techniques determine concentration profiles by solving the integral equations that describe the simultaneous interactions of all atoms. In this section, the terminology of these approaches is defined, then we discuss the Ornstein–Zernike equation and its solution, as well as the resulting conclusions as to the value of distribution functions.

DISTRIBUTION AND CORRELATION FUNCTIONS

This section describes terminology to describe spatial distributions of molecules. We use *distribution functions* and *correlation functions* to help describe these distributions. Although these definitions are unnecessary for systems in the ideal-solution approximation, these definitions describe both interacting and noninteracting systems.

Although we define many different functions, which seem similar and have potentially confusing names, they are distinguished by their definitions and use.[3] Our goal is to predict or understand the *distribution function* f_d, which describes the spatial distribution of atoms relative to a particular object (say, the distribution of ions in an EDL with respect to a charged wall). To predict this function, we use the *direct correlation function* f_dc, which describes the potential energy landscape caused for the atoms by an object (for example, the potential energy of the ion induced by a nonzero charge density at the wall). The direct correlation function is combined with the *Ornstein–Zernike equation* to generate the *total correlation function* f_tc, which gives a more correct energy landscape that accounts for the combined influence of all atoms, rather than just the one object (for example, the potential energy landscape induced by a nonzero charge density at the wall *combined with* the cloud of ions that are close to the wall). This total correlation function can then be directly related to the distribution function. Conceptually, the key distinction is that the distribution function describes the spatial variation of *atomic concentration*, whereas the direct and total correlation functions describe the *potential energy landscapes* for atoms from, respectively, (1) a consideration of one pair interaction alone or (2) consideration of all pair interactions in a system.

Distribution functions in the ideal-solution limit. We start by considering a solution with species i with bulk concentration $c_{i,\infty}$ and consider the effect of a wall on the spatial distribution of a species as a function of its distance from the wall Δr. We use Δr (rather than a Cartesian coordinate) because these same relations apply for spatial distributions of atoms with respect to each other, thus a radial coordinate is most general.

If the wall had no effect on species distributions, the concentration of species i would be $c_{i,\infty}$ everywhere. If there is some effect of the wall, we describe the spatial variations of concentration by using *distribution functions* as follows. Define the distribution

[3] The notation used here is not the most common in the field. Correlation and distribution functions are commonly denoted using g, h, and c. We define symbols as we do in this section to avoid confusion with other symbols in this text, such as electrochemical potential and electrolyte concentration.

function $f_{d,i}(\Delta r)$ as the concentration of species i at a distance Δr from the first object, normalized by the bulk concentration:

distribution function for a chemical species

$$f_{d,i}(\Delta r) = \frac{c_i(\Delta r)}{c_{i,\infty}} . \tag{H.8}$$

So, if the wall has no effect, $f_{d,i}$ is given by $f_{d,i} = 1$ everywhere.

We also define an *adjusted distribution function* f_{ad}:

definition of adjusted distribution function

$$f_{ad,i}(\Delta r) = -1 + f_{d,i}(\Delta r) , \tag{H.9}$$

which is a measure of the effect the object at $\Delta r = 0$ has on the concentration of ions at Δr. When $f_{ad} > 0$, ions are more likely to be at a distance Δr because of the ion at $\Delta r = 0$; when $f_{ad} < 0$, ions are less likely to be at Δr because of the molecule at $\Delta r = 0$.[4] If the wall has no effect, f_{ad} is zero everywhere.

For example, in Chapter 9 we wrote that the equilibrium concentration of an ion in an EDL with a spatially varying potential is given in the ideal-solution approximation by

$$c_i = c_{i,\infty} \exp\left(-\frac{z_i F \varphi}{RT}\right) . \tag{H.10}$$

In this case, the distribution function is given by

$$f_{d,i}(\Delta r) = \frac{c_i(\Delta r)}{c_{i,\infty}} = \exp\left(-\frac{z_i F \varphi(\Delta r)}{RT}\right) . \tag{H.11}$$

MAYER f FUNCTIONS AND THE POTENTIAL OF MEAN FORCE

Two definitions are of further use in relating potential energies and distribution functions. The Mayer f function defines a Boltzmann-equivalent distribution function for a given electrical potential function, whereas the *potential of mean force* defines a Boltzmann-equivalent potential function for a given distribution function.

Given a potential field $e_1(\Delta r)$, the *Mayer f function* f_M is given by

definition of Mayer f function

$$f_M = -1 + \exp(-e_1/k_B T) . \tag{H.12}$$

The Mayer f function can be thought of as the total correlation function that $e_1(\Delta r)$ would generate in the ideal-solution approximation. It is a way to express a potential energy field in terms of a correlation function. A plot of a potential distribution and the corresponding Mayer f function is shown in Fig. H.2. The inverse of the Mayer f

[4] f_{ad} and f_d, of course, contain the same information. The adjusted distribution function is much easier to implement mathematically than the distribution function, because its integral over an infinite domain is bounded.

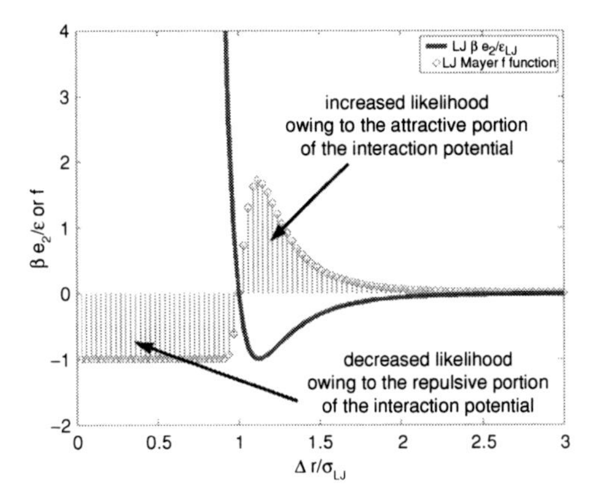

Fig. H.2 A Lennard–Jones potential distribution and the corresponding Mayer f function. The f function is temperature dependent; this example is for $T^* = 1$, i.e., $T = \varepsilon_{\mathrm{LJ}}/k_{\mathrm{B}}$.

function is the *potential of mean force.* The potential of mean force is a way to express a distribution function in terms of a potential energy field. Given a radial distribution function $f_{\mathrm{d}}(\Delta r)$, the potential of mean force e_{mf} is given by

definition of potential of mean force

$$e_{\mathrm{mf}} = -k_{\mathrm{B}} T \ln f_{\mathrm{d}} \,. \tag{H.13}$$

The definition of the potential of mean force is just the inverse of the Mayer f function – it defines the potential that would be required for the distribution function (in the Boltzmann approximation) to be equal to f_{d}.

CORRELATION FUNCTIONS

Distribution functions can be calculated directly from the Mayer f function in the Boltzmann approximation; however, for interacting systems, the process is more complicated, and we need to use *correlation functions* to describe the effects of atomic interactions on distribution functions.

Direct correlation function. The *direct correlation* function $f_{\mathrm{dc}}(\Delta r)$ is minus one plus the likelihood of the state, assuming *no other objects have any effect on the system*:

definition of direct correlation function; uses *only* singlet energy

$$f_{\mathrm{dc}}(\Delta r) = -1 + \exp(-e_1/k_{\mathrm{B}} T) \,. \tag{H.14}$$

The direct correlation function is simply the Mayer f function of the potential energy function.

In sparse or noninteracting systems, in which three-body interactions are absent, the adjusted distribution function, the direct correlation function, and the Mayer f function are the same:

$$f_{\mathrm{ad}}(\Delta r) = f_{\mathrm{dc}}(\Delta r) = f_{\mathrm{M}}(\Delta r) = -1 + \exp(-e_1/k_{\mathrm{B}} T) \,. \tag{H.15}$$

Total correlation function. In interacting systems, the adjusted distribution function is *not* equal to the direct correlation function. Because of this, we use the *total correlation function*, which takes into account all molecules in the system. Any calculation of the total correlation function typically requires an approximation to limit the need for integrodifferential equations. The total correlation function is a function both of the direct correlation function and the number density of all molecules in the system. Most important, the total correlation function that we calculate is an approximation of the adjusted distribution function, i.e., if we properly calculate f_{tc}, then $f_{\text{tc}} = f_{\text{ad}}$ and can be used to determine the distribution f_{d}. Before we get into details, though, we briefly discuss the reasons why f_{dc} and f_{tc} are different.

H.2.2 Why the direct correlation function does not describe concentration profiles

The direct correlation function gives the likelihood of a state based on its energy in the absence of third-party perturbers. In the absence of three-body interactions, the direct correlation function correctly predicts the likelihood of a state.

However, the presence of other objects changes the interaction between two bodies. Consider a macroscopic example. Consider a ball of mass m at a height Δr acted on by gravity with gravitational acceleration g. The potential energy of the ball is given by $mg\Delta r$, and thus the direct correlation function for the ball is given by $f_{\text{d}}(\Delta r) = \exp(-mg\Delta r/k_{\text{B}}T)$. For such a macroscopic system, one can show that $k_{\text{B}}T$ is tiny compared with $mg\Delta r$, and our result is that a ball acted on solely by gravity has a finite distribution function only at $\Delta r = 0$, i.e., the ball is always located on the ground.

What happens if we have a styrofoam ball submerged in water? Gravity still acts on this system, and the direct correlation function between the ground and the ball is still $f_{\text{d}}(\Delta r) = \exp(-mg\Delta r/k_{\text{B}}T)$. However, we know that the styrofoam ball rises. Macroscopically, we describe this using buoyancy arguments and we use a *mass difference*. Fundamentally, though, the equilibrium position of the styrofoam ball *is not* given by the direct correlation function, owing to the fact that the water gets in the way (in this case, by being more dense). Here, the water acts as a third party to make the statistical distribution of styrofoam ball locations *not* equal to the direct correlation function.

Now consider metal balls of diameter d dropped into a tube with a diameter just barely larger than d. All of the balls have a direct correlation function given by $f_{\text{d}}(\Delta r) = \exp(-mg\Delta r/k_{\text{B}}T)$; however, they cannot fall to the bottom of the tube because the others get in the way. Thus, although the direct correlation function is $f_{\text{d}}(\Delta r) = \exp(-mg\Delta r/k_{\text{B}}T)$, the balls are all located at $\Delta r = d/2$, $\Delta r = 3d/2$, $\Delta r = 5d/2$, etc.

H.2.3 Total correlation functions and the Ornstein–Zernike equation

Determining the total correlation function from the direct correlation function is the key step required for correctly predicting the distribution function. To determine the total correlation function, we must solve the *Ornstein–Zernike* equation:

$$f_{\text{tc}}(\Delta r_{12}) = f_{\text{dc}}(\Delta r_{12}) + n\int_{-\infty}^{\infty} f_{\text{tc}}(\Delta r_{12}) f_{\text{dc}}(\Delta r_{32}) d\Delta\vec{r}_{13}\,, \tag{H.16}$$

where the integration is over all space. Here the total correlation function between atoms 1 and 2 is a function of the distance Δr_{12} between them, as well as the integrated effect of all two-body interactions between atom 1 and atom 3 and between atom 3 and atom 2. This equation essentially says that the total correlation function is given by the direct correlation function adjusted to account for the effect (integrated over all space) of the effect of third parties. For a number density (n) of zero, this says that $f_{tc} = f_{dc}$. As the number density increases, the effects of these third-party interactions increase.

CLOSURE RELATIONS

The Ornstein–Zernike equation is one equation with two unknowns (f_{tc} and f_{dc}) and cannot be solved without an additional equation. Many such *closure relations* have been proposed. These include hypernetted-chain closures (with varying levels of bridge diagrams), the Percus–Yevick closure, and potential of mean force techniques. We discuss only the hypernetted-chain closure, in which we assume that

approximate hypernetted-chain closure relation

$$f_{tc}(\Delta r_{12}) = -1 + \exp\left[-e_1/k_B T + f_{tc}(\Delta r_{12}) - f_{dc}(\Delta r_{12})\right] . \tag{H.17}$$

This relation approximates[5] the total correlation function as the likelihood of the state in isolation plus an additional correction factor that is proportional to the difference between the total and direct correlation functions.

SOLUTIONS OF THE ORNSTEIN–ZERNIKE EQUATION

We typically solve the Ornstein–Zernike equation by Fourier-transforming the Ornstein–Zernike equation and iterating between the Ornstein–Zernike equation (in frequency space) and the closure relation (in physical space). At low temperatures (where attractive forces are important) and high densities (where atomic interactions are prevalent), the distribution function has multiple peaks and valleys (Fig. H.3).

H.3 EXCLUDED VOLUME CALCULATIONS

It often becomes important to calculate the volume taken up by a molecule. This is relevant for modified Poisson–Boltzmann models of ion distributions in EDLs (Chapter 9), as well as excluded volume models to properly predict DNA conformation in nanochannels (Chapter 14).

The *excluded volume* V_{ex} of an atom is given by the integral of the Mayer f function for that atom over all space. For a spherically symmetric atom, this is given by

$$V_{ex} = \int_0^{\infty} f_M \, 4\pi \Delta r^2 d\Delta r . \tag{H.18}$$

This relation can approximate the effective hard-sphere diameter λ_{HS} of an ion or (in cylindrical form) the excluded volume of a DNA polymer.

H.4 ATOMISTIC SIMULATIONS

Although liquid-state theories generate *equilibrium* distributions for interacting atoms, they do not inform our understanding of the dynamics of systems out of equilibrium.

[5] Technically, an infinite number of additional terms should be added inside the exponential; these are called *bridge functions* or *bridge diagrams*. For our purposes, we note only that these functions are usually small and always hard to compute. The hypernetted-chain closure essentially ignores the small, difficult terms, and can be considered a perturbation expansion of the total correlation function around the direct correlation function.

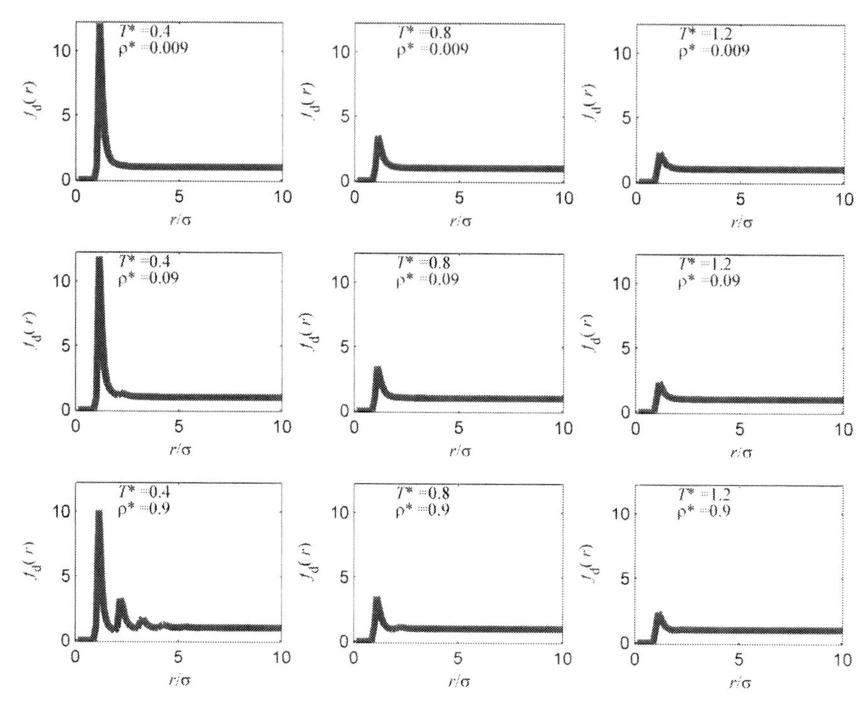

Fig. H.3 Solutions of the Ornstein–Zernike equation for the radial distribution function of Lennard–Jones atoms for several different ρ^* and T^* values. The hypernetted-chain closure in Eq. (H.17) was used for these calculations.

Atomistic simulations use the same interaction potentials to describe the velocities and accelerations of atoms, and this approach is useful when the continuum approximation in a *dynamic* fluid-mechanical system breaks down.

The continuum approximation is a fundamental and powerful aspect of most engineering analysis, simplifying analysis by discussing field properties such as velocity, density, and viscosity. Roughly speaking, we can assess the validity of the continuum approximation by comparing the characteristic dimensions of the flow with the characteristic mean free path of atoms or molecules in the system. Unless the dimensions of the system are 10 or 50 times bigger than the characteristic mean free path of the system, the continuum approximation cannot be used without significant error. For liquid flows, the mean free path of molecules is of the order of 0.2 times the diameter of the molecule. Thus the continuum approximation is quite sound for length scales greater than about ten molecular diameters, approximately 2–5 nm for water. However, some key applications for which continuum modeling does not work well and for which atomistic simulations are preferred include (1) interfacial phenomena such as fluid slip, flow in nanochannels and nanopores, and interactions of macromolecules with these nanoscale structures.

One atomistic simulation approach is molecular dynamics (MD). MD is the technique that most dramatically adds to the tools we have to understand fluid flow at small scales, because it has the potential to overcome the key limitations of continuum mechanics, namely the inability of continuum mechanics to speak to the effects of boundaries in any way other than defining empirical or phenomenological boundary conditions. MD is a deterministic approach and the one best suited to handle situations in which the continuum approximation is invalid.

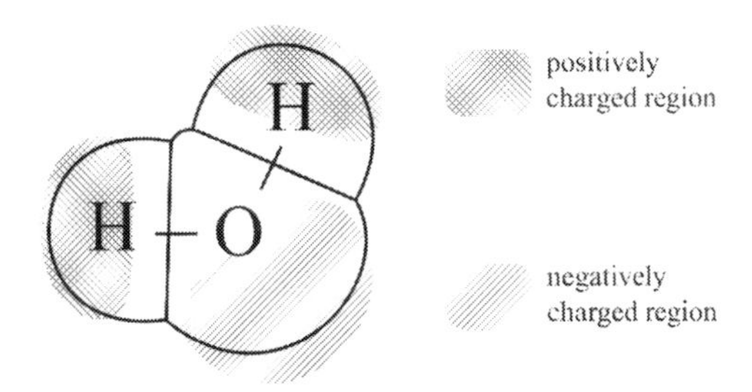

Fig. H.4 A space-filling depiction of a water molecule, with the positively and negatively charged regions specified.

MD is simple in concept but challenging in practice. Every molecule in the system is tracked by defining its potential energy as a function of its environment and integrating Newton's equations of motion in time. MD's limitations and challenges fall into two general categories: (a) defining the potential energies is difficult, especially for water; and (b) the computing power required for tracking every molecule as a function of time is enormous, greatly limiting the systems that can be studied.

H.4.1 Defining atomic forces and accelerations

Defining the potential energy of the system (and therefore the forces and accelerations) is quite complicated. To be precise, the potential energy is a function of the locations of all of the molecules of the system, and because this system is not linear, it cannot simply be expressed as a sum of the potential energy terms of individual molecules.[6] Typically, the energy of the system is defined as being a function of two components, singlet energies and pair potentials. To be mathematically precise, the potential energy of the system consists of an infinite number of components, including singlet potentials, pair potentials, triplet potentials, etc. However, it is intractable to account for so many interactions. MD simulations approximate things by using only singlet and pair potentials, usually adjusting the pair potentials a bit to make the overall system behave as if the triplet and higher order potentials were included.

H.4.2 Water models

The Lennard–Jones potential discussed in Section H.1 has a spherically symmetric potential with an attraction proportional to $\Delta r^{1/6}$ stemming from correlations between dipole fluctuations and a spherically symmetric repulsion stemming from orbital repulsions. The LJ potential is well designed for spherically symmetric, neutral molecules such as liquid argon, for which these are the only two forces.

For water, though, the situation is much more complicated. Water is not spherically symmetric, and the dipole of the OH bonds leads to an effective partial charge on each of the component atoms (Fig. H.4). We can split the water molecule up into several elements (for example, the atoms) and treat the interactions between all of these elements, or we can define a complex interaction potential that describes the potential in terms of the both the relative orientations of and the distance between the molecular centers.

[6] Note a contrast here with our use of mean-field, point-charge Boltzmann statistics in the derivation of the electrical double-layer equations. When using Boltzmann statistics for EDLs in Chapter 9, we modeled ions as point charges in a mean field, which treated all ions as independent from one another and affected only by the mean electrical field. Such an approximation is valid only with sparse ions in a solvent field. Here, we are deriving equations of motion for densely packed atoms, and such an approximation would not be valid.

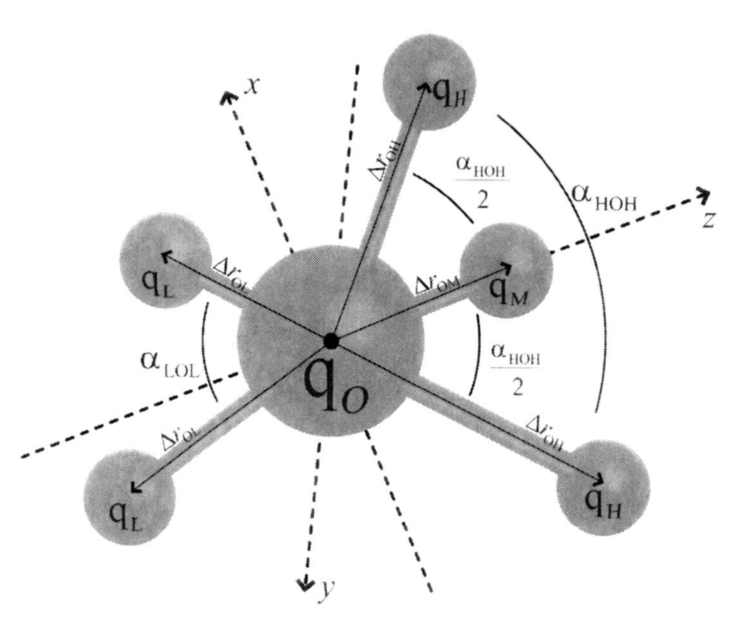

Fig. H.5 A general depiction of the six points used in water models. Most models use only a subsection of these points, combining several points into one.

In both cases, the approach is essentially a pragmatic one. To resolve every detail of water's classical electrostatic interaction is currently not close to possible in any system of physical or engineering significance, and to incorporate the quantum mechanical effects in play is even farther from possible. However, if a simplified molecular mechanics model can be designed with enough detail so that certain crucial components of water's interaction (its equation of state, diffractive measurements of radial distribution function, etc.) are well approximated by a simulation that is computationally tractable, then we can analyze systems of interest with that model and compare the results with experiment. These models all use tradeoffs between computational efficiency and physical accuracy – ignoring certain effects that are assumed small but hard to compute, and focusing on other effects that are assumed large but easier to compute. These models typically have a number of parameters that have been adjusted to best match experimental data, and the adjustment of these parameters, in part, corrects for features that are absent from the model.

MULTIPOINT MODELS

We describe several multipoint models, which combine a Lennard–Jones potential with point charges stemming from partial charges on the oxygen and hydrogen atoms. The multipoint models (usually three-point to six-point models) break the water molecule into up to six points. These locations are denoted with symbols that roughly indicate a chemical part of the molecule; however, in terms of the model, their meaning most precisely refers to a physical location at which we evaluate forces and accelerations:

1. One O: the location of the "oxygen atom" – the location at which we calculate distances for Lennard–Jones potentials.
2. One M: the location of the "center of mass," which is the point used to keep track of where the molecule is in the equation of motion for the molecule.

Table H.1 Water model parameters, focusing on static point charge models.

Parameter	SPC	SPC/E	TIP3P	TIP4P	TIP5P
Δr_{OH}	1 Å	1 Å	0.9572 Å	0.9572 Å	0.9572 Å
α_{HOH}	109.47°	109.47°	104.52°	104.52°	104.52°
Δr_{OL}	n/a	n/a	n/a	n/a	0.7Å
Δr_{OM}	n/a	n/a	n/a	0.15Å	n/a
α_{LOL}	n/a	n/a	n/a	n/a	109.47°
$A\left[\text{nm}\left(\frac{\text{kJ}}{\text{mol}}\right)^{1/6}\right]$	0.37122	0.37122	0.343	0.354	0.321
$B\left[\text{nm}\left(\frac{\text{kJ}}{\text{mol}}\right)^{1/12}\right]$	0.3428	0.3428	0.326	0.334	0.377
q_L	n/a	n/a	n/a	n/a	$-.241e$
q_O	$-0.82e$	$-0.8476e$	$-0.834e$	n/a	n/a
q_M	n/a	n/a	n/a	$-1.04e$	n/a
q_H	$+0.41e$	$+0.4238e$	$+0.417e$	$+0.52e$	$+0.241e$

3. Two H's: the location of the "hydrogen atoms," typically treated as positive point charges.
4. Two L's: the location of "lone pair" electrons, typically treated as negative point charges.

Models with fewer points usually assume that some of these points coincide. For example, five-point models usually treat O and M as the same point, four-point models usually treat the O and the L's as the same point, and three-point models treat O, L's, and M as all occurring at the same point. A general geometry is shown in Fig. H.5. Most models treat the structure of a water molecule as being infinitely rigid, meaning that the bond lengths and angles and the location of electron clouds are not affected by any external perturbation. This assumption greatly reduces the computation time required to simulate the system.

The structure of the orbitals surrounding the oxygen atom in a water molecule is approximately tetrahedral, where the oxygen is at the center of the tetrahedron, the hydrogens are at two of the corners of the tetrahedron, and the two pairs of unbonded electrons are at the other corner. The bond lengths (i.e., the distances from the oxygen atom to the corners of the tetrahedron) are approximately 1 Å. If the molecule were perfectly symmetric (as a methane molecule is), the angle between each of these bonds would be 109.47°, the tetrahedral angle.[7] The multipoint models all choose some geometry (some examples of these geometric specifications for several models are specified in Table H.1) that is approximately tetrahedral, with 3–6 points denoting specific elements of the molecule, and calculate intermolecular pair potentials with an equation of the following form:

$$e_2 = e_2^{\mathrm{LJ}} + e_2^{\mathrm{C}}, \tag{H.19}$$

where e_2^{LJ} is the Lennard–Jones component of the pair potential, which is evaluated as a function of the distance between the oxygen atoms, and e_2^{C} is the Coulomb component of the pair potential, which is evaluated as a sum of all of the Coulomb interactions

[7] Spectroscopic measurements of gas-phase water give the HOH angle as 104.5° and the bond length as 0.95 Å.

between all of the partial charges on the two molecules.[8] Thus the Lennard–Jones component is written as

$$e_2^{\mathrm{LJ}}(\Delta r) = 4\varepsilon_{\mathrm{LJ}}\left[\left(\frac{\Delta r}{\sigma_{\mathrm{LJ}}}\right)^{-12} - \left(\frac{\Delta r}{\sigma_{\mathrm{LJ}}}\right)^{-6}\right] \tag{H.20}$$

or

$$e_2^{\mathrm{LJ}}(\Delta r) = \left[\left(\frac{\Delta r}{B}\right)^{-12} - \left(\frac{\Delta r}{A}\right)^{-6}\right], \tag{H.21}$$

where Δr now specifically denotes the distance between oxygen atoms. Equation (H.21) is a physically equivalent but notationally different specification of the Lennard–Jones potential.

The Coulomb potential felt by one molecule caused by another is written as

$$e_2^{C} = \sum_i \sum_j \frac{1}{4\pi\varepsilon_0} \frac{q_i q_j}{\Delta r_{ij}}, \tag{H.22}$$

where indices $i = 1, 2, 3$ correspond to point charges on the first molecule and indices $j = 1, 2, 3$ correspond to point charges on the second atom. q_i and q_j correspond to the values of these point charges, and Δr_{ij} corresponds to the distance between these point charges.

The force two molecules impart on each other is simply the derivative of the pair potential with respect to their distance Δr, and the torque they impart on each other is the derivative of the pair potential with respect to an orientational coordinate.

Simple point charge (SPC) model. The SPC model is a three-point model that collocates the oxygen atom, the negative charge, and the center of mass. It assumes that the hydrogen atoms each have a charge of $+0.41e$, where e is the magnitude of the electron charge, and the oxygen has a charge of $-0.82e$. The bond lengths are 1 Å and the HOH angle is 109.47°. A modified Lennard–Jones model is used, with $A = 0.37122\ \mathrm{nm}\left(\frac{\mathrm{kJ}}{\mathrm{mol}}\right)^{1/6}$ and $B = 0.3428\ \mathrm{nm}\left(\frac{\mathrm{kJ}}{\mathrm{mol}}\right)^{1/12}$.

Effective simple point charge (SPC/E) model. The SPC/E model is a three-point model that collocates the oxygen atom, the negative charge, and the center of mass. It assumes that the hydrogen atoms each have a charge of $+0.4238e$, where e is the magnitude of the electron charge, and the oxygen has a charge of $-0.8476e$. The bond lengths are 1 Å and the HOH angle is 109.47°. A modified Lennard–Jones model is used, with $A = 0.37122\ \mathrm{nm}\left(\frac{\mathrm{kJ}}{\mathrm{mol}}\right)^{1/6}$ and $B = 0.3428\ \mathrm{nm}\left(\frac{\mathrm{kJ}}{\mathrm{mol}}\right)^{1/12}$.

[8] Such a model essentially treats the pair potential as a perturbation around the basic (spherically symmetric) state, assuming that the primary difference between real water interactions and spherically symmetric water interactions is the existence of isolated partial point charges. Effects of water structure on the spatial distribution of fluctuating dipoles (i.e., spatial variation in the LJ coefficients) is ignored, and only the spatial distribution of static monopoles (i.e., point charges) is retained. The model is then fit to experimental data using parameters such as LJ coefficients, charge magnitudes, bond angles, and bond distances. This approach is taken simply because the computing power required to simulate all effects in their entirety is not currently available.

EXAMPLE PROBLEM H.1

Calculate the magnitude of the dipole moment (in debyes) for a water molecule given the geometry of the SPC/E model.

SOLUTION: The magnitude of the dipole moment is given by the charge multiplied by the distance between charges along the axis of symmetry. For the SPC/E model, the distance in angstroms along the axis of symmetry is given by

$$d = \left(\cos \frac{109.47^\circ}{2}\right), \tag{H.23}$$

which gives $d = 0.577$ Å or $d = 0.577 \times 10^{-10}$ m. The charge magnitude is $0.8476e$ or 1.36×10^{-19} C. Thus the dipole moment is given by $p = 7.83 \times 10^{-30}$ Cm or, because 1 D $= 3.33 \times 10^{-30}$ Cm,

$$p = 2.35 \text{ D} . \tag{H.24}$$

Transferable intermolecular potential three-point model. The TIP3P model is a three-point model that collocates the oxygen atom, the negative charge, and the center of mass. It assumes that the hydrogen atoms each have a charge of $+0.417e$, where e is the magnitude of the electron charge, and the oxygen has a charge of $-0.834e$. The bond lengths are 0.9572 Å and the HOH angle is 104.52°. A Lennard–Jones model is used, with $\sigma_{LJ} = 3.12$ Å and $\varepsilon_{LJ} = 0.636$ kJ/mol.

Transferable intermolecular potential four-point model. The TIP4P model separates the oxygen atom and the center of mass by a distance of 0.15 Å. It assumes that the hydrogen atoms each have a charge of $+0.52e$, where e is the magnitude of the electron charge, and the *center of mass* has a charge of $-1.04e$. The bond lengths are 0.9572 Å and the HOH angle is 104.52°. A Lennard–Jones model is used, with $\sigma_{LJ} = 3.15$ Å and $\varepsilon_{LJ} = 0.649$ kJ/mol.

Transferable intermolecular potential five-point model. The TIP5P model collocates the oxygen atom and the center of mass but treats the negative charge as residing on lone pair electron sites. It assumes that the hydrogen atoms each have a charge of $+0.241e$, where e is the magnitude of the electron charge, and the center of mass has a charge of $-1.04e$. The bond lengths are 0.9572 Å, the HOH angle is 104.52°, the lone pair electron distance is -0.7 Å, and the LOL angle is 109.47°. A Lennard–Jones model is used, with $\sigma_{LJ} = 3.12$ Å and $\varepsilon_{LJ} = 0.669$ kJ/mol.

Model applicability. Applicability of these models varies depending on the problem. These models include only static charge and work best for systems in which the potential energy landscape of the water is relatively homogeneous. These models work relatively well for predicting water properties, given their relative simplicity. The performance of these models degrades for inhomogeneous mixtures, for example, for mixtures

Table H.2 Fundamental scales for nondimensionalizing liquid-state and MD parameters.

Fundamental scale	Value in terms of LJ units
Mass	m
Length	σ_{LJ}
Time	$\sigma_{LJ}\sqrt{m/\varepsilon_{LJ}}$
Temperature	ε_{LJ}/k_B

Note: For electrolyte solutions, the nondimensionalizations are usually based on the properties of the solvent.

of solvents or for solutions with dissolved electrolytes, owing to the larger variation of potential energy landscapes that the water molecules see and the failure of these models to account for atomic or electronic polarization.

H.4.3 Nondimensionalization in MD simulations

Understanding the typical nondimensionalization in MD simulations is critical, because MD papers often do not define their nondimensionalization owing to the fact that it has been standard in the field for years. To an outsider trying to interpret an MD paper without knowledge of these standards, this nondimensionalization can impede interpretation of data. MD simulations are typically normalized by m (mass of one molecule), ε_{LJ} (depth of the attractive energy well), σ_{LJ} (distance at which attractive and repulsive energies cancel), and k_B (Boltzmann constant). These lead to four fundamental scales for mass, length, time, and temperature. Table H.2 shows these fundamental scales, and Table H.3 shows how variables are often nondimensionalized. Further, authors in the atomistic literature often omit the $\frac{1}{4\pi\varepsilon_0}$ from electrostatic energy expressions such as Eq. (H.22) – in those cases, the $\frac{1}{4\pi\varepsilon_0}$ is implied or non-SI units are used so that the permittivity of free space has a value of $1/4\pi$. We explicitly write the $\frac{1}{4\pi\varepsilon_0}$ here to maintain consistent units, for clarity, and to remind the reader that, when treating water atomistically, we resolve all of the electrostatic interactions and thus the permittivity of free space is used, not the permittivity of water.

Table H.3 Some nondimensional units used for MD.

Quantity	Nondimensional quantity definition		
Radial position	r^*	=	r/σ_{LJ}
Density (usually number density)	ρ^*	=	$n\sigma_{LJ}^3/V$
Force	F^*	=	$F\sigma_{LJ}/\varepsilon_{LJ}$
Pressure	p^*	=	$p\sigma_{LJ}^3/\varepsilon_{LJ}$
Temperature	T^*	=	$k_B T/\varepsilon_{LJ}$
Velocity	v^*	=	$v\sqrt{m/\varepsilon_{LJ}}$

Notes: n is the number of molecules, and V is the volume of the studied region. Other nondimensional units can be straightforwardly derived by using the fundamental scales in the text.

H.5 SUMMARY

This appendix summarizes interaction potentials between solvents and solutes. These interaction potentials lead to an integro-differential formulation of solvent properties such as distribution functions, analysis of excluded volume interactions of macromolecules, and atomistic simulation in nanosystems.

The Lennard–Jones potential,

$$e_2(\Delta r_{12}) = 4\varepsilon_{LJ}\left[\left(\frac{\Delta r_{12}}{\sigma_{LJ}}\right)^{-12} - \left(\frac{\Delta r_{12}}{\sigma_{LJ}}\right)^{-6}\right], \tag{H.25}$$

is presented as a typical pair potential for interactions between molecules. Use of these sorts of functions motivates the Mayer f function

$$f_M = -1 + \exp(-e_1/k_B T), \tag{H.26}$$

which describes distribution functions in terms of interaction potentials, and the potential of mean force,

$$e_{mf} = -k_B T \ln f_d, \tag{H.27}$$

which describes interaction potentials in terms of distribution functions. The Mayer f function is shown to be related to the excluded volume around a molecule, and these pair potentials are combined with the Ornstein–Zernike equation

$$f_{tc}(\Delta r_{12}) = f_{dc}(\Delta r_{12}) + n\int_{-\infty}^{\infty} f_{tc}(\Delta r_{12})\, f_{dc}(\Delta r_{32})\, d\bar{r}_{13}, \tag{H.28}$$

combined with a closure relation to predict the equilibrium properties of condensed matter. Nonequilibrium behavior of liquids is predicted by use of atomistic simulations, which integrate Newton's second law at the molecular level. For water, which is highly polar and cannot be modeled with only a Lennard–Jones interaction potential, a number of models are presented that account for the distribution of charge and more accurately simulate molecular interactions.

H.6 SUPPLEMENTARY READING

This chapter gives some preliminary information about interaction potentials and their relation to atomistic simulations, excluded volume descriptions of polymers, and liquid-state theory descriptions; however, the treatment is sparse and does not constitute preparation to pursue these endeavors in detail. For those interested in atomistic simulations of liquids, the texts of Haile [298] and Karniadakis et al. [3] each provide an accessible introduction. Haile, in particular [298], has a concise description of nondimensionalization in MD simulations. Allen and Tildesley [299] is a standard reference text and is more well suited for those engaged in MD research.

This appendix omits discussion of direct simulation Monte Carlo techniques, lattice-Boltzmann techniques, and Brownian dynamics approaches. Boltzmann equation techniques treat statistical representations of properties (specifically, atomic or molecular motions) and derive transport equations for the distribution functions for these properties. Lattice-Boltzmann techniques are designed specifically to simplify and approximate Boltzmann techniques by using a lattice representation. Brownian dynamics is a means for simplifying molecular dynamics simulations by coarse-graining the system.

For those interested specifically in applications of molecular dynamics simulations to EDLs, some recent work includes Refs. [120, 121, 300, 301, 302]. For those interested in integro-differential approaches using Ornstein–Zernike formulations, Attard [123, 303] has presented detailed work in this area. Rubinstein and Colby [36] discuss interaction potentials in the context of excluded volume effects for polymers.

The description of water models here is restricted to static charge models and ignores charge-on-spring models, inducible dipole models, and fluctuating charge models, as well as any aspect of quantum treatment of the water molecule. Sources that discuss water models, a description of water properties, or the relation between these include [304, 305, 306, 307].

H.7 EXERCISES

H.1 Consider a pair potential given by

$$e_2 = \infty \quad \text{if} \quad r < a \tag{H.29}$$

$$e_2 = k_B T \quad \text{if} \quad a < r < 2a \tag{H.30}$$

$$e_2 = 0 \quad \text{if} \quad 2a < r \tag{H.31}$$

Calculate and plot the Mayer f function f_M for this potential.

H.2 Write a numerical routine to solve the Ornstein–Zernike equation with hypernetted-chain closure to find the radial distribution function for a homogeneous Lennard–Jones fluid.

Proceed as follows:

(a) Use an iterative technique that, in turn, uses the hypernetted-chain closure in Eq. (H.17) to solve for f_{tc} and the Ornstein–Zernike equation (H.16) to solve for f_{dc}.

(b) Start by setting $f_{tc}(r) = f_{dc}(r) = 0$ on a domain that ranges from $r = 0$ to $r = 512\sigma$.

(c) In each step, define a new f_{tc} by using the hypernetted-chain relation:

$$f_{tc,new}(r) = -1 + \exp\left[-e_1(r)/k_B T + f_{tc,old}(r) - f_{dc}(r)\right] . \tag{H.32}$$

Note that $e_1(r)$ in this case is the Lennard–Jones potential.

(d) In each step, define a new $f_{\rm dc}$ by Fourier-transforming $f_{\rm tc}$ and $f_{\rm dc}$, applying the Fourier-transformed Ornstein–Zernike equation to get a new $\hat{f}_{\rm dc}$, and inverse Fourier-transforming $\hat{f}_{\rm dc}$ to get a new $f_{\rm dc}$. We do this because the Fourier-transformed Ornstein–Zernike equation is much easier to deal with (the spatial integral becomes a product when Fourier transformed):

$$\hat{f}_{\rm tc}(k) = \hat{f}_{\rm dc}(k) + \rho\, \hat{f}_{\rm dc}(k)\, \hat{f}_{\rm tc}(k)\,. \tag{H.33}$$

Here k is the frequency variable and $\hat{f}$ is the Fourier transform of f. This can be rearranged to give

$$\hat{f}_{\rm dc,new}(k) = \frac{\hat{f}_{\rm tc,new}(k)}{1 + \rho\, \hat{f}_{\rm tc,new}(k)} \tag{H.34}$$

So we Fourier-transform $f_{\rm tc}$ and $f_{\rm dc}$ to get $\hat{f}_{\rm tc}$ and $\hat{f}_{\rm dc}$, apply Eq. (H.34), and then transform $\hat{f}_{\rm dc}$ back.[9]

(e) The two previous steps are repeated until the solutions for $f_{\rm tc}$ and $f_{\rm dc}$ are no longer changing. Some attention to numerical stability is needed, especially if ρ^* is high and T^* is low.

Plot your results for nine cases as follows: three values of ρ^* (0.1, 0.4, 0.8) and three values of T^* (0.5, 1.0, 1.5).

H.3 Given the results from Exercise H.2, calculate the potential of mean force that atoms see in this case. Plot the $e_{\rm mf}$ for the nine cases from Exercise H.2.

H.4 Calculate the magnitude of the dipole moment (in debyes) for a water molecule given the geometry of the SPC model.

H.5 Consider hard spheres of radius a.

(a) What is the closest approach of the centers of two spheres?

(b) What is the doublet potential for the interaction between the two spheres?

(c) Calculate the excluded volume. How does the excluded volume compare with the volume of one of the spheres?

H.6 Show that the nonintegrability of the Coulomb pair potential guarantees that physical systems must be overall electroneutral.

H.7 Using the equation for monopole interaction potentials, explain why sodium chloride might be expected to be a crystalline solid when dry but dissolves when exposed to water.

---

[9] Note that one must be careful about how one transforms $f_{\rm tc}$ and $f_{\rm dc}$. The FFT and IFFT algorithms in this case should be symmetric with regards to the number of points. That is, for Eq. (H.34) to apply, the definition of the FFT must be

$$\hat{f}(k) = N^{-1/2} \sum_{n=1}^{N} f(n) \exp(2\pi i/N)^{(n-1)(k-1)}\,, \tag{H.35}$$

and the IFFT must be

$$f(n) = N^{-1/2} \sum_{k=1}^{N} \hat{f}(k) \exp(2\pi i/N)^{-(n-1)(k-1)}\,. \tag{H.36}$$

Many software packages define the fast Fourier transform function slightly differently.

Bibliography

[1] Becker, F. F., Wang, X. B., Huang, Y., Pethig, R., Vykoukal, J., and Gascoyne, P. R. C. *Journal of Physics D-Applied Physics* **27**(12), 2659–2662 Dec 14 (1994).

[2] Yasukawa, T., Suzuki, M., Sekiya, T., Shiku, H., and Matsue, T. *Biosensors & Bioelectronics* **22**(11), 2730–2736 May 15 (2007).

[3] Karniadakis, G., Beskok, A., and Aluru, N. *Microflows and Nanoflows*. Springer, (2005).

[4] Madou, M. *Fundamentals of Microfabrication: The Science of Miniaturization*. CRC, (2002).

[5] Brodie, I. and Murray, J. *The Physics of Microfabrication*. Springer, (1982).

[6] Tabeling, P. *Introduction to Microfluidics*. Oxford, (2005).

[7] Nguyen, N.-T. and Wereley, S. *Fundamentals of Microfluidics*. Artech House, (2006).

[8] Peyret, R. and Taylor, T. *Computational Methods for Fluid Flow*. Springer-Verlag, (1983).

[9] Hirsch, C. *Numerical Computation of Internal and External Flows*. Wiley, (1988).

[10] Hirsch, C. *Numerical Computation of Internal and External Flows Volume 2: Computational Methods for Inviscid and Viscous Flows*. Wiley, (1990).

[11] Fletcher, C. *Computational Techniques for Fluid Dynamics, volumes 1 & 2*. Springer, (1991).

[12] Greibel, M., Dornsheifer, T., and Neunhoeffer, T. *Numerical Simulation in Fluid Dynamics: A Practical Introduction*. SIAM, (1998).

[13] Ferziger, J. and Peric, M. *Computational Methods for Fluid Dynamics*. Springer, (2001).

[14] Tu, J., Yeoh, G., and Liu, C. *Computational Fluid Dynamics: A Practical Approach*. Butterworth-Heinemann, (2007).

[15] Prosperetti, A. and Tryggvason, G. *Computational Methods for Multiphase Flow*. Cambridge University Press, (2007).

[16] Anna, S. and Mayer, H. *Physics of Fluids* **18**, 121512 (2006).

[17] Fox, R., Pritchard, P., and McDonald, A. *Introduction to Fluid Mechanics*. Wiley, (2008).

[18] Munson, B., Young, D., and Okiishi, T. *Fundamentals of Fluid Mechanics*. Wiley, (2006).

[19] White, F. *Fluid Mechanics*. Wiley, (2006).

[20] Bird, R., Stewart, W., and Lightfoot, E. *Transport Phenomena*. Wiley, (2006).

[21] Panton, R. *Incompressible Flow*. Wiley, (2005).

[22] White, F. *Viscous Fluid Flow*. McGraw-Hill, (2005).

[23] Kundu, P. and Cohen, I. *Fluid Mechanics*. Academic Press, (2008).

[24] Batchelor, G. *Introduction to Fluid Dynamics*. Cambridge University Press, (2000).

[25] Born, M. and Green, H. *A General Kinetic Theory of Liquids*. University Press, (1949).

[26] Frenkel, Y. *Kinetic Theory of Liquids*. Dover, (1955).

[27] Lauga, E. and Stone, H. in *Springer Handbook of Experimental Fluid Mechanics*, Microfluidics: The no-slip boundary condition. Springer (2007).

[28] Edwards, D., Brenner, H., and Wasan, D. *Interfacial Transport Processes and Rheology*. Butterworth-Heinemann, (1991).

[29] Probstein, R. *Physicochemical Hydrodynamics*. Wiley, (1994).

[30] Leal, L. *Advanced Transport Processes: Fluid Mechanics and Convective Transport Processes*. Cambridge University Press, (2007).

[31] Chen, T., Chiu, M.-S., and Wen, C.-N. *Journal of Applied Physics* **100**, 074308 (2006).

[32] Russel, W., Saville, D., and Schowalter, W. *Colloidal Dispersions*. Cambridge University Press, (1989).

[33] Bruus, H. *Theoretical Microfluidics*. Oxford, (2007).
[34] Van Dyke, M. *Perturbation Methods in Fluid Mechanics*. Parabolic Press, (1964).
[35] Stroock, A., Dertinger, S., Ajdari, A., Mezic, I., Stone, H., and Whitesides, G. *Science* **295**, 647–651 (2002).
[36] Rubinstein, M. and Colby, R. *Polymer Physics*. Oxford, (2003).
[37] Taylor, R. and Krishna, R. *Multicomponent Mass Transfer*. Wiley, (1993).
[38] Ottino, J. *The Kinematics of Mixing: Stretching, Chaos, and Transport*. Cambridge University Press, (1989).
[39] Strogatz, S. *Nonlinear Dynamics and Chaos: With Applications to Physics, Biology, Chemistry, and Engineering*. Westview, (2001).
[40] Ottino, J. and Wiggins, S. *Philosophical Transactions A: Mathematics, Physics, Engineering, and Science* **362**, 923–35 (2004).
[41] Nguyen, N.-T. and Wu, Z. *Journal of Micromechanics and Microengineering* **15**, R1–R16 (2005).
[42] Hessel, V., Lowe, H., and Schonfeld, F. *Chemical Engineering Science* **60**, 2479–2501 (2005).
[43] Song, H., Bringer, M., Tice, J., Gerdts, C., and Ismagilov, R. *Applied Physics Letters* **83**, 4664–4666 (2003).
[44] Simonnet, C. and Groisman, A. *Physical Review Letters* **94**, 134501 (2005).
[45] Takayama, S., Ostuni, E., LeDuc, P., Naruse, K., Ingber, D., and Whitesides, G. *Nature (London)* **411**, 1016 (2001).
[46] Beebe, D., Moore, J., Yu, Q., Liu, R., Kraft, M., Jo, B.-H., and Devadoss, C. *Proceedings of the National Academy of Science of the United States of America* **97**, 13488–13493 (2000).
[47] Raynal, F., Plaza, F., Beuf, A., and Carriere, P. *Physics of Fluids* **16**, L63–L66 (2004).
[48] McQuain, M., Seale, K., Peek, J., Fisher, T., Levy, S., Stremler, M., and Haselton, F. *Analytical Biochemistry* **325**, 215–226 (2004).
[49] Wei, C.-W., Cheng, J.-Y., Huang, C.-T., Yen, M.-H., and Young, T.-H. *Nucleic Acids Research* **33**, e78 (2005).
[50] Stremler, M. and Cola, B. *Physics of Fluids* **18**, 011701 (2006).
[51] Hertzsch, J.-M., Struman, R., and Wiggins, S. *Small* **3**, 202–218 (2007).
[52] Chang, H.-C. and Yeo, L. *Electrokinetically Driven Microfluidics and Nanofluidics*. Cambridge University Press, (2010).
[53] Stone, H. *Physics of Fluids A* **1**, 1112–1122 (1989).
[54] Ismagilov, R., Stroock, A., Kenis, P., Whitesides, G., and Stone, H. *Applied Physics Letters* **76**, 2376–2378 (2000).
[55] Griffiths, D. J. *Introduction to Electrodynamics*. Prentice-Hall, 3rd edition, (1981).
[56] Jackson, J. *Classical Electrodynamics*. Wiley, 3rd edition, (1999).
[57] Haus, H. and Melcher, J. *Electromagnetic Fields and Energy*. Prentice Hall, (1989).
[58] Bockris, J. and Reddy, A. *Modern Electrochemistry*. Plenum, (1970).
[59] Bard, A. and Faulkner, L. *Electrochemical Methods*. Wiley, (1980).
[60] Morgan, H. and Green, N. *AC Electrokinetics: Colloids and Nanoparticles*. Research Studies Press, (2002).
[61] Jones, T. B. *Electromechanics of Particles*. Cambridge University Press, (1995).
[62] Pethig, R. *Dielectric and Electric Properties of Biological Materials*. Wiley, (1979).
[63] Smith, R. and Dorf, R. *Circuits, Devices, and Systems: A First Course in Electrical Engineering, 5th edition*. Wiley, (1991).
[64] Oesterle, J. *Journal of Applied Mechanics* **31**, 161–164 (1964).
[65] Zeng, S., Chen, C., Mikkelsen, J., and Santiago, J. *Sensors and Actuators B* **79**, 107–114 (2001).
[66] Reichmuth, D., Chirica, G., and Kirby, B. *Sensors and Actuators B* **92**, 37–43 (2003).
[67] Reichmuth, D. and Kirby, B. *Journal of Chromatography A* **1013**, 93–101 (2003).
[68] Levich, V. *Physicochemical Hydrodynamics*. Prentice Hall, (1962).
[69] Hunter, R. *Zeta Potential in Colloid Science*. Academic, (1981).
[70] Hunter, R. J. *Introduction to Modern Colloid Science*. Oxford, (1994).

[71] Lyklema, J. *Fundamentals of Interface and Colloid Science: Volume II: Solid-Liquid Interfaces*. Academic Press, (1995).

[72] Li, D. *Electrokinetics in Microfluidics*. Elsevier, (2004).

[73] Overbeek, J. *Colloid Science, Volume I: Irreversible Systems*. Elsevier (1952).

[74] Cummings, E., Griffiths, S., Nilson, R., and Paul, P. *Analytical Chemistry* **72**, 2526–2532 (2000).

[75] Santiago, J. *Analytical Chemistry* **73**, 2353–2365 (2001).

[76] Oh, J. and Kang, K. *Journal of Colloid and Interface Science* **310**, 607–616 (2007).

[77] Santiago, J. *Journal of Colloid and Interface Science* **310**, 675–677 (2007).

[78] Soderman, O. and Jonsson, B. *Journal of Chemical Physics* **105**, 10300–10311 (1996).

[79] Herr, A., Molho, J., Santiago, J., Mungal, M., Kenny, T., and Garguilo, M. *Analytical Chemistry* **72**, 1053–1057 (2000).

[80] Min, J., Hasselbrink, E., and Kim, S. *Sensors and Actuators B* **98**, 368–377 (2004).

[81] Griffiths, S. and Nilson, R. *Electrophoresis* **26**, 351–361 (2005).

[82] Griffiths, S. and Nilson, R. *Analytical Chemistry* **73**, 272–278 (2001).

[83] Molho, J., Herr, A., Mosier, B., Santiago, J., Kenny, T., Brennen, R., Gordon, G., and Mohammadi, B. *Analytical Chemistry* **73**, 1350–1360 (2001).

[84] Currie, I. *Fundamental Mechanics of Fluids*. Marcel Dekker, (2002).

[85] Kuethe, A., Schetzer, J., and Chow, C.-Y. *Foundations of Aerodynamics*. Wiley, (1987).

[86] Anderson, J. *Fundamentals of Aerodynamics*. McGraw-Hill, (2006).

[87] Chapra, S. and Canale, R. *Numerical Methods for Engineers: With Software and Programming Applications*. McGraw-Hill, (2001).

[88] Happel, J. and Brenner, H. *Low Reynolds Number Hydrodynamics: With Special Applications To Particulate Media*. Kluwer, (1983).

[89] Kim, S. and Karrila, S. *Microhydrodynamics: Principles and Selected Applications*. Dover, (2005).

[90] Doi, M. and Edwards, S. *The Theory of Polymer Dynamics*. Oxford, (1986).

[91] Onsager, L. *Physical Review* **37**, 405–426 (1931).

[92] Onsager, L. *Physical Review* **38**, 2265–2279 (1931).

[93] Batchelor, G. *Journal of Fluid Mechanics* **74**, 1–29 (1976).

[94] Chwang, A. and Wu, T. *Journal of Fluid Mechanics* **75**, 677–689 June (1976).

[95] Brenner, H. *Chemical Engineering Science* **16**, 242–251 (1961).

[96] Goldman, A., Cox, R., and Brenner, H. *Chemical Engineering Science* **21**, 1151–1170 (1966).

[97] Green, N. G. and Jones, T. B. *Journal of Physics D-Applied Physics* **40**(1), 78–85 Jan 7 (2007).

[98] Raffel, M. *Particle-Image Velocimetry: A Practical Guide*. Springer, (1998).

[99] Beatus, T., Tlusty, T., and Bar-Ziv, R. *Nature Physics* **2**, 743–748 (2006).

[100] Segre, G. and Silverberg, A. *Journal of Fluid Mechanics* **14**, 115–136 (1962).

[101] Segre, G. and Silverberg, A. *Journal of Fluid Mechanics* **14**, 137–157 (1962).

[102] Gouy, M. *Journal de Physique* **9**, 457–468 (1910).

[103] Chapman, D. *Philosophical Magazine* **25**, 475–481 (1913).

[104] Israelachvili, J. *Intermolecular and Surface Forces*. Academic, (1992).

[105] Anderson, J. *Annual Review of Fluid Mechanics* **21**, 61–99 (1989).

[106] Lyklema, J. and Overbeek, J. *Journal of Colloid Science* **16**, 501–512 (1961).

[107] Lyklema, J. *Colloids and Surfaces A* **92**, 41–49 (1994).

[108] Stern, O. *Z. Electrochemical* **30**, 508–516 (1924).

[109] Bikerman, J. *Philosophical Magazine* **33**, 384 (1942).

[110] Borukhov, I., Andelman, D., and Orland, H. *Physical Review Letters* **79**, 435 (1997).

[111] Kilic, M., Bazant, M., and Ajdari, A. *Physical Review E* **75**, 021502 (2007).

[112] Kilic, M., Bazant, M., and Ajdari, A. *Physical Review E* **75**, 021503 (2007).

[113] Carnahan, N. and Starling, K. *Journal of Chemical Physics* **51**, 635 (1969).

[114] Boublik, T. *Journal of Chemical Physics* **53**, 471 (1970).

[115] Mansoori, G., Carnahan, N., Starling, K., and Leland, T. *Journal of Chemical Physics* **54**, 1523 (1971).

[116] Hansen, J.-P. and McDonald, J. *Theory of Simple Liquids*. Academic, (1986).
[117] di Caprio, D., Borkowska, Z., and Stafiej, J. *Journal of Electroanalytical Chemistry* **572**, 51–59 (2004).
[118] Chakraborty, S. *Physical Review Letters* **100**, 09801 (2008).
[119] Marcelja, S. *Langmuir* **16**, 6081–6083 (2000).
[120] Qiao, R. and Aluru, N. *Journal of Chemical Physics* **118**, 4692–4701 (2003).
[121] Joly, L., Ybert, C., Trizac, E., and Bocquet, L. *Physical Review Letters* **93**, 257805 (2004).
[122] Vlachy, V. *Annual Review of Physical Chemistry* **50**, 145–165 (1990).
[123] Attard, P. *Advances in Chemical Physics* **92**, 1–159 (1996).
[124] Biesheuvel, P. and van Soestbergen, M. *Journal of Colloid and Interface Science* **316**, 490–499 (2007).
[125] Grochowski, P. and Trylska, J. *Biopolymers* **89**, 93–113 (2008).
[126] Kirby, B. and Hasselbrink, E. *Electrophoresis* **25**, 187–202 (2004).
[127] Gaudin, A. and Fursteneau, D. *Trans. ASME* **202**, 66–72 (1955).
[128] Atamna, I., Issaq, H., Muschik, G., and Janini, G. *Journal of Chromatography* **559**, 69–80 (1991).
[129] Scales, P., Greiser, F., and Healy, T. *Langmuir* **8**, 965–974 (1992).
[130] Kosmulski, M. and Matijevic, E. *Langmuir* **8**, 1060–1064 (1992).
[131] Caslavska, J. and Thormann, W. *Journal of Microcolumn Separations* **13**, 69–83 (2001).
[132] Kirby, B. and Hasselbrink, E. *Electrophoresis* **25**, 203–213 (2004).
[133] Tandon, V., Bhagavatula, S., Nelson, W., and Kirby, B. *Electrophoresis* **29**, 1092–1101 (2008).
[134] Tandon, V. and Kirby, B. *Electrophoresis* **29**, 1102–1114 (2008).
[135] Delgado, A., Gonazalez-Caballero, F., Hunter, R., Koopal, L., and Lyklema, J. *Pure and Applied Chemistry* **77**, 1753–1805 (2005).
[136] Kirby, B., Wheeler, A., Zare, R., Freutel, J., and Shepodd, T. *Lab on a Chip* **3**, 5–10 (2003).
[137] Grodzinsky, A. *Fields, Forces, and Flows in Biological Systems*. Garland Science, (2008).
[138] Atkinson, G. *Institute of Physics Handbook, 3rd edition*, Electrochemical information. (1972).
[139] Paul, P., Garguillo, M., and Rakestraw, D. *Analytical Chemistry* **70**, 2459–2467 (1998).
[140] Devasenathipathy, S. and Santiago, J. *Microscale Diagnostic Techniques*, chapter Electrokinetic Flow Diagnostics, 113–154. Springer (2005).
[141] Mosier, B., Molho, J., and Santiago, J. *Experiments in Fluids* **33**, 545 (2002).
[142] Lempert, W., Ronney, P., Magee, K., Gee, K., and Haugland, R. *Experiments in Fluids* **18**, 249–257 (1995).
[143] Dahm, W., Su, L., and Southerland, K. *Physics of Fluids A* **4**, 2191–2206 (1992).
[144] Ramsey, J., Jacobson, S., Culbertson, C., and Ramsey, J. *Analytical Chemistry* **75**, 3758–3764 (2003).
[145] Shadpour, H., Hupert, M., Patterson, D., Liu, C., Galloway, M., Stryjewski, W., Goettert, J., and Soper, S. *Analytical Chemistry* **79**, 870–878 (2007).
[146] Paegel, B., Hutt, L., Simpson, P., and Mathies, R. *Analytical Chemistry* **72**, 3030–3037 (2000).
[147] Fiechtner, G. and Cummings, E. *Analytical Chemistry* **75**, 4747–4755 (2003).
[148] Shackman, J., Munson, M., and Ross, D. *Analytical Chemistry* **79**, 565 (2007).
[149] Reichmuth, D., Shepodd, T., and Kirby, B. *Analytical Chemistry* **77**, 2997–3000 (2005).
[150] Skoog, D., Holler, F., and Crouch, S. *Principles of Instrumental Analysis*. Brooks Cole, (2006).
[151] Giddings, J. *Unified Separation Science*. Wiley, (1991).
[152] Righetti, P. *Isoelectric Focusing – Theory, Methodology and Applications*. Elsevier, (1983).
[153] Shadpour, H. and Soper, S. *Analytical Chemistry* **78**, 3519–3527 (2006).
[154] Herr, A., Molho, J., Drouvalakis, K., Santiago, J., and Kenny, T. *Analytical Chemistry* **75**, 1180–1187 (2003).
[155] Wang, Y., Choi, M., and Han, J. *Analytical Chemistry* **76**, 4426–4431 (2004).
[156] Gottschlich, N., Jacobson, S., Culbertson, C., and Ramsey, J. *Analytical Chemistry* **73**, 2669–2674 (2001).

[157] Kirby, B., Reichmuth, D., Renzi, R., Shepodd, T., and Wiedenman, B. *Lab on a Chip* (2004).
[158] Ottewill, R. and Shaw, J. *Journal of Electroanalytical Chemistry* **37**, 133 (1972).
[159] O'Brien, R. and White, L. *Journal of the Chemical Society, Faraday Transactions* **2**, 1607–1626 (1978).
[160] Ohshima, H., Healy, T., and White, L. *Journal of the Chemical Society, Faraday Transactions 2* **79**, 1613 (1983).
[161] Henry, D. *Proceedings of the Royal Society Series A* **133**, 106–129 (1931).
[162] Henry, D. *Transactions of the Faraday Society* **44**, 1021–1026 (1948).
[163] Booth, F. *Transactions of the Faraday Society* **44**, 955–959 (1948).
[164] Booth, F. *Proceedings of the Royal Society Series A* **203**, 514–533 (1950).
[165] Wiersema, P., Loeb, A., and Overbeek, J. *Journal of Colloid and Interface Science* **23**, 78–99 (1966).
[166] Ohshima, H., Healy, T., White, L., and O'Brien, R. *Journal of the Chemical Society, Faraday Transactions 2: Molecular and Chemical Physics* **80**, 1299–1317 (1984).
[167] Dukhin, S. *Advances in Colloid and Interface Science* **61**, 17–49 (1995).
[168] Ohshima, H., Healy, T., and White, L. *Journal of the Chemical Society, Faraday Transactions 2: Molecular and Chemical Physics* **80**, 1643–1667 (1984).
[169] Strychalski, E., Stavis, S., and Craighead, H. *Nanotechnology* **19**, 315301 (2008).
[170] Balducci, A., Mao, P., Han, J., and Doyle, P. *Macromolecules* **39**, 6273–6281 (2006).
[171] Strychalski, E., Levy, S., and Craighead, H. *Macromolecules* **41**, 7716–7721 (2008).
[172] Ott, A., Magnasco, M., Simon, A., and Libchaber, A. *Physical Review E* **48**, R1642 (1993).
[173] Nkodo, A., Garnier, J., Tinland, B., Ren, H., Desruisseaux, C., McCormick, L., Drouin, G., and Slater, G. *Electrophoresis* **22**, 2424–2432 (2001).
[174] Stellwagen, N., Gelfi, C., and Righetti, P. *Biopolymers* **42**, 687–703 (1997).
[175] Stellwagen, E. and Stellwagen, N. *Electrophoresis* **23**, 1935–1941 (2002).
[176] de Gennes, P. *Scaling Concepts in Polymer Physics*. Cornell University Press, (1979).
[177] Flory, P. *Principles of Polymer Chemistry*. Cornell University Press, (1971).
[178] Han, J. and Craighead, H. *Science* **288**, 1026–1029 (2000).
[179] Viovy, J. *Reviews of Modern Physics* **72**, 813–872 (2000).
[180] Ugaz, V. and Burns, M. *Philosophical Transactions of the Royal Society of London Series A-Mathematical Physical and Engineering Sciences* **362**, 1105–1129 (2004).
[181] Fu, J., Yoo, J., and Han, J. *Physical Review Letters* **97**, 018103 (2006).
[182] Huang, L., Tegenfeldt, J., Kraeft, J., Sturm, J., Austin, R., and Cox, E. *Nature Biotechnology* **20**, 1048–1051 (2002).
[183] Paegel, B., Blazej, R., and Mathies, R. *Current Opinion in Biotechnology* **14**, 42–50 (2003).
[184] Doi, M. *Introduction to Polymer Physics*. Oxford, (2001).
[185] Reccius, C., Mannion, J., Cross, J., and Craighead, H. *Physical Review Letters* **95**, 268101 (2005).
[186] Reisner, W., Beech, J., Larsen, N., Flyvbjerg, H., Kristensen, A., and Tegenfeldt, J. *Physical Review Letters* **99**, 058302 (2007).
[187] Reisner, W., Morton, K., Riehn, R., Wang, Y., Yu, Z., Rosen, M., Sturm, J., Chou, S., Frey, E., and Austin, R. *Physical Review Letters* **94**, 196101 (2005).
[188] Odijk, T. *Physical Review E* **77**, 060901 (2008).
[189] Bonthuis, D., Meyer, C., Stein, D., and Dekker, C. *Physical Review Letters* **101**, 108303 (2008).
[190] Edel, J. and de Mello, A., editors. *Nanofluidics: Nanoscience and Nanotechnology*. Royal Society of Chemistry, (2009).
[191] Lehninger, A., Nelson, D., and Cox, M. *Principles of Biochemistry*. Freeman, (2008).
[192] Schoch, R., Han, J., and Renaud, P. *Reviews of Modern Physics* **80**, 839–883 (2008).
[193] Rubinstein, I. and Zaltzman, B. *Physical Review E* **62**, 2238–2251 (2000).
[194] Pennathur, S. and Santiago, J. *Analytical Chemistry* **77**, 6772–6781 (2005).
[195] Pennathur, S. and Santiago, J. *Analytical Chemistry* **77**, 6782–6789 (2005).
[196] Baldessari, F. and Santiago, J. *Journal of Nanobiotechnology* **4**, 12 (2006).
[197] Karnik, R., Cuan, C., Castelino, K., Daiguji, H., and Majumdar, A. *Nano Letters* **7**, 547–551 (2007).

[198] Outhwaite, C. and Bhuiyan, L. *Journal of the Chemical Society–Faraday Transactions* **76**, 1388–1408 (1980).
[199] Outhwaite, C. and Bhuiyan, L. *Journal of the Chemical Society–Faraday Transactions* **78**, 707–718 (1983).
[200] Bhuiyan, L. and Outhwaite, C. *Physical Chemistry Chemical Physics* **6**, 3467–3473 (2004).
[201] Liu, Y., Liu, M., Lau, W., and Yang, J. *Langmuir* **24**, 2884–2891 (2008).
[202] Bazant, M. and Squires, T. *Physical Review Letters* **92**, 066101 (2004).
[203] Green, N., Ramos, A., Gonzalez, A., and Morgan, H. *Physical Review E* **66**, 026305 (2002).
[204] Friese, V. *Zeitschrift fur Electrochemie* **56**, 822–827 (1952).
[205] Squires, T. and Bazant, M. *Journal of Fluid Mechanics* **509**, 217–252 (2004).
[206] Chu, K. and Bazant, M. *Physical Review E* **74**, 011501 (2006).
[207] Bazant, M., Thornton, K., and Ajdari, A. *Physical Review E* **70**, 021506 (2004).
[208] Kilic, M. and Bazant, M. *Physical Review E* **75**, 021503 (2007).
[209] Green, N. G., Ramos, A., Gonzalez, A., Castellanos, A., and Morgan, H. *Journal of Physics D-Applied Physics* **33**(2), L13–L17 Jan 21 (2000).
[210] Ramos, A., Morgan, H., Green, N., and Castellanos, A. *Journal of Colloid and Interface Science* **217** (1000).
[211] Gonzales, A., Ramos, A., Green, N., Castellanos, A., and Morgan, H. *Physical Review E* **61**, 4019 (2000).
[212] Brown, A., Smith, C., and Rennie, A. *Physical Review E* **63**, 016305 (2000).
[213] Ramos, A., Morgan, H., Green, N., and Gonzalez, A. *Journal of Applied Physics* **97**, 084906 (2005).
[214] di Caprio, D., Borkowska, A., and Stafiej, J. *Journal of Electroanalytical Chemistry* **540**, 17–23 (2003).
[215] Urdaneta, M. and Smela, E. *Electrophoresis* **28**(18), 3145–3155 Sep (2007).
[216] James, C. D., Okandan, M., Galambos, P., Mani, S. S., Bennett, D., Khusid, B., and Acrivos, A. *Journal of Fluids Engineering-Transactions of the Asme* **128**(1), 14–19 Jan (2006).
[217] Voldman, J. *Annual Review of Biomedical Engineering* **8**, 425–454 (2006).
[218] Cummings, E. B. *IEEE Engineering in Medicine and Biology Magazine* **22**(6), 75–84 Nov-Dec (2003).
[219] Hawkins, B. G., Smith, A. E., Syed, Y. A., and Kirby, B. J. *Analytical Chemistry* **79**(19), 7291–7300 Oct 1 (2007).
[220] Srinivasan, V., Pamula, V., and Fair, R. *Lab on a Chip* **4**, 310–315 (2004).
[221] Moon, H., Cho, S., Garrell, R., and Kim, C. *Journal of Applied Physics* **92**, 4080–4087 (2002).
[222] Pohl, H. A. *Dielectrophoresis: The behavior of neutral matter in nonuniform electric fields*. Cambridge University Press, (1978).
[223] Wang, X. B., Huang, Y., Becker, F. F., and Gascoyne, P. R. C. *Journal of Physics D-Applied Physics* **27**(7), 1571–1574 Jul 14 (1994).
[224] Wang, X. J., Wang, X. B., and Gascoyne, P. R. C. *Journal of Electrostatics* **39**(4), 277–295 Aug (1997).
[225] Kang, K. H. and Li, D. Q. *Journal of Colloid and Interface Science* **286**(2), 792–806 Jun 15 (2005).
[226] Liu, H. and Bau, H. H. *Physics of Fluids* **16**(5), 1217–1228 May (2004).
[227] Rosales, C. and Lim, K. M. *Electrophoresis* **26**(11), 2057–2065 Jun (2005).
[228] Al-Jarro, A., Paul, J., Thomas, D. W. P., Crowe, J., Sawyer, N., Rose, F. R. A., and Shakesheff, K. M. *Journal of Physics D-Applied Physics* **40**(1), 71–77 Jan 7 (2007).
[229] Jones, T. B., Wang, K. L., and Yao, D. J. *Langmuir* **20**(7), 2813–2818 Mar 30 (2004).
[230] Liu, Y., Liu, W. K., Belytschko, T., Patankar, N., To, A. C., Kopacz, A., and Chung, J. H. *International Journal for Numerical Methods in Engineering* **71**(4), 379–405 Jul 23 (2007).
[231] Singh, P. and Aubry, N. *Physical Review E* **72**(1), 016612 Jul (2005).
[232] Washizu, M. and Jones, T. B. *Journal of Electrostatics* **33**(2), 187–198 Sep (1994).
[233] Castellarnau, M., Errachid, A., Madrid, C., Juarez, A., and Samitier, J. *Biophysical Journal* **91**(10), 3937–3945 Nov (2006).

[234] Ehe, A. Z., Ramirez, A., Starostenko, O., and Sanchez, A. *Cross-Disciplinary Applied Research in Materials Science and Technology* **480–481**, 251–255 (2005).

[235] Gimsa, J. *Bioelectrochemistry* **54**(1), 23–31 Aug (2001).

[236] Gimsa, J., Schnelle, T., Zechel, G., and Glaser, R. *Biophysical Journal* **66**(4), 1244–1253 Apr (1994).

[237] Maswiwat, K., Wachner, D., Warnke, R., and Gimsa, J. *Journal of Physics D-Applied Physics* **40**(3), 914–923 Feb 7 (2007).

[238] Rivette, N. J. and Baygents, J. C. *Chemical Engineering Science* **51**(23), 5205–5211 Dec (1996).

[239] Archer, S., Morgan, H., and Rixon, F. J. *Biophysical Journal* **76**(5), 2833–2842 May (1999).

[240] Bakirov, T. S., Generalov, V. M., Chepurnov, A. A., Tyunnikov, G. I., and Poryavaev, V. D. *Doklady Akademii Nauk* **363**(2), 258–259 Nov (1998).

[241] Becker, F. F., Wang, X. B., Huang, Y., Pethig, R., Vykoukal, J., and Gascoyne, P. R. C. *Proceedings of the National Academy of Sciences of the United States of America* **92**(3), 860–864 Jan 31 (1995).

[242] Chan, K. L., Gascoyne, P. R. C., Becker, F. F., and Pethig, R. *Biochimica et Biophysica Acta-Lipids and Lipid Metabolism* **1349**(2), 182–196 Nov 15 (1997).

[243] Egger, M. and Donath, E. *Biophysical Journal* **68**(1), 364–372 Jan (1995).

[244] Falokun, C. D. and Markx, G. H. *Journal of Electrostatics* **65**(7), 475–482 Jun (2007).

[245] Falokun, C. D., Mavituna, F., and Markx, G. H. *Plant Cell Tissue and Organ Culture* **75**(3), 261–272 Dec (2003).

[246] Gascoyne, P., Mahidol, C., Ruchirawat, M., Satayavivad, J., Watcharasit, P., and Becker, F. F. *Lab on a Chip* **2**(2), 70–75 (2002).

[247] Gimsa, J., Marszalek, P., Loewe, U., and Tsong, T. Y. *Biophysical Journal* **60**(4), 749–760 Oct (1991).

[248] Huang, Y., Wang, X. B., Becker, F. F., and Gascoyne, P. R. C. *Biochimica et Biophysica Acta-Biomembranes* **1282**(1), 76–84 Jun 13 (1996).

[249] Huang, Y., Wang, X. B., Holzel, R., Becker, F. F., and Gascoyne, P. R. C. *Physics in Medicine and Biology* **40**(11), 1789–1806 Nov (1995).

[250] Simeonova, M. and Gimsa, J. *Journal of Physics-Condensed Matter* **17**(50), 7817–7831 Dec 21 (2005).

[251] Castellanos, A., Ramos, A., Gonzalez, A., Green, N. G., and Morgan, H. *Journal of Physics D-Applied Physics* **36**(20), 2584–2597 Oct 21 (2003).

[252] Mietchen, D., Schnelle, T., Muller, T., Hagedorn, R., and Fuhr, G. *Journal of Physics D-Applied Physics* **35**(11), 1258–1270 Jun 7 (2002).

[253] Morgan, H., Sun, T., Holmes, D., Gawad, S., and Green, N. G. *Journal of Physics D-Applied Physics* **40**(1), 61–70 Jan 7 (2007).

[254] Holmes, D. and Morgan, H. *Electrostatics 2003* **178**, 107–112 (2004).

[255] Holmes, D., Morgan, H., and Green, N. G. *Biosensors & Bioelectronics* **21**(8), 1621–1630 Feb 15 (2006).

[256] Huang, Y., Wang, X. B., Gascoyne, P. R. C., and Becker, F. F. *Biochimica et Biophysica Acta-Biomembranes* **1417**(1), 51–62 Feb 4 (1999).

[257] Kim, Y., Hong, S., Lee, S. H., Lee, K., Yun, S., Kang, Y., Paek, K. K., Ju, B. K., and Kim, B. *Review of Scientific Instruments* **78**(7), 074301 Jul (2007).

[258] Labeed, F. H., Coley, H. M., Thomas, H., and Hughes, M. P. *Biophysical Journal* **85**(3), 2028–2034 Sep (2003).

[259] Hughes, M. P. and Hoettges, K. F. *Biophysical Journal* **88**(1), 172A–172A Jan (2005).

[260] Gascoyne, P. R. C., Pethig, R., Burt, J. P. H., and Becker, F. F. *Biochimica et Biophysica Acta* **1149**(1), 119–126 Jun 18 (1993).

[261] Docoslis, A., Kalogerakis, N., Behie, L. A., and Kaler, K. V. I. S. *Biotechnology and Bioengineering* **54**(3), 239–250 May 5 (1997).

[262] Docoslis, A., Kalogerakis, N., and Behie, L. A. *Cytotechnology* **30**(1–3), 133–142 (1999).

[263] Labeed, F. H., Coley, H. M., and Hughes, M. P. *Biochimica et Biophysica Acta-General Subjects* **1760**(6), 922–929 Jun (2006).

[264] Lapizco-Encinas, B. H., Simmons, B. A., Cummings, E. B., and Fintschenko, Y. *Analytical Chemistry* **76**(6), 1571–1579 Mar 15 (2004).
[265] Kaler, K. V. I. S., Xie, J. P., Jones, T. B., and Paul, R. *Biophysical Journal* **63**(1), 58–69 Jul (1992).
[266] Pethig, R., Talary, M. S., and Lee, R. S. *IEEE Engineering in Medicine and Biology Magazine* **22**(6), 43–50 Nov-Dec (2003).
[267] Hu, X. Y., Bessette, P. H., Qian, J. R., Meinhart, C. D., Daugherty, P. S., and Soh, H. T. *Proceedings of the National Academy of Sciences of the United States of America* **102**(44), 15757–15761 Nov 1 (2005).
[268] Markx, G. H., Rousselet, J., and Pethig, R. *Journal of Liquid Chromatography & Related Technologies* **20**(16–17), 2857–2872 (1997).
[269] Huang, Y., Wang, X. B., Becker, F. F., and Gascoyne, P. R. C. *Biophysical Journal* **73**(2), 1118–1129 Aug (1997).
[270] Kang, K. H., Kang, Y. J., Xuan, X. C., and Li, D. Q. *Electrophoresis* **27**(3), 694–702 Feb (2006).
[271] Markx, G. H. and Pethig, R. *Biotechnology and Bioengineering* **45**(4), 337–343 Feb 20 (1995).
[272] Li, J. Q., Zhang, Q., Yan, Y. H., Li, S., and Chen, L. Q. *IEEE Transactions on Nanotechnology* **6**(4), 481–484 Jul (2007).
[273] Voldman, J., Gray, M. L., Toner, M., and Schmidt, M. A. *Analytical Chemistry* **74**(16), 3984–3990 Aug 15 (2002).
[274] Taff, B. M. and Voldman, J. *Analytical Chemistry* **77**(24), 7976–7983 Dec 15 (2005).
[275] Shih, T. C., Chu, K. H., and Liu, C. H. *Journal of Microelectromechanical Systems* **16**(4), 816–825 Aug (2007).
[276] Albrecht, D. R., Underhill, G. H., Wassermann, T. B., Sah, R. L., and Bhatia, S. N. *Nature Methods* **3**(5), 369–375 May (2006).
[277] Fair, R. B., Khlystov, A., Tailor, T. D., Ivanov, V., Evans, R. D., Griffin, P. B., Srinivasan, V., Pamula, V. K., Pollack, M. G., and Zhou, J. *IEEE Design & Test of Computers* **24**(1), 10–24 Jan-Feb (2007).
[278] Franks, F., editor. *Water: A Comprehensive Treatise*. Plenum, (1973).
[279] *CRC Handbook of Chemistry and Physics*. CRC Press, (2008).
[280] Gubskaya, A. and Kusalik, P. *Journal of Chemical Physics* **117**, 5290–5302 (2002).
[281] Tu., Y. and Laaksonen, A. *Chemical Physics Letters* **329**, 283–288 (2000).
[282] Coutinho, K., Guedes, R., Cabral, B., and Canuto, S. *Chemical Physics Letters* **369**, 345–353 (2003).
[283] Murrell, J. and Jenkins, A. *Properties of Liquids and Solutions*. Wiley, (1994).
[284] Hasted, J. *Aqueous Dielectrics*. Chapman and Hall, (1973).
[285] Arnold, W. M., Gessner, A. G., and Zimmermann, U. *Biochimica et Biophysica Acta* **1157**(1), 32–44 May 7 (1993).
[286] Akerlof, G. *Journal of the Americal Chemical Society* **54**, 4125 (1932).
[287] Galin, M., Chapoton, J.-C., and Galin, J.-C. *Journal of the Chemical Society Perkin Chem.* **74**, 2623 (2002).
[288] Segel, I. *Biochemical calculations*. Wiley, (1976).
[289] Wilson, E. *Vector Analysis*. Yale University Press, (1902).
[290] Aris, R. *Vectors, Tensors, and the Basic Equations of Fluid Mechanics*. Prentice Hall, (1962).
[291] Greenberg, M. *Advanced Engineering Mathematics*. Prentice-Hall, (1998).
[292] Pope, S. *Turbulent Flows*. Cambridge University Press, (2000).
[293] Squires, T. and Quake, S. *Reviews of Modern Physics* **77**, 977 (2005).
[294] Marsden, J. and Hoffman, M. *Basic Complex Analysis*. W. H. Freeman, (1998).
[295] R., S. and Laurra, P. *Conformal Mapping: Methods and Applications*. Dover, (2003).
[296] Sun, T., Morgan, H., and Green, N. G. *Physical Review E* **76**(4), 046610 Oct (2007).
[297] Sun, T., Green, N., and Morgan, H. *Applied Physics Letters* **92**, 173901 (2008).
[298] Haile, J. *Molecular Dynamics Simulation: Elementary Methods*. Wiley, (1992).
[299] Allen, M. and Tildesley, D. *Computer Simulation of Liquids*. Oxford University Press, (1987).

[300] Freund, J. *Journal of Chemical Physics* **116**, 2194–2200 (2002).
[301] Thompson, A. *Journal of Chemical Physics* **119**, 7503–7511 (2003).
[302] Lorenz, C., Crozier, P., Anderson, J., and Travesset, A. *Journal of Physical Chemistry* **112**, 10222–10232 (2008).
[303] Attard, P. *Thermodynamics and Statistical Mechanics*. Academic Press, (2002).
[304] Dougherty, R. and Howard, L. *Journal of Chemical Physics* **109**, 7379–7392 (1998).
[305] Errington, J. and Debenedetti, P. *Nature (London)* **409**, 318–321 (2001).
[306] Guillot, B. *Journal of Molecular Liquids* **101**, 219–260 (2002).
[307] Schropp, B. and Tavan, P. *J. Physical Chemistry B* **112**, 6233–6240 (2008).

Index

CPSIA information can be obtained at www.ICGtesting.com
Printed in the USA
LVOW10s0456130215
426908LV00003B/6/P